**CHEMICAL
ANALYSIS**

McGRAW-HILL SERIES IN ADVANCED CHEMISTRY

**McGRAW-HILL
BOOK COMPANY**
New York
St. Louis
San Francisco
Auckland
Düsseldorf
Johannesburg
Kuala Lumpur
London
Mexico
Montreal
New Delhi
Panama
Paris
São Paulo
Singapore
Sydney
Tokyo
Toronto

HERBERT A. LAITINEN

*Graduate Research Professor
University of Florida*

WALTER E. HARRIS

*Professor of Analytical Chemistry
Chairman, Department of Chemistry
University of Alberta*

Chemical Analysis

AN ADVANCED TEXT AND REFERENCE

SECOND EDITION

This book was set in Times Roman.
The editors were Robert H. Summersgill and Shelly Levine Langman;
the production supervisor was Charles Hess.
The drawings were done by Glen R. Johanson.
Kingsport Press, Inc., was printer and binder.

Library of Congress Cataloging in Publication Data

Laitinen, Herbert August, date
    Chemical analysis.

    (McGraw-Hill series in advanced chemistry)
    Includes bibliographies.
    1. Chemistry, Analytic—Quantitative. I. Harris,
Walter Edgar, date joint author. II. Title.
[DNLM: 1. Chemistry, Analytical. QD75 L189c]
QD101.2.L34   1975   545   74-11497
ISBN 0-07-036086-3

**CHEMICAL
ANALYSIS**

234567890KPKP798765

to

IZAAK MAURITS KOLTHOFF

"theory guides, experiment decides"

CONTENTS

PREFACE

This edition, like the first, primarily is intended to serve as a textbook at the advanced undergraduate and beginning graduate level for courses stressing fundamental principles of analytical chemistry. In addition, it is intended to serve as a reference work and guide to the literature for instructors in quantitative analysis and for practicing analytical chemists. Again, the aim is to cover the principles rigorously.

To keep the length of the first edition within the bounds of a single volume, it was necessary to avoid detailed descriptions of analytical procedures and to describe only the features that contribute to understanding of basic principles. It was noted that most chapters could have been written more easily at twice their length. The introduction of new topics, as well as the expansion and increased activity in most areas, intensified this problem for the second edition. To keep it at substantially the same length as the first, we have had to delete some admittedly important topics covered in the first edition; inevitably, some readers will disagree with certain of our choices.

The manuscript for this edition was completed in August 1973; literature covered therefore includes references of which we were aware through the first part of 1973.

For most topics it was possible to examine the research literature in addition to reference books and monographs. In some areas the literature is so extensive that it was not possible to survey the literature completely. For instance, in the field of complexation several thousand publications, and in chromatography tens of thousands of papers and hundreds of books, have appeared.

Significant new material is included in all major topics. There is a new chapter on standards, another on ion-selective electrodes, and the single separations chapter has been expanded to four. The major condensation is in the precipitations chapters, and throughout the book there is somewhat less emphasis on applications. Although not by design, there are again 27 chapters.

We have adopted insofar as seemed feasible the nomenclature adopted or tentatively agreed upon by members of IUPAC committees. Nevertheless, troublesome questions of symbolism and nomenclature persist because of the broad range of subjects covered. For example, in the complexation field the symbol α is often used to represent the reciprocal of a fraction rather than a fraction. For consistency with several other fields, we have chosen to use α in the sense of fraction throughout. Even within a single field, misunderstanding is sometimes possible. In chromatography the symbol D_m has been used elsewhere to denote both mass distribution ratio and diffusion coefficient in the mobile phase. The infinite-dilution activity coefficient, familiar to gas chromatographers, has remarkable similarities to the transfer activity coefficient used in nonaqueous chemistry. The recent agreement to assign a negative sign to cathodic currents disturbs many analytical chemists; against our inclination, we have adopted this convention in Chapter 14.

The generosity of colleagues and associates has been invaluable. First, not last as is usual, indebtedness must be acknowledged to Phyllis Harris, whose untiring assistance with every phase of the writing and production, in matters of style, clarity, and consistency, have greatly enhanced the readability. The assistance of the members of the Analytical Division at the University of Alberta is appreciated: Professors J. A. Plambeck and D. L. Rabenstein reviewed an early draft of the entire manuscript; Professor Byron Kratochvil reviewed all but the separations chapters; Professor Gary Horlick reviewed all but the redox chapters, and Professor W. F. Allen reviewed the first eleven chapters. Special thanks are due also to Professor D. N. Hume of Massachusetts Institute of Technology, who reviewed the entire manuscript and made many valuable suggestions, and to Professor H. Pardue of Purdue University, who reviewed two drafts of the chapter on kinetics and made important suggestions about several items, including organization. We wish to thank also those who reviewed several chapters and made helpful suggestions: Dr. P. C. Kelly of Canada Packers, the separations, statistics, and sampling chapters; Dr. H. W. Habgood of the Alberta Research Council, the four separations chapters; Dr. Josef Novák of the Institute for Instrumental Analytical Chemistry, Brno, Czechoslovakia, the two chapters on chromatography; Dr. Hasuk Kim, of the University of Illinois, the first seven chapters and Chapters 17 and 20; Dr. R. R. Gadde, of the University of Illinois, Chapters 12, 13, 16, and 25; Mr. J. M. Conley, of the University of Illinois, Chapters 14 and 21; Professor Saul Zalik of the University of Alberta, the chapter on statistics; Professor H. B. Dunford of the University of Alberta, the chapter on kinetics; and Professor J. F. Coetzee of the University of Pittsburgh, and Professor I. M. Kolthoff of the University of Minnesota, Chapter 4 on nonaqueous chemistry. Mrs. Barbara Burrows, who is a librarian as well as a chemist, collected and checked all references. With care and patience, Mrs. Carol Reed typed many of the earlier chapters, Mrs. Lu Ziola typed the later chapters and the final manuscript, and Mr. Glen Johanson prepared the illustrations. To all of these and others who played a role in the preparation of the manuscript, we wish to express our appreciation. None of the people named is in any way responsible for such faults as remain; we will be grateful if readers who detect errors would call them to our attention.

A book such as this could not be completed without the use and adaptation of ideas, tables, and diagrams from innumerable copyrighted sources. A large debt to the creative thoughts of many is gratefully acknowledged.

HERBERT A. LAITINEN
WALTER E. HARRIS

1

THE OPERATIONS OF ANALYSIS

Chemical analyses have their origins in problems needing reliable measurements for their solution or interpretation; therefore, the first step in chemical analysis is *definition of the goal*. A problem may involve, for instance, control of a manufacturing process, gypsy moth devastation, fluoridation of a water supply, or monitoring motor oil for engine wear. If at least part of the problem involves analysis, these aspects need to be defined, isolated, or recognized by someone who may or may not be a chemist. After the analytical aspects have been adequately recognized, formal evaluation of the analytical chemistry can be undertaken.

To solve the analytical problems facing him, a chemist must be prepared to use any part of the arsenal of chemical information and tools. This includes knowledge not only of methods for measuring chemical and physical properties of atoms, ions, and molecules, but also of modern methods of separation, sampling, and handling of statistical data. Part of the fascination of analytical chemistry is its enormous range of application, extending from time-tested precipitations and titrations to sophisticated modern theory, instrumentation, and technology. Inadequate knowledge of any aspect cripples the ability to attack problems successfully. An analytical chemist can be judged in part by his skill in the critical selection of methods. Therefore an investment in competence in all aspects of modern analytical chemistry is worthwhile.

In multidisciplinary research the special contribution of the analytical chemist is his recommendation as to the best methods for making measurements of chemical significance. A sound recommendation must be based on both breadth and depth of knowledge of a host of methods and also on the ability to adapt them intelligently to particular needs. We are concerned here primarily with the noninstrumental aspects of theory underlying the practice of analytical chemistry. This area is crucial to the performance of competent analytical work.

The sample analyzed must validly represent the composition of the material of interest. *Sampling* (Chapter 27) is a complex topic only partly amenable to theoretical description. Because by its very nature the sampling operation requires statistical treatment, it requires a sound understanding of statistics.

Another major step in many analyses is *separation* (Chapters 22 to 25). When, because of the method chosen or the nature of the sample, this unit operation is not required, much effort can be saved. For example, if a masking agent will complex an interfering metal ion in an EDTA titration, a separation step may be avoided. Where a separation is essential, a choice among several techniques is usually available. In general, separation involves the formation of two phases, physically separated, one containing the material of interest and the other the interference. Either phase may be a gas, liquid, or solid. Thus six major types of separation processes are possible. Once separation has been effected, the quantitative determination by physical means is often straightforward.

Still another major operation in analysis is *measurement*, which may be carried out by physical, chemical, or biological means. In each of these three areas a wide range in techniques is available. For example, titrimetry is the most common of the chemical methods of measurement, and spectroscopy the most widely used of the available physical methods. In most analytical studies the bulk of the effort is directed to an examination of the theoretical background, experimental limitations, and applications of various techniques of measurement. Since methods of analysis are usually defined in terms of the final measurement step, the impression is often given that this stage constitutes the entire subject of analytical chemistry. Even though the measurement aspect deserves much attention, it should be remembered that the preliminary steps of definition of the problem, sampling, and separation are also critical to the overall process.

Classical methods of final measurement will long continue to be important. In the first place, they are inherently simple. For an occasional determination or standardization the use of a titration or gravimetric determination often will require the least time and effort and will involve no investment in expensive equipment. Second, classical methods are accurate. Many instrumental methods are designed for speed or sensitivity rather than accuracy, and often must be calibrated by classical methods.

Much of the theory of analytical measurements is concerned with nonquantitativeness arising from unfavorable equilibria, undesirable side reactions, and catalyzed or induced reactions. In the precipitation of a sparingly soluble salt, loss of precipitate may occur through solubility or peptization, and coprecipitation and postprecipitation of foreign materials may introduce error. In acid-base titrations, lack of coincidence of indicator change with the equivalence point may be a problem. In redox titrations, unfavorable equilibria may introduce error, but more often the lack of exact stoichiometry is traceable to slow reactions, side reactions, and unexpected occurrences such as induced reactions. In chromatography, irreversible adsorption on the stationary phase and inadequate resolution often present problems.

A frequent complication is that several simultaneous equilibria must be considered (Section 3-1). Our objective is to simplify mathematical operations by suitable approximations, without loss of chemical precision. An experienced chemist with sound chemical instinct usually can handle several solution equilibria correctly. Frequently, the greatest uncertainty in equilibrium calculations is imposed not so much by the necessity to approximate as by the existence of equilibria that are unsuspected or for which quantitative data for equilibrium constants are not available. Many calculations can be based on concentrations rather than activities, a procedure justifiable on the practical grounds that values of equilibrium constants are obtained by determining equilibrium concentrations at finite ionic strengths and that extrapolated values at zero ionic strength are unavailable. Often the thermodynamic values based on activities may be less useful than the practical values determined under conditions comparable to those under which the values are used. Similarly, thermodynamically significant standard electrode potentials may be of less immediate value than formal potentials measured under actual conditions.

Knowledge of statistics (Chapter 26) is basic to effective handling of the last unit operation, *evaluation of data*. Many chemists rarely go beyond a calculation of the standard deviation for a set of determinations and have the erroneous notion that the use of more advanced statistical methods is restricted to enormous bodies of data. A point often overlooked is that a relatively small number of systematically planned observations may yield more information than a larger number of repeated identical observations. For example, in the simple matter of running triplicate analyses of a sample, it may be best to weigh three samples of substantially different sizes. The results obtained may reveal determinate errors that would be unsuspected with samples of equal size.

In the analysis of real materials any one of the operations may be the most important. One general objective of an analysis is to obtain decisive results quickly and at low cost. When sufficient sample is available and no interferences are present, analysis usually presents few problems, and it is often possible to go directly from sampling to measurement. Complex mixtures of closely related substances can be approached in two ways. One is through use of a measurement technique of high specificity; this permits the sampling and separation stages of analysis to be deemphasized, and often the separation stage can be omitted completely. The other approach is to isolate the materials of interest and then measure each one by simple techniques. For complex samples it may be necessary first to separate classes of compounds and then to examine them by techniques of high specificity.

In summary, the thesis of this book is that knowledge of chemical reactions is important, first because it is needed for direct application to classical methods, and second because it is essential in instrumental methods where chemical reactions are involved in operations preceding the use of an instrument in the final measurement. We focus on the processes occurring in sampling, separation, and measurement, emphasizing the chemistry, rather than the physics, of analysis. The chemical aspects of analytical chemistry are alive with challenging unsolved problems of a fundamental nature. By searching for answers, the analytical chemist can make important contributions not only to the field of chemical and instrumental analysis, but to the larger subject of chemistry as a whole.

EQUILIBRIUM AND ACTIVITY

Much of the work of analytical chemists involves reactions that take place at appreciable concentrations; yet the equilibrium constants of fundamental importance are the thermodynamic values obtained by extrapolation to infinite dilution or to zero ionic strength. The purpose in this chapter is to: examine some of the basic thermodynamic concepts that apply to solutions; estimate the magnitude of the errors introduced by neglecting effects of ionic strength in aqueous solutions; consider the extent to which these errors can be minimized by suitable corrections; and examine the behavior of nonelectrolytes in solution.

2-1 THE CONDITION OF EQUILIBRIUM

A chemical reaction is at equilibrium when the sum of the chemical potentials[1] of the reactants is equal to that of the products. Thus for the reaction

$$A + B \rightleftharpoons C + D \qquad (2\text{-}1)$$

the condition of equilibrium is

$$\mu_A + \mu_B = \mu_C + \mu_D \qquad (2\text{-}2)$$

where μ is the chemical potential. At constant temperature and pressure the chemical potential is the partial molal free energy,[1] and (2-2) is equivalent to stating that at constant temperature and pressure the free-energy change of a reaction is zero at equilibrium.

Lewis[2] defined chemical activity in terms of chemical potential by the equation

$$\mu_i = k_i + RT \ln a_i \qquad (2\text{-}3)$$

where a_i is the chemical activity of species i, R is the gas constant, T is absolute temperature, and k_i is the value of μ_i at $a_i = 1$. Substituting (2-3) in (2-2), we have

$$\frac{k_A + k_B - k_C - k_D}{RT} = \ln \frac{a_C a_D}{a_A a_B} \qquad (2\text{-}4)$$

The equilibrium constant is defined by

$$K_{eq} = \frac{a_C a_D}{a_A a_B} \qquad (2\text{-}5)$$

For the general reaction

$$mM + nN \rightleftharpoons pP + rR \qquad (2\text{-}6)$$

Equation (2-4) becomes

$$\frac{mk_M + nk_N - pk_P - rk_R}{RT} = \ln \frac{a_P{}^p a_R{}^r}{a_M{}^m a_N{}^n} = \ln K_{eq} \qquad (2\text{-}7)$$

2-2 ACTIVITY AND THE STANDARD STATE

The numerical value of k_i in (2-3) depends on how activity is defined and on the units in which concentration is expressed (molarity, mole fraction, partial pressure). Measurement of the absolute activity, or chemical potential, of an individual ion is one of the classical unsolved problems.[3] Since we cannot measure absolute ion activity, we are then necessarily interested in the next best—comparative changes in activities with changing conditions. To obtain comparative values numerically, we measure activity with respect to an *arbitrarily* chosen standard state under a given set of conditions of temperature and pressure, where the substance is *assigned* unit activity. The value of k_i in (2-3) thus depends on the arbitrary standard state chosen; accordingly, the value of the equilibrium constant also depends on the choice of standard states.

Although the standard state could be based on any reference behavior, for simplicity the choices are conventionally limited to one of two main types (Figure 2-1). One is the limiting behavior of a substance as it approaches zero mole fraction (condensed phase) or zero partial pressure (gas phase); this is called henryan reference behavior. The other is the limiting behavior of a substance as it approaches unit

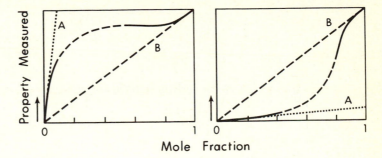

FIGURE 2-1 Variation of a property (schematic) as a function of mole fraction, to illustrate henryan (*A*) and raoultian (*B*) reference behavior.

mole fraction; this is called raoultian reference behavior. Thus activity is assumed to approach concentration as a limit, in one case as concentration approaches zero, and in the other as mole fraction approaches unity. In practice we choose a raoultian standard state whenever possible and a henryan one when we must. For gases the reference behavior is that of the gas at a pressure sufficiently low that there is no interaction, and the behavior under this reference condition is extrapolated so that an "ideal" gas has unit activity (unit fugacity) at unit partial pressure (1 atm). Real gases usually have an activity less than unity at unit partial pressure. For pure liquids the reference behavior is that of the pure substance at a specified temperature and pressure. To avoid the problem of requiring knowledge of molecular weights (water, for example, is an associated solvent), we express concentration for pure liquids in mole fractions, and therefore a pure liquid has unit activity by definition. Similarly, pure solids are assigned a value of unit activity in their most stable crystalline state.

For solutions the standard state usually is defined differently for solvents and for solutes. For solvents the standard state is the pure solvent, whose activity is given by

$$a_i = N_i \gamma_i \qquad (2\text{-}8)$$

where N_i is the mole fraction and γ_i is the activity coefficient.† Equation (2-8) constitutes a definition of activity coefficient γ_i, and since $a_i = 1$ when $N_i = 1$ (pure substance), $\gamma_i = 1$ for pure solvent. If the activity is proportional to the mole fraction (Raoult's law), $\gamma_i = 1$. When dealing with mixtures containing several solvents or with solid solutions, it is convenient to employ (2-8) to compare the activity in a mixture with that of a pure substance. Thus, for example, the activity of water in a sodium chloride solution is best compared with that of pure water at the same temperature.

† The symbol γ_i can be used when concentrations are expressed as molality m, y_i when expressed as molarity M, and f_i when expressed as mole fraction. In dilute aqueous solutions the difference between γ_i and y_i is small and usually negligible.

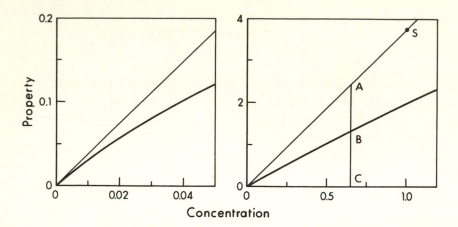

FIGURE 2-2 Relation between concentration and a physical property of a solution. Upper lines, the value extrapolated from the lowest concentrations. Left, an expansion of the initial part of the curve, to make clear the asymptotic behavior. (This example involves the lowering of the freezing point of magnesium sulfate solutions as a function of molality.)

For solutes the standard state and the activity usually must be defined in terms of behavior under conditions of infinite dilution, where by definition the activity of a solute is set equal to its concentration. Thus at infinite dilution the ratio of activity to concentration (in whatever units) is unity, and $\gamma_i = 1$. When the value of some physical property of a solution is plotted as a function of concentration, a curve like those in Figure 2-2 is obtained. If the asymptote passing through the origin on the concentration scale is extrapolated to higher concentrations, we obtain the standard state of unit activity for the property in question. This hypothetical solution, labeled S, of unit concentration exhibits the same type of behavior as the infinitely dilute solution. The extent to which the real value of the physical property measured differs from the hypothetical value at a specific concentration is expressed by the activity coefficient, a coefficient that is simply the ratio between two measurable quantities. In Figure 2-2 the activity coefficient γ_i is the ratio BC/AC and is defined by

$$a_i = C_i\gamma_i \qquad (2\text{-}9)$$

where C_i is the concentration. For a solution of magnesium sulfate in water, the solute would have unit activity at a concentration of about 2 m. (Extrapolate the asymptote of the lower line of Figure 2-2 from the lowest concentration to higher values.) There is no simple relation between this unit activity of magnesium sulfate and the unit activity defined for the pure solid. Similarly, for a substance such as hexane several different standard states might easily by employed—one when it is used as a solvent, another when its vapor pressure is of interest, and still another when it is present as a solute. The concentration of hexane in these three situations probably would be expressed as mole fraction, partial pressure, and molarity. When equilibrium constants are used, the standard states must be either specified or clearly understood.

Activity is given in the same units as concentration—molarity, mole fraction, and so forth. In (2-9) the concentration C_i usually is expressed either in moles per liter of solution (molarity M) or in moles per kilogram of solvent (molality m). In dilute aqueous solutions the molarity and molality are nearly equal; in nonaqueous solutions molarity is usually larger than molality, since the density of the solvent is usually less than unity. Analytical chemists ordinarily find it more convenient to express concentration in molarity, even though it varies slightly with temperature. The *analytical concentration* is represented by the symbol C, to indicate the moles of solute added per liter of solution. Analytical concentration should be distinguished from the *equilibrium concentration*, which is indicated by enclosure in square brackets.

2-3 THE DEBYE-HÜCKEL THEORY

In 1923 Debye and Hückel[4] made an important contribution to our understanding of solutions by deriving a theoretical expression for the activity coefficients of individual ions and for the mean activity coefficients of strong electrolytes. Their derivation is based on two laws that describe interactions among the ions of an electrolyte. These are (1) Coulomb's law, the inverse-square law of interaction for particles of unlike charge and of repulsion for particles of like charge, and (2) the Boltzmann distribution law, which describes the tendency for thermal agitation to counteract the effects of electrical attraction and repulsion. In the simplest form of the Debye-Hückel derivation, ions are assumed to be point charges and their finite sizes are neglected. We first examine the consequences of the simplest treatment, the Debye-Hückel limiting law (DHLL), and then consider the more exact treatments, which take ion size and ion hydration into account.

According to the Boltzmann distribution law

$$C_i = C_i^0 \exp \frac{-Z_i e \Psi}{kT} \qquad (2\text{-}10)$$

where Ψ is electrical potential of a point in solution with respect to an electrically neutral point (Ψ is positive around a cation and negative around an anion), $Z_i e \Psi$ is electrical potential energy of the ith ion with respect to a neutral point, where Z_i is a positive or negative integer and e is the charge of a proton (4.803×10^{-10} esu), C_i^0 is concentration of the ith ion at an electrically neutral point, C_i is concentration of an ion at potential Ψ, k is Boltzmann's constant (1.38×10^{-16} erg/K), and T is absolute temperature (kT is a measure of thermal energy). According to the Boltzmann expression the concentration of like-charged ions (Z_i and Ψ both positive or both negative) is diminished in the vicinity of a particular ion, whereas unlike-charged ions are concentrated to form an *ion atmosphere*. Since the solution as a whole is electrically neutral, the total charge in the ion atmosphere surrounding a particular ion is equal to the charge of the central ion. The distribution of charge around an ion falls off exponentially with distance and depends on the temperature, because at higher temperatures thermal agitation tends to counteract the electrical attraction of unlike ions.

In the simplest derivation, in which the ions are assumed to be point charges, the activity coefficient γ_i of an ion of charge $Z_i e$ in a solvent of dielectric constant D is given by

$$\ln \gamma_i = \frac{-Z_i^2 e^2 \kappa}{2DkT} \tag{2-11}$$

For a single electrolyte a quantity κ may be defined as the reciprocal of the radius of the ionic atmosphere and is proportional to the square root of the concentration. In other words, the charge in the ionic atmosphere may be visualized as being uniformly distributed over the surface of a sphere of radius $1/\kappa$.

The relation between κ (in units of 1/cm) and concentration is given by

$$\kappa = \sqrt{\frac{8\pi e^2 N}{1000 DkT}} \sqrt{\mu} = 0.33 \times 10^8 \sqrt{\mu} \qquad \text{for water at 25°C} \tag{2-12}$$

where N is Avogadro's number and μ is ionic strength,†

$$\mu = \frac{\sum C_i Z_i^2}{2} \tag{2-13}$$

For a single electrolyte the ionic strength is proportional to the concentration. Thus, for molarity C of the following electrolytes we have

$$A^+ B^- \qquad \mu = \frac{C_A Z_A^2 + C_B Z_B^2}{2} = \frac{C_A + C_B}{2} = C$$

$$A^{++} B_2^- \qquad C_A = C \qquad C_B = 2C \qquad \mu = \frac{4C + 2C}{2} = 3C$$

$$A^{++} B^= \qquad C_A = C_B = C \qquad \mu = \frac{4C + 4C}{2} = 4C$$

$$A^{3+} B_3^- \qquad C_A = C \qquad C_B = 3C \qquad \mu = \frac{9C + 3C}{2} = 6C$$

$$A_m^{p+} B_n^{q-} \qquad C_A = mC \qquad C_B = nC \qquad \mu = \frac{C(mp^2 + nq^2)}{2}$$

† The symbol μ frequently is used for both chemical potential and ionic strength. Since we refer to chemical potential but little, no confusion should result from the use of a boldface $\boldsymbol{\mu}$ for chemical potential and an ordinary μ for ionic strength.

The radius $1/\kappa$ of the ionic atmosphere depends on the charge type of the electrolyte as well as on the concentration. Thus from (2-12) for an A^+B^- type of electrolyte in 0.1, 0.001, and 10^{-5} M solutions, $1/\kappa$ is 9.5×10^{-8}, 9.5×10^{-7}, and 9.5×10^{-6} cm. For $A_2^+B^=$ in 0.1 M solution, $1/\kappa$ is 5.5×10^{-8} cm.

From (2-11) and (2-12), for the ions of point charge the DHLL takes the form

$$-\log \gamma_i = AZ_i^2\sqrt{\mu} \qquad (2\text{-}14)$$

where the constant A is proportional to the $-\frac{3}{2}$ power of both the dielectric constant of the solvent and the absolute temperature and contains the factor $(1/2.303)$ to convert natural to base-10 logarithms. The constant A for water at $0°$, $25°$, and $100°C$ is 0.492, 0.511, and 0.596. Therefore, for aqueous solutions at room temperature, we may write the approximation

$$-\log \gamma_i = 0.5Z_i^2\sqrt{\mu} \qquad (2\text{-}15)$$

For an ionic electrolyte A_mB_n the *mean activity coefficient* $\gamma_\pm$ is defined by the equation

$$(m + n) \log \gamma_\pm = m \log \gamma_A + n \log \gamma_B \qquad (2\text{-}16)$$

Combining (2-15) and (2-16), we can write the DHLL in the form

$$-\log \gamma_\pm = 0.5\sqrt{\mu} \frac{mZ_A^2 + nZ_B^2}{m + n} = 0.5Z_AZ_B\sqrt{\mu} \qquad (2\text{-}17)$$

in which Z_A and Z_B are taken *without regard to sign*.

Interestingly, Lewis[5] introduced the concept of ionic strength 2 years before the derivation of the Debye-Hückel equation; he had shown that in dilute solutions the logarithm of the activity coefficient of a strong electrolyte is in general a linear function of the square root of the ionic strength. It is a triumph of the Debye-Hückel theory that, not only is the linear dependence predicted without special assumptions, but even the slope of the curve is quantitatively predicted[6] in solutions of both high and low dielectric constant (Figure 2-3). Thus ionic strength, first introduced as an empirical quantity, was given sound theoretical justification.

EXAMPLE 2-1 Using the DHLL, calculate the activity coefficients of each of the ions and the mean activity coefficient of the salt in a 10^{-4} m solution of potassium sulfate.

ANSWER The ionic strength is $3C = 3 \times 10^{-4}$, and $-\log \gamma_{K^+} = 0.5(3 \times 10^{-4})^{\frac{1}{2}}$, or $\gamma_{K^+} = 0.98$. Similarly, $-\log \gamma_{SO_4^=} = 0.5 \times 2^2 \times (3 \times 10^{-4})^{\frac{1}{2}}$, or $\gamma_{SO_4^=} = 0.923$; and $-\log \gamma_\pm = 0.5 \times 2 \times (3 \times 10^{-4})^{\frac{1}{2}}$, or $\gamma_\pm = 0.961$. ////

According to the DHLL the activity coefficient of an ion is determined by its charge and by the total ionic strength of the solution, which may be due primarily to

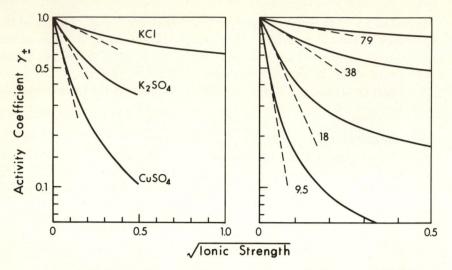

FIGURE 2-3 Activity coefficients calculated by the limiting Debye-Hückel equation (dotted lines) and those observed experimentally. Left, electrolytes of three charge types in water. Right, hydrochloric acid in water-dioxane mixtures with bulk dielectric constants as indicated. (*Adapted from Harned and Owen.*[6])

other electrolytes in solution. For example, in a saturated solution of silver chloride in 0.01 M potassium chloride, the concentration of silver chloride is only about 10^{-8} M and so makes no appreciable contribution to the total ionic strength. Yet the activity coefficient of the silver ion is equal to that of potassium or chloride ions (0.89, from the DHLL).

Recall that the DHLL does not take into account the finite sizes of ions. If the radii of hydrated ions were equal to a, the nearest approach of their centers would be $2a$, and in the calculation of the distribution of ions in the ion atmosphere it would be incorrect to include any approach closer than $2a$. The DHLL, therefore, tends to overcorrect for the effects of interionic attraction and repulsion. Nonetheless, it is useful at low ionic strengths because of its simplicity, and no resort to empiricism need be made to fit experiment to theory. But, at electrolytic concentrations of more practical interest, some empiricism must be introduced to make practice and theory agree. In their original paper Debye and Hückel did take into account the finite sizes of ions and introduced the parameter a, described as "the mean distance of approach of the ions, positive or negative." They derived the equation

$$-\log \gamma_i = \frac{AZ_i^2\sqrt{\mu}}{1 + \kappa a} \qquad (2\text{-}18)$$

in which the constants A and κ vary with temperature and are slightly different depending on whether concentration is given in terms of molarity or molality. In view of (2-12), (2-18) becomes

$$-\log \gamma_i = \frac{AZ_i^2\sqrt{\mu}}{1 + 0.33 \times 10^8 a\sqrt{\mu}} \qquad (2\text{-}19)$$

Although a correction for ion size is necessary, the correction to be used in terms of (2-19) can be predicted only roughly. For many ions the ion-size parameter a is of the order of 3×10^{-8} cm, and hence

$$-\log \gamma_i \simeq \frac{AZ_i^2\sqrt{\mu}}{1 + \sqrt{\mu}} \qquad (2\text{-}20)$$

For an electrolyte A_mB_n the mean activity coefficient in water at room temperature is given by

$$-\log \gamma_{\pm} = \frac{0.5Z_AZ_B\sqrt{\mu}}{1 + 0.33 \times 10^8 a\sqrt{\mu}} \simeq \frac{0.5Z_AZ_B\sqrt{\mu}}{1 + \sqrt{\mu}} \qquad (2\text{-}21)$$

Either (2-20) or (2-21) is referred to[7] as the extended Debye-Hückel equation (EDHE); this pair of equations gives results appreciably different from the DHLL when $\mu > 0.01$ (that is, $\sqrt{\mu} > 0.1$). For comparison, some ionic activity coefficients calculated from (2-15) and (2-20) are listed in Table 2-1.

Unfortunately, the calculated values of γ_i cannot be confirmed by direct experiment, because in principle all experimental methods yield the mean activity coefficient $\gamma_{\pm}$ rather than the individual ionic values.[5] By use of the definition given in (2-16), the experimentally determined value $\gamma_{\pm}$ can be apportioned to give γ_A and γ_B. This procedure is theoretically justified only at high dilution, where the DHLL is valid because the limiting slope of $\log \gamma_{\pm}$ plotted against $\sqrt{\mu}$ is found experimentally to be $0.5Z_AZ_B$, as required by (2-17). At higher values of μ the ion-size parameter a must be introduced.

Kielland[8] assigned to each ion an empirical value of a parameter a and used (2-19) to calculate its activity coefficient. A selection of Kielland's values is listed in

Table 2-1   CALCULATED ACTIVITY COEFFICIENTS OF IONS IN WATER AT 25°C

Ion charge	From DHLL, for $\mu =$				From EDHE, for $\mu =$			
	0.005	0.01	0.05	0.1	0.005	0.01	0.05	0.1
1	0.92	0.89	0.78	0.70	0.93	0.90	0.81	0.76
2	0.73	0.63	0.36	0.23	0.74	0.65	0.43	0.33
3	0.48	0.36	0.10	0.039	0.50	0.39	0.15	0.083
4	0.28	0.17	0.017	0.003	0.30	0.18	0.035	0.013

Table 2-2. As a justification of his procedure the values of mean activity coefficients calculated for various electrolytes from the individual ionic values are in satisfactory agreement with the experimental values up to an ionic strength of about 0.1.

With the present feasibility of evaluating experimental data with the aid of a computer, more refined or optimum values of the ion-size parameter can be obtained. For example, Paabo and Bates,[9] in a study of the dissociation of deuteriophosphoric acid or deuteriocarbonate ion, selected the ion-size parameter as that which gave the smallest standard deviation for the least-squares intercept for the ionization constant.

In brief, an assumption of the Debye-Hückel theory is that activity coefficients in aqueous solution for ordinary neutral molecules may be assigned a value of unity. For ions of ± 1 charge as a group, the activity coefficients will deviate from unity somewhat; for ions of ± 2 charge, still more deviation is expected. In each group the activity coefficient increases with increasing radius of the hydrated ion. Further, in a given solution ions of a given charge have approximately the same activity coefficient without regard to their individual concentrations.

Table 2-2   INDIVIDUAL ION ACTIVITY COEFFICIENTS IN WATER AT 25°C

Ion	Ion-size parameter a, cm $\times 10^8$	Ionic strength[†]			
		0.005	0.01	0.05	0.1
H^+	9	0.933	0.914	0.86	0.83
$(C_3H_7)_4N^+$	8	0.931	0.912	0.85	0.82
$(C_3H_7)_3NH^+$, $\{OC_6H_2(NO_3)_3\}^-$	7	0.930	0.909	0.845	0.81
Li^+, $C_6H_5COO^-$, $(C_2H_5)_4N^+$	6	0.929	0.907	0.835	0.80
$CHCl_2COO^-$, $(C_2H_5)_3NH^+$	5	0.928	0.904	0.83	0.79
Na^+, IO_3^-, HSO_3^-, $(CH_3)_3NH^+$, $C_2H_5NH_3^+$	4–4.5	0.927	0.901	0.815	0.77
K^+, Cl^-, Br^-, I^-, CN^-, NO_2^-, NO_3^-	3	0.925	0.899	0.805	0.755
Rb^+, Cs^+, NH_4^+, Tl^+, Ag^+	2.5	0.924	0.898	0.80	0.75
Mg^{++}, Be^{++}	8	0.755	0.69	0.52	0.45
Ca^{++}, Cu^{++}, Zn^{++}, Mn^{++}, Ni^{++}, Co^{++}	6	0.749	0.675	0.485	0.405
Sr^{++}, Ba^{++}, Cd^{++}, $H_2C(COO)_2^=$	5	0.744	0.67	0.465	0.38
Hg_2^{++}, $SO_4^=$, $CrO_4^=$	4	0.740	0.660	0.445	0.355
Al^{3+}, Fe^{3+}, Cr^{3+}, La^{3+}	9	0.54	0.445	0.245	0.18
$\{Co(en)_3\}^{3+}$	6	0.52	0.415	0.195	0.13
$Citrate^{3-}$	5	0.51	0.405	0.18	0.115
PO_4^{3-}, $Fe(CN)_6^{3-}$, $\{CO(NH_3)_6\}^{3+}$	4	0.505	0.395	0.16	0.095
Th^{4+}, Zr^{4+}, Ce^{4+}	11	0.35	0.255	0.10	0.065
$Fe(CN)_6^{4-}$	5	0.31	0.20	0.048	0.021

† Kielland's table[8] is given in terms of ionic concentrations.

2-4 ACTIVITY COEFFICIENTS AT HIGH IONIC STRENGTHS

For solutions of *single electrolytes* of the 1:1 or 1:2 charge types, several theoretical approaches[10–16] have proved useful in interpreting the variation of the *mean* activity coefficient up to relatively high concentrations. In an early modification due to Hückel,[11] a term similar in form to Equation (2-29) was added to (2-21), yielding the empirical expression

$$\log \gamma_{\pm} = \frac{-AZ_AZ_B\sqrt{\mu}}{1 + \kappa a} + BC \qquad (2\text{-}22)$$

The term BC was included to correct for the change in dielectric constant of the solvent upon addition of electrolyte. Equations of this form, containing two empirical constants, are satisfactory for 1:1 electrolytes up to an ionic strength of the order of unity.

Stokes and Robinson[12] achieved remarkable success with a one-parameter equation, the single parameter being a hydration number. For a salt of the type AB or AB_2, they wrote the equation

$$\log \gamma_{\pm} = \frac{-0.51Z_AZ_B\sqrt{\mu}}{1 + 0.329 \times 10^8 a\sqrt{\mu}} - \frac{n}{v} \log a_w - \log\left[1 - 0.018(n - v)m\right] \qquad (2\text{-}23)$$

where n is number of water molecules bound by one "molecule" of solute, v is number of ions per "molecule" of solute, a_w is activity of water, and m is molality. The first term on the right side is recognized as the Debye-Hückel term, which is always negative in sign. The second term, or *solvent term*, corrects for the decreased activity of water in the salt solution and is always positive in sign because a_w is less than unity. The third term, or *scale term*, takes into account the hydration of the ions, with the consequent binding of n water molecules with v ions to remove them from acting as solvent molecules. In principle, the ion-size parameter a should be related to the hydration parameter n. If the water of hydration is assumed to be associated largely with the cations of the salt, the ion-size parameter can be estimated[12] from the crystallographic radii of the unhydrated ions and the hydration number n. If a semi-empirical relation between a and n is introduced into (2-23), the hydration number n becomes the only parameter, which is then found by trial and error.

The remarkable agreement between experiment and the fitted one-parameter equation up to an ionic strength of the order of 4 is evident from Figure 2-4. In general, the equation breaks down when the product nm (hydration number times molality) exceeds 10 or 15. Since only 55.5 moles of water are available for m moles of salt, it is reasonable to expect that n would begin to decrease with concentration when neighboring ions begin to compete for the available water. The hydration numbers quoted in Figure 2-4 are reasonable in magnitude and show the expected trends with the nature of the cation. They should not, however, be taken too literally.[13–15]

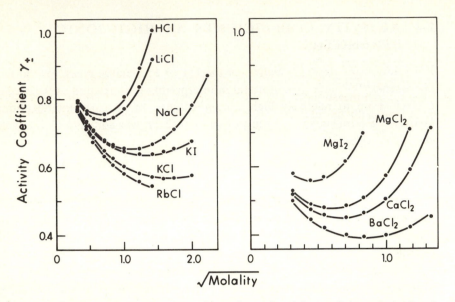

FIGURE 2-4 Comparison of experimental activity coefficients (points) and those predicted (solid lines) by a one-parameter equation. The hydration numbers n are HCl, 7.3; LiCl, 6.5; NaCl, 3.5; KI, 2.45; KCl, 1.9; RbCl, 1.25; MgI$_2$, 20.0; MgCl$_2$, 13.9; CaCl$_2$, 11.9; and BaCl$_2$, 8.4. (*From Stokes and Robinson.*[12])

Ion association has long been recognized as a complicating factor; it was considered in detail first by Bjerrum[17] and later by Fuoss and Kraus.[18] The formation of ion pairs, ion triplets, and higher-ion clusters becomes increasingly important as the charges of the ions increase and the dielectric constant of the medium decreases (Section 4-4). For particularly large ions, such as the complex ions of heavy metals, the forces of interionic attraction are only moderately large, and even electrolytes of higher charge types can be handled.[19,20] Alkaline earth halides act as strong electrolytes,[21] but alkali metal sulfates and alkaline earth nitrates show some evidence of association, as do the sulfates of divalent metals.[22]

2-5 ACTIVITY COEFFICIENTS IN MIXED ELECTROLYTES

A practical problem frequently encountered by the analytical chemist is that of maintaining as nearly as possible a constant ionic environment while varying the concentration of one component of a mixture of electrolytes. For example, in evaluating the successive formation constants of halide complexes of a metal ion, it is necessary to vary the halide ion concentration over wide limits. The practice generally adopted is to use an electrolyte such as sodium perchlorate, which is believed not to form competing complexes, and to compensate for the increased concentration of, say, sodium chloride by a corresponding decrease in sodium perchlorate concentration. In this manner a constant ionic strength is maintained at a high level, about 1 or 2 M. Even though measurements of equilibria taken under these conditions often give the most practically useful data, and in fact may give the only useful and consistent data, some question arises as to the validity of the assumption that ionic activity coefficients are maintained at a constant level in this procedure.

By measurement of solubilities of sparingly soluble electrolytes or by appropriate cell emf measurements, the *mean* activity coefficient of one electrolyte can be determined as a function of composition in solutions of mixed electrolytes. It can be concluded[6] that for mixed electrolytes of the 1:1 charge type the mean activity coefficient of one component, for instance silver chloride, is accurately determined by the total ionic strength as expected from the Debye-Hückel theory, at least up to ionic strengths of the order of 0.1. For mixtures of unlike charge types, for example 1:1 and 2:1, the situation is less satisfactory.[23,24] Maintenance of constant ionic strength by electrolyte substitution is best carried out with electrolytes of the same charge type, and the most consistent results are to be expected when both electrolytes are of the 1:1 charge type, as in the sodium chloride–sodium perchlorate example cited above. However, no procedure should be used blindly. Thus, because of the varying charge types of the successive chloride complexes, the formation constants evaluated in sodium chloride–sodium perchlorate mixtures cannot be expected to be valid in solutions alike in ionic strength but widely different in composition. Nor is it valid, as has sometimes been suggested, to evaluate equilibrium constants only at ionic strengths of 1 and 2 M and extrapolate the values thus obtained to zero ionic strength.

2-6 SALT EFFECTS ON EQUILIBRIUM CONSTANTS

The equilibrium-constant expression Equation (2-7), written in terms of activities of the reactants, shows no salt effect. In practice, however, an equilibrium constant usually is determined experimentally in terms of concentrations. Thus we can write

$$K' = \frac{[P]^p[R]^r}{[M]^m[N]^n} \qquad (2\text{-}24)$$

and substituting concentrations times activity coefficients in (2-7),

$$K = \frac{[P]^p[R]^r}{[M]^m[N]^n} \frac{\gamma_P^{\,p}\gamma_R^{\,r}}{\gamma_M^{\,m}\gamma_N^{\,n}} = K' \frac{\gamma_P^{\,p}\gamma_R^{\,r}}{\gamma_M^{\,m}\gamma_N^{\,n}} \qquad (2\text{-}25)$$

We distinguish here between the simple K, written in terms of activities (sometimes called the thermodynamic equilibrium constant), and K', written in terms of concentrations. Since for ionic solutes the γ's in the equation approach unity at infinite dilution, the usual practice is to extrapolate the experimental values of K' at various ionic strengths to obtain K, which is the limiting value at infinite dilution.

An estimate of the magnitude of the salt effect can be obtained from the Debye-Hückel theory. Defining

$$pK = -\log K \qquad (2\text{-}26)$$

and

$$pK' = -\log K' \qquad (2\text{-}27)$$

we may rewrite (2-25)

$$pK' = pK + p \log \gamma_P + r \log \gamma_R - m \log \gamma_M - n \log \gamma_N \qquad (2\text{-}28)$$

and the values of the activity coefficients can be estimated from the DHLL or EDHE.

EXAMPLE 2-2   Estimate the effect of ionic strength on the two successive ionization constants of a dibasic acid composed of neutral molecules H_2A.

ANSWER   For the first ionization

$$H_2A \rightleftharpoons H^+ + HA^-$$

$$K_1 = \frac{a_{H^+}a_{HA^-}}{a_{H_2A}} = K_1' \frac{\gamma_{H^+}\gamma_{HA^-}}{\gamma_{H_2A}}$$

$$pK_1' = pK_1 + \log \gamma_{H^+} + \log \gamma_{HA^-} - \log \gamma_{H_2A}$$

From the DHLL

$$\log \gamma_{H^+} = \log \gamma_{HA^-} = -0.5\sqrt{\mu}$$

$$\log \gamma_{H_2A} = 0$$

and

$$pK_1' = pK_1 - \sqrt{\mu}$$

Or from the EDHE

$$pK_1' = pK_1 - \frac{\sqrt{\mu}}{1 + \sqrt{\mu}}$$

For the second ionization

$$HA^- \rightleftharpoons H^+ + A^=$$

$$pK_2' = pK_2 + \log \gamma_{H^+} + \log \gamma_{A^=} - \log \gamma_{HA^-}$$

From the DHLL

$$pK_2' = pK_2 - 2\sqrt{\mu}$$

Or from the EDHE

$$pK_2' = pK_2 - \frac{2\sqrt{\mu}}{1 + \sqrt{\mu}} \qquad ////$$

2-7 NONELECTROLYTES AND ACTIVITY

Going beyond solutions of electrolytes in water, several other possibilities need consideration: electrolytes in nonaqueous solvents, nonelectrolytic behavior in solutions, and nonelectrolytes in nonaqueous solvents. None of the theories proposed for the quantitative prediction of solution behavior has been as successful as that of Debye and Hückel for dilute ionic aqueous solutions. Nevertheless, general trends can be predicted.

Electrolytes in nonaqueous solvents The most significant work for analytical chemists in this area has been concerned with acids and bases; discussion of this topic is reserved for Chapter 4.

Nonelectrolytic behavior in solutions containing electrolytes A fundamental assumption of the Debye-Hückel theory is that molecular solutes have activity coefficients of unity at all ionic strengths when the standard state is defined in terms of henryan reference behavior. Experimentally, the deviations of activity coefficients from unity are much smaller for molecular than for ionic solutes. Qualitatively, however, when addition of electrolyte to an aqueous solution has a *salting-out effect*, that is, the solubility of a solute is lowered, the activity coefficient must increase with increasing ionic strength. Suppose that the solubility of a nonelectrolyte in water S_w is greater than its solubility in a salt solution S_s. By definition both saturated solutions exist in equilibrium with the pure solid; hence the activities of the nonelectrolyte are equal in the two solutions. Thus $a_w = \gamma_w S_w = a_s = \gamma_s S_s$, leading to $\gamma_s > \gamma_w$. This situation is much more commonly encountered than the reverse (*salting in*),[6] in which $\gamma_w > \gamma_s$.

Other than specific effects that result from conventional chemical interactions (such as acid-base or complex formation), the main factors to be considered are hydration of ions, electrostatic effects, and change in dielectric constant of the solvent. For example, hydration of ions of added salt effectively removes some of the free solvent, so that less is available for solution of the nonelectrolyte. The Setschenow[25] equation probably best represents the activity coefficient of dilute solutions (less than 0.1 M) of nonelectrolytes in aqueous solutions of salts up to relatively high concentrations (about 5 M):

$$\log \gamma_0 = k\mu \qquad (2\text{-}29)$$

According to (2-29) the logarithm of the activity coefficient should be proportional to the ionic strength. The proportionality constant k is positive if the solute has a lower dielectric constant than the solvent, as is usually the case for aqueous solutions. As a first approximation an equation of this form is valid for many solutes up to ionic strengths of the order of unity.[26] The quantity k, called the *salting coefficient*, depends on the nature of the solute and of the electrolyte and usually has a value of 0.01 to 0.10.

From a practical viewpoint we may conclude that molecular solutes have activity coefficients near unity up to an ionic strength of 0.1 and that deviations are moderate even at ionic strengths of the order of unity. In contrast to those of ionic solutes, activity coefficients of molecular solutes usually are slightly greater than unity.

Nonelectrolytes in nonaqueous solvents   Activity coefficients of dilute solutions of solutes can be studied experimentally by liquid-liquid chromatography[27] as well as techniques such as solvent extraction, light scattering, vapor pressure, and freezing point depression.

Significant developments from the point of view of analysis have taken place in gas chromatography, which has made feasible the study of the thermodynamics of a wide variety of solutions involving volatile solutes. On the basis of several assumptions, such as that the vapor pressure of a volatile solute in dilute solution is proportional to its mole fraction (Raoult's law) and that adsorption is insignificant at the several phase boundaries, expressions relating the activity coefficient to measurable gas-chromatographic quantities have been developed by Keulemans[28] and Porter, Deal, and Stross.[29]

For a solution of a volatile solute the partial vapor pressure P_i of the solute in equilibrium with the solution is

$$P_i = \gamma_i X_i P_i^0 \qquad (2\text{-}30)$$

where γ_i is activity coefficient, X_i is mole fraction of solute in solution, and P_i^0 is vapor pressure of pure solute. With γ_i equal to unity, Equation (2-30) is simply a statement of Raoult's law. It is important to recognize that the activity coefficient is thus defined with reference to pure solute, with activity and concentration approaching each other as mole fraction approaches unity (Figure 2-1). If there are no interactions in the gas phase,

$$P_i = M_{ig}RT \qquad (2\text{-}31)$$

where M_{ig} is the concentration of solute in the gas phase in moles per liter, R is the gas constant, and T is temperature. For dilute solutions the mole fraction X_i, that is, $N_i/(N_i + N_s)$, reaches a limiting value X_i^0 equal to the mole ratio N_i/N_s, where N_i is moles of solute and N_s is moles of solvent. The quantity N_s, in turn, is equal to the ratio of the volume of solvent to its molar volume. For a dilute solution

$$X_i^0 = \frac{N_i}{N_s} = \frac{N_i V_{\text{molar}}}{V_s} = M_{is} V_{\text{molar}} \qquad (2\text{-}32)$$

where V_{molar} is molar volume and M_{is} is concentration of solute in the solution in moles per liter. Substituting (2-31) and (2-32) in (2-30) and rearranging, we obtain

$$\frac{M_{is}}{M_{ig}} = \frac{RT}{\gamma_i^0 V_{\text{molar}} P_i^0} \qquad (2\text{-}33)$$

where γ_i^0 corresponds to a solution with mole fraction X_i^0. The quantity γ_i^0 is called the *infinite-dilution activity coefficient*. [It is analogous to the term *transfer activity coefficient* used in nonaqueous chemistry (Section 4-1).]

The ratio of concentration of solute in the solution to that in the gas phase, M_{is}/M_{ig}, that is, $RT/\gamma_i^0 V_{molar}P_i^0$, is a distribution constant K_D analogous to that in Equation (23-1). Hence

$$\gamma_i^0 = \frac{RT}{K_D V_{molar}P_i^0} \qquad (2\text{-}34)$$

The quantity K_D is related to measurable quantities through the expression

$$K_D = \frac{V_i - V_m}{V_s} \qquad (2\text{-}35)$$

where V_i is the retention volume (Section 24-1) of component i, V_m the volume occupied by the mobile gas phase in the column, and V_s the volume occupied by non-volatile solvent in the column. Combining (2-34) and (2-35), we obtain

$$\gamma_i^0 = \frac{RTV_s}{(V_i - V_m)V_{molar}P_i^0} = \frac{RTw_s}{(V_i - V_m)P_i^0 M_s} \qquad (2\text{-}36)$$

where w_s is the weight of solvent in the chromatographic column and M_s its molecular weight. If $V_i - V_m$ is expressed as net retention volume *per gram* of solvent, $\overline{V}_i - \overline{V}_m$, we obtain

$$\gamma_i^0 = \frac{RT}{(\overline{V}_i - \overline{V}_m)P_i^0 M_s} = \frac{6.24 \times 10^4 T}{(\overline{V}_i - \overline{V}_m)P_i^0 M_s} \qquad (2\text{-}37)$$

Here the net retention volume per gram of solvent is expressed at temperature T and at a pressure expressed in millimeters of mercury. If net retention volume is calculated at 0°C, T then can be set equal to 273, and (2-37) becomes

$$\gamma_i^0 = \frac{1.704 \times 10^7}{(\overline{V}_i - \overline{V}_m)P_i^0 M_s} \qquad (2\text{-}38)$$

Figure 2-5 shows infinite-dilution activity coefficients for alkanes as measured for solvents ranging in polarity from water to *n*-heptane.[30] As can be seen, activity coefficients can have truly enormous values. This figure illustrates the striking advantage of being able to use a single standard state dependent on only the properties of the pure solute, in that activities then become directly relatable to one another in the different solvents. Thus, decane in a dilute solution with triethylene glycol would have an activity about 100 times that of the same concentration of decane in *n*-heptane as solvent. When use of a single standard state is not possible, two solutions of a solute in different solvents may be assigned the same value of activity coefficient and behave in a sharply different manner (Section 4-1). If the conventional henryan standard state for ionic solutes were adopted, all activity coefficients at infinite dilution would be unity by definition.

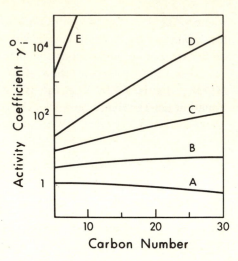

FIGURE 2-5   Infinite-dilution activity coefficients $\gamma_i{}^0$ as a function of carbon number for n-alkanes. A, n-heptane at 90°C; B, 2-butanone at 90°C; C, phenol at 90°C; D, triethylene glycol at 90°C; E, water at 25°C. (*From Pierotti, Deal, Derr, and Porter.*[30])

EXAMPLE 2-3   Suppose the corrected net retention volume at 90°C for a limiting small sample of dodecane is 74.1 ml/g of triethylene glycol at stationary phase. Similarly, the net retention volume for a sample of moderate size is 78.4 ml/g.

(*a*)   Calculate the value of $\gamma_i{}^0$ in the first case and an apparent value of γ_i in the second case, assuming $X_i \simeq N_i/N_s$. Take the vapor pressure of dodecane at 90°C to be 10.0 mm.

ANSWER

$$\gamma^0_{\text{dodecane}} = \frac{6.24 \times 10^4 \times T}{(\overline{V}_{\text{dodecane}} - \overline{V}_m)\, P^0_{\text{dodecane}}\, M_{\text{triethylene glycol}}}$$

$$= \frac{6.24 \times 10^4 \times 363}{74.1 \times 10.0 \times 150.2} = 204$$

Similarly, γ_{dodecane} for moderate sample size is 192.

(*b*)   Suppose that, instead of a standard state of fugacity that is equal to unity, the standard state is based on activities and concentrations being equal in solution at infinite dilution. What would be the values of the activity coefficients in each of the two cases?

ANSWER   For the limiting small sample the activity coefficient would be 1.0 by definition. For a sample of moderate size, it would be $\frac{192}{204} = 0.94$. For an extremely large sample, in this case it would approach a small value as a limit (about 0.005). ////

The deviation of the activity coefficient γ_i from unity results from two factors—one arising from differences in the sizes of the solute and solvent molecules and the other from the interaction of molecules with each other. Ashworth and Everett[31] expressed the relation by

$$\ln \gamma_i = \ln \gamma_i^s + \ln \gamma_i^e \qquad (2\text{-}39)$$

where γ_i^s denotes a factor arising from size differences and γ_i^e one arising from the energy of interaction. The first of these terms can be considered an entropy term and the second an energy term. When molecules of different sizes are mixed, the resulting system has increased entropy; this effect would be expected to lead to lowering of the activity coefficient. The contribution of the size factor[32,33] has been expressed by[34]

$$\ln \gamma_i^s = \ln \frac{1}{r} + \left(1 - \frac{1}{r}\right) \qquad (2\text{-}40)$$

where r is the ratio of partial molar volumes of solvent and solute. This term has been estimated[34] for several solutes in hydrocarbons of high molecular weight, and negative values in the region of a few tenths to about 1 unit in $\ln \gamma_i$ have been obtained. Equation (2-40) indicates that the size effect is small for solute and solvent molecules of the same size.

The molecular-interaction parameter can give rise to either positive or negative contributions to the term $\ln \gamma_i$. If the solvent tends to be self-associated, then the addition of a nonpolar solute disrupts solvent structure. Figure 2-5 indicates that n-decane in water has an activity coefficient of about 100,000. This high activity coefficient in solution is typical of the weak interaction of solute and solvent. On the other hand, where the molecular interaction between solute and solvent is strong, the activity coefficient would be expected to be low, and solute-solvent interactions and association constants may be measurable.[35]

REFERENCES

1   J. W. GIBBS in "Collected Works of J. Willard Gibbs," vol. 1, p. 93, Yale University Press, New Haven, 1928.

2   G. N. LEWIS: *Proc. Amer. Acad. Arts Sci.*, **37**:49 (1901); **43**:259 (1907).

3   E. A. GUGGENHEIM: *J. Phys. Chem.*, **33**:842 (1929).

4   P. DEBYE and E. HÜCKEL: *Phys. Z.*, **24**:185 (1923).

5   G. N. LEWIS and M. RANDALL: *J. Amer. Chem. Soc.*, **43**:1112 (1921).

6   H. S. HARNED and B. B. OWEN: "The Physical Chemistry of Electrolytic Solutions," 3d ed., app. A, Reinhold, New York, 1958.

7   E. A. GUGGENHEIM and T. D. SCHINDLER: *J. Phys. Chem.*, **38**:533 (1934).

8   J. KIELLAND: *J. Amer. Chem. Soc.*, **59**:1675 (1937).

9   M. PAABO and R. G. BATES: *J. Phys. Chem.*, **74**:706 (1970); **73**:3014 (1969).

10   T. H. GRONWALL, V. K. LA MER, and K. SANDVED: *Phys. Z.*, **29**:358 (1928); V. K. LA MER, T. H. GRONWALL, and L. J. GRIEFF: *J. Phys. Chem.*, **35**:2245 (1931).

11 E. HÜCKEL: *Phys. Z.*, **26**:93 (1925).

12 R. A. ROBINSON and H. S. HARNED: *Chem Rev.*, **28**:419 (1941); R. H. STOKES and R. A. ROBINSON: *J. Amer. Chem. Soc.*, **70**:1870 (1948).

13 E. GLUECKAUF: *Trans. Faraday Soc.*, **51**:1235 (1955).

14 D. G. MILLER: *J. Phys. Chem.*, **60**:1296 (1956).

15 R. M. DIAMOND: *J. Amer. Chem. Soc.*, **80**:4808 (1958).

16 K. S. PITZER: *J. Phys. Chem.*, **77**:268 (1973).

17 N. BJERRUM: *Kgl. Dan. Vidensk. Selsk. Mat.-Fys. Medd.*, (9)**7**:1 (1926).

18 R. M. FUOSS and C. A. KRAUS: *J. Amer. Chem. Soc.*, **55**:1019, 2387 (1933); **57**:1 (1935); R. M. FUOSS: **56**:1027 (1934).

19 J. N. BRØNSTED and V. K. LA MER: *J. Amer. Chem. Soc.*, **46**:555 (1924).

20 C. H. BRUBAKER, JR.: *J. Amer. Chem. Soc.*, **78**:5762 (1956).

21 H. S. HARNED and C. M. MASON: *J. Amer. Chem. Soc.*, **54**:1439 (1932).

22 E. C. RIGHELLATO and C. W. DAVIES: *Trans. Faraday Soc.*, **26**:592 (1930); C. W. DAVIES: *J. Chem. Soc.*, 2093 (1938).

23 J. R. PARTINGTON and H. J. STONEHILL: *Phil. Mag.*, (7)**22**:857 (1936).

24 W. J. ARGERSINGER, JR., and D. M. MOHILNER: *J. Phys. Chem.*, **61**:99 (1957).

25 J. SETSCHENOW: *Z. Phys. Chem.* (Leipzig), **4**:117 (1889). See also P. Debye and J. McAulay, *Phys. Z.*, **26**:22 (1925).

26 HARNED and OWEN,[6] p. 531.

27 D. C. LOCKE, in "Advances in Chromatography," J. C. Giddings and R. A. Keller (eds.), vol. 8, p. 47, Dekker, New York, 1969.

28 A. I. M. KEULEMANS: "Gas Chromatography," 2d ed., p. 170, Reinhold, New York, 1959.

29 P. E. PORTER, C. H. DEAL, and F. H. STROSS: *J. Amer. Chem. Soc.*, **78**:2999 (1956).

30 G. J. PIEROTTI, C. H. DEAL, E. L. DERR, and P. E. PORTER: *J. Amer. Chem. Soc.*, **78**:2989 (1956).

31 A. J. ASHWORTH and D. H. EVERETT: *Trans. Faraday Soc.,* **56**:1609 (1960).

32 P. J. FLORY: *J. Chem. Phys.*, **10**:51 (1942).

33 H. C. LONGUET-HIGGINS: *Discuss. Faraday Soc.*, **15**:73 (1953).

34 Y. B. TEWARI, D. E. MARTIRE, and J. P. SHERIDAN: *J. Phys. Chem.*, **74**:2345 (1970).

35 H. LIAO, D. E. MARTIRE, and J. P. SHERIDAN: *Anal. Chem.*, **45**:2087 (1973); E. F. MEYER, K. S. STEC, and R. D. HOTZ: *J. Phys. Chem.*, **77**:2140 (1973).

36 G. N. LEWIS and M. RANDALL: "Thermodynamics," 2d ed., p. 338, revised by K. S. Pitzer and L. Brewer, McGraw-Hill, New York, 1961.

PROBLEMS

2-1 The solubility of AgCl in water is 1.278×10^{-5} M. Using the DHLL, calculate the following: (*a*) the solubility product; (*b*) the solubility in 0.01 m and 0.03 m KNO_3; (*c*) the solubility in 0.01 m K_2SO_4.

   Answer (*a*) 1.620 (not 1.63) $\times 10^{-10}$. (*b*) 1.43×10^{-5}; 1.55×10^{-5}. (*c*) 1.55×10^{-5}.

2-2 The solubility product written in terms of concentrations of $CaSO_4$ at 10°C is 6.4×10^{-5}. (a) Using the DHLL and EDHE, estimate the solubility product, written in terms of activities. (b) Using the EDHE, calculate by successive approximations the solubility in 0.01 M $MgCl_2$ and in 0.01 M $MgSO_4$.

> Answer   (a) 1.2×10^{-5}; 1.6×10^{-5}. (b) In $MgCl_2$, $\mu = 0.07$ and $S = 1.05 \times 10^{-2}$; in $MgSO_4$, $\mu = 0.065$ and $S = 6.2 \times 10^{-3}$.

2-3 Using the data given in Table 2-2, estimate the solubility product (in terms of activities) of $BaMoO_4$ at an ionic strength of 0.1. The solubility is 58 mg/100 ml. Assume the ion-size parameter for molybdate to be the same as for chromate.

> Answer 5.1×10^{-7}.

2-4 Using the EDHE, write expressions for the equilibrium constants of the following reactions as a function of the ionic strength: (a) $Ca_3(PO_4)_2$ (solid) $\rightleftharpoons$ $3Ca^{++} + 2PO_4^{3-}$; (b) $HPO_4^{=} \rightleftharpoons H^+ + PO_4^{3-}$; (c) $Fe(C_2O_4)_3^{3-} \rightleftharpoons Fe(C_2O_4)_2^{-} + C_2O_4^{=}$; (d) $I_3^{-} + 2S_2O_3^{=} \rightleftharpoons 3I^{-} + S_4O_6^{=}$.

> Answer $pK' = pK - [\sqrt{\mu}/(1 + \sqrt{\mu})]y$, where $y = $ (a) 15; (b) 3; (c) -2; (d) -1.

2-5 At 20°C the vapor pressure of water with a H_2SO_4 solution of density 1.3947 containing 49.52% H_2SO_4 is 6.6 mm. Pure water at the same temperature has a vapor pressure of 17.4 mm. Calculate the activity coefficient of water in the H_2SO_4 solution.

> Answer   0.45.

2-6 Suppose the corrected net retention volume at 25°C for a limiting small sample of a solute is 10.0 ml/g of stationary phase. Similarly, the net retention volume for a sample of moderate size is 11.0 ml/g. (a) What are γ_i^{0} and γ_i (assume $X_i \simeq N_i/N_s$) in the two samples if the molecular weight of the solvent is 100 and the vapor pressure of the solute P_i^{0} at column operating temperature is 2.0 mm? (b) What would be the values of the activity coefficients in these two cases if the standard state were based on activities and concentrations being equal in solution at infinite dilution?

> Answer   (a) $\gamma_i^{0} = 9300$ and $\gamma_i = 8460$; (b) $\gamma_i = 1.00$ and 0.91.

2-7 Assuming that the DHLL applies to all solvents including those containing no water, use Equation (2-14) to calculate the activity coefficient $\gamma_{\pm}$ of a 1:1 electrolyte at a concentration of 0.01 m and 25°C in water, ethanol, and acetic acid, whose dielectric constants are 78.3, 24.3, and 6.2. From Equations 2-11 and 2-12 the value of A is obtainable,[36] from

$$A^2 = \frac{2\pi N}{1000 \times (2.303)^2} \left(\frac{e^2}{DkT} \right)^3$$

> Answer   Activity coefficients in water, ethanol, and acetic acid are 0.89, 0.505, and 0.005.

2-8 Assume that the DHLL applies to dilute solutions of $1:1$ electrolytes in dioxane ($D = 2.21$). What would be the maximum permissible ionic strength of such an electrolyte if $\gamma_{\pm}$ is not to fall below 0.9 at 25°C?

Answer 2×10^{-7}.

2-9 Using the DHLL, derive an expression for the solubility product of a precipitate A_mB_n, composed of ions of charge $+Z_A$ and $-Z_B$ units, as a function of ionic strength.

Answer $pK'_{sp} = pK_{sp} - 0.5\,(mZ_A^2 + nZ_B^2)\sqrt{\mu}$.

2-10 Show that for a salt A_mB_n the quantity $(mZ_A^2 + nZ_B^2)/(m + n)$ appearing in Equation (2-17) is equal in absolute magnitude to Z_AZ_B. *Hint:* Using the electroneutrality condition $mZ_A = nZ_B$, express the numerator and denominator to contain the factor $Z_A + Z_B$, which can be canceled.

2-11 If the value of the equilibrium constant for the reaction $H_2O + D_2O \rightleftharpoons 2\,HOD$ is 3.3 at 25°C, calculate the weight percent of H_2O, HOD, and D_2O: (a) in natural water, if natural hydrogen contains 99.985% H and 0.015% D by weight; (b) in reactor-grade heavy water with a purity of 99.75 mole % in terms of deuterium.

Answer (a) 99.98%, 0.016%, 7.6×10^{-7}%; (b) 6.8×10^{-4}%, 0.48%, 99.52%.

3

ACID-BASE EQUILIBRIA IN WATER

Acids and bases were defined and described by early chemists, including Boyle, Lavoisier, Davy, Berzelius, Liebig, and Arrhenius. At the present time, depending on objectives, one of two definitions of acids and bases is likely to be accepted. These two definitions, by Brønsted and Lowry[1,2] and by Lewis,[3] were proposed about the same time. According to the Brønsted definition acids are substances having a tendency to lose a proton, and bases are those having a tendency to accept a proton. Thus, for an acid HA the *acid-base half-reaction* is

$$HA \rightleftharpoons H^+ + A^- \qquad (3\text{-}1)$$

where HA and A^- constitute a *conjugate pair* and A^- is the *conjugate base* of the acid. Acids therefore include substances such as the ammonium cation as well as those more conventionally regarded as acids, such as hydrogen chloride. The Lewis definition broadly defines an acid as a substance that can accept an unshared pair of electrons. In analytical work this definition is often awkward; silver ion, for instance, is defined as an acid (electron acceptor), whereas only with difficulty can hydrogen chloride be included. We do not further consider acids and bases in the Lewis sense or the extension of the definition to hard and soft acids and bases.[4]

3-1   METHODS OF DEALING WITH EQUILIBRIA

One method of obtaining equilibrium concentrations of the various species involves no simplifying assumptions. In this method as many independent equations relating the unknown quantities are obtained as there are unknowns, in which a mathematical statement for each concentration and a mathematical equation for each equilibrium reaction can be written. These are then solved. For complex systems the calculations may be cumbersome. Sillen[5] developed a computer program HALTAFALL (available in FORTRAN) by which equilibrium concentrations of species can be calculated without simplifying assumptions. Precise values of equilibrium constants and activity coefficients are required for best results.

In another method the behavior of equilibrium systems can often be visualized and clarified through one of several graphical approaches.[6-9] Such approaches enable one to tell quickly which species are significant and which are not (for example, see Figures 7-1, 11-2, and 11-8). For simple systems, and for experimentalists who have well-developed chemical intuition, such diagrams may be unnecessary; for complicated systems or for those who have not developed the intuition, graphical presentations allow rapid evaluation of certain chemical systems. One type of graphical presentation increasingly being used involves so-called logarithmic diagrams with a master variable. They consist mostly of a series of straight lines with slopes of 0, $+1$, or -1. In acid-base equilibria the master variable is usually pH; in redox equilibria, it is usually an electrochemical potential. Generally some simplifying assumptions are made, such as that the total concentration and the activity coefficients are constant with change in the master variable. In acid-base equilibria a *system point* is defined as the point whose abscissa value is pK_a and ordinate is $\log C_0$, where C_0 is initial concentration. Figure 3-1 illustrates a logarithmic diagram for aqueous 10^{-2} M acetic acid ($pK_a = 4.7$). Such a diagram is constructed[7] by first selecting two axes, one for pH from 0 to pK_w and the other for $\log C$ corresponding to a change of several orders of magnitude of C. Lines of slope ± 1 are drawn for hydrogen ion and hydroxyl ion corresponding to $\log [OH^-] = pH - pK_w$. Next, the system point at $pH = 4.7$ and $\log C_0 = -2$ is located. A line of slope 0 and downward lines of $+1$ and -1 are drawn with origins at the system point. Finally, a point 0.3 unit below the system point is located, and the linear parts of each species are joined with short curves passing through this point. The value 0.3 (or log 2) arises from the fact that at $pH = pK_a$ the value of $\log [CH_3COOH] = \log [CH_3COO^-] = \log (C_0/2) = \log C_0 - \log 2$.

A third approach, the one we emphasize, involves the use of judgment or chemical intuition to introduce simplifying approximations at early stages in the equilibrium calculations. Our aim is to make the operating principles clear and un-ambiguous without introducing significant error. Thus, equilibrium calculations are not treated simply as mechanical operations. In most practical cases a simple equation

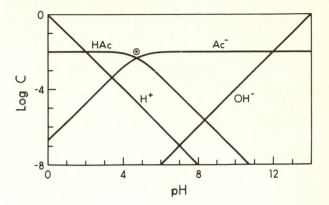

FIGURE 3-1 Logarithmic diagram for 10^{-2} M acetic acid ($pK_a = 4.7$). The system point is denoted by $\odot$.

gives a result as precise as is justified by either the reliability of the equilibrium constant or its application to the system in question including control of the temperature. Butler[10] summarized information on the precision and accuracy of equilibrium data and concluded that, while the ion product for water is known to a precision of about 1 ppt, most other constants have a relative error in the range of 1 to 10%. Therefore, the calculation of equilibrium concentrations to within a range of a few percent is all that normally is warranted.

3-2 EQUILIBRIUM IN PURE WATER

The autoionization of water can be represented by

$$2H_2O \rightleftharpoons H_3O^+ + OH^- \qquad (3-2)$$

or conventionally by

$$H_2O \rightleftharpoons H^+ + OH^- \qquad (3-3)$$

with the understanding that the symbol H^+ represents the *hydrated proton*, which is written in the most general way as $(H_2O)_nH^+$. From (3-2) the ionization constant K_i of water in terms of activities is

$$K_i = \frac{a_{H_3O^+}a_{OH^-}}{a_{H_2O}^2} \qquad (3-4)$$

The usual standard state for water as solvent (Section 2-2) is defined as pure water having unit activity. Hence the equilibrium constant is generally written as the ion activity product of water:

$$K_w = a_{H^+}a_{OH^-} \qquad (3-5)$$

The activity of water when solutes are present is less than that of pure water; but for all practical purposes the activity of water in dilute aqueous solutions may still be regarded as unity, because even in solutions of electrolytes as concentrated as 1 M, the vapor pressure is diminished by only 2 to 4%.[11]

3-3   DEFINITION OF THE pH SCALE

The negative logarithm of hydrogen ion *concentration* was defined by Sørensen[12] as the pH.† Sørensen did not actually measure hydrogen ion concentrations, but something more nearly related to activities. He measured the emf of galvanic cells such as

$$H_2, Pt \left| buffer\ X \left| \frac{KCl}{salt\ bridge} \right| 0.1\ M\ calomel\ electrode \right. \qquad (3\text{-}6)$$

and attempted to eliminate the liquid-junction, or diffusion, potentials by an extrapolation procedure developed by Bjerrum.[14] At the time of Sørensen's original work the difference in cell emf of two such cells with different hydrogen ion concentrations was assumed to be a measure of the ratio of hydrogen ion concentrations in the solutions, according to the Nernst equation:

$$E_1 - E_2 = \frac{RT}{F} \ln \frac{[H^+]_2}{[H^+]_1} \qquad (3\text{-}7)$$

To establish a pH scale, Sørensen chose a dilute hydrochloric acid solution for a standard. He took the concentration of hydrogen ions in such a solution to be given by αC, where C is the concentration of hydrochloric acid and α is a *degree of dissociation* determined from conductance measurements. His procedure had drawbacks: first, there is evidence that the extrapolation procedure does not actually reduce the liquid-junction potential of Cell (3-6) to zero; second, the hydrochloric acid is completely dissociated (dissociation constant about 1.6×10^6),[15] and therefore the concentration of H^+ is C rather than a somewhat smaller quantity.

The Sørensen pH unit, often designated psH, is neither the negative logarithm of concentration pc_H nor the activity of hydrogen ion pa_H. It does, however, closely resemble modern "operational" pH scales.

When the concept of thermodynamic activity became established, Sørensen and Linderstrøm-Lang[16] defined the term pa_H as the negative logarithm of hydrogen ion activity. Thus

$$pa_H = -\log a_{H^+} = -\log [H^+]\gamma_{H^+} \qquad (3\text{-}8)$$

† By international agreement[13] the quantity pH is written on line in roman type.

Unfortunately, since neither liquid-junction potentials nor individual ion activity coefficients can be evaluated rigorously, the pa_H of this definition cannot be accurately related to experimental quantities. Several attempts have been made to determine pH scales that are not subject to the theoretical limitations of psH or pa_H and yet are capable of experimental measurement; these have been described elsewhere.[17]

From the viewpoint of analytical measurements, a practical scale of acidity should permit interpretation of the most important experimental measurements, namely, those made with the glass electrode (Section 13-2) and the saturated calomel electrode, represented by the cell

$$\begin{array}{c|c|c} \text{glass} & \text{sample} & \text{saturated} \\ \text{electrode} & \text{solution} & \begin{array}{c}\text{calomel}\\ \text{electrode}\end{array} \end{array} \qquad (3\text{-}9)$$

This cell closely resembles the original Sørensen cell, with a glass electrode substituted for the hydrogen electrode. Since glass electrodes are subject to an unpredictable *asymmetry potential*, pH measurements are made in practice by substituting a standard buffer for the sample solution and then comparing the pH of the sample with the pH of the standard, pH_s, according to the equation

$$pH = pH_s + \frac{(E - E_s)F}{2.303RT} \qquad (3\text{-}10)$$

Similar calibrations can also be performed conveniently when any pH electrode, such as the hydrogen or the quinhydrone electrode, is substituted for the glass electrode in Cell (3-9).

Equation (3-10) is the present operational definition of pH and is recommended by the International Union of Pure and Applied Chemistry.[18,19] The value of pH_s is determined once and for all by careful measurements of cells such as

$$H_2, \text{Pt} \mid \text{solution S, AgCl} \mid \text{Ag}$$

without liquid junction, with reasonable assumptions about activity coefficients so that pH_s represents as nearly as possible $-\log a_{H^+} = -\log [H^+]\gamma_\pm$, where $\gamma_\pm$ is the mean activity coefficient of a typical 1:1 electrolyte in the standard buffer.

A pH scale could be defined in terms of a single standard buffer.[18] Every practical pH reading, however, involves a liquid-junction potential, and the variation in this potential between the readings of Cell (3-9) for the unknown and the standard is tacitly included in Equation (3-10). The liquid-junction potential is essentially constant for solutions of intermediate pH values, say 3 to 11, but beyond these limits it varies considerably because the concentrations of the unusually mobile H^+ or OH^- ions become appreciable. Although this variation does not affect the definition of pH by (3-10), the measured value of pH deviates appreciably from the known best values of $-\log a_{H^+}$ for solutions of high acidity or alkalinity. For this reason, a series of standards covering a range of values of pH_s has been adopted. For an experimental determination it is recommended that the pH standard chosen be the one nearest the sample. By use of two standard buffers, preferably one on each side of the sample, the response of the pH indicator electrode can be standardized, and comparative results of high reliability can be attained.

A different set of pH_s values must be adopted at each temperature; in effect, there is a different standard state for hydrogen ions, and therefore a different pH scale, for each temperature. Selected values of pH_s for several standards recommended by the U.S. National Bureau of Standards[20] are shown in Table 3-1.

The uncertainty in the pH_s values for the seven primary standards in Table 3-1 (not including tetroxalate or calcium hydroxide) is judged[21] to be about ± 0.006 pH unit at 25°C. At pH values much less than 3 or more than 11, or at temperatures other than 25°C, the uncertainties become greater. As to the significance of measured pH values, it may be concluded that, at best, pH may be regarded as an estimate of $-\log a_{H+}$ or $-\log [H^+]\gamma_{H+}$. The validity of the estimate depends on how constant the liquid-junction potential remains during measurement of standard and unknown

Table 3-1 pH_s VALUES OF NBS STANDARDS†

Temperature, °C	2.303 RT/F	0.05 m KH₃(C₂O₄)₂·2H₂O (tetroxalate)‡	KHC₄H₄O₆ (tartrate) (saturated, 25°C)	0.05 m KH₂C₆H₅O₇ (citrate)	0.05 m KHC₈H₄O₄ (phthalate)	0.025 m KH₂PO₄, 0.025 m Na₂HPO₄	0.008695 m KH₂PO₄, 0.03043 m Na₂HPO₄	0.01 m Na₂B₄O₇·10H₂O (borax)	0.025 m NaHCO₃, 0.025 m Na₂CO₃	Ca(OH)₂‡ (saturated, 25°C)
0	0.05420	1.666	...	3.863	4.003	6.984	7.534	9.464	10.317	13.423
15	0.05717	1.672	...	3.802	3.999	6.900	7.448	9.276	10.118	12.810
20	0.05816	1.675	...	3.788	4.002	6.881	7.429	9.225	10.062	12.627
25	0.05916	1.679	3.557	3.776	4.008	6.865	7.413	9.180	10.012	12.454
30	0.06015	1.683	3.552	3.766	4.015	6.853	7.400	9.139	9.966	12.289
40	0.06213	1.694	3.547	3.753	4.035	6.838	7.380	9.068	9.889	11.984
50	0.06411	1.707	3.549	3.749	4.060	6.833	7.367	9.011	9.828	11.705
60	0.06610	1.723	3.560	...	4.091	6.836	...	8.962	...	11.449
90	0.07205	1.792	3.650	...	4.205	6.877	...	8.850	...	...

† From Bates.[20] See also B. R. Staples and R. G. Bates, *J. Res. Nat. Bur. Stand.*, **73A**:37 (1969). Note that the concentrations are given in terms of molality *m* rather than molarity *M*. See "Handbook of Chemistry and Physics," 53d ed., p. D-105, Chemical Rubber, Cleveland, Ohio, 1972–1973, for additional details of preparation.

‡ Secondary standard.

and on the validity of the assumptions made concerning activity coefficients when the standards are set; hence a_{H^+} cannot be exactly known, but neither is it completely unknown. Probably few pH measurements are more accurate than 0.01 pH unit when all sources of uncertainty are included. *Changes* in pH, however, of 0.001 or less in closely similar solutions can be measured and may be of practical significance (in blood plasma, for example). Absolute values of pH are not possible because single-ion activities cannot be obtained. Interpretations of pH values are soundest when associated with dilute aqueous solutions of simple solutes; they become unsound when associated with nonaqueous solvents, colloids, or solutions of high ionic strength. For most applications the pH value may be regarded as a practical and comparative measure of acidity, and no attempt need be made to interpret it rigorously in terms of single-ion activity.

In many calculations the hydrogen ion concentration is more accessible than the activity. For example, the electroneutrality condition is written in terms of concentrations rather than activities. Also, from stoichiometric considerations, the concentrations of solution components are often directly available. Therefore, the hydrogen ion concentration is most readily calculated from equilibrium constants written in terms of concentration. When a comparison of hydrogen ion concentrations with measured pH values is required (in calculation of equilibrium constants, for example), an estimate of the hydrogen ion activity coefficient can be made by application of the Debye-Hückel theory; if necessary, an estimate of liquid-junction potentials also can be made.[22,23] Alternatively, the glass electrode can be calibrated with solutions of known hydrogen ion concentration and constant ionic strength.[23a, 23b]

3-4 EQUILIBRIA IN A SINGLE ACID-BASE SYSTEM IN WATER[24,25]

The following general equations are developed and then modified to suit particular needs. If a Brønsted acid HA or its conjugate base A^-, or both, are added to water, we have the equilibrium

$$HA + H_2O \rightleftharpoons H_3O^+ + A^- \qquad (3\text{-}11)$$

or conventionally,

$$HA \rightleftharpoons H^+ + A^- \qquad (3\text{-}12)$$

In a more general way, HA can be written HA^n, where the charge n may be an integer with a positive, negative, or zero value. The conjugate base is then A^{n-1}. Similarly, for a base designated B the conjugate acid would be BH^+.

For the ionization of water,

$$H_2O \rightleftharpoons H^+ + OH^- \qquad (3\text{-}13)$$

If C_{HA} and C_{A^-} are the analytical concentrations of HA and A^-, and if [HA] and $[A^-]$ are the equilibrium concentrations, the relations

$$[HA] = C_{HA} - ([H^+] - [OH^-]) \qquad (3-14)$$

and
$$[A^-] = C_{A^-} + ([H^+] - [OH^-]) \qquad (3-15)$$

result because the analytical concentration of HA is diminished by the amount of hydrogen ion produced in Reaction (3-12), which in turn is the total hydrogen ion concentration minus the hydroxyl ion concentration.

The equilibrium constant K_{eq} of (3-12) in terms of concentration is

$$K_{eq} = \frac{[H^+][A^-]}{[HA]} = K_a \qquad (3-16)$$

where K_a is by definition the acid-dissociation constant, or ionization constant, of the acid HA. Solving for hydrogen ion concentration gives

$$[H^+] = K_a \frac{[HA]}{[A^-]} = K_a \frac{C_{HA} - [H^+] + [OH^-]}{C_{A^-} + [H^+] - [OH^-]} \qquad (3-17)$$

This is the general equation that may be used, in conjunction with Equations (3-11) to (3-16), to calculate hydrogen ion concentrations in various situations that involve a single acid-base system in water. To illustrate, we consider several examples.

Solution of strong acid A strong acid, by definition, is one for which (3-11) is complete to the right, so that [HA] is essentially zero. From (3-14)

$$[H^+] = C_{HA} + [OH^-] \qquad (3-18)$$

The total concentration of hydrogen ion is that from the strong acid plus that from the water (equal to $[OH^-]$). Unless the solution is extremely dilute ($C_{HA} < 10^{-6}\ M$), the second term may be neglected. Consider, for example, $C_{HA} = 10^{-6}\ M$. As a first approximation $[H^+] = 10^{-6}\ M$, and therefore $[OH^-] = 10^{-14}/10^{-6} = 10^{-8}\ M$; this represents only a 1% correction on the first approximation.

Solution of strong base In a solution of strong base the equilibrium concentration of base other than hydroxyl ion is zero. Thus, from (3-15)

$$[OH^-] = C_{A^-} + [H^+] \qquad (3-19)$$

an expression analogous to (3-18).

Solution of weak acid If HA but not A^- is added to water, $C_{A^-} = 0$, and since hydrogen ions are formed, $[H^+] \gg [OH^-]$ for solutions that are not highly dilute. Equation (3-17) then becomes

$$[H^+] = K_a \frac{C_{HA} - [H^+]}{[H^+]} \qquad (3-20)$$

which can be written

$$[H^+] = \sqrt{K_a(C_{HA} - [H^+])} \qquad (3\text{-}21)$$

As a first approximation, if the degree of dissociation is small,

$$[H^+] \ll C_{HA}$$

and

$$[H^+] \simeq \sqrt{K_a C_{HA}} \qquad (3\text{-}22)$$

Thus from (3-22) the hydrogen ion concentration of a 10^{-3} M solution of an acid HA of ionization constant 10^{-7} is calculated to be 10^{-5} M. Since $[H^+]$ is only 0.01 C_A (see 3-21), the approximation is valid for this case.

EXAMPLE 3-1   Calculate the hydrogen ion concentration of a 10^{-3} M solution of acid HA, of ionization constant $K_a = 10^{-5}$.

ANSWER   From (3-22), $[H^+] \simeq \sqrt{10^{-5} \times 10^{-3}} = 10^{-4}$. From (3-21), $[H^+]'' = \sqrt{10^{-5}(10^{-3} - 10^{-4})} = 9.5 \times 10^{-5}$. (The second and succeeding approximations are indicated by primes.)

$$[H^+]''' = \sqrt{10^{-5}(10^{-3} - 9.5 \times 10^{-5})} = 9.5 \times 10^{-5}$$

Therefore the second approximation was adequate.   ////

EXAMPLE 3-2   Calculate the hydrogen ion concentration of a 10^{-3} M solution of acid HA, of ionization constant $K_a = 10^{-3}$.

ANSWER   From (3-22), $[H^+] \simeq \sqrt{10^{-3} \times 10^{-3}} = 10^{-3}$. From (3-21) no answer is possible as a first approximation. From (3-20), $[H^+]^2 + K_a[H^+] - K_a C_{HA} = 0$. From quadratic solution, $[H^+] = 6.2 \times 10^{-4}$.   ////

Note that nothing has been stated about the charge of the acid HA, which can be a neutral molecule HA or a cation BH^+ such as the ammonium ion. The acid cannot, however, be an anion HA^-, because in general such an anion is also a weak base and so two acid-base systems in addition to water are involved. For a cationic acid BH^+ the ionization constants of the acid and of the conjugate base are related. Thus the ionization constant of BH^+ as an acid is

$$K_a = \frac{[H^+][B]}{[BH^+]} \qquad (3\text{-}23)$$

For the conjugate base B the ionization constant K_b is

$$K_b = \frac{[BH^+][OH^-]}{[B]} \qquad (3\text{-}24)$$

Multiplying (3-23) by (3-24), we obtain

$$K_a K_b = [H^+][OH^-] = K_w \qquad (3\text{-}25)$$

The ionization constant K_a for the cationic acid is thus K_w/K_b, the classical value for the hydrolysis† constant of the cation (conjugate acid) of a weak base. Thus for a solution containing ammonium ions the acid-base reaction can be written

$$NH_4^+ + H_2O \rightleftharpoons NH_3 + H_3O^+$$

Solution of weak base   If A^- but not HA is added to water, $C_{HA} = 0$, and with hydrogen ions removed from water to form HA, we have as an approximation $[H^+] \ll [OH^-]$. Equation (3-17) becomes

$$[H^+] = K_a \frac{[OH^-]}{C_{A^-} - [OH^-]} = \frac{K_w}{[OH^-]} \qquad (3\text{-}26)$$

Solving for $[OH^-]$, we have

$$[OH^-] = \sqrt{\frac{K_w}{K_a} (C_{A^-} - [OH^-])} \qquad (3\text{-}27)$$

If $[OH^-] \ll C_{A^-}$,

$$[OH^-] \simeq \sqrt{\frac{K_w}{K_a} C_{A^-}} \qquad (3\text{-}28)$$

In (3-27) or (3-28), K_w/K_a is the classical hydrolysis constant if A^- is the anion of a weak acid undergoing the reaction $A^- + H_2O \rightleftharpoons HA + OH^-$.

For a classical molecular weak base, such as ammonia, the reaction $NH_3 + H_2O \rightleftharpoons NH_4^+ + OH^-$ leads to equations identical to (3-27) or (3-28) with K_w/K_a replaced by K_b, the classical ionization constant of a weak base.

EXAMPLE 3-3   Calculate the pH of 0.05 M sodium acetate if $K_a = 2 \times 10^{-5}$.

ANSWER   From (3-28)

$$[OH^-] = \sqrt{(10^{-14}/2 \times 10^{-5}) \times 0.05} = 5.0 \times 10^{-6}$$
$$pOH = 6 - \log 5 = 5.3$$
$$pH = 14 - 5.3 = 8.7$$

Compare these results with (3-27) to verify the validity of the approximation.   ////

Solution of weak acid and its conjugate base   If both HA and A^- of a conjugate acid-base pair are added to water, hydrogen ions may be either added or removed.

---

† Since the classical concept of hydrolysis is not essential in the quantitative treatment of acid-base equilibria, we use the concept but little.

Therefore we have one of two situations, depending on whether the solution at equilibrium is acidic or alkaline.

For an acidic solution, $[H^+] \gg [OH^-]$, and Equation (3-17) becomes

$$[H^+] = K_a \frac{C_{HA} - [H^+]}{C_{A^-} + [H^+]} \qquad (3\text{-}29)$$

This often can be replaced with the approximate expression

$$[H^+] \simeq K_a \frac{C_{HA}}{C_{A^-}} \qquad (3\text{-}30)$$

if C_{HA} and C_{A^-} are both much greater than $[H^+]$. Equation (3-30) is sometimes known as the *Henderson equation*.[26]

For an alkaline solution, $[H^+] \ll [OH^-]$, and (3-17) becomes

$$[H^+] = K_a \frac{C_{HA} + [OH^-]}{C_{A^-} - [OH^-]} \qquad (3\text{-}31)$$

which again is reduced to (3-30) if C_{HA} and C_A are both much greater than $[OH^-]$.

EXAMPLE 3-4   Calculate the hydrogen ion concentration of a solution containing 0.10 M HA and 0.002 M A$^-$ if $K_a = 10^{-5}$.

ANSWER   From (3-30)

$$[H^+] = 10^{-5} \frac{0.10}{0.002} = 5 \times 10^{-4}$$

From (3-29)

$$[H^+]'' = 10^{-5} \frac{0.10}{0.0025} = 4.0 \times 10^{-4}$$

and

$$[H^+]''' = 10^{-5} \frac{0.10}{0.0024} = 4.2 \times 10^{-4} \qquad ////$$

3-5 EQUILIBRIA IN MULTIPLE ACID-BASE SYSTEMS

When more than one conjugate acid-base pair is in equilibrium with water, the exact mathematical relations for calculation of $[H^+]$ become complex, especially if a single equation is to represent all possible initial conditions.[27,28] The derivation of such equations can clarify the nature of approximations made in practical applications. In many cases, however, simplification may be achieved at the outset by using approximate calculations to estimate the concentrations of the major species concerned and then testing the validity of the approximations. If the concentration levels or equilibrium constants for a system are so unusual that the simple equations are not valid, exact equations can be used.[28-31]

Solution of polybasic acid A solution containing only a polybasic acid H_nA and water can almost always be treated by considering only the first step of ionization. To illustrate, if the dibasic acid H_2A is added to water, we have the equilibria

$$H_2A \rightleftharpoons H^+ + HA^- \qquad K_1 = \frac{[H^+][HA^-]}{[H_2A]} \qquad (3\text{-}32)$$

and

$$HA^- \rightleftharpoons H^+ + A^= \qquad K_2 = \frac{[H^+][A^=]}{[HA^-]} \qquad (3\text{-}33)$$

These, together with the stoichiometric equation

$$C_{HA} = [H_2A] + [HA^-] + [A^=] \qquad (3\text{-}34)$$

and the electroneutrality condition

$$[H^+] = [HA^-] + 2[A^=] \qquad (3\text{-}35)$$

give four equations and four unknowns. Rigorous solution of the equations can be awkward; fortunately, it can be avoided by simple approximations.

As a first approximation, ignore the second ionization step. Thus

$$[H^+]' \simeq [HA^-]' \qquad (3\text{-}36)$$

and from Equation (3-21)

$$[H^+]' = \sqrt{K_1(C_{H_2A} - [H^+]')} \qquad (3\text{-}37)$$

as for a monobasic acid. To test the validity of the approximation, consider (3-33) and (3-36); whence

$$[A^=] \simeq K_2 \qquad (3\text{-}38)$$

which states that the divalent anion concentration is given approximately by the second ionization constant and is *independent of the acid concentration*.

If necessary, second and further approximations may be made by adding (3-35) and (3-36) to give

$$[H^+]'' = [HA^-]' + [A^=]' \qquad (3\text{-}39)$$

and

$$[HA^-]'' = [HA^-]' - [A^=]' \qquad (3\text{-}40)$$

From the new values of $[H^+]''$ and $[HA^-]''$ a better value of $[A^=]$, namely $[A^=]''$, is calculated, which can be substituted for $[A^=]'$ in (3-39) and (3-40). These additional steps, however, are rarely necessary, as is demonstrated by the following examples.

EXAMPLE 3-5 Calculate the hydrogen ion concentration of a 10^{-2} M solution of acid H_2A, of ionization constants $K_1 = 10^{-6}$ and $K_2 = 10^{-7}$.

ANSWER From (3-37), $[H^+]' = [HA^-]' = 10^{-4}$. From (3-38), $[A^=]' = 10^{-7}$. Therefore, the second ionization is negligible (Section 3-1), and the approximation for $[H^+] = 10^{-4}$ M is valid. ////

EXAMPLE 3-6   Calculate the hydrogen ion concentration of a 10^{-3} M solution of acid H_2A, of ionization constants $K_1 = 10^{-3}$ and $K_2 = 10^{-4}$.

ANSWER   From (3-37), by quadratic solution,

$$[H^+]' = [HA^-]' = 6.2 \times 10^{-4}$$

From (3-38), $[A^=]' = 10^{-4}$

From (3-39), $[H^+]'' = (6.2 + 1.0)10^{-4} = 7.2 \times 10^{-4}$

From (3-40), $[HA^-]'' = (6.2 - 1.0)10^{-4} = 5.2 \times 10^{-4}$

From (3-33), $[A^=]'' = \frac{5.2}{7.2}10^{-4} = 7.2 \times 10^{-5}$

From (3-39), $[H^+]''' = (6.2 + 0.7)10^{-4} = 6.9 \times 10^{-4}$

From (3-40), $[HA^-]''' = (6.2 - 0.7)10^{-4} = 5.5 \times 10^{-4}$

Observe that, even in this extreme case of a dilute solution of an acid with two closely adjoining ionization steps, the second approximation is essentially correct. In almost all practical cases the first approximation is sufficient.   ////

Solution of ampholyte   The equilibria involved in a solution of the monohydrogen salt HA^- of a dibasic acid H_2A are considered here. In general, the equations are independent of the charge type of the acid; therefore, they are valid also for solutions of a salt BH^+A^-, where BH^+ is the cationic acid of a weak base B, as well as for intermediate states of neutralization of polybasic acids (H_2A^-, $HA^=$, and so on). If HA^- is added to water, it can act as an acid or as a base, and the water is subject to autoionization:

$$HA^- \rightleftharpoons A^= + H^+ \qquad K_2 = \frac{[H^+][A^=]}{[HA^-]} \qquad (3\text{-}41)$$

$$HA^- + H^+ \rightleftharpoons H_2A \qquad \frac{1}{K_1} = \frac{[H_2A]}{[HA^-][H^+]} \qquad (3\text{-}42)$$

$$H_2O \rightleftharpoons H^+ + OH^- \qquad K_w = [H^+][OH^-] \qquad (3\text{-}43)$$

From stoichiometry

$$[H^+] = [A^=] + [OH^-] - [H_2A] \qquad (3\text{-}44)$$

Or, with the use of the equilibrium constants,

$$[H^+] = K_2 \frac{[HA^-]}{[H^+]} + \frac{K_w}{[H^+]} - \frac{[H^+][HA^-]}{K_1} \qquad (3\text{-}45)$$

Clearing fractions and rearranging give

$$[H^+]^2(K_1 + [HA^-]) = K_1 K_2 [HA^-] + K_1 K_w$$

or

$$[H^+] = \sqrt{\frac{K_1(K_2[HA^-] + K_w)}{K_1 + [HA^-]}} \qquad (3\text{-}46)$$

Equation (3-46) usually may be simplified by suitable approximations. If C is the analytical concentration of salt, $[HA^-] \simeq C$. In the numerator of (3-46), K_w is often negligible compared with $K_2 C$; then

$$[H^+] \simeq \sqrt{\frac{K_1 K_2 C}{K_1 + C}} \qquad (3\text{-}47)$$

If C is large compared with K_1, (3-47) becomes

$$[H^+] \simeq \sqrt{K_1 K_2} \qquad (3\text{-}48)$$

so the hydrogen ion concentration to a first approximation is independent of salt concentration. If $K_1 K_2 > K_w$, the solution is acidic; if $K_1 K_2 < K_w$, it is alkaline. The approximation in going from (3-46) to (3-47) is valid particularly if the solution is acidic, for then K_2 cannot be comparable to K_w. If the solution is alkaline and $[HA^-] \simeq C$, (3-46) can be rearranged to

$$[H^+]^2 = \frac{K_1 K_2 C}{K_1 + C} + \frac{K_1 K_w}{K_1 + C} \qquad (3\text{-}49)$$

At reasonable values of C, K_1 can be neglected in comparison with C, and

$$[H^+] \simeq \sqrt{K_1 K_2 + \frac{K_1 K_w}{C}} \qquad (3\text{-}50)$$

This is reduced to (3-48) if $K_1 K_w/C$ is negligible compared with $K_1 K_2$, as it will be for usual values of C (where K_1/C is small), unless K_2 is smaller than K_w. Thus the limiting forms of (3-47) and (3-50) are both (3-48), though the effects of concentration in these two cases differ somewhat.

EXAMPLE 3-7   Calculate the hydrogen concentration of 0.01 M sodium hydrogen sulfide taking $K_1 = 9 \times 10^{-8}$ and $K_2 = 10^{-15}$.

ANSWER   Since $K_1 K_2 < K_w$, the solution is alkaline. Because of the small value of K_2, (3-50) is used, giving $[H^+] \simeq \sqrt{K_1 K_w/C} \simeq 3 \times 10^{-10}$ M. Note that the approximation (3-48) would give $[H^+] \simeq 10^{-11}$ M.   ////

According to (3-48), $\sqrt{K_1 K_2}$ is an approximate expression for the hydrogen ion concentration at the first end point of the titration of a dibasic weak acid. It can be used in a more general way to describe the successive intermediate end points of the titration curve of a polybasic acid; that is,

$$[H^+] \simeq \sqrt{K_2 K_3} \qquad \text{or} \qquad [H^+] \simeq \sqrt{K_3 K_4}$$

provided that the concentration is not too low and that the other ionization constants are sufficiently different from the two in question for their effects to be ignored.

Equation (3-46) and its simplifications can also be used to describe the hydrogen ion concentration of a solution of the salt of a weak acid and a weak base. According to the Brønsted concept a solution of the salt BH^+A^- may be regarded as the half-neutralized solution of an equimolar mixture of the acids HA and BH^+. For the salt BH^+A^- to form, HA must be a stronger acid than BH^+. Thus the ionization constants K_1 and K_2 may be replaced by the constants K_a and K_w/K_b, where K_b is the classical dissociation constant of a base. Thus (3-46) becomes

$$[H^+] = \sqrt{\frac{K_w K_a (C + K_b)}{K_b (C + K_a)}} \qquad (3\text{-}51)$$

which is valid if $K_a K_b \gg K_w$. If the concentration C is large compared with K_a and K_b, (3-51) is reduced to

$$[H^+] = \sqrt{\frac{K_w K_a}{K_b}} \qquad (3\text{-}52)$$

which is analogous to (3-48).

Solution of multivalent anions   In the Brønsted sense a multivalent anion A^{n-} is a multiacidic base, which can add protons stepwise. If the protons originate from water, and for an ion $A^=$, we have two stages of hydrolysis:

$$A^= + H_2O \rightleftharpoons HA^- + OH^- \qquad K_{h_1} = K_{b_1} = \frac{K_w}{K_2} \qquad (3\text{-}53)$$

$$HA^- + H_2O \rightleftharpoons H_2A + OH^- \qquad K_{h_2} = K_{b_2} = \frac{K_w}{K_1} \qquad (3\text{-}54)$$

Equations (3-53) and (3-54) bear a relation to each other similar to that of (3-32) and (3-33) for the successive ionization steps of multibasic acids. Therefore, in general, the *second and higher stages of hydrolysis may be ignored* except in situations involving extreme dilution and close proximity of the constants K_1 and K_2.

A simple equation [(3-27) or (3-28)] may therefore be used in almost all practical situations involving titration end points. But, in the calculation of solubilities of sparingly soluble salts of polybasic weak acids, the concentration of anion may be so low that successive hydrolysis steps must be considered (Section 7-4).

3-6   CALCULATION OF CONCENTRATIONS OF SPECIES PRESENT AT A GIVEN pH

In many situations, calculation of $[H^+]$ or of a dominant species is not the objective, and it is necessary to calculate the concentrations of the minor as well as major constituents of a solution at a given pH value. Such computations are involved in finding the relative concentrations of the two forms of an acid-base indicator at a given pH, in calculating the ionic strength of a buffer, and in calculating the concentration of complexing agents in buffers where the actual complexing species is not the dominant one in solution. Consider first a single acid–base system:

$$HA \rightleftharpoons A^- + H^+ \qquad K_a = \frac{[A^-][H^+]}{[HA]} \qquad (3\text{-}55)$$

Let α_{HA} and α_{A^-} be the fractions of the total concentration $[HA] + [A^-]$ in the forms HA and A^-. Since $[HA] = [A^-][H^+]/K_a$,

$$\alpha_{HA} = \frac{[HA]}{[HA] + [A^-]} = \frac{[H^+]}{[H^+] + K_a} \qquad \text{and} \qquad \alpha_{A^-} = \frac{K_a}{[H^+] + K_a} \qquad (3\text{-}56)$$

Consider next a dibasic acid system, H_2A and its ions, and let us calculate the fractions α_0, α_1, and α_2 in the forms H_2A, HA^-, and $A^=$. The total concentration is

$$C_{H_2A} = [H_2A] + [HA^-] + [A^=] \qquad (3\text{-}57)$$

which, by substitution from (3-32) and (3-33), can be written in terms of $[H_2A]$ as follows:

$$C_{H_2A} = [H_2A] + \frac{K_1[H_2A]}{[H^+]} + \frac{K_1 K_2[H_2A]}{[H^+]^2} \qquad (3\text{-}58)$$

so

$$\alpha_0 = \frac{[H_2A]}{C_{H_2A}} = \frac{[H^+]^2}{[H^+]^2 + K_1[H^+] + K_1 K_2} \qquad (3\text{-}59)$$

Similarly,

$$\alpha_1 = \frac{[HA^-]}{C_{H_2A}} = \frac{K_1[H^+]}{[H^+]^2 + K_1[H^+] + K_1 K_2} \qquad (3\text{-}60)$$

and

$$\alpha_2 = \frac{[A^=]}{C_{H_2A}} = \frac{K_1 K_2}{[H^+]^2 + K_1[H^+] + K_1 K_2} \qquad (3\text{-}61)$$

For the general case of a polybasic acid a similar result is obtained. Here the denominator becomes a polynomial,

$$[H^+]^n + K_1[H^+]^{n-1} + K_1 K_2[H^+]^{n-2} + \cdots + K_1 K_2 \cdots K_n \qquad (3\text{-}62)$$

containing $n + 1$ terms. The fractions $\alpha_0, \alpha_1, \ldots, \alpha_n$ are obtained by taking each of the $n + 1$ terms in turn as the numerator of the fraction:

$$\alpha_0 = \frac{[H^+]^n}{[H^+]^n + K_1[H^+]^{n-1} + K_1 K_2[H^+]^{n-2} + \cdots + K_1 K_2 \cdots K_n} \qquad (3\text{-}63)$$

$$\alpha_1 = \frac{K_1[H^+]^{n-1}}{[H^+]^n + K_1[H^+]^{n-1} + K_1 K_2[H^+]^{n-2} + \cdots + K_1 K_2 \cdots K_n} \qquad (3\text{-}64)$$

$$\alpha_2 = \frac{K_1 K_2[H^+]^{n-2}}{[H^+]^n + K_1[H^+]^{n-1} + K_1 K_2[H^+]^{n-2} + \cdots + K_1 K_2 \cdots K_n} \qquad (3\text{-}65)$$

$$\alpha_n = \frac{K_1 K_2 \cdots K_n}{[H^+]^n + K_1[H^+]^{n-1} + K_1 K_2[H^+]^{n-2} + \cdots + K_1 K_2 \cdots K_n} \qquad (3\text{-}66)$$

EXAMPLE 3-8   Calculate the concentration of $HPO_4^=$ in a 0.1 M phosphate buffer of pH $= 7$, given $pK_1 = 2.15$, $pK_2 = 7.15$, $pK_3 = 12.40$.

ANSWER   At pH $= 7$ the denominator in (3-65) is given by $10^{-21} + 10^{-16.15} + 10^{-16.30} + 10^{-21.70} = 10^{-15.92}$, and

$$\alpha_2 = \frac{10^{-16.30}}{10^{-15.92}} = 10^{-0.38} = 0.42$$

The concentration of $HPO_4^- = 0.42 \times 0.10 = 0.042$ M.   ////

3-7   TITRATION CURVES

From the viewpoint of analytical applications the titration curves of greatest importance are those involving a *strong* acid or base as titrant, because the sharpest possible end points are thus obtained. According to the Brønsted concept the titration of monobasic weak acids and bases can be described by the same curves as illustrated by the family of curves shown in Figure 3-2, calculated for 0.1 M acid or base, neglecting dilution. Since the necessary equations were presented earlier, details of the calculations are omitted here. One important practical consideration, however, is that the sharpness of the end point varies with the strength of the acid or base being titrated. The sharpness also depends on the concentration, because the pH increases with dilution for acids and decreases with dilution for bases.

According to the calculations of Meites and Goldman,[32] when both the effects of dilution during titration and the ionization of water are taken into account, the inflection point precedes the equivalence point in both the titration of a strong acid with a strong base and the reverse. But only at high dilution (about 10^{-5} M) does the difference between the inflection-point and equivalence-point volumes become significant (1 ppt relative). At extremely high dilution the inflection point vanishes $(C < \sqrt{32K_w})$.

A titration such as that of a monobasic weak acid with a strong base or of the last step of a polybasic weak acid usually shows two inflection points, one where the slope of the curve is at a minimum and the other where it is at a maximum. The first inflection point is usually near the 50% neutralization point, but follows it for very weak acids, precedes it for moderately strong acids, and disappears for the strongest acids. The second inflection point precedes the equivalence point; the difference amounts to as much as 1 ppt only for very weak acids ($K_a < 10^{-9}$ for 0.1 M solutions) or highly dilute solutions. The second inflection point disappears for highly dilute or exceedingly weak acids. Automated techniques for end-point detection normally rely on the inflection point; significant error therefore may be incurred under certain circumstances.

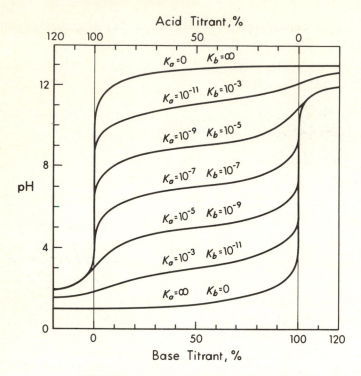

FIGURE 3-2   Titration curves of 0.1 M acid or base.

A titration curve of a dibasic acid is essentially a composite of the titration curves of an equimolar mixture of two weak acids with dissociation constants K_1 and K_2. In the two buffer regions, where Equation (3-30) applies, the shape can be calculated directly if $K_1 \gg K_2$. The hydrogen ion concentration at the first end point is given approximately by $\sqrt{K_1 K_2}$, and at the second end point it is nearly the same as that for the titration of a monobasic weak acid with $K_a = K_2$. If K_1 is not much larger than K_2, the above approximations are no longer valid; then the precision of the intermediate end point is so low that it is no longer of analytical value. The titration curve is too shallow for quantitative use unless K_1/K_2 exceeds about 10^3 or 10^4.

The steepness of the titration curve in the immediate vicinity of the end point is the decisive quantity as to whether determining an end point is feasible. The *relative precision*[33] of location of an end point is the fraction of the stoichiometric quantity of reagent required to traverse the region represented by 0.1 pH unit on either side of the end point. The change of 0.1 pH unit represents about the limit of visual observation of indicator color change using a color-comparison technique. A relative precision of 10^{-3} indicates an expected end-point precision of about 1 ppt. Modern end-point detection techniques require less pH change, and end points can be detected with improved sensitivity.

The relative precision is inversely related to the slope of the titration curve at the end point, and important conclusions from this relation are: for a strong acid–strong base titration the relative precision is approximately $\sqrt{K_w/C}$; for a weak acid–strong base titration the relative precision is approximately $\sqrt{K_w/K_aC}$; and for an intermediate end point in a polybasic acid titration the relative precision is approximately $\sqrt{K_2/K_1}$ (independent of concentration).

3-8 MICROSCOPIC IONIZATION CONSTANTS

For detailed understanding of many chemical reactions, more knowledge about ionization constants is needed than has been considered so far. The macroscopic constants $K_1, K_2, \ldots$ of polybasic acids can have several components arising from the detailed ionization processes. It has been shown[34–38] that, if a dibasic acid has two equivalent and independent ionization processes, the minimum value of K_1/K_2 is 4.

Suppose that a dibasic acid HAH has two independent ionization processes yielding the anions AH^- and HA^- with equilibrium constants k_1 and k_2, where the subscripts 1 and 2 denote the first and second protons in HAH as written. These in turn ionize to yield the same anion $A^=$ with equilibrium constants k_{12} and k_{21}, where the subscript 12 denotes loss of proton 2 following loss of proton 1 and subscript 21 denotes loss of proton 1 following loss of proton 2. The ordinarily defined K_1 and K_2 are written in terms of the total concentration $[HA^-] + [AH^-]$. It follows that $K_1 = k_1 + k_2$ and $K_2 = k_{12}k_{21}/(k_{12} + k_{21})$. In the limiting case there is no interaction involving the two independent and equivalent proton sites, and then $k_1 = k_2 = k_{12} = k_{21}$. Therefore $K_1 = 2k_1$ and $K_2 = k_1/2$, and thus $K_1 = 4K_2$. The factor 4 is statistically reasonable. The neutral molecule HAH can lose a proton at either of two positions and its conjugate base can regain it at only one, whereas the anion $A^=$ can regain a proton at either of two and its conjugate acid can lose it at only one.

From a qualitative point of view, a minimum difference in ionization constants is to be expected, because even though a molecule contains two equivalent ionizable protons, once one proton has been lost, the second is retained more strongly by an energy amounting to e^2/Dr (where e is the electronic charge, D is the dielectric constant, and r is the distance between the proton and the charge on the acid anion). Thus, for oxalic acid, K_1 is about $1000K_2$; for adipic acid, K_1 is only about $5K_2$ (nearly the factor of 4). The titration curve of an acid in which $K_1 = 4K_2$ can be shown[39] to be identical in every respect to the titration curve of a monobasic acid of one-half its molecular weight and with dissociation constant $K_1/2$.

The constants k_1, k_2, k_{12}, and k_{21} are known as specific or *microscopic ionization constants*. Such constants assume more than theoretical significance when the protons are attached to groups that are not equivalent, but whose ionization processes can occur independently, or nearly so, of each other. Of the molecules of this type, cysteine has been studied extensively. Fully protonated cysteine, $HSCH_2CHNH_3^+-COOH$, has three acid groups: sulfhydryl, ammonium, and carboxyl. A proton can be lost from any one of the three groups to give three different ionized species; the loss of a proton from the carboxyl group is most probable, and from the sulfhydryl group least probable. Therefore this concept of microscopic ionization constants k_1, k_2, and k_3 may be applied, where k_1 involving the carboxyl proton is

$$k_1 = \frac{[H^+][HSCH_2CHNH_3^+COO^-]}{[HSCH_2CHNH_3^+COOH]} \qquad (3\text{-}67)$$

k_2 involving the sulfhydryl proton is

$$k_2 = \frac{[H^+][S^-CH_2CHNH_3^+COOH]}{[HSCH_2CHNH_3^+COOH]} \qquad (3\text{-}68)$$

and k_3 involving the ammonium proton is

$$k_3 = \frac{[H^+][HSCH_2CHNH_2COOH]}{[HSCH_2CHNH_3^+COOH]} \qquad (3\text{-}69)$$

At appreciable ionic strength these microscopic constants should be written in terms of activities rather than concentrations. The constants are normally estimated by spectral means. There are three microscopic ionization constants for the loss of the first proton, six for the loss of the second, and three for the loss of the last.[40] In Table 3-2 the subscripts 1, 2, and 3 denote the carboxyl, sulfhydryl, and ammonium protons; thus k_{32} is the microscopic constant for the ion formed by loss of the sulfhydryl proton from the species that has already lost the ammonium proton.

It can be shown that for a tribasic acid the first ionization constant K_1 is the sum $k_1 + k_2 + k_3$, the second ionization constant K_2 is $(k_1k_{12} + k_1k_{13} + k_2k_{23})/K_1$ or $(k_2k_{21} + k_3k_{31} + k_3k_{32})/K_1$, and K_3 is $k_1k_{12}k_{123}/K_1K_2$.

Table 3-2   **MICROSCOPIC IONIZATION CONSTANTS FOR CYSTEINE**

Ionizing group					
Carboxyl		Sulfhydryl		Ammonium	
pk_1	1.7	pk_2	7.4	pk_3	6.8
pk_{21}	2.8	pk_{12}	8.5	pk_{13}	8.9
pk_{31}	3.8	pk_{32}	9.1	pk_{23}	8.4
pk_{231}	4.7	pk_{132}	10.0	pk_{123}	10.4

If a pH-titration curve for a polybasic acid such as fully protonated cysteine is examined, three distinct pH regions are observed, each consuming 1 mole of hydroxyl ion per mole of cysteine. The chemical interpretation of the pH changes is not straightforward, even though from model compounds the carboxyl proton is predicted to be the most acidic and the sulfhydryl proton the least acidic. Calculations involving the microscopic constants indicate that the first sharp change in the titration curve corresponds to removal of the carboxyl proton almost, but not completely, exclusively. On the other hand, the second and the third breaks in pH correspond to removal of the second and third protons from ammonium and sulfhydryl groups in comparable proportions. Thus in Table 3-2 for the second proton the two important constants are pk_{12} of 8.5 and pk_{13} of 8.9, indicating a composite of roughly equal numbers of two species. For the last proton the two most significant constants are pk_{132} and pk_{123} of 10.0 and 10.4, again not greatly different.

A knowledge of the concentrations of each species in solution is necessary in the elucidation of the chemistry of metal ion complexes with molecules such as EDTA and glutathione.

3-9   EFFECT OF TEMPERATURE ON ACID-BASE EQUILIBRIA

The temperature coefficient of an acid dissociation constant is determined by the enthalpy change of the dissociation reaction, and is given by the thermodynamic relation

$$-2.303 \frac{dpK}{dT} = \frac{d \ln K}{dT} = \frac{\Delta H°}{RT^2} \qquad (3\text{-}70)$$

The standard enthalpy change $\Delta H°$ varies with temperature according to

$$\frac{d \Delta H°}{dT} = \Delta C_p \qquad (3\text{-}71)$$

where ΔC_p is the difference in molal heat capacities of the products and reactants. For many acid dissociation reactions, ΔC_p is independent of temperature,[41,42] so it is possible to write

$$\Delta H° = \Delta H_0° + \Delta C_p T \qquad (3\text{-}72)$$

where $\Delta H_0°$ is a hypothetical standard enthalpy change at absolute zero. The quantity $\Delta H_0°$ should be regarded as an extrapolated limiting value that has no physical reality, because ΔC_p does not remain constant down to absolute zero. For many acids, $\Delta H_0°$ is a positive and ΔC_p a negative quantity, so that at some temperature T_m (calculated from $T_m = -\Delta H_0°/\Delta C_p$), $\Delta H°$ is zero. At this temperature, $d \ln K/dT$ is zero, and the ionization constant passes through a maximum (295°C for acetic acid and 564°C

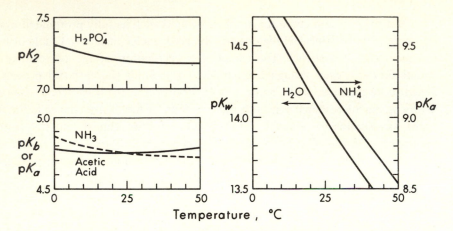

FIGURE 3-3   Variation of ionization constants with temperature. All values
are on the molal scale. (*Data from "Handbook of Chemistry and Physics,"* 53d ed.,
p. D-121, *Chemical Rubber, Cleveland, Ohio, 1972–1973.*)

for water). Figure 3-3 shows the variation of ionization constants with temperature
for several substances.

Qualitatively the maximum ionization constant occurs when the tendency
toward increased proton transfer with rising temperature has been just counterbalanced by the decreasing tendency toward charge separation. For cationic acids
there is no net separation of charged species, with the result that the ionization
constant increases regularly with increasing temperature (see NH_4^+, Figure 3-3).

It is noteworthy that for various uncharged acids the quantity ΔC_p has a relatively large and constant negative value of the order of -160 J/deg. Evidently this
regularity is associated with the formation of a hydrogen ion and an anion, the nature
of the anion exerting only a minor effect. The temperature T_m of maximum ionization
constant falls below 25°C for acids showing negative ΔH_{298}° values. For water, T_m is
so high that a regular increase in ionization constant is observed (Figure 3-3).

3-10   ACID-BASE INDICATORS

The equilibrium between the acid form In_A of an indicator In and the alkaline form
In_B may be written

$$In_A \rightleftharpoons H^+ + In_B \qquad (3\text{-}73)$$

and the corresponding equilibrium constant is

$$K_{In} = \frac{a_{H^+} a_{In_B}}{a_{In_A}} \qquad (3\text{-}74)$$

The color of an indicator, as perceived by the eye, is determined primarily by the ratio of concentrations $[In_A]/[In_B]$, given by

$$\frac{[In_A]}{[In_B]} = \frac{a_{H^+}\gamma_{In_B}}{K_{In}\gamma_{In_A}} \quad (3\text{-}75)$$

where γ_{In_A} and γ_{In_B} are the activity coefficients of the acid and alkaline forms of the indicator. Equation (3-75), written in logarithmic form and rearranged, becomes

$$-\log a_{H^+} = pH = pK_{In} + \log\frac{[In_B]}{[In_A]} + \log\frac{\gamma_{In_B}}{\gamma_{In_A}} \quad (3\text{-}76)$$

The *apparent indicator constant* pK'_{In} is defined by

$$pK'_{In} = pK_{In} + \log\frac{\gamma_{In_B}}{\gamma_{In_A}} \quad (3\text{-}77)$$

The activity coefficient of hydrogen ion is not included in (3-77), so that (3-78) may be expressed as nearly as possible in terms of the pH measured by the operational pH scale.

The color-change equilibrium at any particular ionic strength (where the term for activity coefficient is constant) can be expressed by the equation

$$pH = pK'_{In} + \log\frac{[In_B]}{[In_A]} \quad (3\text{-}78)$$

The ratio $[In_B]/[In_A]$ is determined experimentally either by spectrophotometry or by visual color comparison. If the eye is able to perceive a color change in the interval $0.1 < [In_B]/[In_A] < 10$, the corresponding pH interval according to (3-78) is $pH = pK'_{In} \pm 1$. A wider pH interval can be covered spectrophotometrically. If the color intensity of either form is much greater than that of the other, the transition interval is unsymmetrical with respect to pK'_{In}. For most indicators the transition interval judged visually depends to some extent on the individual and is somewhat less than 2 units, showing that more than 10% must be present in a given form for the color to be perceptible. For a one-color indicator the transition interval is indeterminate unless the total concentration of indicator is defined. Table 3-3 gives transition intervals for a selection of acid-base indicators. A more complete listing of indicators is given by Banyai,[43] including the analytically important color-change reactions for several types of indicators.

The charge types of the acid and alkaline forms of the indicators significantly affect salt and solvent behavior of acid-base indicators. A brief summary of the major effects is as follows.

Salt effects Aside from specific salt effects, the general effects of electrolytes on the indicator equilibrium can be interpreted reasonably well on the basis of (3-77). As a first approximation we have from the DHLL (Section 2-3), at low ionic strengths for three charge types of indicators,

$$HIn^+ \rightleftharpoons H^+ + In \qquad pK'_{In} = pK_{In} + 0.5\sqrt{\mu} \qquad (3\text{-}79)$$

$$HIn \rightleftharpoons H^+ + In^- \qquad pK'_{In} = pK_{In} - 0.5\sqrt{\mu} \qquad (3\text{-}80)$$

$$HIn^- \rightleftharpoons H^+ + In^= \qquad pK'_{In} = pK_{In} - 1.5\sqrt{\mu} \qquad (3\text{-}81)$$

Indicators can exist as dipolar ions, and particularly at high ionic strength and in solvents of relatively low dielectric constant, each species is best regarded as consisting of n univalent anions, where n is the total charge including the dipolar charges.[44]

Temperature effects   The effect of temperature on the color of an indicator in a given solution is determined by the effect of temperature on the pH of the solution and on the response of the indicator. Thus the pH of dilute hydrochloric acid is insensitive to heating, whereas pOH decreases markedly. In contrast, the pOH of dilute sodium hydroxide is insensitive to heating, whereas the pH decreases markedly. Indicators of the HIn or HIn$^-$ type show relatively small changes in K_{In} with temperature.

Colloid effects   An important limitation on the use of indicators is that colloids may cause relatively large errors[45] because, for example, an indicator can complex

Table 3-3   TRANSITION INTERVALS OF SELECTED ACID-BASE INDICATORS

Common name	Chemical name	Acid color	Base color	pH transition interval	pK_{In}
Thymol blue	Thymolsulfonephthalein	Red	Yellow	1.2–2.8	1.6
Methyl yellow	p-Dimethylaminoazo-benzene	Red	Yellow	2.9–4.0	3.3
Methyl orange	p-Dimethylaminoazo-benzenesulfonic acid	Red	Orange-yellow	3.1–4.4	4.2
Bromocresol green	Tetrabromo-m-cresol-sulfonephthalein	Yellow	Blue	3.8–5.4	4.7
Methyl red	Dimethylaminoazoben-zene-o-carboxylic acid	Red	Yellow	4.2–6.2	5.0
Chlorophenol red	Dichlorosulfonephtha-lein	Yellow	Red	4.8–6.4	6.0
Bromothymol blue	Dibromothymolsulfone-phthalein	Yellow	Blue	6.0–7.6	7.1
Phenol red	Phenolsulfonephthalein	Yellow	Red	6.4–8.0	7.4
Cresol purple	m-Cresolsulfonephtha-lein	Yellow	Purple	7.4–9.0	8.3
Thymol blue	Thymolsulfonephthalein	Yellow	Blue	8.0–9.6	8.9
Phenolphthalein	Phenolphthalein	Colorless	Red-violet	8.0–9.8	9.7
Thymolphthalein	Thymolphthalein	Colorless	Blue	9.3–10.5	9.9

with or be adsorbed on protein of opposite charge. Thus the color can be shifted toward the acid or alkaline color, as well as have a different absorption spectrum. Adsorbed hydroxyl ions on the surface of a colloidal hydrous oxide may cause an adsorbed indicator to exhibit its alkaline color on the surface.

Mixed indicators For certain titrations, particularly those carried to a definite end-point pH value, sharpening of the color of an indicator is desirable. This can be accomplished by either of two methods: addition of a pH-insensitive dye that produces a color complementary to one of the indicator colors, or use of two indicators of closely adjoining pK'_{In} values so that the overlapping colors are complementary at an intermediate pH value.[43,46]

3-11 ACIDITY IN HEAVY WATER

Their unique relation to water systems favors the inclusion of acid-base reactions in deuterium oxide with aqueous acid-base equilibria, even though some aspects of the chemistry suggest inclusion with nonaqueous solvents. In studies such as those of deuterium isotope effects, it is desirable to be able to measure pD as an index of acidity in heavy water. Glass electrodes respond in a nernstian way[47,48] to changes in deuterium ion concentration, and therefore the usual combination of glass and calomel electrodes can form the basis of an operational definition[49] of pD:

$$pD = pD_s + \frac{(E - E_s)F}{2.303RT} \qquad (3\text{-}82)$$

where pD_s is the pD value of the reference buffer in deuterium oxide and E and E_s are the potentials with the sample solution and standard buffer in Cell (3-9). The saturated calomel reference electrode in this cell is the usual one with light water; the standard buffer solutions have pa_D values established by means of cells without liquid junction, as have the aqueous standard buffers of Table 3-1. Reference buffers[49,50] for use in connection with Equation (3-82) are shown in Table 3-4. Standardization of the pD

Table 3-4 pa_D VALUES OF REFERENCE BUFFERS

Temperature, °C	0.025 m KD$_2$PO$_4$ and 0.025 m Na$_2$DPO$_4$ in D$_2$O	0.05 m CH$_3$COOH and 0.05 m CH$_3$COONa in D$_2$O
5	7.537	5.265
15	7.475	5.243
20	7.450	5.235
25	7.429	5.230
30	7.411	5.227
40	7.386	5.226
50	7.377	5.235

scale using these buffers is recommended in preference to the earlier empirical technique[47] using aqueous buffers.

At 25°C, pa_D values on the molality scale are 0.043 unit higher than on the molarity scale,[49] since the density of deuterium oxide is higher than that of water. The accuracy of measured pD values depends in part on the variability of the liquid-junction potential between the deuterium oxide and saturated potassium chloride in the reference electrode.

Although solutes ordinarily would not be expected to behave much differently in deuterium oxide than in water, nevertheless noteworthy differences are observed. The value of pK_{D_2O} at 25°C is 14.955 (molality) or 14.869 (molarity).[51] The temperature coefficient of the ion product is large and similar to that of ordinary water (Figure 3-3). Ionization constants for a variety of organic and inorganic acids have been measured,[52] and the ionization constants have been found to be smaller in deuterium oxide, ranging up to about 1 pK unit. For acetic acid the pK values[53] at 25°C are: CH_3COOH in H_2O, 4.756; CD_3COOH in H_2O, 4.772; CH_3COOD in D_2O, 5.312; CD_3COOD in D_2O, 5.325.

REFERENCES

1 J. N. BRØNSTED: *Rec. Trav. Chim.*, **42**:718 (1923); *Chem. Rev.*, **5**:231 (1928).

2 T. M. LOWRY: *Trans. Faraday Soc.*, **20**:13 (1924).

3 G. N. LEWIS: "Valence and the Structure of Atoms and Molecules," American Chemical Soc. Monograph Series, p. 142, ACS, New York, 1923.

4 R. G. PEARSON: *J. Chem. Educ.*, **45**:581, 643 (1968).

5 N. INGRI, W. KAKOLOWICZ, L. G. SILLEN, and B. WARNQVIST: *Talanta*, **14**:1261 (1967); R. EKELUND, L. G. SILLEN, and O. WAHLBERG: *Acta Chem. Scand.*, **24**:3073 (1970).

6 N. BJERRUM: *Samml. Chem. Chem.-Technol. Vortr.*, **21**:69 (1914).

7 L. G. SILLEN in "Treatise on Analytical Chemistry," I. M. Kolthoff and P. J. Elving, (eds.), pt. 1, vol. 1, sec. B, p. 277, Interscience, New York, 1959.

8 H. FREISER: *J. Chem. Educ.*, **47**:844 (1970).

9 W. STUMM and J. J. MORGAN: "Aquatic Chemistry," Wiley, New York, 1970.

10 J. N. BUTLER: "Ionic Equilibrium: A Mathematical Approach," p. 55, Addison-Wesley, Reading, Mass., 1964.

11 H. S. HARNED and B. B. OWEN: "The Physical Chemistry of Electrolytic Solutions," 3d ed., p. 574, Reinhold, New York, 1958.

12 S. P. L. SØRENSON: *Biochem. Z.*, **21**:131 (1909).

13 J. A. CHRISTIANSEN: *J. Amer. Chem. Soc.*, **82**:5517 (1960).

14 N. BJERRUM: *Z. Phys. Chem.* (Leipzig), A**53**:428 (1905). See also R. G. Bates, "Determination of pH," p. 44, Wiley, New York, 1973.

15 R. A. ROBINSON and R. G. BATES: *Anal. Chem.*, **43**:969 (1971).

16 S. P. L. SØRENSEN and K. LINDERSTROM-LANG: *Compt. Rend. Trav. Lab.* (Carlsberg), **15**:(6) (1924).

17 BATES,[14] chap. 2.

18 R. G. BATES and E. A. GUGGENHEIM: *Pure Appl. Chem.*, **1**:163 (1960).

19 M. L. MCGLASHAN: *Pure Appl. Chem.*, **21**:1 (1970).

20 R. G. BATES: *J. Res. Nat. Bur. Stand.*, **66A**:179 (1962).

21 BATES,[14] p. 86.

22 R. G. PICKNETT: *Trans. Faraday Soc.*, **64**:1059 (1968).

23 R. N. GOLDBERG and H. S. FRANK: *J. Phys. Chem.*, **76**:1758 (1972).

23a BATES,[14] p. 268.

23b G. R. HEDWIG and H. K. J. POWELL: *Anal. Chem.*, **43**:1206 (1971).

24 G. CHARLOT: *Anal. Chim. Acta*, **1**:59 (1947).

25 D. D. DEFORD: *J. Chem. Educ.*, **27**:554 (1950).

26 L. J. HENDERSON: *J. Amer. Chem. Soc.*, **30**:954 (1908).

27 D. D. DEFORD: *Anal. Chim. Acta*, **5**:345, 352 (1951).

28 J. E. RICCI: "Hydrogen Ion Concentration," Princeton, Princeton, N.J., 1952.

29 A. CLAEYS: *Bull. Soc. Chim. Belges*, **72**:102 (1963).

30 E. BISHOP: *Anal. Chim. Acta*, **28**:299 (1963).

31 L. MEITES and E. BISHOP: *Anal. Chim. Acta*, **29**:484 (1963).

32 L. MEITES and J. A. GOLDMAN: *Anal. Chim. Acta*, **29**:472 (1963).

33 A. A. BENEDETTI-PICHLER: "Essentials of Quantitative Analysis," Ronald, New York, 1956.

34 E. Q. ADAMS: *J. Amer. Chem. Soc.*, **38**:1503 (1916).

35 M. T. BECK: "Chemistry of Complex Equilibria," p. 146, Van Nostrand Reinhold, New York, 1970.

36 J. GREENSPAN: *Chem. Rev.*, **12**:339 (1933).

37 J. E. PRUE in "International Encyclopedia of Physical Chemistry and Chemical Physics," vol. 3, "Ionic Equilibria," p. 85, Pergamon, New York, 1966.

38 L. MEITES: *J. Chem. Educ.*, **49**:682 (1972).

39 RICCI,[28] p. 95.

40 J. T. EDSALL and J. WYMAN: "Biophysical Chemistry," p. 503, Academic, New York, 1958. See also E. L. Elson and J. T. Edsall, *Biochemistry*, **1**:1 (1962), and E. Coates, C. G. Marsden, and B. Rigg, *Trans. Faraday Soc.*, **65**:3032 (1969).

41 K. S. PITZER: *J. Amer. Chem. Soc.*, **59**:2365 (1937).

42 D. H. EVERETT and W. F. K. WYNNE-JONES: *Trans. Faraday. Soc.*, **35**:1380 (1939).

43 E. BANYAI in "Indicators," E. Bishop (ed.), chap. 3, Pergamon, New York, 1972.

44 L. S. GUSS and I. M. KOLTHOFF: *J. Amer. Chem. Soc.*, **62**:249 (1940).

45 I. M. KLOTZ: *Chem. Rev.*, **41**:373 (1947).

46 C. L. HILTON in "Encyclopedia of Industrial Chemical Analysis," vol. 2, p. 248, F. D. Snell and C. L. Hilton (eds.), Wiley, New York, 1966.

47 P. K. GLASOE and F. A. LONG: *J. Phys. Chem.*, **64**:188 (1960).

48 P. R. HAMMOND: *Chem. Ind.* (London), 311 (1962).

49 R. GARY, R. G. BATES, and R. A. ROBINSON: *J. Phys. Chem.*, **68**:3806 (1964).

50 R. GARY, R. G. BATES, and R. A. ROBINSON: *J. Phys. Chem.*, **69**:2750 (1965).

51 A. K. COVINGTON, R. A. ROBINSON, and R. G. BATES: *J. Phys. Chem.*, **70**:3820 (1966).

52 R. A. ROBINSON, M. PAABO, and R. G. BATES: *J. Res. Nat. Bur. Stand.*, **73A**:299 (1969).

53 M. PAABO, R. G. BATES, and R. A. ROBINSON: *J. Phys. Chem.*, **70**:540, 2073 (1966).

PROBLEMS

3-1 Cell (3-9) is used to follow the kinetics of the reaction $PtCl_4^= + H_2O \rightarrow$ $PtCl_3(OH)^= + H^+ + Cl^-$, starting with a solution of K_2PtCl_4 of initial concentration $0.010 \ M$ in pure water. Assuming that the pH meter reading corresponds to $-\log a_{H^+}$, use the EDHE to calculate the concentration of $PtCl_4^=$ corresponding to a pH meter reading of 3.00.

Answer $\mu = 0.031; \ C = 8.8 \times 10^{-3} \ M.$

3-2 A 50-ml aliquot of $0.100 \ M \ H_2A$ is titrated with $0.100 \ M \ NaOH$. The equilibrium constants K_1 and K_2 are 1.00×10^{-4} and 1.00×10^{-6}. For each of the titration points listed below, calculate the pH (a) assuming activity coefficients of unity and (b) using the EDHE to estimate activity coefficients. Volume of NaOH added $= 0, 10, 50, 75, 100$ ml.

Answer (a) 2.51, 3.41, 5.00, 6.00, 9.26. (b) 2.51, 3.35, 5.00, 5.67, 9.02.

3-3 Assuming activity coefficients of unity, calculate the pH of each of the following solutions: (a) Water in equilibrium with CO_2 of the air; $C_{H_2A} = 1.3 \times 10^{-5}$, $K_1 = 3 \times 10^{-7}$, $K_2 = 6 \times 10^{-11}$. (b) Water as in part (a) brought to $pH = 7.00$ with NaOH and allowed to regain equilibrium with CO_2. (c) Standard $0.100 \ M \ HCl$ in equilibrium with CO_2 of the air, titrated with an equal volume of NaOH, of total alkalinity $0.100 \ M$ but contaminated with carbonate to the extent of 2% (relative).

Answer (a) 5.74. (b) 6.48 (not 6.40). (c) 4.92.

3-4 The titration in Problem 3-3c is continued to (a) pH 7.0 and (b) pH 9.0. Estimate the titration errors, in relative percentages.

Answer (a) 0.76%. (b) 1.06%.

3-5 (a) A H_3PO_4 solution is brought to $pH = 7$ by the addition of NaOH. Calculate the concentrations of the various forms of orthophosphate if the total phosphate concentration in the buffer is $0.200 \ M$; $pK_1 = 2.15$, $pK_2 = 7.15$, $pK_3 = 12.40$. (b) Calculate the ionic strength of the buffer.

Answer (a) $[H_3PO_4] = 1.6 \times 10^{-6}$, $[H_2PO_4^-] = 0.117$, $[HPO_4^=] = 0.083$, $[PO_4^{3-}] = 3.3 \times 10^{-7}$. (b) $\mu = 0.365$.

3-6 Calculate the values of relative precision for the titration of $0.05 \ M \ Na_2CO_3$ with $0.1 \ M \ HCl$, assuming no loss of CO_2; $K_1 = 4 \times 10^{-7}$, $K_2 = 4 \times 10^{-11}$.

Answer First end point 10^{-2}; second end point $\sqrt{K_1/C} = 0.004$.

3-7 Calculate the relative precision of (a) the first end point in the titration of citric acid with strong base and (b) the second end point; $K_1 = 8.7 \times 10^{-4}$, $K_2 = 1.8 \times 10^{-5}$, $K_3 = 4 \times 10^{-6}$. (c) Is either end point feasible for practical use?

Answer (a) 0.14. (b) 0.47. (c) Neither end point is feasible.

3-8 Suppose that a tribasic acid H_3A has three independent and equivalent ionization processes. What would be the minimum value of (a) K_1/K_2 and (b) K_1/K_3?

Answer (a) 3. (b) 9.

3-9 (a) Calculate the ionization constants K_1, K_2, and K_3 for cysteine from the microscopic constants given in Table 3-2. (b) At a pH of 5, calculate the concentration in 0.01 M cysteine of each of the three doubly protonated species. Make reasonable approximations in the calculations.

$\quad$ *Answer* (a) $K_1 = 2.0 \times 10^{-2}$, $K_2 = 4.4 \times 10^{-9}$, $K_3 = 2.8 \times 10^{-11}$.

$\quad$ (b) $[COO^-CHNH_3{}^+CH_2SH] = 10^{-2}$ M,

$\quad$ $[COOHCHNH_2CH_2SH] = 8 \times 10^{-8}$ M,

$\quad$ $[COOHCHNH_3{}^+CH_2S^-] = 2 \times 10^{-8}$ M.

3-10 In 1 M KOH the activity of water is 0.963 at 24°C. If pK_w is 14.0, what is the product of the activities of hydrogen and hydroxyl ions in 1 M KOH?

$\quad$ *Answer* 9.63×10^{-15}.

3-11 Construct a logarithmic diagram of log C against pH for (a) 0.01 M NH_4Cl ($pK_a = 9.26$) and (b) 0.1 M H_3PO_4 with pK values as in Problem 3-5.

4

ACID-BASE EQUILIBRIA IN NONAQUEOUS SOLVENTS

Acid-base reactions in solvents other than water are of both theoretical and practical significance, and their fundamental chemistry is becoming increasingly understood. It should be realized at the outset that solvents play an active rather than a passive role in acid-base reactions and that water as a solvent, though of unique importance, is highly atypical. The important considerations are general dielectric-constant effects, acidic behavior and basic behavior of solvents, and specific interactions of solvent with solute.

The dielectric constant is a property of major concern in understanding acid-base behavior in various solvents. When the dielectric constant of a solvent is low, ion association and homoconjugation can take place, resulting in modification of otherwise simple proton transfer reactions.

Acidic and basic behavior in solvent-solute systems can be complex even though it involves only proton transfer reactions in the Brønsted sense. The acid or base strength of a solvent is a major factor in whether a solute gives an acidic or a basic solution. Thus substances act as bases (proton acceptors) more readily in glacial acetic acid than in ethylenediamine. On the other hand, acetic acid itself in sulfuric acid as solvent acts as a base in that it accepts a proton to give H_2OAc^+. It is important to recognize that different acid-base conjugate pairs are involved when the acid rather than the base strength of a solvent is of concern. Thus the conjugate pair for water acting as a base is H_3O^+/H_2O, whereas for water acting as an acid it is H_2O/OH^-. Consequently, the acid and base strengths of a solvent molecule are not simply the inverse of one another (Section 4-3). Water has considerable strength as both an acid and a base. Acetonitrile is both a weaker acid and a weaker base than water.

In this chapter several general topics important to understanding acid-base systems are considered and then illustrated by acid-base reactions in three typical solvent types. Finally, pH measurements in solvent mixtures and the Hammett acidity function are discussed.

4-1 THE STANDARD-STATE PROBLEM AND NONAQUEOUS SOLVENTS

A study of the acid-base properties of solutes in nonaqueous solvents must include consideration of hydrogen ion activities and in particular a comparison of their activities in different solvents. Attempting to transpose interpretations and methods of approach from aqueous to nonaqueous systems may lead to difficulty. The usual standard state (Section 2-2) for a nonvolatile solute is arbitrarily defined in terms of a reference condition with activity equal to concentration at infinite dilution. Comparisons of activities are unsatisfactory when applied to different solvents, because different standard states are then necessarily involved. For such comparisons it would be gratifying if the standard state could be defined solely with reference to the properties of the pure solute, as it is for volatile nonelectrolytes (Section 2-7). Unfortunately, for ionic solutes a different standard state is defined for every solvent and every temperature.

To compare activities of solutes in different solvents, a single reference state for the solute must be chosen. Although from some points of view it is awkward,[1] water is a logical choice for a single reference solvent in which the behavior of solutes in other solvents can be compared. To make comparisons of solute activities among solvents, it is convenient to consider separately the effect of dilution within a given solvent and the difference in the usual reference states of a solute at infinite dilution in different solvents. The activity coefficient γ_i of a species i in a solvent may be considered the product of two terms:[2]

$$\gamma_i = \gamma\gamma_t \qquad (4\text{-}1)$$

where γ may be called the salt activity coefficient or simply the activity coefficient, which represents the effect of electrostatic ion-ion interactions and which is given by the Debye-Hückel equation (Section 2-3), and γ_t is the *transfer activity coefficient*, which represents the difference in ion-solvent interactions in two solvents. In recent years the transfer activity coefficient has been commonly referred to as the medium activity coefficient, medium effect, or primary medium effect. It has also been called the partition or distribution coefficient[3,4] and the degenerate activity coefficient.[5] Except for differences in reference behavior on which the standard states are chosen, the quantity γ_t is equivalent to the infinite-dilution activity coefficient γ^0 described in Section 2-7.

In Equation (4-1), γ is a measure of the effects of interionic or interparticle interactions, and γ_t a measure of the effect of changing the solvent. As the concentration of solute approaches zero, γ approaches unity and γ_i approaches γ_t, a constant for each solute-solvent pair at a given temperature. The value of γ_t is related to the difference between the free energies of the solute in the usual standard states in the solvent and the reference solvent:

$$G_i^\circ \text{ (solvent)} - G_i^\circ \text{ (reference solvent)} = RT \ln \gamma_t \qquad (4\text{-}2)$$

In the reference solvent, γ_t would be unity by definition.

Differences in transfer activity coefficients arise as a result of the nonspecific effects of differences in dielectric constants and of specific interactions between the solute and solvent. Although these two factors cannot really be considered in isolation from each other, for convenience they are treated separately, in Sections 4-2 and 4-3.

The estimation of transfer activity coefficients was reviewed by Popovych.[6] One method for measurement of transfer activity coefficients for electrolytes or neutral molecules is by measurement of solubility.[1,7,8] When the solubility is low, the effects of contamination by traces of water can be profound. Furthermore, reliable solubility values even in water are difficult to obtain. Nevertheless, if saturated solutions of a substance in water and another solvent are considered, and if each solution can be shown to be in equilibrium with the same solid, the value of $\gamma_{t\pm}{}^n$ is given by the ratio of the solubility products $(K_{sp})_{water}/(K_{sp})_{solvent}$ for an electrolyte producing n ions. For the hypothetical case of totally immiscible solvents, the relative transfer activity coefficient for a solute is equal to the distribution ratio of the substance between the two solvents.

Mean activity coefficients have been evaluated for hydrochloric acid by potential measurements[2,9,10] in alcohols. The salt-effect activity coefficient (left) and its product with the transfer activity coefficient (right) are shown in Figure 4-1. The values of $\gamma_\pm$ are lower than would be calculated from the appropriate modification of the Debye-Hückel equation (2-21) applied in the usual way to account for interionic interactions. The low values result from significant ion pairing due to the low dielectric constant. Thus, 0.1 M hydrochloric acid in 95% ethanol is about half in the form of ion pairs[9] rather than being completely dissociated. As shown in Figure (4-1), at low concentrations the salt-effect activity coefficients approach unity, as they must by definition, whereas at moderate concentrations they are somewhat less than unity. On

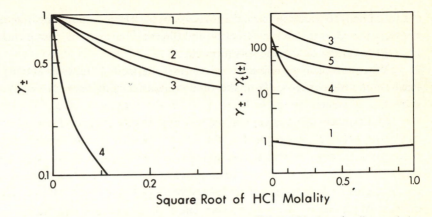

FIGURE 4-1 Comparison of the mean coefficients for the salt effect and the overall activity coefficient for hydrochloric acid in several solvents at 25°C: *1*, water; *2*, 80% 2-methoxyethanol; *3*, ethanol; *4*, 82% dioxane; *5*, methanol. (*From Bates,*[2] *Pool and Bates,*[9] *and Thun, Staples, and Bates.*[10])

the other hand, the transfer activity coefficients are generally large at all values; those in ethanol are about two orders of magnitude greater than those in water.

Transfer activity coefficients for single ions are not rigorously measurable by thermodynamic methods. If γ_t could be determined for any single ion, then a universal scale for all ion activities could be developed. Estimates of relative single-ion values for γ_t for either cations or anions can, however, be made by extrathermodynamic means.[6] One possible assumption is that transfer activity coefficients are equal in an electrolyte consisting of similar, large, symmetrical ions of low charge. One suggestion[11] for such a salt is tetraphenylphosphonium tetraphenyl borate, another[12] is triisoamyl-*n*-butylammonium tetraphenyl borate, and another is tetraphenylarsonium tetraphenyl borate.[1,13] Another assumption is that a large ion and its corresponding uncharged species form a pair whose transfer activity coefficients are the same; ferrocene-ferricinium ion, for example, is a couple[14] whose potential might be assumed to be independent of solvent. Still another assumption might be that a negligible liquid-junction potential exists between two Ag | AgClO$_4$ (0.01 M) half-cells in two solvents when linked by a 0.1 M tetraethylammonium picrate salt bridge.[15,16]

These several assumptions do not lead to the same conclusions. For example, transfer activity coefficients obtained by the tetraphenylarsonium tetraphenyl borate assumption differ[1] in water and polar aprotic solvents by up to 3 log units from those based on the ferrocene assumption. From data compiled by Kratochvil and Yeager[17] on limiting ionic conductivities in many organic solvents, it is clear that no reference salt can serve for a valid comparison of all solvents. For example, the tetraphenylarsonium and tetraphenyl borate ions have limiting conductivities of 55.8 and 58.3 in acetonitrile.[18] Krishnan and Friedman[19] concluded that the solvation enthalpy of the tetraphenyl borate ion in methanol is about 30 kJ/mole more negative than that of the tetraphenylarsonium ion. Coetzee and Sharpe,[20] using pmr, found that these ions undergo a variety of specific solvation effects.

When only a single solvent is of interest, questions of transfer activity coefficient need be of little concern. To make valid comparisons of data for different solvents, however, agreement needs to be reached on the problem of transfer activity coefficients, probably in somewhat the same way as it has on the definition of an operational pH scale. Meanwhile, values of γ_t for single ions are not completely unknown, even though not known exactly in a defined sense. The estimate obtained for any one ion depends to some extent on the assumption adopted. Recent comparative estimates for a number of single ions are given by several authors.[1,20,21] With a standard state arbitrarily defined in terms of activity equal to concentration at infinite dilution in water, the transfer activity coefficient for the sodium ion in methanol[1] has been given as about 30. It can be symbolized† as $^{W}\gamma^{M}_{t(Na^+)} = 30$, where W stands for water (the reference solvent, for which γ_t is unity by definition) and M for methanol. Similarly, the values[1] of transfer activity coefficients for hydrogen ions are as follows: $^{W}\gamma^{M}_{t(H^+)}$ about 80, $^{W}\gamma^{AN}_{t(H^+)}$ about 1.3×10^8, $^{W}\gamma^{DMF}_{t(H^+)}$ about 3×10^{-3}, and $^{W}\gamma^{DMSO}_{t(H^+)}$ about 5×10^{-4}, where AN, DMF, and DMSO stand for acetonitrile, dimethylformamide, and dimethyl sulfoxide. The value for $^{W}\gamma^{M}_{t(Ag^+)}$ is about 20, and for $^{W}\gamma^{AN}_{t(Na^+)}$ about 250. Acetonitrile solvates many ions more weakly[22] than does water, and their resulting high reactivity is reflected in large transfer activity coefficients. In contrast,[1,23] the transfer activity coefficient for silver ions in acetonitrile is small, about 1.3×10^{-4}. For hydrogen ions in pyridine, Mukherjee[24] gives a value of about 1.4×10^{-4}. Values such as these have led to useful correlations and qualitative predictions.

4-2 DIELECTRIC CONSTANT

According to Equation (4-1) the activity coefficient is made up of two terms, one of which is the transfer activity coefficient γ_t. This term may be regarded as resulting from the specific interactions of solute and solvent and the nonspecific (electrostatic) effects from differences in dielectric constant. In this section we consider the electrostatic effects, and in the next section specific interaction effects.

Transfer activity coefficient The electrostatic, or nonspecific, contribution to the transfer activity coefficient can be obtained by estimating the free-energy change involved in transferring a sphere of radius r and charge $Z_i e$ from one solvent to another of different dielectric constant. According to the Born equation,[25–28] when 1 mole

† It can also be symbolized as $\gamma_t(Na^+)_{W\to M}$. J. F. Coetzee, private communication.

of an electrolyte is transferred from the gas phase to a solvent of dielectric constant D, the free-energy change is

$$\Delta G = -\frac{Ne^2}{2}\left(1 - \frac{1}{D}\right)\left(\frac{Z_+^2}{r_+} + \frac{Z_-^2}{r_-}\right) \qquad (4\text{-}3)$$

where N is Avogadro's number, e the electron charge, and r the radius of an ion in solution. If the transfer is from a medium of dielectric constant D_1 to another of dielectric constant D_2, (4-3) becomes

$$\Delta G = -\frac{Ne^2}{2}\left(\frac{1}{D_1} - \frac{1}{D_2}\right)\left(\frac{Z_+^2}{r_+} + \frac{Z_-^2}{r_-}\right) \qquad (4\text{-}4)$$

For electrolytes in which $Z_+ = Z_-$ and $r_+ = r_-$, Equation (4-4) becomes, for 1 mole of electrolyte,

$$\Delta G_\pm = \frac{NZ_\pm^2 e^2}{r}\left(\frac{1}{D_1} - \frac{1}{D_2}\right) \qquad (4\text{-}5)$$

The chemical potential is the partial molal free energy, and by modifying Equation (2-3), we obtain

$$\Delta\mu = -\Delta G_\pm = RT \ln \gamma_t^2$$

and hence

$$\log \gamma_t = \frac{NZ_\pm^2 e^2}{2 \times 2.303 RTr}\left(\frac{1}{D_1} - \frac{1}{D_2}\right) \qquad (4\text{-}6)$$

In (4-6), R is 8.3145×10^7 ergs/deg-mole. The radius r is usually estimated to be equal to the crystal radii and should not be confused with the ion-size parameter used in Debye-Hückel calculations.

EXAMPLE 4-1 Calculate the expected electrostatic contribution to the transfer activity coefficient from differences in dielectric constant for ethanol and water (dielectric constants 24.3 and 78.3) at 25°C for a 1:1 electrolyte with an average ionic radius of 1.5×10^{-8} cm.

ANSWER

$\log \gamma_t$

$$= \frac{6.023 \times 10^{23} \times 1^2 (4.803 \times 10^{-10})^2}{2.303 \times 2 \times 8.3145 \times 10^7 \times 298.1 \times 1.5 \times 10^{-8}}\left(\frac{1}{24.3} - \frac{1}{78.3}\right)$$

$$= 2.30$$

$$\gamma_t = 200 \qquad\qquad ////$$

4-3   SPECIFIC SOLUTE-SOLVENT INTERACTIONS AND PROTON TRANSFER REACTIONS

Conjugate acid-base pairs   Recall from Chapter 3 that the definition of acids and bases most useful to analytical chemistry is that of Brønsted, in which a conjugate acid-base pair is related by the reaction

$$HA \rightleftharpoons H^+ + A^- \qquad (4\text{-}7)$$

Equation (4-7) represents an acid-base half-reaction, which involves protons, analogous to an oxidation-reduction half-reaction (Chapter 15), which involves electrons. Protons, even less than electrons, do not exist in a free state to an appreciable extent. Therefore an acid dissociates to yield protons only when a base is available to accept them; that is, two conjugate pairs are necessary for an acid-base reaction. Several conjugate acid-base pairs, arranged in order of decreasing acidity of HA and therefore increasing basicity of A^-, are listed in Table 4-1.

The equilibrium constant of the acid-base half-reaction (4-7) could be taken to be a measure of the intrinsic acidity of HA, and its reciprocal a measure of the intrinsic basicity of A^-. Thus[29]

$$K_{acidity} = \frac{a_A \cdot a_{H^+}}{a_{HA}} = \frac{1}{K_{basicity}} \qquad (4\text{-}8)$$

For a complete acid-base reaction, two acid-base half-reactions are combined in the correct way,

$$HA' + A''^- \rightleftharpoons HA'' + A'^-$$

and the equilibrium constant for the whole reaction is

$$K_{eq} = \frac{a_{HA''}\, a_{A'^-}}{a_{HA'}\, a_{A''^-}} = (K'_{acidity})(K''_{basicity}) \qquad (4\text{-}9)$$

Table 4-1   SOME CONJUGATE ACID-BASE PAIRS

Acid	Base
$HClO_4 \rightleftharpoons ClO_4^- + H^+$	
$H_2OAC^+ \rightleftharpoons HOAC + H^+$	
$H_3O^+ \rightleftharpoons H_2O + H^+$	
$HOAc \rightleftharpoons OAc^- + H^+$	
$NH_4^+ \rightleftharpoons NH_3 + H^+$	
$H_2O \rightleftharpoons OH^- + H^+$	
$EtOH \rightleftharpoons OEt^- + H^+$	
$NH_3 \rightleftharpoons NH_2^- + H^+$	
$OH^- \rightleftharpoons O^= + H^+$	

A solvent SH can fulfill the role of one of the conjugate pairs, so that the following acid-base reaction is obtained:

$$HA + SH \rightleftharpoons A^- + SH_2^+ \qquad (4\text{-}10)$$

The equilibrium constant for (4-10) is the ionization constant of the acid HA in the solvent SH, which could formally be considered to be

$$K_i = \frac{a_{A^-}a_{SH_2^+}}{a_{HA}a_{SH}} = (K_{acidity})_{HA}(K_{basicity})_{SH} \qquad (4\text{-}11)$$

Equation (4-11) illustrates the role of solvent basicity in determining the strength of a solute acid. Equation (4-11) is, however, of little direct use in analytical acid-base measurements because we have no methods of evaluating the absolute constants $K_{acidity}$ and $K_{basicity}$. But suppose we wish to compare the ionization constants of two acids HA' and HA" in a solvent SH. From (4-11)

$$\frac{K_i'}{K_i''} = \frac{(K_{acidity})_{HA'}}{(K_{acidity})_{HA''}} \qquad (4\text{-}12)$$

and the basicity of the solvent should cancel. Furthermore, for two solvents S'H and S"H the ratio K_i'/K_i'' for a pair of acids should be independent of the nature of the solvents.

Actually, Equation (4-11) does not describe adequately the acid-base reaction (4-10) as we usually determine it experimentally. Reaction (4-10) can be considered to take place in two stages; the first is the proton transfer or *ion-pair generation* reaction, and the second is the *formation of free ions*. These two stages for an acid can be expressed as

$$HA + SH \rightleftharpoons SH_2^+A^- \rightleftharpoons SH_2^+ + A^- \qquad (4\text{-}13)$$

For a base this becomes

$$SH + B \rightleftharpoons BH^+S^- \rightleftharpoons BH^+ + S^- \qquad (4\text{-}14)$$

Usually the concentrations or activities of the free ions such as A^- and SH_2^+ are of primary interest. For most 1:1 electrolytes the equilibrium for both steps is far to the right in solvents of high dielectric constant, but not in those of low dielectric constant. The distinction between ion-pair generation and the formation of free ions then becomes important.[30] We denote the equilibrium constants for the two steps by $K_{i_{IP}}$ for the ion-pair step and $K_{i_{FI}}$ for the free-ion step (Section 4-5).

Solvent classification   In acid-base reactions the solvent plays an active or specific role in two ways: it may react generally with ions and molecules (solvation), and as indicated above, it has acidic and basic properties that are of active concern. Broadly, solute-solvent interactions are studied by electrical and spectral methods.[18,27,31–35] Nmr and gas-phase solvation techniques are especially useful in elucidating interaction phenomena.

Solvents may be classified† broadly according to their proton donor-acceptor properties as either *amphiprotic*, that is, both acidic and basic, or *aprotic*, neither acidic nor basic.

Autoprotolysis Amphiprotic solvents in particular undergo *autoprotolysis*, or self-ionization, as illustrated by the following reactions:

$$2H_2O \rightleftharpoons H_3O^+ + OH^- \qquad (4\text{-}15)$$

$$2EtOH \rightleftharpoons EtOH_2^+ + OEt^- \qquad (4\text{-}16)$$

$$2HOAc \rightleftharpoons H_2OAc^+ + OAc^- \qquad (4\text{-}17)$$

$$2NH_3 \rightleftharpoons NH_4^+ + NH_2^- \qquad (4\text{-}18)$$

or in general,

$$2SH \rightleftharpoons SH_2^+ + S^- \qquad (4\text{-}19)$$

where SH_2^+ denotes the solvated proton and S^- the conjugate base of the solvent. Amphiprotic solvents may be predominantly acidic, basic, or neither. Thus we may regard glacial acetic acid as an acidic solvent, pyridine as a basic solvent, and ethanol as predominantly neither acidic nor basic.

As a result of its acidic and basic properties, an amphiprotic solvent may exert a *leveling effect*[36] on solute acids and bases. Leveling occurs if a solute acid HA is a much stronger proton donor than the protonated solvent SH_2^+. Reaction (4-10) then goes to completion, as far as can be determined experimentally, and the acid is said to be *leveled* to the strength of the solvated proton. In water, acids such as perchloric acid, hydrochloric acid, and nitric acid are stronger proton donors than H_3O^+. Therefore, all are leveled to the same strength, that of the hydronium ion. In a less basic solvent, such as glacial acetic acid, perchloric acid is a stronger acid than hydrochloric acid. Glacial acetic acid acts as a *differentiating solvent* in this instance because it levels acids at a higher ultimate strength, namely that of H_2OAc^+, and so permits differentiation of acids that are both strong in aqueous solution. In liquid ammonia, a solvent more basic than water, solute acids are leveled to the strength of the ammonium ion, which is a weaker acid than the hydronium ion. Therefore, in liquid ammonia, perchloric acid is no stronger than acetic acid.

Analogous specific interactions or leveling effects exist for solute bases. In water any base stronger than hydroxyl ion is leveled to the strength of the hydroxyl ion.

---

† Classification systems (I. M. Kolthoff, private communication) according to proton-donating or -accepting ability (protophilic, protogenic, inert, and so on) and dielectric constant (greater or less than 20 to 30) have been proposed. A continuing problem is that the borderlines between classes are necessarily arbitrary and particular solvents often do not fit well into a classification.

Thus, oxide ions or amide ions cannot exist in measurable concentrations in water because the reactions

$$O^= + H_2O \rightarrow 2OH^- \qquad (4\text{-}20)$$

and

$$NH_2^- + H_2O \rightarrow NH_3 + OH^- \qquad (4\text{-}21)$$

are quantitative. Liquid ammonia can differentiate between stronger bases than can water. In glacial acetic acid, on the other hand, bases stronger than acetate ion will be leveled to its strength.

In solvents of moderately high dielectric constant, the dissociation constant of an acid is determined primarily by the basicity of the solvent. Thus in two solvents S'H and S"H with similar dielectric constants, for a particular acid HA we may write the equilibria

$$HA + S'H \rightleftharpoons S'H_2^+ + A^- \qquad K'_a \qquad (4\text{-}22)$$

$$HA + S''H \rightleftharpoons S''H_2^+ + A^- \qquad K''_a \qquad (4\text{-}23)$$

where the ratio K'_a/K''_a is determined by the relative tendencies for S'H and S"H to act as bases (their relative basicities). Since this observation is true for all acids of the same charge type, it follows that the relative strengths of *any two acids* of the same charge type are approximately the same in solvents having similar dielectric constants. Wooten and Hammett[37] studied a number of uncharged acids in *n*-butyl alcohol ($D = 17.4$) and in water. They found that, in spite of the relatively large difference between the dielectric constants of the solvents, the relative strengths of the acids were essentially independent of the nature of the acid.

Other things being equal, specific solute-solvent interactions resulting from the acidity or basicity of the solvent affect the transfer activity coefficient of the hydrogen ion. Qualitatively, if the basicity of a solvent is greater than that of water, the proton from an acid is attracted to the solvent more strongly than it is to water, and the activity of the acid is smaller than in water at the same concentration. Conversely, if the basicity of the solvent is less, the transfer activity coefficient is larger. Comparisons of this sort must be made with care, however; the dielectric constant of the solvent and the charge type of the acid also exert important nonspecific effects. The conjugate base, too, is significant. For example, the acetate ion associates with cations to varying degrees, and therefore different acetate salts do not exhibit equal basicity.

The autoprotolysis constant   The extent of ionization (4-19) of a pure amphiprotic solvent is measured by the autoprotolysis constant K_{SH}, defined as the product $a_{SH_2^+} a_{S^-}$. Since the autoprotolysis reaction results in the formation of both solvent cations and solvent anions, the autoprotolysis constant is a measure of the differentiating ability of a solvent. If a solvent has a large K_{SH} value, the existence in it of a wide range of strengths of either acids or bases is not possible. In contrast, if the autoprotolysis constant is small, acids and bases of varying strengths show titration curves distinctly different from each other.

Since autoprotolysis is an acid-base reaction in which one molecule of solvent acts as an acid and another as a base, the extent of autoprotolysis is determined by the acid strength, the base strength, and the dielectric constant of the solvent. The acid strength and base strength are not reciprocally related. Hence the acidity of water is the inverse of the basicity of the hydroxyl ion, not the basicity of water; similarly, the basicity of water is the inverse of the acidity of the hydronium ion. For each autoprotolysis reaction we are concerned with two acid-base conjugate pairs, SH_2^+/SH and SH/S^-. Thus, to compare acidities of the solvents S'H and S"H, we must determine the equilibrium constant of the reaction

$$S'H + S''^- \rightleftharpoons S'^- + S''H \qquad (4\text{-}24)$$

whereas to compare basicities, we must determine the equilibrium constant of

$$S'H_2^+ + S''H \rightleftharpoons S'H + S''H_2^+ \qquad (4\text{-}25)$$

For example, comparing ethanol and water as acids in isopropanol solution, Hine and Hine[38] found ethanol to be 0.95 as acidic as water, and Kolthoff[39] found ethanol to be 0.0025 as basic as water. Accordingly, considering acidic and basic properties alone, we might expect the autoprotolysis constant of ethanol to be $0.95 \times 0.0025 \times 10^{-14}$, or 2.4×10^{-17}. The actual value[40] is smaller, 3×10^{-20}, evidently in part because the low dielectric constant of 25 causes more association than expected. In addition, the associated structure of water and alcohols in the liquid state complicates estimates of this type.

The low autoprotolysis constant[41] of liquid ammonia (10^{-33} at $-50°C$) suggests that the strongly basic properties of ammonia in comparison with water are more than counterbalanced by its feeble acidic properties. Once more, a relatively low dielectric constant of 22 contributes to association to a minor extent.

Glacial acetic acid represents the other extreme from ammonia, that of a solvent strongly acidic but weakly basic compared with water. These two characteristics by themselves would cause glacial acetic acid to have a relatively high autoprotolysis constant. Nevertheless, owing to the low dielectric constant (6.13), the autoprotolysis constant ($pK_{SH} = 14.45$) turns out to be about the same as that of water.[42]

Formic acid is more acidic than acetic acid, but has a high dielectric constant (62). The autoprotolysis constant[43] is so large ($pK_{SH} = 6.2$) that formic acid is relatively useless as a titration medium. Similarly, sulfuric acid has a high autoprotolysis constant ($pK_{SH} = 3.85$) and a high dielectric constant.

Calculations such as in Example 4-1 are but a first estimate and do not allow for specific solute-solvent interactions. The calculations cannot be entirely correct because on the molecular level a solvent is not a dielectric continuum; the effective dielectric constant near the intense field of an ion is decreased. Furthermore, the equation assumes that ions are spherical, nonpolarizable entities with the charge located at the center. Latimer, Pitzer, and Slanski[44] modified the Born equation by including correction terms for ionic radii. Noyes[45] and later Hepler[46] modified the equation to allow for the lowering of the dielectric constant that takes place in the immediate vicinity of ions (dielectric saturation). DeLigny and Alfenaar[47] and more recently Popovych[6] reviewed methods of estimating the nonspecific effects of the solvent on the transfer activity coefficient.

In general, the activity coefficient of an electrolyte will be larger in a solvent of lower dielectric constant than in water. Transfer activity coefficients are likely to be large if the dielectric constant is small or if the ions are of high charge or small radius.

4-4 ELECTROLYTIC DISSOCIATION

As stated in Section 2-3, electrostatic interactions among ions in solution are governed by the Coulomb and Boltzmann laws. According to Coulomb's law the force of attraction between two ions of charge Z_+e and Z_-e is proportional to the product of their charges and inversely proportional to the product of the dielectric constant and the square of the distance between them. The Boltzman distribution law describes the tendency for thermal agitation to counteract the effects of electrical attraction. If two ions are near each other, thermal energy may be less than the energy of mutual attraction, and they may persist as an ion pair. If they are farther apart, thermal energy may be greater than the electrostatic attraction, and the ion pair probably will not survive collisions with solvent molecules long enough to be considered a separate entity from the viewpoint of participation in chemical reactions. At some critical intermediate distance the forces of electrical attraction and thermal agitation just counterbalance. This critical distance q, the distance of minimum probability of finding an ion of charge opposite that of a central ion, was defined by Bjerrum[48,49] by

$$q = \frac{Z_+Z_-e^2}{2DkT} \qquad (4\text{-}26)$$

where the absolute values of the ion charges Z_+ and Z_- are taken and k is Boltzmann's constant. This electrostatic expression is limited in that it takes into account neither specific interaction effects nor dielectric saturation. Like the Born equation, (4-26) can be used only to obtain an approximation. Figure 4-2 illustrates[50] the predicted linear relation between the formation constant of a salt (determined by conductance) and the inverse of the dielectric constant of a solvent.

Figure 4-2 illustrates that, in solvents of low dielectric constant, dissociation constants are small. In practice, 1:1 electrolytes are incompletely dissociated in water at 1 M or higher and in other solvents of dielectric constant less than about 40 (Figure 4-2) at reasonable concentrations above about 10^{-3} M. Even in water, 2:2 electrolytes are incompletely dissociated in spite of the strong specific solvent-solute

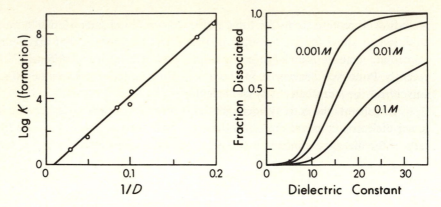

FIGURE 4-2   Left, relation between the observed formation constant for the salt tetra *n*-butylammonium picrate and the dielectric constant of the solvent. (*From Inami, Bodenseh, and Ramsey.*[50]) Right, relation between fraction dissociated and dielectric constant calculated from data in the left part of the figure.

interactions and the high dielectric constant. According to Grunwald and Kirschenbaum,[51] with the usual analytical concentrations the principal species in solution can be assumed to be an ion pair when the dissociation constant is less than about 10^{-3} and to be free ions above this value.

From (4-26), for a 1:1 electrolyte in water the Bjerrum critical distance is 3.6×10^{-8} cm. When it is realized that ions in water are normally highly solvated and that the sum of ionic crystal radii for typical anions and cations often approaches or exceeds 3.6×10^{-8} cm, it is reasonable to find that dissociation constants for ion pairs in water are large; thus for sodium hydroxide the dissociation constant is about 5. On the other hand, for a 2:2 electrolyte in water the critical distance is 14.3×10^{-8} cm, and for a 1:1 electrolyte in ethanol, 11.5×10^{-8} cm. In these cases, even highly solvated ions can readily approach to the distance necessary to form an ion pair. For magnesium sulfate the dissociation constant in water[51] is 6×10^{-3}, and for sodium sulfate, 0.2.

In addition to ion-pair formation, other types of association reactions are important in nonaqueous solvents. Evidence[49,52,53] for triple and quadruple ion formation in nonaqueous solvents is obtained from conductimetric, solvent extraction, calorimetric, or cryoscopic measurements. Self-association reactions, that is, equilibria such as $2HA \rightleftharpoons (HA)_2$, have been reviewed[54] and have been studied by differential-vapor-pressure techniques.[55] Frequently the anion A^- obtained from an acid HA is poorly solvated; stabilization then may occur by interaction (homoconjugation) with a second molecule of acid to give HA_2^-. Homoconjugation can be studied by techniques such as spectroscopy or conductimetry. Thus, the homoconjugation formation constant[56] for 3,5-dinitrobenzoic acid in methyl isobutyl ketone is 2×10^4.

The question of hydrogen bonding has been given insufficient attention.[57] Small anions with localized charge are strongly solvated in water by hydrogen bonding; HF_2^- represents the most stable hydrogen bond. Such solvation is poor in polar aprotic solvents such as acetonitrile, dimethylformamide, and dimethylsulfoxide. Dimerization and polymerization of acids are further examples of hydrogen bonding. The solvation of cations in such polar aprotic solvents becomes relatively more significant than that of anions. With anions such as picrate, in which the charge is not localized, the transfer activity coefficients in polar aprotic solvents are not abnormally large.[1]

4-5   EQUILIBRIUM EXPRESSIONS USED IN pH CALCULATIONS

Water is highly unusual in the extent of its interactions with solutes, but even minimal solvent-solute interactions can play a major role in the nature of chemical reactions. To calculate pH during acid-base titrations in a nonaqueous solvent, we must consider not only the equilibria discussed in Chapter 3 but also reactions discussed in Sections 4-2, 4-3, and 4-4.

When ion-pair formation is significant, ion-pair generation must be differentiated from formation of free ions as indicated by Equations (4-13) and (4-14). The equilibrium constants corresponding to these two equations can be written just as for other equilibria, either in terms of activities (not including transfer activity coefficients) or approximately in terms of concentrations. If uncharged entities (including ion pairs) are assumed to have activity coefficients of unity, the constant for ion-pair generation $SH_2^+A^-$ from an acid HA in a solvent SH, $K_{HA_{IP}}$, is written

$$K_{HA_{IP}} = \frac{[SH_2^+A^-]}{[HA]} \qquad (4\text{-}27)$$

The constant for dissociation of the ion pair $SH_2^+A^-$ to free ions, $K_{HA_{FI}}$, is

$$K_{HA_{FI}} = \frac{a_{SH_2^+}a_{A^-}}{[SH_2^+A^-]} \qquad (4\text{-}28)$$

In terms of ion pairs and free ions the usual overall dissociation constant K_{HA} is

$$K_{HA} = \frac{a_{SH_2^+}a_{A^-}}{[HA] + [SH_2^+A^-]} \qquad (4\text{-}29)$$

and is related to the other two constants, since from (4-27) and (4-28), K_{HA} is equal to $K_{HA_{IP}}K_{HA_{FI}}/(1 + K_{HA_{IP}})$. Corresponding expressions for a base B are:

$$K_{B_{IP}} = \frac{[BH^+S^-]}{[B]} \qquad (4\text{-}30)$$

$$K_{B_{FI}} = \frac{a_{BH^+}a_{S^-}}{[BH^+S^-]} \qquad (4\text{-}31)$$

$$K_B = \frac{a_{BH^+}a_{S^-}}{[B] + [BH^+S^-]} \qquad (4\text{-}32)$$

The salt BHA formed from the reaction of the acid with the base must also be considered

$$BHA \rightleftharpoons BH^+A^- \rightleftharpoons BH^+ + A^- \qquad (4\text{-}33)$$

$$K_{BHA_{IP}} = \frac{[BH^+A^-]}{[BHA]} \qquad (4\text{-}34)$$

$$K_{BHA_{FI}} = \frac{a_{BH^+}a_{A^-}}{[BH^+A^-]} \qquad (4\text{-}35)$$

and the overall dissociation constant is by definition

$$K_{BHA} = \frac{a_{BH^+}a_{A^-}}{[BHA] + [BH^+A^-]} \qquad (4\text{-}36)$$

When solutes are poorly solvated, ion stabilization through homoconjugation must be allowed for in calculations:

$$A^- + HA \rightleftharpoons HA_2^- \qquad (4\text{-}37)$$

Thus the formation constant $K_{f_{HA_2^-}}$ (Section 11-1) for homoconjugation is

$$K_{f_{HA_2^-}} = \frac{a_{HA_2^-}}{[HA]a_{A^-}} \qquad (4\text{-}38)$$

Similar homoconjugate complexes should be considered for a conjugate acid BH^+:

$$BH^+ + B \rightleftharpoons BHB^+ \qquad (4\text{-}39)$$

$$K_{f_{BHB^+}} = \frac{a_{BHB^+}}{a_{BH^+}[B]} \qquad (4\text{-}40)$$

Finally, we must include the autoprotolysis-constant expression from (4-19):

$$K_{SH} = a_{SH_2^+}a_{S^-} \qquad (4\text{-}41)$$

If the dielectric constant is low enough, even "strong" electrolytes are only slightly dissociated (Figure 4-2), and so high ionic strengths are not encountered. Therefore, little error results from writing concentrations rather than activities for the ions as well as the uncharged substances in such solvents.

Unless otherwise indicated, in the use of Equations (4-27) through (4-41) in the remaining sections of this chapter, activity coefficients are used without including the transfer activity coefficient.

4-6   ACID-BASE EQUILIBRIA IN GLACIAL ACETIC ACID

Glacial acetic acid, though like water in being classed as amphiprotic, represents a solvent type distinctly different from water in that it is a much weaker base. This weak basicity makes it a useful solvent for the titration of weakly basic substances. As mentioned in Section 4-3, the autoprotolysis constant has a pK_{SH} value of 14.45 for the reaction

$$2HOAc \rightleftharpoons H_2OAc^+ + OAc^- \qquad (4\text{-}42)$$

Its dielectric constant, 6.13, is low compared with that of water, 78.3. This low value results in a strong tendency toward incomplete dissociation of strong electrolytes, few of which have dissociation constants as large as 10^{-5}. Therefore, in the application of the equations in Section 4-5, ion activities can be written simply as concentrations.

An acid may, rather arbitrarily, be called a strong acid in glacial acetic acid if $K_{HA_{IP}} \geq 1$. Thus perchloric acid is a strong acid, and yet the pK for the overall dissociation constant is only 4.87 because it exists largely as ion pairs.[42] Hydrochloric acid has an overall pK value of 8.55; ammonia, 6.40; pyridine, 6.10; sodium acetate, 6.68; potassium chloride, 6.88; and sodium perchlorate, 5.48. Perchloric acid is the strongest acid and the one used for titration of bases that may be too weak to be titrated in water as solvent. At first it appears that an attempt to titrate a base such as pyridine with perchloric acid would fail, since both have small overall dissociation constants. Critical to the success of such titrations is the small dissociation constant of the salt formed, which results in a large favorable equilibrium constant for the reaction.

In the following sections we consider the equilibria involved in the titration of a base B with perchloric acid, a reaction of practical interest; the titration of acids with acetic acid as solvent is unimportant. The acidity of the solutions can be determined by an indicator or by the change in potential of an electrode responsive to free solvated protons. In water these two methods give the same results. In acetic acid, measurements of potential depend on the extent of formation of free solvated protons, whereas indicators respond to the extent of proton formation whether ion-paired or not; thus two different measures of acidity are possible. The constant for ion-pair generation sometimes may be large, but the constant for dissociation to free ions never is. We discuss here the change in acidity during a titration as it would be obtained potentiometrically and consider the behavior of indicators in Section 4-11.

Acidity of a solution of base B   In a solution of base B in acetic acid, if C_B is the analytical concentration of base,

$$C_B = [B] + [BH^+OAc^-] + [BH^+] \qquad (4\text{-}43)$$

and the concentration of undissociated base is $[B] + [BH^+OAc^-]$. Since at concentrations of analytical interest the extent of dissociation of BH^+OAc^- is small, we can approximate C_B to be equal to $[B] + [BH^+OAc^-]$. Neglecting the contribution to acetate concentration arising from the autoprotolysis reaction (4-42), we then can set $[BH^+]$ equal to $[OAc^-]$, and using concentrations instead of activities and the symbol $[OAc^-]$ instead of $[S^-]$ in (4-32), we obtain

$$K_B = \frac{[OAc^-]^2}{C_B} \qquad (4\text{-}44)$$

Substituting this value of $[OAc^-]$ in the autoprotolysis expression (4-41) yields

$$[H_2OAc^+] = \frac{K_{SH}}{\sqrt{K_B C_B}} \qquad (4\text{-}45)$$

Thus the initial acidity, or hydrogen ion concentration, varies inversely with the square root of the concentration of the base just as it does for a weak base in water. For a 100-fold increase in base concentration the pH increase is 1 unit.[42, 58, 59]

Solution of base B and its perchlorate   At the points in a titration between the initial and equivalence points, it is not possible either to reason directly by analogy with titrations in water or to use buffer formulas derived for weak acids in aqueous solutions. The dissociation equilibria of the ion pair $BH^+ClO_4{}^-$, the acid $HClO_4$, the base B, and the solvent HOAc all must be considered.

The electroneutrality expression is

$$[H_2OAc^+] = [OAc^-] + [ClO_4{}^-] - [BH^+] \qquad (4\text{-}46)$$

Here the H_2OAc^+ ions in solution come from autoprotolysis of the solvent and from dissociation of perchloric acid. From (4-32) and (4-42), again with the assumption that $[B] + [BH_2{}^+OAc^-]$ approximately equals C_B,

$$[BH^+] = \frac{K_B C_B}{[OAc^-]} = \frac{K_B C_B [H_2OAc^+]}{K_{SH}} \qquad (4\text{-}47)$$

Applying (4-36) to the perchlorate salt, setting $[BHClO_4] + [BH^+ClO_4{}^-]$ equal to C_{BHClO_4}, and combining with (4-47) results in

$$[ClO_4{}^-] = \frac{K_{BHClO_4} C_{BHClO_4}}{[BH^+]} = \frac{K_{BHClO_4} C_{BHClO_4} K_{SH}}{K_B C_B [H_2OAc^+]} \qquad (4\text{-}48)$$

Combining Equations (4-42), (4-46), (4-47), and (4-48) gives

$$[H_2AOc^+] = \left\{ \frac{K_{SH}(1 + K_{BHClO_4} C_{BHClO_4}/K_B C_B)}{1 + K_B C_B/K_{SH}} \right\}^{\frac{1}{2}} \qquad (4\text{-}49)$$

Further simplifying approximations can be made. If the base is strong enough to give a detectable potential break at the end point, $K_B C_B / K_{SH}$ is much greater than 1, so (4-49) becomes

$$[H_2OAc^+] = \frac{K_{SH}}{K_B C_B} (K_{BHClO_4} C_{BHClO_4} + K_B C_B)^{\frac{1}{2}} \qquad (4\text{-}50)$$

Kolthoff and Bruckenstein[60] pointed out a useful relation for the case in which K_B and K_{BHClO_4} are approximately equal. For example, the value of pK for diethylaniline, both as a base and as a salt, is 5.78. If the titrant is relatively concentrated so that the sum $C_B + C_{BHClO_4}$ remains sensibly constant and equal to C, (4-50) is reduced to

$$[H_2OAc^+] = \frac{K_{SH}}{C_B} \sqrt{\frac{C}{K_B}} = \frac{K'}{C_B} \qquad (4\text{-}51)$$

where $K' = K_{SH}\sqrt{C/K_B}$, a quantity approximately constant during titration. The hydrogen ion concentration is thus inversely proportional to the concentration of untitrated base. If X is the fraction titrated, a plot of log $(1 - X)$ against pH is a straight line, as is observed for a strong acid–strong base titration in water. Such a relation has been experimentally observed[58] for guanidine and diethylaniline. It appeared paradoxical because, despite the apparent strong acid–strong base titration behavior, the pH change upon dilution of a solution of the pure base was still that of a typical weak base.

For an equimolar mixture of base and its salt, such that $C_B = C_{BHClO_4} = C$, (4-50) becomes

$$[H_2OAc^+] = K_{SH} \left(\frac{K_{BHClO_4} + K_B}{K_B{}^2 C} \right)^{\frac{1}{2}} \qquad (4\text{-}52)$$

and therefore the hydrogen ion concentration increases 10-fold for a 100-fold dilution. Hall and Werner[58] obtained a similar result. When the base is so weak that K_{BHClO_4} is much greater than K_B, (4-50) may be simplified to

$$[H_2OAc^+] = \frac{K_{SH}}{K_B C_B} \sqrt{K_{BHClO_4} C_{BHClO_4}} \qquad (4\text{-}53)$$

Solution of a pure salt. The equivalence point   To calculate the acidity at the equivalence point, we have only BHClO$_4$ to consider. During the titration the net reaction is

$$B + HClO_4 \rightleftharpoons BHClO_4 \qquad (4\text{-}54)$$

The salt dissociates in part to give equal numbers of BH$^+$ and ClO$_4{}^-$ ions. At the equivalence point, to the extent that the salt also reacts with solvent, equal amounts of base $B + BH^+OAc^-$ and of acid $HClO_4 + H_2{}^+OAcClO_4{}^-$ are produced. Combining Equations (4-29), (4-32), and (4-41) to give the expression for $K_{SH}K_{HClO_4}/K_B$, and canceling terms of the same magnitude results in

$$\frac{K_{SH}K_{HClO_4}}{K_B}$$

$$= \frac{[H_2{}^+OAc][OAc^-][H_2{}^+OAc][ClO_4{}^-]([B] + [B^+HOAc^-])}{[OAc^-][BH^+]([HClO_4] + [H_2{}^+OAcClO_4{}^-])}$$

$$= [H_2{}^+OAc]^2 \tag{4-55}$$

or

$$[H_2{}^+OAc] = \sqrt{\frac{K_{SH}K_{HClO_4}}{K_B}} \tag{4-56}$$

Solution of salt plus excess acid After the equivalence point the solution contains the salt $BHClO_4$ and excess acid, $HClO_4$. From (4-54) the formation constant of $BHClO_4$ is equal to $C_{BHClO_4}/C_B C_{HClO_4}$. Combining (4-29), (4-32), (4-36), and (4-41) gives

$$\frac{C_{BHClO_4}}{C_B C_{HClO_4}} = \frac{K_{HClO_4}K_B}{K_{SH}K_{BHClO_4}} \tag{4-57}$$

By substituting $K_B C_B/K_{SH}$ from (4-57) in (4-49) we can obtain[60]

$$[H_2OAc^+]^2 = \frac{K_{SH} + K_{HClO_4}C_{HClO_4}}{1 + K_{BHClO_4}C_{BHClO_4}/K_{HClO_4}C_{HClO_4}} \tag{4-58}$$

which holds at all points of the titration. In the presence of an appreciable excess of acid, $K_{HClO_4}C_{HClO_4}$ is much greater than K_{SH}, and (4-58) becomes

$$[H_2OAc^+] = \frac{K_{HClO_4}C_{HClO_4}}{\sqrt{K_{HClO_4}C_{HClO_4} + K_{BHClO_4}C_{BHClO_4}}} \tag{4-59}$$

In the vicinity of the end point, C_{BHClO_4} is much greater than C_{HClO_4}. Either of the two terms under the square-root sign may be negligible, depending on the relative magnitudes of K_{HClO_4} and K_{BHClO_4}.

4-7 ACID-BASE EQUILIBRIA IN ETHYLENEDIAMINE

Ethylenediamine (en), $NH_2C_2H_4NH_2$, a strongly basic substance, may be considered to represent solvents that are weakly acidic compared with water. Ethylenediamine is therefore useful as a solvent for the titration of weakly acidic substances.[61] It is a leveling solvent for acids whose ionization constants are larger than about 10^{-5} in water; thus acetic acid and hydrochloric acid are leveled to about equal strength. The titrant base normally used in en is sodium ethanolamine. The autoprotolysis constant[62] of en is 5×10^{-16} for the equilibrium

$$2NH_2C_2H_4NH_2 \rightleftharpoons NH_2C_2H_4NH_3{}^+ + NH_2C_2H_4NH^- \tag{4-60}$$

The dielectric constant for en of 12.5 means that ion pairing is extensive, though less so than for acetic acid. Although en, like acetic acid, contains no completely dissociated electrolytes, as a result of the larger dielectric constant the dissociation constants for ion pairing are larger, typically of the order of 10^{-3} for the strongest electrolytes. This means that, in the application of Equations (4-29), (4-32), and (4-36), approximations regarding concentrations of the type made for acetic acid as solvent frequently are invalid. For example, for many of the calculations the concentration of undissociated electrolyte cannot be assumed to be equal to the analytical concentration. Again, ion pairing in the salt product of neutralization reactions is important in determining the position of equilibrium. Homoconjugation is of some importance because anions of acids generally are weakly solvated in en, and accordingly, Equation (4-37) often becomes significant. Thus[62] the activity ratio for phenol $a_{HA_2^-}/a_{HA}a_{A^-}$ is 15. In the calculations, concentrations are assumed to be low enough that homoconjugation effects can be neglected. Another complicating factor is that activity-coefficient effects must be taken into account.

Activity coefficients in en   Schaap and others[63] reviewed the application of equations of the Debye-Hückel type to solutions of electrolytes in en. The appropriate modification of the limiting law (DHLL) becomes

$$-\log \gamma_\pm = 8.02\sqrt{\mu} \qquad (4\text{-}61)$$

where μ is the ionic strength resulting from the dissociated ions. This equation gives useful results up to an ionic strength of about 10^{-4} for $1:1$ electrolytes. Marshall and Grunwald[64] used an extended equation that for en becomes

$$-\log \gamma_\pm = \frac{8.02\sqrt{\mu}}{(1 + 18.1\sqrt{\mu})^{\frac{3}{2}}} \qquad (4\text{-}62)$$

On the basis of this equation a $1:1$ electrolyte with a dissociation constant of 10^{-4} has an activity coefficient[63] appreciably less than unity when the total salt concentration is in excess of about 10^{-4} M.

Acidity in a solution of weak acid   In the determination of acidity of a solution of acid HA in en, Equation (4-29) can be written to include activity coefficients of the ionic species:

$$K_{HA} = \frac{[SH_2^+][A^-]\gamma_{SH_2^+}\gamma_{A^-}}{[HA] + [SH_2^+A^-]} \qquad (4\text{-}63)$$

With the assumption that autoprotolysis and homoconjugation are negligible,

$$[HA] + [SH_2^+A^-] = C_{HA} - [SH_2^+] \qquad (4\text{-}64)$$

Since $[SH_2^+]$ is equal to $[A^-]$, and if we assume $\gamma_{SH_2^+} = \gamma_{A^-} = \gamma_\pm$, (4-63) becomes

$$K_{HA} = \frac{[SH_2^+]^2\gamma_\pm^2}{C_{HA} - [SH_2^+]} \qquad (4\text{-}65)$$

This equation can be solved for $[SH_2^+]$ by successive approximations: obtain an initial value for $[SH_2^+]$ assuming $\gamma_\pm^2$ to be unity, then use (4-62) and $[SH_2^+]$ to obtain a second $\gamma_\pm^2$ value and recycle.

EXAMPLE 4-2   Calculate the hydrogen ion concentration and pH in en of 0.1 M HA with a dissociation constant of 10^{-4}. Assume negligible dissociation of the solvent and homoconjugation of the solute.

ANSWER   Equation (4-65) can be written

$$K_{HA} = \frac{[SH_2^+]^2\gamma_\pm^2}{0.1 - [SH_2^+]} = 10^{-4}$$

First approximation:

$$[SH_2^+] \ll 0.1 \qquad \text{and} \qquad \gamma_\pm = 1$$

$$[SH_2^+]' = \sqrt{K_{HA} \times 0.1} = 3.2 \times 10^{-3} \ M$$

Second approximation: Use the Marshall-Grunwald equation (4-62) to estimate $\gamma_\pm$ for an ionic strength of 3.2×10^{-3}. A value of about 0.5 is obtained. Substitute 3.2×10^{-3} for $[SH_2^+]$ in the denominator and 0.5 for $\gamma_\pm$ and recalculate:

$$[SH_2^+]'' = \frac{\sqrt{(0.1 - 3.2 \times 10^{-3})1 \times 10^{-4}}}{0.5} = 6.2 \times 10^{-3} \ M$$

Third approximation:

$$(\gamma_\pm = 0.44), [SH_2^+]''' = 7.0 \times 10^{-3} \ M$$

Fourth approximation:

$$(\gamma_\pm = 0.43), [SH_2^+]'''' = 7.1 \times 10^{-3} \ M$$

and

$$pH = -\log 7.1 \times 10^{-3} \times 0.43 = 2.5$$

The first approximation gives a seriously low result.   ////

If the acid is only slightly dissociated at moderate concentrations, the quantity $[SH_2^+]$ can be neglected in comparison with C_{HA}; further, the ionic strength is sufficiently low that ionic activity coefficients are near unity. Equation (4-65) then becomes simplified to $[SH_2^+] = \sqrt{C_{HA}K_{HA}}$.

Solution of an acid and its salt   With the assumptions that dissociation of the solvent is negligible and that the anion from the salt represses dissociation of the acid, $C_{HA} \simeq [HA] + [SH_2{}^+A^-]$, and (4-29) becomes

$$K_{HA} = \frac{[SH_2{}^+][A^-]\gamma_\pm{}^2}{C_{HA}} \qquad (4\text{-}66)$$

or

$$[A^-] = \frac{K_{HA}C_{HA}}{[SH_2{}^+]\gamma_\pm{}^2} \qquad (4\text{-}67)$$

Equation (4-36) can be written

$$K_{BHA} = \frac{[BH^+][A^-]\gamma_\pm{}^2}{[BHA] + [BH^+A^-]} \simeq \frac{[BH^+][A^-]\gamma_\pm{}^2}{C_{BHA} - [BH^+]} \qquad (4\text{-}68)$$

The quantity $[BH^+] \simeq [A^-]$, so

$$K_{BHA} = \frac{[A^-]^2\gamma_\pm{}^2}{C_{BHA} - [A^-]} \qquad (4\text{-}69)$$

Substituting $[A^-]$ from (4-67) in (4-69) gives

$$
\begin{aligned}
K_{BHA} &= \frac{(K_{HA}C_{HA}/[SH_2{}^+]\gamma_\pm{}^2)^2\gamma_\pm{}^2}{C_{BHA} - K_{HA}C_{HA}/[SH_2{}^+]\gamma_\pm{}^2} \\[2mm]
&= \frac{(K_{HA}C_{HA}/[SH_2{}^+])^2}{(C_{BHA} - K_{HA}C_{HA}/[SH_2{}^+]\gamma_\pm{}^2)\gamma_\pm{}^2} \qquad (4\text{-}70)
\end{aligned}
$$

Either (4-70) can be solved for $[SH_2{}^+]$ by trial and error, or (4-69) can be used first to obtain $[A^-]$ and $\gamma_\pm$ and then these values can be substituted in (4-67) to obtain $[SH_2{}^+]$.

Solution of a salt at the equivalence point   At the equivalence point the net reaction is $B + HA \rightleftharpoons BH^+A^- \rightleftharpoons BH^+ + A^-$. To the extent that the salt BHA reacts with the solvent SH, equal amounts of base B and acid HA are produced. Again, combining the expressions for K_{SH}, K_{HA}, and K_B as in (4-55) gives

$$\frac{K_{SH}K_{HA}}{K_B} = \frac{[SH_2{}^+][A^-]\gamma_\pm{}^2[SH_2{}^+][S^-]\gamma_\pm{}^2([B] + [BH^+S^-])}{([HA] + [SH_2{}^+A^-])[BH^+][S^-]\gamma_\pm{}^2} \qquad (4\text{-}71)$$

On the assumption that $[B] + [BH^+S^-]$ equals $[HA] + [SH_2{}^+A^-]$ and $[BH^+]$ equals $[A^-]$, Equation (4-71) gives

$$[SH_2{}^+] = \sqrt{\frac{K_{SH}K_{HA}}{K_B\gamma_\pm{}^2}} \qquad \text{or} \qquad a_{SH_2{}^+} = \sqrt{\frac{K_{SH}K_{HA}}{K_B}} \qquad (4\text{-}72)$$

Solution of salt and excess base   After the equivalence point the solution contains the salt BHA and excess base B. The formation constant of BHA, written in terms

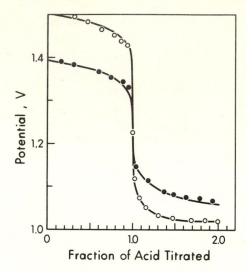

FIGURE 4-3   Comparison of theoretical (lines) and observed (points) titration curves of HBr ($K_{HA} = 4.18 \times 10^{-3}$) with NaOC$_2H_4NH_2$ in en. Open circles, 0.01 M HBr; closed circles, 0.01 M HBr in presence of 1 M NaBr. Dissociation constant of NaBr, 3.95×10^{-3}. (*From Schaap and others.*[63])

of concentrations, is equal to $C_{BHA}/C_B C_{HA}$. Analogous to Equations (4-58) and (4-59), but with activity coefficients included,

$$[S^-]\gamma_\pm = \frac{K_{SH}}{[SH_2{}^+]\gamma_\pm}$$

$$= \left(\frac{K_{SH} + K_B C_B}{1 + K_{BHA}C_{BHA}/K_B C_B}\right)^{\frac{1}{2}} \simeq \frac{K_B C_B}{\sqrt{K_B C_B + K_{BHA}C_{BHA}}} \qquad (4\text{-}73)$$

Schaap and others[63] calculated the shape of neutralization titration curves in en under a variety of conditions. Calculations were undertaken for the case where the only salt present is that formed during neutralization and for the case where a large excess of supporting electrolyte is present. If an acid is titrated in a solution containing electrolyte, activity-coefficient effects reduce the change in hydrogen ion concentration. Figure 4-3 is a comparison of theoretical and observed potentiometric titration curves with a hydrogen stainless-steel indicator electrode for hydrogen bromide with and without sodium bromide present.

Another effect of electrolyte is the formation of double salts of moderate stability. These would be expected to enhance the change in hydrogen ion concentration in the vicinity of the end points.

Figure 4-4 illustrates comparative effects of the dissociation constants of the acid titrated and the salt produced on the shapes of the titration curves.

Clearly, if the salt is comparatively highly dissociated and the acid weak, titrations are less satisfactory. The analytical utility of an acid titration in en depends not only on the dissociation constant of the acid but also on the formation constant of the salt produced.

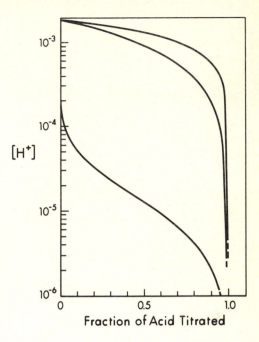

FIGURE 4-4 Theoretical titration curves of acids in en. Top curve, $K_{HA} = 3.3 \times 10^{-4}$ and $K_{BHA} \ll K_{HA}$; middle curve, $K_{HA} = K_{BHA} = 3.3 \times 10^{-4}$; lowest curve, $K_{HA} = 3.3 \times 10^{-6}$ and $K_{BHA} = 3.3 \times 10^{-4}$. (*From Schaap and others.*[63])

4-8   ACID-BASE EQUILIBRIA IN ACETONITRILE

Acetonitrile, CH_3CN, represents solvents that are neither strongly basic nor acidic; acetonitrile itself is nearly aprotic. Its autoprotolysis constant[65] is 6×10^{-33} or even smaller.[57] Acetonitrile is a differentiating solvent, since it is not strong enough as a base to cause appreciable leveling of acids and at the same time not strong enough as an acid to interfere in reactions of weak acids. A similar argument holds for bases. Since acetonitrile is a poor proton acceptor, dissociation constants of acids in it tend to be smaller than in water. The dielectric constant is 36, appreciably larger than that of acetic acid or en, but smaller than that of water. Although ion pairing does occur to some extent, perchloric acid, for example, is extensively dissociated in acetonitrile,[66,67] in contrast to its behavior in acetic acid.

Probably the most important characteristic of acetonitrile is its poor ability to stabilize anions by solvation. The nature of the anion is therefore critical in the determination of solubility, ion pairing, and homoconjugation of salts. The lack of solvation of anions results in high values for transfer activity coefficients [Equation (4-1)]; chloride ion, for instance,[68,69] has a transfer activity coefficient of about 10^6 (Figure 4-5). High transfer activity coefficients mean that solubilities of many electrolytes are low;[68] sodium chloride, for instance, has a solubility of only 3×10^{-5} mole/l at 25°C, and sodium hydroxide is sparingly soluble. Conversely, salts with univalent anions that are either large or highly polarizable tend to be more soluble;

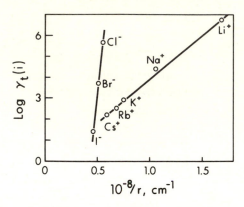

FIGURE 4-5 Transfer activity coefficients for halide and alkaline-metal
ions in acetonitrile as solvent. (*From
Coetzee and Campion.*[69])

thus iodides, perchlorates, and picrates are moderately soluble, as are the tetraalkylammonium halides.

Because acetonitrile is a poor proton donor, the transfer activity coefficient of
the hydrogen ion also is large. The solvated proton has been described[68] as behaving
like a superacid in acetonitrile. For most solvents, dissociation of acids occurs through
a reaction of the type shown in Equation (4-13), with the anion stabilized through
solvation. In acetonitrile, dissociation to ions is often a result of anion stabilization
through hydrogen bonding with one or more molecules of the acid (homoconjugation):

$$2HA + SH \rightleftharpoons SH_2^+ + AHA^-$$
$$3HA + SH \rightleftharpoons SH_2^+ + A(HA)_2^-$$

(4-74)

If other substances in solution can serve the same function, heteroconjugation may
occur, as in the formation of $BrHI^-$. Owing to the poor solvating ability of acetonitrile, many possible chemical species may be present in an acetonitrile solution of a
base B and an acid HA. Among them are SH_2^+ (solvated proton), $(HA)_n$ or $(B)_n$
(self-association), $SH_2^+A^-$, $A(HA)_n^-$ (polyhomoconjugation), BH^+A^-, BH^+AHA^-,
and BHB^+A^-, as well as the simple ions A^- and BH^+.

Activity coefficients for ions in acetonitrile may be calculated[68] from

$$-\log \gamma_{\pm} = \frac{1.64 Z_i^2 \sqrt{\mu}}{1 + 0.48 a \sqrt{\mu}}$$

(4-75)

For rigorous calculations of equilibrium concentrations of the various species in
solution, values of the constants for all the equilibria involved are needed; use of a
computer may be indicated. For example, Sellers, Eller, and Caruso[70] developed a
theoretical model that assumes the existence of a wide variety of species such as can
be found in acetonitrile solution. Eleven equations in eleven unknowns are solved
with the aid of a computer to give information about conductance, potentiometric,
and photometric titration curves in nonaqueous media.

In the following discussions of expressions relating hydrogen ion concentration to various points in the titration of a weak base with an acid, several simplifying assumptions are made:

1 Since the autoprotolysis constant of acetonitrile is small, the SH_2^+ contribution from dissociation of the solvent may be neglected.

2 The acid HA used as titrant is strong, or completely dissociated. In practice the strongest acids, such as perchloric acid and 2,4,6-trinitrobenzenesulfonic acid, are completely ionized and have large dissociation constants (of the order of 10^{-2}).

3 The salt BHA is completely dissociated to free ions. This assumption introduces only moderate error for many 1:1 electrolytes in acetonitrile.

4 The effects of self-association and polyhomoconjugation may be neglected.

For some systems these assumptions will be valid and reasonable.

Solution of a weak base   Acetonitrile is so weakly acidic that proton transfer from it to most bases is negligible; the hydrogen ion concentration therefore is exceedingly low and is likely determined by traces of impurity in the solvent. Nevertheless, on the assumption that slight reaction occurs we may write

$$B + SH \rightleftharpoons BH^+ + S^- \qquad (4\text{-}76)$$

$$BH^+ + B \rightleftharpoons BHB^+ \qquad (4\text{-}77)$$

$$SH + SH \rightleftharpoons SH_2^+ + S^- \qquad (4\text{-}78)$$

The electroneutrality expression is

$$[SH_2^+] = [S^-] - [BH^+] - [BHB^+] \qquad (4\text{-}79)$$

If the base is only slightly dissociated, the ionic strength is low, and activity coefficients for univalent ions may be assumed to be unity. Equations (4-32), (4-40), and (4-41) can be simplified to the following three equilibrium-constant expressions:

$$K_B = \frac{[BH^+][S^-]}{[B]} \qquad (4\text{-}80)$$

$$K_{f_{BHB^+}} = \frac{[BHB^+]}{[BH^+][B]} \qquad (4\text{-}81)$$

$$K_{SH} = [SH_2^+][S^-] \qquad (4\text{-}82)$$

These three equations and (4-79) can be combined to give:

$$[SH_2^+] = \frac{K_{SH}}{[SH_2^+]} - \frac{K_B[B][SH_2^+]}{K_{SH}} - \frac{K_{f_{BHB^+}}[B]K_B[B][SH_2^+]}{K_{SH}} \qquad (4\text{-}83)$$

and

$$[SH_2^+] = \frac{K_{SH}}{(K_{SH} + K_B[B] + K_{f_{BHB^+}}K_B[B]^2)^{\frac{1}{2}}} \qquad (4\text{-}84)$$

Solution of a base B and its salt BHA   When an appreciable concentration of salt is present, activity-coefficient effects may not be negligible. To the extent that $\gamma_+ = \gamma_- = \gamma_\pm$ the activity coefficients largely cancel out in expressions for the calculation of hydrogen ion concentration. The overall reactions considered are:

$$B + H^+ + A^- \rightleftharpoons BH^+ + A^- \qquad (4\text{-}85)$$

and

$$BH^+ + B \rightleftharpoons BHB^+ \qquad (4\text{-}86)$$

With the assumption that ionization of the base (4-76) and of the solvent (4-78) contribute negligibly to the cation concentration,

$$C_{BHA} = [BH^+] + [BHB^+] \qquad (4\text{-}87)$$

and

$$C_B{}^0 = [B] + [BH^+] + 2[BHB^+] \qquad (4\text{-}88)$$

where $C_B{}^0$ is the initial analytical concentration of base B. Combining (4-87) and (4-88) with (4-40) to eliminate $[BHB^+]$, and assuming γ_{BHB+} equals γ_{BH+}, we obtain

$$C_B{}^0 = [B] + [BH^+] + 2K_{f_{BHB+}}[BH^+][B] \qquad (4\text{-}89)$$

Similarly, combining (4-40) and (4-87) to eliminate $[BHB^+]$ gives

$$[B] = \frac{C_{BHA} - [BH^+]}{K_{f_{BHB+}}[BH^+]} \qquad (4\text{-}90)$$

From (4-32) and (4-41)

$$[BH^+] = \frac{K_B[B][SH_2{}^+]\gamma_\pm}{K_{SH}\gamma_\pm} = \frac{K_B[B][SH_2{}^+]}{K_{SH}} \qquad (4\text{-}91)$$

When we combine (4-89), (4-90), and (4-91),

$$C_B{}^0 = [B] + \frac{K_B[B][SH_2{}^+]}{K_{SH}} + \frac{2K_{f_{BHB+}}K_B[B]^2[SH_2{}^+]}{K_{SH}} \qquad (4\text{-}92)$$

Now substitute (4-91) in (4-90) and solve for $[SH_2{}^+]$:

$$[SH_2{}^+] = \frac{K_{SH}C_{BHA}}{K_B[B] + K_BK_{f_{BHB+}}[B]^2} \qquad (4\text{-}93)$$

Substitute (4-93) in (4-92) to obtain

$$K_{f_{BHB+}}[B]^2 + \{2K_{f_{BHB+}}C_{BHA} - C_B{}^0K_{f_{BHB+}} + 1\}[B]$$
$$+ (C_{BHA} - C_B{}^0) = 0 \qquad (4\text{-}94)$$

Consolidation of (4-93) and (4-94) to solve for $[SH_2{}^+]$ yields a complex expression. Therefore the numerical values of analytical concentrations of salt C_{BHA} and initial base $C_B{}^0$ along with the formation constant for the homoconjugate complex are inserted in (4-94), and $[B]$ is obtained at a specified point in a titration. This value of $[B]$ is substituted in (4-93) to obtain $[SH_2{}^+]$. Fortunately, the activity coefficients drop out in this method of calculating hydrogen ion concentration.

EXAMPLE 4-3 What is the hydrogen ion concentration (dilution neglected) after 10% reaction of 0.01 M base with a K_B value of 10^{-15}?

ANSWER Take $K_{SH} = 6 \times 10^{-33}$ and $K_{f_{BHB^+}}$ for homoconjugation = 20, and assume the titrant acid is strong. From Equations (4-93) and (4-94),

$$20[B]^2 + (2 \times 20 \times 10^{-3} - 0.01 \times 20 + 1)[B] + (10^{-3} - 10^{-2}) = 0$$

$$[B] = 8.8 \times 10^{-3} \; M$$

$$[SH_2^+] = \frac{6 \times 10^{-33} \times 0.001}{10^{-15} \times 8.8 \times 10^{-3} + 10^{-15} \times 20 \times (8.8 \times 10^{-3})^2}$$

$$= 5.8 \times 10^{-19} \qquad\qquad ////$$

At the half-neutralization point, $C_{BHA} = C_B = C_B^0/2$. Equation (4-94) can be simplified to

$$K_{f_{BHB^+}}[B]^2 + [B] - \frac{C_B^0}{2} = 0 \qquad (4-95)$$

and (4-93) can be rearranged to

$$[SH_2^+] = \frac{K_{SH}}{K_B} \frac{C_{BHA}}{K_{f_{BHB^+}}[B]^2 + [B]} \qquad (4-96)$$

Therefore

$$[SH_2^+] = \frac{K_{SH}}{K_B} \qquad (4-97)$$

Thus the extent of formation of BHB^+ at this point in the titration does not influence the hydrogen ion concentration because equimolar amounts of B and BH^+ are required to form the complex.

Solution of salt, BHA   The significant reaction is

$$BH^+ + SH \rightleftharpoons SH_2^+ + B \qquad (4-98)$$

$$[SH_2^+] = [B] + [BHB^+] \qquad (4-99)$$

Combine (4-32) and (4-41), eliminating a_{S^-}, to obtain

$$[B] = \frac{[BH^+]\gamma_\pm K_{SH}}{[SH_2^+]\gamma_\pm K_B} \qquad (4-100)$$

Substitute (4-40) in (4-99) and combine with (4-100):

$$[SH_2^+] = \frac{[BH^+]K_{SH}}{[SH_2^+]K_B} + \frac{K_{f_{BHB^+}}[BH^+]^2 K_{SH}}{[SH_2^+]K_B} \qquad (4-101)$$

$$[SH_2^+] = \left(\frac{[BH^+]K_{SH}}{K_B} \{1 + K_{f_{BHB^+}}[BH^+]\} \right)^{\frac{1}{2}} \qquad (4-102)$$

Solution of salt and excess acid   After the equivalence point the solution contains the salt BHA and excess HA. If both are assumed to be completely dissociated,

$[SH_2^+]$ will be equal to the concentration of excess acid (except close to the equivalence point), and

$$a_{SH_2+} = C_{HA}\gamma_\pm \qquad (4\text{-}103)$$

In general, for a titration of a weak acid with a strong base, considerations similar to those outlined above apply.

In practice, detailed calculations involving equilibria in acetonitrile using a K_{SH} value as small as 6×10^{-33} may be misleading because of the difficulty in obtaining and maintaining the solvent in high purity. Traces of water usually are present, as well as ammonia or ammonium acetate from solvent decomposition. In titrations of weak acids the tetraalkylammonium hydroxide often contains some alcohol. Kolthoff and Chantooni[71] reported that 0.1 M water in acetonitrile caused a decrease of 1.5 units in the pa_H value of a sulfate-bisulfate mixture, and 0.1 M p-bromophenol caused a decrease of 7.5 units. In this instance the doubly charged sulfate ion has a strong tendency to form heteroconjugates with proton-donor substances.

4-9 ACIDS AND BASES IN ALCOHOL-WATER MIXTURES

Although acid-base titrations in alcohol-water mixtures have been studied extensively, we do not consider them in detail since titration curves and indicator equilibria in ethanol-water and methanol-water mixtures can be calculated in the same way as in water. Values of K_{SH}, the autoprotolysis constants of the mixtures, are close to K_w for mixtures containing only a moderate amount of alcohol. On the other hand, even a trace of water in ethanol causes a large increase in K_{SH}. According to Gutbezahl and Grunwald,[72] pK_{SH} is 14.33, 14.88, 15.91, and 19.5 for ethanol-water mixtures containing 20, 50, 80, and 100 wt % ethanol.

Dissociation constants for acids and bases in alcohol-water mixtures have long been a subject of great interest. One approach to relating dissociation constants of acids and bases in different solvents, that of correlating values only with the dielectric constant (Figure 4-2), is inadequate in that it fails to account for specific interactions between solutes and solvents. Another approach, developed by Grunwald,[5,72,73] relates the ΔpK values of an acid (or base) in two solvents by equations with empirical constants whose values depend on the nature of the solvent, the nature of the acid or base, and particularly on the charge type. According to this empirical approach, acids of a given charge type should have similar ΔpK_a values for any two alcohol-water mixtures taken as solvents. The comparisons in Table 4-2 show this relation in a striking manner (Column 4).

Acid-base indicators can be characterized[74] according to the charges present on their acid and alkaline forms.

1 Sulfonephthaleins (for example, bromophenol blue, bromocresol green, phenol red), which are transformed from singly to doubly charged negative ions,

Table 4-2 COMPARISON OF ΔpK_a VALUES OF ACIDS IN
WATER† AND IN ANHYDROUS ETHANOL AT 25°C

Acid	pK_a in ethanol	pK_a in water	ΔpK_a
Formic	9.15	3.75	5.40
Acetic	10.32	4.76	5.56
Cyanoacetic	7.49	2.47	5.02
Benzoic	10.25	4.20	6.05
Salicylic	8.68	3.00	5.68
Anilinium ion	5.70	4.63	1.07
o-Toluidinium ion	5.55	4.46	1.09
m-Toluidinium ion	5.78	4.77	1.01
p-Toluidinium ion	6.24	5.10	1.14
N-Methylanilinium ion	4.86	4.84	0.02

† From Gutbezahl and Grunwald[73] and Grunwald and Berkowitz.[5]

undergo a relatively large change in pK_{In} in going from water to ethanol.

2 Cationic acid indicators (for example, methyl yellow and neutral red) undergo a small change in pK_{In} because of the small change in charge separation in the reaction $InH^+ \rightleftharpoons In + H^+$.

3 Indicators that exist as dipolar ions in the acid form and as singly charged anions in the alkaline form (for example, methyl orange) undergo a small change in pK_{In}. Here the color-change reaction may be represented by $^+HIn^- \rightleftharpoons In^- + H^+$, which indicates once more only a small change in charge separation.

Methyl red represents a special case in which the intermediate form (red 2) acts as a neutral molecule rather than a dipolar ion. Consequently, its first transition (at low pH) is characteristic of a cationic acid, whereas its second transition is characteristic of a molecular acid. The salt effects with methyl red in ethanol and methanol also point to a neutral molecular species as the intermediate form.[74] Bear in mind that the degree of dipolar ion character is expected to diminish with decreasing dielectric constant of the solvent.

4-10 pH MEASUREMENTS IN NONAQUEOUS SOLVENTS

Bates[75] indicated that the soundest procedure for experimentally establishing a practical scale for pH in a given solvent requires that the hydrogen gas electrode and the silver–silver chloride electrodes be thermodynamically reversible and stable in the solvent system, the glass (or other) electrode respond in a Nernstian way, and the liquid-junction potential be little affected by change in acidity of the solution. A reference value for pH should be selected that is close to that of the solution to be measured and that gives rational meaning to pH values for the solutions examined.

To measure pH in a nonaqueous solvent, the most satisfactory procedure involves calibration of a practical pH scale in terms of buffers established for this specific purpose in the particular solvent. Bates's recommendation extends to solvents as similar to water as deuterium oxide. An operational pH scale for deuterium oxide (heavy water) is described in Section 3-11.

Fortunately, glass electrodes have wide applicability in solvents other than water. Even in a solvent as different from water as acetonitrile,[76,77] glass electrodes respond reversibly to changes in hydrogen ion activity, in agreement with the Nernst equation. In setting up pH values for reference buffers, it is not always possible to use internal reference electrodes without liquid junction. For example, Ag–AgX electrodes are unstable in acetonitrile owing to slow formation of species of the type $Ag_nX_{n+1}^-$. When external reference electrodes must be substituted, the reliability of the measurements is reduced because uncertainties in the liquid junction are always present.

In general, operational definitions of pH in nonaqueous solvents are similar to Equation (3-10) for water:

$$pH = pH_s + \frac{(E - E_s)F}{2.303RT} \qquad (4\text{-}104)$$

where pH_s is the pH value of a reference buffer in the nonaqueous solvent, and E and E_s are the potentials observed for the sample solution and standard buffer in the cell used for making the measurement.

In acetonitrile the glass electrode can be calibrated with a buffer of picric acid and tetraethylammonium picrate to give a calibration for the pH scale within a narrow region. The dissociation constant of 10^{-11} for picric acid is the only one that has been determined with the care necessary for calibration purposes.[77,78] Picric acid was chosen because it undergoes little homoconjugation (homoconjugation formation constant[79] of 2.4) and the tetraalkylammonium salt is essentially completely dissociated (Figure 4-2) up to moderate concentrations. For pH measurement the reference-electrode half of the cell is Ag/0.01 M AgNO$_3$ in acetonitrile, and the salt bridge is 0.1 M tetraethylammonium perchlorate.

For 50% methanol-water mixtures three series of buffers involving acetate, succinate, and phosphate have been established[80] in the temperature range from 10 to 40°C by emf data obtained without liquid junction. With 50% methanol solutions an aqueous saturated calomel electrode is used as reference electrode, since liquid-junction potentials are adequately reproducible[81] for operational pH values.

For most solvents, standard buffers have not been established for standardizing the pH scale. Many measurements of pH in nonaqueous solvents have therefore been made by extension of the pH scale in water by use of aqueous buffers. Even for hydrogen-bonding solvents similar to water, extrapolation procedures are uncertain since the effect of both liquid-junction potential and transfer activity coefficient are present. Fundamentally, aqueous standard buffers are applicable only if the liquid-junction potential E_j arising from a change in solvent can be determined and if transfer activity coefficients can be taken into account.

Gutbezahl and Grunwald[72,73] considered liquid-junction potentials between a solution of aqueous potassium chloride and solutions of acids in ethanol-water mixtures both theoretically and experimentally. They concluded that for mixtures containing up to 35% ethanol the liquid-junction potential should be 6 mV or less. For solvents containing higher percentages of alcohol, the liquid-junction potential increases rapidly—25 mV for 50%, 44 mV for 65%, and 75 mV for 80% ethanol. These numerical values should not be interpreted too literally, particularly as the composition approaches 100% ethanol. Calculated liquid-junction potentials contain an indeterminate term that involves all quantities other than those arising from unequal transfer activity coefficients (such as dipole orientation effects).

In practical terms the operational scale for acidity in a nonaqueous solvent standardized with aqueous buffers may be written

$$pH = pH_s^* + \frac{(E - E_s)F}{2.303RT} \qquad (4\text{-}105)$$

Here the pH is the value for acidity referred to the standard state defined from the limiting reference behavior for the hydrogen ions in the solvent in question,[82] and pH_s^* is the pH value of the aqueous reference buffer corrected for liquid junction and transfer activity coefficient as follows:

$$pH_s^* = pH - (E_j - \log \gamma_t) \qquad (4\text{-}106)$$

Bates, Paabo, and Robinson[81] expressed the net effect of the terms E_j and $\log \gamma_t$ in units of pH. The quantity $E_j - \log \gamma_t$ has been tabulated for several alcohol-water mixtures.[83] For example, in 50 and 100% methanol the quantity amounts to 0.13 and -2.34 pH units, and in 50 and 100% ethanol, 0.21 and about -2.9 pH units. Although the values of E_j and γ_t cannot be obtained independently of each other, the subtraction of a tabulated constant involving their difference from the pH value of the aqueous reference standard can yield useful values for pH.

The meaning to be attached to pH values obtained in different solvents and under different conditions needs interpretation. For example, solutions of pH 5 in the three solvents water, ethanol, and acetonitrile do not denote the same acidity (acidity defined as a measure of the tendency for protons to be donated to basic substances). In view of what is known about transfer activity coefficients (Section 4-1), an acetonitrile solution of pH 5 is far more acidic (higher absolute activity) than an ethanol solution of pH 5 and, in turn, one in ethanol is more acidic than one in water. The significant point is that a solution of, say, pH 5 in a given solvent has an acidity 10 times the acidity of another solution of pH 6 in *that same solvent*. A single universal scale of pH for all solvents does not exist; instead there is a different scale for every solvent of differing composition. To denote pH values in nonaqueous solvents, the symbol pH* is often used instead of the symbol pH in Equation (4-105); this serves as a reminder that the standard state for hydrogen ions is defined in terms of activity equal to concentration at infinite dilution in the solvent in question rather than in water. We chose to omit the asterisk.

4-11   ACID-BASE INDICATORS IN SOLVENTS OF LOW DIELECTRIC CONSTANT

In aqueous solution, and in solvents of high dielectric constant, the behavior of indicators generally can be related directly to pH changes (Equation 3-76). With solvents of low dielectric constant the color of an indicator depends on the extent of its ion formation, whether as ion pairs or free ions. The extent of ion formation may bear only an obscure relation to the hydrogen ion activity, which depends on the extent of formation of dissociated protonated solvent. Analogous to Equation (4-13), the reaction between an indicator base In and an acid HA may be represented in its simplest form by

$$In + HA \rightleftharpoons InH^+A^- \rightleftharpoons InH^+ + A^- \qquad (4\text{-}107)$$

where In represents the alkaline form and InH^+A^- and InH^+ represent the acid forms of the indicator. For example,[30] the absorption spectra of the various acid forms (including higher-ion aggregates) of p-naphtholbenzein are identical and independent of the nature of the anion X^-.

The first equilibrium step of (4-107) is described by the expression

$$K_{In_{IP}} = \frac{[InH^+A^-]}{[In][HA]} \qquad (4\text{-}108)$$

and the second equilibrium by

$$K_{In_{FI}} = \frac{[InH^+][A^-]}{[InH^+A^-]} \qquad (4\text{-}109)$$

The ratio of acid form to alkaline form is

$$\frac{\Sigma\,[InH^+]}{[In]} = \frac{[InH^+] + [InH^+A^-]}{[In]} \qquad (4\text{-}110)$$

No simple relation exists between this ratio and the hydrogen ion concentration, as is shown in the following discussion.

Indicators and glacial acetic acid   Kolthoff and Bruckenstein[30] found that for perchloric acid in glacial acetic acid the ratio of acid to alkaline forms of p-naphtholbenzein is proportional to the concentration of perchloric acid. On the other hand, the hydrogen ion concentration should vary with the square root of the concentration of perchloric acid [Equation (4-59)]. For weak acids such as hydrochloric acid the relation between indicator color and acid concentration is more complicated and will not be considered in detail here. Note, however, that the color varies with the concentration of anion. With increasing chloride ion concentration, generation of ion pairs and formation of free ions from the indicator salt is repressed, and the ratio of acid to basic form of the indicator decreases. But, when the chloride ion concentration reaches a certain level, higher-ion aggregates such as $Cl^-InH^+Cl^-$ begin to form and the indicator once more shifts toward the acidic color.

From the viewpoint of titrimetry in glacial acetic acid, the behavior of indicators during the titration of weak bases with perchloric acid is most important. It at first seems that Equation (4-108) could be used in a straightforward way; nevertheless, this is not an experimentally useful form because the concentration of molecular perchloric acid, $[HClO_4]$, is unknown.

For the reaction of the indicator base with perchloric acid,

$$In + HClO_4 \rightleftharpoons InHClO_4 \qquad (4\text{-}111)$$

The expression for the formation constant is

$$K_{f_{InHClO_4}} = \frac{C_{InHClO_4}}{C_{In}C_{HClO_4}} \qquad (4\text{-}112)$$

where C denotes analytical concentration and implies nothing about chemical form.

During the titration the salt $BH^+ClO_4^-$ is present in large excess and suppresses dissociation of the indicator salt $InH^+ClO_4^-$; consequently, $[InH^+]$ can be considered negligible. The indicator salt can be considered completely ion-paired, so that $[InH^+ClO_4^-] \simeq C_{InHClO_4}$. Since the indicator base is weak, it undergoes little ionization, and $[In] \simeq C_{In}$. Hence from (4-112)

$$\frac{[InH^+ClO_4^-]}{[In]} = K_{f_{InHClO_4}}C_{HClO_4} \qquad (4\text{-}113)$$

For the reaction

$$B + HClO_4 \rightleftharpoons BHClO_4 \qquad (4\text{-}114)$$

$$K_{f_{InHClO_4}} = \frac{C_{BHClO_4}}{C_B C_{HClO_4}} \quad \text{or} \quad C_{HClO_4} = \frac{C_{BHClO_4}}{C_B K_{f_{InHClO_4}}} \qquad (4\text{-}115)$$

Equations (4-113) and (4-115) can be combined to obtain the expression giving the ratio of the two colored forms of the indicator:

$$\frac{[InH^+ClO_4^-]}{[In]} = \frac{K_{f_{InHClO_4}}}{K_{f_{BHClO_4}}} \frac{C_{BHClO_4}}{C_B} \qquad (4\text{-}116)$$

Equation (4-116) shows that the color ratio of the indicator is determined by the *ratio* of concentrations of salt to base and therefore is independent of dilution at a given point in the titration (neglecting solvolysis). As previously indicated, in Equation

(4-51) the hydrogen ion concentration varies inversely with the square root of the concentration; therefore the indicator does not respond to changes in hydrogen ion concentration in acetic acid in the same manner as it does in water.

If the ratio C_{BHClO_4}/C_B is replaced by $X/(1 - X)$, (4-116) becomes

$$\frac{[InH^+ClO_4^-]}{[In]} = \frac{K_{f_{InHClO_4}}}{K_{f_{BHClO_4}}} \frac{X}{1 - X} \qquad (4\text{-}117)$$

where X is the fraction of the stoichiometric quantity of titrant. Thus, the indicator color ratio changes during the course of the titration in the same way as in aqueous solution for a titration of a weak base with a strong acid. If an indicator is chosen so that $K_{f_{InHClO_4}}$ is of the same order of magnitude as $K_{f_{BHClO_4}}$, the spectrophotometric measurement of indicator color ratio is equivalent to a determination of the ratio $X/(1 - X)$, and a graphical method can be used to detect the end point.[84]

At the equivalence point, corresponding to a solution of the salt BHClO$_4$, C_B equals C_{HClO_4}, and from (4-115)

$$C_{HClO_4}^2 = \frac{C_{BHClO_4}}{K_{f_{BHClO_4}}} \qquad (4\text{-}118)$$

which, with (4-113), gives

$$\frac{[InH^+ClO_4^-]}{[In]} = K_{f_{InHClO_4}} \left(\frac{C_{BHClO_4}}{K_{f_{BHClO_4}}}\right)^{\frac{1}{2}} \qquad (4\text{-}119)$$

The color ratio of the indicator therefore varies with the square root of salt concentration. The hydrogen ion concentration at the equivalence point has been shown in (4-56) to be independent of salt concentration.

Beyond the equivalence point the ratio of acid to alkaline forms of the indicator is proportional to the concentration of excess perchloric acid. Thus, upon 10-fold dilution the color ratio decreases 10-fold, while the hydrogen ion concentration decreases by a factor of $\sqrt{10}$. Again, the indicator color is not a straightforward index of acidity.

4-12   ACIDITY AND STRONGLY ACID SOLUTIONS. HAMMETT ACIDITY FUNCTION H_0

To obtain a comparative measure of acidities in strongly acid solutions or to measure the strengths of exceedingly weak bases, Hammett[43,85] defined acidity in terms of the degree of ionization of a series of indicators that are weak organic bases. To set up an acidity scale he chose a group of indicators, the nitroanilines, that were of the same charge type and also resembled each other closely in molecular structure. The first members of the series, m- and p-nitroaniline, called the *reference bases*, are strong enough that the thermodynamic dissociation constants of the conjugate acids BH^+ have been measured in water. These two reference bases have pK_a values of 2.50 and 0.99. The ratio $[B]/[BH^+]$ for these indicators can be measured reliably over about a 100-fold change in acidity; for p-nitroaniline this includes both dilute aqueous solutions of strong acid and solutions of up to 15% strong acid. In solutions more acidic than this, the ratio $[B]/[BH^+]$ is too small to be measurable with accuracy. The weaker base o-nitroaniline ($pK_a = -0.29$) can be used in the same range as p-nitroaniline and in more strongly acid solutions as well, again over about a 100-fold change in acidity. In turn, 4-chloro-2-nitroaniline ($pK_a = -1.03$) overlaps with o-nitroaniline and extends still farther into strongly acidic solutions. Sixteen nitroaniline indicators were chosen that completely cover acidities up to and beyond that of 100% sulfuric acid.

If two nitroaniline indicators B′ and B″ are introduced into separate portions of a strongly acidic solvent and the ratio of $[B]/[BH^+]$ is measured in both, then

$$pK_a' - pK_a'' = -\log \frac{[B'][B''H^+]}{[B'H^+][B'']} - \log \frac{\gamma_{B'}\gamma_{B''H^+}}{\gamma_{B'H^+}\gamma_{B''}} \qquad (4\text{-}120)$$

The Hammett postulate in essence is that the quantity $\gamma_{B'}\gamma_{B''H^+}/\gamma_{B'H^+}\gamma_{B''}$ is unity, independent of the medium and of the particular indicator in the series. For solvents of high dielectric constant and for the closely similar nitroaniline indicators, both theoretical expectation and experimental justification uphold this postulate. By the use of (4-120), values of pK_a for the various nitroaniline indicators were measured up to a pK_a value of -10.10 for 2,4,6-trinitroaniline.

Having determined values of K_a, Hammett then expressed the acidity of the solvent by defining the function H_0:

$$H_0 = pK_a + \log \frac{[B]}{[BH^+]} \qquad (4\text{-}121)$$

This value of H_0 is characteristic of a particular solvent, because the ratio γ_B/γ_{BH^+} was found to have the same value for all the nitroanilines in the solvent. For *dilute aqueous solutions* the value of H_0 approaches pH as the ionic strength approaches zero. For all other solvents, H_0 is an empirical quantity that decreases with increasing acidity and measures on a logarithmic scale the tendency for the solvent to donate protons to uncharged molecules of base. In 100% sulfuric acid, H_0 has a value of -12.2. Since all acidities on the H_0 scale relate to the reference bases in dilute aqueous acidic solutions, the acidities are defined with reference to the standard state for hydrogen ions in water as solvent. Gillespie[86] measured acidities as high as $H_0 \simeq -20$ in systems such as $SbF_5 \cdot 3SO_3$ and $HSO_3F \cdot SbF_5 \cdot SO_3$. Tables of H_0 values have been compiled by Boyd.[87]

As the acidity of the strongly acidic solution being measured increases, errors in the determination of the K_a values leading up to the indicator in question become statistically additive, and the accuracy of the H_0 values is lowered. Nevertheless, internally consistent values can be obtained, and by means of the H_0 function it has been possible to evaluate the pK values of weak uncharged bases and interpret the effects of acid catalysis in numerous acid-water mixtures.

The acidity function does not and cannot fully account for effects of specific interactions (Section 4-3) between solute and solvent except as they reflect the behavior of the nitroanilines. The quantity H_0 therefore should be used as permissive rather than conclusive evidence for mechanisms of reactions or even for relative acidities.

Similar acidity functions H_- and H_+ for anionic and cationic bases have been proposed. As might be predicted, the specific interaction effects with cationic, anionic, and neutral indicators are different, and the acidity values H_- and H_+ may be either higher or lower than H_0. The nitroanilines apparently are exceptionally well-behaved substances in terms of setting up an acidity scale, and Hammett later stated that the choice was fortunate.[85]

Acidity scales for highly acidic solutions in media of low dielectric constant have been proposed, such as for ethanol-water mixtures[88] and glacial acetic acid–acetic anhydride.[89] Ion pairing is a complicating factor in these solvents. The extent of formation of ion pairs and of free ions differs for various indicators and salts; in particular, dissociation constants are sensitive to the size of the anion.[90,91] The Hammett postulate therefore cannot easily be extended to include media of low dielectric constant.

REFERENCES

1 I. M. KOLTHOFF and M. K. CHANTOONI: *J. Amer. Chem. Soc.*, **93**:7104 (1971); *J. Phys. Chem.*, **76**:2024 (1972).

2 R. G. BATES: "Determination of pH," 2d ed., p. 215, Wiley, New York, 1973.

3 N. BJERRUM and E. LARSON: *Z. Phys. Chem.*, **127**:358 (1927).

4 I. M. KOLTHOFF, J. J. LINGANE, and W. D. LARSON: *J. Amer. Chem. Soc.*, **60**:2512 (1938).

5 E. GRUNWALD and B. J. BERKOWITZ: *J. Amer. Chem. Soc.*, **73**:4939 (1951).

6 O. POPOVYCH: *Crit. Rev. Anal. Chem.*, **1**:73 (1970).

7 I. M. KOLTHOFF and M. K. CHANTOONI, JR.: *Anal. Chem.*, **44**:194 (1972).

8 I. M. KOLTHOFF and M. K. CHANTOONI, JR.: *J. Phys. Chem.*, **77**:1, 523, 527 (1973).

9 K. H. POOL and R. G. BATES: *J. Chem. Thermodyn.*, **1**:21 (1969).

10 H. P. THUN, B. R. STAPLES, and R. G. BATES: *J. Res. Nat. Bur. Stand.*, **74A**:641 (1970).

11 E. GRUNWALD, G. BAUGHMAN, and G. KOHNSTAM: *J. Amer. Chem. Soc.*, **82**:5801 (1960).

12 O. POPOVYCH: *Anal. Chem.*, **38**:558 (1966).

13 R. ALEXANDER and A. J. PARKER: *J. Amer. Chem. Soc.*, **89**:5549 (1967).

14 H. M. KOEPP, H. WENDT, and H. STREHLOW: *Z. Elektrochem.*, **64**:483 (1960).

15 R. ALEXANDER, A. J. PARKER, J. H. SHARP, and W. E. WAGHORNE: *J. Amer. Chem. Soc.*, **94**:1148 (1972).

16 B. G. COX, A. J. PARKER, and W. E. WAGHORNE: *J. Amer. Chem. Soc.*, **95**:1010 (1973).

17 B. KRATOCHVIL and H. L. YEAGER: *Top. Current Chem.*, **27**:1 (1972).

18 C. H. SPRINGER, J. F. COETZEE, and R. L. KAY: *J. Phys. Chem.*, **73**:471 (1969).

19 C. V. KRISHNAN and H. L. FRIEDMAN: *J. Phys. Chem.*, **75**:3606 (1971).

20 J. F. COETZEE and W. R. SHARPE: *J. Phys. Chem.*, **75**:3141 (1971).

21 O. POPOVYCH, A. GIBOFSKY, and D. H. BERNE: *Anal. Chem.*, **44**:811 (1972); D. H. BERNE and O. POPOVYCH: *Anal. Chem.*, **44**:817 (1972).

22 J. F. COETZEE, J. J. CAMPION, and D. R. LIBERMAN: *Anal. Chem.*, **45**:343 (1973).

23 B. G. COX and A. J. PARKER: *J. Amer. Chem. Soc.*, **94**:3674 (1972).

24 L. M. MUKHERJEE: *J. Phys. Chem.*, **76**:243 (1972).

25 M. BORN: *Z. Phys.*, **1**:45 (1920).

26 J. BRONSTED: *Chem. Rev.*, **5**:231 (1928).

27 J. F. COETZEE and J. J. CAMPION: *J. Amer. Chem. Soc.*, **89**:2513 (1967).

28 S. GOLDMAN and R. G. BATES: *J. Amer. Chem. Soc.*, **94**:1476 (1972).

29 J. BRONSTED: *Z. Phys. Chem.* (Leipzig), **A169**:52 (1934).

30 I. M. KOLTHOFF and S. BRUCKENSTEIN: *J. Amer. Chem. Soc.*, **78**:1 (1956).

31 J. F. COETZEE and J. J. CAMPION: *J. Amer. Chem. Soc.*, **89**:2517 (1967).

32 R. G. BATES: *J. Electroanal. Chem.*, **29**:1 (1971).

33 M. HERLEM and A. I. POPOV: *J. Amer. Chem. Soc.*, **94**:1431 (1972).

34 D. R. COGLEY, M. FALK, J. N. BUTLER, and E. GRUNWALD: *J. Phys. Chem.*, **76**:855 (1972).

35 P. KEBARLE in "Ions and Ion Pairs in Organic Reactions," M. Swzarc (ed.), vol. 1, p. 27, Wiley, New York, 1972.

36 A. HANTZSCH: *Z. Elektrochem.*, **29**:221 (1923).

37 L. A. WOOTEN and L. P. HAMMETT: *J. Amer. Chem. Soc.*, **57**:2289 (1935).

38 J. HINE and M. HINE: *J. Amer. Chem. Soc.*, **74**:5266 (1952).

39 I. M. KOLTHOFF: *J. Phys. Chem.*, **35**:2732 (1931).

40 S. KILPI and H. WARSILA: *Z. Phys. Chem.* (Leipzig), **A177**:427 (1936).

41 V. A. PLESKOV and A. M. MONOSSON: *J. Phys. Chem. USSR*, **6**:513 (1935).

42 S. BRUCKENSTEIN and I. M. KOLTHOFF: *J. Amer. Chem. Soc.*, **78**:2974 (1956).

43 L. P. HAMMETT and A. J. DEYRUP: *J. Amer. Chem. Soc.*, **54**:4239 (1932).

44 W. M. LATIMER, K. S. PITZER, and C. M. SLANSKY: *J. Chem. Phys.*, **7**:108 (1939).

45 R. M. NOYES: *J. Amer. Chem. Soc.*, **84**:513 (1962).

46 L. G. HEPLER: *Aust. J. Chem.*, **17**:587 (1964).

47 C. L. DELIGNY and M. ALFENAAR: *Rec. Trav. Chim. Pays-Bas*, **84**:81 (1965); **86**:929. 952 (1967).

48 N. BJERRUM: *Kgl. Dan. Vidensk. Selsk. Mat.-Fys. Medd.*, (9)**7**:1 (1926).

49 R. A. ROBINSON and R. H. STOKES: "Electrolyte Solutions," 2d ed., p. 392, Butterworths. London, 1959.

50 Y. H. INAMI, H. K. BODENSEH, and J. B. RAMSEY: *J. Amer. Chem. Soc.*, **83**:4745 (1961).

51 E. GRUNWALD and L. J. KIRSCHENBAUM: "Introduction to Quantitative Analysis," p. 109, Prentice-Hall, Engelwood Cliffs, N.J., 1972.

52 S. BRUCKENSTEIN and L. D. PETTIT: *J. Amer. Chem. Soc.*, **88**:4790 (1966).

53 S. GOLDMAN and G. C. B. CAVE: *Can. J. Chem.*, **49**:4096 (1971).

54 G. ALLEN and E. F. CALDIN: *Quart. Rev.* (London), **7**:255 (1953).

55 J. F. COETZEE and R. M. LOK: *J. Phys. Chem.*, **69**:2690 (1965).

56 J. JUILLARD and I. M. KOLTHOFF: *J. Phys. Chem.*, **75**:2496 (1971).

57 I. M. KOLTHOFF: private communication.

58 N. F. HALL and T. H. WERNER: *J. Amer. Chem. Soc.*, **50**:2367 (1928).

59 T. HIGUCHI, M. L. DANGUILAN, and A. D. COOPER: *J. Phys. Chem.*, **58**:1167 (1954).

60 I. M. KOLTHOFF and S. BRUCKENSTEIN: *J. Amer. Chem. Soc.*, **79**:1 (1957).

61 M. L. MOSS, J. H. ELLIOT, and R. T. HALL: *Anal. Chem.*, **20**:784 (1948).

62 S. BRUCKENSTEIN and L. M. MUKHERJEE: *J. Phys. Chem.*, **66**:2228 (1962).

63 W. B. SCHAAP, R. E. BAYER, J. R. SIEFKER, J. Y. KIM, P. W. BREWSTER, and F. C. SCHMIDT: *Rec. Chem. Progr.*, **22**:197 (1961).

64 H. P. MARSHALL and E. GRUNWALD: *J. Chem. Phys.*, **21**:2143 (1953).

65 I. M. KOLTHOFF and M. K. CHANTOONI, JR.: *J. Phys. Chem.*, **72**:2270 (1968).

66 J. F. COETZEE and I. M. KOLTHOFF: *J. Amer. Chem. Soc.*, **79**:6110 (1957).

67 J. F. COETZEE and D. K. MCGUIRE: *J. Phys. Chem.*, **67**:1810 (1963).

68 J. F. COETZEE: *Progr. Phys. Org. Chem.*, **4**:45 (1967).

69 J. F. COETZEE and J. J. CAMPION: *J. Amer. Chem. Soc.*, **89**:2517 (1967).

70 N. G. SELLERS, P. M. P. ELLER, and J. A. CARUSO: *J. Phys. Chem.*, **76**:3618 (1972).

71 I. M. KOLTHOFF and M. K. CHANTOONI, JR.: *J. Amer. Chem. Soc.*, **91**:25 (1969).

72 B. GUTBEZAHL and E. GRUNWALD: *J. Amer. Chem. Soc.*, **75**:565 (1953).

73 B. GUTBEZAHL and E. GRUNWALD: *J. Amer. Chem. Soc.*, **75**:559 (1953).

74 L. S. GUSS and I. M. KOLTHOFF: *J. Amer. Chem. Soc.*, **62**:249 (1940).

75 R. G. BATES: *Pure Appl. Chem.*, **18**:421 (1969).

76 J. F. COETZEE and G. R. PADMANABHAN: *J. Phys. Chem.*, **66**:1708 (1962).

77 I. M. KOLTHOFF and M. K. CHANTOONI, JR.: *J. Amer. Chem. Soc.*, **87**:4428 (1965).

78 I. M. KOLTHOFF and M. K. CHANTOONI, JR.: *J. Amer. Chem. Soc.*, **91**:6907 (1969).

79 I. M. KOLTHOFF and M. K. CHANTOONI, JR.: *J. Phys. Chem.*, **73**:4029 (1969).

80 M. PAABO, R. A. ROBINSON, and R. G. BATES: *J. Amer. Chem. Soc.*, **87**:415 (1965).

81 R. G. BATES, M. PAABO, and R. A. ROBINSON: *J. Phys. Chem.*, **67**:1833 (1963).

82 C. L. DELIGNY, P. F. M. LUYKX, M. REHBACH, and A. A. WIENEKE: *Rec. Trav. Chim. Pays-Bas*, **79**:699, 713 (1960).

83 BATES,[2] p. 245.

84 T. HIGUCHI, C. REHM, and C. BARNSTEIN: *Anal. Chem.*, **28**:1506 (1956). See also C. Rehm and T. Higuchi, *Anal. Chem.*, **29**:367 (1957).

85 L. P. HAMMETT and A. J. DEYRUP: *J. Amer. Chem. Soc.*, **54**:2721 (1932). Also see L. P. Hammett, "Physical Organic Chemistry," 2d ed., p. 267, McGraw-Hill, New York, 1970.

86 R. J. GILLESPIE: *Endeavour*, **32**(115):3 (1973).

87 R. H. BOYD in "Solute-Solvent Interactions," p. 97, Dekker, New York, 1969.

88 M. A. PAUL and F. A. LONG: *Chem. Rev.*, **57**:1 (1957).

89 O. W. KOLLING and T. L. STEVENS: *Anal. Chem.*, **34**:1653 (1962).

90 S. BRUCKENSTEIN: *J. Amer. Chem. Soc.*, **82**:307 (1960).

91 R. G. BATES in "The Chemistry of Nonaqueous Solutions," J. J. Lagowski (ed.), vol. 1, Academic, p. 97, New York, 1966.

PROBLEMS

4-1 Taking K_{SH} for ethanol to be 3×10^{-20}, calculate the pH of ethanol for 0, 50, 90, 99, 99.9, 100, 100.1, and 101% of the stoichiometric amount of reagent added for the titration of 10^{-3} M $HClO_4$ with 0.01 M sodium ethoxide in 100% ethanol.
Answer 3.0, 3.32, 4.03, 5.04, 6.04, 9.76, 13.48, 14.48.

4-2 Calculate the pH at 0, 1, 50, 90, 99, 100, and 101% neutralization of 0.01 M acetic acid ($K_a = 4.8 \times 10^{-11}$) in 100% ethanol ($K_{SH} = 3 \times 10^{-20}$) with 0.1 M sodium ethoxide. Would neutral red be a suitable indicator (see Table 6-2) for this titration?
Answer 6.15, 8.32, 10.32, 11.27, 12.31, 13.90, 15.48.

4-3 Calculate the magnitude of the change in pH in going from 10^{-3} M strong acid to 10^{-3} M strong base in water, ethanol, ethylenediamine, formic acid ($pK_{SH} = 6.2$), and acetic acid (see Table 6-1). In acetic acid, consider that ion-pair formation of the strong acid and strong base is complete but that the constants for dissociation of ion pairs to free ions are both 10^{-5}.
Answer 8.0, 13.5, 9.3, 0.52, 6.4.

4-4 For the titration of N,N-diethylaniline ($pK_B = 5.78$) with $HClO_4$ ($pK_a = 4.87$) in glacial acetic acid ($pK_{SH} = 14.45$), calculate the pH at the following percentages of the stoichiometric amount of $HClO_4$ added: 0, 50, 99, 100, 101. Assume $pK_{BH+ClO_4-} = 5.78$. Take the initial concentration of base to be 0.01 M and neglect dilution.
Answer 10.56, 10.26, 8.56, 6.77, 5.00.

4-5 For the titration of an acid HA in en ($pK_{SH} = 15.3$) with overall dissociation constant of 10^{-6} with sodium aminoethoxide, calculate the pH at the following percentages of the stoichiometric amount of reagent added: 0, 50, 90, 100, 110. Take the initial concentration of acid to be 0.01 M, neglect dilution, and assume that the salt has an overall dissociation constant of 10^{-3} and that pK_b for sodium aminoethoxide[63] is 5.84.
Answer 4.00, 5.47, 6.33, 7.73, 9.46.

4-6 Calculate as in Problem 4-5, but with the salt having an overall dissociation constant of 10^{-6} instead of 10^{-3}.
Answer 4.00, 5.47, 6.33, 7.73, 10.43.

4-7 For the titration of a base in acetonitrile with $K_B = 10^{-15}$ and a homoconjugation formation constant of 20, calculate the hydrogen ion concentration for each of the following percentages of the stoichiometric amount of $HClO_4$ added: 0, 50, 90, 100, 110. Take the initial concentration of base to be 0.001 M and K_{SH} to be 10^{-32}, and neglect dilution.
Answer 1×10^{-23}, 1.0×10^{-17}, 9.1×10^{-17}, 1.0×10^{-10}, 1.0×10^{-4}.

4-8   Assuming that Equation (4-6) applies to single ions, calculate γ_t in acetonitrile compared to a transfer activity coefficient of unity in water for Cl^-, Br^-, I^-, and Na^+, whose ionic radii are assumed to be 2.5, 2.37, 2.33, and 1.35×10^{-8} cm. Dielectric constants of water and acetonitrile are 78.3 and 36.0.

Answer $\gamma_t = 5.38, 5.90, 6.08$, and 22.6. (Compare with Figure 4-5.)

4-9   The 2,6-dinitroaniline cation in a particular solvent has a K_a value of 3.47×10^5. If the ratio $[B]/[BH^+]$ determined with a spectrophotometer is 0.28 in 71.9% H_2SO_4, what is the value of H_0?   *Answer* -6.09.

CHEMICAL STANDARDS

After an analytical problem has been defined, a method of analysis is selected or developed. If it is a new method, materials should be obtained for testing its reliability, precision, and accuracy. Even if it is a well-tested method, materials should be available for standardizing the response in terms of known materials. Instrumental methods, for example, are mainly empirical; standard substances are necessary for their calibration.

In the matter of standards, intelligent skepticism is essential for reliable work; every experienced chemist is cautious about accepting claims regarding the reliability of methods and applicability of materials.

5-1 HISTORY

The determination of atomic weights has been preeminent in the study of the reliability of analytical methods and in the definition and development of reference materials. In the history of chemistry no significant problem has been under more continuous investigation than that of atomic weights. Determinations of atomic weights by chemical methods require highly refined experimental techniques, and only a few people of each generation possessed the combination of intelligence, persistence, and dexterity that produced work of such caliber as to be generally accepted.

The question of standards is an integral part of atomic-weight determinations. Dalton used hydrogen equal to 1 as standard, Berzelius (whose atomic weights are the most significant in the earliest work) oxygen equal to 100, Wollaston oxygen equal to 10, and Thompson oxygen equal to 1. Then, in 1860 Stas suggested the value of exactly 16 for oxygen, the standard that served science for an entire century. The contributions of Stas, who spent most of his life working in this field, were outstanding for the time. He took elaborate precautions in purifying materials and in carrying out the chemical reactions. His use of relatively concentrated solutions, however, resulted in unrecognized serious errors due to coprecipitation.

The monumental work of Richards[1] on this subject stands alone; his work in the determination of atomic weights by chemical means was definitive. He found that even the most carefully purified potassium chlorate always contains some potassium chloride, that precipitated silver chloride carries down traces of potassium chloride that cannot be removed by washing, and that metallic silver always contains some oxygen unless stringent precautions are taken. Richards's first goal was revision of the atomic weights of silver, chlorine, and nitrogen. Once these weights were known, those of most of the other metallic elements could be determined by preparation of their pure chlorides, followed by conversion to the equivalent amount of silver chloride. He made certain critical measurements with the aid of a nephelometer that he developed (marking the beginning of instrumental analysis). Richards was awarded the Nobel prize in chemistry in 1915. Three of the most important atomic weights he obtained were 107.880, 35.458, and 14.008, representing the atomic weights of silver, chlorine, and nitrogen. These values were remarkably accurate; for example, the 1960 mass-spectrometric value[2] for silver was found to be 107.8731 $\pm$ 0.0018.

With the discovery of isotopes and the realization that atoms of oxygen are not all identical, the standard of oxygen exactly equal to 16 became unsatisfactory. Because the atomic weight of any element cannot be known more precisely than the standard to which it is compared, dissatisfaction grew as natural variations in the isotopic composition of oxygen were found to make its atomic weight uncertain by ± 0.0001 atomic-weight units. For a brief period, both chemical (oxygen equal to 16) and physical (oxygen 16 equal to 16) scales of atomic weight existed. In 1959, agreement was reached that the isotope of carbon 12 should be assigned a mass of exactly 12. All atomic weights are now referred to this standard.[3]

Although we can never know atomic weights absolutely, values generally are becoming more and more accurate. For some elements, especially sulfur (32.06) and boron (10.81), natural variations in isotopic composition limit the precision of the weights adopted. The accepted atomic weight for aluminum has undergone striking changes within the last few decades. For instance, in 1921, 1925, and 1961, the values were 27.1, 26.97, and 26.9815. Lead is an interesting example of an element for which the precision of its accepted atomic weight has historically gone through a maximum. Its atomic weight (converted to the carbon-12 scale) has been given as 206 (Dalton, 1810), 207.11 (Berzelius, 1826), 206.905 (Stas, 1890), 207.19 (Richards, 1923), and 207.2 (present).[4]

5-2 OPERATIONAL STANDARDS: SILVER AND THE COULOMB

Although carbon 12 serves admirably as the ultimate standard for determination of atomic weights, it is unsuitable as a working standard for chemical analyses. One of two operational standards, metallic silver or the coulomb, is best for this purpose. Each has its proponents. Silver has a venerable history as the key element in practical atomic-weight determinations, at least since the time of Stas. As a *defined quantity* the coulomb can justifiably be claimed to be a more fundamental basis for an operational standard, being defined, through the ampere, in terms of the fundamental units of mass, length, and time. In addition, the number of electrons equal to the number of atoms in 12 g of carbon 12 exactly defines Avogadro's number, and the number of coulombs equal to 1 mole of electrons is the electrochemical equivalent, the faraday. Thus, the faraday also is an exactly defined quantity. Experimentally determining its value to the highest precision is another matter. In practice the faraday usually has been measured[5-8] in terms of the amount of electricity required for an electrode reaction for unit weight of silver (or other pure chemical such as iodine or benzoic acid). Evidence for complete quantitativeness of the electrode reaction with silver can be tested experimentally only to a certain tolerance. That the reaction is quantitative has been substantiated to an uncertainty of probably no better than 10 to 20 ppm.[9]

Experimentally, then, the two operational standards are intimately related. Silver of high purity (99.999+%) is obtainable at the present time. Also available are techniques and equipment for measuring current and time to high precision as well as conditions for generation of reagents at electrodes with high current efficiency. Thus either silver or the coulomb can be used as the operational standard with justification.

Using silver, systems have been worked out for evaluating the purity and reactions of other supposedly pure substances and for standardizing solutions of chemicals. One such system[10] develops sodium carbonate, sodium chloride, and iodine as intermediate standards. For example, a solution of hydrochloric acid can be standardized by dispensing, with a weight buret, slightly less of the hydrochloric acid solution than is required to react with a weighed quantity of silver that has been brought into solution. The silver chloride precipitate formed is filtered, washed, and discarded, and then the slight excess of silver in solution is determined by precipitation with an excess of hydrochloric acid. If all operations are performed carefully, the concentration can be calculated to high precision from the weights of silver, the hydrochloric acid solution, and the silver chloride formed with the slight excess. Solutions of sodium chloride can be standardized by similar procedures. By extension, standard solutions for acid-base, precipitation, and redox titrations can be related directly or indirectly to silver.

Coulometry as an analytical technique, which began with Szebellédy and Somogyi,[11] initially was accurate enough only for routine analysis. Tutundžić[12] suggested that the coulomb could be used practically as a standard and thereby replace several pure chemicals. Now that the faraday has been measured[6,9] to high precision, coulometry makes standardizations using current-time measurements reasonable. The reagent, electrons, can be added or removed at any desired rate provided that current efficiency differs insignificantly from 100%. A highly accurate constant-current generator is required (Figure 5-1).

In coulometric analysis the product of current in amperes and time in seconds that are required to complete a reaction or generate a reagent gives the number of coulombs. To bring about 1 equivalent of electrode reaction, 96,487 (96,487.0 ± 1.6) C is required, and hence the number of equivalents E of reaction is

$$E = \frac{it}{96,487} \qquad (5\text{-}1)$$

Usually, after a reaction is about 98% complete, a current of about 0.1 of the normal value is used. In the apparatus developed for coulometry of high accuracy, the voltage of the regulated power supply seldom changes by more than 0.001% in a day, the current is accurately measurable through the use of a standard resistor (known to better than 1 ppm) of high precision, and adjustment of the iR drop is made equal to that of a Weston cell. The time can be measured to an accuracy of 1 ppm or better.[13]

In coulometry the stoichiometry of the electrode process should be known and should proceed with 100% current efficiency, and the product of reaction at any other electrode must not interfere with the reaction at the electrode of interest. If there are intermediate reactions, they too must proceed with the desired accuracy. In practice the electrolytic cell is designed to include isolation chambers. Losses of solute through diffusion, through ionic or electrical migration, and simply through bulk transfer must be minimal. Finally, the end point has to be determined by one of the many techniques used in titrations generally, whether coulometric or not. Both indeterminate and determinate end-point errors limit the overall accuracy achieved. Cooper and Quayle[14] critically examined errors in coulometry, and Lewis[15] reviewed coulometric techniques.

Acids and bases[13] have been standardized coulometrically with standard deviations of 0.003%; halides[16] by generation of silver ions to 0.005%; potassium dichromate by generation of iron to 0.003%; and conditions established under which

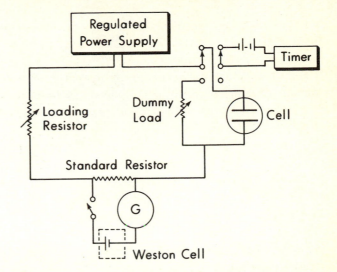

FIGURE 5-1 Constant-current and timing circuits for coulometric analysis. (*From Marinenko and Taylor.*[7])

iodine is generated[7] with a current efficiency of at least 99.9999%. Knoeck and Diehl[17] assayed potassium hydrogen phthalate, potassium dichromate, and ammonium hexanitratocerate by coulometric generation of reagent. Coulometric titrations are discussed in Section 14-10.

5-3 PURE MATERIALS

Chemicals are obtainable in a range of purity, including commercial chemicals for industrial use, chemically pure (CP) chemicals purified according to certain procedures, reagent-grade chemicals purified and tested with specified maximum limits of stated impurities, and primary-standard chemicals mainly for use in titrimetric standardizations. The publications[18] of the ACS Committee on Analytical Reagents can be referred to for specific information about certain analytical reagents. Data often are provided for impurities that can be easily tested for, and major impurities such as water that are not easy to measure are ignored.

In a titrimetric standardization the solution to be standardized is compared directly or indirectly against a standard substance. Some substances can be weighed accurately, but do not give a solution of known concentration. Others cannot be weighed accurately, but give stable solutions. A *primary standard* is a pure substance (element or compound) that is stable enough to be stored indefinitely without decomposition, can be weighed accurately without special precautions when exposed to

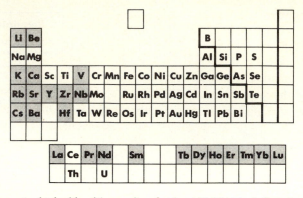

FIGURE 5-2   Elements obtainable with a purity of at least 99.9% (shaded) and at least 99.99% (unshaded). (*From ASTM.*[21])

laboratory air, and will undergo an accurate stoichiometric reaction in a titration. It should be readily available in a state of high purity, and simple tests should be available to detect common impurities. Ideally, to minimize weighing errors, it should possess a high equivalent weight.

Wichers[19] reviewed the preparation and evaluation of pure-substance standards; only a few are suitable for use as primary standards. Purity is often difficult to describe quantitatively. Terms such as extremely pure, triple-distilled, zone-refined, and chromatographically pure often are informative only to a few and may mislead some. Absolute purity is experimentally unattainable—although a gram of water with 1 part of impurity in 10^9 parts is extremely pure, it contains about 10^{13} particles of foreign species, with probably every radiochemically stable element present. A quantitative description of purity can be arrived at by (*1*) determining the concentration of each foreign constituent, or (*2*) determining the sum of all impurities through observation of how much a selected physical property, such as freezing point, deviates from the expected behavior of the pure substance. When impurity levels are low, refined techniques for such determinations as the measurement of freezing point depression have been developed in which temperature changes as small as 0.0001°C can be observed.

From the viewpoint of primary standards for ordinary analysis, acceptable purity depends on the substance. Many contain at least 99.9+% of the major component, for example, potassium hydrogen phthalate, potassium chloride, or calcium carbonate; but cholesterol[20] of high purity is 99.4%, and bilirubin 99%.

Most of the elements are obtainable in a state of high purity. Those available with a purity of at least 99.9% or 99.99% are shown in Figure 5-2.[21] Many, but certainly not all, of these elements can be used as primary standards to prepare standard solutions determinately. Smith and Parsons[22] listed suitable choices for preparation of standard solutions for 72 of the elements.

The preparation and examination of high-purity iron is as follows.[23-27] The preparation begins with low-impurity iron obtained from an ore of favorable composition. This is refined by several combinations of chemical reactions and zone melting. Analysis by spark-source mass spectrometry is commonly used for measuring the impurity level of nearly all elements present, even though the relative precision[26] of the measurements is only about 25%. Other analytical techniques involve vacuum fusion, gas chromatography, colorimetry, atomic absorption, and combustion. Some elements that cause difficulty in assessment of the quality of highly purified iron are hydrogen, carbon, oxygen, nitrogen, aluminum, silicon, and mercury.

5-4 STANDARD SAMPLES

Many analyses depend on the comparison of a sample with a standard whose composition is as much like that of the sample as possible. Such standards are used in the evaluation of new methods, techniques, and procedures. A new procedure for nickel in an ore, for example, cannot be evaluated by a comparison of results from the sample with those from pure nickel; the sample probably contains interferences, whereas the standard does not. The evaluation should be made with a standard containing the same kind and amount of impurities as the sample or through the use of the standard addition technique. A crucial problem in the preparation of standard samples is recognition of the significance of *matrix effects*. These effects result from often subtle influences of the chemical and physical composition of the sample on the magnitude of a measurement for the component to be assayed. Ideal standards eliminate some determinate errors in analyses by resembling as closely as possible the chemical and physical composition of the sample. This matrix problem is met in a serious form in spectroscopy.

Standard samples are prepared by many groups; best known are the standard reference materials prepared under the direction of the Office of Standard Reference Materials of the National Bureau of Standards.[28-30] The NBS standards comprise more than 600 different materials that range from pure chemicals such as metals, benzoic acid, and potassium chloride through standards for clinical chemistry to such complex substances as orchard leaves, beef liver,[20] and gelatin as a carrier for mercury reference materials.[31] In the samples of orchard leaves, 10 trace elements have their concentration certified, and many others have preliminary values established or are to be studied. The sources for about 3000 standard samples are listed by the American Society for Testing and Materials.[21] Most of these were developed because of the needs for spectrochemical analysis; 1743 are available for spectrochemical analysis of aluminum-containing materials alone. In the preparation of standard samples the problem of homogeneity and sampling variance needs to be recognized (Section 27-3).

Water as solvent deserves special mention. Hughes, Mürau, and Gundersen[32] compared the results of various techniques for water purification. They concluded that the nearly universal practice of assessing the quality of water by measurement of conductivity should not be relied upon. Also, metals are unsuitable as materials for construction of systems for producing highly purified water. Deionization with cation or mixed-bed resins removes only ionized impurities and is ineffective for the removal of either suspended or nonionized material.

More extensive sources[33–36] should be consulted for detailed information associated with the preparation, characterization, storage, and use of pure materials.

REFERENCES

1   T. W. RICHARDS: *Chem. Rev.*, **1**:1 (1924). G. S. FORBES: *J. Chem. Educ.*, **9**:453 (1932).

2   NBS OFFICE OF TECHNICAL INFORMATION: *J. Chem. Educ.*, **37**:531 (1960).

3   REPORT OF THE COMMISSION ON ATOMIC WEIGHTS (1961): *Pure Appl. Chem.*, **5**:255 (1962).

4   TABLES OF ATOMIC WEIGHTS, SELECTED RADIOACTIVE ISOTOPES, AND ATOMIC MASSES OF SELECTED ISOTOPES, 1969: *J. Amer. Chem. Soc.*, **93**:2579 (1971).

5   D. N. CRAIG, J. I. HOFFMAN, C. A. LAW, and W. J. HAMER, *J. Res. Nat. Bur. Stand.*, **64A**:381 (1960).

6   W. J. HAMER: *J. Res. Nat. Bur. Stand.*, **72A**:435 (1968).

7   G. MARINENKO and J. K. TAYLOR: *Anal. Chem.*, **39**:1568 (1967); **40**:1645 (1968).

8   V. E. BOWER in "U.S. National Bureau of Standards, Special Publication No. 343," p. 147, NBS, Washington D.C., 1971.

9   A. HORSFIELD in "U.S. National Bureau of Standards, Special Publication No. 343," p. 137, NBS, Washington, D.C., 1971.

10   ANALYTICAL CHEMISTS' COMMITTEE OF IMPERIAL CHEMICAL INDUSTRIES: *Analyst*, **75**:577 (1950).

11   L. SZEBELLÉDY and Z. SOMOGYI: *Z. Anal. Chem.*, **112**:313, 323, 332, 385, 391, 395, 400 (1938).

12   P. S. TUTUNDŽIĆ: *Anal. Chim. Acta*, **18**:60 (1958).

13   J. K. TAYLOR and S. W. SMITH: *J. Res. Nat. Bur. Stand.*, **63A**:153 (1959).

14   F. A. COOPER and J. C. QUAYLE: *Analyst*, **91**:355, 363 (1966).

15   D. T. LEWIS: *Analyst*, **86**:494 (1961).

16   G. MARINENKO and J. K. TAYLOR: *J. Res. Nat. Bur. Stand.*, **67A**:31 (1963).

17   J. KNOECK and H. DIEHL: *Talanta*, **16**:181, 567 (1969).

18   AMERICAN CHEMICAL SOCIETY, COMMITTEE ON ANALYTICAL REAGENTS: "Reagent Chemicals," American Chemical Society Specifications, 1960. ACS, Washington, D.C., 1961.

19   E. WICHERS: *Anal. Chem.*, **33**(4):23A (1961). See also *Anal Chem.*, **36**(12):23A (1964).

20   W. W. MEINKE: *Anal. Chem.*, **43**(6):28A (1971).

21   R. E. MICHAELIS: "Report on Available Standard Samples, Reference Samples, and High-Purity Materials for Spectrochemical Analysis," ASTM Special Technical Publication No. 58-E, pp. 125–147, ASTM, Philadelphia, 1963.

22   B. W. SMITH and M. L. PARSONS: *J. Chem. Educ.*, **50**:679 (1973).

23   G. W. P. RENGSTORFF: *Mem. Sci. Rev. Met.*, **65**:85 (1968).

24   P. ALBERT: *Mem. Sci. Rev. Met.*, **65**:3(1968).

25   O. KAMMORI: *Trans. Iron Steel Inst. Jap.*, **9**:76 (1969).

26   J. M. MCCREA: *Appl. Spectrosc.*, **23**:55 (1969).

27   P. J. PAULSEN, R. ALVAREZ, and C. W. MUELLER: *Anal. Chem.*, **42**:673 (1970).

28   THE NBS STANDARD REFERENCE MATERIALS PROGRAM: *Anal. Chem.*, **38**(8):27A (1966).

29   "Standard Reference Materials. Price and Availability List," National Bureau of Standards Special Publication No. 260, NBS, Washington, D.C., 1974.

30   W. W. MEINKE: *Mater. Res. Stand.*, **9**(10):15 (1969).

31   D. H. ANDERSON, J. J. MURPHY, and W. W. WHITE: *Anal Chem.*, **44**:2099 (1972).

32   R. C. HUGHES, P. C. MÜRAU, and G. GUNDERSEN: *Anal. Chem.*, **43**:691 (1971).

33   F. H. STROSS and J. H. BADLEY in "Treatise on Analytical Chemistry," I. M. Kolthoff and P. J. Elving (eds.), part I, vol. 9, p. 5863, Wiley, New York, 1971.

34   M. ZIEF and R. SPEIGHTS (eds.), "Ultrapurity," Dekker, New York, 1972.

35   E. C. KUEHNER, R. ALVAREZ, P. J. PAULSEN, and T. J. MURPHY: *Anal. Chem.*, **44**:2050 (1972).

36   K. HICKMAN, I. WHITE, and E. STARK: *Science*, **180**:15 (1973).

PROBLEMS

5-1   The anode compartment in a coulometric titration cell contains 250 ml of 1 M Na_2SO_4, with the CO_2 removed. Suppose that 1.2463 g of pure Na_2CO_3 is added and electrolyzed with platinum electrodes to completion of the neutralization. How long will this procedure take if the current is 194.36 mA?

Answer   3.2430 h.

5-2   A 50.006-ml sample of HCl is introduced into the cathode chamber of a coulometric titration cell. The end point is reached in 1.4632 h. What is the concentration of the HCl if the Weston cell used has a voltage of 1.01865 V (temperature maintained at 20°C) and the standard resistor in Figure 5-1 a resistance of 5.1463 Ω?

Answer   0.21610 M.

5-3   A sample of "pure" $(NH_4)_2Ce(NO_3)_6$ indicates a purity (for cerium) of 99.98% when it reacts with primary standard arsenic(III). Examination of the $(NH_4)_2 \cdot Ce(NO_3)_6$ by emission spectroscopy indicates that it contains 0.06% of the element thorium [presumably as $(NH_4)_2Th(NO_3)_6$]. Such results can be rationalized by assuming a deficiency of NH_3 and that the sample contains $H_2Ce(NO_3)_6$. (*a*) What is the percentage of the theoretical cerium content? (*b*) Calculate the percentage of $H_2Ce(NO_3)_6$ in the sample. (*c*) Calculate the percentage of the theoretical NH_3 content.

Answer   (*a*) 99.98%. (*b*) 2.20%. (*c*) 97.78%.

5-4   As a literature-search problem, find out what is known about the development of a selected standard substance. One such problem could involve how a metallo-organic standard (say, one of Al, Ag, Cr, Cu, Mg, Fe, Ni, Ti or Si) has been prepared for oil analysis (for example, in routine maintenance of aircraft engines). Other standards might be: a reference standard for fluoride for use with a fluoride ion-selective electrode; a SO_2 standard for use in connection with air pollution analyses; measurement of the "purity" of 99.99+% pure silver. Consider particularly the problems that had to be surmounted. For example, as a fluoride standard, NaF is often unsuitable because of difficulties with ion pairing and activity coefficients.

APPLICATIONS OF ACID-BASE TITRATIONS

The purpose of this chapter is to consider the preparation and standardization of acids and bases and to review some of the important applications of acid-base titrations in aqueous and nonaqueous systems. For end points to be detected most precisely, the pH in the vicinity of the equivalence point should change sharply. For this reason a solution of strong acid or base is chosen as titrant whenever possible.

More than brief discussion of the numerous ways in which end points can be taken other than by visual methods is beyond our scope. For example, end-point techniques may involve photometry, potentiometry, amperometry, conductometry, and thermal methods. In principle, many physical properties can be used to follow the course of a titration; in acid-base titrations, use of the pH meter is common. In terms of speed and cost, visual indicators are usually preferred to instrumental methods when they give adequate precision and accuracy for the purposes at hand. Selected instrumental methods may be used when a suitable indicator is not available, when higher accuracy under unfavorable equilibrium conditions is required, or for the routine analysis of large numbers of samples.

It should be mentioned here that automation of repetitive analyses is also possible. This includes all steps, including taking the sample and its preparation, titrant delivery, end-point detection, readout, and computation.[1-5] Automation can include closed-loop process control and continuous analysis. A detailed treatment of automation is not undertaken here. The development of instrumental techniques and automated analyses has by no means, however, rendered indicators and pH methods unimportant or the chemist obsolete.

6-1   PREPARATION OF STANDARD ACID

Hydrochloric acid is the most frequently used titrant in analytical acid-base work. According to Kolthoff and Stenger,[6] 0.1 M solutions of hydrochloric acid can be boiled for 1 h without loss of acid if the evaporated water is replaced. Even 0.5 M hydrochloric acid can be boiled for 10 min without appreciable loss. Sulfuric acid has the disadvantage of a relatively weak second step of ionization ($pK_a \simeq 2.0$). Moreover, a number of metallic and basic sulfates are sparingly soluble. Nitric acid is relatively unstable, though useful in special procedures such as the alkalimetric method for phosphorus.

It is possible to prepare a standard solution of hydrochloric acid determinately by weighing constant-boiling hydrochloric acid,[7,8] often using some form of weight buret. Composition of the constant-boiling acid varies slightly with barometric pressure; at pressures of 760, 750, 740, and 730 mm, constant-boiling hydrochloric acid contains 20.221, 20.245, 20.269, and 20.293 wt % hydrogen chloride (a nearly linear relation). A determinate solution of hydrochloric acid can also be prepared by the use of pure sodium or potassium chloride as a primary standard. The chloride is dissolved in water and passed through a well-rinsed cation-exchange column (Section 25-2) in the hydrogen form. An equivalent amount of hydrochloric acid is produced and is rinsed from the column into a volumetric flask and diluted to volume.[9,10]

Although the determinate preparation of standard acids is possible, the use of a standardization method is usual practice.

6-2   PREPARATION OF STANDARD BASE

An important consideration in the preparation and storage of standard solutions of strong bases is the possible interference of carbonate. When carbonate is present, the titrant solution behaves chemically as though it were a mixture of the strong base with moderately strong base (carbonate ion) and weak base (bicarbonate ion formed from carbonate ion during titration). Normally, the only two substances considered for the preparation of aqueous standard solutions are sodium hydroxide and barium hydroxide. Barium hydroxide has the advantage of being automatically free of carbonate, since the solubility of barium carbonate is negligible. The low solubility of barium sulfate and phosphate is a disadvantage in some applications.

Sodium hydroxide is almost universally used as the titrant base. The solid usually contains appreciable amounts of such substances as water and carbonate. The classical method[11] of removing carbonate contamination is to prepare a concentrated solution (1:1 or 4:5 sodium hydroxide to water by weight), in which sodium carbonate is only slightly soluble. After filtration, centrifugation, or settling,[11] the solution is diluted with freshly boiled and cooled water or with distilled water in equilibrium with the carbon dioxide of the air. Such "equilibrium water," prepared by bubbling with air that has been passed successively through wash bottles containing dilute acid and water, contains only 1.5×10^{-5} mole/l of carbon dioxide. The analogous procedure with potassium hydroxide is not successful because potassium carbonate is relatively soluble in concentrated potassium hydroxide.

An ion-exchange method is convenient and efficient for the removal of carbonate from either sodium or potassium hydroxide.[12] The solution is passed through a strong base anion-exchange column (Section 25-2) in the chloride form. Since the first portions of hydroxide convert the resin to the hydroxide form, the first portions of effluent contain chloride. If either the presence of chloride or dilution of the standard base is objectionable, the effluent is discarded until the resin has been converted to the hydroxide form. Carbonate is retained on the anion-exchange column, and carbonate-free base is collected. When the column becomes saturated with carbonate, it is reconverted to the chloride form by passing through it dilute hydrochloric acid, followed by water to remove the excess acid. Anion exchangers have been used to prepare standard solutions of sodium hydroxide from weighed amounts of primary-standard sodium chloride[13] or potassium chloride.[14]

Solutions of strong base attack glass and therefore should be stored in polyethylene containers. Polyethylene is slightly permeable to carbon dioxide[15] (permeability constant about 1×10^{-9} ml-cm/sec-cm^2-cm Hg), but no significant carbonate contamination occurs within a reasonable time (see Problem 6-6).

When carbonate contamination is undesirable, protection from the carbon dioxide of the atmosphere should be provided. The usual method is to use a storage bottle with a protective tube containing either sodium hydroxide and asbestos (ascarite) or sodium and calcium hydroxides (soda lime). A siphon is used to deliver the clear supernatant liquid to the buret without exposure to carbon dioxide of the air.

6-3 STANDARDIZATION OF ACIDS AND BASES

Under the auspices of the International Union of Pure and Applied Chemistry, an Analytical Methods Committee carried out a critical examination of some primary standards suggested for use in acid-base titrations,[16] with a view to making recommendations concerning practical applications. Some of the findings of this committee were as follows: (1) Benzoic acid as a primary standard is not recommended because of difficulty in controlling loss through volatility, even though it can be prepared with a purity[17] of 99.997% or better. (2) The use of constant-boiling hydrochloric acid is not recommended because of difficulties in preparation and storage. (3) Sulfamic acid is suitable as a primary standard for acidimetric work, but no commercial source of adequate purity was found. The laboratory preparation of this primary standard requires care, and a detailed procedure involving recrystallization has been tested.[16] (4) Sodium carbonate is the most generally suitable primary standard for the standardization of solutions of acid. Material with a purity within the limits of $100 \pm 0.02\%$ is available commercially and is therefore directly acceptable as primary standard.

Sodium carbonate as a primary standard for acids   The Analytical Methods Committee[16] endorsed the common practice of using sodium carbonate as the primary standard of choice for standardizing solutions of acids. For the laboratory preparation of primary-standard sodium carbonate, the committee recommended preparing a sodium bicarbonate solution containing 769 g per 3 l of water at 86°C, allowing it to cool to 75°C, and filtering it to remove the precipitated material. Then the solution is cooled to 20°C, filtered, and washed with ice water. The product is dried at 100°C, ground, and converted to sodium carbonate by heating at 270°C to constant weight. A serious loss of carbon dioxide occurs above 300°C in air, 1 h of heating at 310 to 315°C giving an error of more than 1%.[18]

For accurate results in the use of sodium carbonate, several precautions are necessary. Kolthoff[19] warned against storage of ignited samples in a desiccator, because moisture in amounts up to 0.1% can be taken up during the opening and closing of the container. According to Richards and Hoover[20] even a freshly ignited sample contains up to 0.05% moisture. The last trace of moisture can be removed by fusing the sample in a stream of pure carbon dioxide, which is gradually displaced by air as the sample is cooled.

For the titration of sodium carbonate with acid, some procedures specify direct titration at room temperature to a methyl orange or a mixed-indicator end point. To achieve accurate results requires precautions, particularly in the standardizing of solutions that are 0.1 M or more dilute. The concentration of carbon dioxide in solution at the end point is not reproducible, because the solution becomes supersaturated as the titration proceeds, and because variable amounts of carbon dioxide are lost near the end point, depending in part on the time elapsed and the amount of shaking. The equilibrium $CO_2 + H_2O \rightleftharpoons H_2CO_3$ is not rapid; elimination of carbon dioxide can be accelerated by the use of the enzyme[21] carbonic acid anhydrase or by raising the temperature. The color of the indicator in a saturated solution of carbon dioxide is sensitive to the concentration of sodium chloride (Section 3-10). With increasing salt concentration the color of methyl orange, for example, is shifted to the acid side. Therefore, for accurate results a comparison solution should be used,

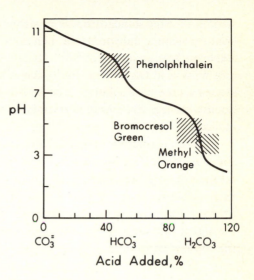

FIGURE 6-1 Titration curve of 0.05 M Na$_2$CO$_3$ with 0.1 M HCl.

saturated with carbon dioxide and containing the same concentration of sodium chloride and indicator as the titrated solution at the end point.

The two-indicator method[22] gives a preliminary estimate of the end point and a sharp end point following the boiling step. First, the titration is carried out at room temperature to a colorless phenolphthalein end point (pH $\simeq$ 8), which lies a little past the first equivalence point (Figure 6-1). Then bromocresol green is added, and the titration continued to the first indication of green color (pH $\simeq$ 5), corresponding to the addition of about 98% of the theoretical quantity of acid. At this point the solution is boiled so that carbon dioxide is eliminated and the buffer capacity of the solution reduced. If the titration is stopped noticeably before the end point, a purple color corresponding to the red of phenolphthalein and the blue of bromocresol green in a solution of bicarbonate (pH $=$ 8.4) is observed after cooling. The titration is continued through the blue to the green end-point color of bromocresol green. The titration error may be estimated as follows: For a standardization involving 5 millimoles of hydrochloric acid and an end-point volume of 100 ml, the end-point concentration of carbonic acid is 0.025 M, with the assumption of no loss of carbon dioxide. At pH 5, $[HCO_3^-] = K_1[H_2CO_3]/[H^+] \simeq 10^{-3} M$, with $K_1 = 3.5 \times 10^{-7}$. This corresponds to 4% of the carbonic acid concentration, or 2% of the total acid used. After the solution is boiled to remove the carbon dioxide and the titration continued, this time to pH 4, 0.4% of the *remaining* bicarbonate will be untitrated, corresponding to an error of less than 0.01% for the entire titration.

Potassium hydrogen phthalate as a primary standard for bases Potassium hydrogen phthalate, C$_6$H$_4$COOKCOOH, is used almost exclusively as the primary standard for the standardization of strong bases. It is readily available in pure form, nonhygroscopic, anhydrous, of high equivalent weight, readily soluble in water, and stable on heating up to 135°C.[23] Its only real disadvantage is that, since it is the salt of a weak acid ($K_2 = 4 \times 10^{-6}$), the end point occurs in the alkaline region. Consequently, the base solution must be free of carbonate contamination if a sharp end point with negligible titration error is to be achieved. Phenolphthalein is used as the indicator.

When a solution of standard base is used only for titration of strong acids, a small amount of carbonate is not a serious source of error provided the end point is taken with an indicator that changes color at a pH of about 4 or 5. For standardizing such carbonate-containing solutions, potassium hydrogen phthalate is an unsuitable primary standard. An alternative[24] is pure potassium chloride, which is passed through a cation-exchange column, converted to hydrochloric acid, and titrated with the sodium hydroxide.

6-4   APPLICATIONS OF AQUEOUS ACID-BASE TITRATIONS

Determination of acids   For accurate results in the titration of weak acids with an indicator, one should be chosen that shows a transition color in the alkaline range coinciding as nearly as possible with the pH at the equivalence point. For best results a comparison solution of the indicator in a solution of the salt of the weak acid should be used. If the salt is not available, a buffer solution of the same pH may be substituted.

The theoretical precision of a titration may be estimated from the slope of the titration curve in the vicinity of the end point (Section 3-7). In practice, however, precision and accuracy often are limited by the presence of carbonate in the strong base. The effect of carbonate can be deduced from the titration curve of carbonic acid, which is the reverse of the curve of a carbonate with a strong acid (Figure 6-1). If we start with a solution of a weak acid and use carbonate-containing base, a dilute solution of carbonic acid forms during the titration. Below a pH of about 4 the carbonic acid formed has no effect on the effective molarity, but between values 4 and 8.5 it is converted to bicarbonate by the further addition of strong base. But, since the rate of reaction between the carbonic acid formed and hydroxyl ion is slow,[25] an indicator such as phenolphthalein or thymol blue exhibits its alkaline color temporarily as titrant base is added and then fades slowly back to its acid color on standing. The permanent alkaline color of phenolphthalein is reached only after the original carbonate has been converted to bicarbonate. Thus the hydroxide converted to carbonate in the standard base loses half its effective molarity as a reagent if the titration is carried to a phenolphthalein end point, whereas no error is caused if a methyl orange or a bromocresol green end point is used.

For the titration of strong acids of concentration 0.1 M or higher with base containing some carbonate, the error involved in titrating to a pH of 4 is negligible. For the titration of more dilute solutions, however, it is advisable (Section 6-3) to titrate to the first perceptible color change of methyl red (pH range 4.4 to 6.0), boil to remove carbon dioxide, cool, and continue to the yellow color of the indicator.

Determination of bicarbonate and carbonate in mixtures The classical method of determining carbonate in the presence of bicarbonate is to titrate in the cold to a bicarbonate end point. Since the pH of sodium bicarbonate is nearly constant at 8.3 to 8.4 over a wide range of concentrations, this method is in principle applicable over a wide range of sample compositions. The titration curve of Figure 6-1, however, exhibits a low slope. The relative precision (Section 3-7) is $\sqrt{K_2/K_1} = 1.2 \times 10^{-2}$, indicating that if an uncertainty of ± 0.1 pH unit is allowed a relative error of $\pm 1.2\%$ may be expected. Even this precision can be attained only by use of a comparison standard of a pure bicarbonate solution containing the same concentration of indicator. Phenolphthalein, the most common indicator, does not become decolorized until a pH of 8.0, corresponding to an error of 3 to 5%. Color comparison is therefore essential. If thymol blue, a two-color indicator, is used, less attention need be paid to attaining identical indicator concentrations in the two solutions. By using a mixed indicator composed of thymol blue and cresol red, Simpson[26] showed that results accurate to within 0.5% can be obtained without a comparison solution.

In the determination of bicarbonate or total alkalinity, titration to a low-pH end point is common practice, with boiling to remove carbon dioxide just prior to the final end point. The two-indicator procedure outlined in connection with the standardization of acids using sodium carbonate may be advantageous. Many practical determinations of total carbonates, however, are based on direct titration procedures involving no boiling step. For example, an official procedure for the determination of carbonates in water[27] calls for the determination of the carbonic acid end point by using a mixed bromocresol green–methyl red indicator. Such a procedure may lead to errors in dilute solutions (below about 0.001 M) because the pH at the equivalence point varies with the concentration of carbonic acid [Equation (3-37)]. Titration curves given by Cooper[28] (Figure 6-2) show the constancy of pH at the bicarbonate end point and the decrease in pH at the carbonic acid end point as the concentration increases. After comparing various indicators, Cooper recommended a mixture of bromocresol green and methyl red. To allow for the change in equivalence-point pH with concentration, he selected different shades of color for various total carbonate contents.

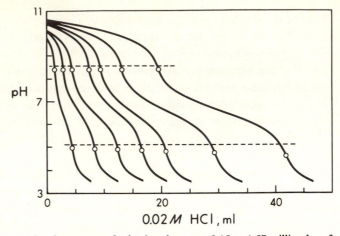

FIGURE 6-2   Titration curves of mixed carbonates: 0.15 to 1.57 millimoles of Na_2CO_3 and 0.05 to 0.21 millimoles of $NaHCO_3$ per liter. (*From Cooper.*[28])

Determination of hydroxyl ion in the presence of carbonate   Owing to the small second ionization constant of carbonic acid ($K_2 = 5 \times 10^{-11}$), the titration of hydroxyl ion alone is not feasible in the presence of an appreciable concentration of carbonate ion. For example, a 0.01 M solution of sodium hydroxide has a pH of about 12, and a 0.01 M solution of sodium carbonate a pH of about 11.2. Thus the pH change in a titration of 0.01 M sodium hydroxide in the presence of 0.01 M sodium carbonate would be only about 0.8 pH unit.

Winkler[29] showed that, if carbonate is precipitated by adding an excess of barium chloride, hydroxyl ion can be successfully titrated in the presence of precipitated carbonate. Thus, if barium chloride is added to give 0.1 M excess, the concentration of carbonate is $K_{sp}/[Ba^{++}] = 5 \times 10^{-9}/0.1 = 5 \times 10^{-8}\ M$. Actually the situation is more complicated, because barium carbonate coprecipitates some carbonate if an excess of barium chloride is added to a cold solution of sodium carbonate.[30]   On the other hand, if an alkali metal hydroxide is present, the barium carbonate is contaminated with hydroxide.[31] For the most accurate results the excess hydroxide should be approximately neutralized by adding an amount of hydrochloric acid estimated from a preliminary test, and the barium carbonate precipitated from a warm solution.[30]   Sulfate tends to cause error by occlusion of hydroxide with the precipitated barium sulfate.[32]

As in the determination of carbonate and bicarbonate, a two-indicator method[33] may be used. It consists of titrating the hydroxide plus one-half the carbonate alkalinity to a phenolphthalein end point and then carrying out the usual determination of total alkalinity.

Kjeldahl method for determination of nitrogen in organic compounds   The
simplest form of the Kjeldahl[34] method calls for decomposing the sample by boiling
with concentrated sulfuric acid to convert the nitrogen to ammonium sulfate. Potassium sulfate is usually added to raise the boiling point, and mercury, mercury(II)
oxide, selenium, or iron(II) sulfate are often used to hasten the decomposition of
organic matter. After the addition of excess sodium hydroxide, the ammonia is
distilled into an excess of acid, which is then titrated.

The usual Kjeldahl procedure works for a large variety of nitrogen compounds
such as amines, amino acids, and alkaloids, but fails for nitrates, nitrites, azo compounds, cyanides, and derivatives of hydrazine. For nitrates a method involving the
addition of salicylic acid may be used.[35] Azo compounds give quantitative results if
first reduced by refluxing with tin(II) chloride and hydrochloric acid. Hydrazine
derivatives are decomposed by heating with formaldehyde, zinc dust, and hydrochloric acid, followed by tin(II) chloride, prior to the usual Kjeldahl digestion.[36]

The ammonia can be distilled into an excess of standard sulfuric acid, and the
excess determined by back titration. Preferably, the distilled ammonia can be absorbed into a solution of boric acid or other weak acid. Direct titration of the boric
acid distillate with standard acid, using bromophenol blue, with a blank determination
gives excellent results. As a routine method for determining traces of nitrogen,
Milner and Zahner[37] distilled the ammonia into a dilute boric acid solution. Urban[38]
suggested an aqueous solution of p-hydroxybenzoic acid (ionization constant
2.9×10^{-5}) as an ammonia-absorbing solution.

The Kjeldahl nitrogen procedure has been automated[39,40] with a colorimetric
method of final measurement.

Titrations in concentrated salt solutions   Critchfield and Johnson[41] found that
bases with ionization constants as low as about 10^{-11} can be titrated with indicator
methods in aqueous solutions containing a high concentration of a neutral salt. The
acidity of the strong acid titrant is increased by the high salt concentration. Thus in
$7\ M$ sodium iodide the pH shortly after the equivalence point in a titration of a weak
base has an effective value of about zero instead of about 2 as in the usual titration.
Since the pH before the equivalence point is essentially unchanged, the pH change
in the equivalence-point region is enhanced by about 2 units.

6-5 APPLICATIONS OF NONAQUEOUS ACID-BASE
TITRATIONS

If feasible, acid-base titrations are best carried out in aqueous media. When, however,
the acid or base to be determined is too weak to be titrated in water or is insoluble,
then nonaqueous titrations may be attractive. An enormous literature has developed

in connection with applications of nonaqueous titrimetry, in marked contrast to the paucity of data on fundamental equilibrium constants in many of the solvents. Since adequate description of these applications is beyond the scope of this work, the reader is referred to books[42-46] and reviews[47,48] on the subject; selected aspects are discussed in the following treatment. Literature such as that in the organic, petrochemical, and pharmaceutical fields can be examined for recent practical developments. For practical details and advice, two early books[42,43] and a booklet[49] on the subject are recommended. Much work has been devoted to establishing the stoichiometry of acid-base reactions that cannot be carried out conveniently or accurately in water and to comparing various indicator and electrometric methods for end-point detection. Even though quantitative equilibrium data are often lacking, particularly for mixed solvents, a qualitative understanding of acid-base principles and a knowledge of the species present can be valuable guides to the application of solvents and titrants in a given problem.

From the practical point of view, nonaqueous solvents can be described broadly as acidic, basic, or neutral. For carrying out titrations, properties such as dielectric constant, melting point, boiling point, and (for amphiprotic solvents, Section 4-3) autoprotolysis constant are important (Table 6-1).

Mixed solvents often are used to modify the acidic or basic strength of the parent solvent or, what is more important, to change its dielectric constant and hydrogen-bonding characteristics. The solubility of the sample can thus be enhanced, and the titration equilibrium can be improved by decreasing the degree of dissociation and the solubilities of ionic products. One empirical guide[47,50] to the applicability of a solvent is the difference between the potential of a pH indicating electrode at half-neutralization of a strong base such as a tetraalkylammonium hydroxide and the potential at half-neutralization of a strong acid such as perchloric acid. A difference of 200 to 300 mV in half-neutralization potential is usually sufficient to permit selective titrations, provided the concentration range is suitable.

Solvent purification has been discussed by many, and recommended procedures have been listed by Kratochvil.[48] Molecular and volatilizable impurities are best measured by gas chromatography, and ionic impurities by electrical conductivity. The commonest and most troublesome impurity, because of its leveling action, is water; it can be measured by the Karl Fischer method (Section 19-8) among others.

In general, a solvent selected for the titration of a weak acid should itself be a much weaker acid (or neutral) and usually should be basic enough compared with water as solvent to increase the extent of dissociation. Comparably, for a weak base the solvent should be a much weaker base and acidic enough to increase the extent of dissociation. For the possible differentiation of acids that are all strong in water, the choice of a solvent less basic than water is appropriate. The solvent should also dissolve the sample, and the reaction products should be either soluble or rapidly form crystalline precipitates.[51] Usually, bases are most easily titrated in an acidic solvent such as acetic acid, and acids in a basic solvent such as pyridine.

The titrant acid used most often is perchloric acid—the strongest common acid. Potassium hydrogen phthalate is the usual primary standard for standardizing perchloric acid in nonaqueous media; in contrast to its use as an acid in water, it is used here as a base. One of the acids in the benzenesulfonic series also can be used as titrant. Of these, 2,4-dinitrobenzenesulfonic acid is almost as strong as perchloric acid.[52] Although somewhat weaker, p-toluenesulfonic acid has several advantages: It is available in high purity, it can be used in the direct preparation of standard solutions, and (unlike stronger acids) solutions of it are stable[53] in acetonitrile.

The titrant base is generally a tetraalkylammonium hydroxide[54] or an alkali metal alkoxide. Hiller[55] used the sodium salt of methylsulfinyl carbanion as titrant base.

When adequate electrical conductivity can be obtained, the coulometric

Table 6-1   PROPERTIES OF SELECTED NONAQUEOUS SOLVENTS

Solvent	Dielectric constant	Dipole moment esu $\times$ 10^{18}	Auto-protolysis constant pK_{SH}	Boiling point, °C	Melting point, °C	Density at 20°C, g/ml
Acetic acid	6.1	0.83	14.45	118	16.5	1.050
Acetic anhydride	20.7	. . .	. . .	140	−73	1.082
Acetonitrile	36.0	3.96	32.2	82	−45	0.777
Chloroform	4.8	1.00	. . .	61	−64	1.498 (15°)
Dimethylfor- mamide	36.7	. . .	. . .	153	−61	0.944
Dimethyl- sulfoxide	46.7	. . .	. . .	189	18	1.101
p-Dioxane	2.21	0	. . .	101	12	1.036
Ethanol	24.3	. . .	19.5	78	−117	0.789
Ethylene carbonate	81	. . .	. . .	248	36	1.322 (40°)
Ethyl acetate	6.02	1.78	. . .	77	−84	0.902
Ethylene glycol	37.7	. . .	. . .	197	−12	1.113
Ethylene- diamine	12.9	. . .	15.3	117	8.5	0.900
Formamide	109	. . .	. . .	193	−5	1.002
n-Hexane	1.91	0	. . .	69	−94	0.659
Methanol	32.6	. . .	16.7	65	−98	0.792
Nitromethane	35.8	3.46	. . .	101	−29	1.131 (25°)
Pyridine	12.3	2.15	. . .	115	−42	0.987
Sulfolane	43.3	. . .	. . .	285	28	1.262 (30°)
Tetrahydrofuran	7.6	. . .	. . .	66	−108	0.888
Toluene	2.38	0.37	. . .	111	−95	0.868
Water	78.3	1.87	14.0	100	0	0.9982

generation of reagents would seem to be especially advantageous in nonaqueous chemistry.[56-58]

A matter of practical importance in nonaqueous titrimetry is that, when volumetric equipment is used, errors should be prevented that arise from solvent volatility and from characteristics of viscosity and surface tension that differ from those of water. Temperature coefficients of expansion are often about six times that of water, so careful control of temperature is needed when volumes are being measured. Gravimetric titration techniques are recommended,[53,59] since they avoid most of these volumetric problems. Details of a gravimetric technique using a syringe have been given.[53]

End-point detection   Most nonaqueous titrations are carried out with a glass indicating electrode, which responds satisfactorily to changes in hydrogen ion activity in many solvents (Section 4-10). Visual indicators of a wide variety have been developed and are often chosen empirically. Crystal violet and methyl violet have long been used in the determination of bases, with glacial acetic acid as solvent. Fritz and Gainer[60] listed indicators for the titration of acids in pyridine with tetrabutylammonium hydroxide. Kolthoff, Chantooni, and Bhowmik[61] studied a series of indicators whose pK values for dissociation cover the approximate range from 2 to 30 in acetonitrile. In alcohol or alcohol-water mixtures the usual indicators for aqueous solution can be used if their shifted pH range is recognized (Section 4-9). Higuchi, Feldman, and Rehm[62] studied the behavior of 13 indicators in glacial acetic acid. Table 6-2 summarizes pK values for selected indicators in various solvents.[61,63,64]

Table 6-2   pK VALUES FOR ACID-BASE INDICATORS IN VARIOUS SOLVENTS

Indicator	Color change	Water	Methanol	Ethanol	Acetonitrile	Dimethylsulfoxide
o-Nitrodiphenylamine	Colorless to yellow	−2.9	...	...	2.0	...
Tropeolin 00	Red to yellow	~1.8	2.2	2.3	8.0	...
Methyl orange	Red to yellow	3.5	3.8	3.4	10.6	...
Neutral red	Red to yellow	7.5	8.2	8.2	15.6	...
Bromophenol blue	Yellow to blue	4.0	8.9	9.5	17.5	...
Bromocresol green	Yellow to blue	4.7	9.8	10.6	18.5	7.3
Bromothymol blue	Yellow to blue	7.1	12.4	13.2	22.3	11.3
Phenol red	Yellow to red	7.9	12.8	13.6	25.0	13.7
Phenolphthalein	Colorless to red	9.5	...	...	29.2	16.3

Several indicators have two or more color-change transition intervals (not shown in the table).

Conductance titrations often are employed that involve an electrode system that is simpler, though less specific, than for pH measurements. Coetzee and Bertozzi[65] titrated bases as weak as acetonitrile and nitrobenzene with perchloric acid in sulfolane by conductance measurements. Watson and Eastham[66] used high-frequency titration end points. Bruckenstein and Vanderborgh[67] used cryoscopic methods incorporating measurement of freezing point depression during titration. Thermometric titrations, described by Jordan,[68] Tyrrell and Beezer,[69] and Bark and Bark,[70] are universally applicable to rapid reactions with an adequate enthalpy change on neutralization. Since the heat capacities of most nonaqueous solvents are lower than that of water, the temperature rise observed is often larger than in aqueous solvents, leading to a sensitivity advantage. For thermometric titrations the titrant normally must be in the same solvent, or the heat of mixing leads to significant thermal effects. Table 6-3 lists pK_a values for the dissociation of acids in several solvents.[64, 71-73]

Acidic solvents The most important acidic solvent is glacial acetic acid with perchloric acid in acetic acid as titrant. This system is generally applicable to the determination of bases whose pK_b values in water are 12 or lower. Salts of carboxylic acids, for instance acetates, citrates, and benzoates are strongly basic and show good titration behavior. Varying degrees of ion association can cause the effective basic strengths to be different even for a closely related series such as the alkali metal

Table 6-3 pK_a VALUES FOR OVERALL DISSOCIATION OF ACIDS IN WATER AND SEVERAL NONAQUEOUS SOLVENTS

Acid	Water	Glacial acetic acid	Ethanol	Ethylene-diamine	Aceto-nitrile	Dimethyl-formamide	Dimethyl-sulfoxide
Perchloric	...	4.87	...	...	...	...	...
Sulfuric	...	7.24 (K_1)	...	...	7.25 (K_1)	3.0 (K_1) 17.2 (K_2)	... ...
p-Toluene-sulfonic	...	8.46	...	...	...	...	...
Hydrobromic	...	...	...	3.62	5.5	...	...
Hydrochloric	...	8.55	...	3.98	8.9	3.4	...
Benzoic	4.18	...	10.25	...	20.7	12.3	11.1
Salicylic	2.97	...	8.68	...	16.7	8.2	6.8
Picric	0.80	...	3.8	...	8.9	1.6	−1
Acetic	4.76	...	10.32	...	22.3	13.5	12.6
Anilinium ion	4.63	...	5.70	...	10.6	4.4	3.6
Phenol	10.0	...	...	...	27.2	...	16.4
Ammonium ion	9.26	...	...	...	16.5	9.4	10.5

acetates. According to the data of Pifer and Wollish,[74] the water content of glacial acetic acid should not exceed about 1%, or else the change in pH in the region of the equivalence point will be seriously impaired as a result of leveling. Concentrated perchloric acid contains 28% water. In glacial acetic acid as solvent the titrant perchloric acid is made anhydrous by the use of acetic anhydride. To prepare[42] 0.1 M perchloric acid, add 8.5 ml of concentrated perchloric acid to 250 ml of glacial acetic acid and 20 ml of acetic anhydride. Dilute to a liter with glacial acetic acid, and allow it to stand several hours until the anhydride reacts completely with the water in the 72% perchloric acid.

An important variation in the application of nonaqueous solvents has been to add various *neutral* (see below) solvents to the sample solvent, the reagent, or both, or to use mixtures of solvents. Fritz[75] dissolved weakly basic samples in a variety of solvents such as chloroform, nitrobenzene, ethyl acetate, and acetonitrile and then titrated with perchloric acid in acetic acid using a glass electrode or methyl violet as indicator. Dioxane has been substituted for acetic acid as a solvent for titrant perchloric acid, and sharper titration curves often are observed. Such a reagent is no longer strictly anhydrous when no provision is made to remove water from the 72% perchloric acid used to prepare the reagent.

Basic solvents   A basic solvent can be used to enhance the acidic properties of weak acids. To take full advantage of a basic solvent requires a strongly basic titrant. Moss, Elliott, and Hall[76] were able to observe separate end points for carboxylic acids and phenols using sodium aminoethoxide as titrant in ethylenediamine as solvent.

Sodium methoxide dissolved in benzene-methanol mixture was introduced as a titrant by Fritz and Lisicki.[77] Various applications have since been made with a number of basic solvents, notably ethylenediamine, butylamine, and dimethyl-formamide. Other titrants that have found use in basic solvents are potassium methoxide in benzene-methanol, potassium hydroxide in ethanol or isopropyl alcohol, and tetrabutylammonium hydroxide in ethanol, isopropyl alcohol, or benzene-methanol mixtures. Benzene-methanol has been recommended especially[78] for selective titrations of mixtures of acids in pyridine as the solvent.

Neutral solvents   The term *neutral solvent* applies here to solvents not predom-inately either acidic (protogenic) or basic (protophilic) in character. Some are weakly basic but not appreciably acidic (ethers, dioxane, acetone, acetonitrile, esters), some aprotic (benzene, carbon tetrachloride, 1,2-dichloroethane), and some amphiprotic solvents (ethanol, methanol). Aprotic solvents are used mainly in mixed solvents to alter the solubility characteristics of the reactants.

Many titrations have been made in such solvents as mixtures of alcohol or acetone with water. As pointed out in Section 4-9, the main effect of adding alcohol to water is to weaken molecular and anionic acids. Only as the composition approaches anhydrous alcohol does the autoprotolysis constant decrease markedly. Therefore, the main advantage of avoiding the leveling effect of water and permitting the use of acids stronger than hydronium ion or bases stronger than hydroxyl ion can be gained *only by using essentially anhydrous solvents.*

Solvents called *G-H mixtures,*[79] where G is a glycol and H is a hydrocarbon solvent, are exceptionally effective for titration of alkali metal salts of organic acids. These salts can be titrated with perchloric acid in the same solvent, with either a visual indicator or a glass electrode.

Mention should be made of lithium aluminum hydride, introduced as a titrant by Higuchi, Lintner, and Schleif[80] in a method of determining alcohols by titration of the hydroxyl hydrogen in tetrahydrofuran as solvent. The same reagent was applied by Higuchi and Zuck[81] to the determination of various functional groups (alcohols, esters, phenols, ketones, aldehydes) that react with lithium aluminum hydride. The excess reagent was determined by titration with alcohol. To avoid the strongly reducing properties of lithium aluminum hydride, Higuchi, Concha, and Kuramoto[82] introduced lithium aluminum amides, prepared by the reaction of amines with lithium aluminum hydride.

Solvent choices for nonaqueous acid-base titrations have been reviewed by Lagowski.[83]

REFERENCES

1 A. JOHANSSON: *Analyst,* **95**:535 (1970); A. JOHANSSON and L. PEHRSSON: *Analyst,* **95**:652 (1970).

2 J. A. FIFIELD and R. G. BLEZARD: *Analyst,* **96**:213 (1971).

3 K. A. MUELLER and M. F. BURKE: *Anal. Chem.,* **43**:641 (1971).

4 G. W. NEFF, W. A. RADKE, C. J. SAMBUCETTI, and G. M. WIDDOWSON: *Clin. Chem.,* **16**:566 (1970).

5 G. M. HIEFTJE and B. M. MANDARANO: *Anal. Chem.,* **44**:1616 (1972).

6 I. M. KOLTHOFF and V. A. STENGER: "Volumetric Analysis," 2d ed., vol. 2, p. 64, Interscience, New York, 1947.

7 G. A. HULETT and W. D. BONNER: *J. Amer. Chem. Soc.,* **31**:390 (1909).

8 J. A. SHAW: *Ind. Eng. Chem.,* **18**:1065 (1926).

9 J. SCHUBERT: *Anal. Chem.,* **22**:1359 (1950).

10 C. J. KEATTCH: *Lab. Pract.,* **5**:208 (1956).

11 R. G. BATES: *Chemist-Analyst,* **50**:117 (1961).

12 C. W. DAVIES and G. H. NANCOLLAS: *Nature,* **165**:237 (1950).

13 J. STEINBACH and H. FREISER: *Anal. Chem.,* **24**:1027 (1952).

14 B. W. GRUNBAUM, W. SCHÖNIGER, and P. L. KIRK: *Anal. Chem.,* **24**:1857 (1952).

15 S. A. STERN, J. T. MULLHAUPT, and P. J. GAREIS: *Amer. Inst. Chem. Eng. J.,* **15**:64 (1969). See also R. W. ROBERTS and K. KAMMERMEYER, *J. Appl. Polym. Sci.,* **7**:2183 (1963).

16 ANALYTICAL METHODS COMMITTEE: *Analyst,* **90**:251 (1965); **92**:587 (1967).

17   E. WICHERS: *Anal. Chem.*, **33**(4):23A (1961).

18   G. F. SMITH and G. F. CROAD: *Ind. Eng. Chem., Anal. Ed.*, **9**:141 (1937).

19   I. M. KOLTHOFF: *J. Amer. Chem. Soc.*, **48**:1447 (1926).

20   T. W. RICHARDS and C. R. HOOVER: *J. Amer. Chem. Soc.*, **37**:95 (1915).

21   A. L. UNDERWOOD: *Anal. Chem.*, **33**:955 (1961).

22   I. M. KOLTHOFF, E. B. SANDELL, E. J. MEEHAN, and S. BRUCKENSTEIN: "Quantitative Chemical Analysis," 4th ed., p. 778, Macmillan, New York, 1969.

23   E. R. CALEY and R. H. BRUNDIN: *Anal. Chem.*, **25**:142 (1953).

24   W. E. HARRIS and B. KRATOCHVIL: "Chemical Separations and Measurements," chap. 11, Saunders, Philadelphia, 1974.

25   D. VORLÄNDER and W. STRUBE: *Ber. Deut. Chem. Ges.*, **46**:172 (1913); A. THIEL and R. STROHECKER: *Ber. Deut. Chem. Ges.*, **47**:945, 1061 (1914).

26   S. G. SIMPSON: *Ind. Eng. Chem.*, **16**:709 (1924).

27   "Standard Methods for the Examination of Water and Wastewater," 12th ed., p. 50, American Public Health Association, New York, 1965.

28   S. S. COOPER: *Ind. Eng. Chem., Anal. Ed.*, **13**:466 (1941).

29   C. WINKLER: "Practische Uebungen in der Maassanalyse," 3d ed., Akademische Verlagsgesellschaft, Leipzig, 1902.

30   S. P. L. SÖRENSON and A. C. ANDERSEN: *Z. Anal. Chem.*, **45**:217 (1906); **47**:279 (1908).

31   J. LINDNER: *Z. Anal. Chem.*, **72**:135 (1927); **78**:188 (1929).

32   E. P. PARTRIDGE and W. C. SCHROEDER: *Ind. Eng. Chem., Anal. Ed.*, **4**:271 (1932).

33   R. B. WARDER: *Amer. Chem. J.*, **3**:55, 232 (1881).

34   J. KJELDAHL: *Z. Anal. Chem.*, **22**:366 (1883).

35   H. C. MOORE: *J. Ind. Eng. Chem.*, **12**:669 (1920); A. L. PRINCE: *J. Assoc. Offic. Agr. Chem.*, **8**:410 (1925).

36   I. K. PHELPS and H. W. DAUDT: *J. Assoc. Offic. Agr. Chem.*, **3**:306 (1920); **4**:72 (1921).

37   O. I. MILNER and R. J. ZAHNER: *Anal. Chem.*, **32**:294 (1960).

38   W. C. URBAN: *Anal. Chem.*, **43**:800 (1971).

39   A. FERRARI: *Ann. N.Y. Acad. Sci.*, **87**:792 (1960).

40   D. G. KRAMME, R. H. GRIFFEN, C. G. HARTFORD, and J. A. CORRADO: *Anal. Chem.*, **45**:405 (1973).

41   F. E. CRITCHFIELD and J. B. JOHNSON: *Anal. Chem.*, **30**:1247 (1958); **31**:570 (1959).

42   J. S. FRITZ: "Acid-Base Titrations in Nonaqueous Solvents," G. F. Smith Chemical, Columbus, Ohio, 1952; "Acid-Base Titrations in Nonaqueous Solvents," Allyn and Bacon, Boston, 1973.

43   J. S. FRITZ and G. S. HAMMOND: "Quantitative Organic Analysis," Chap. 3, Wiley, New York, 1957.

44   W. HUBER: "Titrations in Nonaqueous Solvents," Academic, New York, 1967.

45   M. R. F. ASHWORTH: "Titrimetric Organic Analysis," Interscience, New York, 1964, 1965.

46   M. M. DAVIS: "Acid-Base Behavior in Aprotic Organic Solvents," National Bureau of Standards Monograph 105, NBS, Washington, D.C., 1968.

47   A. P. KRESHKOV: *Talanta*, **17**:1029 (1970).

48   B. KRATOCHVIL: *Crit. Rev. Anal. Chem.*, **1**:415 (1971).

49   "Reagents for Nonaqueous Titrimetry," Kodak Publication No. JJ-4, Eastman Kodak, Rochester, New York, 1970.

50 H. B. VAN DER HEIJDE and E. A. M. F. DAHMEN: *Anal. Chim. Acta*, **16**:378 (1957).

51 J. S. FRITZ: *Anal. Chem.*, **26**:1701 (1954).

52 D. J. PIETRZYK and J. BELISLE: *Anal. Chem.*, **38**:969 (1966).

53 B. KRATOCHVIL, E. J. FINDLAY, and W. E. HARRIS: *J. Chem. Educ.*, **50**:629 (1973).

54 R. H. CUNDIFF and P. C. MARKUNAS: *Anal. Chem.*, **28**:792 (1956).

55 L. K. HILLER, JR.: *Anal. Chem.*, **42**:30 (1970).

56 W. B. MATHER, JR., and F. C. ANSON: *Anal. Chim. Acta*, **21**:468 (1959).

57 V. J. VAJGAND, F. F. GAÁL, and S. S. BRUSIN: *Talanta*, **17**:415 (1970).

58 W. L. JOLLY and E. A. BOYLE: *Anal. Chem.*, **43**:514 (1971).

59 E. A. BUTLER and E. H. SWIFT: *J. Chem. Educ.*, **49**:425 (1972).

60 J. S. FRITZ and F. E. GAINER: *Talanta*, **13**:939 (1966).

61 I. M. KOLTHOFF, M. K. CHANTOONI, JR., and S. BHOWMIK: *Anal. Chem.*, **39**:315 (1967).

62 T. HIGUCHI, J. A. FELDMAN, and C. R. REHM: *Anal. Chem.*, **28**:1120 (1956).

63 L. S. GUSS and I. M. KOLTHOFF: *J. Amer. Chem. Soc.*, **62**:249 (1940).

64 I. M. KOLTHOFF, M. K. CHANTOONI, JR., and S. BHOWMIK: *J. Amer. Chem. Soc.*, **90**:23 (1968).

65 J. F. COETZEE and R. J. BERTOZZI: *Anal. Chem.*, **45**:1064 (1973).

66 S. C. WATSON and J. F. EASTHAM: *Anal. Chem.*, **39**:171 (1967).

67 S. BRUCKENSTEIN and N. E. VANDERBORGH: *Anal. Chem.*, **38**:687 (1966).

68 J. JORDAN in "Treatise on Analytical Chemistry," I. M. Kolthoff and P. J. Elving (eds.), pt. 1, vol. 8, p. 5175, Interscience, New York, 1968.

69 H. J. V. TYRRELL and A. E. BEEZER: "Thermometric Titrimetry," chap. 5, Chapman & Hall, London, 1968.

70 L. S. BARK and S. M. BARK: "Thermometric Titrimetry," chap. 7, Pergamon, London, 1969.

71 S. BRUCKENSTEIN and I. M. KOLTHOFF: *J. Amer. Chem. Soc.*, **78**:2974 (1956); I. M. KOLTHOFF, S. BRUCKENSTEIN, and M. K. CHANTOONI, JR.: *J. Amer. Chem. Soc.*, **83**:3927 (1961); I. M. KOLTHOFF, M. K. CHANTOONI, JR., and H. SMAGOWSKI: *Anal. Chem.*, **42**:1622 (1970).

72 E. GRUNWALD and B. J. BERKOWITZ: *J. Amer. Chem. Soc.*, **73**:4939 (1951); B. GUTBEZAHL and E. GRUNWALD: *J. Amer. Chem. Soc.*, **75**:559 (1953).

73 W. B. SCHAAP, R. E. BAYER, J. R. SIEFKER, J. Y. KIM, P. W. BREWSTER, and F. C. SCHMIDT: *Rec. Chem. Progr.*, **22**:197 (1961).

74 C. W. PIFER and E. G. WOLLISH: *Anal. Chem.*, **24**:300 (1952).

75 J. S. FRITZ: *Anal. Chem.*, **22**:1028 (1950).

76 M. L. MOSS, J. H. ELLIOTT, and R. T. HALL: *Anal. Chem.*, **20**:784 (1948).

77 J. S. FRITZ and N. M. LISICKI: *Anal. Chem.*, **23**:589 (1951).

78 R. H. CUNDIFF and P. C. MARKUNAS: *Anal. Chem.*, **30**:1447, 1450 (1958).

79 S. R. PALIT: *Ind. Eng. Chem., Anal. Ed.*, **18**:246 (1946).

80 T. HIGUCHI, C. J. LINTNER, and R. H. SCHLEIF: *Science*, **111**:63 (1950); *Anal. Chem.*, **22**:534 (1950).

81 T. HIGUCHI and D. A. ZUCK: *J. Amer. Chem. Soc.*, **73**:2676 (1951).

82 T. HIGUCHI, J. CONCHA, and R. KURAMOTO: *Anal. Chem.*, **24**:685 (1952).

83 J. J. LAGOWSKI: *Anal. Chem.*, **42**:305R (1970); **44**:524R (1972).

PROBLEMS

6-1 A standard NaOH solution is contaminated with Na_2CO_3. With phenolphthalein in the cold, 30.50 ml is required to titrate 50.00 ml of 0.5010 M HCl. With boiling to remove CO_2, 30.00 ml is required for the same amount of acid to the methyl orange end point. Calculate the moles of (*a*) NaOH and (*b*) Na_2CO_3 per liter. *Answer* (*a*) 0.808. (*b*) 0.0137.

6-2 A sample of wheat weighing 1.2461 g was analyzed for protein nitrogen by the Kjeldahl procedure, with excess H_3BO_3 in the distillate receiving flask. Calculate the percentage of nitrogen if 13.19 ml of 0.0962 M HCl was required to the bromocresol green end point. *Answer* 1.426%.

6-3 A 0.3126-g sample of pure KCl was passed through a cation-exchange column in the acid form. If 41.63 ml of a NaOH solution was required to titrate the effluent to the methyl red end point, what was the molarity of the NaOH? *Answer* 0.1007 M.

6-4 A sample weighing 1.6321 g containing Na_3PO_4 and NaCN was passed through a cation-exchange column in the acid form. The effluent required 34.14 ml of 0.1041 M NaOH to reach the bromocresol green end point. What was the percentage of Na_3PO_4 in the sample? *Answer* 35.70%.

6-5 If 0.3636 g of glycine, H_2NCH_2COOH, is dissolved in glacial acetic acid and titrated with 0.1063 M $HClO_4$, what volume of acid is required for neutralization? *Answer* 45.57 ml.

6-6 If the permeability constant of CO_2 to polyethylene is 1×10^{-9} ml-cm/sec-cm²-cm Hg, and if air consists of 0.033% CO_2, estimate how much carbonate would be formed in 1 week at room temperature in a closed liter bottle (thickness 1 mm; area 500 cm²) of polyethylene containing 0.1 M NaOH. *Answer* 3×10^{-6} moles.

SOLUBILITY OF PRECIPITATES

Broadly considered, solubilities depend in part on nonspecific electrolyte effects and in part on specific effects. The nonspecific effects can be considered in terms of activity coefficients (Chapter 2). But activity-coefficient effects often are negligible compared[1] with the uncertainties arising from disregarded or unknown side reactions and also with uncertainties arising from the crystalline state, the state of hydration, the extent of aging of the precipitate, and intrinsic solubility, all of which may contribute to the solubility of the precipitate. To the extent that each can be identified and measured, each can be accounted for. Nevertheless, the magnitude of unsuspected effects makes it expedient to assume activity coefficients of unity unless otherwise specifically indicated for relatively soluble salts or solutions containing moderate amounts of electrolytes.

In this chapter the specific chemical and physical factors affecting the solubility of a precipitate are examined. Solubilities of precipitates are considered first, since an understanding of this aspect is basic to their properties generally. In Chapters 8 and 9, other properties of precipitates are described—their mode of formation, physical properties, and chemical purity.

7-1   SOLUBILITY, INTRINSIC SOLUBILITY, AND SOLUBILITY PRODUCT

In Chapters 3, 4, and 6, dissociation equilibria are discussed and are applied in detail principally to acidic or basic substances. In general, for a substance MA formed from reaction of a cation M^+ with an anion A^-, the dissociation equilibrium expression can be written

$$MA \rightleftharpoons M^+ + A^- \qquad K_d = \frac{[M^+][A^-]\gamma_\pm^2}{[MA]\gamma_0} \qquad (7\text{-}1)$$

A modified and simplified dissociation constant can be obtained if the constraint is imposed that the solution is saturated (solid MA present):

$$MA \text{ (solid)} \rightleftharpoons MA \text{ (soln)} \rightleftharpoons M^+ + A^- \qquad (7\text{-}2)$$

where MA (soln) denotes both the uncharged molecule MA and the ion pair M^+A^-. In a saturated solution, $[MA]\gamma_0$ is constant for a particular solute in a solvent at a given temperature, where $[MA]$ includes both MA and M^+A^-. Equation 7-1 can be rearranged to

$$[M^+][A^-]\gamma_\pm^2 = K_d[MA]\gamma_0 \qquad (7\text{-}3)$$

which is applicable to unsaturated as well as saturated solutions. For a saturated solution the value of $[MA]$ is defined[2] as the *molecular or intrinsic solubility* S^0. Equation (7-3) is the basis for a definition of the *solubility product* K_{sp}:

$$K_{sp} = [M^+][A^-]\gamma_\pm^2 = K_d S^0 \qquad (7\text{-}4)$$

where γ_0 is assumed to be unity. If no other equilibria are involved, the *solubility S* for this 1:1 electrolyte is the sum of $[M^+]$ or $[A^-]$, and S^0. The relation between solubility product and solubility is therefore not necessarily straightforward.[2-4]

EXAMPLE 7-1   In water the acid 2,4,6 trichlorophenol, $Cl_3C_6H_2OH$, has a solubility[2] of 4.0×10^{-3} M and a dissociation constant of 1.0×10^{-6}. Calculate the intrinsic solubility and the solubility product, assuming activity coefficients of unity.

ANSWER

$$Cl_3C_6H_2OH \text{ (solid)} \rightleftharpoons Cl_3C_6H_2OH \text{ (soln)} \rightleftharpoons H^+ + Cl_3C_6H_2O^-$$

$$\frac{[H^+][Cl_3C_6H_2O^-]}{[Cl_3C_6H_2OH]} = 1.0 \times 10^{-6}$$

Solve for $[H^+]$ and $[Cl_3C_6H_2OH]$.

$$[H^+] = [Cl_3C_6H_2O^-] = 6.3 \times 10^{-5} \ M$$

and $$S^0 = [Cl_3C_6H_2OH] - [H^+] \simeq 3.94 \times 10^{-3} \ M$$

$$K_{sp} = [H^+][Cl_3C_6H_2O^-] = (6.3 \times 10^{-5})^2 = 4.0 \times 10^{-9}$$

Thus the solubility of this substance, $([Cl_3C_6H_2O^-] + [Cl_3C_6H_2OH])$, is far larger than $\sqrt{K_{sp}}$ and is due mainly to intrinsic solubility.   ////

Intrinsic solubilities and dissociation constants have been measured for only a small fraction of the slightly soluble substances of interest to analysis. A frequently made tacit assumption is that dissociation is complete and that S^0 contributes negligibly to the total solubility. This assumption is frequently invalid with water and probably is rarely applicable with organic precipitants or with solvents of low dielectric constant. The relation between dissociation, dissociation constant, and concentration of a 1:1 solute is illustrated by Figure 7-1. Only when conditions are such that MA in solution approaches complete dissociation is it valid to ignore S^0. These conditions vary. Note that complete dissociation is to be expected only when the concentration is low and the dissociation constant large. Thus for calcium sulfate the contribution of S^0 to total solubility cannot be ignored[3] even though it has a moderately large dissociation constant (5.2×10^{-3}), since its concentration in saturated solution is $0.015 \ M$. According to Figure 7-1 it is less than half-dissociated in saturated solution. The dissociation constants for silver halides[5] are even smaller: 3.9×10^{-4} for silver chloride, 2.1×10^{-5} for silver bromide, and 2.6×10^{-7} for silver iodide. The fraction dissociated is large for these substances, however, because of the low concentrations in their saturated solutions, and S^0 can be neglected safely in these cases.

For determination of total solubility of a material, equilibria in addition to those represented by Equation 7-2 need consideration:

$$\text{MA (solid)} \rightleftharpoons \text{MA (soln)} \rightleftharpoons \text{M}^+ \ + \ \text{A}^-$$

$$\qquad\qquad \updownarrow X \qquad\qquad \updownarrow Y \quad \updownarrow Z \qquad\qquad (7\text{-}5)$$

$$\qquad\qquad \text{MA(X)} \qquad \text{M}^+(\text{Y}) \quad \text{A}^-(\text{Z})$$

In (7-5) the three sets of vertical reversible arrows represent reactions of MA (including the ion pair M^+A^-), M^+, and A^- with other entities in solution. The effect of these additional equilibria, which is to increase solubility, is the principal subject treated in this chapter. In Sections 7-2 to 7-5 the simplifying assumption is made that solubilities and dissociation constants are such that $[MA]$ is negligible. When the additional equilibria of Equation (7-5) are included, solubility becomes equal to the sum of $[MA] + [MA(X)] + [M^+] + [M^+(Y)]$, or $[MA] + [MA(X)] + [A^-] + [A^-(Z)]$.

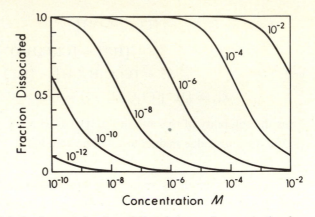

FIGURE 7-1   Relation between fraction of dissociation and concentration for 1:1 solutes with dissociation constants from 10^{-12} to 10^{-2}.

7-2   SOLUBILITY OF A PRECIPITATE IN PURE SOLUTION OR WITH A COMMON ION

If activity coefficients are assumed to be unity, and if a substance is virtually completely dissociated (S^0 = zero) and its ions undergo no secondary reactions, then $[M^+] = [A^-] = S$, the molar solubility, and hence in this case $S = \sqrt{K_{sp}}$. For a precipitate M_mA_a

$$M_mA_a \text{ (solid)} \rightleftharpoons M_mA_a \text{ (soln)} \rightleftharpoons mM^{n+} + aA^{b-} \qquad (7\text{-}6)$$

and
$$K_{sp} = [M^{n+}]^m[A^{b-}]^a = (mS)^m(aS)^a \qquad (7\text{-}7)$$

Equation (7-7) can be solved for the value of molar solubility.

EXAMPLE 7-2   The solubility product of calcium fluoride is 4×10^{-11}. Calculate its solubility S, neglecting hydrolysis of fluoride and assuming S^0 equal to zero.

ANSWER $[Ca^{++}] = S$; $[F^-] = 2S$; $4S^3 = 4 \times 10^{-11}$; $S = 2.2 \times 10^{-4}$ mole/l.   ////

In the presence of a common ion of the precipitate, the contribution of the solubility to the total common ion concentration often may be neglected. If excess M^{n+} is present to the extent C_M, the solubility S may be calculated from

$$K_{sp} = (C_M + mS)^m(nS)^n \simeq (C_M)^m(nS)^n$$

In accurate calculations employing activity coefficients, the total ionic strength for all solutes, including the contribution of the slightly soluble substance, determines the solubility (Section 2-6).

7-3 EFFECT OF HYDROGEN ION CONCENTRATION

Frequently the ions in equilibrium with a precipitate are involved in additional equilibria, as explained in Section 7-1. For example, if the anion of the precipitate is that of a weak acid, the solubility will vary with acidity.

Consider first the salt of a monobasic weak acid HA:

$$MA_n \rightleftharpoons M^{n+} + nA^- \qquad (7\text{-}8)$$

$$A^- + H^+ \rightleftharpoons HA \qquad (7\text{-}9)$$

If C_A is the total concentration of A, or $C_A = [A^-] + [HA]$, and if α_1 is the fraction of the total A in the ionized form [Equation (3-56)], then

$$K_{sp} = [M^{n+}][A^-]^n = [M^{n+}]\alpha_1{}^n C_A{}^n \qquad (7\text{-}10)$$

where

$$\alpha_1 = \frac{[A^-]}{[A^-] + [HA]} = \frac{K_a}{[H^+] + K_a} \qquad (7\text{-}11)$$

If $[H^+]$ is known, α_1 may be calculated and substituted in (7-10) to give a *conditional solubility product* K'_{sp}:

$$K'_{sp} = [M^{n+}]C_A{}^n = \frac{K_{sp}}{\alpha_1{}^i} \qquad (7\text{-}12)$$

which varies with pH and is used conveniently to calculate solubility.

Consider now the salt of a dibasic acid. The procedure is the same except that the divalent anion concentration $[A^=]$ is given by $\alpha_2 C_A$, where α_2 is the fraction of the total anion in the $A^=$ form [comparable to Equation (3-61)]:

$$\alpha_2 = \frac{K_1 K_2}{[H^+]^2 + K_1[H^+] + K_1 K_2} \qquad (7\text{-}13)$$

EXAMPLE 7-3   Calculate the solubility of calcium oxalate in a solution of pH 3.0 that contains a total oxalate excess of 0.01 M.

ANSWER $\alpha_2 = 0.056$, $S = [Ca^{++}]$, $C_{ox} = S + 0.01$, $S(S + 0.01) = K_{sp}/\alpha_2 = 3.6 \times 10^{-8}$, $S = 3.6 \times 10^{-6}$ mole/l. ////

7-4 EFFECT OF HYDROLYSIS OF THE ANION IN UNBUFFERED SOLUTION

The anion A^- of a sparingly soluble salt MA may undergo hydrolysis in water:

$$A^- + H_2O \rightleftharpoons HA + OH^- \qquad (7\text{-}14)$$

With the ion product for water, the solubility product for MA, the equilibrium expression K_w/K_a for (7-14), and the electroneutrality relation

$$[M^+] + [H^+] = [A^-] + [OH^-] \qquad (7\text{-}15)$$

four equations in four unknowns are obtainable. The system is even more complex when two or more stages of hydrolysis of a polyvalent anion must be considered. A simplification is usually possible at the outset. If K_{sp} is small, the hydroxyl ion contribution from (7-14) is negligible compared with that already present in the water, and the pH is near 7. Corresponding to this pH, the fraction of anion in the unhydrolyzed form can be calculated from the ionization constant K_a.

EXAMPLE 7-4   Calculate the solubility S of silver carbonate in water, considering the hydrolysis of the carbonate ion. The K_{sp} of Ag_2CO_3 is 8×10^{-12}; for H_2CO_3, $K_1 = 4 \times 10^{-7}$ and $K_2 = 5 \times 10^{-11}$.

ANSWER   From the relatively large K_{sp} and small K_2 we conclude that the reaction is best represented by

$$Ag_2CO_3 + H_2O \rightleftharpoons 2Ag^+ + HCO_3^- + OH^- \qquad K = \frac{K_{sp}K_w}{K_2}$$

Then $[HCO_3^-] = [OH^-] = S$; $[Ag^+] = 2S$; $4S^4 = 8 \times 10^{-12} \times 10^{-14}/5 \times 10^{-11}$; and $S = 1.4 \times 10^{-4}$ mole/l. More exactly,

$$[Ag^+] = 2([CO_3^=] + [HCO_3^-] + [H_2CO_3]) = \frac{2[CO_3^=]}{\alpha_2}$$

But we calculate at $[OH^-] = 1.4 \times 10^{-4}$,

$$\alpha_2 = 0.41$$

From K_{sp}

$$4S^3 = \frac{K_{sp}}{\alpha_2}$$

$$S = 1.7 \times 10^{-4} \text{ mole/l} \qquad ////$$

7-5   EFFECT OF HYDROLYSIS OF THE CATION

Many cations of heavy metals act as acids and undergo hydrolysis, with a resultant increase in the solubility of precipitates involving these metals.

In the simplest type of hydrolysis reactions, the steps

$$M^{n+} + H_2O \rightleftharpoons MOH^{(n-1)+} + H^+ \qquad K_1$$

$$MOH^{(n-1)+} + H_2O \rightleftharpoons M(OH)_2^{(n-2)+} + H^+ \qquad K_2 \qquad (7\text{-}16)$$

$$\cdots$$

may be regarded as the successive acid dissociation constants of the aquated cation

M^{n+}. Correspondingly, the fraction α_M of metal in the aquated form may be calculated from equations of the type

$$\alpha_M = \frac{[H^+]^n}{[H^+]^n + K_1[H^+]^{n-1} + K_1K_2[H^+]^{n-2} + \cdots + K_1K_2 \cdots K_n} \tag{7-17}$$

This expression is identical to Equation (3-63) for the fraction of undissociated acid in a solution of a polybasic acid and its salts.

Hydrolysis reactions of many cations are more complicated than (7-16) implies. Metal ions usually cannot be treated as simple polybasic acids. In particular, metal ions of charge 2 or greater may have a pronounced tendency to form ionic species containing more than one metal atom. Thus Hedström[6] deduced that the hydrolysis of iron(II) could be represented by the single reaction

$$Fe^{++} + H_2O \rightleftharpoons FeOH^+ + H^+ \qquad K = 3 \times 10^{-10}$$

whereas that of iron(III) involved the reactions

$$Fe^{3+} + H_2O \rightleftharpoons FeOH^{++} + H^+ \qquad K_1 = 9 \times 10^{-4}$$

$$FeOH^{++} + H_2O \rightleftharpoons Fe(OH)_2^+ + H^+ \qquad K_2 = 5 \times 10^{-7} \tag{7-18}$$

$$2Fe^{3+} + 2H_2O \rightleftharpoons Fe_2(OH)_2^{4+} + 2H^+ \qquad K = 1.2 \times 10^{-3}$$

The equilibrium constants of Reactions (7-18) are such that under certain conditions the dimeric species $Fe_2(OH)_2^{4+}$ is the major form of iron(III) present.

The hydrolysis of aluminum(III) is still more complicated.[7] The mononuclear species $Al(OH)^{++}$ and $Al(OH)_2^+$ and the polynuclear species $Al_2(OH)_2^{4+}$ and $Al_3(OH)_6^{3+}$, if they exist, are not the main products. The experimental data can be explained on the assumption of either a series of complexes $Al[(OH)_5Al_2]_n^{(n+3)+}$ or a single large complex $Al_6(OH)_{15}^{3+}$.

Similarly, the hydrolysis of thorium(IV) can be accounted for on the assumption of a series of polynuclear complexes $Th[(OOH)Th]_n^{(n+4)+}$, where n has values exceeding 6 in certain cases.[8]

With cadmium ion, only the simple product $CdOH^+$ is formed,[9] whereas with copper(II) a binuclear reaction predominates:[10]

$$2Cu^{++} + 2H_2O \rightleftharpoons Cu_2(OH)_2^{++} + 2H^+ \qquad K = 2.5 \times 10^{-11}$$

For the cation $CH_3HgH_2O^+$ the hydrolysis reactions are[11]

$$CH_3HgH_2O^+ \rightleftharpoons CH_3HgOH + H^+$$

$$2CH_3HgOH \rightleftharpoons (CH_3Hg)_2OH^+ + OH^-$$

If calculations of the solubilities of heavy-metal salts are to be accurate, careful attention must be paid to the nature of the hydrolytic species present. As the following example illustrates, however, in an absolute sense the effect of cation hydrolysis is often of minor importance.

EXAMPLE 7-5   Calculate the solubility of cadmium sulfide in water, considering hydrolysis of Cd^{++} and $S^=$. $K_{sp} = 10^{-28}$. The first acid dissociation constant of $Cd^{++} = 10^{-9}$.

ANSWER   From the small K_{sp} we conclude that $[H^+] = 10^{-7}$, and therefore that $\alpha_2 = 4.7 \times 10^{-9}$ [Equation (7-13)] and $\alpha_M = [H^+]/([H^+] + K_H) = 0.99$. If $S =$ solubility, $[Cd^{++}] = \alpha_M S$, $[S^=] = \alpha_2 S$, and $S = 1.5 \times 10^{-10}$ mole/l.   ////

7-6   EFFECT OF FOREIGN COMPLEXING AGENTS

If a slightly soluble salt MA is treated with a complexing agent L, the following equilibria are involved:

$$MA \rightleftharpoons M + A \qquad K_{sp} \qquad (7\text{-}19)$$

$$M + L \rightleftharpoons ML \qquad K_1 = \frac{[ML]}{[M][L]} \qquad (7\text{-}20)$$

$$ML + L \rightleftharpoons ML_2 \qquad K_2 = \frac{[ML_2]}{[ML][L]}$$

$$= \frac{[ML_2]}{K_1[M][L]^2} \qquad (7\text{-}21)$$

$$ML_{n-1} + L \rightleftharpoons ML_n \qquad K_n = \frac{[ML_n]}{[ML_{n-1}][L]} = \frac{[ML_n]}{(K_1 K_2 \cdots K_{n-1})[M][L]^n} \qquad (7\text{-}22)$$

where both M and A may be univalent, divalent, or so on, and L may be molecular or ionic; therefore, charges are omitted for simplicity. The constants $K_1, K_2, \ldots, K_n$ are *successive formation constants* or *stepwise formation constants* of the complexes $ML, ML_2, \ldots, ML_n$ (Section 11-1). The product of the successive formation constants is the *overall formation constant*, or

$$\beta_i = K_1 K_2 \cdots K_i \qquad (7\text{-}23)$$

The fraction α_M of the metal ion in the uncomplexed form M may be calculated. The total metal ion concentration C_M is given by

$$\begin{aligned} C_M &= [M] + [ML] + [ML_2] + \cdots + [ML_n] \\ &= [M]\{1 + K_1[L] + K_1 K_2[L]^2 + \cdots + (K_1 K_2 \cdots K_n)[L]^n\} \\ &= \frac{[M]}{\alpha_M} \end{aligned} \qquad (7\text{-}24)$$

where $1/\alpha_M$ represents the terms within the braces. The solubility product expression is

$$K_{sp} = [M][A] = \alpha_M C_M[A] \qquad (7\text{-}25)$$

In the absence of a common ion the solubility S is given by

$$S = C_M = [A] = \sqrt{\frac{K_{sp}}{\alpha_M}} \qquad (7\text{-}26)$$

The effect of complex formation is to increase the solubility in proportion to the square root of the decreasing fraction of metal ion in the uncomplexed form. Similar considerations may be applied to more involved types of precipitates.

EXAMPLE 7-6 Calculate the solubility of silver iodide in 0.01 M ammonia, if the solubility product of silver iodide is 9×10^{-17} and the logarithms of the successive formation constants of the silver-ammonia complexes are 3.2 and 3.8.

ANSWER   From (7-24), $1/\alpha_M = 1 + 10^{3.2} \times 10^{-2} + 10^{7.0} \times 10^{-4} = 10^3$. From (7-26)

$$S = \sqrt{9 \times 10^{-17} \times 10^3} = 3 \times 10^{-7} \text{ mole/l}$$

Note that the concentration of $Ag(NH_3)^+$ is small compared with that of $Ag(NH_3)_2{}^+$.

////

7-7 EFFECT OF COMPLEX FORMATION WITH PRECIPITATING ANION

If a metal ion forms a sparingly soluble salt with an anion, it frequently forms a complex with excess anion. The solubility passes through a minimum as the suppressing effect of the common ion is balanced by the increasing solubility due to complex formation.

Consider first a 1:1 salt MA, which dissolves in excess reagent to form a series of complexes. M and A both may be univalent, divalent, or so on; therefore we omit charges. The precipitation reaction may be written as the reverse of (7-2). The intrinsic solubility S^0 of MA is a constant at a given temperature. The successive complexes are formed by the reactions

$$MA + A \rightleftharpoons MA_2 \qquad K_2 = \frac{[MA_2]}{[MA][A]} = \frac{[MA_2]}{S^0[A]} \qquad (7\text{-}27)$$

$$MA_2 + A \rightleftharpoons MA_3 \qquad K_3 = \frac{[MA_3]}{[MA_2][A]} \qquad (7\text{-}28)$$

$$MA_{n-1} + A \rightleftharpoons MA_n \qquad K_n = \frac{[MA_n]}{[MA_{n-1}][A]} \qquad (7\text{-}29)$$

where the constants $K_2, K_3, \ldots, K_n$ are again the successive formation constants.

The first formation constant K_1 is defined by the inverse of Equation (7-3),

$$K_1 = \frac{[MA]_{soln}}{[M][A]} = \frac{S^0}{[M][A]} = \frac{S^0}{K_{sp}} \tag{7-30}$$

The solubility of the precipitate is given by the total concentration of all forms of M, or

$$S = [M] + [MA] + [MA_2] + [MA_3] + \cdots + [MA_n] \tag{7-31}$$

Substituting from (7-4), (7-27), (7-28), and (7-29), we obtain

$$S = \frac{K_{sp}}{[A]} + S^0 + S^0 K_2[A] + S^0 K_2 K_3[A]^2 + \cdots$$
$$+ S^0(K_2 K_3 \cdots K_n)[A]^{n-1} \tag{7-32}$$

and from (7-30)

$$S = \frac{K_{sp}}{[A]} + K_1 K_{sp} + K_1 K_2 K_{sp}[A] + K_1 K_2 K_3 K_{sp}[A]^2 + \cdots$$
$$+ K_{sp}(K_1 K_2 K_3 \cdots K_n)[A]^{n-1} \tag{7-33}$$

In terms of the overall formation constants of (7-23),

$$S = K_{sp}\left(\frac{1}{[A]} + \beta_1 + \beta_2[A] + \beta_3[A]^2 + \cdots + \beta_n[A]^{n-1}\right) \tag{7-34}$$

or
$$S = \frac{K_{sp}}{[A]} + K_{sp}\sum_{i=1}^{n}\beta_i[A]^{i-1} \tag{7-35}$$

The solubility given by the polynomial (7-34) or (7-35) passes through a minimum at some value of [A] and, beyond that concentration, increases with increasing concentration of A.

If no higher complexes than MA_2 are formed, (7-34) or (7-35) becomes

$$S = K_{sp}\left(\frac{1}{[A]} + K_1 + K_1 K_2[A]\right)$$
$$= \frac{K_{sp}}{[A]} + S^0 + \beta_2 K_{sp}[A] \tag{7-36}$$

showing that the solubility increases linearly with [A] at sufficiently high values of [A]. The minimum solubility is determined by setting $dS/d[A]$ equal to zero. For the condition described by (7-36) this corresponds to

$$\frac{K_{sp}}{[A]_{min}^2} = \beta_2 K_{sp} \quad \text{or} \quad [A]_{min} = \frac{1}{\sqrt{\beta_2}}$$

and
$$S_{min} = S^0 + 2K_{sp}\sqrt{\beta_2} \tag{7-37}$$

EXAMPLE 7-7 Johnston, Cuta, and Garrett[12] found that the solubility of silver oxide in sodium hydroxide solutions could be expressed by the equation $S = 5 \times 10^{-8}/[OH^-] + 1.95 \times 10^{-4}[OH^-]$. Comparing this equation with (7-36), we find $S^0 = 0$; $K_{sp} = 5 \times 10^{-8}$; $\beta_2 K_{sp} = 1.95 \times 10^{-4}$, or $\beta_2 = 3.9 \times 10^3$. From (7-37) the condition of minimum solubility corresponds to $[OH^-]_{min} = 0.016$ M, $S_{min} = 6.2 \times 10^{-6}$ M. ////

If a succession of complexes is formed, a corresponding number of terms from (7-34) must be added. For example, Forbes and Cole[13] expressed the solubility of silver chloride by

$$S = \frac{2 \times 10^{-10}}{[Cl^-]} + b + 3.4 \times 10^{-5}[Cl^-]$$

where the constant 2×10^{-10} is the solubility product. They ascribed the constant b to undissociated silver chloride plus colloidally dispersed solid. From the reproducible value of b (6.3×10^{-7} M for sodium chloride solutions, 6.1×10^{-7} M for hydrochloric acid solutions), b appears to represent the intrinsic solubility S^0. From $S^0 = K_1 K_{sp}$ we calculate the first formation constant $K_1 = 3 \times 10^3$, and from $\beta_2 K_{sp} = 3.4 \times 10^{-5}$ we estimate the second formation constant $\beta_2 = 1.7 \times 10^5$.

At chloride concentrations above 2.5 M, Forbes[14] expressed the solubility of silver chloride by $S = 10^{-4}[Cl^-]^2 + 4.5 \times 10^{-5}[Cl^-]^3$, from which we deduce $\beta_3 = 10^{-4}/K_{sp} = 5 \times 10^5$, and $\beta_4 = 4.5 \times 10^{-5}/K_{sp} = 2 \times 10^5$.

EXAMPLE 7-8 From the above equilibrium constants, estimate the solubility of silver chloride in solutions containing chloride at the following concentrations: 10^{-3}, 10^{-2}, 10^{-1}, 1, and 2 M. What is the minimum solubility, and at what chloride ion concentration does it occur?

ANSWER From (7-34) we calculate the following solubilities at the specified chloride concentrations: 8.3×10^{-7}, 9.7×10^{-7}, 5×10^{-6}, 1.8×10^{-4}, and 8×10^{-4} M. Taking only the first three terms, we calculate that the minimum solubility of 7.6×10^{-7} M occurs at a chloride ion concentration of 2.4×10^{-3} M. ////

Figure 7-2 represents data for solubility of silver chloride plotted to illustrate[15] the minimum solubility, the common ion effect and intrinsic solubility S^0, and the formation of the chloro complex. As Figure 7-2B indicates, at low concentrations of excess chloride an essentially linear relation is obtained between solubility and the reciprocal of the product of chloride ion concentration and the square of the activity coefficient. The zero intercept corresponds to S^0 (in this case an intrinsic solubility

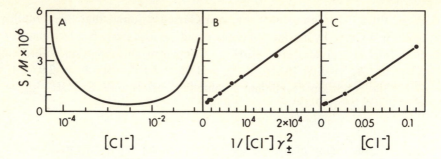

FIGURE 7-2   Solubility of AgCl as a function of excess chloride: *A*, on a logarithmic scale; *B*, at low concentrations; *C*, at moderate concentrations. (*Adapted from Ramette.*[15])

value somewhat smaller than that obtained by Forbes and Cole[13]). Since complex formation is inappreciable at low chloride ion concentrations, only the first two terms in Equation (7-33) are significant. At higher concentrations an essentially linear relation should result between solubility and chloride ion concentration if only the one chloro complex $AgCl_2^-$ is formed [Equation (7-36)]. The slight curvature in Figure 7-2*C* suggests the presence of higher complexes.

Consider now a precipitate of the 1:2 type, such as MA_2, where M^{++} is a divalent cation and A^- a univalent anion. The solubility is given by

$$S = [M^{++}] + [MA^+] + [MA_2] + [MA_3^-] + \cdots \tag{7-38}$$

$$= K_{sp}\left(\frac{1}{[A^-]^2} + \frac{\beta_1}{[A^-]} + \beta_2 + \beta_3[A^-] + \cdots + \beta_n[A^-]^{n-2}\right) \tag{7-39}$$

In (7-39) the solubility passes through a minimum at some value of $[A^-]$. Since experimental data are unfortunately often lacking, especially with respect to the intrinsic solubility, few applications of equations such as (7-39) can be made.

EXAMPLE 7-9   For the reaction of Th^{4+} with F^- the logarithms of the successive formation constants are $\log K_1 = 7.8$, $\log K_2 = 6.1$, $\log K_3 = 4.7$; $K_{sp} = 4 \times 10^{-28}$. Calculate the solubility at a fluoride ion concentration of 10^{-2} *M*, neglecting the intrinsic solubility.

ANSWER

$$S = \frac{10^{-27.4}}{10^{-8}} + \frac{10^{7.8} \times 10^{-27.4}}{10^{-6}} + \frac{10^{13.9} \times 10^{-27.4}}{10^{-4}}$$

$$+ \frac{10^{18.6} \times 10^{-27.4}}{10^{-2}}$$

$$= 10^{-6.8} = 1.6 \times 10^{-7} \; M$$

due mainly to ThF_3^+.

////

7-8 SOLUBILITY OF METASTABLE FORMS OF PRECIPITATES

A precipitate often comes down initially in a metastable form, which gradually reverts on standing to a stable form. Hydrous oxides, for example, often are precipitated in an amorphous form or as exceedingly small crystallites that adsorb enormous quantities of water.[16] Calcium oxalate, precipitated at room temperature, comes down as a mixture of dihydrate and trihydrate, which are metastable with respect to the monohydrate.[17] Mercury(II) sulfide, precipitated from acid solution in the black cubic form, is gradually transformed during aging to red trigonal cinnabar.[18] Because the transformation from a metastable to a stable form is a spontaneous process, it follows that under any given set of conditions *the metastable form is more soluble than the stable form.*

7-9 SOLUBILITY AND PARTICLE SIZE

As early as 1878, Gibbs[19] concluded that the breakdown or growth of a crystal was not a continuous transformation, as the gas-liquid transition was considered to be. Thomson[20] derived what has come to be known as the Gibbs-Thomson equation, relating the vapor pressure of liquid droplets to the size of the droplets. Ostwald[21] extended the concept to the problem of solubility, but made a numerical error later corrected by Freundlich.[22] Similar to the Gibbs-Thomson equation, the Ostwald-Freundlich equation was expressed by

$$\frac{RT}{M} \ln \frac{S_2}{S_1} = \frac{2\sigma}{\rho} \left(\frac{1}{r_2} - \frac{1}{r_1} \right) \tag{7-40}$$

in which S_2 and S_1 are the solubilities of spherical particles of radii r_2 and r_1, M is the molecular weight, σ is the surface tension of the solid-liquid interface, and ρ is the density of the solid.

According to (7-40) a substance of large surface tension σ would show a pronounced dependence of solubility on particle size. A noticeable increase in solubility occurs only for particles of small radius (of the order of 1 μm or less). Surface tension (or surface free energy) can therefore be used as a measure of the tendency of a finely divided solid to decrease its total surface area by growing to larger particle size. This tendency is reflected in increased solubilities for particles of small size. If r_1 is large, $1/r_1$ is negligible and S_1 may be replaced by S, the solubility of large crystals. Equation (7-40) then becomes

$$\frac{RT}{M} \ln \frac{S_r}{S} = \frac{2\sigma}{\rho r} \tag{7-41}$$

where S_r is the solubility of particles of radius r. This equation, however, is valid only for a nonionic solid composed of spherical particles whose surface tension does not change with r. The effect of ionization was considered by Jones[23] and also by Dundon and Mack,[24] who introduced the van't Hoff factor $1 - \alpha + v\alpha$ on the left side of (7-41). The degree of ionization is α, and v is the number of ions resulting from the dissociation of one "molecule" of solute. May and Kolthoff presented a derivation[25] assuming the solute to be a strong electrolyte. The result is identical to (7-41) except that the solubility S is replaced by the solubility product:

$$\frac{RT}{M} \ln \frac{(K_{sp})_r}{K_{sp}} = \frac{2\sigma}{\rho r} \qquad (7\text{-}42)$$

Even though the experimental fact that particle size increases with time in contact with a solution is unquestionable, experimental verification of (7-41) and (7-42) has proved to be extraordinarily difficult.

Hulett[26] found from conductivity measurements that, when barium sulfate was freshly precipitated, its concentration gradually decreased from 4.6 to 2.9 mg/l. The use of conductivity measurements for determining the solubility of finely divided solids was questioned by Cohen and Blekkingh[27] on the grounds that the suspended material could contribute to the conductance, both by motion of charged particles and by motion of ions in the double layer at the surface of the particles. They found no evidence for increased solubility of fine particles, but did not actually measure particle size. Balarev[28] similarly found no increase in the solubility of barium sulfate particles of 0.1-μm diameter. May and Kolthoff[25] measured the solubilities of various preparations of lead chromate. To avoid the questionable features of conductance measurements, they analyzed the solution for both lead and chromate ions and also measured the solubilities with varying amounts of solid present. They showed that supersaturation was relieved rapidly (within 20 s) by shaking the supersaturated solution with aged lead chromate. Particle radii were calculated from the specific surface (area per gram) measured by dye adsorption and radioactive-isotope exchange. All precipitate preparations gave the same final solubility, though fresh precipitate was found to be about 1.6 times as soluble as an aged preparation. Growth of particles took place with time, so the excess solubility was a temporary phenomenon. Enüstün and Turkevich[29] synthesized fine particles of strontium sulfate by a reproducible procedure, determined solubilities by a radiotracer technique, and measured particle sizes by electron microscopy. They found the solubility of their smallest particles to be 1.5 times that of large particles. They also clearly pointed out that the size of the smallest particles controls the observed solubilities; that is, the effective particle size is that of the smallest.

An enhanced solubility of small particles could not be expected to be an equilibrium condition. Even if all particles were uniform in size, the system would be metastable. If a slight dissolution of any particle occurred, the solution would become supersaturated with respect to all the other particles and unsaturated with respect to the particle that had decreased in size, so that ultimately this particle would dissolve. Thus solids in contact with their saturated solution are not totally at equilibrium unless the solid is present as a single particle.

In conclusion, precipitates of high surface energy (such as barium sulfate) show qualitatively a greater trend to increasing solubility with decreasing particle size than precipitates of low surface tension (such as silver chloride). Evidence of this behavior is found, for example, in the much greater tendency toward stability of slightly super-saturated solutions of barium sulfate. It does not appear valid, however, to apply the Ostwald-Freundlich equation quantitatively when comparing the properties of different precipitates.

7-10 THE DIVERSE ION EFFECT

Equation (7-4) indicates that the solubility product includes an activity-coefficient term, a term which has been assumed to be unity up to this time. The introduction to this chapter pointed out that errors arising from neglect of the effects of the activity coefficient are usually small when compared with several uncertainties or side reactions. The activity coefficient in Equation (7-4) depends on the kind and concentration of all electrolytes in solution, not merely those involved directly with the precipitate. The correction to solubility calculations that must be made to account for the activity-coefficient effect is known as the *diverse ion effect*. The appropriate background is discussed in Chapter 2, and Problems 2-1, 2-2, and 2-3 are examples of the calculations. For 1:1 electrolytes in solution, activity coefficients can usually be assumed to be unity when concentrations are much less than 0.1 M. Common ion and diverse ion effects can be significant at the same time, for example, when a large excess of common ion is added in a precipitation. The diverse ion effect is one of the reasons that the haphazard addition of a large excess of precipitant should be avoided.

REFERENCES

1 A. RINGBOM: *J. Chem. Educ.*, **35**:282 (1958).

2 H. M. IRVING and R. J. P. WILLIAMS: *Analyst*, **77**:813 (1952).

3 L. MEITES, J. S. F. PODE, and H. C. THOMAS: *J. Chem. Educ.*, **43**:667 (1966).

4 M. L. LYNDRUP, E. A. ROBINSON, and J. N. SPENCER: *J. Chem. Educ.*, **49**:641 (1972).

5 D. L. LEUSSING in "Treatise on Analytical Chemistry," I. M. Kolthoff and P. J. Elving (eds.), part 1, vol. 1, p. 705, Interscience, New York, 1959.

6 B. O. A. HEDSTRÖM: *Ark. Kemi*, **5**:457 (1953); **6**:1 (1953).

7 C. BROSSET, G. BIEDERMANN, and L. G. SILLÉN: *Acta Chem. Scand.*, **8**:1917 (1954).

8 S. HIETANEN: *Acta Chem. Scand.*, **8**:1626 (1954).

9 Y. MARCUS: *Acta Chem. Scand.*, **11**:690 (1957).

10 C. BERECKI-BIEDERMANN: *Ark. Kemi*, **9**:175 (1956).

11 S. LIBICH and D. L. RABENSTEIN: *Anal. Chem.*, **45**:118 (1973).

12   H. L. JOHNSTON, F. CUTA, and A. B. GARRETT: *J. Amer. Chem. Soc.*, **55**:2311 (1933). See also C. A. Reynolds and W. J. Argersinger, Jr., *J. Phys. Chem.*, **56**:417 (1952).

13   G. S. FORBES and H. I. COLE: *J. Amer. Chem. Soc.*, **43**:2492 (1921).

14   G. S. FORBES: *J. Amer. Chem. Soc.*, **33**:1937 (1911).

15   R. RAMETTE: *J. Chem. Educ.*, **37**:348 (1960).

16   H. B. WEISER and W. O. MILLIGAN: *Chem. Rev.*, **25**:1 (1939).

17   E. B. SANDELL and I. M. KOLTHOFF: *J. Phys. Chem.*, **37**:153 (1933).

18   R. MOLTZAU and I. M. KOLTHOFF: *J. Phys. Chem.*, **40**:637 (1936).

19   "The Collected Works of J. Willard Gibbs," vol. 1, p. 324, Longmans, New York, 1928.

20   W. THOMSON: *Phil. Mag.*, (4)**42**:448 (1881).

21   W. OSTWALD: *Z. Phys. Chem.* (Leipzig), **34**:495 (1900).

22   H. FREUNDLICH: "Kapillarchemie," Akademische Verlagsgesellschaft, Leipzig, 1909.

23   W. J. JONES: *Z. Phys. Chem.* (Leipzig), **82**:448 (1913).

24   M. L. DUNDON and E. MACK, JR.: *J. Amer. Chem. Soc.*, **45**:2479 (1923); M. L. DUNDON: *J. Amer. Chem. Soc.*, **45**:2658 (1923).

25   D. R. MAY and I. M. KOLTHOFF: *J. Phys. Colloid. Chem.*, **52**:836 (1948).

26   G. A. HULETT: *Z. Phys. Chem.* (Leipzig), **37**:385 (1901).

27   E. COHEN and J. J. A. BLEKKINGH, JR.: *Z. Phys. Chem.* (Leipzig), **A186**:257 (1940).

28   D. BALAREV: *Kolloid-Z.*, **96**:19 (1941).

29   B. V. ENÜSTÜN and J. TURKEVICH: *J. Amer. Chem. Soc.*, **82**:4502 (1960).

PROBLEMS

7-1   The solubility product of $CaCO_3$ is 1×10^{-8}. Assuming activity coefficients of unity and considering hydrolysis of the carbonate ion ($pK_1 = 6.5$, $pK_2 = 10.3$), calculate (a) the solubility of $CaCO_3$ in water, (b) the pH of a saturated solution of $CaCO_3$, (c) the solubility of $CaCO_3$ at a pH of 7.00.

Answer   (a) 1.6×10^{-4} M.   (b) 10.18.   (c) 0.005 M.

7-2   The solubility product of CdS is 10^{-28}. The successive stepwise formation constants of $Cd(NH_3)_4{}^{++}$ are $K_1 = 300$, $K_2 = 100$, $K_3 = 20$, $K_4 = 6$. Taking the base dissociation constant of NH_3 to be 2×10^{-5} and the acid dissociation constants of H_2S to be $K_1 = 10^{-7}$ and $K_2 = 10^{-15}$, calculate the molar solubility of CdS in (a) 0.1 M NH_3 and (b) a buffer of pH 9 containing a total concentration of $NH_3 + NH_4Cl = 0.1$ M.

Answer   (a) 3.0×10^{-11}.   (b) 8.5×10^{-11}.

7-3   The solubility product of CaC_2O_4 is 2×10^{-9}. For oxalic acid, $K_1 = 6 \times 10^{-2}$, and $K_2 = 6 \times 10^{-5}$. Assuming activity coefficients of unity calculate the solubility of CaC_2O_4 at pH values of 1, 2, 3, 4, 5, 6 and depict the results graphically.

Answer $S = 3.0 \times 10^{-3}$, 6.2×10^{-4}, 1.9×10^{-4}, 7.3×10^{-5}, 4.8×10^{-5}, 4.5×10^{-5} M.

7-4 The solubility product of $MgNH_4PO_4$ is 2.5×10^{-12}. Calculate its solubility in buffers of pH 8, 9, and 10 containing $NH_4^+ + NH_3$ at a total concentration of $0.2\ M$ and phosphate at a total concentration of $0.01\ M$. The formation constant of $Mg(OH)^+$ is 300, K_a of NH_4^+ is 5×10^{-10}; pK values of H_3PO_4 are 2.15, 7.15, 12.4.

Answer $S = 3.8 \times 10^{-5}, 4.8 \times 10^{-6}, 1.9 \times 10^{-6}\ M.$

7-5 The value of pK_{sp} is 18.09 for BiI_3, and the overall formation constants of the higher complexes between Bi^{3+} and I^- are $\log \beta_4 = 14.95$, $\log \beta_5 = 16.8$, $\log \beta_6 = 18.8$. The lower complexes make a negligible contribution. From these data calculate the solubility of BiI_3 in $0.1\ M$ KI.

Answer $4.0 \times 10^{-3}\ M$ (first approximation 5.7×10^{-3}).

7-6 The solubility of mercury(I) chloride at chloride ion concentrations of 0.1 to $1.0\ M$ follows the empirical equation $S = 5.9 \times 10^{-6} + 7.9 \times 10^{-5}[Cl^-]$. Taking $K_{sp} = 1.1 \times 10^{-18}$, evaluate the formation constants of the complexes Hg_2Cl_2 and $Hg_2Cl_3^-$.   *Answer* $\beta_2 = 5.4 \times 10^{12}, \beta_3 = 7.2 \times 10^{13}.$

7-7 The formation constant of $Zn(OH)^+$ is 2×10^4. The solubility product of $Zn(OH)_2$ is 1.3×10^{-17}. The equilibrium constant of the reaction $Zn(OH)_2$ (solid) $\rightleftharpoons HZnO_2^- + H^+$ is 3×10^{-17}, and that of the reaction $Zn(OH)_2$ (solid) $\rightleftharpoons ZnO_2^= + 2H^+$ is 10^{-29}. (a) Evaluate β_3 and β_4, the formation constants of $HZnO_2^-$ and $ZnO_2^=$. (b) Calculate the solubility of $Zn(OH)_2$ at pH = 6, 7, 8, 9, 10, 11, 12, 13, assuming intrinsic solubility is negligible.

Answer   (a) $\beta_3 = 2.3 \times 10^{14}, \beta_4 = 7.7 \times 10^{15}$. (b) $S = 0.13, 1.3 \times 10^{-3}$,
$1.3 \times 10^{-5}, 1.9 \times 10^{-7}, 3 \times 10^{-7}, 3 \times 10^{-6}, 4 \times 10^{-5}, 1.3 \times 10^{-3}\ M.$

7-8 Assuming activity coefficients of unity, calculate the solubility of a 1:1 substance in water if the solubility product is 10^{-10} and the dissociation constant is $10^{-2}, 10^{-4}, 10^{-6},$ or 10^{-8}.

Answer $10^{-5}\ M, 1.1 \times 10^{-5}\ M, 1.1 \times 10^{-4}\ M, 10^{-2}\ M.$

7-9 Dimethylglyoxime, H_2Dx, reacts as follows: $2H_2Dx + Ni^{++} \rightleftharpoons Ni(HDx)_2 + 2H^+$. The solubility product of $[Ni^{++}][HDx^-]^2$ is 4.3×10^{-24}; K_1 for H_2Dx is 2.6×10^{-11}. Calculate the solubility at a pH of 5 in a solution containing $3 \times 10^{-3}\ M$ excess H_2Dx if S^0 is 9.7×10^{-7}. Assume activity coefficients of unity.   *Answer* $1.04 \times 10^{-6}\ M.$

7-10 If at 30°C KCl in glacial acetic acid has a solubility of 0.031 mole/l and a dissociation constant of 1.3×10^{-7}, calculate the intrinsic solubility and the solubility product.

Answer   Intrinsic solubility equals $0.031 - 6.3 \times 10^{-5} \simeq 0.031$.
$K_{sp} = 4.0 \times 10^{-9}.$

7-11 How would you determine whether the increase in solubility of MgC_2O_4 in excess $(NH_4)_2C_2O_4$ is due primarily to an activity effect or to complex formation?

8

FORMATION AND PROPERTIES OF PRECIPITATES

Several factors must be considered when deciding whether a given precipitation reaction is a feasible basis for an analytical method; solubility certainly is one, another relates to physical properties, and still another to chemical purity and stability. The physical properties of a precipitate are influenced by the mode of its formation as well as by the specific compound involved. Broadly speaking, precipitates and their properties depend on the processes of nucleation, crystal growth, and aging. This chapter deals with these topics and also the properties of colloidal precipitates.

8-1 NUCLEATION

Nucleation is the process occurring in a supersaturated solution that results in the formation of the smallest particles that are capable of growth to larger ones. This process is vitally important to many aspects of disciplines such as meteorology, physiology, metallurgy, and chemistry; analytical chemists in particular have been concerned about understanding the process. By controlling the nucleation process, the analyst can largely control the physical properties and to some extent the chemical composition of his materials. The question of how nucleation occurs is extraordinarily difficult to study experimentally;[1-7] as a result speculation, controversy, and interpretive hypotheses abound, typical of situations where reliable experimental data are scarce.

One difficulty in the study of nucleation is that it is a transient phenomenon in a usually dynamic system. Essentially no observable changes occur prior to nucleation, but once nucleation occurs, it is over in a short time, and the processes of growth and aging take over. (An interesting exception[8] involves polar fishes, in which nucleation of ice crystals occurs in their blood at one temperature, and crystal growth about 1°C lower.) Literature on the subject is extensive, often with conflicting observations. In the remainder of this section a few of the more significant experimental observations are summarized, followed by a section in which the mechanism is considered. The discussion by Walton[7] gives many additional details on the subject of formation and properties of precipitates.

When considering nucleation and crystal growth, it is convenient to use the term *supersaturation ratio*, defined[1] by $(Q - S)/S$, where Q is concentration of solute and S is equilibrium solubility.

Some experimental observations Perhaps the most striking, and rather general, phenomenon is the induction period that occurs after the mixing of reagents and before the appearance of a visible precipitate. When a solution is dilute, usually no precipitation occurs for a period of time. Thus Enüstün and Turkevich[9] found the minimum strontium sulfate concentration that initiates precipitation in about 10 h to be 0.015 M, far above the equilibrium solubility of 6×10^{-4} M. The duration of the induction period varies greatly for different precipitates, being long for barium sulfate and short for silver chloride. These two salts have been interesting cases for extensive study and comparison because of their similar molar solubility.

Nielsen[10] compared the induction periods observed by various investigators for several precipitates. He reported the induction period for silver chloride to be inversely proportional to the fifth power of the initial concentration after mixing. Similarly, induction periods for silver chromate, calcium fluoride, calcium oxalate, and potassium perchlorate are inversely proportional to the 4.7, 9, 3.3, and 2.6 powers of the initial concentrations. For barium sulfate, on the other hand, a variety of discordant values has been observed. La Mer[11] summarized the data of two groups of observers, covering a range of concentrations (Figure 8-1).

The duration of the induction period seems to be independent of the method of observation, whether by direct visual observation, more sensitive optical means, or measurement of electrical conductance. In addition,[12] the conductivity remains nearly constant during the induction period, indicating that during this period only a small fraction of the solute exists as ion pairs or higher aggregates.

The induction period varies in length according to the experimental technique as well as with the chemical nature of the solid being formed. La Mer and Dinegar,[13]

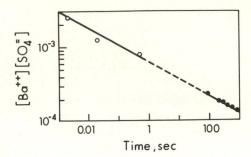

FIGURE 8-1 Induction period for BaSO$_4$ as a function of the product of ion concentrations. (*From La Mer.*[11])

by generating sulfate ion in homogeneous solution in the presence of barium ion, found that a solution of a given supersaturation ratio persists without precipitation for a longer time than when the same solution is prepared by direct mixing. They concluded that, with the homogeneous method, precipitation occurs at a specific limiting value for $(a_{Ba^{++}} a_{SO_4^=}/K_{sp})^{\frac{1}{2}} = 22$, no matter what the barium ion concentration or the rate of generation of sulfate. Collins and Leineweber[14] found that the limiting supersaturation ratio is dependent on the purity of the reagents; after repeated recrystallization and filtration of the reagents they observed ratios as high as 32. Nielsen[15] found that, if the walls of the precipitation vessel were thoroughly cleaned by prolonged steaming, the number of barium sulfate crystals formed per unit volume could be decreased by a factor of 10 or more. A decreasing number and increased size of barium sulfate particles have been observed when aged and filtered barium chloride solutions are used.[16–18] Enüstün and Turkevich[9] found a critical supersaturation ratio of 24 for strontium sulfate. Walton[19] found a value of 5.5 for silver chloride, typical of a softer type of precipitate.

In an attempt to eliminate solid-liquid interfaces such as are present when container walls or microscopic clay particles are involved, Vonnegut[20] devised the drop method of studying the precipitation process. In this method the solution is subdivided into microscopic droplets, which are then suspended in an oil matrix. The technique of precipitation from homogeneous solution (Section 10-2) has been combined[21–23] with the drop method to generate a sparingly soluble substance slowly in solution and thereby continually increase the supersaturation to the point of precipitation. Different droplets are observed to become turbid at widely varying times. A plot of turbid-droplet count against time (or concentration) gives an S-shaped curve.

Another significant observation, beginning with von Weimarn,[1] concerns the variation in size and number of precipitate particles with the concentration of precipitating reagents. Von Weimarn postulated that a maximum would occur in the curve of particle size as a function of concentration of reactants and also that average particle size would increase with time. He studied precipitation of the alkaline earth sulfates and silver sulfate and acetate, but did not distinguish clearly between forma-

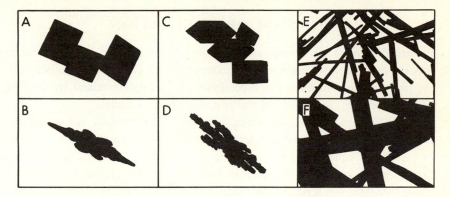

FIGURE 8-2 Electron photomicrographs of precipitates: $BaSO_4$ from a 0.0026 M solution (A), a 0.026 M solution (B), a solution of pH 0.5 (C), and a solution of pH 2.0 (D); nickel dimethylglyoxime prepared by conventional precipitation (E) and by homogeneous precipitation (F). (*From Fischer*[16] *and Takiyama and Gordon.*[26])

tion of precipitate particles and later changes that are now classified as aging. He observed that for barium sulfate, as the initial concentration was increased from 10^{-3} to 1 M, the freshly formed crystals became less perfect, and more needles and skeletal crystals appeared. Above 1 M, jellylike masses were formed first, later changing to voluminous fine-grained precipitates. Interestingly, von Weimarn believed early in the century that the presumably amorphous precipitates were actually crystalline but composed of extremely small and imperfect particles. This view has been confirmed by x-ray investigations, which generally show that freshly formed ionic precipitates have the same powder diffraction patterns as the aged products but with more diffuse lines. Even precipitated metal "hydroxides" should be regarded as highly imperfect hydrous oxides.[24] Van't Riet and Kolthoff[25] obtained the maximum particle size for lead sulfate at a value of $a_{Pb^{++}}a_{SO_4^{=}} = 4 \times 10^{-6}$. This is about 200 times the value of the solubility product, corresponding to a supersaturation of $\sqrt{200}$, or about 15. At higher concentrations the maximum particle size decreases rapidly.

Figure 8-2 shows a selection of electron photomicrographs of barium sulfate[16] and nickel dimethylglyoxime[26] precipitates formed under a variety of conditions. The size and shape of precipitate particles clearly varies widely with the nature of the precipitate and the mode of precipitation. The perfection of crystals of the same size also varies with conditions. Particle size as reflected by a linear dimension may be misleading.

The effect of temperature is complex. Solubility usually increases so that the supersaturation ratio decreases with increasing temperature. On the other hand, the size of the critical agglomerate for nucleation decreases with increasing temperature [Equation (8-5), Section 8-4]. Therefore an optimum temperature exists for precipitate formation.

8-2   THE MECHANISM OF NUCLEATION

Presented here is an abbreviated, possibly oversimplified view of the nucleation process under conditions normally encountered in analytical chemistry. In general, the particle-size distribution of a precipitate must be determined by the relative rates of two processes: formation of nuclei and growth of nuclei.

Distinction should be made between homogeneous (or spontaneous) and heterogeneous (or induced) nucleation. In *homogeneous nucleation* a cluster of ions or molecules equal to that for a nucleus forms in a supersaturated solution. In *heterogeneous nucleation* the formation of the critical cluster is aided by a second phase (for instance, a particle of dust). In analytical precipitation reactions, homogeneous nucleation probably is rarely, if indeed it is ever, the predominant mechanism and occurs only when the supersaturation ratio is large. Unfortunately, most of the hypotheses concerning the nucleation mechanism relate to the homogeneous process; those for the heterogeneous type are not easily developed.

According to Ostwald[2] a supersaturated solution may be metastable; that is, it can remain indefinitely homogeneous unless suitably inoculated with nuclei for crystal formation. Beyond a certain degree of supersaturation (the metastable limit) he regarded the solution as labile, or subject to spontaneous nucleation and crystallization. In most analytical precipitation reactions the number of nuclei formed in a primary process depends on the number of nucleation sites provided by the reagents, solvent, and reaction vessel, the effectiveness of these sites, and the extent of supersaturation. Fischer[27] indicated that even highly purified reagents may contain astronomical numbers of small insoluble particles (probably silicates), far more than are needed for nucleation sites. Solutions made from chemically pure reagents probably contain at least 10^6 such sites per milliliter, and hence the number of nucleation sites is not easily amenable to experimental control. Effectiveness of the sites varies enormously; the most effective are those having the same chemical composition and physical arrangement as the solid in supersaturated solution. Solid phases generally are more effective sites for nucleation of solids than are gases or liquids. Similarly, gas phases are effective sites for gases in supersaturated solution. When the number and effectiveness of nucleation sites are relatively constant, the number of nuclei formed depends mainly on the supersaturation ratio. Meehan and Chiu[28] measured the number of colloidal particles of silver bromide formed over a range of five orders of magnitude of supersaturation ratio. They found a linear relation between the number of particles and the supersaturation ratio; at a ratio of 100 about 10^{10} particles per milliliter were present 2 min after reaction.

Heterogeneous nucleation of an ionic precipitate can be viewed as a sequence[29] involving the diffusion of ions or ion pairs to a surface and their adsorption and surface diffusion to form a two-dimensional cluster or island. The critical nucleus so formed probably consists of relatively few ions.[19] If the lattice spacings of the

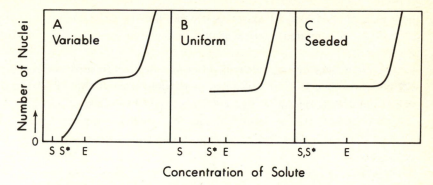

FIGURE 8-3 Schematic relation between the number of nuclei formed and the concentration of solute at the instant of nucleation. Here S is equilibrium solubility and S^* is critical or limiting supersaturation for precipitate formation. A, variable effectiveness of nucleation sites; B and C, uniform effectiveness of nucleation sites. In C the sites and solute are of the same composition. The vertical scales are arbitrary.

nucleation site and of the material in supersaturated solution are closely similar, then the energy barrier to nucleation decreases and nucleation is favored. Thus water in the presence of powdered silver iodide can be supercooled by only 2.5°C, whereas in the presence of powdered Teflon it can be supercooled by at least 16°C. The lattice spacing of silver iodide compares closely to that of water. When some of the iodine atoms in silver iodide are replaced by bromine atoms, the disregistry (noncoincidence of lattice spacings) with ice is even smaller and the supercooling of water is less.[30]

Figure 8-3 is a schematic representation of the number of nuclei formed during a precipitation when the effectiveness of nucleation sites is variable and when it is uniform. In this figure homogeneous nucleation is indicated by the sharply rising curve at high concentrations. If, at some temperature, solutions of reagents are added to increase the concentration of a solute gradually, no precipitate can form until the concentration becomes equal to or greater than the value of S^*. At or above this concentration, for example at E, the solution is said to be labile, and nucleation can occur after a suitable induction period. The number of nuclei formed depends on the extent of supersaturation and the effectiveness of the nuclei. If the nuclei are uniform, their number will be equal to the number of nucleation sites (Figure 8-3B). Once nucleation occurs, the excess material in solution is removed and the concentration drops to S, the solubility value at equilibrium. Because solid of the same composition as the solute is present, further addition of reagent does not increase the concentration beyond S for a substantial period of time. New nuclei are not formed because the instantaneous local concentration does not reach high supersaturation values. Thus Fischer[31] found that, for precipitation of barium sulfate from homogeneous solution, there were about 1.5×10^7 particles per milliliter at early stages in the precipitation (about 2% precipitated) and essentially the same number (1.6×10^7) at 90% precipitated.

A precipitate formed under conditions where additional nuclei are not formed will be relatively coarse and uniform in particle size. If the initial concentration is equal to or greater than E in Figure 8-3A, the number of particles will be independent of concentration over a certain range. If concentration becomes high enough, homogeneous as well as heterogeneous nucleation occurs, and large numbers of particles form.

Figure 8-4 schematically represents particle size as a function of initial concentration of solute prior to nucleation, with the number of particles assumed equal to the number of nucleation sites present. Generally, particle size can be expected to increase with concentration until homogeneous nucleation takes place, after which particle size will decrease sharply with concentration. In Figure 8-4 the difference between the concentrations S and S^* can be large or small. For barium sulfate it is larger than for silver chloride, with the result that barium sulfate tends to be nucleated much less extensively than silver chloride and consequently forms larger particles.

On the assumption that heterogeneous nucleation is the important process, precipitation of barium sulfate from a supersaturated solution can be viewed as beginning by a series of steps *on a nucleation site:*

$$Ba^{++} + SO_4^{=} \rightleftharpoons Ba^{++}SO_4^{=} \quad \text{ion pair} \tag{8-1}$$

$$Ba^{++}SO_4^{=} + Ba^{++}(\text{or } SO_4^{=}) \rightleftharpoons Ba_2SO_4^{++}[\text{or } Ba(SO_4)_2^{=}] \tag{8-2}$$

$$Ba_2SO_4^{++} + SO_4^{=} \rightleftharpoons (Ba^{++}SO_4^{=})_2 \tag{8-3}$$

$$(Ba^{++}SO_4^{=})_2 + Ba^{++}(\text{or } SO_4^{=}) \rightleftharpoons Ba_3(SO_4)_2^{++}[\text{or } Ba_2(SO_4)_3^{=}] \tag{8-4}$$

During the induction period this process continues on the surface of the nucleation site until the critical cluster has collected; the next ion to be added triggers nucleation. Crystal growth then can follow. For barium sulfate, La Mer[11] concluded that the slope of the line in Figure 8-1 is six and therefore the nucleation of barium sulfate is a seventh-order reaction overall. The critical cluster is then $(Ba^{++}SO_4^{=})_3$, and the addition of the seventh ion, either Ba^{++} or $SO_4^{=}$, constitutes the final step of the nucleation process. The question of the number of ions in the critical cluster, however, is by no means settled. Christiansen and Nielsen[6] concluded that for barium sulfate the number is 8. Johnson and O'Rourke[12] also concluded that the nucleation rate of this salt is proportional to the fourth power of the concentration. The concept of a small critical nucleus is intuitively satisfying in that the nucleus then requires only a small number of steps for its formation. On the other hand, application of the

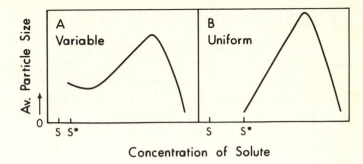

FIGURE 8-4 Particle size as a function of concentration: variable (A) and uniform (B) effectiveness of nucleation sites.

Becker-Döring[4] nucleation hypothesis indicates a much larger number, of the order of 100, for the critical cluster. Klein and Driy,[21] in nucleation studies combining the drop method and homogeneous precipitation, found the rate of nucleation of strontium sulfate to depend on the 27th power of the concentration, indicating a nucleus containing 52 ions.

For the onset of homogeneous in addition to heterogeneous nucleation, the supersaturation ratio must be high. Walton[32] compiled estimates of the critical size of cluster for several substances as shown in Table 8-1. Serious difficulties in making estimates of this type arise from the dearth of knowledge about the surface tension of solids with extremely small particles and the assumption that the particles are spheres with equivalent surface sites.

Table 8-1 CRITICAL RATIOS, SURFACE TENSIONS, AND CRITICAL SIZES FOR SPHERICAL NUCLEI IN HOMOGENEOUS NUCLEATION†

Precipitate	Critical ratio S_{crit}/S	Surface tension σ, ergs/cm^2	Critical radius r_{crit}, cm $\times$ 10^8
BaSO$_4$	1000	116	5.5
PbSO$_4$	28	74	6.5
SrCO$_3$	30	86	6.0
AgCl	5.5	72	7.5
AgBr	3.7	56	7.5
CaC$_2$O$_4$	31	67	6.5

† From Walton.[32]

8-3 CRYSTAL GROWTH

Growth of nuclei to larger particles consists mainly of diffusion of material to the surface, followed by deposition. If diffusion in solution were the rate-controlling step, growth rates (except under conditions of a high supersaturation ratio) would be higher than generally observed. Marc,[33] in an extensive series of investigations, found that most growth processes are second order, rather than first order as required by a single-ion diffusion mechanism. Beyond a certain stirring rate the rate of crystallization becomes independent of stirring rate. He observed that crystal growth is often retarded or inhibited by adsorbed dyes that do not appreciably influence the rate of dissolution. Such observations point to some factor other than diffusion that contributes to the rate, particularly for small particles, which have the highest diffusional flux. For many metal ions the rate of solvent release may be small compared with the rate of diffusion,[34] and the rate-determining step may be loss of solvent from the solvated ions. Near the end of growth, diffusion, which undoubtedly always plays a role, may become rate-controlling.[35]

If foreign substances are present in solution, the induction period, the rate of crystal growth, and even the crystal form may be changed. For example, the addition of gelatin greatly increases the induction period and decreases the growth rate for barium sulfate precipitation. Davies and Nancollas[36] found that many organic substances such as potassium benzoate, even at low concentrations, decrease the rate of recrystallization of silver chloride. Eosin can prevent crystal growth completely.

Broadly speaking, four principal types of crystal growth are postulated, involving screw dislocation, two-dimensional nucleation, dendrite formation, and amorphous precipitation. In special cases postprecipitation is important.

Screw dislocation At low supersaturation values, rates of crystal growth may be explained in terms of screw dislocations.[37,38] This is probably the most important type of growth for analytical precipitates formed under conditions giving large, well-formed particles. Initiation of a dislocation in a crystal surface may result from occlusion of an impurity. Once started, such an imperfection can grow with a spirally advancing face, and nucleation on a plane surface is not required. In an angular sense, the relative rate of growth at the center of a dislocation is much greater than farther away, so that the growing dislocation winds itself into a spiral, as indicated schematically by Figure 8-5. Somorjai[39] stated that dislocation densities of the order of $10^6/cm^2$ are usually observable on ionic crystal surfaces. Since the surface concentration of atoms is about $10^{15}/cm^2$, one dislocation in about 10^9 atoms may be present on the average in a well-crystallized material. The step height can be more than one molecular dimension, permitting direct microscopic observations of the spiral growth patterns. Many substances on condensation from a supersaturated vapor show whisker formation by screw dislocation. Walton[40] showed that, when

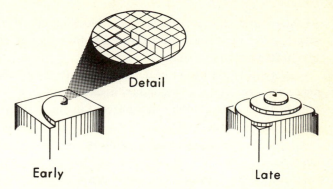

Detail

Early   Late

FIGURE 8-5 Schematic representation of crystal growth by screw dislocation. Left, an early stage; right, growth spiral at a later stage; center, detail at the dislocation site.

supersaturation is low and a spiral has developed, the growth rate should be proportional to the square of the supersaturation, and when high, directly proportional to the supersaturation.

Two-dimensional surface nucleation As early as 1878, Gibbs[41] suggested that crystal growth proceeds by nucleation on a crystal face, forming monolayer islands that grow rapidly to the boundaries of the crystal face. Theoretical rate expressions for mechanisms involving two-dimensional nucleation have been derived by Volmer,[42] Kossel,[43] and Stranski.[44] Such nucleation would require a critical supersaturation level, because energy is associated with the edge of a surface cluster. With this mechanism the nucleation process would have to be repeated for each new crystal layer. The mechanism for crystal growth would be operative, then, only when the concentration exceeded some critical supersaturation level, and at low supersaturation values it would become unsatisfactory. For example, Volmer and Schultze[45] found that, at a supersaturation of only 1%, crystals of naphthalene, mercury, and phosphorus could be grown from the vapor. If two-dimensional surface nucleation were the mechanism for growth, then growth could occur only adjacent to atoms brought out of the plane surface by thermal agitation, at a rate 10^{1000} times lower[46] than the rate actually observed! As indicated in the preceding subsection, at low supersaturation values growth can occur by a screw-dislocation mechanism.

Dendritic growth At high supersaturation values, where crystal growth is limited by diffusion, dendritic growth predominates. This growth involves irregular or branched aggregates, as in snowflakes. For an ionic precipitate, there is diffusion of solvated ions to the growing crystal surface, deposition of the ions, and release of solvent molecules followed by diffusion of the solvent away from the growing surface.

At edges, and especially corners, the blocking effect of the released solvent is less severe, and so growth is most favored at those points. This process has been called the "traffic-jam" mechanism.[47]

An important aspect of dendritic growth is that crystals fragment[48] easily, and hence what can be called secondary nucleation results. Thus, the number of particles obtained in a precipitation can far exceed the number of nucleation sites even though homogeneous nucleation does not occur. In meteorological seeding, each silver iodide nucleation site can result in thousands of rain drops through fragmentation of dendritic ice crystals. Nielsen[15] found that stirring during crystal growth leads to smaller final particle size, presumably again because of the fragmentation of dendritic crystals during the early stages of precipitation. Ultrasonic vibration during precipitation also results in small particle size. Walton[49] suggested that fragmentation of dendritic crystals may sometimes be an alternative to the onset of homogeneous nucleation.

Amorphous precipitation   The fourth possibility for crystal growth, amorphous precipitate formation, is important at extremely high supersaturation ratios. Such ratios are hard to avoid when dealing with substances with exceedingly small solubility products. The form of a precipitate depends in part on the rates of the aggregation and orientation velocities.[50] In certain salts, when the supersaturation is high enough, aggregates of ions may come together rapidly in random arrangement and then tend to gain stability by rearrangement to a regular crystal lattice. When the orientation velocity is low compared with the aggregation velocity, the precipitate is amorphous. Hydrous metal oxides are an example.

Postprecipitation   Postprecipitation[51] involves the formation of a second insoluble substance on a precipitate already formed, as a result of differences in rates of precipitation. For example, in the separation of calcium from magnesium by oxalate precipitation of calcium, the solubility of magnesium oxalate may be exceeded. But, since magnesium oxalate has a pronounced tendency to remain in supersaturated solution,[52,53] it slowly precipitates on the calcium oxalate over a period of many hours.

Especially pronounced postprecipitation has been observed in mixtures of metal sulfides. Zinc sulfide, for example, has a definite tendency to postprecipitate on other metal sulfides. Although in dilute strong acids (0.1 to 0.3 M) zinc sulfide is actually insoluble, it will remain indefinitely in supersaturated solution unless nuclei are present for crystallization.[54] This characteristic is evidently due to the low concentration of sulfide and even of bisulfide ions present in acidic solution. Crystals of another sulfide, say copper or bismuth, induce the postprecipitation of zinc sulfide[55] because hydrogen sulfide is adsorbed on the solid sulfide as HS^- or $S^=$, owing to the strong attraction of the metal sulfide lattice for sulfide ions. Postprecipitation begins slowly on copper(II) or bismuth(III) sulfides, probably because these sulfides are not efficient surfaces for crystallization. On the other hand, on mercury(II) sulfide,[56] which is isomorphous with zinc sulfide, no induction period exists, and postprecipitation curves begin steeply from the origin. That no coprecipitation (Chapter 9) is taking place is shown by the fact that, when not all the mercury is precipitated, the mercury sulfide contains no appreciable zinc. The postprecipitated zinc sulfide evidently enters solid solution (Section 9-4) with the mercury(II) sulfide, since it cannot be extracted completely by continuous extraction with 3 M hydrochloric acid, whereas it was extracted readily from bismuth(III) sulfide or copper(II) sulfide.

8-4 AGING

Aging has been defined[57] to include all irreversible structural changes that occur in a precipitate after it has formed. These changes may include: (*1*) recrystallization of primary particles; (*2*) cementing of primary particles by recrystallization in an agglomerated state; (*3*) thermal aging, or perfection by thermal agitation of ions to form a more nearly perfect structure; (*4*) transformation of a metastable modification into another, more stable form; (*5*) chemical aging, including processes involving changes in composition.

As indicated in Section 8-1, at a critical supersaturation concentration and after a suitable induction period, clusters of critical size form in a solution to constitute nucleation. In a simple way, this critical size[58] can be calculated for a given temperature from Equation (7-41) rearranged to

$$r_{\text{crit}} = \frac{2\delta M}{\rho RT \ln (S^*/S)} \quad (8\text{-}5)$$

where r_{crit} is "radius" of the nucleus, δ is surface tension, M/ρ is molar volume (molecular weight divided by density), S^* is the supersaturation at which nucleation occurs, and S is equilibrium solubility. While nuclei are growing as a result of precipitation, supersaturation decreases, and according to (8-5), r_{crit} must become correspondingly larger. The smaller particles will have a solubility exceeding the solute concentration and will tend to dissolve. Immediately following the onset of precipitation, then, a continuous race begins between the larger and smaller particles and between the inactive and active surfaces, with the smaller particles and the more active surfaces the losers. At equilibrium, which is never attained in practice, there would be only a single particle of precipitate. At early stages of precipitation, recrystallization is fast, with the number of particles dropping rapidly.

Undoubtedly, the number of particles of a precipitate experimentally measured gives little indication of the number of effective nucleation sites. Primary particles normally have large surface energy, dendritic crystals in particular having many corners, prominences, and edges. Greater stability and reduction of the surface area and energy are attained by various aging processes.

Experimental studies of aging usually employ radioactive-tracer techniques for observation of exchange of ions between precipitate and solution, the adsorption of suitable dyes on the surface of a precipitate to measure the extent of the specific surface, microscopic or x-ray observation of the precipitate, or a combination of these.

Recrystallization of primary particles (Ostwald ripening) Ostwald ripening is aging that involves transfer of solute from small to large particles by way of the solution. At the time of Ostwald, when it first became generally realized that the solubility of small particles should be greater than normal, it was logical to deduce that a freshly formed precipitate would undergo a process of ripening by dissolution of small particles and growth of large ones. But the recrystallization of highly imperfect particles immediately following precipitation is so rapid that lattice ions have little opportunity to go from the surface to the bulk of the solution. Rather, the ions move mainly within a thin film around the particles or by surface diffusion, while the active surface disappears and the crystal perfects itself. Therefore, Ostwald ripening is probably of subordinate importance in aging,[59–61] at least during the early stages when recrystallization is rapid and under conditions of low solubility of the solid. For example, the rate of aging is not increased by increased agitation for lead sulfate,[60] lead chromate,[59] or barium sulfate,[62] although with silver bromide[63] in excess bromide appreciable Ostwald ripening does occur. In general, it may be expected to be noticeable under conditions of increased solubility.[59,61]

The rate of recrystallization is high for a fresh precipitate, but diminishes gradually as the particles become perfected. Figure 8-6 illustrates the exchange of lead-212 (half-life 10.6 h) ions in solution with lead ions in lead sulfate or lead chromate precipitates.[59,64] In both systems the rate of aging decreases with age, but a distinctly measurable rate is still observable with a lead sulfate precipitate aged for 3 h. With lead chromate[59] it was found that under certain conditions the amount of radioactive lead taken up by the solid exceeded the equilibrium amount corresponding to a uniform distribution between solid and solution. A product only 15 s old reaches the homogeneous composition rapidly through recrystallization. A product 10 min old is recrystallizing more slowly; it temporarily extracts a relatively large quantity of radioactive lead, because the surface originally exposed to rich radioactive solution is buried by recrystallization, so that the precipitate in effect performs a multiple extraction. Eventually the abnormally rich solid approaches equilibrium by continuous recrystallization. These experiments provide striking evidence for a severalfold recrystallization process, occurring at a gradually diminishing rate as the aging proceeds. Conti, d'Alessandro, and di Napoli[65] correlated the exchange of chromium 51 with lead chromate with the kinetics of the monoclinic–orthorhombic phase transition.

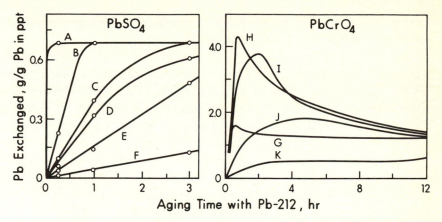

FIGURE 8-6 Exchange of lead-212 ions with lead sulfate or lead chromate precipitates. Left: *A*, fresh lead sulfate; *B*, aged 12 min; *C*, 20 min; *D*, 45 min; *E*, 57 min; *F*, 3 h. Right: *G*, lead chromate aged 15 s; *H*, 10 min; *I*, 20 min; *J*, 30 min; *K*, 1 h. (*From Kolthoff and Eggertsen*[59] *and Kolthoff and von Fischer.*[64])

They concluded that exchange exceeding the equilibrium value is due to phase transition. During aging the total number of precipitate particles decreases strongly. Meehan and Chiu[28] found that for colloidal silver bromide the number of particles decreased 80-fold during the aging that occurred after 2 min and up to 1 day.

The rate of aging is strongly influenced by other solutes in solution and thus can be increased or decreased by the presence of excess lattice ions in solution. Barium sulfate ages more slowly in barium ion solution than in sulfate and more slowly in sulfate than in water.[62] The aging of silver chloride is impeded by silver ion, but speeded by chloride ion;[61] a similar effect exists for silver bromide.[63] For lead chromate no particular lattice ion effect was noticed.[59] Apparently the rate of aging does not parallel solubility, which is decreased by the common ion effect. Kolthoff and others postulated that the solubility in the adsorbed water layer may be different from that in the bulk of the solution. For example, in the case of silver chloride in the presence of adsorbed chloride ion, the solubility may be increased owing to the formation (Section 7-7) of $AgCl_2^-$ in the immediate vicinity of the surface. It appears likely that the adsorbed lattice ion also has a pronounced effect on the rate of recrystallization, which is not necessarily parallel with solubility even in the adsorbed water layer.

In general, other factors that affect solubility may also change the rate of aging. For example, barium sulfate ages more rapidly in 0.1 M nitric acid and more slowly in ethanol than it does in water.[62,66] Similarly, the aging of silver chloride is promoted by ammonia and inhibited by ethanol.[61] It is striking that lead chromate ages much more rapidly in the flocculated, or agglomerated, state than in a colloidal state.[67] In contrast, a colloidal dispersion of silver bromide forms homogeneous solid solutions almost instantaneously with added chloride ion.[68]

Thermal aging   *Thermal aging* may be defined as a process in which perfecting of crystal structure occurs through thermal agitation within a crystal rather than through recrystallization. The rate of such a process increases rapidly with rising temperature, but at any temperature is independent of the nature of the solvent or even whether solvent is present. The rate of recrystallization processes also increases rapidly with rising temperature, and so temperature dependence is not the criterion of thermal aging.

Silver bromide apparently undergoes appreciable thermal aging even at room temperature;[63,69] complete homogeneous exchange with radioactive bromine occurs within a few seconds.[70] The exceptional thermal aging properties of silver bromide have been attributed to lattice defects that give rise to a high mobility of ions, at least in the layers near the surface. Compressed pellets of freshly precipitated silver bromide have considerable electrical conductivity, an effect that also has been attributed to highly mobile surface ions.[71,72] Lead chromate undergoes no thermal aging at room temperature, but shows a pronounced effect upon heating at 355°C.[67] Barium sulfate shows no thermal aging in 24 h at 300°C or in 1 h at 400°C, but shows a marked sintering at 500°C. Thermal aging becomes more rapid above 700°C, and occluded sodium chloride is volatilized at temperatures above 800°C.[73] Particles of silica gel undergo two types of aging upon heating: a low-temperature perfecting of individual particles below 700°C and a sintering process above 700°C with a pronounced decrease in porosity.[74]

In general, the critical temperature at which thermal aging becomes appreciable corresponds to Tammann's *relaxation temperature*,[75] the temperature at which thermal agitation begins to overcome lattice forces. This occurs at about half the melting point on the absolute temperature scale.

Other aging effects   Aging due to cementing of primary particles is difficult to observe directly because of other simultaneous changes. Nevertheless, it may be inferred that such processes going on in the flocculated state render impossible the subsequent peptization, or dispersal, of an aged product. Kolthoff and others have discussed the cementing process for barium sulfate,[62] lead chromate,[59] and silver bromide.[76]

A special type of aging occurs in calcium oxalate,[77] which at room temperature is precipitated as a mixture of dihydrate and trihydrate. Upon digestion at higher temperatures these products become metastable with respect to the monohydrate. As a result of the drastic recrystallization, coprecipitated impurities are largely removed by digestion.

Another type of aging occurs with hydrous oxides. Hydrous iron(III) oxide precipitated at room temperature is at first amorphous to x-rays, but after standing at room temperature for several weeks, it shows the band diffraction pattern of hematite. After several months the pattern has changed to give sharp lines. A similar sharp pattern is observed after a few hours at the boiling point. The x-ray results give no evidence of hydrates such as α-$Fe_2O_3 \cdot H_2O$. X-ray evidence[24] indicates that the aging process should be regarded as a growth of fine crystallites of iron(III) oxide to form crystals large enough to give a sharp diffraction pattern. The rate of aging of hydrous iron(III) oxide has been found to be negligible at room temperature in water or dilute acid but to increase markedly with hydroxyl ion concentration (ammoniacal or sodium hydroxide solution),[78,79] even though the solubility decreases with increasing alkalinity. Digestion at 98°C greatly speeds the aging process. A striking observation was that the process is inhibited by adsorbed divalent metal ions such as zinc, nickel, cobalt, or magnesium but not by calcium. The inhibiting effect was attributed to the replacement of hydroxyl hydrogen by metals (ferrite formation) to prevent polymerization. In support of this view was the finding that the precipitate removed increased amounts of zinc, nickel, or cobalt from solution upon digestion at 98°C and (slowly) even on standing at room temperature.

8-5   PHYSICAL PROPERTIES OF COLLOIDS

With recognition that chemical differences cannot be ignored, the physical properties of two precipitates are likely to be similar if they are formed under comparable conditions of supersaturation. If few nuclei form, relatively few large particles result; if many nuclei, a large number of small particles of a precipitate of large surface area result. From several points of view, coarsely crystalline precipitates are preferable when a filtration operation is involved. Handling of a precipitate is a practical problem, and for finely divided substances, knowing how to manipulate conditions to advantage is a decided asset. When filtering finely divided precipitates, knowledge of how to coagulate, or flocculate, a colloidal dispersion of the solid to permit filtration and to prevent peptization upon washing and passage through the filtering medium is important. At other times, such as in titrimetric operations, the problem may be to keep a precipitate in dispersed form. An understanding of the basic principles of the colloidal or surface chemistry of precipitates is therefore desirable.

Colloids are generally considered to comprise particles having diameters in the range 10^{-7} to 10^{-4} cm or those containing 10^3 to 10^9 atoms. Some colloidal particles are single macromolecules and others are needle-shaped, disk-shaped, or spherical aggregates for which a single linear dimension has little meaning. Two types of colloids are: *lyophobic* (also called suspensoids or hydrophobic, inorganic, irreversible, or irresoluble colloids) and *lyophilic* (also called emulsoids or hydrophilic, organic, reversible, or resoluble colloids). Lyophobic colloids are generally dispersions of insoluble inorganic substances in a liquid medium, usually aqueous solution, and have little attraction for the solvent. They are characterized by sensitivity to coagulation by electrolytes and by the fact that the flocculation process cannot be reversed completely upon dilution. The dispersions (sols) are of relatively low viscosity, and the flocculated solid contains relatively little water. Typical examples are colloidal sulfur, gold, silver iodide, and arsenic(III) sulfide. Lyophilic colloids, on the other hand, have strong affinity for the solvent and are relatively viscous and insensitive to electrolytes. On addition of a high concentration of salt, the system tends to break into two liquid phases (coacervation). This process is reversed upon dilution. The suspended solid is highly hydrated and, if dried, hygroscopic. Typical examples are proteins and starches. Most precipitates of analytical interest are lyophobic in character, although certain precipitates, such as freshly formed silicic acid, behave like lyophilic colloids. Hydrated metal oxides exhibit lyophilic behavior to a lesser degree.

Colloidal dispersions owe their stability to a surface charge and the resultant electrical repulsion of charged particles. This charge is acquired by adsorption of cations or anions on the surface. For example, an ionic precipitate placed in pure water will reach solubility equilibrium as determined by its solubility product, but the solid may not have the same attraction for both its ions. Solid silver iodide has greater attraction for iodide than for silver ions, so that the zero point of charge (the isoelectric point) corresponds to a silver ion concentration much greater than iodide, rather than to equal concentrations of the two ions. The isoelectric points of the three silver halides are:[80–82] silver chloride, pAg = 4, pCl = 5.7; silver bromide, pAg = 5.4, pBr = 6.9; silver iodide, pAg = 5.5, pI = 10.6. For barium sulfate the isoelectric point seems to be dependent on the source of the product and its degree of perfection.[83]

Lange and Berger[84] studied the adsorption of potential-determining ions, or ions that carry a charge to the solid phase. For many substances, including silver iodide, they found that adsorption follows the equation

$$\Delta X = k \, \Delta \ln C \qquad (8\text{-}6)$$

where ΔX is the change in the amount of adsorbed ion brought about by a change $\Delta \ln C$ in the concentration of potential-determining ion in solution. They gave the following equation for the resulting potential difference ΔE between solid and solution:

$$\Delta E = \frac{RT}{ZF} \Delta \ln C \qquad (8\text{-}7)$$

where Z is the charge of the potential-determining ion. If (8-7) is applied between the concentration C and the concentration at the point of zero charge C^0, we have, using the solubility-product relation,

$$E = \frac{RT}{Z_+ F} \ln \frac{C_+}{C_+^0} = \frac{RT}{Z_- F} \ln \frac{C_-^0}{C_-} \qquad (8-8)$$

where C_+ and C_- refer to the concentrations of cation and anion. When $C_+ = C_+^0$ or $C_- = C_-^0$, E is zero and corresponds to the potential of the solid with respect to the solution, if any residual boundary potential for a neutral solid is ignored.

Although the potential difference between solid and solution cannot be measured directly, it is interesting that Kolthoff and Sanders[85] found the potentials of silver halide membrane electrodes to vary with halide ion (or silver ion) activity nearly as expected from (8-8). There are some complications, however. Overbeek[86] pointed out that Equation (8-7) states that the potential across the double layer is proportional to the charge transferred across it and thus implies that the electrical capacity of the double layer is a constant. More accurate adsorption experiments[87] indicate that the capacity of the double layer does not remain constant with varying amounts of adsorbed lattice ion (varying potential of the solid). Another difficulty is that the colloidal behavior of precipitates is influenced strongly by the solvent. A small amount of acetone shifts the point of zero charge of silver iodide far toward higher silver ion concentrations,[87] probably because of the effect of an oriented adsorbed layer of acetone, which changes the phase-boundary potential. Still another complication in attempts to calculate the potential change of the solid is that foreign ions other than lattice ions can be potential-determining to a greater or lesser extent. For example, Reyerson, Kolthoff, and Coad[88] found that lead ions act as potential-determining ions for barium sulfate. Citrate ion was actually more effective than sulfate ion in causing the precipitate to assume a negative charge. Similarly, silver iodide is peptized as a negative colloid by various ions other than iodide, such as bromide, chloride, cyanide, thiocyanate, and phosphate.[89]

8-6 THE ELECTRICAL DOUBLE LAYER AND COLLOID STABILITY

If lattice ions or other potential-determining ions are adsorbed on a solid surface, they may be regarded as belonging to the solid and imparting an electrical charge to it. For the sake of overall electrical neutrality, an equivalent number of oppositely charged ions (counterions) exist in solution, drawn to the charged surface by electrical attraction. The counterions and the adsorbed lattice ions form an electrical double layer. The closest counterions cannot be nearer the surface than a finite distance (inner Helmholtz plane[90]) that depends on the ionic radius.

To account for the behavior of colloids, several hypotheses have been suggested;[91-93] one simple, reasonable concept is schematically depicted in Figure (8-7).

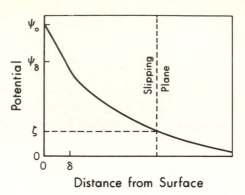

FIGURE 8-7   Potential as a function of distance from a charged surface.

As indicated in Figure (8-7), at the surface the potential is given as ψ_0, an unmeasurable quantity. A layer of counterions adjacent to the surface causes the potential to drop linearly with distance to a value ψ_δ at their outer boundary δ. Beyond the distance δ the counterions form a diffuse layer causing the potential to fall exponentially to zero. The theory of the diffuse double layer[92,93] takes into account the effect of thermal agitation; it is similar in principle to the calculation of the ionic atmosphere in the Debye-Hückel theory, the early developments of which it antedated by a decade. In fact, the quantity $1/\kappa$, which in the Debye-Hückel theory has the significance of the radius of the ionic atmosphere (Section 2-3), is a measure of the thickness of the double layer in the Gouy-Chapman theory.[92,93] The extent of the diffuse ion layer varies inversely with approximately the square root of the concentration of a particular electrolyte. Because the force of attraction increases with the square of the ionic charge, the double layer is much more compressed for a counterion of high charge than for one of low charge.

When an electric field is applied to a solution containing charged colloidal particles, the rate of motion of the particles is proportional to neither ψ_0 nor ψ_δ but to a smaller, measurable value called the *electrokinetic* or *zeta potential* (Figure 8-7). In the region near the particle surface there is a strongly oriented[94] layer of water a few molecules thick, or solvent molecules that tend to move as though they were part of the solid. A *slipping plane* is defined as the outer boundary of the attached water sheath, and the zeta potential is the potential between the slipping plane and the bulk of the solution. For a given surface potential the double layer can be increasingly compressed by increasing the electrolyte concentration and by using counterions of higher charge. Several possibilities are shown in Figure 8-8.

Figure 8-8 indicates schematically that the zeta potential decreases with decreasing surface charge, increasing electrolyte concentration, and increasing charge of the counterion. In some instances the sign of the zeta potential can be reversed. Strongly adsorbed substances, such as dyes of charge opposite that of the solid, often cause such a reversal. This effect is due to adsorption of more than an equivalent amount of

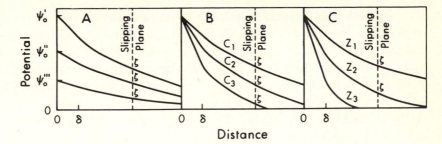

FIGURE 8-8 Schematic diagram of potential as a function of distance from a charged surface: A, effect of surface charge; B, effect of electrolyte concentration $(C_1 > C_2 > C_3)$; C, effect of charge of counterion $(Z_1 < Z_2 < Z_3)$.

oppositely charged material, resulting in a triple layer of charge. The diffuse layer of counterions is now of the same charge as the original surface, and the zeta potential undergoes a reversal in sign.

Coagulation of lyophobic colloids: Secondary stability   Verwey[95] distinguished between primary stability of colloids, which is imparted by the surface charge, and secondary stability, which pertains to the effective repulsion of colloidal particles. The primary stability is governed by the total potential ψ_0 of the solid, whereas the secondary stability is determined primarily by the zeta potential. For the existence of a stable sol the solid phase must have a primary charge, and the diffuse double layer must extend beyond the slipping plane. If most of the counterions are within a distance less than that of the slipping plane, then two colloidal particles can approach each other with little repulsion until they are near enough that their water sheaths coalesce and coagulation results.

Schulze[96] first showed that inorganic colloids are especially sensitive to electrolytes of high charge, and Hardy[97] pointed out that their stability is closely related to mobility in an electric field. The Schulze-Hardy rule states that the sensitivity of lyophobic colloids to coagulating electrolytes is governed by the charge of the ion opposite that of the colloid and that the sensitivity increases more rapidly than the charge of the ion.

The effects of various electrolytes are usually compared in terms of the *flocculation values*, or minimal concentrations (expressed in millimoles per liter), required to bring about coagulation. Flocculation values for nonspecific electrolytes can be interpreted in terms of electrostatic repulsion and van der Waals attraction. The van der Waals attractive forces vary with size and shape of the particles, but roughly speaking the force is appreciable between colloidal particles at distances of the order of magnitude of their own radii.[98] The significant feature is that flocculation occurs before the zeta potential reaches zero, that is, when it reaches a small critical value.

In some cases unusual flocculation behavior is observed as increasing concentrations of electrolyte are added. A small concentration of electrolyte brings about coagulation, a higher concentration causes peptization to a colloid of opposite charge, and this is finally flocculated by a still higher electrolyte concentration. When potential-determining ions are added (silver ions to negatively charged silver halide or hydroxide ions to a positively charged hydrous iron(III) oxide sol), the effect is due to the reversal of surface charge. The reversal can also be brought about by highly adsorbed counterions such as laurylpyridinium bromide on a negative silver iodide sol.[99]

Colloids can be protected from, or sensitized to, flocculation. If a lyophilic colloid is added to a lyophobic one, it tends to take on the characteristics of the lyophilic one and resists flocculation. One measure of the protective action has been defined by a *gold number*, the number of milligrams of hydrophilic colloid required to prevent flocculation of 10 ml of a colloidal solution of gold by the addition of 1 ml of 10% sodium chloride. The gold number for gelatin is 0.01, and for dextrin 20. The opposite effect, sensitization, is observed upon the addition of small amounts of lyophilic colloids to certain lyophobic colloids. When the lyophilic and lyophobic colloids are oppositely charged, for example in the addition of positively charged gelatin to negatively charged silver iodide, the effect is simply a mutual coagulation.

Sensitization occurs when an exclusively negative colloid like gum arabic or starch is added to a negative silver iodide sol.[100] Apparently when the repulsion of the two colloids is lowered by addition of electrolyte, an agglomeration occurs in which a lyophilic particle is attached to two or more lyophobic ones. If enough lyophilic colloid is added, the agglomerates are unstable, because the entire surfaces of the lyophobic particles are covered.

A practical problem sometimes arises when a flocculated colloid is filtered and then washed with water. During washing the concentration of ions in the double layer decreases, and the flocculated particles may regain enough zeta potential to repel each other and hence become dispersed once more, or peptized. A simple preventive often applied is to wash with a dilute electrolyte that is volatile upon ignition (such as nitric acid or ammonium nitrate). Peptization does not always occur. In the first place, recrystallization of the primary particles (Section 8-4) causes a diminution and perfecting of the surface, with the result that adsorbed potential-determining ions are released. In addition, coarse particles may grow at the expense of fine ones, or colloidal particles may merge through partial recrystallization.

REFERENCES

1   P. P. VON WEIMARN: *Chem. Rev.*, **2**:217 (1926).

2   W. OSTWALD: *Z. Phys. Chem.* (Leipzig), **22**:289 (1897).

3   A. C. ZETTLEMOYER (ed.): "Nucleation," Dekker, New York, 1969.

4 R. BECKER and W. DÖRING: *Ann. Phys.* (Leipzig), **24**:719 (1935).

5 W. F. PFANN in "Solid State Physics," F. Seitz and D. Turnbull (eds.), vol. 4, p. 502, Academic, New York, 1957.

6 J. A. CHRISTIANSEN and A. E. NIELSEN: *Z. Elektrochem.*, **56**:465 (1952); *Acta Chem. Scand.*, **5**:673, 674 (1951).

7 A. G. WALTON: "The Formation and Properties of Precipitates," Wiley, New York, 1967.

8 A. R. HARGENS: *Science*, **176**:184 (1972).

9 B. V. ENÜSTÜN and J. TURKEVICH: *J. Amer. Chem. Soc.*, **82**:4502 (1960).

10 A. E. NIELSEN: *J. Colloid Sci.*, **10**:576 (1955).

11 V. K. LA MER: *Ind. Eng. Chem.*, **44**:1270 (1952).

12 R. A. JOHNSON and J. D. O'ROURKE: *J. Amer. Chem. Soc.*, **76**:2124 (1954).

13 V. K. LA MER and R. H. DINEGAR: *J. Amer. Chem. Soc.*, **73**:380 (1951).

14 F. C. COLLINS and J. P. LEINEWEBER: *J. Phys. Chem.*, **60**:389 (1956).

15 A. E. NIELSEN: *Acta Chem. Scand.*, **11**:1512 (1957).

16 R. B. FISCHER and T. B. RHINEHAMMER: *Anal. Chem.*, **25**:1544 (1953); **26**:244 (1954).

17 E. J. BOGAN and H. V. MOYER: *Anal. Chem.*, **28**:473 (1956).

18 J. D. O'ROURKE and R. A. JOHNSON: *Anal. Chem.*, **27**:1699 (1955).

19 A. G. WALTON: *Mikrochim. Acta*, 422 (1963).

20 B. VONNEGUT: *J. Colloid Sci.*, **3**:563 (1948).

21 D. H. KLEIN and J. A. DRIY: *Talanta*, **13**:289 (1966).

22 D. MEALOR and A. TOWNSHEND: *Talanta*, **13**:1191 (1966).

23 S. THOMPSON and L. GORDON: *Talanta*, **14**:137 (1967).

24 H. B. WEISER and W. O. MILLIGAN: *Chem. Rev.*, **25**:1 (1939), "Advances in Colloid Science," vol. 1, p. 227, E. O. Kraemer (ed.), Interscience, New York, 1942.

25 B. VAN'T RIET and I. M. KOLTHOFF: *J. Phys. Chem.*, **64**:1045 (1960).

26 K. TAKIYAMA and L. GORDON: *Talanta*, **10**:1165 (1963).

27 R. B. FISCHER: *Anal. Chim. Acta*, **22**:501 (1960).

28 E. J. MEEHAN and G. CHIU: *J. Amer. Chem. Soc.*, **86**:1443 (1964).

29 WALTON,[7] p. 9.

30 B. VONNEGUT and H. CHESSIN: *Science*, **174**:945 (1971); *J. Amer. Chem. Soc.*, **93**:4964 (1971).

31 R. B. FISCHER: *Anal. Chem.*, **32**:1127 (1960).

32 WALTON,[7] p. 30.

33 R. MARC: *Z. Phys. Chem.* (Leipzig), **79**:71 (1912); *Z. Elektrochem.*, **18**:161 (1912); also earlier papers.

34 M. EIGEN: *Pure Appl. Chem.*, **6**:97 (1963).

35 D. TURNBULL: *Acta Met.*, **1**:684 (1953).

36 C. W. DAVIES and G. H. NANCOLLAS: *Trans. Faraday Soc.*, **51**:823 (1955).

37 I. M. DAWSON: *Proc. Roy. Soc.*, Ser. A, **214**:72 (1952).

38 F. C. FRANK: *Discuss. Faraday Soc.*, **5**:48 (1949).

39 G. A. SOMORJAI: "Principles of Surface Chemistry," p. 4, Prentice-Hall, Englewood Cliffs, N.J., 1972.

40 WALTON,[7] p. 50.

41 J. W. GIBBS: "The Collected Works of J. Willard Gibbs," vol. 1, p. 325, Longmans, New York, 1928.

42 M. VOLMER: *Z. Phys. Chem.* (Leipzig), **102**:267 (1922); *Z. Elektrochem.*, **35**:555 (1929).

43 W. KOSSEL: *Nachr. Akad. Wiss. Göttingen, Math.-Phys. Kl.*, 135 (1927).

44 I. N. STRANSKI: *Z. Phys. Chem.* (Leipzig), **136**:259 (1928).

45 M. VOLMER and W. SCHULTZE: *Z. Phys. Chem.* (Leipzig), **156A**:1 (1931).

46 R. L. FULLMAN: *Sci. Amer.*, **192**(3):74 (1955). Perhaps the all-time record for discrepancy between hypothesis and experiment.

47 A. G. WALTON in "Dispersion of Powders in Liquids," G. D. PARFITT (ed.), p. 156, Elsevier, Amsterdam, 1969.

48 K. A. JACKSON: *Ind. Eng. Chem.*, **57**(12):28 (1965).

49 WALTON,[7] p. 165.

50 I. M. KOLTHOFF, E. B. SANDELL, E. J. MEEHAN, and S. BRUCKENSTEIN: "Quantitative Chemical Analysis," 4th ed., p. 214, Macmillan, London, 1969.

51 I. M. KOLTHOFF and D. R. MOLTZAU: *Chem. Rev.*, **17**:293 (1935).

52 W. M. FISCHER: *Z. Anorg. Allg. Chem.*, **153**:62 (1926).

53 T. HOLTH: *Anal. Chem.*, **21**:1221 (1949).

54 I. M. KOLTHOFF and E. A. PEARSON: *J. Phys. Chem.*, **36**:549 (1932).

55 I. M. KOLTHOFF and F. S. GRIFFITH: *J. Phys. Chem.*, **42**:531 (1938).

56 I. M. KOLTHOFF and R. MOLTZAU: *J. Phys. Chem.*, **40**:779 (1936).

57 I. M. KOLTHOFF: *Analyst* (London), **77**:1000 (1952).

58 M. KAHLWEIT in "Proceedings of the 6th International Symposium on the Reactivity of Solids, 1968," J. W. MITCHELL, R. C. DEVRIES, R. W. ROBERTS, and P. CANNON (eds.), p. 93, Wiley, New York, 1969.

59 I. M. KOLTHOFF and F. T. EGGERTSEN: *J. Amer. Chem. Soc.*, **63**:1412 (1941).

60 I. M. KOLTHOFF and C. ROSENBLUM: *J. Amer. Chem. Soc.*, **58**:121 (1936).

61 I. M. KOLTHOFF and H. C. YUTZY: *J. Amer. Chem. Soc.*, **59**:1215, 1634 (1937).

62 I. M. KOLTHOFF and G. E. NOPONEN: *J. Amer. Chem. Soc.*, **60**:499, 505 (1938).

63 I. M. KOLTHOFF and A. S. O'BRIEN: *J. Amer. Chem. Soc.*, **61**:3414 (1939).

64 I. M. KOLTHOFF and W. VON FISCHER: *J. Amer. Chem. Soc.*, **61**:191 (1939).

65 L. G. CONTI, R. D'ALESSANDRO, and V. DI NAPOLI: *J. Phys. Chem.*, **75**:350 (1971).

66 I. M. KOLTHOFF and W. M. MACNEVIN: *J. Amer. Chem. Soc.*, **58**:499, 725 (1936).

67 I. M. KOLTHOFF and F. T. EGGERTSEN: *J. Phys. Chem.*, **46**:458 (1942).

68 I. M. KOLTHOFF and F. T. EGGERTSEN: *J. Amer. Chem. Soc.*, **61**:1036 (1939).

69 I. M. KOLTHOFF and A. S. O'BRIEN: *J. Chem. Phys.*, **7**:401 (1939).

70 I. M. KOLTHOFF and R. C. BOWERS: *J. Amer. Chem. Soc.*, **76**:1503 (1954).

71 I. SHAPIRO and I. M. KOLTHOFF: *J. Chem. Phys.*, **15**:41 (1947).

72 I. SHAPIRO and I. M. KOLTHOFF: *J. Phys. Colloid Chem.*, **52**:1319 (1948).

73 I. M. KOLTHOFF and W. M. MACNEVIN: *J. Phys. Chem.*, **44**:921 (1940).

74 I. SHAPIRO and I. M. KOLTHOFF: *J. Amer. Chem. Soc.*, **72**:776 (1950).

75 G. TAMMANN and A. SWORYKIN: *Z. Anorg. Allg. Chem.*, **176**:46 (1928).

76 I. M. KOLTHOFF and R. C. BOWERS: *J. Amer. Chem. Soc.*, **76**:1510 (1954).

77 E. B. SANDELL and I. M. KOLTHOFF: *J. Phys. Chem.*, **37**:153 (1933).

78 I. M. KOLTHOFF and B. MOSKOVITZ: *J. Phys. Chem.*, **41**:629 (1937).

79 I. M. KOLTHOFF and L. G. OVERHOLSER: *J. Phys. Chem.*, **43**: 909 (1939).

80 J. A. W. VAN LAAR quoted in "Colloid Science," H. R. KRUYT (ed.), vol. 1, p. 161, Elsevier, Amsterdam, 1952.

81 E. J. W. VERWEY and H. R. KRUYT: *Z. Phys. Chem.* (Leipzig), **A167**:149 (1934); A. BASINSKI: *Rec. Trav. Chim. Pays-Bas*, **60**:267 (1941).

82 E. LANGE and P. W. CRANE: *Z. Phys. Chem.* (Leipzig), **A141**:225 (1929).

83 R. RUYSSEN and R. LOOS: *Nature*, **162**:741 (1948).

84 E. LANGE and R. BERGER: *Z. Elektrochem.*, **36**:171 (1930).

85 I. M. KOLTHOFF and H. L. SANDERS: *J. Amer. Chem. Soc.*, **59**:416 (1937).

86 J. TH. G. OVERBEEK: in Kruyt.[80]

87 E. L. MACKOR: *Rec. Trav. Chim. Pays-Bas*, **70**:747, 763 (1951).

88 L. H. REYERSON, I. M. KOLTHOFF, and K. COAD: *J. Phys. Colloid Chem.*, **51**:321 (1947).

89 H. B. WEISER: "Inorganic Colloid Chemistry," vol. 3, p. 115, Wiley, New York, 1938.

90 H. R. THIRSK and J. A. HARRISON: "A Guide to the Study of Electrode Kinetics," p. 2, Academic, London, 1972.

91 J. PERRIN: *J. Chim. Phys.*, **2**:601 (1904).

92 G. GOUY: *J. Phys.* (Paris), (4)**9**:457 (1910); *Ann. Phys.* (Paris), (9)7:129 (1917).

93 D. L. CHAPMAN: *Phil. Mag.*, (6)**25**:475 (1913).

94 D. C. GRAHAME: *J. Amer. Chem. Soc.*, **79**:2093 (1957).

95 E. J. W. VERWEY: *Chem. Rev.*, **16**:363 (1935).

96 H. SCHULZE: *J. Prakt. Chem.* (2)**25**:431 (1882); **27**:320 (1883).

97 W. B. HARDY: *Proc. Roy. Soc.* (London), **66**:110 (1900).

98 J. TH. G. OVERBEEK in Kruyt,[80] p. 264.

99 A. LOTTERMOSER and R. STEUDEL: *Kolloid-Z.*, **82**:319 (1938); **83**:37 (1938).

100 H. R. KRUYT and C. W. HORSTING: *Rec. Trav. Chim. Pays-Bas*, **57**:737 (1938); J. TH. G. OVERBEEK: *Chem. Weekbl.*, **35**:117 (1938).

9

CONTAMINATION OF PRECIPITATES

Two types of precipitate contamination have been defined:[1] (*1*) *coprecipitation*, in which the main precipitate and the impurity come down together; and (*2*) *post-precipitation*, in which the main precipitate may be initially pure, but is contaminated by a second substance later. Postprecipitation usually occurs from supersaturated solutions. It is dealt with in Section 8-3 and is not considered further here.

It should be noted that coprecipitation, by definition, includes only the contamination of a precipitate by *normally soluble substances*. Thus the fact that two substances are carried down together is insufficient to classify the phenomenon as coprecipitation. If a trace of beryllium oxide is brought down quantitatively with a large amount of hydrous aluminum oxide under conditions such that both are insoluble, we speak of *gathering* rather than coprecipitation; when the relative proportions of the two substances are not extreme, the term *simultaneous precipitation* is appropriate. Two broad classes of coprecipitation are recognized: (*1*) *Adsorption* is the carrying down of impurities on the surface of particles. Even when impurities on the surfaces of primary colloidal particles become a sort of internal surface of coagulated colloids (Section 8-6), the phenomenon is still considered an adsorption process. (*2*) *Occlusion*, the second class, is used here to denote the carrying down of impurities in the interior of primary particles, by whatever mechanism it may occur. Two such mechanisms are solid-solution formation and growth of precipitate around adsorbed ions. *Inclusion*, or entrainment, of pockets of mother liquor occurs frequently in the crystallization of soluble salts, but is of relatively little importance in analytical precipitates. Inclusion of mother liquor in the interstices of colloidal precipitates is another type of gross entrainment; it can be minimized by proper choice of precipitation conditions to yield a more compact form of precipitate.

9-1 ADSORPTION

Several types of adsorption processes have been classified by Kolthoff.[2]

Adsorption of potential-determining ions   Ideally, as seen in Section 8-5, lattice ions held at the surface of a precipitate[3] containing the same ions are adsorbed in accordance with an equation of the form

$$d\frac{X}{m} = k\,d\ln C_i$$

or, in integrated form,

$$\frac{X}{m} = k\ln\frac{C_i}{C_i^0} \qquad (9\text{-}1)$$

where X is the amount adsorbed (in milligrams, grams, or moles), m is weight of precipitate, C_i is concentration of lattice ion in solution, C_i^0 is isoelectric concentration (corresponding to $X = 0$ when $C_i = C_i^0$), and k is a constant.

Actually, adsorption of lattice ions is far from ideal (Section 8-5), owing to the nonequivalence of perfect and imperfect crystal surfaces and the interference of nonlattice ions that may also be potential-determining.

Adsorption by ion exchange   Exchange adsorption[2] can take place through exchange of either lattice ions or counterions. An example[4] of exchange of lattice ions is

$$\underset{\text{surf}}{\text{BaSO}_4} + \underset{\text{soln}}{\text{Pb}^{++}} \rightleftharpoons \underset{\text{surf}}{\text{PbSO}_4} + \underset{\text{soln}}{\text{Ba}^{++}} \qquad (9\text{-}2)$$

which is described by the equilibrium constant

$$K_{eq} = \frac{[\text{Ba}^{++}]_{\text{soln}}\,[\text{PbSO}_4]_{\text{surf}}}{[\text{BaSO}_4]_{\text{surf}}\,[\text{Pb}^{++}]_{\text{soln}}} \qquad (9\text{-}3)$$

Hence X/m is proportional to $[\text{Pb}^{++}]/[\text{Ba}^{++}]$. An example of exchange of counterions is

$$\text{AgI}\cdot\text{I}^-\,\vdots\,\text{Na}^+ + \text{NH}_4^{\,+} \rightleftharpoons \text{AgI}\cdot\text{I}^-\,\vdots\,\text{NH}_4^{\,+} + \text{Na}^+$$

which is described by an equilibrium expression similar to (9-3).

With the aid of the information in Section 8-5, rules may be formulated for predicting which of several ions in solution will be preferentially adsorbed as counterions.

1 Of two ions present at equal concentration, the ion of higher charge is preferentially adsorbed.

2 Of two ions of equal charge, the ion present at higher concentration is preferentially adsorbed.

3 Of two ions of equal charge at the same concentration, the ion most strongly attracted by the lattice ions is preferentially adsorbed (Paneth-Fajans-Hahn adsorption rule).[5-9] Stronger interionic attraction between adsorbed lattice ions and counterions is indicated by (*a*) lower solubility, (*b*) lesser degree of dissociation, (*c*) greater covalency, or (*d*) greater electrical polarizability of the anion and greater polarizing character of the cation.

Thus, calcium ion is adsorbed preferentially over magnesium ion on negatively charged barium sulfate, as expected because calcium sulfate is less soluble than magnesium sulfate. Silver acetate is more strongly adsorbed on silver iodide than is silver nitrate, an observation consistent with the lower solubility and greater covalency of silver acetate. Dye anions are strongly adsorbed on positively charged silver halides, in line with the covalency of the silver dye salts. Hydrogen sulfide is strongly adsorbed on metal sulfides as a result of its weakly ionized character. It should be mentioned here that the Paneth-Fajans-Hahn rule often is applied erroneously to situations involving solid solution rather than adsorption.

Molecular or ion-pair adsorption The adsorption of potassium bromate on the surface of barium sulfate appears to involve the simultaneous occupation of adjacent sites on the crystal surface by potassium ions and bromate ions; this is called an *equivalent adsorption*[10] because equal amounts of the two ions are adsorbed. Such adsorptions, and in fact many molecular adsorption processes, follow an empirical equation known as the *Freundlich adsorption isotherm*:

$$\frac{X}{m} = kC^{1/n} \qquad (9\text{-}4)$$

where k and n are constants. The constant n is usually of the order of 2 to 4. Equation (9-4) represents a parabola of order n and so does not describe those saturation processes in which X/m reaches a limiting value at high concentrations.

Monolayer adsorption If the adsorbed substance can occupy only a limited number of surface sites, and if the process does not continue far enough to form several molecular layers, a condition of surface saturation is reached at high concentrations of adsorbed substance. This behavior is described by the *Langmuir adsorption isotherm*,

$$\frac{X}{m} = \frac{k_1 C}{1 + k_2 C} \qquad (9\text{-}5)$$

where X/m reaches the saturation value k_1/k_2 as C approaches infinity. For small values of C ($k_2 C \ll 1$), $X/m \simeq k_1 C$; that is, the amount adsorbed is proportional to concentration. For intermediate values of C the Langmuir isotherm expresses variation of X/m with a fractional power of concentration, and over a limited range the adsorption can be described equally well by the Freundlich isotherm. An example of monolayer adsorption is the adsorption of dyes, which reaches a saturation value at low concentrations and thus is useful in measuring the specific surface of precipitates.

Several other types of adsorption isotherms have been described; one applicable to fractional monolayers in which there is linear variation between the amount adsorbed and log C is called *Temkin adsorption*.

9-2   CONTAMINATION BY ADSORPTION

Adsorption is the principal source of contamination of precipitates that have large surfaces, for example, flocculated colloids (metal sulfides, silver halides, hydrous oxides). The extent of adsorption may be relatively small, as it usually is with silver halides, or severe, as it often is with hydrous oxides.

A silver halide precipitate brought down with excess alkali halide will carry down halide ions adsorbed as lattice ions and alkali metal ions as counterions. Washing with dilute nitric acid can largely displace the alkali metal ions by counterion exchange. The adsorbed hydrogen halide volatilizes upon ignition of the precipitate. If silver ion is the adsorbed lattice ion, washing is not effective for removing the adsorbed silver salt, which then does not volatilize upon ignition. The amount of adsorbed material can be significantly decreased by aging during digestion, when recrystallization causes a marked decrease in total surface and also creates a more nearly perfect surface structure that has a smaller tendency to adsorb lattice ions.

Coprecipitation with the hydrous oxides, such as of iron(III) and aluminum, occurs by adsorption and possibly also by compound formation. The precipitates, coming down in either amorphous or finely crystalline form with extensive surface, adsorb large amounts of water and adsorb hydroxide ions as potential-determining ions. Figure 9-1 illustrates[11] the effect of varying the concentration of ammonium chloride and ammonium hydroxide on the amount of coprecipitation of divalent metal ions with hydrous iron(III) oxide. When the concentration of ammonium chloride is increased at a constant ammonia concentration, the adsorption is decreased

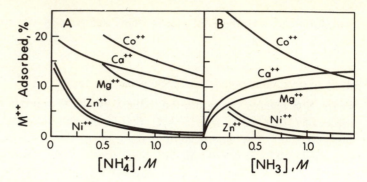

FIGURE 9-1 Adsorption of divalent metal ions on hydrous Fe(III) oxide: *A*, as a function of NH_4Cl concentration with a NH_3 concentration of 0.9 *M*, except 0.7 *M* for zinc; *B*, as a function of NH_3 concentration with a NH_4Cl concentration of 1 *M*. (*From Kolthoff and Overholser.*[11])

because (*1*) the primary adsorption of hydroxyl ion is less at the lower pH value and (*2*) ammonium ion competes with the foreign metal ions for counterion adsorption. When the concentration of ammonia is increased at a constant ammonium chloride level, adsorption of calcium and magnesium increases, owing to greater primary adsorption of hydroxyl ion at the higher pH. On the other hand, the adsorption of cobalt, nickel, and zinc ions decreases, because these metals form complexes with ammonia. It was shown experimentally that zinc did not carry down an appreciable amount of ammonia and was therefore adsorbed as the aquated ion. The adsorption of cobalt(II) is unusually extensive, perhaps because its coordination sphere is not completely occupied by ammonia at the concentrations used or it is partly oxidized to cobalt(III) in the presence of ammonia. As mentioned in Section 8-4, coprecipitation of copper, zinc, nickel, or cobalt actually increases upon digestion at 98°C. The formation of divalent metal ferrites was suggested.

Recent studies of coprecipitation with the hydrous oxides have presented further evidence for definite 1:1 compound formation between Zn^{++} or Cu^{++} and iron(III) hydroxide.[12, 13] X-ray information indicated a chemical compound with Cu^{++} and tin(IV) hydrous oxide, presumably with the formula $Cu_3SnO_5 \cdot nH_2O$.[14] In each case the first step was suggested to be adsorption, followed by reaction.

Specific reversible adsorption of up to 0.28 mole of Pb^{++} per mole of iron in colloidal hydrous iron(III) oxide has been found[15] to be accompanied by the release of 1.6 hydrogen ions per mole of Pb^{++} adsorbed at pH 6.

Evidently a low ammonia concentration (0.003 *M*) and a high ammonium chloride concentration (1.5 to 2 *M*) are favorable for efficient separation of metal ions that do not form ammine complexes (Ca^{++}, Mg^{++}). In fact, a single precipitation has been found to be effective under these conditions.[11] To separate iron(III) from copper, zinc, and nickel, high concentrations of both ammonia and ammonium chloride produced excellent results in a single precipitation.

9-3   OCCLUSION BY ADSORPTION

Kolthoff[1] proposed the concept that entrapment of foreign ions, involving growth of precipitate around adsorbed ions, is an important source of contamination, particularly of crystalline precipitates such as barium sulfate and calcium oxalate. Essential to this concept is that occlusion is not an equilibrium process and that recrystallization during aging can effect purification. The foreign ions represent *lattice imperfections* unless they are actually held in solid solution.

Surfaces on which adsorption of substances in solution occurs include those of growing crystals. If as a result of continuing growth the adsorbed material is not desorbed, it will be retained inside a crystal in a position where it is no longer on a surface. This type of coprecipitation is generally more important than adsorption in that it affects all precipitates, not only those with large surfaces. Since such occlusion occurs by the mechanism of adsorption, it can be qualitatively predicted by the same rules. There is one additional factor: the order in which reagents are mixed can be expected to influence the extent of occlusion of particular types of ions. When a salt of the cation of a precipitate is added slowly to a solution of a salt of the anion, foreign cation coprecipitation predominates. Meanwhile, occlusion of foreign anions is minimized, because under these circumstances the concentration of precipitate anions during crystal growth is relatively high. Similarly, occlusion of foreign cations can be minimized if the order of addition is selected so that precipitate cations in solution are in excess during precipitation and crystal growth.

The effect of the order of mixing on anion occlusion is illustrated by the data of Weiser and Sherrick,[16] who found that the occlusion of chloride was decreased from 15.8 to 1.25 millimoles per mole of barium sulfate by adding barium to sulfate rather than the reverse. Cation occlusion can be decreased by reverse precipitation, that is, by adding the sulfate to the acidified barium chloride solution,[17] thereby increasing the occlusion of the chloride. Rieman and Hagen[18] made a comparison of various methods of precipitating barium sulfate; they concluded that the method of Hintz and Weber,[19] involving the rapid addition of barium chloride to the sulfate solution, gave the best results for sulfate, particularly in the presence of sodium chloride. The good results obtained by this method may be partly due to a compensation of errors, high results following from some types of occlusion and low results from others.

9-4   OCCLUSION BY SOLID-SOLUTION FORMATION

One of the ions of a precipitate can be replaced by a foreign ion of the same charge, provided the ions do not differ in size by more than about 5% and provided the two salts crystallize in the same system.[20] Thus silver chloride and silver bromide form a complete series of solid solutions[21] by isomorphous replacement of bromide by chloride.

If pure silver bromide is shaken with a solution containing chloride ions, an equilibrium is set up:

$$Ag(Br, Cl) \rightleftharpoons Ag^+ + (Br^- \text{ or } Cl^-)$$

This may be regarded as a distribution equilibrium of the two halide ions between solid and solution:

$$AgBr + Cl^- \rightleftharpoons AgCl + Br^- \qquad (9\text{-}6)$$
$$\text{solid} \quad\quad \text{soln} \quad\quad \text{solid} \quad\quad \text{soln}$$

The equilibrium constant for (9-6) may be called a distribution ratio D:

$$K_{eq} = \frac{[Br^-]_{soln} N_{AgCl}}{[Cl^-]_{soln} N_{AgBr}} = D \qquad (9\text{-}7)$$

where N stands for mole fraction in the solid.

To express theoretically the relations involved, we follow the treatment given by Flood[22] and by Flood and Bruun.[23] The work of Vaslow and Boyd[24] may be consulted for a more complete thermodynamic consideration of silver chloride–silver bromide equilibria. An equation for the distribution ratio D corresponding to Equilibrium (9-6) can be obtained in terms of solubility products. First, consider the solid to be an impure silver bromide, with chloride as a foreign ion. Thus,

$$\frac{a_{Ag^+} a_{Br^-}}{a_{AgBr}} = K_{eq,1} \qquad \text{or} \qquad a_{Ag^+} a_{Br^-} = K_{eq,1} a_{AgBr} \qquad (9\text{-}8)$$

But, since in the limit this equation should hold for pure silver bromide ($a_{AgBr} = 1$), we see that $K_{eq.1} = K_{sp,AgBr}$. Similarly, consider the precipitate to be an impure silver chloride. Then

$$a_{Ag^+} a_{Cl^-} = K_{eq,2} a_{AgCl} \qquad (9\text{-}9)$$

where $K_{eq,2} = K_{sp,AgCl}$. Dividing (9-8) by (9-9) to eliminate the silver ion activity, we have

$$\frac{a_{Br^-}}{a_{Cl^-}} = \frac{K_{sp,AgBr}}{K_{sp,AgCl}} \frac{a_{AgBr}}{a_{AgCl}} \qquad (9\text{-}10)$$

The activities of the solids may be replaced by mole fractions multiplied by activity coefficients:

$$\frac{a_{Br^-}}{a_{Cl^-}} = \frac{K_{sp,AgBr}}{K_{sp,AgCl}} \frac{N_{AgBr} \gamma_{AgBr}}{N_{AgCl} \gamma_{AgCl}} \qquad (9\text{-}11)$$

The activity ratio in solution may be replaced to a close approximation by the concentration ratio, because the activity coefficients of two ions of the same charge and similar size will be nearly equal.[25]

For an *ideal solid solution*, in which the activity coefficient ratio $\gamma_{AgBr}/\gamma_{AgCl}$ is unity, (9-11) may be rewritten

$$\frac{K_{sp,AgBr}}{K_{sp,AgCl}} = \frac{[Br^-] N_{AgCl}}{[Cl^-] N_{AgBr}} \qquad (9\text{-}12)$$

which is identical to the distribution-ratio expression (9-7), provided that the distribution ratio D is equal to the ratio $K_{sp,AgBr}/K_{sp,AgCl}$ (or 2.75×10^{-3} as written).

Although experimental distribution ratios are sometimes of the same order of magnitude as the ratio of solubility products, they often disagree widely; moreover, D usually varies with the composition of the solid phase, indicating that activities are not directly proportional to concentration in solid solution. Thus with 92 mole % silver chloride the value[26] of D was found to be 2×10^{-3}, with about 15 mole % silver chloride it was 4×10^{-3}, and with[24] 99.9+ mole % silver chloride it was about 5×10^{-3}. Nevertheless, in the absence of experimental data an expression similar to (9-12) serves as a useful guide for estimating the possible extent of coprecipitation due to solid-solution formation.

Another important factor that governs the extent of contamination is the rate of attainment of solid-solution equilibrium. Thus the equilibrium described by (9-12) may be attained rapidly or slowly. At one extreme the precipitate could undergo recrystallization so rapidly that, as the solution composition changes during reaction, the precipitate is at all times homogeneous and in equilibrium with the particular composition at that instant. At the other extreme it could be imagined that every infinitesimal increment of solid is in equilibrium with the composition at the instant of its formation and that it does not undergo recrystallization. At this extreme the continually changing composition of the solid solution would result in its being heterogeneous. In this situation the first bit of precipitate, having come down from a solution rich in bromide, would be relatively pure silver bromide. Therefore, the average composition of the heterogeneous precipitate would be purer in silver bromide than the homogeneous precipitate.

The quantitative relations involved in heterogeneous solid-solution formation were considered by Doerner and Hoskins,[27] who derived the equilibrium expression. For silver bromide and chloride the two extreme situations are represented by

$$\frac{[Br^-]_f}{[Cl^-]_f} = D \frac{N_{AgBr}}{N_{AgCl}} \qquad \text{homogeneous} \qquad (9\text{-}13)$$

$$\lambda \log \frac{[Br^-]_f}{[Br^-]_i} = \log \frac{[Cl^-]_f}{[Cl^-]_i} \qquad \text{heterogeneous} \qquad (9\text{-}14)$$

where the subscripts f and i refer to final and initial concentrations in solution. The coefficient λ is identical to D; the different symbols are commonly used to distinguish between the two extreme types of equilibrium. As (9-13) and (9-14) are written, D and λ are smaller than unity for the example of chloride and bromide. In this situation, less chloride is precipitated under heterogeneous than homogeneous conditions. On the other hand, for carrier precipitations in which a trace component is to be collected, D or λ should be greater than unity. The enrichment in this instance will be greater under heterogeneous than under homogeneous conditions. Doerner and Hoskins[27]

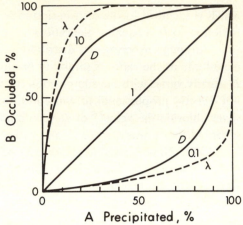

FIGURE 9-2   Occlusion of B by solid-solution formation. Solid lines, homogeneous solid solution with $D = 0.1$ or 10. Dotted lines, heterogeneous solid solution with $\lambda = 0.1$ or 10. The straight line corresponds to λ or $D = 1$. (*From Klein.*[28])

also discussed the collection of radium by fractional crystallization. Figure 9-2 indicates generally[28] the extent of occlusion to be expected under conditions of heterogeneous or homogeneous solid-solution formation.

Table 9-1 shows the experimental results for precipitation of bromide from equimolar mixtures of bromide and chloride determined by Kolthoff and Eggertsen[29] compared with the values calculated on the assumption of homogeneous and heterogeneous solid solutions. It appears that in the absence of aluminum, when the precipitate remained colloidally dispersed, homogeneous distribution equilibrium was

Table 9-1   EFFECT OF SOLID-SOLUTION FORMATION ON PRECIPITATION OF BROMIDE FROM EQUIMOLAR BROMIDE-CHLORIDE SOLUTION†

| | Bromide left in solution, % | | | |
| | Calculated | | Found | |
Ag^+ added, %‡	Homogeneous§	Heterogeneous¶	Al(III) absent	Al(III) present (3.7 min)
98	5.90	...	5.82	...
99	5.33	1.99	5.30	...
100	4.80	1.1	4.81	2.18
101	4.33	0.40	4.26	1.52
102	3.90	0.03	3.87	1.19

† From Kolthoff and Eggertsen.[29]
‡ Based on bromide equivalence point.
§ $D = 2.54 \times 10^{-3}$.
¶ $\lambda = 2.54 \times 10^{-3}$.

reached, whereas in the presence of aluminum ions added as a coagulant the theoretical heterogeneous composition was approached. This limit cannot be reached in practice, where finite rather than infinitesimal increments of precipitate are involved. Recrystallization through aging of the precipitate also prevents its isolation from the solution. Interestingly, when the precipitation was carried out in the presence of a dye that inhibited aging, coprecipitation of chloride was more pronounced; local excesses of silver carried down chloride, which was trapped because dye adsorption inhibited recrystallization.

In contrast to this rapid attainment of equilibrium, Kolthoff and Noponen[30] found equilibrium to be slowly attained between solid solutions of barium and lead sulfates and the aqueous solution. Therefore, the amount of contamination depended on the order of mixing of the reagents. When sulfate was added to an equimolar mixture of barium and lead ions in the presence of acetate to increase the solubility of lead sulfate, the amount of coprecipitation was relatively low. With addition of barium to a mixture of lead and sulfate in the presence of acetate, coprecipitation was almost 10 times as high. These observations would be expected on the basis of the relative concentrations of barium ion in solution under the two conditions of precipitation. On aging, the two types of precipitate slowly approached the same composition from opposite sides. Other examples of solid-solution formation of analytical interest are $MgNH_4PO_4$ and $MgKPO_4$; $MnNH_4PO_4$ and $ZnNH_4PO_4$; $ZnHg(SCN)_4$ and $CuHg(SCN)_4$; $BaSO_4$ and $SrSO_4$; HgS and ZnS, MnS, CdS, or PbS.

EXAMPLE 9-1 The influence of order of addition on the extent of occlusion can be demonstrated by precipitating potassium perchlorate in the presence of permanganate as follows. First, slowly add 1.5 M perchloric acid to an equal volume of 1.5 M potassium nitrate containing a few drops of 0.1 M potassium permanganate. Then destroy the excess potassium permanganate in solution by adding 1 or 2 ml of sodium hydrogen sulfite solution. A precipitate formed in this manner will be intensely colored by occluded permanganate. Second, add 1.5 M potassium nitrate to 1.5 M perchloric acid containing permanganate (reverse order of addition), again followed by addition of sodium hydrogen sulfite to destroy the permanganate in solution. The potassium perchlorate precipitate now will be almost white, since permanganate adsorption during crystal growth was largely prevented. ////

Another type of solid-solution formation is encountered when two ions as a pair can replace two other ions in a crystal lattice. Grimm and Wagner[31] pointed out that such a twofold replacement can occur if the two salts have the same type of chemical structure and crystallize in the same type of crystal with lattice dimensions not too dissimilar. An apt example is barium sulfate, which can form solid solutions with potassium permanganate; x-ray diffraction has revealed that solid-solution formation actually occurs.[32] As has also been shown, solid-solution equilibrium can be reached from both directions.[33] Other examples of ion-pair replacements are $BaSeO_4 + KMnO_4$, $BaCrO_4 + KMnO_4$, $BaSO_4 + KBF_4$.

In conclusion, it may be stated that solid-solution formation represents an exceptionally troublesome situation in analysis in that substantial contamination may be present even at equilibrium. Aging can help to reach homogeneous equilibrium, but whether the purity of the precipitate is better or worse after aging depends on the direction from which equilibrium is approached.

REFERENCES

1 I. M. KOLTHOFF: *J. Phys. Chem.*, **36**:860 (1932).

2 I. M. KOLTHOFF: *J. Phys. Chem.*, **40**:1027 (1936).

3 K. FAJANS and W. FRANKENBURGER: *Z. Phys. Chem.* (Leipzig), **105**:255 (1923).

4 I. M. KOLTHOFF and W. M. MACNEVIN: *J. Amer. Chem. Soc.*, **58**:499 (1936).

5 F. PANETH: *Phys. Z.*, **15**:924 (1914); K. HOROWITZ and F. PANETH: *Z. Phys. Chem.* (Leipzig), **89**:513 (1915).

6 K. FAJANS and K. VON BECKERATH: *Z. Phys. Chem.* (Leipzig), **97**:478 (1921).

7 K. FAJANS and G. JOOS: *Z. Phys.* **23**:1 (1924).

8 O. HAHN: *Ber Deut. Chem. Ges.*, **59B**:2014 (1926); *Naturwissenschaften*, **14**:1196 (1926).

9 O. HAHN: "Applied Radiochemistry," Cornell University Press, Ithaca, N.Y., 1936.

10 L. H. REYERSON, I. M. KOLTHOFF, and K. COAD: *J. Phys. Colloid Chem.*, **51**:321 (1947).

11 I. M. KOLTHOFF and B. MOSKOVITZ: *J. Phys. Chem.*, **41**:629 (1937); I. M. KOLTHOFF and L. G. OVERHOLSER: *J. Phys. Chem.*, **43**:767, 909 (1937).

12 N. A. RUDNEV, G. I. MALOFEEVA, N. P. ANDREEVA, and T. V. TIKHONOVA: *Zh. Anal. Khim.*, **26**:697 (1971); *J. Anal. Chem. USSR*, **26**:604 (1971).

13 I. L. TEODOROVICH, KH. Z. MECHOS, and R. I. GUTNIKOVA: *Zh. Anal. Khim.*, **25**:526 (1970); *J. Anal. Chem. USSR*, **25**:451 (1970).

14 G. I. MALOFEEVA, V. S. RASSKAZOVA, N. P. ANDREEVA, and N. A. RUDNEV: *Zh. Anal. Khim.*, **26**:703 (1971); *J. Anal. Chem. USSR*, **26**:609 (1971).

15 R. R. GADDE and H. A. LAITINEN: *Environ. Lett.*, **5**:223 (1973).

16 H. B. WEISER and J. L. SHERRICK: *J. Phys. Chem.*, **23**:205 (1919).

17 S. POPOFF and E. W. NEUMAN: *Ind. Eng. Chem., Anal. Ed.*, **2**:45 (1930).

18 W. RIEMAN, III, and G. HAGEN: *Ind. Eng. Chem., Anal. Ed.*, **14**:150 (1942).

19 E. HINTZ and H. WEBER: *Z. Anal. Chem.*, **45**:31 (1906).

20 H. G. GRIMM: *Z. Phys. Chem.* (Leipzig), **98**:353 (1921); *Z. Elektrochem.*, **28**:75 (1922); **30**:467 (1924).

21 R. B. WILSEY: *J. Franklin Inst.*, **200**:739 (1925).

22 H. FLOOD: *Z. Anorg. Allg. Chem.*, **229**:76 (1936).

23 H. FLOOD and B. BRUUN: *Z. Anorg. Allg. Chem.*, **229**:85 (1936).

24 F. VASLOW and G. E. BOYD: *J. Amer. Chem. Soc.*, **74**:4691 (1952).

25 A. P. RATNER: *J. Chem. Phys.*, **1**:789 (1933).

26 H. C. YUTZY and I. M. KOLTHOFF: *J. Amer. Chem. Soc.*, **59**:916 (1937).

27 H. A. DOERNER and W. M. HOSKINS: *J. Amer. Chem. Soc.*, **47**:662 (1925).

28 D. H. KLEIN in "Encyclopedia of Industrial Chemical Analysis," vol. 2, p. 204, F. D. SNELL and C. L. HILTON (eds.), Interscience, New York, 1966.

29 I. M. KOLTHOFF and F. T. EGGERTSEN: *J. Amer. Chem. Soc.*, **61**:1036 (1939).

30 I. M. KOLTHOFF and G. E. NOPONEN: *J. Amer. Chem. Soc.*, **60**:508 (1938).

31 H. G. GRIMM and G. WAGNER: *Z. Phys. Chem.* (Leipzig), **132**:131 (1928).

32 G. WAGNER: *Z. Phys. Chem.* (Leipzig), **B2**:27 (1929).

33 I. M. KOLTHOFF and G. E. NOPONEN: *J. Phys. Chem.*, **42**:237 (1938).

PROBLEMS

9-1 If $BaCO_3$ ($K_{sp} = 8 \times 10^{-9}$) and $SrCO_3$ ($K_{sp} = 2 \times 10^{-9}$) form an ideal, homogeneous solid solution, what is the theoretical mole fraction of $BaCO_3$ in the precipitate if half the Sr^{++} is precipitated from a solution initially equimolar in Ba^{++} and Sr^{++}? *Answer* 0.286.

9-2 The distribution ratio for the system $MgNH_4PO_4 + AsO_4^{3-} \rightleftharpoons MgNH_4AsO_4 + PO_4^{3-}$ is 0.18. (*a*) Which of the two pure solids has the higher solubility? (*b*) If a liter of 10^{-7} *M* arsenate is treated with phosphate to bring its concentration to 0.1 *M*, and if 99.9% of the phosphate is then precipitated as $MgNH_4PO_4 \cdot 6H_2O$, what percentage of the arsenate remains in solution? *Answer* 0.56%.

9-3 The solubility products of CaC_2O_4 and SrC_2O_4 are 1.8×10^{-9} and 5.4×10^{-8}. A liter of solution containing 0.1 mole of Ca^{++} and 10^{-6} mole of Sr^{++} is treated with 0.11 mole of oxalate ion. (*a*) Calculate the concentrations of Ca^{++} and Sr^{++} remaining in solution, assuming no interaction between solid CaC_2O_4 and SrC_2O_4 and no coprecipitation by ion entrapment. (*b*) Calculate the concentration of Sr^{++} remaining if CaC_2O_4 and SrC_2O_4 form an ideal, homogeneous solid solution. (*c*) If solid-solution equilibrium is slow and a heterogeneous solid is obtained, which order of mixing would give less coprecipitation of Sr^{++}?
Answer (*a*) 1.8×10^{-7} *M*, 10^{-6} *M*. (*b*) 5.4×10^{-11} *M*.
(*c*) Add oxalate to $Ca^{++} + Sr^{++}$.

9-4 A liter of solution containing 10^{-6} mole of chromate and 0.2 mole of molybdate is treated with enough lead ions to precipitate half the molybdate. Calculate the fraction of the chromate precipitated if *D* or λ is 0.004 ($PbCrO_4$ more soluble than $PbMoO_4$), assuming (*a*) homogeneous solid-solution formation and (*b*) Doerner-Hoskins heterogeneous solid-solution formation.
Answer (*a*) 0.004. (*b*) 0.0028.

9-5 Recalculate the calculated values in Table 9-1 for 99 and 101% of Ag^+ added. (For additional background see I. M. Kolthoff and V. A. Stenger, "Volumetric Analysis," 2d ed., vol. 1, p. 203, Interscience, New York, 1942.)

10

PRACTICAL ASPECTS OF PRECIPITATION REACTIONS

Precipitation reactions have several applications in analysis: in gravimetric methods, in precipitation titrations, and in separations. Gravimetry, which used to be a major part of analytical chemistry, has expanded less rapidly than other aspects of analysis and does not now occupy a prominent place. Precipitation titrimetry always has been restricted in application because most precipitation reactions fail to meet the requirements of rapid reaction rate and adequate stoichiometry. In separations, precipitation reactions are used in two ways; in one the precipitate involved is of direct concern, and in the other it acts as a carrier for another substance of interest. The application of precipitation reactions to separations is described in Chapter 22.

10-1 GRAVIMETRIC METHODS

Gravimetry normally consists of separation of a substance by precipitation and then isolation and weighing in a form of known composition. Gravimetry also includes other processes, such as volatilization and electrodeposition. Although gravimetric methods are usually time-consuming, they have the advantage that for macro amounts of sample their limitations are few. Gravimetric methods have been developed for nearly every element; for details of gravimetric procedures, the reader is referred to other sources.[1]

A precipitant can be added by direct mixing, by formation in a homogeneous chemical reaction (Section 10-2), or by electrolytic generation (Chapter 14). Organic precipitants are often useful (discussed in Section 22-4). The precipitation form and the weighing form are frequently not the same. For example, even though calcium can be precipitated and weighed as the oxalate, the oxalate salt is usually weighed either after conversion to the oxide by heating strongly, or to the sulfate by reaction with ammonium sulfate.

A precipitate normally must be washed. The wash solution is chosen with a view to minimizing both solubility losses and the effects of coprecipitation or peptization.

10-2 PRECIPITATION FROM HOMOGENEOUS SOLUTIONS

Conditions during precipitate formation are often critical. Most precipitation procedures specify slow addition of precipitant with efficient stirring. Such a technique tends to produce large filterable crystals with minimal coprecipitation. Tedium can be avoided by use of the ultimate technique for slow addition, precipitation from homogeneous solution (PFHS). In PFHS the precipitant is generated homogeneously in an unsaturated solution, usually by chemical reaction. Two general methods are used: either slowly changing the pH of a solution in which a substance has a solubility dependent on pH, or slowly increasing the concentration of one of the reactants. The rate of generation of reagent can be decreased to the point that hours or even days are required for quantitative precipitation. Such slow precipitation produces a precipitate with highly favorable physical and chemical properties, because homogeneous nucleation (Section 8-2) is avoided; also, heterogeneous nucleation probably occurs on only a small proportion of the nucleation sites.

Although PFHS has been used for more than a century, only during the last decade or two has much attention been given to improving classical gravimetric methods. The background and techniques for PFHS have been reviewed by Willard[2] and more recently by Cartwright, Newman, and Wilson.[3]

Nucleation Fischer[4] found that, in PFHS, nucleation takes place early and only growth occurs thereafter. In the homogeneous precipitation of barium sulfate by sulfate generation from sulfamic acid in a solution containing 0.01 M barium ion, he found (Section 8-2) that the number of particles was sensibly constant from the beginning to the end of precipitation. Evidently nucleation was complete within the first small fraction of the total precipitation time. From the rate of sulfate generation it was calculated that the product of sulfate ion and barium ion concentrations in a 0.01 M barium ion solution exceeds the solubility product in only a second or two. Furthermore, for a 0.01 M solution of barium ion, reagent-grade sulfamic acid probably contains several thousand times the sulfate needed to reach the saturation value. Fischer therefore concluded that nucleation often occurs on direct mixing of the reagents but at a low supersaturation ratio. Under these conditions only the most efficient nucleation sites[5,6] are operative. On the other hand, Haberman and Gordon[7] pointed out that, for sulfamic acid, reaching the limiting supersaturation ratio (Section 8-1) may require several minutes. In the PFHS precipitation of hydrated metal oxides by raising the pH, the solubility product is never exceeded on direct mixing.

pH control   Willard and Tang[8] introduced the hydrolysis of urea,

$$CO(NH_2)_2 + H_2O \rightarrow CO_2\uparrow + 2NH_3 \qquad (10\text{-}1)$$

as a method for gradually neutralizing acid and increasing the pH by the ammonia generated within the solution. Urea has the advantage of being a weak base ($K_b = 1.5 \times 10^{-14}$) that does not hydrolyze at an appreciable rate at room temperature but does so at a convenient rate at 90 to 100°C. The final pH attainable depends on the initial concentration of ammonium ion, but is about 9.3 in its absence. Willard and Tang used urea hydrolysis in the PFHS of aluminum. In general, urea is an excellent reagent for the precipitation of insoluble hydrous oxides and has been applied to several, including those of iron and thorium. Control of pH has been used also as a general method for precipitation of metal chelates,[9] for instance, nickel dimethylglyoxime.[10] Other examples of precipitations based on urea hydrolysis are the precipitation of calcium as the oxalate and of barium as the chromate.[11] Willard and Tang[8] also made the important observation that the nature of the anion is important in the formation of more compact (larger particle size) precipitates and that certain basic salts are more favorable than hydroxides in the precipitation of aluminum. Gordon[12] compared the compactness of several basic salt precipitates prepared from homogeneous solution with the usual hydrous oxides prepared by addition of ammonia. The apparent volume of basic iron(III) formate was 20 times smaller than that of hydrous iron(III) oxide that had settled for 2 months. Similarly, basic tin(IV) sulfate was 20 times and basic thorium(IV) formate 9 times more compact than the hydrous oxide. As a consequence of the larger particle size, filtration and washing were facilitated.

PFHS often markedly improves the efficiency of separation. For example, in the precipitation of aluminum hydroxide by addition of ammonia, narrow pH limits must be maintained to get quantitative precipitation, because of the amphoteric properties of aluminum hydroxide. Consequently, it is impossible to regulate the ratio of ammonium ion and ammonia to attain minimum coprecipitation. Copper, when present to the extent of 50 mg in the presence of 0.1 g of aluminum, is extensively coprecipitated (approximately 40%) by the direct addition of ammonia. With the urea-succinate method only 0.05 mg was coprecipitated from 1.0 g of copper initially present.[8]

The pH can also be controlled by evaporation of ammonia. For example, barium chromate can be precipitated in the presence of strontium and lead by first complexing the metal ions at a pH of 10.4, followed by evaporation of ammonia. The pH decreases slowly, and barium chromate precipitates since the barium ion is no longer complexed at the lower pH.[13]

Anion or cation generation Anions can be generated slowly in solution to bring about homogeneous precipitation. Swift and Butler[14] reviewed precipitation of the metal sulfides by use of thioacetamide or thiourea. PFHS of sulfides of cadmium, mercury, zinc, and nickel have been studied more recently by Swift and others.[15] Phosphate can be generated by hydrolysis of triethyl phosphate,[16] oxalate by hydrolysis of methyl oxalate,[17] and sulfate by hydrolysis of diethyl sulfate or sulfamic acid.

Cations can be generated slowly in solution by releasing them from their EDTA complexes by oxidative destruction of the EDTA.[18,19]

Reagent synthesis Slow reaction of biacetyl and hydroxylamine in solution will generate dimethylglyoxime for the precipitation of nickel and palladium as their dimethylglyoximates[20] by PFHS. The reaction of phenylhydroxylamine and sodium nitrite can be used for the PFHS of the cupferron complex of copper[21] or titanium;[22] cobalt can be precipitated as the 1-nitroso-2-naphtholate by the reaction of 2-naphthol and nitrous acid in the presence of cobalt;[23] zirconium as the tetramandelate by hydrolysis of acetylmandelic acid;[24] nickel and copper by their salicylaldimine complexes through slow reaction of salicylaldehyde and ammonia[25] or hydroxylamine;[26] thorium as the iodate by slow photochemical reduction of periodate.[27,28]

Solvent evaporation Precipitation reactions can be controlled by use of mixed solvents.[29] Slow evaporation of acetone from a buffered water-acetone solution produces results similar to PFHS in the precipitation of aluminum 8-hydroxyquinolate.

10-3 IGNITION OF PRECIPITATES

Many precipitates are formed with a composition unsuitable for weighing, or containing varying amounts of water (or other solvent) that need removal; most precipitates therefore need to be heated to convert them to compounds of known composition. Water may be present as: superficial adherent water clinging to the moist precipitate; inclusions within crystals; adsorbed water on the surface; imbibed water (lyophilic colloids); and essential water, either as water of hydration or as water of constitution. The effects of several contaminants may differ: some give positive errors, others negative; some volatilize, others do not; incomplete removal of water may compensate for the negative effect of a lighter ion replacing a lattice ion. The ignition of precipitates often entails thermal decomposition reactions involving the dissociation of salts into acidic and basic components. As examples we may cite the decomposition of carbonates to give basic oxides and of sulfates to give acidic oxides. Knowing that the decomposition temperature is related to the acidic and basic properties of the oxides thus produced, we can predict some important trends in the stabilities of related compounds. Accordingly, since the basicity of alkali and alkaline earth oxides increases in moving down in a periodic group, the stabilities of alkali metal carbonates and sulfates also increase in the same order. Likewise, since sulfur trioxide is more acidic than carbon dioxide, the thermal stability of a particular metal sulfate is usually greater than that of the carbonate. Predictions of this nature are generally valid if more deep-seated changes, such as changes in oxidation state, are not involved. Other acid-base reactions, for instance combination or displacement reactions, also can occur during ignition.

The weight of an ignited precipitate is itself not necessarily represented accurately by the balance reading; there may be equilibration of the precipitate and container with atmospheric moisture (the atmosphere in a desiccator probably is not dry), and absorption of carbon dioxide or ammonia.

The study of precipitates by differential thermal analysis or thermogravimetric analysis[30,31] can give detailed information about their properties. In differential thermal analysis, transformations in a substance on heating (phase changes, decomposition) that involve thermal effects (with or without weight changes) are recorded as a function of temperature. In thermogravimetric analysis, loss of weight as a function of temperature is measured. A combination of these two techniques is more powerful than either alone. Duval[32] and collaborators prepared and examined the thermal behavior of over 1000 precipitates proposed for use in gravimetric analysis and worked out a large number of detailed heating procedures.

For the determination of single constituents, thermogravimetric methods are applicable as rapid control methods in which the weighing operation has been rendered automatic. Accuracy is limited to about 1 part in 300. Determinations of more than one component also are possible. For example, a mixture of calcium and magnesium oxalates can be analyzed by heating and weighing calcium carbonate and magnesium oxide at 500°C and calcium oxide and magnesium oxide at 900°C.[33] Similarly, a mixture of silver and copper(II) nitrates yields silver nitrate and copper oxide at 280 to 400°C and silver and copper oxide at temperatures above 529°C. Potassium perchlorate can be determined[34] in the presence of barium nitrate by taking advantage of the fact that barium nitrate catalyzes the thermal decomposition of potassium perchlorate.

Important to our present purposes is that Duval's compilation of data[32] gives the kind of information needed to guide the choice of drying or ignition temperatures for precipitates. Correct choice of conditions varies widely. For instance, silver chloride is easily dried[32] within temperature limits of 70 to 600°C. For careful gravimetric analyses, heating to 130 to 150°C is normally recommended. It should be realized that about 0.01% of adsorbed water remains at this temperature and the last traces are lost only upon fusion, which occurs at 455°C. A precipitate of barium sulfate contains not only adsorbed water, which is given off by 115°C, but also occluded water in solid solution.[35] For example, Fales and Thompson[36] found that, in the ignition of relatively uncontaminated barium sulfate that had been dried at 115°C, further weight losses of 0.1 and 0.3% occurred upon heating for two 2-h periods at 300 and 600°C. An additional weight loss of 0.05% occurred in 1 h of heating at 800°C. Ordinarily an ignition temperature of 800 to 900°C is recommended; according to Duval the thermolysis curve becomes essentially horizontal at 780°C. The upper temperature limit is imposed by decomposition of barium sulfate to give barium oxide and sulfur trioxide, a reaction that becomes appreciable at about 1400°C. But, if impurities such as iron oxide or silica are present, a loss of sulfur trioxide may begin at 1000°C, no doubt because the iron oxide or silica acts as an acidic oxide and reacts with the strongly basic barium oxide to favor the loss of volatile sulfur trioxide. Filter paper, if used, should be removed at a temperature not exceeding 600°C and in an oxidizing atmosphere, as a precaution against reduction of barium sulfate by carbon to give barium sulfide.

Hydrous oxides contain relatively large quantities of adsorbed and imbibed water and, in some instances, water of constitution (hydroxide) as well. Duval observed that the minimum temperature necessary for quantitative dehydration of hydrous oxides often depends on the method of precipitation. Alumina precipitated by gaseous ammonia was dried at 475°C, a product precipitated by the urea-succinate method was dried at 611°C, and material precipitated with aqueous ammonia required 1031°C. These temperature limits were determined by recording the weight as a function of temperature while heating to give a continuously rising temperature. The resulting thermolysis curves showed horizontal regions corresponding to the attainment of constant weight. The results of such continuous-heating experiments are not necessarily valid for the ordinary (static) conditions of gravimetric analysis. Milner and Gordon[37] clearly showed that, with a conventional ignition and weighing procedure, above 800°C a weight loss amounting to several percent occurs on heating hydrous aluminum oxide precipitated by these several techniques. They recommended a temperature of 1200°C. In thermogravimetric analysis, since the temperature is continuously increased at an arbitrary rate, equilibrium is not necessarily achieved. Moreover, the precipitate is not exposed to a cool and humid atmosphere because the weighing is carried out at elevated temperatures. Because alumina that has been ignited at 900 to 1000°C is hygroscopic,[38] during the first few minutes of exposure to moist air it takes up a large proportion of the water that it will absorb in 24 h. On the other hand, if alumina is ignited at 1200°C, the hygroscopic γ-aluminum oxide is transformed into α-aluminum oxide, which is nonhygroscopic and can be weighed at leisure.[39]

Silica reaches a constant weight at 358°C in a recording thermobalance, but in ordinary gravimetry is ignited at high temperatures to lower its hygroscopicity. Miehr, Koch, and Kratzert[40] found the following percentages of water in silica that had been heated for 1 h at the indicated temperature and cooled for 30 min in a desiccator: 900°C, 0.9%; 1000°C, 0.5%; 1100°C, 0.2%; 1200°C, 0.1%.

Potassium tetraphenylborate, normally dried at 105 to 120°C, is stable up to 265°C, according to the thermolysis curve of Wendlandt.[41] At 715 to 825°C the metaborate KBO_2 is formed.

Calcium oxalate normally is precipitated as the monohydrate from hot solution. The thermolysis curve[32] shows several plateaus, corresponding to the monohydrate from room temperature to 100°C, anhydrous calcium oxalate from 226 to 398°C, calcium carbonate from 420 to 660°C, and calcium oxide above 840 to 850°C. Sandell and Kolthoff[42] concluded that the monohydrate is not a reliable weighing form because of its tendency to retain excess moisture. Coprecipitated ammonium oxalate also remains undecomposed, so the results are usually 0.5 to 1.0% too high when the precipitate is dried at 105 to 110°C. Anhydrous calcium oxalate also is unsuitable as a weighing form because of its hygroscopicity.

Willard and Boldyreff[43] concluded that calcium carbonate is an excellent weighing form if the oxalate is ignited at a temperature of 500 $\pm$ 25°C. The necessity of this closeness of temperature control is evident from the following consideration. The minimum temperature is determined by the *rate* of the irreversible decomposition

$$CaC_2O_4 \rightarrow CaCO_3 + CO \qquad (10\text{-}2)$$

which is too slow to reach completion in a reasonable time at 450°C, but becomes rapid at 475°C. The maximum temperature limit is determined by the *equilibrium* pressure of carbon dioxide at a given temperature, no matter what the ratio of calcium carbonate to calcium oxide. Therefore, if the equilibrium pressure exceeds the partial pressure of carbon dioxide in the atmosphere, the carbonate should decompose completely. The dissociation pressure at 500°C is 0.15 mm, compared with a partial pressure of 0.23 mm in normal atmospheric air. The dissociation pressure reaches 0.23 mm at 509°C, but the dissociation rate does not become appreciable until the temperature is above 525°C. Thus calcium carbonate is the ideal weighing form, provided that means are available either to control the temperature at a value near 500°C or to heat the oxalate in an atmosphere of carbon dioxide below a temperature of 880°C. The dissociation pressure of calcium carbonate reaches 760 mm at 882°C, and it can be ignited to calcium oxide above that temperature. Calcium oxide, being hygroscopic, is a less desirable weighing form than calcium carbonate.

10-4   TITRATION METHODS

By analogy to pH titration curves of acids and bases, it is customary in precipitation titrations to plot the quantity pM (defined by either $- \log [M^{m+}]$ or $- \log a_{M^{m+}}$) against titration volume. For certain metals that form reversible electrodes with their ions, the measured electrode potential is a linear function of the logarithm of ion activity, so the titration curve can be realized experimentally in a potentiometric titration. In any case, the curve gives a useful indication of the sharpness of an end-point break.

As it was for acid-base titrations, the concept of relative precision (Section 3-7) is useful for comparing the steepness of titration curves in the immediate vicinity of the end point. For the formation of a precipitate of symmetrical charge type ($m = n$), with activity coefficients assumed to be unity,

$$M^{m+} + A^{n-} \rightleftharpoons MA \qquad K_{sp} = [M^{m+}] [A^{n-}] \qquad (10\text{-}3)$$

the titration curve is analogous to that of a titration of strong acid with strong base. The relative precision, defined as the fraction of the stoichiometric quantity of reagent required to traverse the region $\pm$ 0.1 pM unit on either side of the end point, is given approximately by $\sqrt{K_{sp}}/C$, where C is the hypothetical molar end-point concentration of MA that would exist if no precipitation had occurred. Thus the steepness of the titration curve in the end-point region decreases in proportion to the ratio of the solubility of precipitate to the hypothetical end-point concentration. Carrying out titrations in nonaqueous media often improves end-point precision, and titration procedures can be extended to substances that are not otherwise applicable (see Problem 10-6).

It is often assumed that in a 1:1 ion combination titration the point of maximum slope (inflection point) coincides with the equivalence point. This assumption would be correct only if no dilution occurred during titration. Meites and Goldman[44] and Carr[45] showed that the inflection point slightly precedes the equivalence point when dilution is taken into account. This effect of dilution is rarely the limiting factor in end-point precision and is an unnecessary complication under virtually all practical conditions. It becomes significant when the concentration is low or the solubility product is large; thus the error is about 1 ppt for the titration of a 10^{-3} M solution of a substance with a 10^{-3} M titrant when the solubility product is 10^{-10}. The effect can be minimized by use of a titrant with a high concentration compared with that of the substance titrated.

For precipitates of asymmetrical charge types, such as MA_2 and M_2A, expressions for relative precision are more complicated. Christopherson,[46] excluding the effect of dilution, indicated that the relative titration error, defined by $(V_{inflection} - V)/V$ (where $V_{inflection}$ is the titrant volume to the inflection point and V is the equivalence-point volume), is generally

$$\text{TE} = \frac{m^2 - n^2}{m^3(C_i/n)} \sqrt[m+n]{K_{sp}\left(\frac{m}{n}\right)^{3n}} \tag{10-4}$$

where C_i is the initial molar concentration of the substance titrated. When $m > n$, the inflection point follows the equivalence point; when $m < n$, it precedes it. According to (10-4) the titration error becomes negligible for $0.1\ M$ solutions when the solubility of a precipitate is less than about $10^{-6}\ M$. Meites and Goldman,[44] however, pointed out that, when dilution is taken into account, the inflection point can be made to coincide with the equivalence point by titrating with a solution of the correct concentration containing the ion of lower oxidation number. Thus, for a precipitate of the type MA_2 the titrant concentration of the solution containing A should be $7.6\sqrt[3]{K_{sp}}$. No inflection point occurs if the concentration is sufficiently low.

For detailed explanations of the chemistry of specific precipitation titrations the reader is referred to other sources.[47] These include methods such as the Mohr method for halides using the silver chromate end point, the Fajans adsorption indicator method, the Volhard method for $FeSCN^{++}$ end point, the titration of halide mixtures with the attendant solid-solution and adsorption effects, the titration of fluoride with thorium(IV), and the titration of sulfate with barium ion.

REFERENCES

1   A. I. VOGEL: "Textbook of Quantitative Inorganic Analysis," 3d ed., chap. 5, Longmans, London, 1961; I. M. KOLTHOFF, E. B. SANDELL, E. J. MEEHAN, and S. BRUCKENSTEIN: "Quantitative Chemical Analysis," 4th ed., pt. 3, Macmillan, London, 1969; W. F. HILLEBRAND, G. E. F. LUNDELL, H. A. BRIGHT, and J. I. HOFFMAN: "Applied Inorganic Analysis," 2d ed., Wiley, New York, 1953; L. ERDEY: "Gravimetric Analysis," 3 vols., Pergamon, New York, 1964–1965; F. E. BEAMISH and W. A. E. MCBRYDE in "Comprehensive Analytical Chemistry," C. L. Wilson and D. W. Wilson (eds.), vol. 1A, chap. VI, Elsevier, New York, 1959; J. F. COETZEE in "Treatise on Analytical Chemistry," I. M. Kolthoff and P. J. Elving (eds.), pt. 1, vol. 1, chap. 19, Interscience, New York, 1959; "Standard Methods of Chemical Analysis," 6th ed., 3 vols., Van Nostrand Reinhold, New York, 1962–1966; American Society for Testing and Materials, "Book of ASTM Standards," ASTM, Philadelphia, annual.

2   H. H. WILLARD: *Anal Chem.*, **22**:1372 (1950).

3   P. F. S. CARTWRIGHT, E. J. NEWMAN, and D. W. WILSON: *Analyst*, **92**:663 (1967).

4   R. B. FISCHER: *Anal. Chem.*, **32**:1127 (1960); **33**:1802 (1961).

5   D. MEALOR and A. TOWNSHEND: *Talanta*, **13**:1069 (1966).

6   D. H. KLEIN and J. A. DRIY: *Talanta*, **13**:289 (1966).

7   N. HABERMAN and L. GORDON: *Anal. Chem.*, **33**:1801 (1961).

8   H. H. WILLARD and N. K. TANG: *J. Amer. Chem. Soc.*, **59**:1190 (1937); *Ind. Eng. Chem.*, *Anal. Ed.*, **9**:357 (1937).

9   F. H. FIRSCHING: *Talanta*, **10**:1169 (1963).

10   E. L. BICKERDIKE and H. H. WILLARD: *Anal. Chem.*, **24**:1026 (1952).

11   L. GORDON and F. H. FIRSCHING: *Anal. Chem.*, **26**:759 (1954).

12   L. GORDON: *Anal. Chem.*, **24**:459 (1952).

13   F. H. FIRSCHING and P. H. WERNER: *Talanta*, **19**:790 (1972).

14   E. H. SWIFT and E. A. BUTLER: *Anal. Chem.*, **28**:146 (1956).

15   D. V. OWENS, E. H. SWIFT, and D. M. SMITH: *Talanta*, **11**:1521 (1964); D. H. KLEIN and E. H. SWIFT: *Talanta*, **12**:349, 363 (1965); D. H. KLEIN, D. G. PETERS, and E. H. SWIFT: *Talanta*, **12**:357 (1965); D. C. TAYLOR, D. M. SMITH, and E. H. SWIFT: *Anal. Chem.*, **36**:1924 (1964).

16   H. H. WILLARD and H. FREUND: *Ind. Eng. Chem.*, *Anal. Ed.*, **18**:195 (1946).

17   L. GORDON and A. F. WROCZYNSKI: *Anal. Chem.*, **24**:896 (1952).

18   R. GRZESKOWIAK and T. A. TURNER: *Talanta*, **16**:649 (1969); **20**:351 (1973).

19   P. F. S. CARTWRIGHT: *Analyst*, **86**:688, 692 (1961); **87**:163 (1962).

20   L. GORDON, P. R. ELLEFSEN, G. WOOD, and O. E. HILEMAN, JR.: *Talanta*, **13**:551 (1966).

21   A. H. A. HEYN and N. G. DAVE: *Talanta*, **13**:27 (1966).

22   A. H. A. HEYN and N. G. DAVE: *Talanta*, **13**:33 (1966).

23   A. H. A. HEYN and P. A. BRAUNER: *Talanta*, **7**:281 (1961).

24   H. SINGH and N. K. MATHUR: *Indian J. Chem.*, **8**:1143 (1970).

25   L. ERDEY and L. PÓLOS: *Talanta*, **17**:1218 (1970).

26   B. S. K. RAO and O. E. HILEMAN, JR.: *Talanta*, **14**:299 (1967).

27   J. M. FITZGERALD, R. J. LUKASIEWICZ, and H. D. DREW: *Anal. Lett.*, **1**:173 (1967).

28   M. DAS, A. H. A. HEYN, and M. Z. HOFFMAN: *Talanta*, **14**:439 (1967).

29   L. C. HOWICK and J. L. JONES: *Talanta*, **9**:1037 (1962); **10**:189, 197 (1963); J. L. JONES: *J. Chem. Educ.*, **45**:433 (1968).

30   W. W. WENDLANDT: "Thermal Methods of Analysis," Wiley, New York, 1964.

31   F. PAULIK, J. PAULIK, and L. ERDEY: *Talanta*, **13**:1405 (1966).

32   C. DUVAL: "Inorganic Thermogravimetric Analysis," 2d ed., translated by R. E. Oesper, Elsevier, Amsterdam, 1963.

33   S. PELTIER and C. DUVAL: *Anal. Chim. Acta*, **1**:408 (1947).

34   V. D. HOGAN, S. GORDON, and C. CAMPBELL: *Anal. Chem.*, **29**:306 (1957).

35   G. WALTON and G. H. WALDEN, JR.: *J. Amer. Chem. Soc.*, **68**:1750 (1946).

36   H. A. FALES and W. S. THOMPSON: *Ind. Eng. Chem.*, *Anal. Ed.*, **11**:206 (1939).

37   O. I. MILNER and L. GORDON: *Talanta*, **4**:115 (1960).

38   HILLEBRAND, LUNDELL, BRIGHT, and HOFFMAN,[1] p. 503.

39   J. N. FRERS: *Z. Anal. Chem.*, **95**:113 (1933).

40   W. MIEHR, P. KOCH, and J. KRATZERT: *Z. Angew. Chem.*, **43**:250 (1930).

41   W. W. WENDLANDT: *Anal. Chem.*, **28**:1001 (1956).

42   E. B. SANDELL and I. M. KOLTHOFF: *Ind. Eng. Chem.*, *Anal. Ed.*, **11**:90 (1939).

43   H. H. WILLARD and A. W. BOLDYREFF: *J. Amer. Chem. Soc.*, **52**:1888 (1930).

44 L. MEITES and J. A. GOLDMAN: *Anal. Chim. Acta*, **30**:18, 200 (1964).
45 P. W. CARR: *Anal. Chem.*, **43**:425 (1971).
46 H. L. CHRISTOPHERSON: *J. Chem. Educ.*, **40**:63 (1963).
47 VOGEL[1]; KOLTHOFF, SANDELL, MEEHAN, and BRUCKENSTEIN[1]; I. M. KOLTHOFF and V. STENGER, "Volumetric Analysis," vol. II, chaps. 8 and 9, Interscience, New York, 1947; A. A. BENEDETTI-PICHLER in "Handbook of Analytical Chemistry," L. Meites (ed.), pp. 3–46, McGraw-Hill, New York, 1963; H. A. LAITINEN: "Chemical Analysis," chap. 12, McGraw-Hill, New York, 1960.

PROBLEMS

10-1 A mixed precipitate of CaC_2O_4 and MgC_2O_4 is to be analyzed by the thermo-gravimetric method. After heating to form $CaCO_3$ and MgO the precipitate weighs 0.4123 g, and after ignition to CaO and MgO it weighs 0.2943 g. Calculate the weight of CaO in the sample.

Answer 0.1504 g.

10-2 A sample weighing 0.9876 g and containing only $CaCO_3$ and $MgCO_3$ is ignited to a mixture of CaO and MgO weighing 0.5123 g. (*a*) Calculate the weight of $CaCO_3$ in the mixture. (*b*) Calculate the weight of the residue if the sample is ignited to MgO and $CaCO_3$.

Answer (*a*) 0.4887 g. (*b*) 0.7272 g.

10-3 A 0.4987-g sample containing only ZnS and CdS is dissolved, and the metals are quantitatively precipitated as $MNH_4PO_4 \cdot 6H_2O$, which is ignited to yield 0.6987 g of mixed pyrophosphates $(M_2P_2O_7)$. Calculate the weight of ZnS in the sample.

Answer 0.0567 g.

10-4 For a mixture of 10^{-3} *M* iodide and 10^{-1} *M* bromide, calculate the theoretical percentage error in titration of the iodide and the bromide with $AgNO_3$, neglecting dilution and solid-solution formation. Take $K_{sp,AgI} = 10^{-16}$, $K_{sp,AgBr} = 4 \times 10^{-13}$.

Answer − 2.5 and 0.025%.

10-5 Calculate the theoretical titration error for the determination of 0.1 *M* bromide in the presence of 0.01 *M* iodide and chloride, neglecting coprecipitation and solid-solution formation. Take $K_{sp,AgCl} = 2 \times 10^{-10}$.

Answer 0.005%.

10-6 Calculate pNa for the precipitation titration of a solution of 0.1 *M* $NaClO_4$ in acetonitrile with 0.1 *M* tetraethylammonium chloride at 0, 50, 90, 100, and 110% of the equivalence point. Include the effect of dilution. Use formation constants of 10, 10, and 0 for $Na^+ClO_4^-$, $Et_4N^+Cl^-$, and $Et_4N^+ClO_4^-$, and a solubility product of 10^{-9} for NaCl.

Answer 1.21, 1.63, 2.46, 4.50, 6.50.

COMPLEXATION EQUILIBRIA: THEORY AND APPLICATIONS

In 1945 Schwarzenbach[1] published the first of a series of fundamental studies of polyaminopolycarboxylic acids as analytical chelating reagents. Of these, ethylene-diaminetetraacetic acid, usually abbreviated EDTA, has become one of the most important of all the reagents used in titrimetry. Another important contribution made by Schwarzenbach and his coworkers was the development of indicators that respond to changes in metal ion activity.

Since 1945 nearly 3000 publications dealing with complexation analysis have appeared. For more complete treatment of this subject than is possible here, monographs such as those by Beck,[2] Perrin,[3] Ringbom,[4] Schwarzenbach and Flaschka,[5] and West[6] may be consulted.

Like other types of titrimetric reactions, complex formation should be rapid, stoichiometric, and quantitative. Most reactions involving the formation of complexes fail to fulfill one or more of these requirements. EDTA is the most important exception. The classical complexing reactions which are analytically useful are those of mercury(II) with halides and of cyanide with silver(I). For discussion of the background chemistry and typical applications the reader is referred to other sources.[7]

In this chapter neither detailed coordination chemistry nor rates of complex formation are considered. Although both topics have broad fundamental significance, adequate description of them is beyond our scope. Nevertheless, it should be borne in mind that complexation does not always involve rapid, reversible processes.

11-1   DEFINITIONS AND GENERAL BACKGROUND

A complexation reaction involves a reaction between a metal ion M and another molecular or ionic entity (a *ligand*) L containing at least one atom with an unshared pair of electrons:

$$M + L \rightleftharpoons ML \qquad (11\text{-}1)$$

where charges on ions are omitted for the sake of generality. It is important to recognize that all the coordination positions on the metal ion M are occupied either by solvent or by other electron-pair donors, so (11-1) simply involves replacement of a molecule of solvent by the ligand L. Again for generality, solvent molecules are omitted in writing the reactions. Ligands are called *unidentate* when they can donate one pair of electrons (ammonia, for example) and *multidentate* when two or more pairs can be donated to form coordinate bonds. When a multidentate ligand forms two or more coordinate bonds to the same acceptor atom, a ring structure results, and the compound is said to be a *chelate*. When a single multidentate ligand coordinates with (bridges) two or more central metal atoms, a *polynuclear* complex results. The *coordination number*, or number of electron pairs a metal ion shares with donor atoms, usually is 4 or 6 and less frequently may be 2 or 8.

N. Bjerrum[8] showed that complexes form in a stepwise manner. Thus Reaction (11-1) can be followed by

$$ML + L \rightleftharpoons ML_2$$
$$ML_2 + L \rightleftharpoons ML_3 \qquad (11\text{-}2)$$
$$ML_{n-1} + L \rightleftharpoons ML_n$$

An important early quantitative study of complex equilibria was that of J. Bjerrum,[9] who showed that unidentate ligands invariably are added in a succession of steps. Unless one particular step is extraordinarily stable, there will be no extended range of concentration of complexing agent over which a single species of complex is formed (except for the last, or highest, complex). The most important examples of analyses based on unidentate ligands are those involving halide complexes of mercury-(II) and the cyanide complex of silver. The importance of the polydentate ligand EDTA as a titrimetric reagent is principally that it forms only 1:1 complexes of high stability with many metal ions. The problem of a succession of complexes overlapping each other therefore does not occur.

The completeness of a complexation titration reaction is governed by the stability of the complex, which in turn is usually favorable when the ligand is poly-dentate. The term *chelate effect* has been used by Schwarzenbach[10] to describe the enhanced stability of a complex involving formation of a ring structure compared with that of a similar complex involving no ring formation. The chelate effect becomes more pronounced as the number of rings per ligand molecule increases and is most pronounced for five-membered rings. This increased stability is illustrated by the values of the stability constants[11] for the ammine complexes of nickel: with ammonia (no chelate ring), $Ni(NH_3)_6^{++}$, 5.5×10^8; and with ethylenediamine (five-membered rings), $Ni(en)_3^{++}$, 2×10^{18}.

Symbolism In contrast to the practice in acid-base and solubility equilibria of using dissociation or simplified dissociation constants (Section 7-1), equilibrium constants in complexation reactions usually are expressed as formation or stability constants. Thus, for Equations 11-1 and 11-2

$$K_1 = \frac{a_{ML}}{a_M a_L} \qquad (11\text{-}3)$$

$$K_2 = \frac{a_{ML_2}}{a_{ML} a_L} \qquad (11\text{-}4)$$

and

$$K_n = \frac{a_{ML_n}}{a_{ML_{n-1}} a_L} \qquad (11\text{-}5)$$

where $K_1, K_2, \ldots, K_n$ are *stepwise formation constants* written as activity constants. These formation constants can be written also as concentration constants or in the form of mixed constants when hydrogen ion activity is involved.[4]

The *overall formation constant*, denoted by the symbol β, is the equilibrium constant for the overall formation reaction $M + nL \rightleftharpoons ML_n$. The relations between stepwise and overall formation constants are as follows: $\beta_1 = K_1$, $\beta_2 = K_1 K_2$, $\beta_3 = K_1 K_2 K_3$, and $\beta_n = K_1 K_2 \cdots K_n$, where n is the maximum coordination number. For example, for copper tetraammine the values of K_1, K_2, K_3, and K_4 are 1.3×10^4, 3.2×10^3, 8×10^2, and 1.3×10^2, and

$$\beta_n = \frac{a_{Cu(NH_3)_4^{++}}}{a_{Cu^+} + a_{NH_3}^4} = K_1 K_2 K_3 K_4 = 4 \times 10^{12}$$

The determination of stability constants has been reviewed by Ringbom and Harju.[12]

11-2 ETHYLENEDIAMINETETRAACETIC ACID (EDTA)

The tetraanion of EDTA, $(—OOCCH_2)_2NCH_2CH_2N(CH_2COO—)_2$, is an especially effective complexer in that it can form five five-membered chelate rings with a single metal ion by coordination through the electron-pair donors of the four (or sometimes

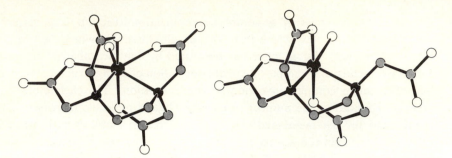

FIGURE 11-1   Computer diagram in perspective of the solid-state structures of the EDTA complex with Fe(III) (left) and with Ni(II) (right). Heavily shaded, central metal atom; solid, nitrogen atoms; lightly shaded, carbon atoms; unshaded, oxygen atoms; hydrogen atoms absent. In each structure an oxygen atom from a molecule of water is included.

three) carboxylate groups and the two nitrogen atoms. The structure of the complexes varies somewhat with the nature of the metal atom; Figure 11-1 depicts the structure of the EDTA complex with iron and nickel ions as determined by x-ray analysis[13] of the solids. The structure of most EDTA complexes in solution is not completely clear. Nickel-EDTA complexes have been examined most extensively, and studies involving nmr[14,15] have shown that at a low pH the complex in solution exists as $Ni(H_2O)HEDTA^-$ with one carboxylate group protonated. At pH 6 or higher[14] about 75% of the EDTA in the nickel complex is hexacoordinate and 25% pentaco-ordinate, with one carboxylate group unbonded (Figure 11-1). Other nmr data[15] indicate that less than 25% is pentacoordinate; on the other hand, studies of absorp-tion and formation constants[16] are in agreement with the 75 to 25% ratio.

Dissociation of EDTA   The successive macroscopic acid dissociation constants of EDTA, H_4Y, at 20°C and an ionic strength of 0.1, are[17] $pK_1 = 2.0$, $pK_2 = 2.67$, $pK_3 = 6.16$, $pK_4 = 10.26$. The EDTA molecule has six basic sites—four carboxylate oxygens and two nitrogens. The distribution of the acidic protons among these sites as a function of the degree of ionization has been studied by nmr.[18] The microscopic dissociation constants (Section 3-8) indicate that, for example, in the divalent anion $H_2Y^=$ the two nitrogen atoms are 96% protonated.[19] This structure accounts for the third and fourth ionization steps being much weaker than the first two. In HY^{3-} the proton is essentially completely associated with the nitrogen atoms.

The fraction of EDTA as H_4Y, H_3Y^-, and so on can be calculated at any pH value by the method described in Section 3-6. Figure 11-2 shows graphically the results of such a calculation neglecting formation of H_5Y^+ (dissociation constant 0.11) and H_6Y^{++} (dissociation constant 0.55).[20] The fraction α_Y present as the tetravalent

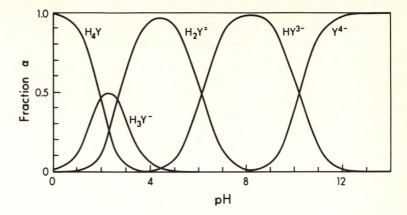

FIGURE 11-2   Fraction of EDTA present in various ionic forms.

anion Y^{4-} is of particular importance in equilibrium calculations.† Its magnitude at various pH values can be calculated by

$$\alpha_Y = \frac{K_1 K_2 K_3 K_4}{[H^+]^4 + K_1[H^+]^3 + K_1 K_2[H^+]^2 + K_1 K_2 K_3[H^+] + K_1 K_2 K_3 K_4} \qquad (11\text{-}6)$$

where $K_1, K_2, \ldots$ are stepwise dissociation constants. Values for α_Y are given in Table 11-1.

Table 11-1   FRACTION OF EDTA AS Y^{4-}

pH	Fraction	pH	Fraction	pH	Fraction
2.0	3.7×10^{-14}	5.0	3.5×10^{-7}	9.0	5.2×10^{-2}
2.5	1.3×10^{-12}	6.0	2.2×10^{-5}	10.0	0.35
3.0	2.5×10^{-11}	7.0	4.8×10^{-4}	11.0	0.85
4.0	3.6×10^{-9}	8.0	5.4×10^{-3}	12.0	0.98

Formation of metal-EDTA complexes   The formation constants of EDTA complexes of various metals are listed in Table 11-2. Written as concentration constants, they involve reactions of the type

$$M^{n+} + Y^{4-} \rightleftharpoons MY^{n-4} \qquad K_{MY} = \frac{[MY^{n-4}]}{[M^{n+}][Y^{4-}]} \qquad (11\text{-}7)$$

† In complexation studies, α and β coefficients are frequently used to represent the *reciprocal* of the fraction.[5] With a broader range of topics the symbol α is nearly universally used to denote fraction; for consistency with other topics in this book we have therefore chosen to continue the use of α to denote fraction.

In terms of the major species present the formation of metal-EDTA complexes for divalent metals may be represented by the equations

$$M^{++} + H_2Y^{=} \rightleftharpoons MY^{=} + 2H^{+} \qquad \text{pH 4 to 5} \qquad (11\text{-}8)$$

$$M^{++} + HY^{3-} \rightleftharpoons MY^{=} + H^{+} \qquad \text{pH 7 to 9} \qquad (11\text{-}9)$$

and at pH greater than 9, $M^{++} + Y^{4-} \rightleftharpoons MY^{=}$, as in (11-7). Since $H_2Y^{=}$ is unreactive mechanistically, Reaction (11-8) is slow as written; a proton dissociation reaction precedes the complexation reaction. In low pH solutions the metal-EDTA complex may exist also as HMY^{-} or H_2MY.

The values for formation constants in Table 11-2 are somewhat misleading in that reactions of Y^{4-} with protons and of metal ions with other substances in solution represent competition for the metal ion–EDTA reaction. As Ringbom[4] pointed out, we rarely need to know the concentration of every species in solution; what is of most concern is the completeness of the reaction of interest. To simplify calculations when there are side reactions, a quantity $[Y']$ is defined as the concentration of all forms of EDTA that are not coordinated to the metal, and a quantity $[M']$ as the concentration of the metal ion that has not reacted with EDTA. With these quantities a *conditional formation constant* $K'_{M'Y'}$ may be defined as

$$K'_{M'Y'} = \frac{[MY]}{[M'][Y']} \qquad (11\text{-}10)$$

where charges on ions are omitted for generality. For EDTA

$$[Y'] = [Y^{4-}] + [HY^{3-}] + [H_2Y^{=}] + [H_3Y^{-}] + [H_4Y] \qquad (11\text{-}11)$$

When the concentration of sodium salts in solution is high, (11-11) should also include $[NaY^{3-}]$, a refinement not included here. At constant pH the relative proportions of the various forms of uncomplexed EDTA are fixed, so that $[Y^{4-}]$ is a constant fraction of the uncomplexed EDTA (Table 11-1), or

$$[Y^{4-}] = \alpha_{Y4-} [Y'] \qquad (11\text{-}12)$$

Table 11-2 FORMATION CONSTANTS OF EDTA COMPLEXES AT 20°C, IONIC STRENGTH 0.1

Metal ion	Log K_{MY}	Metal ion	Log K_{MY}	Metal ion	Log K_{MY}
Fe^{3+}	25.1	TiO^{++}	17.3	Mn^{++}	14.0
Th^{4+}	23.2	Zn^{++}	16.5	Ca^{++}	10.7
Hg^{++}	21.8	Cd^{++}	16.5	Mg^{++}	8.7
Ti^{3+}	21.3	Co^{++}	16.3	Sr^{++}	8.6
Cu^{++}	18.8	Al^{3+}	16.1	Ba^{++}	7.8
Ni^{++}	18.6	La^{3+}	15.4	Ag^{+}	7.3
Pb^{++}	18.0	Fe^{++}	14.3	Na^{+}	1.7

In practice an auxiliary complexing agent usually is added during EDTA titrations to prevent the precipitation of heavy metals as hydroxides or basic salts. Therefore, $[M']$ will be in general the sum of the concentrations of the hydrated metal ion and the various metal-hydroxy complexes, in addition to the concentrations of the complexes of the metal with the auxiliary agent. At a fixed pH, and with a concentration of auxiliary complexing agent that is large compared with that of the metal ion, the proportions of the various species containing metal ions will be fixed. Hence, the fraction α_M as simple metal ion is

$$\alpha_M = \frac{[M]}{[M']} \qquad (11\text{-}13)$$

Combining (11-9), (11-10), (11-12), and (11-13), we obtain for EDTA at a pH ≥ 4

$$K'_{M'Y'} = \frac{[MY]}{[M'][Y']} = K_{MY}\alpha_M\alpha_{Y^{4-}} \qquad (11\text{-}14)$$

or in general, for a ligand L

$$K'_{M'L'} = \frac{[ML']}{[M'][L']} = K_{ML}\frac{\alpha_M\alpha_{L^{n-}}}{\alpha_{ML'}} \qquad (11\text{-}15)$$

where $\alpha_{ML'}$ is the fraction of the complex present in a 1:1 ratio with the metal and includes $[ML]$, $[HML]$, and so forth.

Because of protonation and complexing side reactions that interfere with the formation of a complex, conditional formation constants are smaller than the formation constants in Table 11-2. In the case of EDTA, $\alpha_{Y^{4-}}$ in (11-14) depends on the pH. Normally the solution being titrated is sufficiently buffered that hydrogen ions produced in (11-8) or (11-9) do not cause a significant change in pH, and hence $\alpha_{Y^{4-}}$ is constant throughout the titration. The value of α_M also is dependent on pH, since most metals forming complexes of analytical value also form hydroxy complexes. Most metals also form complexes with anions such as chloride, nitrate, or sulfate normally present in solution, as well as forming complexes with auxiliary complexing agents.

Figure 11-3 shows the relation[4] between conditional formation constants and pH for a number of metal-EDTA complexes, taking into account the effect of pH on $\alpha_{Y^{4-}}$ for EDTA and the effect of hydroxy complexes on α_M (Equation 7-17). As can be seen, the conditional constants are smaller than the thermodynamic constants listed in Table 11-2. In general, there is an optimum[21] pH for each metal ion. In the case of zinc, for example, the conditional constant approaches the K_{MY} value of 16.5 in only one pH region. The curves in Figure 11-3 in general represent maximum values for the conditional constants because auxiliary complexing agents usually are added. The concentration of an auxiliary agent is usually high compared with that of the metal ion, and therefore the fraction α_M is decreased and is a function of the pH and the nature and concentration of the auxiliary agent. When an ammine complex

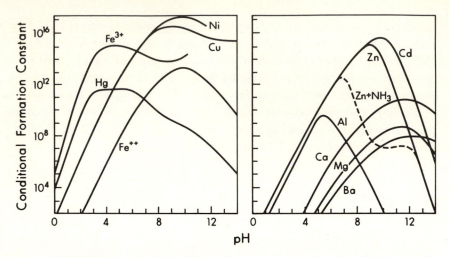

FIGURE 11-3 Conditional formation constants as a function of pH for metal-EDTA complexes. The dotted curve, $Zn + NH_3$, represents zinc in the presence of $[NH_3] + [NH_4^+] = 1$ M. (*Adapted from Ringbom.*[4])

can be formed, the curves of Figure 11-3 for conditional formation constants are lowered further. The dotted curve labeled $Zn + NH_3$, for example, indicates how the curve for zinc is changed as a result of having in solution $[NH_3] + [NH_4^+] = 1$ M. Many EDTA titrations are carried out in ammonia–ammonium chloride buffers, which serve not only to adjust pH but also to provide ammonia as an auxiliary complexing agent. Figure 11-4 shows how α_M depends on ammonia concentration for several metals. In general, the fraction α_M can be calculated from part of Equation (7-24):

$$\alpha_M = \frac{1}{1 + \beta_1[NH_3] + \beta_2[NH_3]^2 + \cdots + \beta_n[NH_3]^n} \qquad (11\text{-}16)$$

where $\beta_1, \beta_2, \beta_n$ are the successive overall formation constants of the metal-ammine complexes.

EXAMPLE 11-1 Calculate the concentration conditional formation constant of $NiY^=$ in a buffer containing 0.05 M ammonia and 0.10 M ammonium chloride. Take values for log K_1, log K_2, ... for the ammine complexes to be 2.75, 2.20, 1.69, 1.15, 0.71, and -0.01, and take K_a for NH_4^+ to be 5 × 10^{-10}.

ANSWER From (11-16), $\alpha_M = 1/(1 + 28 + 223 + 546 + 385 + 99 + 5) = 7.8 \times 10^{-4}$. From K_a we calculate hydrogen ion concentration to be 1.0×10^{-9} M, and from Equation (11-6), $\alpha_Y = 0.052$. From Table 11-2, take log $K_{NiY^=} = 18.6$, and $K'_{Ni'Y'} = 10^{18.6} \times 7.8 \times 10^{-4} \times 0.052 = 1.6 \times 10^{14}$. ////

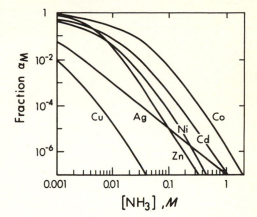

FIGURE 11-4 Fraction α_M for several metals forming ammine complexes as a function of ammonia concentration. Metal ion concentration is assumed to be small compared with ammonia. (*Adapted from Ringbom,*[4] *p. 39.*)

11-3 EDTA TITRATION CURVES

Logarithmic diagrams (Section 3-1) are helpful devices for drawing conclusions about the predominant species in solution at various stages in titrations. Johansson[22] described their application to complexation reactions.

In analogy to a pH titration curve, pM ($-\log[M]$) may be plotted against the fraction titrated. Under the usual titration conditions, in which the concentration of metal ions is small compared with the concentrations of the buffer and the auxiliary complexing agents, the fractions α_Y and α_M are essentially constant during the titration. The titration curve then can be calculated directly from the conditional formation constant since it also remains constant.

EXAMPLE 11-2 Taking the initial concentration of nickel to be 10^{-3} M and neglecting dilution, calculate pNi at the following percentages of the stoichiometric amount of EDTA added under the conditions of Example 11-1: 0, 50, 90, 99, 99.9, 100, 100.1, 101, and 110%.

ANSWER Before the end point, $[Ni'] = C_{Ni}(1 - X)$, where C_{Ni} is the initial concentration of nickel and X is the fraction of the stoichiometric quantity of EDTA added. At the equivalence point, $[Ni'] = [Y'] = \sqrt{C_{Ni}/K'_{Ni'Y'}}$. Beyond the equivalence point, $[Y'] = C_{Ni}(X - 1)$, and from Equation (11-10), $[Ni'] = 1/(X - 1)K'_{Ni'Y'}$. At all points, pNi $= -\log[Ni^{++}] = -\log \alpha_{Ni}[Ni'] = 3.11 - \log[Ni']$. The results are: pNi $= 6.1, 6.4, 7.1, 8.1, 8.4, 11.7, 14.3, 15.3,$ and 16.3 ////

Figure 11-5 shows plots of calculated titration curves of 10^{-3} M nickel(II), iron(III), and calcium(II) at various pH values in buffers of ammonia–ammonium ion at a total buffer concentration of 0.1 M. *Before the end point,* the nickel curves at

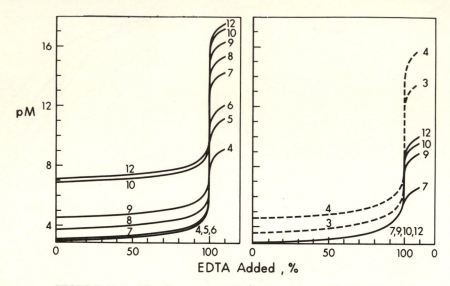

FIGURE 11-5 Titration curves of 10^{-3} M metal ions with EDTA in a solution with $[NH_3] + [NH_4^+] = 0.1$ M at various pH values as indicated. Left, Ni(II); right, calcium (solid lines) and iron (dotted lines).

low pH values (4 to 6) coincide, because no appreciable complexation between nickel(II) and ammonia occurs. At higher pH values the pNi values are higher in an amount determined by the stabilities of the various nickel-ammonia complexes. Much above pH 10 essentially all the buffer is present as ammonia, so little additional pH effect exists. *After the end point* the positions of the curves depend not on the fraction α_M but only on K_{MY} and α_Y. Therefore the titration curves beyond the end point for a given metal depend only on the pH and not on the nature of the auxiliary complexing agent.

The curves for calcium ions differ from those for nickel in two ways. (*1*) Before the end point the curves are essentially independent of pH because calcium does not form ammine complexes. (*2*) After the end point the pM value is smaller than for nickel because of the smaller value of the formation constant $K_{CaY^=}$. For the same reason, at low pH values, α_Y and $K'_{CaY'}$ are so small that no pM break occurs; at high pH, α_Y approaches unity, so it is advantageous to perform calcium titrations at pH 10 to 12.

For titrations of iron(III) with EDTA, a low pH range (3 to 4) is necessary to prevent the hydrolysis of the iron(III); here a low pH value is permissible because the high stability of the complex FeY^- offsets increased ligand protonation.

Titration curves like those of Figure 11-5 can be generalized (Figure 11-6) to a single set based only on conditional formation constants if the quantity pM' is used as ordinate instead of pM. When the conditional formation constant is much less

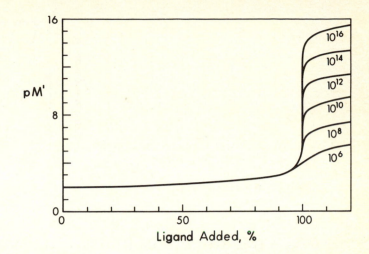

FIGURE 11-6   Titration curves of 10^{-2} M metal ion solutions with a ligand forming 1:1 complexes, with conditional formation constants ranging from 10^6 to 10^{16}.

than 10^8, the change in pM' is too small for precise detection of end points. With the effects of dilution by the titrant neglected, the titration curve would be symmetrical about the equivalence point for a 1:1 complex, and the inflection and equivalence points would coincide. With the effect of dilution taken into account, reasoning similar to that in Sections 3-7 and 10-4 would indicate that the inflection point should slightly precede the equivalence point.

11-4   TITRATION ERROR

For a 1:1 complex of metal and ligand, Ringbom[4] computed the titration error that results from under- or overtitrating the equivalence point by an amount $\pm \Delta$pM (or $\pm \Delta$pL, since ΔpM $= - \Delta$pL). The titration error in percentage is

$$\% \text{ TE} = 100 \frac{[\text{L}]_{\text{end}} - [\text{M}]_{\text{end}}}{C_{\text{M}}} \qquad (11\text{-}17)$$

where C_{M} is the total analytical concentration of metal ion in solution. Thus

$$\Delta \text{pM} = \text{pM}_{\text{end}} - \text{pM}_{\text{eq}} \qquad (11\text{-}18)$$

or

$$[\text{M}]_{\text{end}} = [\text{M}]_{\text{eq}} 10^{-\Delta \text{pM}} \qquad (11\text{-}19)$$

and

$$[\text{L}]_{\text{end}} = [\text{L}]_{\text{eq}} 10^{-\Delta \text{pL}} \qquad (11\text{-}20)$$

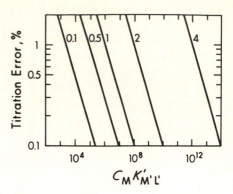

FIGURE 11-7 Titration error as a function of conditional formation constant for ΔpM values from 0.1 to 4. (*Adapted from Ringbom.*[4])

Since $[M]_{eq} = [L]_{eq}$, (11-17), (11-18), (11-19), and (11-20) give

$$\% \ TE = 100[M]_{eq} \frac{10^{\Delta pM} - 10^{-\Delta pM}}{C_M}$$

$$= 100 \frac{10^{\Delta pM} - 10^{-\Delta pM}}{\sqrt{C_M K'_{M'L'}}} \qquad (11\text{-}21)$$

When the Δ values are sufficiently small ($\Delta pM \simeq$ or < 0.4), they can be substituted for differentials to give

$$\% \ TE \simeq 100 \frac{4.6 \ \Delta pM}{\sqrt{C_M K'_{M'L'}}} \qquad (11\text{-}22)$$

Figure 11-7 depicts the relation between the titration error and the product $C_M K'_{M'L'}$. Figure 11-3 indicates that conditional formation constants of at least 10^{10} are possible for most metal ions in a selected pH region. According to Figure 11-7 a titration error of 0.1% should be easily attainable for 0.01 M solutions of such metals. For a 0.1% relative excess of reagent in the titration of 0.01 M metal ion, $[Y']$ of the reagent then equals $10^{-5} \ M$. A value of $K'_{M'Y'} = 10^8$ then corresponds to a 99.9% conversion of metal ion to EDTA complex since $[MY]/C_M = 1000$.

11-5   METAL ION INDICATORS

An important factor in the application of EDTA titration methods has been the development of suitable metal ion indicators, which permit visual titrations to be carried out in dilute solutions. A metal ion indicator is usually a dyestuff that forms metal ion complexes of a color different from that of the uncomplexed indicator. The complex forms over some characteristic range of values of pM, exactly as an acid-base indicator forms a hydrogen ion complex over a characteristic range of pH values. Since metal ion indicators are in general also hydrogen ion indicators, the acid-base equilibria, as well as the metal ion equilibria, must be considered for each indicator. The structures, properties, and classification of complexation indicators are given in publications of West,[6] and Ringbom and Wanninen.[23]

Eriochrome black T was one of the first and most widely used of the metal indicators.[24] Unfortunately, it is unstable in solution, probably because the molecule has both an oxidizing (nitro) and a reducing (azo) group. Lindstrom and Diehl[25] developed as a replacement the indicator 1-(1-hydroxy-4-methyl-2-phenylazo)-2-naphthol-4-sulfonic acid, called calmagite, with the structure

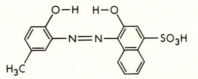

It is stable in aqueous solution, has a sharper color change than eriochrome black T, and can be substituted for it without requiring changes in procedure. For the sake of brevity we shall consider in detail only this indicator; the principles involved are similar for most other indicators.

Calmagite is a tribasic acid, which we designate H_3In. The first proton originating from the sulfonic acid group has a large dissociation constant and is ignored. Thus the following acid-base behavior is significant:

$$H_2In^- \underset{\text{red}}{\overset{pK\ 8.1}{\rightleftharpoons}} HIn^= \underset{\text{blue}}{\overset{pK\ 12.4}{\rightleftharpoons}} In^{3-} \quad (11\text{-}23)$$
$$\text{red} \qquad\qquad \text{blue} \qquad\qquad \text{red-orange}$$

In the pH range 9 to 11, in which the dye itself exhibits a blue color, many metal ions form red 1:1 complexes as a result of the o, o'-dihydroxyazo grouping. Calmagite, having a high molar absorptivity (about 20,000 at pH 10), is a sensitive detector for metals with which it reacts; for example, 10^{-6} to 10^{-7} M solutions of magnesium ion give a distinct red color with this indicator.

With magnesium ions the color-change reaction is represented by

$$\underset{\text{blue}}{HIn^=} + Mg^{++} \rightleftharpoons \underset{\text{red}}{MgIn^-} + H^+ \quad (11\text{-}24)$$

According to (11-23) and (11-24) the indicator color change is affected by the hydrogen ion concentration. From a practical viewpoint a *conditional indicator constant* $K'_{In'}$, which depends on the pH of the buffered solution, may be conveniently defined by

$$K'_{In'} = \frac{[MgIn^-]}{[Mg^{++}][In']} = K_{MgIn}\alpha_{In} \quad (11\text{-}25)$$

Here $[MgIn^-]$ is concentration of metal ion–indicator complex; $[Mg^{++}]$ is concentration of aquated metal ion, which is equal to $\alpha_{Mg}[Mg']$; and $[In']$ is concentration

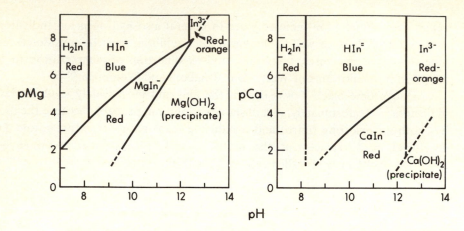

FIGURE 11-8   Predominance-region diagrams illustrating calmagite behavior as a function of pH and pMg (left) and pCa (right). (Color transitions occur over a range; the lines are not meant to imply that they take place at unique values.)

of indicator not complexed with metal ion, which is $[H_3In] + [H_2In^-] + [HIn^=] + [In^{3-}]$. $[In^{3-}] = \alpha_{In}[In']$, where α_{In} is the fraction of the indicator in the completely ionized form.

$$\alpha_{In} = \frac{K_1K_2K_3}{[H^+]^3 + K_1[H^+]^2 + K_1K_2[H^+] + K_1K_2K_3}$$

$$\simeq \frac{K_2K_3}{[H^+]^2 + K_2[H^+] + K_2K_3} \tag{11-26}$$

when K_1 is large compared with K_2 and K_3. From (11-25)

$$\log K'_{MgIn'} = pMg^{++} + \log \frac{[MgIn^-]}{[In']} = \log K_{MgIn}\alpha_{In} \tag{11-27}$$

where K_{MgIn} is the formation constant. The visual transition range will be seen while $[MgIn^-]$ goes from about 0.1 to about $10 \times [In']$. This range can be estimated from the K_2 and K_3 values for calmagite and the value of $\log K_{MgIn}$ of 8.1. Thus at pH 10, $\alpha_{In^{3-}}$ is calculated to be 3.9×10^{-3}, and $\log K'_{MgIn'}$ is then 5.7. At this pH a visible color change of calmagite occurs in the pM range from about 4.7 to 6.7. At pH values 8, 9, 10, 11, and 12, $\log K'_{MgIn'}$ is 3.34, 4.75, 5.69, 6.68, and 7.55. Figure 11-8 summarizes the behavior of calmagite at various pH values as a function of the magnesium and calcium ion concentrations.

The logarithm of the formation constant of calmagite with calcium is 6.1, a smaller value than with magnesium. The effect of this difference is that the conditional formation constant is not large enough at a pH of about 10 for calmagite to act as a sensitive indicator for calcium (Figure 11-8). Consequently, for a sharp end point in calcium titrations a small amount of magnesium must be present. The magnesium-EDTA complex has a smaller conditional formation constant (Figure 11-3) than the calcium, and therefore the magnesium-indicator-EDTA reaction and color change do not occur until the calcium-EDTA reaction is completed.

The indicator titration error can be estimated from values of conditional formation constants.

EXAMPLE 11-3   Estimate the theoretical titration error for the titration of 10^{-3} M magnesium ion at pH 10, using calmagite as the indicator. Assume that end points are taken at (a) 9% and (b) 91% conversion of the indicator from $MgIn^-$ to $HIn^=$. Neglect the effects of dilution and of hydrolysis of magnesium ion.

ANSWER   From Tables 11-1 and 11-2 we have log $K'_{MgY'} = 8.7 - 0.46 = 8.2$. At pH 10, log $K'_{MgIn'} = 5.69$. At the equivalence point, $[Mg^{++}] = \sqrt{C_{Mg}/K'_{MgY'}}$; hence pMg = 5.62. (a) At 9% indicator conversion, from (11-27), pMg = 4.69; so the end point occurs before the equivalence point. Setting $[Mg^{++}][Y'] = C_{Mg}/K'_{MgY'}$ and $[Mg^{++}] = C_{Mg}(1 - X) + [Y']$, we have $1 - X = [Mg^{++}]C_{Mg}/ - 1/[Mg^{++}]K'_{MgY'} = 0.020$. Titration error is -2.0%. (b) At 91% indicator conversion, pMg = 6.69; so the end point occurs after the equivalence point. Setting $[Y'] = C_{Mg}(X - 1) + [Mg^{++}]$, and $[Mg^{++}][Y'] = C_{Mg}/K'_{MgY'}$, we have $X - 1 = 1/[Mg^{++}]K'_{MgY'} - [Mg^{++}]/C_{Mg} = 0.031$. Titration error is $+3.1\%$.   ////

Conditions favorable[26] to a small titration error, with practical limits that are useful as guides, are:

1   Large conditional indicator constant: $K'_{MIn'} > 10^4$.
2   Large ratio of $K'_{M'Y'}$ to $K'_{MIn'}$: $K'_{M'Y'}/K'_{MIn'} > 10^4$.
3   Small ratio of indicator concentration to metal ion concentration: $C_{In}/C_M < 0.01$.

Detailed discussions of end-point detection by metal ion indicators have been presented by several.[23,27-29]

11-6   END-POINT DETECTION BY THE MERCURY-EDTA ELECTRODE

It might at first seem that the direct measurement of electrode potential would, by analogy to pH measurement, afford a simple means for determining pM during the course of EDTA titrations. Unfortunately, many metal electrodes do not behave reversibly, particularly at the extremely low metal ion concentrations involved near the end point in EDTA titrations. Not only might the exchange current densities be prohibitively small (for the transition metals, reversible behavior cannot be observed even at relatively high metal ion concentrations), but electrode side reactions often lead to a *mixed potential* phenomenon (Section 12-9). This is particularly true of the more active metals in solutions containing low concentrations of their ions, which have strongly reducing potentials and therefore are subject to many interferences. In particular, hydrogen evolution often would intervene.

On the other hand, for metals for which ion-selective electrodes have been developed, pM values can be more generally determined through the measurement of membrane potentials. The electrode involving mercury-mercury(II) and EDTA is almost uniquely suitable and broadly applicable as an indicator electrode for several reasons:

1 The equilibrium between metallic mercury and mercury(II) to give mercury(I) is rapid.

2 The mercury-mercury(I) electrode has an exceptionally high exchange current density.[30,31]

3 Mercury is relatively unreactive, so that in spite of the stable EDTA complex of mercury(II) (log K_{HgY} = 22.1), the equilibrium potential is not excessively negative.

4 The high hydrogen overpotential of mercury prevents mixed potential behavior due to hydrogen evolution.

Reilley and Schmid[32] made the important observation that the mercury-mercury (II)-EDTA electrode can be used indirectly as an indicator electrode for various metal ions, and this method was applied[33] to the determination of 29 different metal ions by either direct- or back-titration procedures. Under some conditions,[34] such as when a high concentration of buffer is present, formation of a mercury(I) precipitate or a complex with mercury(II) may result in incorrect end points.

If a metal ion–EDTA complex $MY^{(n-4)+}$ is present in equilibrium with mercury and the mercury(II) complex $HgY^=$, we have the half-cell

$$M^{n+}, MY^{(n-4)+}, HgY^=, Hg^{++} | Hg \qquad (11\text{-}28)$$

It can be shown that the potential at equilibrium is given by

$$E = E^{0'}_{Hg^{++},Hg} - \frac{RT}{2F} \ln \frac{K_{HgY}}{[HgY^=]K_{MY}} - \frac{RT}{2F} \ln \frac{[MY^{(n-4)+}]}{[M^{n+}]} \qquad (11\text{-}29)$$

The first two terms on the right side of (11-29) are essentially constant during a titration. In the region of the end point, $[MY^{(n-4)+}]$ in the third term changes little, so that the measured potential of the mercury electrode becomes a linear function of pM.

EXAMPLE 11-4 Calculate the relation between the potential of an indicator electrode of the type (11-28) and pNi near the end point of an EDTA titration of nickel(II).

ANSWER With $E^{0\prime}_{Hg^{++},Hg} = E^0 = 0.858$, $HgY^= = NiY^= = 10^{-3}$. From Table 11-2, $\log K_{HgY}/K_{NiY} = 21.8 - 18.6 = 3.2$. We then calculate, from (11-29), $E = 0.763 - 0.0296$ pNi. ////

Next consider in more detail a specific example of the use of the mercury-EDTA electrode in the titration of calcium with standard EDTA. In such a titration a small amount of a solution containing mercury-EDTA complex $HgY^=$ is added to the calcium solution, and the indicating electrode is a mercury-coated gold wire. The following equilibria are involved:

$$Hg' + Y' \rightleftharpoons HgY^= \qquad K'_{Hg'Y'} = \frac{[HgY^=]}{[Hg'][Y']} \qquad (11\text{-}30)$$

and

$$Ca' + Y' \rightleftharpoons CaY^= \qquad K'_{Ca'Y'} = \frac{[CaY^=]}{[Ca'][Y']} \qquad (11\text{-}31)$$

where $[Y']$ is the concentration of all forms of EDTA not coordinated to metal (Section 11-2) and $[Hg']$ and $[Ca']$ are the concentrations of all forms of these metal ions not coordinated to EDTA. The potential E of the indicating electrode is given by

$$E = E^0 - \frac{RT}{2F} \ln \frac{1}{[Hg^{++}]} \qquad (11\text{-}32)$$

From (11-31) and (11-12), however,

$$[Y^{4-}] = \frac{[CaY^=]\alpha_Y}{[Ca']K'_{Ca'Y'}} \qquad (11\text{-}33)$$

and

$$[Hg^{++}] = \frac{[HgY^=]\alpha_{Hg^{++}}}{[Y^{4-}]K'_{Hg'Y'}} = \left(\frac{[HgY^=]\alpha_{Hg^{++}}K'_{Ca'Y'}}{\alpha_Y K'_{Hg'Y'}}\right)\frac{[Ca']}{[CaY^=]} \qquad (11\text{-}34)$$

The quantity enclosed in parentheses is essentially constant. Therefore, from (11-32)

$$E = E^0 - \frac{RT}{2F} \ln \frac{\alpha_Y K'_{Hg'Y'}}{[HgY^=]\alpha_{Hg^{++}}K'_{Ca'Y'}} - \frac{RT}{2F} \ln \frac{[CaY^=]}{[Ca']}$$

$$= E^{0\prime} - \frac{RT}{2F} \ln \frac{[CaY^=]}{[Ca']} \qquad (11\text{-}35)$$

From (11-35), and as illustrated schematically by Figure 11-9A, at constant $[CaY^=]$ the electrode potential decreases as calcium ion concentration increases. During an EDTA titration, pCa increases sharply near the equivalence point (Figure 11-9B). If we consider now the entire titration and take into account the sharply increasing value of the ratio $[CaY^=]/[Ca']$ during the early parts of the titration, the potential

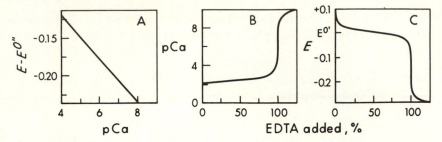

FIGURE 11-9 *A*, Schematic illustration of change in mercury electrode potential with pCa at constant $[CaY^=]$; *B*, relation between pCa and percentage reacted with EDTA for 0.01 *M* Ca^{++} (formation constant $10^{10.7}$); *C*, relation between potential and percentage of EDTA added. In *C* the ratio $[CaY^=]/[Ca']$ is variable, as in a titration.

will decrease during the titration and decrease sharply in the region of the equivalence point, as indicated by Figure 11-9*C*. In general, the absolute value of the change in potential near the equivalence point depends in part on the stability constant of the metal-EDTA complex formed.

A silver indicating electrode [35] can be used in a manner analogous to the mercury electrode for EDTA titrations. In contrast to the mercury electrode, the success of the silver electrode depends on the formation of a silver-EDTA complex that is weak compared with virtually all other metal ions. A trace of silver(I) is added, and after the EDTA has reacted with other metal ions, the first excess of EDTA reacts with silver and sharply decreases the potential of the silver electrode. Owing to the low stability constant of the silver-EDTA complex, titrations must be carried out at high pH.

Another type of potentiometric indication, of less general interest, is based on an electrode reaction involving two different oxidation states of the same metal. Such an electrode reaction has been used[36,37] to follow the titration of iron(III)-iron(II) mixtures with EDTA, measuring the potential of the iron(III)-iron(II) couple. At pH 3, iron(II) remains uncomplexed with EDTA during the titration, and a large potential change [Equation (11-36)] accompanies the abrupt change in iron(III) concentration in the end-point region. Redox indicators also can be used[5] as visual indicators to detect the change in potential.

The scope of the present treatment does not include details of the various instrumental methods for the detection of EDTA titration end points. Nevertheless, we may mention spectrophotometric detection methods, which are of two types. The first is based on instrumental observation of the color changes of metal ion indicators. The second is based on the absorption of radiation in the visible or ultraviolet regions of the spectrum by the metal-EDTA complex.[38,39] For example, $MgY^=$ shows appreciable absorbance at a wavelength of 222 nm, whereas the reagent $H_2Y^=$

is nearly transparent. When the metal concentration is low (10^{-4} to 10^{-5} M), indicator in an amount approximately equivalent to the metal ion is used and extrapolated photometric end points do not always give correct results.[40]

11-7 APPLICATIONS OF EDTA TITRATIONS

Direct titrations with EDTA are commonly carried out by the use of disodium dihydrogen ethylenediaminetetraacetate, Na_2H_2Y, which is available[41] in pure form and can be dried as $Na_2H_2Y \cdot 2H_2O$. A multitude of such titration procedures has been described by Schwarzenbach and Flaschka,[5] West,[6] Welcher,[42] Přibil,[43] and Reilley and Barnard.[44] Extensive tables of fundamental equilibrium data relating to ligands, metals, and indicators also are available.[11]

In direct titrations auxiliary complexing agents such as citrate, tartrate, or triethanolamine are added, if necessary, to prevent the precipitation of metal hydroxides or basic salts. When a pH range of 9 to 10 is suitable, a buffer of ammonia and ammonium chloride often is added in relatively concentrated form, both to adjust the pH and to supply ammonia as a complexing agent for those metal ions that form ammine complexes. A few metals, notably iron(III), bismuth, and thorium, are titrated in acid solution.

Back titrations based on the addition of excess EDTA followed by back titration of the excess reagent are useful when reactions are slow or a suitable direct indicator is not available. The excess generally is determined by titration with standard solutions of magnesium or zinc ion. These titrants are chosen because they form EDTA complexes of relatively low stability, thereby preventing the possible titration of EDTA bound by the sample metal ion. Examples of the indirect method are the following:

1 Analysis for chromium(III) and cobalt [after oxidation to cobalt(III)]. CrY^- and CoY^- complexes are too slowly formed to be titrated directly, but once formed are sufficiently inert that back titration of excess EDTA can be carried out at low pH with a solution of iron(III), even though FeY^- is a thermodynamically more stable complex. The exchange reaction $CrY^- + Fe^{3+} \rightleftharpoons Cr^{3+} + FeY^-$ is negligible in the time necessary for a titration.

2 The determination of metals in precipitates (lead in lead sulfate, magnesium in magnesium ammonium phosphate, calcium in calcium oxalate).

3 Determination of thallium, which forms a stable EDTA complex but does not respond to the usual metal ion indicators.

Replacement titrations are based on a reaction of the type

$$M^{n+} + MgY^{=} \rightleftharpoons MY^{(n-4)+} + Mg^{++}$$

in which the metal ion M^{n+} displaces magnesium ion from its relatively weak EDTA complex. The magnesium-EDTA solution can be added in excess, and the resulting magnesium ion titrated with EDTA with calmagite as the indicator. In principle, ions other than magnesium could be used equally well if they form a weaker EDTA complex than the metal ion being determined. Examples of replacement titrations include the following:

1 Determination of calcium by displacement of magnesium, which uses to advantage the superior magnesium-calmagite color reaction.
2 Determination of barium by displacement of zinc ions in a strongly ammoniacal solution. In general, replacement titrations seem to offer little advantage over back titrations.

Redox titrations can be carried out in the presence of excess EDTA. Here EDTA acts to change the ease of oxidation of a couple, usually by forming a more stable complex with the higher oxidation state. If $M_{ox}Y$ and $M_{red}Y$ represent the EDTA complexes of the oxidized and reduced forms of the metal, and if $K_{M_{ox}Y}$ and $K_{M_{red}Y}$ are the formation constants, we have the following equation for the electrode potential [see Equation (12-62)]:

$$E = E^{0'}_{M_{ox}M_{red}} + \frac{RT}{nF} \ln \frac{K_{M_{red}Y}}{K_{M_{ox}Y}} - \frac{RT}{nF} \ln \frac{[M_{red}Y]}{[M_{ox}Y]} \qquad (11\text{-}36)$$

Usually $K_{M_{ox}Y} > K_{M_{red}Y}$, and hence the couple becomes a stronger reducing agent in the presence of excess EDTA. The effect is pronounced with the cobalt(III)-cobalt(II) couple, whose potential is shifted[45] about 1.2 V, so that cobalt(II) can be titrated with cerium(IV).[46] Alternatively, cobalt(III) can be titrated to cobalt(II) with chromium (II) as the reducing agent.[45,47]

Manganese(II) can be titrated directly to manganese(III) with ferricyanide as the oxidant. Alternatively, manganese(III), prepared by oxidation of the manganese-(II)-EDTA complex with lead dioxide, can be determined by titration with standard iron(II) sulfate.[48] Note here that iron(II) becomes a much stronger reducing agent in the presence of EDTA. The formal potential (Section 12-3) of the iron(III)-iron(II) couple is 0.117 V at pH values of 4 to 6, where both form stable EDTA complexes.[49] Iron(II) becomes such a strong reductant that air must be excluded. It has been used as a reducing reagent for iodine,[50] silver(I),[51] copper(II),[52] and iron(III).[53] For the iron(III) titration the iron(II)-EDTA complex was generated electrolytically by reduction of the iron(III)-EDTA complex.

Indirect determinations of several types can be carried out. Sulfate has been determined by adding an excess of standard barium(II) solution and back-titrating the excess.[54] By titrating the cations in moderately soluble precipitates, other ions can be determined indirectly. Thus sodium has been determined by titration of zinc in sodium zinc uranyl acetate,[55] and phosphate by determination of magnesium in magnesium ammonium phosphate.[56] Quantitative formation of tetracyano nickelate(II) has been used for the indirect determination of cyanide.[57]

Instead of using a standard solution of EDTA for titrations, Reilley and Porter-field[58] devised an electrolytic method of generation of EDTA to replace the usual titrant solution. By means of the cathode reaction

$$HgY^= + 2e^- \rightarrow Hg + Y^{4-} \qquad (11\text{-}37)$$

the EDTA anion Y^{4-} can be generated at a rate proportional to the current and used for the titration of metal ions such as calcium, copper, zinc, and lead.

It is noteworthy that no procedures involving the direct reaction of EDTA with molybdenum(VI) or tungsten(VI) have been developed for the measurement of these metals.

11-8 MASKING

In complexation analysis it is appropriate to comment on masking as a separate topic. Perrin[59] defined a masking reagent as one that decreases the concentration of a free metal ion or a ligand to a level where certain of its chemical reactions are prevented. For example, fluoride can be used as a masking reagent to prevent reaction of aluminum with EDTA,[60] ammonia as a masking reagent to prevent precipitation of copper hydroxide at high pH, and EDTA to mask iron(III) so that it no longer forms interfering red thiocyanate complexes. A simple change in pH can change the reactivity of an ion, as in the formation of the hydroxy complexes of aluminum, $Al(OH)_4^-$, at high pH. Similarly, a change in oxidation number can be classed broadly as a masking phenomenon.

A *masking index*[61] is defined by $-\log \alpha M^{n+}$ [Equation (11-13), where M' includes the free metal ions, the various hydroxy complexes, and also ML, ML_2, ... from reactions of the metal ion with masking ligands]. The masking index can be calculated from the same type of data used to derive Figure 11-4.

Both because most ligands undergo acid-base reactions so that in general their masking ability increases at high pH, and because hydroxy complexes form more extensively with increasing pH, the masking index usually trends upward with increasing pH, as illustrated in Figure 11-10.

Ordinarily, if the masking index is much larger than the logarithm of the conditional formation constant of a metal ion and reagent, the masking agent will prevent a reaction, and if the masking index is much smaller, there will be no interference with the reaction. In contrast to titration reagents, where the stoichiometry of the reaction must be known and favorable, masking reagents are useful even though they may form a variety of complexes with similar equilibrium constants. For this reason the variety of masking agents is wider.[62]

Although not highly selective, amines and ammonia are commonly used as masking agents with masking indices in the range of 5 to 25 for metal ions such as

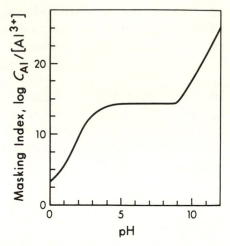

FIGURE 11-10   Masking index as a function of pH for aluminum in 0.1 M fluoride. (*From Perrin,*[3] *p. 27.*)

mercury(II), copper(II), silver, zinc, nickel, and cadmium (Figure 11-4). Acetic acid buffers can be used to mask lead ions to prevent lead sulfate precipitation (masking index about 3 or 4). Citrate in 0.5 M solution at pH 13 gives masking indices of 26 for aluminum and 22 for iron(III); the formation of the soluble complexes of oxalate, citrate, and tartrate are useful in preventing precipitation of hydroxides of many metals. At lower pH, oxalate is a better masking agent for these ions than is citrate. Cyanide masks such ions as silver, cadmium, cobalt, copper, iron, mercury, nickel, and zinc against EDTA reaction at high pH, but does not affect aluminum, bismuth, magnesium, manganese, lead, or calcium. It is therefore useful in differential EDTA titrations of combinations of these metals. Thiols are often proposed in place of cyanide for masking because they are less hazardous at low pH. Triethanolamine is most frequently used to mask iron and aluminum in EDTA titrations. EDTA itself and similar reagents are powerful masking agents for most metal ions; they are used, for example, to prevent precipitation or to eliminate their interference with reactions of anions in solution. A summary of titrations involving masking agents has been made.[44]

Substoichiometric masking[63] involves an amount of masking agent less than that for complete reaction but enough to bring a substance below its interference threshold. Such restricted amounts are used when a larger amount would interfere with the main reaction of interest.

Demasking is the process[64] by which a masked substance is made capable of undergoing its usual reactions. Demasking can be brought about by a displacement reaction involving addition of, for instance, another cation that reacts more strongly with the masking ligand and liberates the masked ion. Since conditional formation constants depend strongly on pH, a common mode of demasking is through change in pH. Thus, in Figure 11-10 aluminum can be largely demasked from fluoride by decreasing the pH to a low value. Other possibilities are destroying or physically removing the masking ligand and changing the oxidation state of the masked ion.

11-9   OTHER CHELATING REAGENTS

Of the many chelating reagents studied, comparatively few offer appreciable advantages over EDTA as analytical reagents. Many of these other titrants have been reviewed by West.[65]

Historically, nitrilotriacetic acid (NTA) is next in importance to EDTA. It forms stable complexes with most metals, even to some extent with the alkali metals (the pK for sodium is 2.1). Unfortunately, its tendency to form 2:1 complexes causes difficulty with the stoichiometry.

A particular reagent may have an advantage of selectivity when used with given combinations of metal ions. An example is ethylene glycol-bis(2-aminoethyl ether)-N,N,N',N'-tetraacetic acid (EGTA). Its log K value with calcium is 11.0, whereas that with magnesium is only 5.2. The difference of 5.8 units means that in principle it should be possible to titrate calcium in the presence of magnesium without interference from the magnesium. With EDTA the log K for calcium is only 2 units larger than that for magnesium, and therefore in water-hardness titrations for calcium and magnesium no differentiation between the two is ordinarily possible. With titration at a pH of 12, magnesium hydroxide is precipitated (Figure 11-8), and calcium alone can be measured with EDTA.[66] Coprecipitation of calcium, however, makes the results low. Replacement of EDTA with EGTA permits direct titration of calcium in the presence of magnesium, but no good visual end-point methods have been developed. An end point can be obtained with the mercury indicator electrode.[67] With the aid of masking agents, pH control, and organic solvents the titration with EGTA of combinations of several metal ions in lubricating oil has been described.[68]

Přibil and Vesely[69] and Harju and Ringbom[70] measured stability constants for a large number of metal ions with triethylenetetraamine-hexaacetic acid (TTHA) and suggested the use of this unique reagent in combination with other complexing agents for the analysis of multicomponent systems. Because TTHA is decadentate, it can form not only 1:1 but also binuclear complexes. For example, thorium, zirconium, indium, and manganese are measurable as 1:1 ML complexes with this reagent, whereas aluminum, iron, and gallium form M_2L complexes. A solution, say one containing thorium and iron, therefore can be resolved by carrying out two titrations, one with EDTA and a second with TTHA. From the difference in the amounts of the two reagents used, the amounts of both thorium and iron could be calculated.

REFERENCES

1   G. SCHWARZENBACH: *Schweiz. Chem. Ztg. Tech. Ind.*, **28**:377 (1945).

2   M. T. BECK: "Chemistry of Complex Equilibria," Van Nostrand Reinhold, London, 1970.

3   D. D. PERRIN: "Masking and Demasking of Chemical Reactions," Wiley, New York, 1970.

4   A. RINGBOM: "Complexation in Analytical Chemistry," Wiley, New York, 1963.

5   G. SCHWARZENBACH and H. FLASCHKA: "Complexometric Titrations," 5th ed., translated by H. M. N. H. Irving, Methuen, London, 1969.

6   T. S. WEST: "Complexometry," 3d ed., BDH Chemicals, London, 1969.

7   I. M. KOLTOFF and V. A. STENGER: "Volumetric Analysis," 2d ed., vol. 2, Interscience, New York, 1947; H. A. LAITINEN, "Chemical Analysis," McGraw-Hill, New York, 1960.

8   N. BJERRUM: *Z. Anorg. Allg. Chem.*, **118**:131 (1921); **119**:39, 51, 179 (1921).

9   J. BJERRUM: *Kgl. Dan. Vidensk. Selsk. Mat.-Fys. Medd.*, **11**:(5) 58 pp. (1931); **11**:3 (1932); **12**(15) 67 pp. (1934). (See also "Metal Ammine Formation in Aqueous Solution," P. Haase and Son, Copenhagen, 1941, reprinted 1957).

10   G. SCHWARZENBACH: *Helv. Chim. Acta*, **35**:2344 (1952).

11   L. G. SILLÉN and A. E. MARTELL: "Stability Constants of Metal-Ion Complexes," Special Publication Nos. 17 and 25, Chemical Society, London, 1964, 1971.

12   A. RINGBOM and L. HARJU: *Anal. Chim. Acta*, **59**:33, 49 (1972).

13   M. D. LIND, M. J. HAMOR, T. A. HAMOR, and J. L. HOARD: *Inorg. Chem.*, **3**:34 (1964); G. S. SMITH and J. L. HOARD: *J. Amer. Chem. Soc.*, **81**:556 (1959).

14   M. W. GRANT, H. W. DODGEN, and J. P. HUNT: *J. Amer. Chem. Soc.*, **93**:6828 (1971).

15   N. A. MATWIYOFF, C. E. STROUSE, and L. O. MORGAN: *J. Amer. Chem. Soc.*, **92**:5222 (1970).

16   W. C. E. HIGGINSON and B. SAMUEL: *J. Chem. Soc.*, A, **1970**:1579.

17   G. SCHWARZENBACH and H. ACKERMANN: *Helv. Chim. Acta*, **31**:1029 (1948).

18   D. E. LEYDEN: *Crit. Rev. Anal. Chem.*, **2**:383 (1971).

19   J. L. SUDMEIER and C. N. REILLEY: *Anal. Chem.*, **36**:1698 (1964).

20   D. C. OLSON and D. W. MARGERUM: *J. Amer. Chem. Soc.*, **82**:5602 (1960).

21   B. W. BUDESINSKY: *Anal. Chem.*, **42**:928 (1970).

22   A. JOHANSSON: *Talanta*, **20**:89 (1973).

23   A. RINGBOM and E. WANNINEN in "Indicators," E. Bishop (ed.), chap. 6, Pergamon, New York, 1972.

24   G. SCHWARZENBACH and W. BIEDERMANN: *Helv. Chim. Acta*, **31**:678 (1948).

25   F. LINDSTROM and H. DIEHL: *Anal. Chem.*, **32**:1123 (1960).

26   H. FLASCHKA and S. KHALAFALLAH: *Fresenius' Z. Anal. Chem.*, **156**:401 (1957); H. FLASCHKA: *Talanta*, **1**:60 (1958).

27   C. N. REILLEY and R. W. SCHMID: *Anal. Chem.*, **31**:887 (1959).

28   RINGBOM,[4] p. 82.

29   SCHWARZENBACH and FLASCHKA,[5] p. 77.

30   H. GERISCHER and K. STAUBACH: *Z. Phys. Chem.* (Leipzig), **6**:118 (1956).

31   K. B. OLDHAM: *Trans. Faraday Soc.*, **53**:80 (1957).

32   C. N. REILLEY and R. W. SCHMID: *Anal. Chem.*, **30**:947 (1958).

33   C. N. REILLEY, R. W. SCHMID, and D. W. LAMSON: *Anal. Chem.*, **30**:953 (1958).

34 U. HANNEMA, G. J. VAN ROSSUM, and G. DEN BOEF: *Fresenius' Z. Anal. Chem.*, **250**:302 (1970).

35 J. S. FRITZ and B. B. GARRALDA: *Anal. Chem.*, **36**:737 (1964).

36 R. PŘIBIL, Z. KOUDELA, and B. MATYSKA: *Collect. Czech. Chem. Commun.*, **16**:80 (1951).

37 A. HULANICKI and R. KARWOWSKA: *Talanta*, **18**:239 (1971).

38 P. B. SWEETSER and C. E. BRICKER: *Anal. Chem.*, **25**:253 (1953).

39 P. B. SWEETSER and C. E. BRICKER: *Anal Chem.*, **26**:195 (1954).

40 J. KRAGTEN: *Fresenius' Z. Anal. Chem.*, **264**:356 (1973); *Talanta*, **20**:937 (1973).

41 W. J. BLAEDEL and H. T. KNIGHT: *Anal. Chem.*, **26**:741 (1954).

42 F. J. WELCHER: "The Analytical Uses of Ethylenediaminetetraacetic Acid," Van Nostrand, Princeton, N.J., 1958.

43 R. PŘIBIL: *Talanta*, **12**:925 (1965); **13**:1223 (1966); **14**:613 (1967).

44 C. N. REILLEY and A. J. BARNARD, JR.: in "Handbook of Analytical Chemistry," L. Meites (ed.), pp. 3–77, McGraw-Hill, New York, 1963.

45 R. PŘIBIL: *Collect. Czech. Chem. Commun.*, **14**:320 (1949).

46 R. PŘIBIL and V. MALICKY: *Collect. Czech. Chem. Commun.*, **14**:413 (1949).

47 R. PŘIBIL and L. SVESTKA: *Collect. Czech. Chem. Commun.*, **15**:31 (1950).

48 R. PŘIBIL and V. SIMON: *Collect. Czech. Chem. Commun.*, **14**:454 (1949); R. PŘIBIL and J. HORAČEK: *Collect. Czech. Chem. Commun.*, **14**:626 (1949).

49 G. SCHWARZENBACH and J. HELLER: *Helv. Chim. Acta*, **34**:576 (1951).

50 R. PŘIBIL, V. SIMON, and J. DOLEŽAL: *Collect. Czech. Chem. Commun.*, **16**:573 (1951).

51 R. PŘIBIL, J. DOLEŽAL, and V. SIMON: *Chem. Listy*, **47**:1017 (1953).

52 K. L. CHENG: *Anal. Chem.*, **27**:1165 (1955).

53 R. W. SCHMID and C. N. REILLEY: *Anal. Chem.*, **28**:520 (1956).

54 G. ANDEREGG, H. FLASCHKA, R. SALLMANN, and G. SCHWARZENBACH: *Helv. Chim. Acta*, **37**:113 (1954).

55 H. FLASCHKA: *Mikrochem. Mikrochim. Acta*, **39**:391 (1952).

56 H. FLASCHKA and A. HOLASEK: *Mikrochem. Mikrochim. Acta*, **39**:101 (1952).

57 F. HUDITZ and H. FLASCHKA: *Fresenius' Z. Anal. Chem.*, **136**:185 (1952).

58 C. N. REILLEY and W. W. PORTERFIELD: *Anal. Chem.*, **28**:443 (1956).

59 PERRIN,[3] p. 1.

60 R. PŘIBIL and V. VESELY: *Talanta*, **9**:23 (1962).

61 PERRIN,[3] p. 25.

62 PERRIN,[3] p. 30.

63 H. FLASCHKA and J. GARRETT: *Talanta*, **15**:589, 595 (1968).

64 PERRIN,[3] p. 48.

65 WEST,[6] p. 123.

66 A. D. KENNEY and V. H. COHN: *Anal. Chem.*, **30**:1366 (1958).

67 R. W. SCHMID and C. N. REILLEY: *Anal. Chem.*, **29**:264 (1957).

68 J. N. WILSON and C. Z. MARCZEWSKI: *Analyst*, **98**:42 (1973).

69 R. PŘIBIL and V. VESELY: *Talanta*, **9**:939 (1962); **10**:899 (1963); **11**:1319 (1964); **18**:395 (1971); R. PŘIBIL: *Talanta*, **12**:925 (1965); **13**:1223 (1966).

70 L. HARJU and A. RINGBOM: *Anal. Chim. Acta*, **49**:205, 221 (1970).

71 G. WESTÖÖ: *Acta Chem. Scand.*, **20**:2131 (1966); **21**:1790 (1967); **22**:2277 (1968).

PROBLEMS

11-1 Using values for the logarithms of the successive formation constants K_1, K_2, K_3, and K_4 of 2.27, 2.34, 2.40, and 2.05 for the zinc ammines, calculate the fraction of Zn^{++} in the uncomplexed form in solutions containing 0.01, 0.1, and 1 M free NH_3.

Answer 0.035, 8.0 × 10^{-6}, 8.6 × 10^{-10}.

11-2 For a series of buffers containing NH_3 (pK_b = 4.7) and ammonium ion at a total concentration of 0.1 M, calculate the fraction of zinc in the uncomplexed form at pH values of 8, 9, 10, and 11. Starting with a total Zn^{++} concentration of 10^{-3} M, calculate the titration curves of zinc with EDTA at these pH values. Use points corresponding to 0, 50, 91, 99, 99.9, 100, 100.1, 101, and 110% of the stoichiometric amounts of EDTA.

Answer At 99.9, pZn = 6.7, 9.3, 10.79, and 11.06.
At 100.0, pZn = 8.93, 10.74, 11.92, and 12.25.
At 100.1, pZn = 11.21, 12.22, 13.04, and 13.43.

11-3 Using the values given in Section 11-5 for conditional indicator constants for calmagite and the titration error curves calculated in Figure 11-7, estimate the titration errors with calmagite for a 0.01 M magnesium titration at pH values of 8 and 9.

Answer For ΔpMg = 0.5: At pH 8, > 10%;
at pH 10, $\simeq$ 4%; at pH 12, $\simeq$ 0.5%.

11-4 A solution saturated with both H_2S and CdS at a pH of 1 has a Cd^{++} concentration of 0.001 M. Another solution that is 1 M in cyanide and saturated with both H_2S and CdS at a pH of 10 has a $CdCN_4^=$ concentration of 0.001 M. What is the Cd^{++} concentration in this second solution? (Numerical values of the various equilibrium constants are not needed.) What is β_4 for $CdCN_4^=$?

Answer $[Cd^{++}]$ = 10^{-21} M; β_4 = 10^{18}.

11-5 Assume that log β_4 for chloride, bromide, and iodide with Zn^{++} is -1, -0.74, and -1.25 and with Cd^{++} is 0.9, 2.53, and 6.1. (a) Calculate the masking index for Zn^{++} and Cd^{++} in 1.0 M chloride, 1.0 M bromide, and 0.1 M and 1.0 M iodide. Assume activity coefficients of unity and that the metal ions are present predominately as their highest complexes. (b) Assuming that the conditional formation constants of Zn^{++} and Cd^{++} with EDTA are the same (Figure 11-3), which condition is most favorable for the titration of zinc in the presence of cadmium?

Answer. (a) Masking index for Zn is less than 0.1 in all cases; masking index for Cd is 0.9, 2.5, 2.1, and 6.1.
(b) Most favorable condition is 1 M iodide.

11-6 Allowing an error in ΔpM of 0.5 for selection of the end point, combine the information in Figures 11-3 and 11-7 to estimate the pH range over which a titration precise to 0.1% can be achieved (a) with a 0.01 M solution of aluminum and (b) with a 0.01 M solution of cadmium.

Answer (a) pH about 5 to 6. (b) pH about 4 to 13.

11-7 Calculate $\alpha_{Y^{4-}}$ for EDTA at a pH of 12 in a solution containing a buffer giving $[Na^+] = 0.1$ M. Assume activity coefficients of unity.

Answer 0.17.

11-8 A 4.013-g sample containing aluminum and indium compounds is dissolved and diluted to 100.08 ml. A 10.014-ml aliquot after suitable adjustment of conditions is titrated with 0.01036 M EDTA, requiring 36.32 ml. A second aliquot requires 18.43 ml of 0.01142 M TTHA. Calculate the percentage of aluminum and of indium in the sample.

Answer Al $= 2.228\%$, In $= 1.277\%$.

11-9 A solution of 0.001 M iron(III) is masked by the addition of 0.05 M EDTA. (a) Calculate the masking index at pH 2, assuming formation of hydroxy complexes is negligible. (b) Calculate the masking index for aluminum(III) under the same conditions.

Answer (a) 10.4. (b) 1.4.

11-10 Mercury is present in foodstuffs in two forms: as mercury(II) and as organic mercury such as CH_3Hg^+. In both cases, presumably the mercury is bonded to ionized sulfhydryl groups (RS^-) including the sulfhydryl groups of amino acids, proteins, and other thiol compounds. In the determination of methylmercury, it needs to be separated from the foodstuffs and from RSHgSR compounds. In one procedure[71] the CH_3Hg^+ can be converted to CH_3HgCl with HCl and then extracted into benzene along with RSH and RSHgSR compounds. The benzene extract is shaken with an aqueous solution of $HgCl_2$ to complex thiol groups as stable, benzene-soluble RSHgSR compounds. The benzene extract is then shaken with 2 M NH_3 solution to extract the methylmercury into the aqueous phase as CH_3HgOH. The success of this method depends in part on the completeness of the conversion of methylmercury to CH_3HgCl in HCl solution and on the effectiveness of the masking of RSH by mercury(II) in ammoniacal solution. (a) Calculate the ratio $[CH_3HgCl]/[CH_3HgSR]$ in 2 M aqueous HCl, assuming the total concentration of sulfhydryl compounds to be 10^{-4} M and of methylmercury compounds to be 10^{-7} M. (b) Calculate the ratio $[CH_3HgOH]/[CH_3HgSR]$ in the aqueous NH_3 solution, assuming the hydroxyl ion concentration to be 0.1 M, the total concentration of methylmercury compounds to be 10^{-7} M, and of sulfhydryl compounds to be 10^{-6} M. Assume that the ionization constant K_a for $RSH \rightleftharpoons RS^- + H^+$ is 3.0×10^{-10} and that the following formation constants apply: $CH_3Hg^+ + Cl^- \rightleftharpoons CH_3HgCl$, $K = 1.8 \times 10^5$; $CH_3Hg^+ + RS^- \rightleftharpoons CH_3HgSR$, $K = 1.3 \times 10^{16}$; $CH_3Hg^+ + OH^- \rightleftharpoons CH_3HgOH$, $K = 4.3 \times 10^{10}$.

Answer   (*a*) 1.8×10^3. (*b*) 0.36.

11-11   Construct a predominance-area diagram for calmagite and zinc ions, given that log K for the zinc calmagite complex is 12.5 and that the solubility product of zinc hydroxide is 5×10^{-18}.

11-12   Construct[22] a logarithmic diagram (Section 3-1) of log C as a function of pY for a solution of 10^{-2} M of a metal whose log K_{MY} is 10.0.

ELECTRODE POTENTIALS

Electroanalytical chemistry has been defined[1] as the application of electrochemistry to analytical chemistry. For the determination of the composition of samples, the three most fundamental measurements in electroanalytical chemistry are those for potential, current, and time. In this chapter several aspects relating to electrode potentials are considered; current and time as well as further consideration of potentials are treated in Chapter 14. The electrode potentials involved in the classical galvanic cell are of considerable theoretical and practical significance for the under-standing of many aspects not only of electroanalytical chemistry but also of thermodynamics and chemical equilibria, including the measurement of equilibrium constants.

12-1 GENERAL BACKGROUND

The difference in electrical potential between two points is defined as the amount of electrical work required to move a unit of positive charge from one point to the other. If the charge is measured in coulombs and the electrical work in joules (volt-coulombs), the potential difference is measured in volts. If an arbitrary point (say, at infinity) is assigned a potential of zero, the work expended in transporting a positive charge from the arbitrary point to another point is proportional to the electrical potential of the second point, with due regard for the sign of the potential at that point. Thus a region of higher potential is a region of higher density of positive charge, or one with a deficiency of electrons.

Over the decades there has been considerable discussion and confusion about the sign (positive or negative) of potentials of electrodes. A brief history of electrochemical sign conventions has been presented by Licht and de Béthune.[2] A significant feature of the so-called European convention (which can be traced back to Gibbs' "Equilibrium of Heterogeneous Substances," 1878) is that the sign of the potential is invariant and corresponds to the electrostatic charge of the metal. This system was adopted by Ostwald[3] and, after falling into temporary disuse, was revived in 1911 by Abegg, Auerbach, and Luther.[4] It was used throughout Europe and extensively in America by physicists, practical electrochemists, engineers, and many analytical chemists and biochemists. The sign convention used by American physical chemists, on the other hand, had as its essential feature that the sign of the potential depended on the direction of writing the half-reaction. Although de Béthune traced the ambivalent sign back to Nernst (a European!), the usage was first set forth clearly by Lewis and Randall.[5]

After a long period of confusion, agreement was reached on conventions regarding the signs of emfs and electrode potentials at the 17th Conference of the International Union of Pure and Applied Chemistry (IUPAC) in Stockholm in 1953. In essence, this agreement[2] consists in using the word *potential* exclusively to describe the quantity associated with an electrode whose sign corresponds to the electrostatic charge. On the other hand, the *half-cell emf* depends on which of the two possible ways the electrode half-reaction is written. Thus the half-cell electrode potential of the standard silver electrode is $+0.80$ V. The half-cell emf of the standard silver electrode written as a reduction process ($Ag^+ + e \rightarrow Ag$) is $+0.80$ V, and written as an oxidation process ($Ag \rightarrow Ag^+ + e$) is -0.80 V. The electrode potential is invariant, whereas the half-cell emf can be either positive or negative.

12-2 ELECTROMOTIVE FORCE OF A CELL AND POTENTIAL OF A HALF-CELL

In a diagram of a cell a single vertical line conventionally represents a phase boundary at which a potential difference is taken into account. A double vertical line represents a liquid junction at which the potential difference is ignored or is considered to be eliminated by an appropriate salt bridge. For example, a cell consisting of zinc and copper half-cells can be expressed by

$$Zn \mid Zn^{++} \parallel Cu^{++} \mid Cu \qquad (12\text{-}1)$$

According to the IUPAC agreement the emf is equal in sign and in magnitude to the electrical potential of the electrode on the right (metallic phase) when the potential of the electrode on the left is taken to be zero. The cell reaction corresponding to (12-1) is written in the direction corresponding to the passage (whether spontaneous or not) of positive electricity from left to right within the cell (electron flow from right to left). Thus, in the example chosen the cell reaction is

$$Zn + Cu^{++} \rightarrow Zn^{++} + Cu \qquad (12\text{-}2)$$

If this is the direction of the spontaneous current when the cell is short-circuited, the cell emf is positive. For reasonable concentrations of Zn^{++} and Cu^{++}, Reaction (12-2) actually is spontaneous and the emf positive.

If the cell were written

$$Cu \mid Cu^{++} \parallel Zn^{++} \mid Zn \qquad (12\text{-}3)$$

the cell reaction would be

$$Cu + Zn^{++} \rightarrow Cu^{++} + Zn \qquad (12\text{-}4)$$

and the emf negative, corresponding to a negative potential for zinc (the electrode on the right) as compared with copper. The cell emf conforms to the convention that the sign of the emf is positive if the electrode on the right is the positive terminal of the cell. (This is the + right rule; in terms of electronics the left side is the common or ground terminal.)

When we speak of the emf of the half-cell $Zn^{++} \mid Zn$, we mean the electromotive force of the cell

$$Pt,H_2 \mid H^+ \parallel Zn^{++} \mid Zn$$
$$a = 1 \quad a = 1 \qquad (12\text{-}5)$$

which implies the reaction

$$H_2 + Zn^{++} \rightarrow 2H^+ + Zn \qquad (12\text{-}6)$$

where the electrode on the left is the hypothetical[6] standard hydrogen electrode. Such a cell emf may properly be called the *electrode potential*. Thus the half-reaction

$$Zn^{++} + 2e^- \rightleftharpoons Zn \qquad (12\text{-}7)$$

may be written to correspond to the value for the emf of Cell (12-5) of -0.7628 V, which may be called the *standard electrode potential* of the zinc electrode, if both zinc and zinc ions are at unit activity.

If we speak of the half-cell $Zn \mid Zn^{++}$, we imply the reverse of Cell (12-5) and Reaction (12-6). This cell emf, which is $+0.7628$ V, should *not* be called the electrode potential of zinc.

The IUPAC agreement for electrode potentials requires that the *electrostatic sign* of the electrical potential with respect to the standard hydrogen electrode be preserved in the designation of electrode potentials. Thus, metals more active than hydrogen acquire a negative charge, relative to the hypothetical standard hydrogen electrode, and are given negative values of electrode potential.

We adopt here the practice of writing half-reactions with the oxidant on the left:

$$Zn^{++} + 2e^- \rightleftharpoons Zn \qquad E° = -0.7628 \text{ V}$$
$$Ag^+ + e \rightleftharpoons Ag \qquad E° = +0.7994 \text{ V}$$

(12-8)

In general, then,

$$ox + ne^- \rightleftharpoons red \qquad (12\text{-}9)$$

and a positive electrode potential means an oxidant ox is stronger than the hydrogen ion (that is, capable of oxidizing H_2 to hydrogen ions), whereas a negative potential means a reductant red is stronger than hydrogen. Table 12-1 lists a selection of standard electrode potentials.[7,8]

Table 12-1   SELECTED LIST† OF STANDARD
ELECTRODE POTENTIALS‡

Half-reaction	$E°$, V
$Na^+ + e^- \rightleftharpoons Na$	−2.713
$H_2AlO_3^- + H_2O + 3e^- \rightleftharpoons Al + 4OH^-$	−2.35
$Al^{3+} + 3e^- \rightleftharpoons Al$	−1.66
$Sn(OH)_6^= + 2e^- \rightleftharpoons HSnO_2^- + H_2O + 3OH^-$	−0.90
$Zn^{++} + 2e^- \rightleftharpoons Zn$	−0.7628
$AsO_4^{3-} + 3H_2O + 2e^- \rightleftharpoons H_2AsO_3^- + 4OH^-$	−0.67
$U^{4+} + e^- \rightleftharpoons U^{3+}$	−0.63
$Fe^{++} + 2e^- \rightleftharpoons Fe$	−0.44
$Cr^{3+} + e^- \rightleftharpoons Cr^{++}$	−0.38
$Cd^{++} + 2e^- \rightleftharpoons Cd$	−0.403
$V^{3+} + e^- \rightleftharpoons V^{++}$	−0.255
$Sn^{++} + 2e^- \rightleftharpoons Sn$	−0.14
$Pb^{++} + 2e^- \rightleftharpoons Pb$	−0.126
$2H^+ + 2e^- \rightleftharpoons H_2$	0.0000
$TiO^{++} + 2H^+ + e^- \rightleftharpoons Ti^{3+} + H_2O$	0.1
$S_4O_6^= + 2e^- \rightleftharpoons 2S_2O_3^=$	0.09
$S + 2H^+ + 2e^- \rightleftharpoons H_2S$	0.14
$Sn^{4+} + 2e^- \rightleftharpoons Sn^{++}$	0.14
$SO_4^= + 4H^+ + 2e^- \rightleftharpoons H_2SO_3 + H_2O$	0.17
$AgCl + e^- \rightleftharpoons Ag + Cl^-$	0.2223
$BiO^+ + 2H^+ + 3e^- \rightleftharpoons Bi + H_2O$	0.32
$UO_2^{++} + 4H^+ + 2e^- \rightleftharpoons U^{4+} + 2H_2O$	0.33
$Cu^{++} + 2e^- \rightleftharpoons Cu$	0.34
$VO^{++} + 2H^+ + e^- \rightleftharpoons V^{3+} + H_2O$	0.34
$O_2 + 2H_2O + 4e^- \rightleftharpoons 4OH^-$	0.401
$H_2SO_3 + 4H^+ + 4e^- \rightleftharpoons S + 3H_2O$	0.45
$Cu^+ + e^- \rightleftharpoons Cu$	0.52
$MnO_4^= + 2H_2O + 2e^- \rightleftharpoons MnO_2 + 4OH^-$	0.5
$I_3 + 2e^- \rightleftharpoons 3I^-$	0.545
$H_3AsO_4 + 2H^+ + 2e^- \rightleftharpoons HAsO_2 + 2H_2O$	0.56
$MnO_4^- + e^- \rightleftharpoons MnO_4^=$	0.57
$I_2 + 2e^- \rightleftharpoons 2I^-$	0.621
$O_2 + 2H^+ + 2e^- \rightleftharpoons H_2O_2$	0.69

Table 12-1 SELECTED LIST† OF STANDARD
ELECTRODE POTENTIALS‡ (*Continued*)

Half-reaction	$E°$, V
$OBr^- + H_2O + 2e^- \rightleftharpoons Br^- + 2OH^-$	0.76
$Fe^{3+} + e^- \rightleftharpoons Fe^{++}$	**0.771**
$Hg_2^{++} + 2e^- \rightleftharpoons 2Hg$	**0.792**
$Ag^+ + e^- \rightleftharpoons Ag$	**0.7994**
$OCl^- + H_2O + 2e^- \rightleftharpoons Cl^- + 2OH^-$	0.89
$2Hg^{++} + 2e^- \rightleftharpoons Hg_2^{++}$	**0.907**
$VO_2^+ + 2H^+ + e^- \rightleftharpoons VO^{++} + H_2O$	0.999
$2ICl_2^- + 2e^- \rightleftharpoons I_2 + 4Cl^-$	1.06
$Br_2 + 2e^- \rightleftharpoons 2Br^-$	**1.08**
$2IO_3^- + 12H^+ + 10e^- \rightleftharpoons I_2 + 6H_2O$	1.19
$MnO_2 + 4H^+ + 2e^- \rightleftharpoons Mn^{++} + 2H_2O$	1.23
$O_2 + 4H^+ + 4e^- \rightleftharpoons 2H_2O$	1.229
$Cr_2O_7^= + 14H^+ + 6e^- \rightleftharpoons 2Cr^{3+} + 7H_2O$	1.33
$Cl_2 + 2e^- \rightleftharpoons 2Cl^-$	**1.358**
$2HIO + 2H^+ + 2e^- \rightleftharpoons I_2 + 2H_2O$	1.45
$PbO_2 + 4H^+ + 2e^- \rightleftharpoons Pb^{++} + 2H_2O$	1.455
$Mn^{3+} + e^- \rightleftharpoons Mn^{++}$	1.51
$MnO_4^- + 8H^+ + 5e^- \rightleftharpoons Mn^{++} + 4H_2O$	**1.51**
$2BrO_3^- + 12H^+ + 10e^- \rightleftharpoons Br_2 + 6H_2O$	1.5
$2HBrO + 2H^+ + 2e^- \rightleftharpoons Br_2 + 2H_2O$	1.6
$Bi_2O_4 + 4H^+ + 2e^- \rightleftharpoons 2BiO^+ + 2H_2O$	1.59
$H_5IO_6 + H^+ + 2e^- \rightleftharpoons IO_3^- + 3H_2O$	~1.6
$Ce^{4+} + e^- \rightleftharpoons Ce^{3+}$	1.61
$2HClO + 2H^+ + 2e^- \rightleftharpoons Cl_2 + 2H_2O$	**1.63**
$MnO_4^- + 4H^+ + 3e^- \rightleftharpoons MnO_2 + 2H_2O$	1.68
$H_2O_2 + 2H^+ + 2e^- \rightleftharpoons 2H_2O$	**1.77**
$Ag^{++} + e^- \rightleftharpoons Ag^+$	1.98
$S_2O_8^= + 2e^- \rightleftharpoons 2SO_4^=$	2.0
$O_3 + 2H^+ + 2e^- \rightleftharpoons O_2 + H_2O$	**2.07**
$F_2 + 2e^- \rightleftharpoons 2F^-$	**2.87**

† Numbers in boldface are considered to be the most reliable[8] by the Electrochemical Commission of the IUPAC.
‡ From Latimer[7] and Charlot and others.[8]

12-3 THE NERNST EQUATION

The electrode potential of a redox couple varies with the activities of the reduced and oxidized forms of the couple in the sense that increasing activity of oxidant increases the value of the potential. Quantitatively, for the reversible half-reaction (12-9) we have the Nernst equation

$$E = E° + \frac{RT}{nF} \ln \frac{a_{ox}}{a_{red}} = E° - \frac{RT}{nF} \ln \frac{a_{red}}{a_{ox}} \qquad (12\text{-}10)$$

where $E°$ is called the *standard electrode potential* and corresponds to the value of the potential E at *unit activities* of oxidant and reductant or, if both activities are variable (such as for Fe^{3+} and Fe^{++}), to an *activity ratio* of unity. When numerical values of RT at 25°C and F are inserted in (12-10), and logarithms of base 10 are used, we have

$$E = E° - \frac{2.3RT}{nF} \log \frac{a_{red}}{a_{ox}} = E°_{25} - \frac{0.059}{n} \log \frac{a_{red}}{a_{ox}} \qquad (12\text{-}11)$$

or at 30°C

$$E = E°_{30} - \frac{0.060}{n} \log \frac{a_{red}}{a_{ox}} \qquad (12\text{-}12)$$

We may substitute in (12-10) for each activity the product of activity coefficient and concentration and thus obtain

$$E = E° - \frac{RT}{nF} \ln \frac{\gamma_{red}}{\gamma_{ox}} - \frac{RT}{nF} \ln \frac{[red]}{[ox]} \qquad (12\text{-}13)$$

Combining the first two terms gives

$$E = E°' - \frac{RT}{nF} \ln \frac{[red]}{[ox]} \qquad (12\text{-}14)$$

where $E°'$ corresponds to the value of E at *unit concentrations* or *unit concentration ratio* of oxidant and reductant and is called the *formal potential* of the electrode. The formal potential varies with activity coefficients and therefore with ionic strength of the solution; it includes implicitly any liquid-junction potential between the reference electrode and the half-cell in question. A given half-reaction may be assigned any number of formal potentials, each corresponding to a different electrolyte composition. Thus, to describe the half-reaction

$$Fe(III) + e^- \rightleftharpoons Fe(II) \qquad (12\text{-}15)$$

a series of values for different concentrations of H_2SO_4, HCl, and so on are necessary. Accordingly, the standard electrode potential (Table 12-1) is $+0.771$ V, and the formal potential in 1 M H_3PO_4 is $+0.44$ V and in 1 M HF is $+0.32$ V. Table 12-2 lists

Table 12-2 FORMAL POTENTIALS OF IRON
AND DICHROMATE
HALF-REACTIONS

Acid present	$E°'_{Fe(III)/Fe(II)}$, V	$E°'_{Cr_2O_7=/Cr(III)}$, V
0.1 M HCl	0.73	0.93
0.5 M HCl	0.72	0.97
1 M HCl	0.70	1.00
2 M HCl	0.69	1.05
3 M HCl	0.68	1.08
0.1 M H$_2$SO$_4$	0.68	0.92
0.5 M H$_2$SO$_4$	0.68	1.08
4 M H$_2$SO$_4$	0.68	1.15
0.1 M HClO$_4$	0.735	0.84
1 M HClO$_4$	0.735	1.025

formal potentials of the Fe(III)-Fe(II) and $Cr_2O_7^=$-Cr(III) couples, evaluated from titration curves.[9] The potentials listed include a junction potential between the acid medium and a saturated calomel electrode.

The formal potentials of a couple are often of greater practical value to the analytical chemist than standard potentials because they represent quantities subject to direct experimental measurement. The use of formal potentials in a sense accomplishes the same objectives as the use of conditional formation constants in complexation; the effects of substances involved in determining the potential whose concentrations are constant are included not in the log term but in the $E^{o'}$ term. For example, during the course of a titration of Fe(II) with Ce(IV) in 1 M $HClO_4$ the ionic strengths and the activity coefficients of the reactants remain essentially constant, although the concentration ratio [Fe(III)]/[Fe(II)] changes enormously and in a known way in the vicinity of the end point. Equation (12-14) is therefore of greater practical value in calculating the course of the titration curve than an equation written in terms of activities.

For more complicated electrode reactions the Nernst equation includes terms to represent all species of variable activity involved in the reaction, such as hydrogen or hydroxyl ion and complexing molecules or ions. Reactants at invariant unit activity, for instance the solvent, pure metals, and pure solids, are omitted in the Nernst equation. Examples are

$$TiO^{++} + 2H^+ + e^- \rightarrow Ti^{3+} + H_2O \qquad (12\text{-}16)$$

$$E = E° - \frac{RT}{F} \ln \frac{a_{Ti^{3+}}}{a_{TiO^{++}}a_{H^+}^2} \qquad (12\text{-}17)$$

$$AgCl(solid) + e^- \rightarrow Ag + Cl^- \qquad (12\text{-}18)$$

$$E = E° - \frac{RT}{F} \ln a_{Cl^-} \qquad (12\text{-}19)$$

12-4   COMBINATION OF HALF-REACTIONS TO FORM NEW HALF-REACTIONS

It is sometimes desirable to calculate the electrode potential of a half-reaction derived from a combination of two or more other half-reactions. We may combine two half-reactions by addition if we regard the free-energy change, or its negative $nFE°$, as being additive. But, since the faraday is common to all such additions, the summation is simplified if we regard the quantity $nE°$ or the volt-electrons as additive, rather than the joules.

To illustrate, combine

$$Fe(III) + e^- \rightarrow Fe(II) \qquad E° = 0.77 \qquad nE° = 0.77 \qquad (12\text{-}20)$$

$$Fe(II) + 2e^- \rightarrow Fe \qquad E° = -0.44 \qquad nE° = -0.88 \qquad (12\text{-}21)$$

to form

$$Fe(III) + 3e^- \rightarrow Fe \qquad nE° = -0.11 \qquad (12\text{-}22)$$

from which $E° = -0.04$.

It may be noted that, though the final half-reaction is not subject to direct measurement (Section 12-9), its potential can be used in making free-energy calculations.

EXAMPLE 12-1   Combine the following two half-reactions by the subtraction $(a) - (b) = (c)$:

(a) $Cu^{++} + 2e^- \rightarrow Cu \qquad E° = 0.34 \qquad nE° = 0.68$
(b) $Cu^+ + e^- \rightarrow Cu \qquad E° = 0.52 \qquad nE° = 0.52$
(c) $Cu^{++} + e^- \rightarrow Cu^+ \qquad nE° = 0.16$

and therefore $E° = 0.16$

Inspection of the values of $E°$ shows that Cu^+ is a stronger oxidant than Cu^{++}. Therefore, if Cu^{++} is reduced, the product must be Cu, since Cu^+ is reduced even more readily. We also infer that Cu^+ should disproportionate according to the equation

$$2Cu^+ \rightarrow Cu + Cu^{++}$$

as indicated by any of the following operations: $2(b) - (a)$; $(a) - 2(c)$; or $(b) - (c)$. The first two give $E°_{cell} = 0.18$; the third gives $E°_{cell} = 0.36$. All three give $\Delta G° = -0.36 \times 96,487 = -34.7$ kJ. In Section 12-7 it is seen that potential is affected by complexation; by forming a stable complex or an insoluble compound, or by the use of certain solvents, Cu^+ can be stabilized. The effect is to lower the $E°$ of half-reaction (a) and raise that of (c) until their order is reversed.   ////

12-5   COMBINATION OF HALF-REACTIONS TO FORM A WHOLE REACTION

An oxidation-reduction reaction is composed of at least two half-reactions. Any two half-reactions may be combined *by subtraction in such a way as to cancel the electrons* and so yield a whole reaction that corresponds to the cell reaction. The cell emf is the algebraic difference between the electrode potentials.

For example, consider

$$Fe(III) + e^- \rightarrow Fe(II) \qquad E^\circ = 0.77 \text{ V} \qquad (12\text{-}23)$$

$$Sn(IV) + 2e^- \rightarrow Sn(II) \qquad E^\circ = 0.14 \text{ V} \qquad (12\text{-}24)$$

First, multiply the iron half-reaction by 2 to provide the same number of electrons in each half-reaction, and then subtract (12-24) from (12-23) to obtain

$$2Fe(III) + Sn(II) \rightarrow 2Fe(II) + Sn(IV) \qquad E^\circ_{cell} = 0.77 - 0.15 = 0.62 \text{ V} \qquad (12\text{-}25)$$

Note that, when the iron half-reaction is doubled, its electrode potential is unaffected because the potential does not depend on the number of ions involved. Thus a Nernst equation written for a doubled reaction is identical to that for a single reaction. If the subtraction is carried out in such a direction as to produce a *positive* cell emf, the equilibrium constant for the reaction is greater than unity, and the reaction is said to be *spontaneous*. This conclusion follows because the stronger oxidant [in this case Fe(III)] has the higher potential, and hence the subtraction is performed in the direction corresponding to the reduction of the stronger oxidant.

To calculate the cell emf from the electrode potentials given in a cell diagram [Cell (12-3)], one subtracts the potentials in the following way:

$$E_{cell} = E_{right} - E_{left} \qquad (12\text{-}26)$$

According to thermodynamic convention a spontaneous cell reaction has a negative free-energy change numerically equal to the electrical work (in joules) per unit of reaction as written. In general, the total free-energy change is

$$-\Delta G = nFE_{cell} \qquad (12\text{-}27)$$

and, if all reactants are in their standard states of unit activity,

$$-\Delta G^\circ = nFE^\circ_{cell} \qquad (12\text{-}28)$$

where n is the number of electrons canceled in subtracting the two half-reactions. In numerical units

$$-\Delta G \text{ (joules)} = n \times 96,487 \times E_{cell} \text{ (volts)} \qquad (12\text{-}29)$$

A particular cell reaction can sometimes be arrived at by different combinations of half-reactions involving different values of n in Equation (12-28). The free-energy change must be independent of the method of arriving at the cell reaction, but the value of the cell emf may vary.

To illustrate, we can use two different combinations of half-reactions combined in hypothetical half-cells to arrive at the reaction

$$2MnO_4^- + 3Mn^{++} + 2H_2O \rightarrow 5MnO_2 + 4H^+ \qquad (12\text{-}30)$$

by combining either

$$MnO_4^- + 8H^+ + 5e^- \rightarrow Mn^{++} + 4H_2O \qquad E^\circ = 1.51 \text{ V} \qquad (12\text{-}31)$$

$$MnO_2 + 4H^+ + 2e^- \rightarrow Mn^{++} + 2H_2O \qquad E^\circ = 1.23 \text{ V} \qquad (12\text{-}32)$$

or $\qquad MnO_4^- + 4H^+ + 3e^- \rightarrow MnO_2 + 2H_2O \qquad E^\circ = 1.68 \text{ V} \qquad (12\text{-}33)$

$$MnO_2 + 4H^+ + 2e^- \rightarrow Mn^{++} + 2H_2O \qquad E^\circ = 1.23 \text{ V} \qquad (12\text{-}34)$$

In the first instance, $n = 10$, $E_{cell}^\circ = 0.28$, and $\Delta G^\circ = -270 \text{ kJ}$. In the second, $n = 6$, $E_{cell}^\circ = 0.45$, and ΔG° has essentially the same value. Still another way is to cancel 15 electrons by combining the five-electron and the three-electron reductions of MnO_4^-. Each subtraction of half-reactions corresponds to a particular cell, which does not necessarily represent a physically measurable entity. None of the three postulated cells in the example can be measured directly (Section 17-1). Nevertheless, the value of ΔG° for Reaction (12-30) is valid.

12-6 EQUILIBRIUM CONSTANT OF A REDOX CHEMICAL REACTION

Cell emf values and electrode potentials are important, not only because they are significant electroanalytically, but more because they can give information about possible chemical reactions and their equilibria. Equilibrium constants can be calculated by use of either cell emf values or electrode potentials.

Calculation of equilibrium constant from emf of a cell The equilibrium constant of a chemical reaction can be calculated from the standard free-energy change by the equation

$$-\Delta G^\circ = RT \ln K = nFE_{cell}^\circ \qquad (12\text{-}35)$$

For a spontaneous reaction, ΔG° is negative and, correspondingly, the equilibrium constant is greater than unity. With numerical values of RT and F at 25°C and conversion to logarithms of base 10,

$$\log K = \frac{nE_{cell}^\circ}{0.0591} = \frac{n(E_{right}^\circ - E_{left}^\circ)}{0.0591} \qquad (12\text{-}36)$$

To illustrate, consider the chemical reaction at 25°C,

$$2Fe(III) + Sn(II) \rightarrow 2Fe(II) + Sn(IV) \qquad (12\text{-}37)$$

Then consider the (hypothetical) cell with all ions at unit activity,

$$Pt \mid Sn(II), Sn(IV) \parallel Fe(III), Fe(II) \mid Pt \qquad (12\text{-}38)$$

The equilibrium constant is given by

$$\log K = 2[0.77 - 0.15]/0.0591 = 21.0 \qquad (12\text{-}39)$$

This large value of the equilibrium constant indicates that the chemical reaction goes essentially to completion to the right as written.

Calculation of equilibrium constant from electrode potentials of the half-reactions The two half-reactions involved in the chemical reaction (12-37) are

$$Fe(III) + e^- \rightleftharpoons Fe(II) \qquad (12\text{-}40)$$

$$Sn(IV) + 2e^- \rightleftharpoons Sn(II) \qquad (12\text{-}41)$$

Their electrode potentials are

$$E = 0.77 - 0.0591 \log \frac{a_{Fe(II)}}{a_{Fe(III)}} \qquad (12\text{-}42)$$

and

$$E = 0.15 - \frac{0.0591}{2} \log \frac{a_{Sn(II)}}{a_{Sn(IV)}} \qquad (12\text{-}43)$$

A single solution at chemical equilibrium containing the components of more than one half-reaction can have only one electrode potential; hence the value of E in (12-43) must equal that in (12-42). Therefore

$$0.77 - 0.0591 \log \frac{a_{Fe(II)}}{a_{Fe(III)}} = 0.77 - \frac{0.0591}{2} \log \frac{a_{Fe(II)}^2}{a_{Fe(III)}^2}$$

$$= 0.15 - \frac{0.0591}{2} \log \frac{a_{Sn(II)}}{a_{Sn(IV)}} \qquad (12\text{-}44)$$

Collect appropriate activity terms in (12-44) on one side of an equality sign and numerical values on the other to yield the equilibrium constant for (12-37):

$$\log \frac{a_{Fe(II)}^2 a_{Sn(IV)}}{a_{Fe(III)}^2 a_{Sn(II)}} = \frac{0.77 - 0.15}{0.0591/2} = \log K = 21.0 \qquad (12\text{-}45)$$

12-7 EFFECT OF COMPLEX FORMATION ON ELECTRODE POTENTIALS

Consider first the half-reaction

$$M^{n+} + ne^- \rightleftharpoons M \qquad (12\text{-}46)$$

and suppose that the aquated ion M^{n+} forms a series of complexes ML, $ML_2, \ldots,$ ML_q with a complexing agent L, which may be either charged or uncharged. For the sake of simplicity we omit the charge on the complex species.

Recall from Equations (7-24) and (11-16) that

$$\frac{1}{\alpha_M} = \frac{C_M}{[M^{n+}]} = 1 + \beta_1[L] + \beta_2[L]^2 + \cdots + \beta_q[L]^q \qquad (12\text{-}47)$$

where α_M is the fraction of the metal ion in the aquated form, C_M is the total metal ion concentration in solution, and $\beta_1, \beta_2, \ldots$ are overall formation constants. With the Nernst equation for (12-46) written in terms of concentrations,

$$E = E^{o\prime}_{M^{n+},M} - \frac{RT}{nF} \ln \frac{1}{[M^{n+}]} \qquad (12\text{-}48)$$

$$E = E^{o\prime}_{M^{n+},M} - \frac{RT}{nF} \ln \frac{1}{C_M} + \frac{RT}{nF} \ln \alpha_M \qquad (12\text{-}49)$$

If the successive formation constants of ML_q are known, the value of α_M can be calculated from (12-47), and the electrode potential can be calculated from (12-49). Conversely, from measurements of the electrode potential at various concentrations of complexing agent L, the value of α_M can be determined for each value of L. If sufficient data are available, the values of the stepwise formation constants $K_1, K_2, \ldots,$ K_q can be evaluated. If there are q such values, we need q equations; these are available from q determinations of E.

The situation is simpler if only a single species ML_q is formed over a range of concentrations of L (EDTA complexes, for example). Then the electrode half-reaction can be written

$$ML_q + ne^- \rightleftharpoons M + qL \qquad (12\text{-}50)$$

The overall formation constant β_q is

$$\beta_q = \frac{[ML_q]}{[M^{n+}][L]^q} \qquad \text{and} \qquad \frac{1}{\alpha_M} = (1 + \beta_q[L]^q) \simeq \beta_q[L]^q \qquad (12\text{-}51)$$

Accordingly, by substitution in (12-48)

$$E = E^{o\prime}_{M^{n+},M} - \frac{RT}{nF} \ln K_q - q\frac{RT}{nF} \ln [L] - \frac{RT}{nF} \ln \frac{1}{[ML_q]} \qquad (12\text{-}52)$$

The first three terms on the right can be combined to give the formal potential of the half-reaction (12-50):

$$E = E^{o\prime}_{ML_q,M} - \frac{RT}{nF} \ln \frac{1}{[ML_q]} \qquad (12\text{-}53)$$

where

$$E^{o\prime}_{ML_q,M} = E^{o\prime}_{M^{n+},M} - \frac{RT}{nF} \ln K_q - q\frac{RT}{nF} \ln [L] \qquad (12\text{-}54)$$

Equation (12-54) indicates that, if the formal potential of a complex is plotted against log [L], a straight line of slope $-2.3qRT/nF$ is obtained. From the slope the value of q, and hence the value of the formula of the complex, can be determined.

From the intercept, which is $E^{\circ\prime}_{M^{n+},M} - (RT/nF) \ln K_q$, the value of the formation constant can be calculated if the formal potential $E^{\circ\prime}_{M^{n+},M}$ is known.

If a series of complexes is formed, the plot of log $[L]$ against E for a given total metal ion concentration [Equation (12-49)] is a curve, because $1/\alpha_M$ is a polynomial in $[L]$.

Since α_M is < 1 and $\beta_q > 1$, the potential E given by (12-49) or (12-53) corresponding to a given total concentration of metal ions C_M or $[ML_q]$ is shifted in a negative direction by complex formation. Thus the metal becomes a stronger reducing agent, or the complex becomes more difficult to reduce, than the aquated ion.

To evaluate the thermodynamic formation constants in terms of activities, rather than the apparent formation constants in terms of concentrations, apparent values must be extrapolated to zero ionic strength. This procedure necessitates measuring the electrode potential as a function of $[L]$ at several levels of ionic strength. If L is an uncharged species, varying $[L]$ does not change the ionic strength; moreover, the various complexes ML, $ML_2, \ldots$ have the same charge as the aquated ion M^{n+}, and so the activity coefficients of all these species are approximately the same. Under these circumstances it is relatively straightforward to evaluate the formation constants at various levels of ionic strength and thus to evaluate the thermodynamic constants. If L is a charged species, however, constant ionic strength must be maintained to compensate for the addition of L by removal of a noncomplexing electrolyte. This may require the presence of relatively high ionic strength throughout the series. Moreover, each of the various species ML, $ML_2, \ldots$ has a different charge and therefore a different activity coefficient. Finally, the assumption that the activity coefficient is constant for constant ionic strength but varying composition becomes progressively less valid as the ionic strength increases. For these reasons the thermodynamic formation constants can seldom be evaluated if the complexing agent is an ion. Nevertheless, many useful formal constants have been evaluated at relatively high ionic strengths.[10]

Turn now to complex formation and a half-reaction involving two oxidation states of a metal in solution (charges omitted for simplicity):

$$M_{ox} + ne^- \rightleftharpoons M_{red} \qquad (12\text{-}55)$$

An equation analogous to (12-47) can be written in terms of the fractions α_{ox} and α_{red} of M_{ox} and M_{red} in the aquated form:

$$\frac{1}{\alpha_{ox}} = \frac{C_{M_{ox}}}{[M_{ox}]} = 1 + (\beta_{ox})_1[L] + (\beta_{ox})_2[L]^2 + \cdots + (\beta_{ox})_p[L]^p \qquad (12\text{-}56)$$

and
$$\frac{1}{\alpha_{red}} = \frac{C_{M_{red}}}{[M_{red}]} = 1 + (\beta_{red})_1[L] + (\beta_{red})_2[L]^2 + \cdots + (\beta_{red})_q[L]^q \qquad (12\text{-}57)$$

and
$$E = E^{\circ\prime}_{M_{ox},M_{red}} - \frac{RT}{nF} \ln \frac{\alpha_{red}}{\alpha_{ox}} - \frac{RT}{nF} \ln \frac{C_{M_{red}}}{C_{M_{ox}}} \qquad (12\text{-}58)$$

In the same way, an equation analogous to (12-54) can be written if the oxidized and reduced forms are present predominantly as the species ML_p and ML_q. The electrode reaction now becomes

$$ML_p + ne^- \rightleftharpoons ML_q + (p - q)L \qquad (12\text{-}59)$$

and we introduce the overall formation constants

$$(\beta_p)_{ox} = \frac{[ML_p]}{[M_{ox}][L]^p} \qquad \text{and} \qquad (\beta_q)_{red} = \frac{[ML_q]}{[M_{red}][L]^q} \qquad (12\text{-}60)$$

Since $C_{M_{ox}} \simeq [ML_p]$ and $C_{M_{red}} \simeq [ML_q]$, (12-56) and (12-57) can be simplified to

$$\frac{1}{\alpha_{ox}} \simeq (\beta_p)_{ox}[L]^p \qquad \text{and} \qquad \frac{1}{\alpha_{red}} \simeq (\beta_q)_{red}[L]^q \qquad (12\text{-}61)$$

which by substitution in (12-58) yield

$$E = E^{\circ\prime}_{M_{ox},M_{red}} - \frac{RT}{nF} \ln \frac{(\beta_p)_{ox}}{(\beta_q)_{red}} - (p - q)\frac{RT}{nF} \ln [L] - \frac{RT}{nF} \ln \frac{[ML_q]}{[ML_p]} \qquad (12\text{-}62)$$

The first three terms on the right are combined to give the formal potential of the complex system (12-59), or

$$E = E^{\circ\prime}_{ML_p,ML_q} - \frac{RT}{nF} \ln \frac{[ML_q]}{[ML_p]} \qquad (12\text{-}63)$$

where

$$E^{\circ\prime}_{ML_p,ML_q} = E^{\circ\prime}_{M_{ox},M_{red}} - \frac{RT}{nF} \ln \frac{(\beta_p)_{ox}}{(\beta_q)_{red}} - (p - q)\frac{RT}{nF} \ln [L] \qquad (12\text{-}64)$$

If we plot the value of the formal potential $E^{\circ\prime}_{ML_p,ML_q}$ against log $[L]$, the slope gives $-2.3(p - q)RT/nF$, from which $p - q$ is evaluated. If both M_{ox} and M_{red} have the same coordination number, and if ML_p and ML_q are formed quantitatively, then $p = q$, and the electrode potential does not vary with concentration of complexing agent. The relation between the electrode potentials of the complex and aquated systems depends on the *relative* stabilities of the two complexes, or the ratio $(\beta_p)_{ox}/(\beta_q)_{red}$. If the oxidant forms the more stable complex, as is frequently the case, the electrode potential of the metal ion couple is lowered by complexation, that is, shifted in the negative direction, corresponding to a weaker oxidant, as would be expected.

As an illustration, the potential of the half-reaction

$$Fe(CN)_6^{3-} + e^- \rightleftharpoons Fe(CN)_6^{4-} \qquad E^\circ = 0.356 \text{ V} \qquad (12\text{-}65)$$

is lower than that of the Fe(III)-Fe(II) couple because ferricyanide is a more stable complex than ferrocyanide.† Moreover, since $p = q = 6$, no shift in potential with cyanide concentration is anticipated. The acid $H_4Fe(CN)_6$, however, is a weak acid

---

† It must not be inferred that the reduction of ferricyanide actually proceeds by dissociation followed by reduction of Fe(III) and reassociation; the reversible potential is independent of the reaction path.

in its third and fourth steps of ionization[11,12] ($pK_3 = 3$; $pK_4 = 4.3$), whereas $H_3Fe(CN)_6$ is a strong acid. Hence the electrode reaction is represented more precisely by

$$Fe(CN)_6{}^{3-} + H^+ + e^- \rightleftharpoons HFe(CN)_6{}^{3-} \qquad pH = 4$$

or $\qquad Fe(CN)_6{}^{3-} + 2H^+ + e^- \rightleftharpoons H_2Fe(CN)_6{}^{=} \qquad 2 > pH > 0 \qquad (12\text{-}66)$

and the formal potential of (12-65) increases with decreasing pH below a pH of about 5. The effect of hydrogen ion could be considered by means of an equation analogous to (12-58), where the ions $HFe(CN)_6{}^{3-}$ and $H_2Fe(CN)_6{}^{=}$ are interpreted to be hydrogen ion complexes whose formation constants are the reciprocals of the third and fourth ionization constants of the acid $H_4Fe(CN)_6$.

Owing to the high charge of the anions, the effect of ionic strength is also large ($E^{\circ\prime} = 0.46$ V for 0.1 M solutions of both potassium ferrocyanide and ferricyanide), and the potential shift is in a direction opposite that observed for the aquo Fe(III)-Fe(II) couple because the more highly charged species is the reductant rather than the oxidant.

For the 1,10-phenanthroline complexes of iron we have $Fe(o\text{-phen})_3{}^{3+} + e^- \rightleftharpoons Fe(o\text{-phen})_3{}^{++}$, and in 1 M H_2SO_4, $E^{\circ\prime} = 1.06$ V. That this value[13] of $E^{\circ\prime}$ is higher than that of the Fe(III)-Fe(II) couple implies a greater stability of the Fe(II) complex, which is confirmed by experiment.[14] Again, the potential is independent of concentration of complexing agent, as long as enough is present to react with both species of iron.

12-8 DEPENDENCE OF POTENTIAL ON pH

Whenever hydrogen or hydroxyl ions appear in the half-reaction, the electrode potential varies with pH. Consider the reaction

$$\text{ox} + m\text{H}^+ + n\text{e}^- \rightleftharpoons \text{red} \qquad (12\text{-}67)$$

The Nernst equation is of the form

$$E = E^\circ - \frac{RT}{nF} \ln \frac{a_{red}}{a_{ox}a_{H^+}{}^m} \qquad (12\text{-}68)$$

or $\qquad E = E^\circ - \frac{RT}{nF} \ln \frac{a_{red}}{a_{ox}} - \frac{mRT}{nF} \ln \frac{1}{a_{H^+}} \qquad (12\text{-}69)$

If we convert the last term to logarithms of base 10 and use the definition pH = $-\log a_{H^+}$, we have at 25°C

$$E = E^\circ - \frac{RT}{nF} \ln \frac{a_{red}}{a_{ox}} - \frac{m}{n} \times 0.0591 \text{ pH} \qquad (12\text{-}70)$$

The relation between potential and pH often is depicted graphically in the form of Pourbaix or *predominance-region* or *potential-pH* diagrams.[15] To illustrate

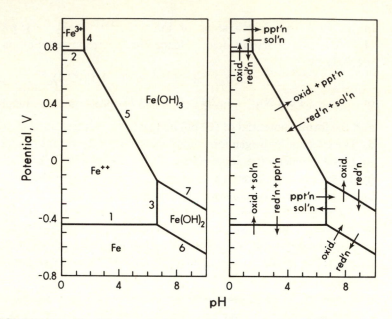

FIGURE 12-1   Simplified predominance-region or potential-pH diagram for the system Fe, Fe(II), Fe(III). Left, predominant species present; right, type of reaction occurring in passing from one region to another.

potential-pH diagrams, the system Fe, Fe(II), Fe(III) is commonly used, as in Figure 12-1. The lines represent the potentials corresponding to unit activity of Fe(II) or Fe(III) at the pH in question.

Line 1 $Fe(II) + 2e^- \rightleftharpoons Fe$; $E° = -0.44$ V (oxidation or reduction independent of pH).

Line 2 $Fe(III) + e^- \rightleftharpoons Fe(II)$; $E° = 0.77$ V (independent of pH).

Line 3 $Fe(OH)_2 \rightleftharpoons Fe(II) + 2OH^-$; $\log K_{sp} = -14.8$; at $a_{Fe(II)} = 1$, pH = 6.6 (precipitation or solution independent of E).

Line 4 $Fe(OH)_3 \rightleftharpoons Fe(III) + 3OH^-$; $\log K_{sp} = -37.4$; at $a_{Fe(III)} = 1$, pH = 1.5 (independent of E).

Line 5 $Fe(OH)_3 + 3H^+ + e^- \rightleftharpoons Fe(II) + 3H_2O$; at $a_{Fe(II)} = 1$, from $1.5 < $ pH < 6.6, $E = 0.77 - 3 \times 0.0591$ (pH $- 1.5$).

Line 6 $Fe(OH)_2 + 2H^+ + 2e^- \rightleftharpoons Fe + 2H_2O$; at pH > 6.6, $E = -0.44 - 0.0591$ (pH $- 6.6$).

Line 7 $Fe(OH)_3 + H^+ + e^- \rightleftharpoons Fe(OH)_2 + H_2O$;   at   pH > 6.6, $E = -0.134 - 0.0591$ (pH $- 6.6$) (oxidation or reduction of precipitate dependent on pH and E).

Potential-pH diagrams are useful graphic devices for concisely and quickly depicting the oxidation-reduction and acid-base behavior of substances. Because such a diagram contains only the information used in its construction, it may present an oversimplified or inexact picture. In the example here, no account was taken of hydrolysis of Fe(III) or Fe(II), which would give lines *1* and *2* a downward slope. In a more complete diagram the existence of species such as magnetite, Fe_3O_4, and ferrate(VI), $FeO_4^=$, would have to be considered and, more important, the reactions of the solvent as well. With water as solvent, at low potentials hydrogen is evolved (0 V at pH $= 0$, that is, before metallic iron forms), and at high potentials oxygen is evolved.

Potential-pH diagrams are often convenient as an aid in predicting the direction and course of redox reactions. If two such diagrams are superimposed, the system of higher potential at any pH will act as an oxidant. If the lines of two systems intersect, the direction of reaction may be reversed by a change in pH. As an example, a comparison of the iodine-iodide and As(V)-As(III) systems (Section 19-5) shows that iodine acts as an oxidant to produce As(V) at higher pH values and As(V) acts as an oxidant to produce iodine at low pH values.

12-9   REVERSIBILITY OF ELECTRODE REACTIONS

A thermodynamically reversible half-reaction can be defined as one that can be made to proceed in either of two opposing directions by an infinitesimal shift in the potential from its equilibrium value. Contrary to general opinion, such reactions are rare. This definition should not necessarily be used as a basis for an experimental test for two reasons: (*1*) a finite potential shift must be made to produce a finite net current, and (*2*) the point of zero current is not always the equilibrium potential. While irreversibility can be revealed in many systems, proof of thermodynamic reversibility at the molecular level in others is virtually impossible.

Regarding the first point, as shown below, if the current drawn in the measurement is kept small compared with the exchange current (Sections 14-2 and 14-4) that passes in both directions at equilibrium, the shift in potential (polarization) can be made small enough to be negligible. (Typically exchange currents are of the order of 10^{-2} A/cm^2 for favorable systems and 10^{-10} A/cm^2 or lower for unfavorable ones.) With modern instrumentation the current drain can be made so low that this condition need not be the limiting factor. When the exchange current is small, however, the rate of attainment of the equilibrium potential may be low.

Consider the half-reaction ox $+ ne^- \rightleftharpoons$ red, which is at equilibrium. The two opposing processes proceed at the same rate, which can be expressed conveniently in terms of the *exchange current density* j_0, in amperes per square centimeter. If the potential is shifted a small amount η, the net current density I is proportional to the potential change [Equation (14-18)]:

$$\eta = \frac{RT}{nF} \frac{I}{j_0} \qquad (12\text{-}71)$$

where η is the overpotential, in volts, and RT/nF is expressed in joules per n faradays, having the value $0.0257/n$ at 25°C. The quantity RT/nFj_0 has the dimensions Ω-cm^2 and is the *polarization resistance* for an electrode 1 cm^2 in area. From (12-71), if $I = 0.01j_0$, $\eta = 0.257/n$ mV. Thus the polarization is negligible if the current drain is small compared with the exchange current.

　　Now, for the electrode to change its potential the electrical double layer (Section 8-6) must be charged or discharged by the passage of current, which must flow through an effective resistance equal to the polarization resistance. The time constant for the charging process is the product of resistance and capacitance, or

$$\tau = \frac{RT}{nFj_0} C_{dl} \qquad (12\text{-}72)$$

where τ is the time constant in seconds and C_{dl} is the differential capacity of the double layer in farads per square centimeter. The decay of overpotential then is given by

$$\eta_t = \eta_0 e^{-t/\tau} \qquad (12\text{-}73)$$

where η_t and η_0 are the overpotentials at times $t = t$ and $t = 0$. For $t = \tau$, η is $1/e$, or 0.36 of its original value. For $t = 2.3\tau$, η is 0.10 of its original value. Thus, if the required accuracy of potential measurement is 0.1 mV, the time required for 1 mV of overpotential (a small value) to decay to 0.1 mV is 2.3 τ, and the same interval is needed for a drift from 10 to 1 mV. For larger values of the overpotential the shift in potential is more rapid, because the overpotential no longer varies linearly with the current but with the logarithm of the current [Tafel equation, (14-21)], and so the effective resistance is lower. In practical work, lengthy electrode equilibration times are to be expected in many systems.

EXAMPLE 12-2 According to Bockris and Huq[16] the exchange current density for the oxygen electrode in acid solution is of the order of 10^{-10} A/cm^2 on a platinum surface. The polarization resistance is $RT/4Fj_0 = 6 \times 10^7$ Ω-cm^2. The double-layer capacity C_{dl} is of the order of 40 μf/cm^2, based on the true area, or perhaps 100 μf/cm^2 based on the geometric or projected area, on which the current density was based. Thus $\tau = 6 \times 10^7 \times 100 \times 10^{-6} = 6 \times 10^3$ s, and the time for overpotential decay from 10 to 1 mV is estimated to be $2.3\tau \simeq 4$ h. Therefore, measurements accurate to within 1 mV would be difficult, but an accuracy of ± 10 mV may reasonably be expected. The actual behavior of an oxygen electrode is complicated by the oxidation of the platinum surface. Even traces of impurities cause currents of the order of magnitude of the exchange current to flow, and the oxygen electrode ordinarily is regarded as irreversible, although in ultrapure solutions it behaves reversibly.   ////

EXAMPLE 12-3   As an example of a reaction normally considered reversible, consider

$$Zn^{++} + Hg + 2e^- \rightleftharpoons Zn(Hg)$$

for which Gerischer[17] found an exchange current density of 5.4×10^{-3} A/cm² for $0.02\ M\ Zn^{++}$ against 1 mole % zinc amalgam. The polarization resistance is $RT/2Fj_0 = 2.4\ \Omega\text{-cm}^2$. With $C_{dl} = 20\ \mu f/cm^2$ for a liquid surface, $\tau = 2.4 \times 20 \times 10^{-6} = 5 \times 10^{-5}$ s. The overpotential would decay from 10 to 0.1 mV in $4.6 \times 5 \times 10^{-5} = 2.3 \times 10^{-4}$ s. So that polarization will not exceed 1 mV, a current density $I \leq \eta j_0 nF/RT = 4.2 \times 10^{-5}$ A/cm² can be used in the measurement.   ////

We now turn to the second problem, which concerns the relation between the point of zero current and the equilibrium potential. For a net current of zero it is necessary only that the total cathodic current be equal to the total anodic current. If the cathodic half-reaction is not just the reverse of the anodic one, a net chemical reaction is proceeding at the electrode, and the potential is not characteristic of either half-reaction.

To take a simple example, consider a solution of Fe(III) containing an electrode of metallic iron. The reactions

$$2Fe(III) + 2e^- \rightarrow 2Fe(II)$$

and
$$Fe \rightarrow Fe(II) + 2e^-$$

proceed at *equal rates*; hence the current is zero. The potential, however, is not characteristic of the Fe(III)-Fe system, but lies somewhere between 0.77 V [the Fe(III)-Fe(II) potential] and -0.44 V [the Fe(II)-Fe(0) potential]. Such a *mixed potential*[18,19] is characteristic of neither reaction and bears no simple relation (such as that of the Nernst equation) to the concentrations or activities of the reacting species. The rates of electron-transfer reactions as well as the rates of transport of the reactants to and from the electrode surface are involved (Section 14-3).

Another example of a mixed potential is encountered in the hypothetical case of a perfectly pure solution of one form of a redox couple. Consider, for instance, a platinum electrode in a solution of Fe(II) containing not even one Fe(III) ion. According to the Nernst equation the potential has a negative value of infinity. Actually, the potential is limited in that a cathodic reaction would occur at some finite potential. In an oxygen-free solution, reduction of hydrogen ions to hydrogen would occur, and the potential would shift to a value such that the rate of reduction would just equal the rate of oxidation of Fe(II) to Fe(III). Fe(III) ions are thus produced in the reaction. Eventually an equilibrium would be reached in which both half-reactions would be at the same potential.

A practical criterion of reversibility is that the Nernst equation be obeyed. From the preceding discussion this cannot be true when two substances that will interact with each other appear in the half-reaction. Thus, a half-reaction such as

$$MnO_4^- + 8H^+ + 5e^- \rightleftharpoons Mn^{++} + 4H_2O \qquad E° = 1.52 \text{ V}$$

cannot follow the Nernst equation because MnO_4^- and Mn^{++} interact to form MnO_2, and the potential must be a mixed one. Suppose, however, that this interaction is so slow as to be negligible. Even then, the Nernst equation cannot be followed unless the reaction can be made to go as written in both directions. In this example both half-reactions proceed through complicated mechanisms. For example, the oxidation of Mn(II) yields adsorbed Mn(III) as the primary product, and solid MnO_2 as the final product.[20] Note that the standard potentials of both the Mn(VII)-Mn(VI) and Mn(III)-Mn(II) couples lie above the theoretical decomposition potential of water. The permanganate potential is probably a mixed one in which the cathodic current due to permanganate reduction is compensated for by an anodic current due to oxidation of water. The potential becomes more positive with increasing permanganate concentration because the cathodic current is increased. The only effect of Mn(II), in the absence of complexing agents that stabilize Mn(III), is to render the permanganate ion less stable.

Another fundamental aspect is that a reversible couple can act as an *electron-transfer catalyst* for an irreversible one. An example is the half-reaction

$$Cr_2O_7^= + 14H^+ + 6e^- \rightleftharpoons 2Cr(III) + 7H_2O$$

which has been described as behaving irreversibly.[21] The potential of a platinum electrode in a solution of dichromate is poorly reproducible and tends to drift slowly. Moreover, the potential is insensitive to changes in Cr(III) and does not vary linearly with pH as expected. On the other hand, certain titration curves of Fe(II) with dichromate are remarkably similar in shape to those expected for a reversible Cr(VI)-Cr(III) couple (Section 15-3). It appears likely that the Fe(III)-Fe(II) couple reaches chemical equilibrium with the Cr(VI)-Cr(III) couple and therefore has the same potential and is the active potential-determining couple.

It should be emphasized that many of the potentials listed in tables of standard electrode potentials are values calculated from thermodynamic data rather than obtained directly from cell emf data. As such they are valuable for calculating equilibrium constants of reactions, but caution should be exercised in using them to predict the behavior of electrodes. A steady value for an electrode potential does not necessarily represent the thermodynamic or equilibrium value.

REFERENCES

1 I. M. KOLTHOFF: *J. Electrochem. Soc.*, **118**:5C (1971).

2 T. S. LICHT and A. J. DE BÉTHUNE: *J. Chem. Educ.*, **34**:433 (1957); A. J. DE BÉTHUNE, *J. Electrochem. Soc.*, **102**:288C (1955).

3 W. OSTWALD: *Z. Phys. Chem.* (Leipzig), **1**:583 (1887).

4 R. ABEGG, F. AUERBACH, and R. LUTHER: Messungen Elektromotorischer Krafter Galvanischer Ketten mit Wässerigen Elektrolyten, *Abh. Deut. Bunsenges. Angew. Phys. Chem.*, No. 5, Halle, 1911.

5 G. N. LEWIS and M. RANDALL: "Thermodynamics and the Free Energy of Chemical Substances," chap. 30, McGraw-Hill, New York, 1923.

6 T. BIEGLER and R. WOODS: *J. Chem. Educ.*, **50**:604 (1973).

7 W. M. LATIMER: "The Oxidation States of the Elements and Their Potentials in Aqueous Solutions," 2d ed., app. 1, Prentice-Hall, Englewood Cliffs, N.J., 1952.

8 G. CHARLOT and others: "Selected Constants," Supplement to *Pure Appl. Chem.*, IUPAC, Butterworth, London, 1971.

9 G. F. SMITH: *Anal. Chem.*, **23**:925 (1951); G. F. SMITH and F. P. RICHTER: "Phenanthroline and Substituted Phenanthroline Indicators," p. 41, G. F. Smith Chemical, Columbus, Ohio, 1944.

10 L. G. SILLÉN and A. E. MARTELL: "Stability Constants of Metal Ion Complexes," Special Publication Nos. 17 and 25, Chemical Society, London, 1964, 1971.

11 I. M. KOLTHOFF and W. J. TOMSICEK: *J. Phys. Chem.*, **39**:945 (1935).

12 B. V. NEKRASOV and G. V. ZOTOV: *J. Appl. Chem. USSR*, **14**:264 (1941).

13 D. N. HUME and I. M. KOLTHOFF: *J. Amer. Chem. Soc.*, **65**:1895 (1943).

14 T. S. LEE, I. M. KOLTHOFF, and D. L. LEUSSING: *J. Amer. Chem. Soc.*, **70**:2348 (1948).

15 R. M. GARRELS and C. L. CHRIST: "Solutions, Minerals, and Equilibria," chap. 7, Harper, New York, 1965. See also L. G. SILLÉN: *J. Chem. Educ.*, **29**:600 (1952); L. G. SILLÉN in "Treatise on Analytical Chemistry," I. M. Kolthoff and P. J. Elving (eds.), pt. 1, vol. 1, chap. 8, Wiley, New York, 1959; W. STUMM and J. J. MORGAN: "Aquatic Chemistry," p. 319, Wiley, New York, 1970.

16 J. O'M. BOCKRIS and A. K. M. S. HUQ: *Proc. Roy. Soc., Ser. A*, **237**:277 (1956).

17 H. GERISCHER: *Z. Elektrochem.*, **59**:604 (1955).

18 C. WAGNER and W. TRAUD: *Z. Elektrochem.*, **44**:391 (1938).

19 I. M. KOLTHOFF and C. S. MILLER: *J. Amer. Chem. Soc.*, **62**:2171 (1940).

20 M. FLEISCHMANN, H. R. THIRSK, and I. M. TORDESILLAS: *Trans. Faraday Soc.*, **58**:1865 (1962).

21 R. LUTHER: *Z. Phys. Chem.* (Leipzig), **30**:628 (1899); I. M. KOLTHOFF: *Chem. Weekbl.*, **16**:450 (1919).

22 GARRELS and CHRIST,[15] p. 195.

PROBLEMS

12-1 From the standard electrode potentials of the S-H_2S and H_2SO_3-S couples, calculate (*a*) the standard potential for the half-reaction

$$H_2SO_3 + 6H^+ + 6e^- \rightarrow H_2S + 3H_2O$$

and (b) the standard free-energy change for the reaction

$$H_2SO_3 + 2H_2S \rightarrow 3S + 3H_2O$$

Answer (a) 0.35 V. (b) -120 kJ.

12-2 From the standard potentials of the Pb^{++}-Pb and PbO_2-Pb^{++} couples, calculate (a) the standard potential of the half-reaction

$$PbO_2 + 4H^+ + 4e^- \rightleftharpoons Pb + 2H_2O$$

and (b) the standard free-energy change of the reaction

$$PbO_2 + Pb + 4H^+ \rightleftharpoons 2Pb^{++} + 2H_2O$$

(c) Sketch the potential-pH diagram, showing regions of stability of Pb, Pb^{++}, $Pb(OH)_2$, PbO_2, and $HPbO_2^-$. For $Pb(OH)_2$, $K_{sp} = 2.5 \times 10^{-16}$; for $Pb(OH)_2$ (solid) $\rightleftharpoons HPbO_2^- + H^+$, $K_a = 10^{-15}$.

Answer (a) 0.665 V. (b) -305 kJ.

12-3 From appropriate data in Table 12-1, calculate (a) the standard potential for the half-reaction

$$VO^{++} + 2H^+ + 2e^- \rightleftharpoons V^{++} + H_2O$$

and (b) the standard free-energy change of the reaction

$$VO^{++} + V^{++} + 2H^+ \rightleftharpoons 2V^{3+} + H_2O$$

Answer (a) 0.042 V. (b) -57 kJ.

12-4 The logarithms of the successive stepwise formation constants of the ammonia complexes of silver are $\log K_1 = 3.2$, $\log K_2 = 3.8$. Calculate the electrode potential of a silver electrode in equilibrium with a solution 10^{-2} M in silver ions, in a NH_3-NH_4NO_3 buffer of total concentration 0.1 M and pH $= 10$. Take K_a of $NH_4^+ = 5 \times 10^{-10}$, and assume activity coefficients of unity.

Answer 0.3950 V.

12-5 Using (1) the cell emf method and (2) the electrode-potential method, calculate the equilibrium constants for the following chemical reactions. (Use the values for standard electrode potentials in Table 12-1.)

(a) $2Cr(II) + Br_2 \rightleftharpoons 2Cr(III) + 2Br^-$
(b) $2Cr(III) + Sn^{++} \rightleftharpoons 2Cr(II) + Sn^{4+}$
(c) $Cr_2O_7^= + 6Fe(II) + 14H^+ \rightleftharpoons 2Cr(III) + 6Fe(III) + 7H_2O$
(d) $5HOBr \rightleftharpoons BrO_3^- + H^+ + 2Br_2 + 2H_2O$

Answer. (a) 2.6×10^{49}. (b) 2.5×10^{-18}. (c) 6.3×10^{56}. (d) 3×10^8.

12-6 Derive an equation for the effect of complex formation on the potential of the half-reaction $M_{ox} + ne^- \rightleftharpoons M_{red}$ for a case in which M_{ox} forms a series of complexes MX, MX_2, ... but M_{red} forms no complex with X.

12-7 Sketch the potential-pH diagram for the system O_2-H_2-H_2O, and superimpose it on Figure 12-1 to illustrate the expected reactions for iron in this solvent.[22]

MEMBRANE POTENTIALS AND ION-SELECTIVE ELECTRODES

An ion-selective electrode consists of a membrane that responds more or less selectively to a single species of ion, in contact on one side with a solution of the ion to be determined and, usually, on the other with a solution having a fixed activity of the ion in contact with a suitable reference electrode. The most familiar example is a glass pH electrode, in which the membrane is glass responsive to hydrogen ions and the reference electrode is a calomel or silver chloride electrode within the glass container.

Although the pH response of a thin glass membrane was recognized[1] long ago, the development[2] of functional glass pH electrodes took a substantial period of time. Similarly, even though electrodes sensitive to ions other than the hydrogen ion were suggested[3] decades ago, few advances were made until recently in the development of functional membrane electrodes for other ions. Probably one cause of the delay was the misleading but common presumption that the glass electrode responds uniquely, on the assumption that the membrane is permeable to hydrogen ions.

After it was recognized that pH glass electrodes were not permeable to hydrogen ions, the systematic development of new electrodes became more intellectually feasible. We have probably only begun to explore the potentialities of membrane electrodes. These electrodes should be thought of as ion-selective rather than ion-specific, since no electrode is entirely specific, but all show mixed response to the activities of several ions.

Reviews on ion-selective electrodes have been presented by Pungor and Toth,[4] Koryta,[5] Buck,[6] and Durst.[7]

13-1 MEMBRANES PERMEABLE TO A SINGLE SPECIES

Consider first a system in which two solutions of an electrolyte A^+B^-, of concentrations C_1 and C_2, are separated by a membrane permeable to both species of ion. For a liquid junction with transference and for dilute solutions, the difference in potential across the membrane is given by[8]

$$\Delta E = (2t_- - 1) \frac{RT}{F} \ln \frac{a_2}{a_1} \simeq (2t_- - 1) \frac{RT}{F} \ln \frac{C_2}{C_1} \qquad (13\text{-}1)$$

in which t_- is the transference number of the anion within the membrane. Now consider two cases, in which the membrane is not permeable to the anion ($t_- = 0$), and not permeable to the cation ($t_- = 1$):

$$\Delta E = \frac{RT}{F} \ln \frac{a_1}{a_2} \quad \text{and} \quad \Delta E = \frac{RT}{F} \ln \frac{a_2}{a_1} \qquad (13\text{-}2)$$

An important practical example of the second case is the lanthanum fluoride single-crystal electrode, permeable to fluoride ion but not to cations. Construct a cell

$$\text{Hg } \text{Hg}_2\text{Cl}_2 \left| (\text{KCl sat}) \right\| \left. \begin{array}{c|c} F^- \text{ test} & \text{LaF}_3 \\ \text{solution} & \text{membrane} \end{array} \right| \left(\begin{array}{c} 0.1\, M\, \text{NaF} \\ 0.1\, M\, \text{NaCl} \end{array} \right) \left| \text{AgCl Ag} \right. \qquad (13\text{-}3)$$

in which the membrane electrode assembly includes the internal silver chloride reference electrode, and in which measurements of fluoride ion activity are made against a saturated calomel electrode. The activity of fluoride ion on one side of the membrane is held constant, and if the potential of the external calomel electrode and the potential of its liquid junction to a series of test solutions can be regarded as constant, then the cell emf becomes

$$E_{\text{cell}} = \text{constant} - \frac{RT}{F} \ln a_{F^-} \qquad (13\text{-}4)$$

In practice, a single crystal of europium-doped lanthanum fluoride ($K_{sp} = 2 \times 10^{-22}$) responds in a matter of seconds[9] according to (13-4) to fluoride ion activity in solutions containing fluoride from about 1 M to 10^{-5} M. Europium doping increases fluoride mobility by introducing lattice disorder. Hydroxide ion, beginning at about 10^{-5} M, is the principal interference, probably[10] the result of competition due to the low solubility of lanthanum hydroxide.

An extensive summary of applications of fluoride ion-selective electrodes to thermodynamic studies of fluoride complexes and analysis by direct potentiometry and potentiometric titrations has been presented by Buck.[6]

13-2   THE GLASS ELECTRODE

In an ion-selective glass electrode the glass membrane is viewed as consisting of at least three distinct regions:

| test solution | hydrated glass layer | dry glass layer | hydrated glass layer | internal solution |

The hydrated layers may be somewhat less than 0.1 μm thick, whereas the dry glass layer is of the order of 50 μm. The hydrated layers on the inner and outer surfaces of the glass are essential to its operation and can be regarded as an immobilized solution[11] in which diffusion coefficients[12] are more than 1000 times those in the dry glass phase. Hydrogen ions diffuse in the hydrated layer[13] at about 3×10^{-3} μm/h.

Haugaard,[14] in attempting to discover whether hydrogen ion transfer occurred, passed a small measured current through a thin glass bulb containing, and immersed in, 0.02 M hydrochloric acid solution. The changes in acidity and the amounts of sodium chloride formed were measured on both sides of the glass, with results that indicated electrolytic conduction by way of sodium ions. Schwabe and Dahms[15] carried out similar electrochemical experiments involving tritium transport. Both studies clearly indicated that hydrogen ions are not transported through a glass membrane.

A conditioned glass electrode contains protons that have replaced sodium or other ions in the hydrated surface, and this reservoir of protons can exchange with cations in solution. The potential at a glass-solution interface is regarded as arising from the cation transfers going on continuously at this interface.[16] Anions appear to play no role in processes occurring within the hydrated layers or the dry glass layer. If a glass membrane separates two solutions of differing pH, equilibrium is quickly established in the surface of each hydrated layer, and an emf is generated because additional protons are either taken up by or lost from the surfaces, depending on the hydrogen ion activities of the two solutions:

$$H^+_{glass} \rightleftharpoons H^+_{soln} \qquad (13-5)$$

This phase-transfer reaction involves a difference in free energy, since the transfer activity coefficient for protons in the glass surface is different from that in water. A small net transfer of ions occurs that gives rise to a potential. A net positive charge on one side of the interface with respect to that on the other is produced in accordance with $-\Delta G = nFE$, where E is potential. The proton pressure and the tendency to acquire a positive potential in the surface layer of the glass change in accordance with the hydrogen ion activity in the solution. In the process of establishing Equilibrium (13-5) a surface potential is developed and, with both interfaces of a glass membrane considered, the net potential across the membrane (*solid-junction potential*) is given by[17]

$$E = \frac{RT}{F} \ln \frac{a_{H^+(test)}}{a_{H^+(ref.\ soln)}} \qquad (13\text{-}6)$$

If the inner surface remains in contact with a solution whose hydrogen ion activity is invariant, its contribution to the solid-junction potential along with the other invariant potentials in the overall cell can be incorporated in a constant, and the potential of a cell with the glass electrode can be written in the usual form of the Nernst equation:

$$E = \text{constant} + \frac{RT}{F} \ln a_{H^+} \qquad (13\text{-}7)$$

where a_{H^+} is that in the test solution.

When the test solution contains sodium and other cations as well as hydrogen ions, the effects of their activities on the membrane potential must be included. For sodium ions the potential is affected by the equilibrium

$$Na^+_{glass} \rightleftharpoons Na^+_{soln} \qquad (13\text{-}8)$$

The higher the sodium ion activity in the test solution, the more positive will be the surface layer of the glass. Sodium and hydrogen ions of equal activity in solution do not undergo exchange with ions in the glass to the same extent, however, and therefore do not affect the potential to the same degree. Because sodium ions affect interfacial potentials less than hydrogen ions, a weighting factor must be applied to obtain the net effect of both ions on the boundary potential. This factor is called a *selectivity coefficient*.

The glass electrode has been treated by Eisenman[18] and coworkers as cation-exchange membranes. Expressions were derived for the membrane potential in relation to the ion-exchange equilibrium constants of the membrane with various ions in solution and in relation to the mobilities of the cation species characteristic of the glass. For a membrane responding to two univalent cations i and j, the potential is given by

$$E = \text{constant} + \frac{RT}{F} \ln \left(a_i + a_j K_{eq_{ij}} \frac{u_i}{u_j} \right) \qquad (13\text{-}9)$$

$$= \text{constant} + \frac{RT}{F} \ln \left(a_i + a_j K_{ij} \right) \qquad (13\text{-}10)$$

where u_i and u_j are the mobilities of ions i and j in the membrane, $K_{eq_{ij}}$ is the equilibrium constant for the reaction

$$i_{soln} + j_{membrane} \rightleftharpoons i_{membrane} + j_{soln} \qquad (13\text{-}11)$$

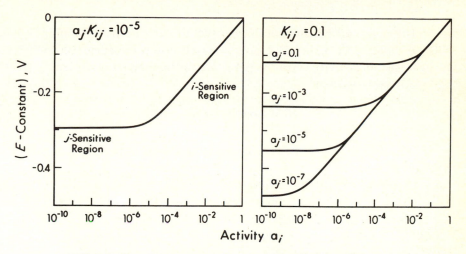

FIGURE 13-1   The potential, $E -$ constant, as a function of activity of an ion i for $a_j K_{ij} = 10^{-5}$ (left) and for $K_{ij} = 0.1$ and a_j variable (right).

and K_{ij} is the selectivity coefficient that is equal to $K_{eq_{i,j}} u_i / u_j$. A value of $K_{ij} = 0.1$ means that the same potential response generated by an activity of i of a_i is generated by $a_j = 10a_i$.

Figure 13-1 shows how, with selected values of K_{ij} and a_j, potential calculated from (13-10) varies with activity of ion i. For a small value of $a_j K_{ij}$ the potential is a linear function of the logarithm of the activity of ion i. When a_i becomes sufficiently small, however, the potential becomes independent of a_i and dependent on a_j. In this region the electrode becomes responsive to a_j. Thus in Figure 13-1 (right), if a_i is 10^{-8}, the potential is linearly dependent on the value of log a_j in the region 0.1 to 10^{-5}, whereas if a_j is less than about 10^{-5}, the potential is proportional to log a_i at values greater than about 10^{-5}. Consequently, an electrode can be used to measure either a_i or a_j, depending on the relative values of a_i, a_j, and K_{ij}.

In general, for several univalent cations $i_1, i_2, \ldots$ in solution the potential given by (13-10) can be generalized to

$$E = \text{constant} + \frac{RT}{F} \ln \left(a_{i_1} + a_{i_2} K_{i_1 i_2} + a_{i_3} K_{i_1 i_3} + \cdots + a_{i_n} K_{i_1 i_n} \right) \quad (13\text{-}12)$$

For a membrane electrode responding to the activity of a cation i with charge Z^+ in the presence of a cation j with charge Y^+, Equation (13-10) becomes

$$E = \text{constant} + \frac{RT}{ZF} \ln \left[a_i^{Z+} + K_{ij}(a_j^{Y+})^{Z/Y} \right] \quad (13\text{-}13)$$

The general equation for an anion-responsive membrane (not a glass electrode) is similar:

$$E = \text{constant} - \frac{RT}{ZF} \ln \left[a_i^{Z-} + K_{ij}(a_j^{Y-})^{Z/Y} \right] \quad (13\text{-}14)$$

The membrane potential of Corning 015 glass (a eutectic containing 72.2 mole % SiO_2, 6.4 mole % CaO, and 21.4 mole % Na_2O) begins to depend on sodium as well as hydrogen ion activity when the pH is greater than about 9. Equation (13-10) for this glass can be written

$$E = \text{constant} + \frac{RT}{F} \ln (a_{H^+} + a_{Na^+} K_{H^+ Na^+}) \quad (13\text{-}15)$$

where $K_{H^+ Na^+}$ is the selectivity coefficient of the glass for hydrogen ions in the presence of sodium ions. If a_{H^+} in (13-15) is much greater than $a_{Na^+} K_{H^+ Na^+}$, the potential depends almost solely on the hydrogen ion activity, and the electrode then functions as a pH electrode. If a_{H^+} is much less than $a_{Na^+} K_{H^+ Na^+}$, the electrode becomes a sodium ion-responsive electrode. Special high pH electrodes are constructed of glasses containing Li_2O, which give better selectivity coefficients in favor of hydrogen ions and less interference from alkali metal ions at high pH. For glass electrodes generally, the value of $K_{H^+ Na^+}$ is small and, for glasses especially suitable for high pH measurements, the value may be as low as 10^{-15}.

Varying the composition of the glass can alter the response to various univalent cations. Eisenman[19] carried out extensive investigations on cation selectivity as a function of glass composition. More recently, Phang and Steel[20] reported on selectivity coefficients, pH, and light responses of several commercially available glass electrodes.

Through the incorporation of Al_2O_3 into the glass, selectivity favoring hydrogen ions can be reduced to yield glasses responding to other cations, providing the pH is relatively high. Thus a glass composed of 15% Li_2O–25% Al_2O_3–60% SiO_2 is useful for measuring lithium ion activity; one of 11% Na_2O–18% Al_2O_3–71% SiO_2, for sodium ion activity; one of 27% Na_2O–5% Al_2O_3–68% SiO_2, for potassium ion activity; and 11% Na_2O–18% Al_2O_3–71% SiO_2, for silver ion activity.[12] For the sodium-selective glass at high pH the membrane potential depends primarily on the activity of sodium ions down to 10^{-5} M. The selectivity coefficient of $K_{Na^+ K^+}$ depends on glass composition and pH, values in the range 10^{-3} to 10^{-5} being possible.[12] For lithium-selective electrodes, $K_{Li^+ Na^+}$ is about 0.3 and $K_{Li^+ K^+}$ about 10^{-3}. For a potassium-selective glass, $K_{K^+ Na^+}$ is about 0.05. Thus the selectivity coefficients between metal ions for alkali metal glasses are many orders of magnitude larger (less selective) than for pH electrodes.

13-3 LIQUID MEMBRANES

A liquid membrane consists of a solvent immiscible with water and a reagent that acts as extractant and complexing agent for an ion. If such a liquid membrane separates two solutions, ion selectivity is achieved through preferential extraction of ions into the membrane phase as well as differences in mobility of the ions within the membrane. Eisenman[17] derived equations for several cases of liquid membranes, depending on the degree of dissociation of the complexes within the membrane, the relative mobilities of the ions, the ion-exchange sites, and the complexes. In each case an expression is derived for the selectivity coefficient K_{ij} in terms of the ion mobilities u_i and u_j, the equilibrium constants k_i and k_j for the extraction of ions into the membrane, the site mobility u_s, and the mobilities u_{is} and u_{js} of the complexes *is* and *js*.

If the complexes *is* and *js* are highly dissociated in the membrane, then

$$K_{ij} = \frac{u_i}{u_j}\frac{k_i}{k_j} = K_{eq\,ij}\frac{u_i}{u_j} \qquad (13\text{-}16)$$

which is the same expression for selectivity coefficient as for a glass electrode. Note that u_i, u_j, k_i, and k_j depend on the solvent but are independent of the exchanger species.

If the complexes *is* and *js* are highly associated and the sites (or complexing agent) are only slightly mobile, then

$$K_{ij} = \frac{u_i + u_s\,k_i}{u_j + u_s\,k_j} \qquad (13\text{-}17)$$

but since $u_s \ll u_i$ or u_j,

$$K_{ij} \simeq \frac{u_i}{u_j}\frac{k_i}{k_j} \qquad (13\text{-}18)$$

which once again is reduced to (13-10).

If the exchange sites are highly mobile, the expression derived by Eisenman is

$$K_{ij} = \frac{u_{is}}{u_{js}}\frac{k_i}{k_j} \qquad (13\text{-}19)$$

in which the relative mobilities of the *complexes* as well as of the extraction constants are involved. Thus not only the nature of the solvent but also the extraction reagent determine the selectivity coefficient.

In one calcium liquid ion-exchanger electrode the calcium salt of didecyl-phosphoric acid is the active component in the solvent di-*n*-acetylphenylphosphonate. In general, alkylamines are suitable carriers for anions, and dinonylnaphthalene-sulfonic acid or di-2-ethylhexylphosphoric acid for cations. The organic liquid is held on a porous, inert support that keeps the liquid ion exchanger in contact with the test and reference solutions with minimum mixing of the two phases. Use of the electrodes is restricted to water and other solvents that do not appreciably dissolve the solvent in the exchanger. A high selectivity coefficient requires that the ion exchanger form a stronger complex with the ion of interest than with interfering ions.

For the calcium electrode the most serious interferences[12] are hydrogen ions (useful pH range 5.5 to 11), strontium ($K_{Ca^{++}Sr^{++}} = 0.014$), magnesium ($K_{Ca^{++}Mg^{++}} = 0.005$), and barium ($K_{Ca^{++}Ba^{++}} = 0.0016$). The selectivity coefficients with sodium and potassium are about 3×10^{-4}. Anions have little effect on the membrane potential.

For a copper-sensitive liquid ion-exchanger electrode,[12] hydrogen ($K_{Cu^{++}H^{+}} = 10$) and iron(II) ($K_{Cu^{++}Fe^{++}} = 140$) interfere seriously, whereas nickel(II) ($K_{Cu^{++}Ni^{++}} = 0.01$) interferes only moderately.

In general, the lower sensitivity limit of these electrodes is governed in large part by the solubility of the ion-organic substrate in the aqueous phase.

Miniature liquid ion-exchanger electrodes[21] have been made of such small size that ion activities in intracellular fluid can be measured.

Another type of liquid membrane involves a neutral substance that acts as a molecular carrier of ions. If the salts $i^{+}X^{-}$ and $j^{+}X^{-}$ exist in solution, and the liquid membrane contains a molecular solute L that complexes the cations, the following equilibria are involved:

$$i^{+} + X^{-} + L(\text{membrane}) \rightleftharpoons iL^{+}(\text{membrane}) + X^{-}(\text{membrane}), k_i \qquad (13\text{-}20)$$

$$j^{+} + X^{-} + L(\text{membrane}) \rightleftharpoons jL^{+}(\text{membrane}) + X^{-}(\text{membrane}), k_j \qquad (13\text{-}21)$$

in which anions are extracted from the solution to preserve electroneutrality in the membrane. The selectivity coefficient is given by [22]

$$K_{ij} = \frac{u_{iL}}{u_{jL}} \frac{k_i}{k_j} \qquad (13\text{-}22)$$

where u_{iL}/u_{jL} is the ratio of mobilities of the species iL^{+} and jL^{+}, and k_i/k_j is the equilibrium constant of the reaction

$$i^{+}(\text{aq}) + jL^{+}(\text{membrane}) \rightleftharpoons j^{+}(\text{aq}) + iL^{+}(\text{membrane}) \qquad (13\text{-}23)$$

Because of the large size of the complexing group, the mobilities of the complexes are approximately equal, and

$$K_{ij} \simeq \frac{k_i}{k_j} \qquad (13\text{-}24)$$

so the selectivity is derived largely from preferential complexation, which is mainly a property of the complexing agent rather than the solvent. An example is valinomycin,[23] an antibiotic with a 36-membered ring, which forms a complex with alkali metal ions. The selectivity coefficient of 2×10^{-4} for K^{+} relative to Na^{+} is about two orders of magnitude more favorable than for glass electrodes. Certain anions affect the measured responses in a predictable manner.[24]

Anion-selective liquid membranes can be made by using a bulky cationic species that will selectively extract anions into an organic solvent. Examples are substituted phenanthroline complexes of iron(III), which selectively extract perchlorate, and similar complexes of nickel(II), which selectively extract nitrate. Selectivity constants are listed by Ross.[25]

13-4   OTHER SOLID-STATE MEMBRANES

Kolthoff and Sanders[3] showed years ago that silver chloride and silver bromide can act as membranes giving a nernstian response to silver ion (and through K_{sp} to halide ion) activity. The silver halides are advantageous in having appreciable ionic conductivity (exclusively due to silver ion), but compressed disks or pellets prepared by fusion have unfavorable physical properties. The Orion mixed Ag_2S-AgX systems[25] take advantage of the favorable properties of Ag_2S in forming stable pellets when mixed with $AgCl$, $AgBr$, AgI, or $AgSCN$. In each case the membrane acts as though it were prepared from pure AgX, because of the low solubility of Ag_2S. The Pungor design[26] involves incorporation of the membrane material in a matrix such as silicone rubber. The proportion of material imbedded in the matrix must be high enough to yield physical contact between particles. The response curves for mixtures of the halides strikingly resemble those of Figure 13-1. The iodide electrode responds to cyanide ion activity at low iodide concentrations, whereas it responds to iodide ion activity at low cyanide concentrations.

Ag_2S can also be used as a matrix for other metal sulfides, notably CuS, PbS, and CdS, giving membranes responsive to the second metal. The electrode can be regarded as responding to the equilibrium activity of Ag^+, which is given by

$$a_{Ag^+}^2 = \frac{K_{sp(Ag_2S)}}{K_{sp(MS)}} a_M \qquad (13\text{-}25)$$

If the solution contains M^{++} but not Ag^+, the electrode obeys the Nernst equation for M^{++}.

Hansen, Lamm, and Růžička[27] described the Selectrode (trademark of Radiometer A/S), which involves a specially treated graphite electrode in direct contact with an electroactive surface. Typical is the halide electrode, involving a layer of silver halide and sulfide in direct contact with a graphite electrode that has been rendered hydrophobic by treatment with Teflon. Another example is the Cu(II) electrode, based on a similar preparation using CuS or CuSe. The Selectrodes couple an electronic conductor directly to the solid-state detector, which is an ionic conductor. Although the mechanism of coupling is not clear, Buck[6] suggested that an interfacial potential can be communicated through solid-state equilibria to the electronic conductor without the intervention of a redox process.

Typical response curves of metal sulfide membrane electrodes in pure aqueous solutions show a linear relation with log C, with the slope expected from the Nernst equation, down to concentrations of 10^{-5} to 10^{-6} M; below this the response flattens out and the electrode response time becomes longer. Significantly, however, in *metal ion buffers* the response time remains linear with ion activity to much lower values (10^{-10} to 10^{-12} M), as long as the analytical concentration is appreciably higher. It appears likely that specific adsorption of the complex species is necessary in appreciable concentration to give thermodynamic response. Care should be taken when such electrodes are calibrated in metal ion buffers and applied to unknowns not containing the same complexing agent.

13-5   ENZYME ELECTRODES

The concept of enzyme substrate electrodes was applied as early as 1962 for the determination of glucose and urea.[28] Guilbault and coworkers[29] expanded the use of such electrodes to the determination of enzymes as well as the substances that interact with them. For example, urea in solution can be converted to ammonium ions through the catalytic action of urease. The enzyme is retained in a gel layer around an ammonium ion–selective electrode, and the ammonium ion produced diffuses to the surface of the electrode. The potential gives a measure of the urea concentration. Similarly, an amygdaline-sensitive electrode can be made by retaining β-glucosidase in a gel layer coupled to a cyanide-sensitive membrane electrode.[30] A highly selective urea and l-tyrosine electrode can be made by coupling a layer of the enzyme urease or tyrosine decarboxylase to a carbon dioxide electrode.[31] A penicillin-selective electrode has been devised by incorporating the enzyme penicillinase in a thin polyacrylamide gel molded around a hydrogen ion glass electrode, which detects an increase in acidity accompanying the hydrolysis of penicillin to penicilloic acid.[32] Automated determination of enzymes including glucose oxidase, β-glucosidase, and rhodanase has been described.[33] The continuing development and application of enzyme electrodes are proceeding at a rapid pace.

13-6   MISCELLANEOUS ION-SELECTIVE ELECTRODES

A carbon dioxide–sensitive electrode[34,35] is a heterogeneous electrode consisting of a pH-sensitive glass electrode around which a gas-permeable membrane retains a layer of sodium bicarbonate solution. The membrane permits carbon dioxide to diffuse into or out of the sodium bicarbonate solution; the glass electrode then responds to the resulting change in pH. Other gas-sensing electrodes have been developed[35a] that respond to ammonia, amines, sulfur dioxide, nitrogen dioxide, hydrogen sulfide, hydrogen cyanide, hydrogen fluoride, acetic acid, or chlorine.

A calcium-sensitive heterogeneous membrane electrode has been made by evaporating an alcohol-ether solution of collodion and calcium dioctylphosphate.[36]

A rugged type of heterogeneous electrode was suggested by Freiser and others[37] in which a slightly soluble salt is deposited as a polymeric matrix simply painted onto a platinum wire, as, for example, a dispersion with an epoxy resin.

Liteanu and Hopirtean[38] described a polyvinylchloride–tricresyl phosphate electrode as being responsive to H^+, Ag^+, Tl^+, Hg_2^{++}, Na^+, and K^+. Although the electrodes composed of polymeric films directly on metals are attractive from the viewpoint of simplicity, the mechanism of their function remains to be elucidated, and the scope of their applications remains to be studied.

An iron-selective electrode based on a glass Fe-1173 ($Ge_{28}Sb_{12}Se_{60}$) has been reported[39] to respond to uncomplexed iron(III) in sulfate solutions. It was used to evaluate formation constants of iron(III)-sulfate complexes and as an indicator for titrations of sulfate with barium. The mechanism of the sensing action is obscure.

13-7 REFERENCE AND STANDARD SOLUTIONS

Reference solutions In an ion-selective electrode assembly the internal reference solution (Figure 13-2) performs a dual function. It must contain one ionic species to provide a stable electrode potential for the internal reference half-cell and another to provide a stable membrane potential at the inner solution–membrane interface. In the usual pH glass electrode assembly the inner reference half-cell is Ag/AgCl; the potential is stable because the internal reference solution is 0.1 M in chloride ion. The potential of the inner membrane surface also is stable because the internal filling solution is 0.1 M in hydrogen ion (or a buffered solution). The single-electrolyte solution 0.1 M HCl serves admirably to provide a high and stable concentration of both ions.

Ion-activity standards In relating membrane potential to concentration of a species being measured, the problem of single-ion activity coefficients arises, just as it does in the measurement of hydrogen ion activity (Section 3-3). In Cell (13-3) exact evaluation is not possible of the liquid-junction potential between the saturated calomel half-cell and test solution, of the asymmetry potential across the membrane, or of the individual ion activities in the test solution. On the assumption that liquid-junction and asymmetry potentials are the same whether an unknown or standard is used as the test solution, a tentative *operational definition* of cation activity is[40]

$$pM = pM_s + \frac{nF(E - E_s)}{2.303RT} \quad (13\text{-}26)$$

where pM is the negative log of the ion activity of a cation in an unknown solution,

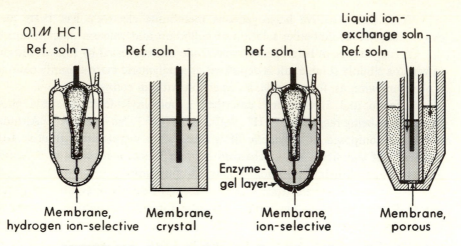

FIGURE 13-2   Schematic diagrams of several membrane electrodes: glass, left; LaF_3; enzyme; and liquid.

pM$_s$ is the negative log of the activity in a standard, and E and E_s are the emf values for a cell such as (13-3) with either the unknown or a standard as the test solution. Similarly for anions,

$$pA = pA_s - \frac{nF(E - E_s)}{2.303RT} \qquad (13\text{-}27)$$

where pA and pA$_s$ are negative logs of ion activities for an anion in unknown and standard solutions.

For cations and anions generally, the assumption that liquid-junction potentials are the same in the measurement of standards and unknowns is less likely to be valid than for pH measurements. It has been suggested[40] that a quantity ΔE_j expressed in pM or pA units be included in (13-26) and (13-27) to correct for changes in junction potential arising from differences in ionic strengths of standard and test solutions. Alternatively, these effects could be eliminated through the use of two reference half-cells composed of electrodes without liquid-junction potentials. For example, if the test solution contained chloride ion, both reference half-cells could be Ag/AgCl, and the liquid-junction potential would be eliminated. In practice, external reference half-cells without liquid junction are not always convenient.

With the ion-selective electrode standardized at an ionic strength of 0.01, and for cells involving silver–silver chloride and saturated calomel half-cells, the residual effect of liquid-junction potential[40] is estimated to amount to about -0.02 and -0.07 pNa units at ionic strengths of 0.1 and 1.0; for chloride, $+0.02$ and $+0.07$ pCl units at 0.1 and 1.0; for calcium, -0.03 and -0.22 pCa units at 0.1 and 1.0.

The potential developed at an ion–selective membrane interface depends on single-ion activities. These potentials cannot be measured in isolation, and only the potential difference between the two interfaces is measurable (with the aid of the two reference electrodes). Furthermore, the activities of single ions cannot be determined exactly. In an operational definition of pM or pA, assumptions regarding ion activities have to be made to obtain pM_s or pA_s. Experimental mean activity coefficients can be split among the ions of a salt according to some convention (Sections 2-3 and 3-3). In an attempt to obtain a consistent set of activity coefficients, it can be assumed that $\gamma_+ = \gamma_- = \gamma_\pm$ for 1:1 electrolytes and $\gamma_{2+} = \gamma_{2-} = \gamma_+^4 = \gamma_-^4$ for 2:1 electrolytes. At low ionic strengths the problem of internal consistency is not serious, though problems begin to arise at ionic strengths of 0.1 or higher. For example, at high ionic strengths the activity coefficient for sodium ion is different in a NaCl solution than in a NaBr solution, and γ_{Br^-} would be different depending on whether the sodium or potassium salt were involved. The activity coefficient for an anion A^- can be estimated[40] by

$$\log \gamma_{A^-} \simeq \log \gamma_{Cl^-} + 2 \log \frac{\gamma_{MA}}{\gamma_{MCl}} \qquad (13\text{-}28)$$

where γ_{MA} and γ_{MCl} are the mean activity coefficients of 1:1 electrolytes involving a cation M^+ and where $-\log \gamma_{Cl^-}$ is $A\sqrt{\mu}/(1 + 1.5\sqrt{\mu})$, where A is the Debye–Hückel constant and μ is the ionic strength [Equation (2-13)]. Similarly, for a 2:1 electrolyte

$$\log \gamma_{A^-} = \log \gamma_{Cl^-} + 1.5 \log \frac{\gamma_{MA_2}}{\gamma_{MCl_2}} \qquad (13\text{-}29)$$

On the basis of (13-28) and (13-29), values of γ_A for several anions are shown in Table 13-1 at an ionic strength of unity.

If ionic hydration theory is used as a basis for obtaining single-ion activity coefficients,[41] somewhat different values result. For example, γ_{Cl^-} values of 0.620 and 0.586 are obtained when NaCl and KCl are used. At an ionic strength of 0.1

Table 13-1 CALCULATED ACTIVITY COEFFICIENTS FOR UNIVALENT ANIONS AT 25°C AND IONIC STRENGTH OF 1.0†

			Anion			
Cation	Cl^-	F^-	Br^-	I^-	ClO_4^-	NO_3^-
Li^+	0.625	...	0.68	0.87	0.81	0.58
Na^+	0.625	0.48	0.69	0.78	0.58	0.44
K^+	0.625	0.71	0.64	0.71	...	0.34
Mg^{++}	0.625	...	0.71	0.81	0.85	0.60
Ca^{++}	0.625	...	0.69	0.79	0.81	0.50
Ba^{++}	0.625	...	0.69	0.81	0.72	0.35

† From Bates and Alfenaar.[40]

or less there is little difference in single-ion activity coefficients obtained by several different and reasonable pathways for splitting mean activity coefficients of electrolytes.

In view of the sometimes widely divergent values of activity coefficients for the same ion at the same ionic strength shown in Table 13-1, agreement needs to be reached on some arbitrary mode of arriving at activity coefficients. It has been proposed[40] that, generally, the activity coefficients of cations in reference solutions for use with cation-selective electrodes be standardized by use of their completely dissociated chloride salts and that anion-selective electrodes be standardized by use of their completely dissociated sodium salts. It has been proposed further that the activity coefficient of chloride ions be obtained from $A\sqrt{\mu}/(1 + 1.5\sqrt{\mu})$ and that of sodium ions from the measured mean activity coefficients of aqueous NaCl solutions (Figure 2-3), and that the following equation be applied:

$$\log \gamma_{Na^+} \simeq 2 \log \gamma_{\pm,NaCl} - \log \gamma_{Cl^-} \qquad (13\text{-}30)$$

For the specific case of a standard for fluoride ion activity KF rather than NaF has been suggested.[42] KF is a better choice because ion pairing is much less. Further, the average hydration number of the fluoride ion is almost the same as that for potassium ion, so that activity coefficients of the two ions are similar. Suggested reference activity values (pM_s or pA_s) for use in the operational definitions for ion-activity measurements [Equations (13-26) or (13-27)] are shown in Table 13-2.[40,42] For the case of fluoride ion, measurements of its activity in NaF-NaCl mixtures up to 1 m and KF-KX mixtures up to 4 m yielded[43] the same values as pure NaF or KF at the same ionic strength.

When concentrations of entities become low, the problem of stable standards becomes serious. Metal ion buffers have been prepared,[26] for example, by complexing

Table 13-2 SUGGESTED VALUES OF REFERENCE STANDARDS AT 25°C IN WATER†

Substance	Molality	pNa_s	pCa_s	pCl_s	pF_s
NaCl	0.001	3.015	...	3.015	...
	0.01	2.044	...	2.044	...
	0.1	1.108	...	1.110	...
	1.0	0.160	...	0.204	...
KF	0.01	...	...	...	2.044
	0.1	...	...	...	1.111
	1.0	...	...	...	0.190
$CaCl_2$	0.000333	...	3.537	3.191	...
	0.00333	...	2.653	2.220	...
	0.0333	...	1.887	1.286	...
	0.333	...	1.105	0.381	...

† From Bates and Alfenaar [40] and Robinson, Duer, and Bates.[42]

a metal ion with a ligand and adjusting the pH. The conditional equilibrium constants are functions of pH, and therefore the metal ion activity can be varied over a wide range of low activities with stability. The systematic development and an investigation of the applicability of such standards should be encouraged.

13-8 APPLICATIONS OF ION-SELECTIVE ELECTRODES

Electrode measurements can be applied in two ways: either the electrode can be used as an end-point detector in a titration or the concentration of a species can be obtained directly from a single measurement of potential. The use of ion-selective electrodes as end-point detectors is the most reliable way to use such electrodes to obtain accurate values of concentrations. During a titration, accurate values of pM or pA are not required, nor is careful standardization. A precision of a few tenths of a percent in end-point volumes is obtainable with even rough potential measurements; that is, potential measurements of low precision can yield highly precise titration results. On the other hand, the direct measurement of activity from a single potential measurement is less precise and furthermore does not give concentration directly, but rather the activity of a species. Careful measurements using ion-selective electrodes taken under highly favorable conditions with precise temperature control have an uncertainty[12,25] of about 0.05 to 0.2 mV; more usually, a precision of a millivolt or two is attained. Relative errors of 2 to 5% have been reported[29] for substrates and enzymes determined with the ammonium ion electrode. Thus, the inherent accuracy attainable is probably in the range of 1 to 10% in most instances. This corresponds to a variation of 0.01 to 0.05 p unit. (A variation of 0.2 mV in emf corresponds to an uncertainty of 0.8% in activity.) Digital readout of potential to four significant figures can therefore be misleading.

Another source of uncertainty in measurements of ion activity arises from lack of nernstian response of the ion-selective electrode. At 25°C, for a univalent ion the change in membrane potential may be less than 59 mV for a 10-fold change in activity. Thus, not all glasses respond to changes in hydrogen ion activity to the same extent: fused quartz has no response; a 2% Na_2O–98% SiO_2 glass shows a change of 15 mV/pH unit; 30% Na_2O–70% SiO_2 glass, a change of 23mV/pH unit. Only glasses of certain compositions approach the response of 59 mV/pH unit. The term *electromotive efficiency* is defined[44] as the ratio of observed to expected change in potential for two solutions containing the ion of interest. For efficient pH-sensitive glass electrodes the electromotive efficiency is near unity (0.995) over a wide range. To minimize error, it is desirable to standardize an ion-selective electrode with a standard reference solution that has an ion activity as close as possible to that of the test solution. Thus, even with an electromotive efficiency of 0.995, an error of 0.03 pH unit occurs if a pH electrode is standardized with a pH 4 buffer and used to measure test solutions with pH values of 10. In ion-responsive glasses the electromotive efficiency is related to the degree of hydration of the membrane surface. As the limit of detection is approached, electromotive efficiency decreases. Errors from this source can be minimized by standardizing the electrode with two solutions, one slightly more and one slightly less concentrated than the test solution. This procedure minimizes errors in liquid-junction potential as well.

Involvement of an ion in a chemical reaction—ion pairing, complexation, precipitation, acid-base—needs to be recognized as a source of misinterpretation of ion activities in solution. Thus, a sulfide-selective electrode in solutions of moderate or low pH should always give small values for sulfide ion activity since sulfide is largely present as HS^- and H_2S.

By the technique of adding small increments and measuring potentials after each addition, low levels of some ions can be determined with better precision than by a single direct measurement. Potential is plotted against concentration on special graph paper (semiantilog) to yield a Gran's plot.[45] The straight line obtained is extrapolated back to the horizontal axis (potential equal to infinity), and the concentration corresponding to this intercept value is the sought-for ion concentration. An example for the microdetermination of fluoride is given by Selig.[46]

REFERENCES

1 M. CREMER: *Z. Biol.*, **47**:562 (1906); F. HABER and Z. KLEMENSIEWICZ: *Z. Phys. Chem.*, **67**:385 (1909).

2 W. S. HUGHES: *J. Amer. Chem. Soc.*, **44**:2860 (1922); *J. Chem. Soc.*, 491 (1928); D. A. MACINNES and M. DOLE, *Ind. Eng. Chem., Anal. Ed.*, **1**:57 (1929); *J. Amer. Chem. Soc.*, **52**:29 (1930).

3 B. LENGYEL and E. BLUM: *Trans. Faraday Soc.*, **30**:461 (1934); H. J. C. TENDELOO, *Proc. Acad. Sci., Amsterdam*, **38**:434 (1935); I. M. KOLTHOFF and H. L. SANDERS, *J. Amer. Chem. Soc.*, **59**:416 (1937).

4 E. PUNGOR and K. TOTH: *Analyst*, **95**:625 (1970).

5 J. KORYTA: *Anal. Chim. Acta*, **61**:329(1972).

6 R. P. BUCK: *Anal. Chem.*, **44**:270R (1972).

7 R. A. DURST (ed.): "Ion-Selective Electrodes," National Bureau of Standards Special Publication 314, NBS, Washington, D.C., 1969.

8 H. S. HARNED and B. B. OWEN: "The Physical Chemistry of Electrolytic Solutions," 3d ed., Reinhold, New York, 1958.

9 R. A. DURST and J. K. TAYLOR: *Anal. Chem.*, **39**:1483 (1967); R. A. DURST, *Anal. Chem.*, **40**:931 (1968).

10 J. N. BUTLER in Durst,[7] chap. 5.

11 K. SCHWABE and H. D. SUSCHKE: *Angew. Chem., Int. Ed. Engl.*, **3**:36 (1964).

12 G. A. RECHNITZ: *Chem. Eng. News*, **45**(25):146 (1967).

13 G. J. MOODY, R. B. OKE, and J. D. R. THOMAS: *Lab. Pract.*, **18**:941 (1969).

14 G. HAUGAARD: *Nature*, **140**:66 (1937); *J. Phys. Chem.*, **45**:148 (1941).

15 K. SCHWABE and H. DAHMS: *Z. Elektrochem.*, **65**:518 (1961).

16 B. P. NIKOLSKI and T. A. TOLMACHEVA, *Zh. Fiz. Khim.*, **10**:495, 504, 513 (1937); F. CONTI and G. EISENMAN: *Biophys. J.*, **5**:247, 511 (1965).

17 G. EISENMAN in Durst,[7] chap. 1.

18 G. KARREMAN and G. EISENMAN, *Bull. Math. Biophys.*, **24**:413 (1962); F. CONTI and G. EISENMAN: *Biophys. J.*, **5**:247, 511 (1965); G. EISENMAN (ed.): "Glass Electrodes for Hydrogen and Other Cations," Dekker, New York, 1967.

19 G. EISENMAN: *Biophys. J.*, **2**:259 (1962).

20 S. PHANG and B. J. STEEL: *Anal. Chem.*, **44**:2230 (1972).

21 J. L. WALKER, JR.: *Anal. Chem.*, **43**(3):89A (1971).

22 G. EISENMAN, S. M. CIANI, and G. SZABO: *Fed. Proc.*, **27**:1289 (1968).

23 W. SIMON: *Proc. Soc. Anal. Chem.*, **9**:250 (1972); L. A. R. PIODA, V. STANKOVA, and W. SIMON: *Anal. Lett.*, **2**:665 (1969).

24 J. H. BOLES and R. P. BUCK: *Anal. Chem.*, **45**:2057 (1973).

25 J. W. ROSS, JR., in Durst,[7] chap. 2.

26 E. PUNGOR: *Anal. Chem.*, **39**(13):28A (1967).

27 J. RŮŽIČKA and C. G. LAMM: *Anal. Chim. Acta*, **54**:1 (1971); E. H. HANSEN, C. G. LAMM, and J. RŮŽIČKA: *Anal. Chim. Acta*, **59**:403 (1972).

28 L. C. CLARK, JR., and C. LYONS: *Anal. N.Y. Acad Sci.*, **102**:29 (1962).

29 G. G. GUILBAULT, R. K. SMITH, and J. G. MONTALVO, JR.: *Anal. Chem.*, **41**:600 (1969); G. G. GUILBAULT: "Enzymatic Methods of Analysis," Pergamon, Oxford, 1970; *Crit. Rev. Anal. Chem.*, **1**:377 (1970).

30 R. A. LLENADO and G. A. RECHNITZ: *Anal. Chem.*, **43**:1457 (1971).

31 G. G. GUILBAULT and F. R. SHU: *Anal. Chem.*, **44**:2161 (1972).

32 G. J. PAPARIELLO, A. K. MUKHERJI, and C. M. SHEARER, *Anal. Chem.*, **45**:790 (1973).

33 R. A. LLENADO and G. A. RECHNITZ, *Anal. Chem.*, **45**:826 (1973).

34 R. W. STOW, R. F. BAER, and B. F. RANDALL, *Arch. Phys. Med. Rehabil.*, **38**:646 (1957).

35 S. R. GAMBINO: *Clin. Chem.*, **7**:336 (1961).

35a J. W. ROSS, J. H. RISEMAN, and J. A. KRUEGER: *Pure Appl. Chem.*, **36**:473(1973).

36 F. A. SCHULTZ, A. J. PETERSEN, C. A. MASK, and R. P. BUCK: *Science*, **162**:267 (1968).

37 R. W. CATTRALL and H. FREISER: *Anal. Chem.*, **43**:1905 (1971); H. JAMES, G. CARMACK, and H. FREISER: *Anal. Chem.*, **44**:856 (1972); B. M. KNEEBONE and H. FREISER: *Anal. Chem.*, **45**:449 (1973).

38 C. LITEANU and E. HOPIRTEAN: *Talanta*, **19**:971 (1972).

39 R. JASINSKI and I. TRACHTENBERG: *Anal. Chem.*, **44**:2373 (1972).

40 R. G. BATES and M. ALFENAAR in Durst,[7] chap. 6.

41 R. G. BATES, B. R. STAPLES, and R. A. ROBINSON: *Anal. Chem.*, **42**:867 (1970).

42 R. A. ROBINSON, W. C. DUER, and R. G. BATES: *Anal. Chem.*, **43**:1862 (1971).

43 J. BAGG and G. A. RECHNITZ: *Anal. Chem.*, **45**:1069 (1973).

44 R. G. BATES: "Determination of pH," p. 342, Wiley, New York, 1973.

45 G. GRAN: *Analyst*, **77**:661 (1952); *Orion Newsl.*, **2**(11, 12):49 (1970).

46 W. SELIG: *Mikrochim. Acta*, **1973**:87.

PROBLEMS

13-1   (a) Assuming activity coefficients of unity and $K_{Na^+,K^+} = 4 \times 10^{-4}$, calculate the expected membrane potential for the following cell at 25°C for a sodium ion–selective electrode: Ag AgCl | 0.1 M KCl, 0.003 M NaNO$_3$ | membrane | 0.1 M NaCl | AgCl Ag. (b) What is the potential if K_{Na^+,K^+} is 4×10^{-3}?

Answer   (a) −0.149 V. (b) −0.146 V.

13-2   Calculate the net potential across a fluoride-sensitive membrane if the test solution is 1.0 M NaF and the reference solution 0.001 M NaF, 0.01 M NaCl. Use the data of Tables 13-1 and 13-2 to estimate activity coefficients.

Answer   0.16 V.

13-3   Calculate the chloride ion activity in a solution if a chloride ion–selective membrane potential of −0.232 V is observed for a test solution and −0.104 V for a 0.01 m NaCl standard.

Answer $a_{Cl} = 6.17 \times 10^{-5}$.

13-4   A solution is 0.1 M in NaF. Using an ionization constant of 7×10^{-4} for HF, calculate and plot the potential to be expected for a fluoride electrode as a function of pH in the region 0 to 7.

ELECTROLYTIC SEPARATIONS AND DETERMINATIONS

Although electrodeposition has long been used for the quantitative separation and determination of metals, much of the work has been highly empirical. Various factors such as current density, concentration, acidity, temperature, stirring rate, and presence of complexing agents or organic additives must be controlled to assure satisfactory results. Our purpose here is to examine the fundamental basis of electrolytic separations, as a guide to the understanding of practical procedures. Not included are electroanalytical techniques involving current-voltage-time interdependence. Thus, most of the major techniques such as polarography, cyclic voltammetry, and chrono-potentiometry are not discussed, nor are such classical techniques as potentiometric end-point detection.

14-1 DEFINITIONS

A *cathode* is an electrode at which reduction occurs. In an electrolytic cell it is the electrode attached to the negative terminal of the source, since electrons leave the source and enter the electrolysis cell at that terminal. Conversely, an *anode* is an electrode at which oxidation occurs. In certain cells a third electrode serving as a nonworking (no current flow) *reference electrode* is added. The potential of a particular working electrode (say, the cathode) may be either positive or negative with respect to the reference electrode, depending on the potential of the latter.

If the potential of an electrode deviates from the reversible or equilibrium value, a current flows in either the anodic or cathodic direction. The deviation of the potential from its equilibrium value is the anodic or cathodic overpotential[1] of the electrode. The terms *emf* and *voltage* are used here to refer to a cell, whereas the term *potential* refers to a single electrode (Section 12-1). *Overvoltage* represents the additional voltage above the reversible cell emf required to permit the passage of a finite current, and *overpotential* refers to the deviation of the potential of a single electrode from its reversible value. In both cases the ohmic voltage drop iR is first subtracted, as seen below.

14-2   BASIC PRINCIPLES

The usual and simplest electrolytic separation is carried out by inserting a pair of electrodes, ordinarily of platinum, into a solution and applying an external source of emf.

If a net chemical change is to be effected, the anodic reaction cannot be simply the reverse of the cathodic reaction. Therefore, owing to the formation of electrolytic products at the electrodes, a galvanic cell is set up when current is caused to flow. The polarity of this galvanic cell is in opposition to the applied emf, giving rise to a *back emf*. The amount of current that flows is given by Ohm's law,

$$E_{appl} - E_{back} = iR \qquad (14\text{-}1)$$

where R is the total resistance of the circuit, mainly the electrolytic resistance of the cell. It should be emphasized that Ohm's law should be applied only after subtraction of the back emf and that *the back emf in general increases with increasing current.*

The back emf can be regarded as being made up of three components: (*1*) a reversible back emf, (*2*) a concentration overvoltage, and (*3*) an activation overpotential.

1   The *reversible back emf* is the reversible emf of the galvanic cell set up by the passage of the electrolytic current, based on concentrations of solutes involved in the electrode reactions in the bulk of the solution. For example, if an acidic solution of copper sulfate is electrolyzed between platinum electrodes, the electrode reactions are

$$Cu^{++} + 2e^- \rightarrow Cu \qquad \text{at the cathode} \qquad E_0 = 0.34$$

$$\text{and} \quad 2H_2O \rightarrow O_2 + 4H^+ + 4e^- \qquad \text{at the anode} \qquad E_0 = 1.229$$

The reversible back emf is the difference in potential between the oxygen electrode and the copper electrode and is calculated from the Nernst equation applied to the prevailing concentrations of copper ion, hydrogen ion, and oxygen in solution. Note that this emf is indeterminate if no oxygen is initially present in solution, but as the oxygen concentration reaches saturation, a definite reversible back emf is attained. Thus for $[Cu^{++}] = [H^+] = 1\ M$ and $p_{O_2} = 1$ atm

$$E_{\text{back, rev}} = E^{o\prime}_{O_2,H^+} - E^{o\prime}_{Cu^{++},Cu} \simeq 1.23 - 0.34 = 0.89\ \text{V}$$

2 The term *concentration overvoltage* is commonly used in electrochemistry to denote the effect of changes in concentration at an electrode surface with reference to the concentration in the bulk of the solution. It is something of a misnomer in that overvoltage implies a deviation of the potentials of the electrode from their reversible values, and in this case the electrodes are presumed to be acting reversibly with respect to the actual solute concentrations at their surfaces. Referring to the above example, we see that changes in concentration occur at both the cathode and the anode. At the cathode, with flow of current, depletion of copper ions occurs near the surface, causing the reversible potential of the copper electrode to shift in the negative direction. At the anode, accumulation of hydrogen ions and perhaps of oxygen (if the solution is not already saturated with it) causes the reversible potential of the oxygen electrode to shift in the positive direction. Both effects tend to increase the back emf. The effect of concentration overvoltage is decreased by stirring and increased by higher current density.

3 The *activation overpotential* is any departure of the potential of an electrode from its reversible value, owing to the passage of electrolytic current other than the above two sources of overpotential. The magnitude of the activation overpotential, as is seen in Section 14-4, is determined primarily by the ratio of the electrolytic current to the *exchange current*, the current passing equally in each direction at the equilibrium potential. Passage of current shifts the cathode potential in the negative and the anode potential in the positive direction, thus once more increasing the back emf. For currents that are small compared with the exchange current, the shift in potential of an electrode is nearly linear with increasing current. For large currents the overpotential is linear with the logarithm of the current.

14-3 CONCENTRATION OVERPOTENTIAL

As pointed out above, changes in concentration occur in the vicinity of both anode and cathode upon passage of electrolytic current. To study these effects more fully, it is convenient to consider the two electrodes separately. This is accomplished experimentally by introduction of a third, reference, electrode by means of a salt bridge. The measuring circuit is arranged so that practically no current passes through the reference electrode (for instance, a silver–silver chloride or saturated calomel electrode); therefore it may be regarded as a point of constant potential against which the anode and cathode potentials are measured. The experimental values of cathode and anode potentials each include a portion of the ohmic iR drop between the two working electrodes. This may often be made negligible by the use of a capillary salt bridge (Luggin capillary) from the reference electrode inserted with its opening close to the surface of the electrode under study or in a region of low current density. Figure 14-1 shows a schematic cathodic polarization curve for a platinum cathode in an acidic air-free solution of the metal ion [here Cu(II)].[2] No appreciable current flows until the equilibrium potential of the metal–metal ion electrode has been reached. As the cathode potential is made increasingly negative, the copper ion concentration at the electrode surface adjusts itself to correspond to the applied potential. We assume here that the Nernst equation can be applied, that activation polarization is negligible, and that the potential has been corrected for iR drop. Then, if $C_{M^{n+}}^{\circ}$ is the concentration of metal ion at the electrode surface,

$$E_{\mathrm{appl}} = E_{M^{n+},M}^{\circ\prime} - \frac{RT}{nF} \ln \frac{1}{C_{M^{n+}}^{\circ}} \qquad (14\text{-}2)$$

For each $0.059/n$ $(=2.3RT/nF)$ volt of increasingly negative potential, the surface concentration is diminished 10-fold. Thus, if the cathode potential is maintained at a value $3 \times 0.059/n$ V more negative than the equilibrium potential, the surface concentration is only 0.1% of the bulk concentration, the concentration at the surface being maintained in accordance with the potential of the electrode. Further increase in cathode (negative) potential can cause no appreciable further increase in the quantity $C_{M^{n+}} - C_{M^{n+}}^{\circ}$, the difference between the bulk concentration and the surface concentration.

Opposing the tendency toward removal of metal ions at the electrode is the transfer of these ions from the bulk of the solution to the surface. At any applied potential, a steady state is reached when the rate of removal by deposition is equal to the rate of *mass transfer* by diffusion, convection, or migration, under the influence of an electric field.[3] If it is assumed that there is an excess of inert electrolyte present in solution, the transference number of the metal ion can be reduced to a negligible value, and migration can be made negligible. If the rate of mass transfer is proportional to the difference in concentration between the bulk of the solution and the electrode surface, we equate the rate of removal to the rate of supply and write

$$\frac{-i}{nFA} = m(C_{M^{n+}} - C_{M^{n+}}^{\circ}) \qquad (14\text{-}3)$$

where $-i/A$ = cathodic current density in amperes per square centimeter, nF = number of coulombs per mole of reduction, $C_{M^{n+}}$ = bulk concentration of M^{n+},

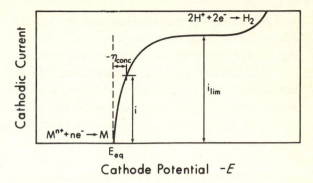

FIGURE 14-1   Schematic representation of cathodic polarization curve of a platinum electrode in acidic solution of metal ion M^{n+}.

$C_{M^{n+}}^{\circ}$ = surface concentration of M^{n+}, and m = mass-transport constant. In accordance with a recent agreement, a negative sign is assigned here to a cathodic current (reduction), and a positive sign is assigned to an anodic current (oxidation). The mass-transport constant m (centimeters per second), when multiplied by the concentration (moles per cubic centimeter), gives the flux (moles per square centimeter per second) at the electrode surface. If diffusion through a boundary layer of effective thickness δ is the sole mode of mass transport, according to Fick's law of diffusion the flux is given by $D(C_{M^{n+}} - C_{M^{n+}}^{\circ})/\delta$, where D is the diffusion coefficient (square centimeters per second). The mass-transport constant is, therefore, given by $m = D/\delta$. When convection is the sole mode of transport, m becomes the convection coefficient[4] (cubic centimeters of solution per square centimeter brought by convection to the surface per second) measured in centimeters per second. In practice, both convection and diffusion usually play a role in mass transport, and the diffusion coefficient enters into m with a fractional exponent instead of the exponent unity for pure diffusion or zero for pure convection.

When the surface concentration $C_{M^{n+}}^{\circ}$ reaches a negligible value compared with the bulk concentration, the current reaches its limiting value, denoted by $-i_{\text{lim}}$.

$$\frac{-i_{\text{lim}}}{nFA} = mC_{M^{n+}} \qquad (14\text{-}4)$$

Substituting from (14-4) into (14-3) then gives the surface concentration:

$$C_{M^{n+}}^{\circ} = C_{M^{n+}}\left(1 - \frac{i}{i_{\text{lim}}}\right) \qquad (14\text{-}5)$$

From the Nernst equation the concentration overpotential η_{conc} is calculated to be

$$\eta_{\text{conc}} = \Delta E = \frac{RT}{nF}\ln\frac{C_{M^{n+}}^{\circ}}{C_{M^{n+}}} = \frac{RT}{nF}\ln\left(1 - \frac{i}{i_{\text{lim}}}\right) \qquad (14\text{-}6)$$

According to (14-6) the concentration overpotential is zero when the current is zero. For $0 < |i| < |i_{lim}|$, the value of η_{conc} is negative, corresponding to cathodic polarization, and increases without limit as i approaches i_{lim}.

The limiting current i_{lim} is determined by the rate of mass transfer to a region of vanishingly small concentration at the surface. Since stirring generally increases the mass-transfer rate, i/i_{lim} decreases with increasing rate of stirring. Correspondingly, the concentration overpotential decreases with increasing stirring rate.

The current extends beyond i_{lim} when the back emf for another cathodic process has been reached. In the present example, hydrogen discharge is the next cathodic process. The back emf for this process, initially indeterminate because of the absence of hydrogen gas, reaches a finite value when saturation is reached. The gradual nature of the increase in current resulting from hydrogen discharge is due largely to the change in activation overpotential of hydrogen at the cathode (Section 14-4).

Concentration overpotential is also observed when the surface concentration is *increased* over the bulk concentration. The most common example is the anodic dissolution of a metal. Suppose that, after part of the metal ion has been plated out from the solution in the above example, the applied emf is decreased to a value below the reversible back emf. The cell now will operate as a galvanic cell, with the metal-plated electrode acting as the anode. The metal ion concentration at the anode surface becomes greater than the bulk concentration of metal ion. As anodic polarization is increased, however, there is no limit to the surface concentration of metal ion except that imposed by the solubility of a salt. Since the surface concentration would have to be 10 times the bulk concentration to produce a concentration overpotential of $0.059/n$ V, the anodic concentration overpotential for metal dissolution is generally small unless the bulk concentration is low.

Another example of concentration overpotential is encountered in the generation of hydrogen and oxygen from unbuffered solutions. The cathode region tends to become alkaline and the anode region to become acidic. In unbuffered solutions the changes in pH can be substantial. For example, for pH 3 near the anode and pH 11 near the cathode the corresponding back emf due to concentration overvoltage is $8 \times 0.059 = 0.47$ V, in addition to the reversible back emf of 1.23 V due to decomposition of water.

14-4   ACTIVATION OVERPOTENTIAL

To cause the passage of a net finite current at an electrode, it is necessary to shift the potential from its equilibrium value. This shift in potential, if no changes in concentration occur in the vicinity of the electrode, is the activation overpotential. The qualitative behavior of *a single electrode* is shown in Figure 14-2. In general, the net

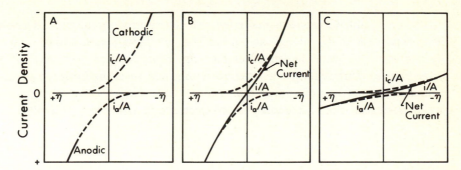

FIGURE 14-2 Schematic diagram of current as a function of polarization η of an electrode. *A*, cathodic and anodic currents; *B*, net electrode current; *C*, net electrode current with lower exchange current density.

current at an electrode is the algebraic sum of two opposing currents. At any potential the net current i is given by the algebraic difference $i = i_a - i_c$. At the equilibrium potential the anodic and cathodic currents are equal, and the net current is zero, or

$$i_c = i_a \qquad (14\text{-}7)$$

For the sake of simplicity we consider here only a half-reaction that proceeds in a single step,

$$\text{ox} + ne^- \underset{k_a}{\overset{k_c}{\rightleftharpoons}} \text{red} \qquad (14\text{-}8)$$

where ox and red are the oxidized and reduced forms of a couple and k_c and k_a are rate constants of the cathodic and anodic half-reactions. We neglect such situations as those in which a chemical reaction precedes or follows the electron-transfer step, which have been reviewed by Vlček.[5] The reactants may be in solution or exist as separate phases as in the case where the metal itself is the reductant.

Consider a half-reaction in which both reactants are in solution at concentrations C_{red}° and C_{ox}° *at the surface of the electrode*. The rates of the forward (cathodic) and backward (anodic) half-reactions are proportional to C_{ox}° and C_{red}°, the proportionality constants being k_c and k_a. The rates are also proportional to the electrode area A. If the rate is expressed in moles per square centimeter per second, it can be equated to i/nF, where i is the current (cathodic or anodic) and nF is the number of coulombs per mole of electrode reaction. Thus

$$i = i_a - i_c = nFA(k_a C_{red}^\circ - k_c C_{ox}^\circ) \qquad (14\text{-}9)$$

If the concentrations are expressed in moles per cubic centimeter, the heterogeneous rate constants k_a and k_c have the units centimeters per second. Thus $k_a C_{red}^\circ$ or $k_c C_{ox}^\circ$ has the dimensions moles per square centimeter per second, or reaction rate per unit of surface.

Both k_a and k_c vary exponentially with the electrode potential.[6] Considering the special case of $C_{red}^{\circ} = C_{ox}^{\circ}$ at equilibrium, we have $i = 0$, and $k_a = k_c = k^{\circ}$ (by definition). Thus k° is defined as *the value of k_a or k_c at the formal potential*† *of the couple.* This definition has the advantage of expressing both the anodic and cathodic rates in terms of a single rate constant.

The dependence of k_c and k_a on the electrode potential may be written[6]

$$k_c = k^{\circ} \exp\left[-\frac{\alpha nF}{RT}(E - E^{\circ\prime})\right] \qquad (14\text{-}10)$$

and

$$k_a = k^{\circ} \exp\left[\frac{(1 - \alpha)nF}{RT}(E - E^{\circ\prime})\right] \qquad (14\text{-}11)$$

where α is the *transfer coefficient*‡ of the electrode reaction. From (14-10) and (14-11) it can be seen that α may be regarded as the fraction of the change in electrode potential in the cathodic direction ΔE that acts to increase the rate of the cathodic reaction. The fraction $1 - \alpha$ acts to decrease the rate of the anodic reaction.

The significance of the transfer coefficient may be amplified by considering the curves for potential energy in Figure 14-3. Curve A represents the potential energy of the reductant; curve B represents the oxidant plus electron in the metal at the equilibrium potential of the couple.[7] For an electron to transfer from metal to oxidant or from reductant to metal, an energy barrier of height $\Delta G^{0\ddagger}$ must be surmounted. $\Delta G^{0\ddagger}$ is the free energy of activation, or the free energy necessary to take a mole of reactant to the activated state in the reaction. The rate of passage of electrons in either direction, and therefore the exchange current density also, are proportional to $\exp(-\Delta G^{0\ddagger}/RT)$. Thus *the exchange current density may be regarded as a measure of the height of the energy barrier* in the transfer of an electron between an electrode and an oxidant or reductant. The higher the barrier, the smaller the exchange current.

Now, if the electrode potential is changed by an amount $-\Delta E$ to favor the cathodic reaction, the free energy of the system oxidant plus electron is changed by an amount $-nF\Delta E$, the effect being to lower curve B to the position represented by curve C. The cathodic free energy of activation now becomes $\Delta G_c^{\ddagger} = \Delta G^{0\ddagger} - \alpha nF\Delta E$, and the anodic value becomes $\Delta G_a^{\ddagger} = \Delta G^0 + (1 - \alpha)nF\Delta E$, where $\alpha = \tan\beta/(\tan\beta + \tan\gamma)$ and $\tan\beta$ and $\tan\gamma$ are the absolute slopes of lines A and C at their intersection. Thus α is the fraction of the total energy $-nF\Delta E$ that acts to

---

† More exactly, at the standard potential if the activity coefficients of oxidant and reductant are included in Equation (14–9). For simplicity we assume activity coefficients of unity.

‡ The treatment here largely follows that of Randles.[7] Bauer[8] critically reviewed the literature on this loosely used term and indicated that the transfer coefficient concept has been employed with various interpretations and in particular the term was introduced for a model that is not now accepted. The assumption that the transfer coefficient is independent of potential is not always valid and much more needs to be done on elucidation of electrode mechanisms.

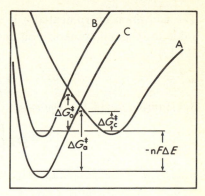

FIGURE 14-3 Change in potential energy for an electrode reaction in which reactants go through an activated complex to products: A, for reductant at equilibrium potential; B, for oxidant at equilibrium potential; C, for oxidant at a potential more negative than the equilibrium potential.

decrease the height of the *cathodic* energy barrier, and $1 - \alpha$ is the fraction that tends to *increase* the height of the *anodic* energy barrier.† *The transfer coefficient thus may be interpreted to be a measure of the symmetry of the energy barrier.* Over a reasonable range of potential, α may be regarded as constant.

From (14-9), (14-10), and (14-11) it follows that

$$i = nFAk^\circ \left\{ C_{red}^\circ \exp\left[\frac{(1-\alpha)nF}{RT}(E - E^{\circ\prime}) \right] - C_{ox}^\circ \exp\left[\frac{-\alpha nF}{RT}(E - E^{\circ\prime}) \right] \right\}$$

(14-12)

At the equilibrium potential, $E = E_{eq}$ and $i = 0$. Because no net current flows, the surface concentrations C_{ox}° and C_{red}° are equal to the bulk concentrations C_{ox} and C_{red}.

$$\frac{C_{ox}}{C_{red}} = \exp\left[\frac{nF}{RT}(E_{eq} - E^{\circ\prime}) \right]$$

(14-13)

or

$$E_{eq} = E^{\circ\prime} - \frac{RT}{nF} \ln \frac{C_{red}}{C_{ox}}$$

shows that the Nernst equation does not involve the transfer coefficient α.

The *exchange current* i_0 is given by either i_c or i_a at equilibrium. From either the first or second terms of (14-12) it follows that

$$i_0 = nFAk^\circ C_{ox}^{1-\alpha} C_{red}^\alpha$$

(14-14)

For the special case of $C_{ox} = C_{red} = C$, (14-14) becomes

$$i_0 = nFAk^\circ C$$

---

† Some authors (Gerischer, for instance) define α in terms of the anodic reaction and $1 - \alpha$ in terms of the cathodic reaction.

The activation overpotential η is defined by $E - E_{eq}$ and therefore has a negative value for cathodic polarization and a positive value for anodic polarization. From (14-13)

$$\eta = E - E_{eq} = E - E^{o'} + \frac{RT}{nF} \ln \frac{C_{red}}{C_{ox}} \qquad (14\text{-}15)$$

Substituting in (14-12) results in

$$i = i_0 \left\{ \frac{C^o_{red}}{C_{red}} \exp\left[\frac{(1 - \alpha)nF}{RT} \eta \right] - \frac{C^o_{ox}}{C_{ox}} \exp\left[\frac{-\alpha nF}{RT} \eta \right] \right\} \qquad (14\text{-}16)$$

where the first exponential term refers to the anodic current and the second to the cathodic current. Since $2.3RT/F = 59.1$ mV at 25°C, at this temperature (14-16) may be written

$$i = i_0 \left[\frac{C^o_{red}}{C_{red}} 10^{(1-\alpha)n\eta/59.1} - \frac{C^o_{ox}}{C_{ox}} 10^{-\alpha n\eta/59.1} \right] \qquad (14\text{-}17)$$

where η is given in millivolts.

For small values of η the overpotential is proportional to the current: $C_{red} \simeq C^o_{red}$ and $C_{ox} \simeq C^o_{ox}$ (negligible concentration polarization). The exponentials in (14-16) can be expanded in a power series, $e^x = 1 + x + x^2/2! + \cdots$, which for $x \ll 1$ yields the approximation $e^x \simeq 1 + x$. Equation (14-16) then becomes

$$i = + \frac{i_0 nF}{RT} \eta \qquad (14\text{-}18)$$

where the positive sign arises from specifying both i and η as negative for cathodic polarization. Note that α cancels out in the approximation. For (14-18) to be valid, $\alpha nF\eta/RT$ and $(1 - \alpha)nF\eta/RT$ must be small compared with unity, or since RT/F has the value 0.0257 J (volt-coulombs) per faraday at 25°C, η must be small compared with $0.0257/\alpha n$ and $0.0257/(1 - \alpha)n$. Taking $\alpha = 0.5$, η must be small compared with 51 mV for $n = 1$ or with 25.7 mV for $n = 2$.

If the cathodic current $-i$ is plotted as ordinate and the cathodic polarization $-\eta$ as abscissa, then the quantity $i_0 nF/RT$ is the slope $(di/d\eta)_{i=0}$ of the current-potential curve at the point of zero current (the equilibrium potential). Its reciprocal RT/nFi_0 has the dimensions of resistance (ohms) and is often called the *polarization resistance*.[9] It is the effective resistance imposed at the electrode surface by the finite rate of the electron-transfer process (Section 12-9).

For sufficiently large absolute values of η, the back reaction can be neglected. Thus, if $\alpha = 0.5$ and $\eta = 59.1/n$ mV, (14-17) becomes $i = \pm i_0(10 - 0.1)$. For this symmetrical case ($\alpha = 0.5$) the back reaction is only 1% of the forward reaction when the current is 10 times the exchange current. For cathodic polarization greater than $-59.1/n\alpha$ mV, (14-16) becomes

$$i_c = i_0 \exp\left(-\frac{\alpha nF}{RT} \eta_c \right) \qquad (14\text{-}19)$$

or in logarithmic form,

$$\eta_c = \frac{2.3RT}{\alpha nF} \log \left| \frac{i_0}{i_c} \right| \qquad (14\text{-}20)$$

An analogous equation can be written for anodic polarization.

The Tafel equation If j_0 and j_c represent the current densities i_0/A and $-i_c/A$, (14-20) can be expressed equally well in the form

$$\eta_c = a + b \log j_c \qquad (14\text{-}21)$$

This is the well-known Tafel[10] equation, expressing overpotential as a linear function of the logarithm of the current density. An equation of this form has long been used to describe hydrogen and oxygen evolution at various electrodes. If the linear logarithmic plot of (14-21) is extrapolated back to zero overpotential, the cathodic component j_c approaches the exchange current density j_0. Thus $\log j_0 = -a/b$.

As shown below, the mechanism of hydrogen discharge cannot be represented by the simple one-step reaction of Reaction (14-8). Nevertheless, even a complex reaction can be resolved into a kinetically equivalent reaction pair[11] that can be represented by a single half-reaction. Whenever the net cathodic or anodic current becomes large compared with the exchange current, a linear *Tafel plot* is to be expected. On the other hand, when the net current is small compared with the exchange current, the current-voltage relation is only approximately linear.

Discharge of hydrogen ions The mechanism of hydrogen discharge has been a subject of controversy for many years. Bockris[12] stressed the importance of stringent purification of electrodes, solutions, and gases in achieving reproducible and significant data. By use of the technique of cyclic sweep voltammetry the deposition[13] on platinum of hydrogen from adequately purified water results in a curve like that of Figure 14-4.

The conclusions presented by Conway and Bockris[14] are summarized briefly. The process of hydrogen discharge consists of two steps, the first being the formation of adsorbed hydrogen atoms (designated MH):

$$H^+ + e^- \xrightarrow{M} MH \qquad \text{acidic solution} \qquad (a_1)$$

or

$$H_2O + e^- \xrightarrow{M} MH + OH^- \qquad \text{alkaline solution} \qquad (a_2)$$

The second step is either the combination of adsorbed atoms

$$2MH \rightarrow 2M + H_2 \qquad (b)$$

or the electrochemical desorption of hydrogen atoms

$$H^+ + MH + e^- \rightarrow H_2(\text{adsorbed}) + M \qquad (c)$$

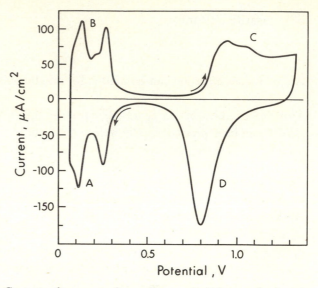

FIGURE 14-4   Current-voltage curve from a linear potential sweep for deposition of hydrogen on platinum from 0.5 M H_2SO_4 in ultrapurified water. A, formation of chemisorbed hydrogen atoms; B, oxidation of hydrogen atoms; C, surface oxidation of platinum; D, reduction of platinum oxide. (*From Conway, Angerstein-Kozlowska, Sharp, and Criddle.*[13])

Rosen and Schuldiner[15] studied hydrogen dissociation and hydrogen atom accumulation on platinum and concluded that only a small fraction of the platinum atoms associated with hydrogen are active dissociation sites.

For hydrogen evolution at various metals, the Tafel slope b is typically about 0.12 V, whereas j_0 varies greatly from one metal to another, being about 10^{-3} A/cm^2 for platinum and 10^{-13} A/cm^2 for mercury[16] in acid solutions.

The Tafel equation also describes the evolution of oxygen at a platinum anode. Bockris and Huq[17] found that, with solutions carefully purified by preelectrolysis, the oxygen electrode exhibits reversible behavior ($E = 1.24$ V, compared with the theoretical 1.23 V). The exchange current density, however, is only of the order of 10^{-9} to 10^{-10} A/cm^2 in dilute sulfuric acid; so polarization occurs readily, and relatively large overpotentials are observed at moderate current densities. In solutions of ordinary chemical purity the Nernst relation fails for the oxygen electrode because of mixed-potential behavior. Criddle,[18] using platinum electrodes in highly purified 1 M KOH, obtained a rest potential of 1.59 V. The potential is reduced by peroxide, which may be formed with impurities such as metals, protein, or carbon.

For metal–metal ion half-reactions, quantitative observations are especially troublesome because of the difficulty in preparing clean and reproducible surfaces. Means for purifying water for electrochemical measurements have been described.[13] With rigorous elimination of organic impurities, reproducible current-voltage profiles can be obtained. The exchange current density is relatively high for metals that readily give reversible or nearly reversible potentials against their ions (Cu, Ag, Zn, Cd, Hg), and so the activation overpotential is small at the current densities used in electro-analysis. On the other hand, transition metals such as Fe, Cr, Ni, and Co have low exchange currents.[19] These metals do not follow Nernst behavior against their ions because the influence of other potential-determining systems gives rise to mixed-potential behavior due to two or more redox couples. Quantitative kinetic studies also are difficult, particularly at solid electrodes, for electron-exchange half-reactions involving oxidants and reductants in solution. There is evidence,[20] such as that shown in Figure 14-4, that oxide films form at platinum surfaces in the presence of strong oxidants or at high positive potentials. These films are removed by electro-reduction or chemical reduction. Such oxide films, as well as adsorbed layers of traces of organic impurities,[13,21] usually have the effect of decreasing the exchange current and therefore of increasing polarization at a given current density.

For the Cl_2/Cl^- system a platinum-coated titanium electrode becomes passive[22] when the anodic overvoltage is more than about 0.2 V, presumably because the rate-determining step $Cl^- - e \rightarrow Cl(adsorbed)$ no longer takes place easily on the electrode surface.

14-5 EFFECTS OF COMPLEXATION AND ADDITION AGENTS

It is often found to be beneficial to carry out electrodeposition from solutions that contain metal ions in the form of a complex rather than the aquated metal ion. A striking example is silver, which forms large, loose crystals when deposited from silver nitrate, but a smooth, adherent, white layer when deposited from a cyanide bath. Nickel cannot be deposited at all from strongly acidic solutions, whereas quantitative deposition can be achieved from ammoniacal solutions.

Complexation has two effects on electrodeposition: a thermodynamic effect, or shift in the equilibrium potential, and a kinetic effect, or alteration of the exchange current. The thermodynamic effect is always in the direction of a more negative potential and so makes deposition more difficult [Equation (12-53)].

The kinetic effect may be in either direction, because the rate of charge transfer between the electrode and the complex species may be greater or less than with the aquated ion. If the discharge of the aquated ion is accompanied by a large activation overpotential because the exchange current is small, the formation of a complex may increase the exchange current so much that the decrease in overpotential more than compensates for the shift in equilibrium potential. In this case, deposition occurs more readily from the complex than from the aquated ion. An example is the aquated nickel ion, which is discharged at a dropping mercury electrode[23] with an over-potential of more than 0.5 V. In the presence of any of several complexing agents, such as thiocyanate ion, pyridine, or high concentrations of chloride, the nickel ion is reduced more readily.

The physical character of plated deposits appears to be governed, at least in part, by the magnitude of the exchange current. Large metallic crystals are characteristic of deposition carried out near the reversible potential. For example, in molten LiCl-KCl eutectic at 450°C even the transition metals (Cr, Mn, Fe, Co, Ni) and noble metals (Pt, Pd, Au) behave reversibly[24] (high exchange current densities), and all are plated in the form of dendritic growths. If the current density is low compared with the exchange current density, a marked recrystallization analogous to the aging of crystalline precipitates (Section 8-4) causes the formation of relatively large crystals. On the other hand, at high current densities the deposition is favored on certain crystal faces so that dendritic growths appear. At high overpotentials the slight energy differences favoring one face over another are counteracted by the large activation energy, and so nucleation of new crystals can readily occur and a microcrystalline deposit is formed.

The striking effects of traces of surface-active compounds can be attributed to the addition of a large energy barrier for deposition, with the result that the exchange current is diminished.[21] This situation does not change the equilibrium potential. A "reversible" electrode reaction then takes on the character of an irreversible one if the current density is appreciable. Thus the addition of a trace of gelatin causes silver to deposit as a smooth coating rather than as a dendritic growth.[2] In electrogravimetric analyses it is advisable to minimize the concentrations of organic additives or to omit them entirely because of the danger of inclusions of organic matter.

In the past the use of complexing and addition agents was largely empirical. As our knowledge of mechanisms of electrode reactions grows, we can hope to develop electrodeposition procedures on a more scientific basis. In the meantime we should be cautious about making arbitrary changes in procedures that have been shown to give satisfactory results.

14-6 SEPARATION AT CONSTANT CURRENT

The classical method for carrying out gravimetric electroanalytical determinations is to adjust the voltage of the source to give the desired electrolytic current, which is maintained at a relatively constant value during the electrolysis.

Suppose we wish copper to be deposited quantitatively from a solution containing copper sulfate and sulfuric acid. The behavior of the cathode is illustrated by Figure 14-5. As electrolysis proceeds, copper ions removed from solution at the cathode are replaced by hydrogen ions produced at the anode, so the electrolytic resistance undergoes little change. Curves A to F represent the behavior of the cathode potential plotted against current during the gradual removal of copper ions. A and B differ only in a displacement of the reversible potential of the copper electrode as the

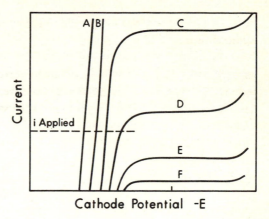

FIGURE 14-5 Current–cathode potential curves. Curves A to F represent decreasing concentrations of metal ion.

concentration of Cu^{++} is decreased. C to F show regions of limiting current in which the current is controlled by the rate of mass transfer. Now, if i_{appl} represents the constant current applied from the source, the cathode potential changes slowly until the limiting current approaches i_{appl}. At this point a sudden change in cathode potential occurs,[25] and both copper and hydrogen ions are discharged. The deposition of copper continues as fast as the rate of mass transfer permits, but the current efficiency for copper discharge gradually approaches zero as the proportion of discharged hydrogen ions increases. Hydrogen evolution unfortunately causes the copper deposit to be rough, spongy, and poorly adherent. Therefore nitric acid generally is added because nitrate ion is reduced to ammonium ion at a copper cathode,

$$NO_3^- + 8e^- + 10H^+ \rightarrow NH_4^+ + 3H_2O$$

at a lower (less negative) cathode potential than hydrogen ion and thus acts to decrease hydrogen evolution.

Similarly, the anode potential can be controlled by adding a suitable reducing agent. For example, if deposition of metallic lead at the cathode is desired, the anodic deposition of lead dioxide can be prevented by use of hydroxylamine or hydrazine in dilute hydrochloric acid. Lingane and Jones[26] found hydrazine to be superior in keeping the anode potential at a lower value and in forming simpler oxidation products (nitrogen instead of mixtures of nitrous oxide, nitrate, and nitrite).

14-7 EXAMPLES OF DETERMINATIONS BY ELECTRODEPOSITION AT CONSTANT CURRENT

The following examples are presented to illustrate some of the sources of error and some of the precautions necessary for accurate results in electrodeposition at constant current.

Copper normally is deposited from a solution that contains both sulfuric and nitric acids and is free of noble metals such as the platinum metals, silver, mercury, and bismuth. To remove nitrous acid, which can oxidize the copper deposit, urea or sulfamic acid sometimes is added. Low temperature and low current density are favored, since they also tend to prevent oxidation of the deposit. Chloride is avoided because in its presence platinum tends to dissolve at the anode and to plate out at the cathode unless a suitable additive such as hydrazine or hydroxylamine is present; also, copper(I) is stabilized as a chloro complex and remains in solution to be oxidized at the anode unless a controlled-cathode-potential method[27] is used. Copper can be separated from ions such as zinc, cadmium, cobalt, nickel, manganese, and aluminum by this classical procedure.[28]

Nickel is deposited quantitatively from ammoniacal solutions, incompletely from weakly acidic solutions, and not at all from strongly acidic ones. (A quantitative separation of copper from nickel requires that the acid concentration be kept high.) Among the important interferences in the determination of nickel are silver, copper, arsenic, and zinc, which can be removed by precipitation with hydrogen sulfide. Iron(II) and chromates are objectionable,[29] but can be removed by precipitation of the hydrous oxides.

Lead is determined by deposition as lead dioxide. The standard potential of the reaction

$$PbO_2 + 4H^+ + 2e^- \rightleftharpoons Pb^{++} + 2H_2O \qquad E^\circ = 1.455 \text{ V}$$

is so high that reductants must be avoided, and advantage must be taken of the high oxygen overpotentials of platinum and lead dioxide. Chloride must be absent, because it is oxidized without appreciable overpotential at a lower potential ($E^\circ = 1.358$ V). A relatively high concentration of nitric acid is useful in preventing cathodic deposition of lead. The use of a copper-plated cathode or the presence of copper in solution is favorable, because the reduction of nitrate proceeds smoothly to ammonia rather than partly to nitrous acid, which would prevent the formation of lead dioxide by oxidation back to nitrate at the anode. Ammonia is not oxidized under these conditions.

Mercury cathode separations at constant current, although not suitable for electrogravimetric determinations, often are useful as adjuncts to other analytical methods. Casto[30] summarized various procedures for the electrolytic removal of metallic impurities from uranium.

Parks, Johnson, and Lykken[31] used a procedure involving several small batches of mercury for the removal of large quantities of heavy metals such as copper, chromium, iron, cobalt, nickel, cadmium, zinc, mercury, tin, and lead, leaving quantitatively in solution even small quantities of aluminum, magnesium, and alkali and alkaline earths for subsequent determination by other methods.

The foregoing examples represent the most important of the classical electrogravimetric procedures. Various other elements, for example, silver, cadmium, zinc, tin, and iron, can be deposited quantitatively on platinum. These methods are impractical for most applications, however, because owing to lack of selectivity in deposition, cumbersome prior separations usually are necessary. Moreover, the more active metals are subject to air oxidation and to loss during washing. Controlled-potential methods have the advantage of being more selective; also, in favorable cases the weighing operation can be avoided entirely, and the determination can be based on measurement of the amount of electricity required for quantitative deposition.

14-8   SEPARATIONS BY ELECTRODEPOSITION AT CONTROLLED POTENTIAL

When electrolysis is carried out under conditions such that the cathode potential corresponds to a point on the limiting-current plateau (see Figure 14-5), the current is limited by the rate of mass transfer. For a given set of conditions (cell, electrode geometry, and stirring) the limiting current represents the maximum current that can be passed at 100% current efficiency at the cathode. Therefore, by using an apparatus that maintains a constant cathode potential regardless of changes in current or in anode potential, one can in principle achieve the *most rapid deposition rate possible* under the given conditions and yet avoid secondary cathode reactions. Because deposition of hydrogen is avoided, the metal is plated in a smooth and compact form that is especially suitable for electrogravimetric determination.

Since the rate of decrease in concentration is proportional to the current (Faraday's law), and since the current is proportional to the concentration, the concentration decreases with time in accordance with a first-order law analogous to the radioactive-decay law,

$$-\frac{dC}{dt} = \lambda C \qquad (14\text{-}22)$$

This can be written also in terms of the current:

$$-\frac{di}{dt} = \lambda i \qquad (14\text{-}23)$$

The proportionality constant λ, which represents the fraction of the solute removed per unit of time at any instant of the electrolysis, can be shown to be given by $\lambda = Am/V$, where A = electrode area, V = volume, and m = mass-transport constant [Equation (14-3)]. The value of λ depends on the particular cell geometry (A/V) and the stirring rate. If the initial current is i^0 and the proportionality holds throughout the electrolysis, after integration we have

$$\ln \frac{i^0}{i} = 2.3 \log \frac{i^0}{i} = \lambda t \qquad (14\text{-}24)$$

Lingane[25] found that (14-24) is obeyed in the deposition of copper onto a platinum cathode from a tartrate solution and in the deposition of lead and reduction of picric acid at a mercury cathode.[32] Lingane and Small[33] found the same type of relation in the anodic formation of silver chloride.

Analogous to the half-life in radioactive decay, the *half-time* of electrolysis is a convenient concept. If $t = t_{\frac{1}{2}}$ when $i/i^0 = \frac{1}{2}$, we find from (14-24) that $t_{\frac{1}{2}} = 2.3 \log 2/\lambda = 0.69/\lambda$ and that $\log i/i^0 = -0.30t/t_{\frac{1}{2}}$. The time required for removal of 0.5, 0.9, 0.99, 0.999, and 0.9999 of the metal in solution corresponds to $\log i/i^0$ values of -0.3, -1.0, -2.0, -3.0, and -4.0 and to electrolysis times (measured in units of half-times) of 1, 3.3, 6.7, 10, and 13.

When the current has decreased to 0.1% of the initial current, the deposition may be regarded as quantitative. As is apparent from the above data, quantitative deposition is reached after a period 10 times as long as the period required for the deposition of half the metal.

The completeness of deposition often can be judged from the magnitude of the electrolytic current. In some practical cases the current no longer decreases after a certain low level has been reached. This residual current is due to traces of extraneous reducible material such as oxygen or to a slow discharge of hydrogen. In such cases, electrolysis is carried out until the current has ceased to diminish for several minutes.

In some instances, Equation (14-24) does not hold at the beginning of the electrolysis, because the initial concentration may be so high that the proportionality between the current and the concentration does not hold. Thus, if the source is unable to supply the total voltage $(E_{back} + iR)$ at the level of current that can be sustained by the rate of mass transfer, the current is determined initially by the back emf and the electrolytic resistance [Equation (14-1)]. This statement amounts to saying that the cathode potential corresponds to a rising portion of the curve for cathodic current against voltage rather than to a point on the plateau.

A fundamental advantage of controlled-potential over constant-current electrolysis is that the theoretical limit of separation efficiency imposed by the electrode potentials can be much more closely approached. The situation is illustrated by the following example.

EXAMPLE 14-1   Suppose that silver and copper are to be separated when both are present in 0.1 M solution. We take the formal potentials to be approximately equal to the standard potentials of 0.80 and 0.34 V. Assume the mass-transfer constant m to be 10^{-2} cm/s. For diffusion control, this corresponds to an effective diffusion-layer thickness $\delta = D/m$ of the order of 10^{-3} cm, since D is usually about 10^{-5} cm^2/s. For unstirred solutions, δ is of the order of 0.04 cm; it decreases rapidly with increasing stirring rate until convection control sets in, when it becomes approximately 10^{-5} cm.[34] The constant $\lambda = Am/V = 20 \times 10^{-2}/100 = 2 \times 10^{-3}$/s for a cell volume of 100 cm^3 and a cathode area of 20 cm^2. This corresponds to a half-time of $0.69/\lambda = 345$ s. Quantitative deposition within 0.1% is reached in 10 half-times, or 3450 s. The initial current is $i^0 = nFAmC = 2$ A; the final current is 2 mA, corresponding to a final silver ion concentration of 10^{-4} M. The equilibrium potential of the silver-plated electrode begins at $0.80 - 0.059 = 0.74$ V; the final equilibrium potential is $0.80 - 4 \times 0.059 = 0.56$ V. The control potential should be $2 \times 0.059 = 0.118$ V more negative than the final equilibrium potential (with overpotential neglected), to correspond to a surface silver ion concentration of 10^{-6} M (1% of the final solution concentration) which assures mass-transfer control. The limits of control potential are, therefore, $0.44 > E_c > 0.31$. The lower limit corresponds to the equilibrium potential of copper.

Now suppose that we are to carry out the same separation at constant current. With the same apparatus, the current cannot exceed 2 mA, the limit imposed by the mass-transfer rate of 10^{-4} M Ag$^+$. This separation would require 5×10^5 s, or 5.8 days! By starting with a higher initial current, say 1 A, which would make it possible to plate out half the silver in 500 s, the time could be shortened, but at the risk of plating out copper accidentally as the rate of mass transfer of silver gradually decreased. Only by interposing a reducible material to consume the excess current without forming solid products could the rate be increased. Such a material should be reducible at the proper current density in just the potential range calculated above. Indeed, such a procedure constitutes an internal form of controlled-potential electrolysis that would permit the same performance as that calculated above for controlled-potential electrolysis.

In principle, if the potential is controlled so as to be just positive to the equilibrium potential of copper (0.31 V), a separation of all but 10^{-8} M Ag$^+$ (calculated from the Nernst equation) is possible. This would require 23.3 half-times, or 8000 s. ////

In favorable cases, in which the metal is plated from a definite oxidation state at 100% current efficiency and in which no secondary oxidation or reduction reactions occur in solution, the determination can be made coulometrically, that is, by application of Faraday's law (Section 5-2). A comprehensive discussion of controlled-potential electrolysis and of coulometric analysis has been given by Lingane.[35]

14-9   EXAMPLES OF ANALYSIS BY CONTROLLED-POTENTIAL ELECTROLYSIS

Controlled-potential electrolysis is valuable for the following purposes: (1) electrogravimetric determinations, (2) coulometric determinations, and (3) selective removal and separation of constituents. For convenience in weighing, platinum electrodes are essential in procedures of the first type. Mercury electrodes are often advantageous in applications where weighing is not involved. For a general discussion of principles, apparatus, and procedures, the monograph by Lingane[35] is recommended.

As an example of an electrogravimetric determination, consider the determination of copper. Torrance[36] and Diehl[37] recommended using a hydrochloric acid solution and controlling the cathode potential at a value sufficiently negative (-0.40 V against the saturated calomel electrode) to avoid forming soluble copper(I) chloro complexes. Lingane,[38] however, found a tartrate buffer of pH 4 to 6 superior to hydrochloric acid solution. Copper can be determined directly in all common alloys containing, for example, antimony, arsenic, lead, tin, nickel, or zinc with an accuracy fully equal to that obtained by more laborious methods.

Lingane and Jones[26] devised an electrogravimetric procedure for the successive determinations of copper, bismuth, lead, and tin in the presence of various other metals. After each deposition the pH and electrode potential are adjusted, and the cathode is replaced in the solution for continued electrodeposition.

Other examples of selective electrodeposition are given in books by Lingane,[35] Diehl,[37] and Sand.[39] As examples may be cited the separation of silver from copper, bismuth from copper, antimony from tin, cadmium from zinc, and rhodium from iridium.

Coulometric determinations of metals with a mercury cathode have been described by Lingane.[35,40] From a tartrate solution, copper, bismuth, lead, and cadmium were successively removed by applying the appropriate cathode potential, which was selected to correspond to a region of diffusion-controlled current determined from current-voltage curves with a dropping mercury electrode. With a silver anode, iodide, bromide, and chloride can be deposited quantitatively as the silver salt. By controlling the anode potential, Lingane and Small[33] determined iodide in the presence of bromide or chloride. The separation of bromide and chloride, however, was not successful because solid solutions were formed (Section 9-4).

Coulometric determinations can be carried out in which no physical separation occurs but simply a quantitative change in oxidation state. For example, MacNevin and Baker[41] determined iron and arsenic by anodic oxidation of iron(II) to iron(III) and arsenic(III) to arsenic(V). The reduction of titanium(IV) to titanium(III) and the reverse oxidation have been used for the analysis of titanium alloys.[42] Conversely, the output current from a cell made from a silver-gauze cathode and a lead anode with potassium hydroxide electrolyte can be used to measure low concentrations of oxygen in inert gases.[43]

Electrolysis at controlled potential can also serve as an elegant method of removing interfering metals from samples to be analyzed by other methods such as spectrophotometry or polarography. The electrogravimetric and coulometric procedures mentioned above represent such separations. The electrolysis can, however, be carried out primarily as a selective separation, with the actual determination being made in the remaining solution. As an example of this technique, Lingane[44] removed copper, together with antimony and bismuth, from hydrochloric acid solutions of copper-base alloys by controlled-potential electrolysis with a mercury cathode. Lead and tin were determined polarographically in the remaining solution and then quantitatively removed by electrolysis at a more negative potential. Finally, nickel and zinc were determined in the residual solution. Other examples of selective deposition with mercury cathodes also have been described by Lingane.[35]

A final example is that of the use of controlled-potential electrolysis for the preparation of carrier-free radioactive silver. Griess and Rogers[45] isolated tracer quantities of radioactive silver, which had been prepared by neutron bombardment of palladium, by selectively depositing the silver onto a platinum surface. Although a small amount of palladium was codeposited, complete separation was achieved by anodic stripping and redeposition.

14-10 TITRATIONS WITH COULOMETRICALLY GENERATED REAGENTS

Another method of applying Faraday's law to analysis is to generate a titration reagent by means of a suitable electrode reaction. If a current efficiency of 100% can be maintained, the rate of addition of reagent can be calculated accurately from the current, and extraordinary precision is attainable (Section 5-2). If the current is maintained accurately at a constant value, a titration can be carried out in which time rather than reagent volume is measured. The end point can be determined by either a visual indicator or instrumental methods.

Two techniques are used, internal or external generation of reagent. Because of the relative complexity of the method and the more stringent requirement for chemical stability of the coulometric intermediate, external generation has found relatively few applications. In the internal generation method, the reagent is generated directly within the titration solution by oxidation or reduction of some component present at relatively high concentration. An appreciable concentration is necessary to maintain 100% current efficiency at high current levels. For example, Swift and coworkers generated bromine, iodine, and chlorine[46] by anodic oxidation of the halide ion and titrated various reducing agents such as arsenic(III), antimony(III), iodide, thallium(I), and thiodiglycol. Similarly, various titrations have been carried out using electrolytically generated iron(II)[47] or cerium(IV).[48] Carter[49] devised a simple, rapid method for sulfur in petroleum products based on combustion to sulfur dioxide, which is titrated with iodine. This principle has been applied to titrations in molten salts, where the addition of titrating reagents poses serious problems of technique.[50] Iron(III), a strong oxidant in molten lithium chloride–potassium chloride eutectic at 450°C, served as a titrant for chromium(II) and vanadium(II).

Whenever the system being titrated forms a reversible redox couple with its reaction product, the second electrode used in the generation reaction must be shielded from the bulk of the sample solution. For example, in the titration of iron(II) with anodically generated cerium(IV),[48] the cathode is placed in a separate compartment to prevent the reduction of iron(III). In this example, iron(II) undergoes *direct* anodic oxidation during the bulk of the titration until the bulk concentration of iron(II) is so low that its rate of mass transfer can no longer sustain the applied current. At this point the intermediate oxidation of cerium(III) permits 100% current efficiency to be maintained to the end point.

Chloride can be determined by titration with anodically generated silver ion.[51,52] Mixtures of bromide and chloride ordinarily cannot be analyzed successfully by this method, but Boyer[53] used a selective oxidation of bromide followed by coulometric titration of chloride, together with a coulometric titration of total halide, to analyze milligram quantities of nonstoichiometric lead bromochloride.

Other examples of coulometric methods are the titration of weak acids in tetrahydrofuran;[54] the titration of iron(III) with electrogenerated dichloro copper(I) ion, $CuCl_2^-$;[55] trace determination of thiols[56] by electrogenerated silver and of cerium(III) by electrogenerated octacyanomolybdate;[57] determination of polythionates by degradation to thiosulfate, which is titrated with electrogenerated iodine;[58] the microdetermination of water by electrogeneration of iodine in a Karl Fischer reagent[59] and of trace amounts of chromium by electrogeneration of iron(II);[60] titration of cyclohexene[61] or furan[62] with electrogenerated bromine; titration of cerium, manganese, chromium, and tin by electrogenerated vanadyl;[63] titration of iridium(IV) with electrogenerated iron(II);[64] and determination of nitrite by controlled-potential coulometry[65] involving direct oxidation to nitrate at pH 4.7. In null-point potentiometric determination of silver,[66] analysis time can be decreased by coulometrically pregenerating a series of standard solutions instead of generating increments in the titration cell.

REFERENCES

1   G. KORTÜM and J. O'M. BOCKRIS: "Textbook of Electrochemistry," vol. 2, p. 395, Elsevier, Amsterdam, 1951.
2   H. A. LAITINEN and I. M. KOLTHOFF: *J. Phys. Chem.*, **45**:1061, 1079 (1941).
3   J. JORDAN: *Anal. Chem.*, **27**:1708 (1955).
4   I. M. KOLTHOFF and J. JORDAN: *J. Amer. Chem. Soc.*, **76**:3843 (1954); I. M. KOLTHOFF, J. JORDAN, and S. PRAGER: *J. Amer. Chem. Soc.*, **76**:5221 (1954).
5   A. A. VLČEK in "Progress in Inorganic Chemistry," F. A. Cotton (ed.), vol. 5, p. 211, Wiley, New York, 1963; A. A. VLČEK in "Progress in Polarography," P. Zuman (ed.), vol. 1, p. 269, Wiley, New York, 1962.

6   B. B. DAMASKIN: "The Principles of Current Methods for the Study of Electrochemical Reactions," translated by G. Mamantov, McGraw-Hill, New York, 1967.

7   For a more complete discussion see J. E. B. Randles, *Trans. Faraday Soc.*, **48**:828 (1952).

8   H. H. BAUER, *J. Electroanal. Chem.*, **16**:419 (1968).

9   K. J. VETTER: *Z. Phys. Chem.* (Leipzig), **194**:199, 284 (1950); "Electrochemical Kinetics," Engl. ed., Academic, New York, 1967.

10 J. TAFEL: *Z. Phys. Chem.* (Leipzig), A**50**:641 (1905).

11 K. B. OLDHAM: *J. Amer. Chem. Soc.*, **77**:4697 (1955).

12 J. O'M. BOCKRIS: *Chem. Rev.*, **43**:525 (1948); *Ann. Rev. Phys. Chem.*, **5**:477 (1954); N. PENTLAND, J. O'M BOCKRIS, and E. SHELDON, *J. Electrochem. Soc.*, **104**:182 (1957).

13 B. E. CONWAY, H. ANGERSTEIN-KOZLOWSKA, W. B. A. SHARP, and E. E. CRIDDLE: *Anal. Chem.*, **45**:1331 (1973).

14 B. E. CONWAY and J. O'M BOCKRIS: *J. Chem. Phys.*, **26**:532 (1957).

15 M. ROSEN and S. SCHULDINER: *J. Electrochem. Soc.*, **117**:35 (1970).

16 J. O'M. BOCKRIS, *J. Electrochem. Soc.*, **99**:366C (1952).

17 J. O'M. BOCKRIS and A. K. M. S. HUQ: *Proc. Roy. Soc., Ser. A*, **237**:277 (1956).

18 E. E. CRIDDLE: *Electrochim. Acta*, **9**:853 (1964).

19 R. PIONTELLI: *J. Chim. Phys.*, **46**:288 (1949); *Z. Elektrochem.*, **55**:128 (1951).

20 F. C. ANSON and J. J. LINGANE: *J. Amer. Chem. Soc.*, **79**:4901 (1957).

21 T. BIEGLER and H. A. LAITINEN: *J. Electrochem. Soc.*, **113**:852 (1966).

22 G. FAITA, G. FIORI, and J. W. AUGUSTYNSKI: *J. Electrochem. Soc.*, **116**:928 (1969).

23 I. M. KOLTHOFF and J. J. LINGANE: "Polarography," 2d ed., vol. 2, p. 486, Interscience, New York, 1952.

24 H. A. LAITINEN and C. H. LIU: *J. Amer. Chem. Soc.*, **80**:1015 (1958).

25 J. J. LINGANE: *Anal. Chim. Acta*, **2**:584 (1948).

26 J. J. LINGANE and S. L. JONES: *Anal. Chem.*, **23**:1798 (1951).

27 H. DIEHL and R. BROUNS: *Iowa State Coll. J. Sci.*, **20**:155 (1945).

28 R. BOCK and H. KAU: *Fresenius' Z. Anal. Chem.*, **217**:401 (1966).

29 G. E. F. LUNDELL and J. I. HOFFMANN: *J. Ind. Eng. Chem.*, **13**:540 (1921).

30 C. C. CASTO in "Analytical Chemistry of the Manhattan Project," C. J. Rodden (ed.), chap. 23, National Nuclear Energy Series, Div. VIII, McGraw-Hill, New York, 1950.

31 T. D. PARKS, H. O. JOHNSON, and L. LYKKEN: *Anal. Chem.*, **20**:148 (1948).

32 J. J. LINGANE: *J. Amer. Chem. Soc.*, **67**:1916 (1945).

33 J. J. LINGANE and L. A. SMALL: *Anal. Chem.*, **21**:1119 (1949).

34 I. M. KOLTHOFF and J. JORDAN: *J. Amer. Chem. Soc.*, **76**:3843 (1954).

35 J. J. LINGANE: "Electroanalytical Chemistry," 2d ed., Interscience, New York, 1958.

36 S. TORRANCE: *Analyst*, **62**:719 (1937); **63**:488 (1938).

37 H. DIEHL: "Electrochemical Analysis with Graded Cathode Potential Control," G. F. Smith Chemical, Columbus, Ohio, 1948.

38 J. J. LINGANE: *Ind. Eng. Chem., Anal. Ed.*, **17**:640 (1945).

39 H. J. S. SAND: "Electrochemistry and Electrochemical Analysis," Blackie, Glasgow, 1939.

40 J. J. LINGANE: *Ind. Eng. Chem., Anal. Ed.*, **16**:147 (1944).

41 W. M. MACNEVIN and B. B. BAKER: *Anal. Chem.*, **24**:986 (1952).

42 L. P. RIGDON and J. E. HARRAR: *Anal. Chem.*, **43**:747 (1971).

43   J. M. IVES, E. E. HUGHES, and J. K. TAYLOR: *Anal. Chem.*, **40**:1853 (1968).

44   J. J. LINGANE: *Ind. Eng. Chem., Anal. Ed.*, **18**:429 (1946).

45   J. C. GRIESS, JR., and L. B. ROGERS: *J. Electrochem. Soc.*, **95**:129 (1949).

46   J. W. SEASE, C. NIEMANN, and E. H. SWIFT: *Ind. Eng. Chem., Anal. Ed.*, **19**:197 (1947); W. J. RAMSEY, P. S. FARRINGTON, and E. H. SWIFT: *Anal. Chem.*, **22**:332 (1950); R. P. BUCK, P. S. FARRINGTON, and E. H. SWIFT: *Anal. Chem.*, **24**:1195 (1952); R. A. BROWN and E. H. SWIFT: *J. Amer. Chem. Soc.*, **71**:2717 (1949); W. S. WOOSTER, P. S. FARRINGTON, and E. H. SWIFT: *Anal. Chem.*, **21**:1457 (1949).

47   W. D. COOKE and N. H. FURMAN: *Anal. Chem.*, **22**:896 (1950).

48   N. H. FURMAN, W. D. COOKE, and C. N. REILLEY: *Anal. Chem.* **23**:945 (1951).

49   J. M. CARTER: *Analyst*, **97**:929 (1972).

50   H. A. LAITINEN and B. B. BHATIA: *Anal. Chem.*, **30**:1995 (1958).

51   B. H. PRISCOTT, T. G. HAND, and E. J. YOUNG: *Analyst*, **91**:48 (1966).

52   A. CEDERGREN and G. JOHANSSON: *Talanta*, **18**:917 (1971).

53   K. W. BOYER: Ph.D. thesis, University of Illinois, 1974.

54   C. E. CHAMPION and D. G. BUSH: *Anal. Chem.*, **45**: 640 (1973).

55   J. J. LINGANE: *Anal. Chem.*, **38**:1489 (1966).

56   F. A. LEISEY: *Anal. Chem.*, **26**:1607 (1954).

57   R. CORDOVA-ORELLANA and F. LUCENA-CONDE: *Talanta*, **18**:505 (1971).

58   E. BLASIUS and J. MÜNCH: *Fresenius' Z. Anal. Chem.*, **261**:198 (1972).

59   R. KARLSSON and K. J. KARRMAN: *Talanta*, **18**:459 (1971); R. KARLSSON: *Talanta*, **19**:1639 (1972).

60   C. E. CHAMPION, G. MARINENKO, J. K. TAYLOR, and W. E. SCHMIDT: *Anal. Chem.*, **42**:1210 (1970).

61   D. H. EVANS: *J. Chem. Educ.*, **45**:88 (1968).

62   A. P. ZOZULYA and E. V. NOVIKOVA: *Zh. Anal. Khim.*, **18**:1380 (1963); *J. Anal. Chem. USSR*, **18**:1200 (1963).

63   V. N. BASOV, P. K. AGASYAN, and A. I. KOSTROMIN: *Zavod. Lab.*, **36**:778 (1970); *Ind. Lab. USSR*, **36**:981 (1970).

64   N. I. STEPINA and P. K. AGASYAN: *Zh. Anal. Khim.*, **20**:351 (1965); *J. Anal. Chem. USSR*, **20**:322 (1965).

65   J. E. HARRAR: *Anal. Chem.*, **43**:143 (1971).

66   R. A. DURST, E. L. MAY, and J. K. TAYLOR: *Anal. Chem.*, **40**:977 (1968).

67   H. GERISCHER: *Z. Elektrochem.*, **54**:366 (1950).

PROBLEMS

14-1   Taking the electron-transfer rate constant $k^\circ = 10^{-3}$ cm/s for a one-electron reaction ox $+ \ e^- \rightleftharpoons$ red, and $\alpha = 0.25$, calculate (*a*) the exchange current density for $C_{ox} = C_{red} = 1\ M$; (*b*) the exchange current density for $C_{ox} = 0.01\ M$, $C_{red} = 1\ M$; (*c*) the current density at a cathodic polarization of 47 mV, concentration as in (*b*); (*d*) same as (*c*) except for anodic polarization.
     Answer   (*a*) 0.096. (*b*) 0.00305. (*c*) 0.00405. (*d*) -0.0101 A/cm^2.

14-2 In a constant-potential electrodeposition the current has fallen to 20% of its initial value in 10 min. Estimate the time required for 99.9% deposition.

Answer 43 min.

14-3 A 100-ml sample of 0.1 M Cu^{++} is electrolyzed at a constant current of 1.0 A under conditions such that the mass-transport constant is 10^{-2} cm/s, with a cathode area of 10 cm^2. (*a*) Calculate the concentration of Cu^{++} remaining when the current efficiency has dropped below 100%. (*b*) How long does it take to reach this point? (*c*) How much longer does it take to plate out 99.9% of the original copper? (*d*) What is the overall cathode current efficiency?

Answer (*a*) 0.052 M. (*b*) 930 s. (*c*) 6250 s. (*d*) 27%.

14-4 According to Gerischer[67] the exchange current density of the Pt | Fe^{3+}, Fe^{++} electrode in 1 M H_2SO_4 is 300 mA/cm^2 for $[Fe^{3+}] = [Fe^{++}] = 1$ M. The value of α is 0.42. Calculate (*a*) the polarization resistance of an electrode of 0.1 cm^2 area in a solution in which $[Fe^{3+}] = [Fe^{++}] = 10^{-4}$ M; (*b*) the value of the electron-transfer rate constant $k°$; (*c*) the exchange current density for a solution in which $[Fe^{3+}] = 10^{-2}$, $[Fe^{++}] = 10^{-3}$ M.

Answer (*a*) 8600 Ω. (*b*) 3.1 × 10^{-3} cm/s. (*c*) 1.14 × 10^{-3} A/cm^2.

14-5 Prove the relation $\lambda = Am/V$ given in Section 14-8.

14-6 Derive Equation (14-14) by substitution of (14-13) in (*a*) the first term and (*b*) the second term of (14-12).

14-7 For a metal-deposition reaction $M^{n+} + ne^- \rightleftharpoons M$, write expressions for the cathodic and anodic currents as a function of the potential. Note that the concentration factor is missing from the expression for anodic current, but that the exponential factor remains. Show that the Nernst equation does not involve the transfer coefficient. Write an expression for the exchange current, and show that Equation (14-16) is valid.

15

OXIDATION-REDUCTION REACTIONS

Many parallels can be drawn between acid-base and oxidation-reduction (redox) reactions. Instead of proton interchange and an acid-base conjugate pair, in redox reactions we have electron interchange and a redox conjugate pair

$$\text{ox} + ne^- \rightleftharpoons \text{red} \qquad (15\text{-}1)$$

where ox is an electron acceptor or oxidizing agent and red an electron donor or reducing agent. The electron pressure exerted by a conjugate pair is expressed quantitatively in terms of the electrode potential (Table 12-1). Acid and base strengths are leveled (Section 4-3) or limited by the solvent; similarly, oxidants and reductants are leveled by the solvent. In water, strong reductants are leveled in strength (at equilibrium) to that of hydrogen, and oxidants to that of oxygen. Fortunately, for many reagents redox equilibrium with water is attained so slowly that the restriction does not apply in practice.

For acid-base reactions, which are fast, tables of ionization constants are directly applicable. In contrast, for redox reactions, many of which are slow, a table of electrode potentials can be no more than a guide to equilibrium conditions as it tells nothing of rates of reactions or mechanisms. For example, a table of electrode potentials leads to the expectation of a quantitative reaction between Ce(IV) and

As(III), a reaction that takes place too slowly to be of direct value unless catalyzed. Moreover, the mechanisms are often obscure. For example, the reduction of Mn(VII) to Mn(II) is complex. A further complication in redox reactions has to do with the stoichiometry; frequently, several redox reactions occur simultaneously. For a slow reaction, a catalyst often can be found to accelerate it. For a reaction where the stoichiometry is unsatisfactory, a more complete understanding of the mechanism often can lead to selection of conditions under which it (a single reaction) can be made to proceed quantitatively.

Redox reactions are more common than all other types. For those that are suitably fast and whose stoichiometry is known and satisfactory, the course of a reaction can be followed during a titration by plotting potential against titration volume. Assuming no complications, consider a stoichiometric reaction derived from two reversible half-reactions

$$ox_1 + n_1e^- \rightleftharpoons red_1 \qquad \text{sample} \qquad (15\text{-}2)$$

and
$$ox_2 + n_2e^- \rightleftharpoons red_2 \qquad \text{titrant} \qquad (15\text{-}3)$$

By subtracting n_1 times Reaction (15-3) from n_2 times Reaction (15-2) to cancel n_1n_2 electrons, we obtain the overall reaction

$$n_2red_1 + n_1ox_2 \rightleftharpoons n_2ox_1 + n_1red_2 \qquad (15\text{-}4)$$

At any point in a titration, after each addition of titrant the reaction proceeds until *at equilibrium the electrode potentials of the two systems become equal.*

15-1 TITRATION CURVES

A titration curve may for convenience be considered to consist of three portions: the region before the equivalence point, the equivalence point, and the region beyond the equivalence point. At all points except at the beginning before any titrant has been added, two redox couples are present, corresponding to the sample and the titrant. In the region before the equivalence point, the potential is calculated conveniently from the known concentration ratio of the sample redox couple. After the equivalence point the concentration ratio of the titrant redox couple is known from the stoichiometry. At the equivalence point both the sample and titrant redox couples are present in the stoichiometric ratio.

The Nernst equation applied to Reaction (15-2), written in terms of the formal potential and concentrations, is

$$E = E_1^{\circ\prime} - \frac{RT}{n_1F} \ln \frac{[red_1]}{[ox_1]} \qquad (15\text{-}5)$$

If the sample is initially in the reduced form, and if X is the percent of the stoichiometric amount of oxidant added, for $0 < X < 100$ we have

$$E = E_1^{\circ\prime} - \frac{RT}{n_1 F} \ln \frac{100 - X}{X} \qquad (15\text{-}6)$$

as the equation of the titration curve. At $X = 50$, $(100 - X)/X = 1$, and $E = E_1^{\circ\prime}$, which shows that the formal potential of the system made up of the sample and its oxidation product is reached at the midpoint of the titration curve. At $25°C$, at $X = 91$

$$\frac{100 - X}{X} \simeq 0.1 \qquad E = E_1^{\circ\prime} - \frac{-0.059}{n_1} \qquad (15\text{-}7)$$

and at $X = 99$

$$\frac{100 - X}{X} \simeq 0.01 \qquad E = E_1^{\circ\prime} - 2\frac{-0.059}{n_1} \qquad (15\text{-}8)$$

Thus, in general, before the equivalence point the potential is in the region of $E_1^{\circ\prime}$.

The equivalence point   At the equivalence point of Reaction (15-4), n_1 moles of ox_2 have been added to n_2 moles of red_1. Applying the Nernst equation to the system being titrated gives

$$E_{\text{equiv}} = E_1^{\circ\prime} - \frac{RT}{n_1 F} \ln \left(\frac{[red_1]}{[ox_1]} \right)_{\text{equiv}} \qquad (15\text{-}9)$$

and to the titrant system gives

$$E_{\text{equiv}} = E_2^{\circ\prime} - \frac{RT}{n_2 F} \ln \left(\frac{[red_2]}{[ox_2]} \right)_{\text{equiv}} \qquad (15\text{-}10)$$

At the equivalence point

$$\left(\frac{[red_1]}{[ox_1]} \right)_{\text{equiv}} = \left(\frac{[ox_2]}{[red_2]} \right)_{\text{equiv}} \qquad (15\text{-}11)$$

Multiplying (15-9) by n_1, (15-10) by n_2, and adding, we find that the last terms cancel each other in view of (15-11) and that

$$(n_1 + n_2)E_{\text{equiv}} = n_1 E_1^{\circ\prime} + n_2 E_2^{\circ\prime} \qquad (15\text{-}12)$$

or

$$E_{\text{equiv}} = \frac{n_1 E_1^{\circ\prime} + n_2 E_2^{\circ\prime}}{n_1 + n_2} \qquad (15\text{-}13)$$

The potential at the equivalence point is thus a weighted arithmetic mean of the formal potentials of the two redox couples involved in the titration. If $n_1 = n_2$,

$$E_{\text{equiv}} = \frac{E_1^{\circ\prime} + E_2^{\circ\prime}}{2} \qquad (15\text{-}14)$$

and the titration curve is symmetrical in the vicinity of the equivalence point if the effects of dilution are neglected.

Region beyond the equivalence point   The Nernst equation applied to Reaction (15-3) yields

$$E = E_2^{\circ\prime} - \frac{RT}{n_2 F} \ln \frac{[\text{red}_2]}{[\text{ox}_2]} \qquad (15\text{-}15)$$

At the equivalence point, from the stoichiometry of Reaction (15-4), for each 100 initial millimoles of red_1, $100\, n_1/n_2$ millimoles of red_2 are formed, corresponding to $X = 100$. For $X > 100$, $[\text{red}_2]/[\text{ox}_2] = 100/(X - 100)$, and

$$E = E_2^{\circ\prime} - \frac{RT}{n_2 F} \ln \frac{100}{X - 100} \qquad (15\text{-}16)$$

At $X = 200$, $E = E_2^{\circ\prime}$, and the potential after the equivalence point is generally in the region of $E_2^{\circ\prime}$.

15-2   THEORETICAL PROPERTIES OF REDOX TITRATION CURVES

By differentiating the titration curve twice and then equating the second derivative to zero, it can be shown[1] that for a symmetrical titration curve ($n_1 = n_2$) the point of maximum slope theoretically coincides with the equivalence point. This conclusion is the basis for potentiometric end-point detection methods. On the other hand, if $n_1 \neq n_2$, the titration curve is asymmetrical in the vicinity of the equivalence point, and there is a small titration error if the end point is taken as the inflection point. In practice the error from this source is usually insignificant compared with such errors as inexact stoichiometry, slowness of titration reaction, and slowness of attainment of electrode equilibria.

In the Nernst equation only the ratio of $[\text{red}]/[\text{ox}]$ appears, and so an important property of the calculated titration curve for Reaction (15-4) is that *the theoretical shape is independent of the concentration of reactants*. Thus the sharpness of the titration is theoretically unaffected by dilution. Note, however, that (15-4) is not perfectly general, because the simple relation of n_1 to n_2 for reactants and products is not always valid. Consider, for example, the reaction

$$2\text{Fe}^{++} + \text{Br}_2 \rightleftharpoons 2\text{Fe}^{3+} + 2\text{Br}^- \qquad (15\text{-}17)$$

for which the expression for equivalence potential is

$$E_{\text{equiv}} = \frac{E_1^{\circ\prime} + 2E_2^{\circ\prime}}{3} - \frac{RT}{3F} \ln 2[\text{Br}^-]_{\text{equiv}} \qquad (15\text{-}18)$$

if no bromide other than that produced in the reaction is present. The equivalence potential in this instance varies with dilution.

The rate of attainment of equilibrium also varies with dilution. We should distinguish between the attainment of equilibrium of the reactants and the attainment of equilibrium of the electrode or visual indicator used to detect the end point.

The exchange current density, which governs the rate of attainment of electrode equilibrium, varies enormously from one potential-determining redox couple to another. It varies not only with the initial concentration but also with the ratio of oxidant to reductant [Equation (14-14)]. In titrations performed at great dilution, the equilibrium near the end point may be reached slowly. Therefore, it may be advantageous to select a method of end-point detection that is not dependent on equilibrium near the end point.

EXAMPLE 15-1   Consider the titration of arsenic(III) with bromate.[2] In a hydrochloric acid solution containing excess bromide, the end point can be determined potentiometrically by using the bromine-bromide couple as the potential-determining system. Alternatively, the same titration can be followed amperometrically by measuring the diffusion-controlled current due to excess bromine slightly beyond the end point. At an initial concentration of $5 \times 10^{-4} M$ arsenic(III), the potentiometric titration can barely be carried out, because several minutes are required for electrode equilibrium at each point of the titration. The amperometric method gives a successful end point even at $5 \times 10^{-7} M$ arsenic(III), the whole titration taking only a few minutes.   ////

When visual indicators are used, the rate of attainment of equilibrium depends on the type of reaction leading to color development, which may be slow. For simple electron exchange reactions like that of ferroin, the rate of indicator response is usually rapid. If, however, the indicator undergoes a more deep-seated structural change, one can anticipate kinetic complications. The oxidation of diphenylamine, for example, is induced (Section 15-8) by the iron(II)-dichromate reaction.

15-3   EXPERIMENTAL TITRATION CURVES

If both the half-reactions involved in a redox titration can be made to behave reversibly at a suitable electrode, the shapes of the titration curves should conform closely to the calculated values, though as pointed out above, the electrode potential reaches its equilibrium value more and more slowly with increasing dilution.

The titration of Fe(II) represents an important practical example, which also serves to illustrate several principles. Since the Fe(III)-Fe(II) couple behaves reversibly, the shape of the curve theoretically predicted by Equation (15-6) is closely achieved, provided that the electrolyte composition is essentially constant during the titration except for the concentration ratio [Fe(II)]/[Fe(III)]. This condition is necessary if the formal potential $E_1^{\circ\prime}$ is to be constant during the titration. The value of the formal potential varies with the nature and concentration of acid present. We can distinguish three effects on the activity ratio $a_{Fe(II)}/a_{Fe(III)}$ as the concentration of any particular acid is increased.

1 Increasing ionic strength tends to increase the activity-coefficient ratio $\gamma_{Fe(II)}/\gamma_{Fe(III)}$ and therefore the activity ratio.
2 Increasing hydrogen ion concentration tends to decrease the activity ratio by suppressing hydrolysis of the Fe(III).
3 Increasing anion concentration may tend to increase the activity ratio by forming complexes with Fe(III) in preference to Fe(II).

The first and third effects cause the value of $E_1^{\circ\prime}$ to decrease; the second has the opposite result.

Some of these effects are illustrated in the experimental curves of Figures 15-1 and 15-2. In the titration of Fe(II), before the end point the shapes are independent of the nature of the oxidant. The value of $E_1^{\circ\prime}$ is highest with perchloric acid, because hydrolysis of Fe(III) is largely suppressed and the perchlorate ion has little tendency to form complexes. With sulfuric acid and with Ce(IV) as titrant, the two effects of hydrolysis and complex formation tend to counteract each other. In the presence of phosphoric acid, complex formation predominates, and $E_1^{\circ\prime}$ is distinctly lower. The shape of the curve beyond the end point is determined primarily by the properties of the oxidant. For Ce(IV) in various media the value of $E_2^{\circ\prime}$ is different, and the potential varies correspondingly. The reasons for these effects are qualitatively the same as for the Fe(III)-Fe(II) couple. Note that three of the experimental curves with Ce(IV) in Figure 15-1 closely resemble the expected shapes; the distortion of the curve for hydrochloric acid after the end point is due to the gradual oxidation of chloride by the excess Ce(IV). For further details see the discussion of Ce(IV) as a reagent (Chapter 18).

Figure 15-2 (left) depicts several titration curves of Fe(II) with permanganate. Beyond the end point the experimental curves differ from the theoretical shape, which is nearly flat beyond the end point (5-equivalent reduction). The essential symmetry of the curves suggests that the potential is determined by the Mn(III)-Mn(II) couple beyond the end point. Evidence for this behavior can be seen in solutions containing sulfate or phosphate, which tend to stabilize Mn(III) (Section 17-1). That sulfuric and phosphoric acids have about the same effect before and after the end point is consistent with the similarity of the behavior of the Mn(III)-Mn(II) and the Fe(III)-Fe(II) systems with respect to changes in activity coefficients as well as with respect to hydrolysis and complex formation.

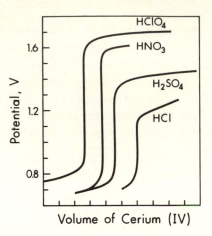

FIGURE 15-1 Potentiometric titration curves of Fe(II) with Ce(IV) in $HClO_4$, HNO_3, H_2SO_4, and HCl. (*From Smith.*[3])

Titration curves for the Fe(II)-dichromate reaction are especially interesting (Figure 15-2, right). In hydrochloric and sulfuric acid solutions, Smith and Richter[3] found that the titration curves had shapes closely resembling those of the theoretical curves for reversible systems, but with a variety of values for the formal potentials $E_1^{\circ\prime}$ and $E_2^{\circ\prime}$ that depended on the nature and concentration of acid. It might be thought that the asymmetry in the titration curves should correspond to $n_1 = 1$ and $n_2 = 6$ because of the 6-equivalent reduction for one dichromate ion. The equilibrium $2HCrO_4^- \rightleftharpoons Cr_2O_7^= + H_2O$, however, is shifted far to the left at the small concentrations of Cr(VI) existing just beyond the end point [Equation (17-11)]; so monomeric Cr(VI) should be regarded as the oxidant.

In Fe(II)-dichromate titrations, Winter and Moyer[4] observed a time dependence of the potential after the end point. When potential readings were taken soon after each addition, an asymmetrical titration curve was observed, but when a time interval of 10 to 15 min was allowed after each addition, the curve approached the theoretical shape. We have noted that automatically recorded titration curves for the Fe(II)-dichromate titration show a considerably smaller potential jump than manually observed curves, the difference being due to lower potentials after the end point. But curves plotted with 15 s of waiting for each point differed only slightly from curves plotted with 150 s of waiting. Ross and Shain[5] also studied the drift in potential of platinum electrodes with time and noted hysteresis effects in recorded potentiometric titration curves. These effects, due to oxidation and reduction of the platinum surface, are discussed below.

Let us first consider the final curves, which closely approximate the theoretical shapes. That they do so may seem surprising, since the Cr(VI)-Cr(III) couple does not behave reversibly. The close approximation makes sense, however, if we assume that the *Fe(III)-Fe(II) couple acts as the potential-determining couple immediately*

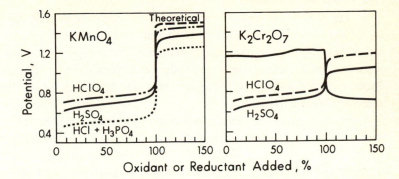

FIGURE 15-2 Titration curves for the Fe(II)-KMnO₄ reaction (left) and the Fe(II)-K₂Cr₂O₇ reaction (right). The downward trending curve is for the reverse titration of dichromate with Fe(II). (*From Smith.*[3])

after the end point. After all, at equilibrium both couples have the same potential, and only because the exchange current density of the Fe(III)-Fe(II) couple becomes low beyond the end point does some other couple with a higher exchange current usually take over as the potential-determining system. From the following example we see that this interpretation is reasonable.

EXAMPLE 15-2 At a potential of 1.0 V the ratio $[\text{Fe(II)}]/[\text{Fe(III)}]$ is 4×10^{-6} in 0.5 M H_2SO_4, calculated from $E^{\circ\prime} = 0.68$. For $[\text{Fe(II)}] = 4 \times 10^{-8}$ $[\text{Fe(III)}] = 0.01$, corresponding to conditions 20% beyond the end point in Figure 15-2 (right), we calculate from the data of Gerischer[6] an exchange current density of 1.5×10^{-5} A/cm², which is sufficient to give a stable potential. On the other hand, at a potential of 1.3 V, corresponding to a point just beyond the permanganate end point (Figure 15-2, left), the ratio $[\text{Fe(II)}]/[\text{Fe(III)}]$ is about 10^{-10}, and the exchange current density is only 1.8×10^{-8} A/cm², which is much more easily swamped out by another potential-determining couple, such as Mn(III)-Mn(II).

This calculation should be regarded as only a rough approximation because the exchange current depends on the state of the electrode surface.   ////

Evidence has long accumulated to show that surfaces of supposedly inert metals readily undergo oxidation. For example, Hickling[7] was able to achieve a cyclic oxidation and reduction of a platinum surface, which could be repeated indefinitely without permanent changes in the metal. (See Figure 14-4.) Kolthoff and Tanaka[8] showed that the oxidation could be carried out either electrochemically or chemically with such strong oxidants as dichromate, Ce(IV), and permanganate.

The present concept of the nature of the oxidation process has been summarized by Vetter and Schultze.[9] Only a small amount of oxygen is bound in a chemisorbed oxygen atom layer. Penetration of oxygen at a rate depending on field strength forms a surface oxide layer, which gradually undergoes aging. The oxide layer permits electron exchange to occur through it, at a rate dependent on its structure and on the redox couple involved.

Surface oxide formation undoubtedly is involved in the Fe(II)-dichromate titration curves, which Smith and Brandt[10] found to be different when the direction of titration was reversed (Figure 15-2, right). Kolthoff and Tanaka[8] found that the rate of oxidation with dichromate was slow, whereas the rate of reduction with Fe(II) was fast. Ross and Shain[5] found the same sort of behavior and noted also that the rates of oxidation and reduction decreased in more dilute solutions. The oxidized surface in a dichromate solution may be largely covered with adsorbed dichromate, as chromium surfaces have been shown to be in some experiments with radio-chromium,[11] so that it is relatively ineffective as an electron-transfer surface for the Fe(III)-Fe(II) system.

The influence of surface oxide and of adsorbed substances on the rates of electron transfer reactions is still often obscure, especially where irreversible half-reactions are involved.

15-4   THEORY OF REDOX INDICATORS

Some redox indicators react specifically with one form of a redox couple to cause a visible color change. Two examples are starch as an indicator for iodine and thiocyanate for Fe(III). Such indicators, being limited in scope, are not amenable to general treatment.

More broadly applicable as redox indicators are substances that undergo oxidation and reduction and exhibit different colors in the two forms. These have been classified[12] into two broad groups, with formal potentials either less or greater than $+0.76$ V, the formal potential for the Fe(III)-Fe(II) couple. Indicators in the group having the higher formal potentials are useful with stronger oxidizing agents—ones that are capable of oxidizing Fe(II) quantitatively. Those in the other group are useful with the stronger reductants—ones that can be quantitatively oxidized by Fe(III).

The general half-reaction applicable to an indicator redox couple is

$$In_{ox} + ne^- \rightleftharpoons In_{red} \qquad (15\text{-}19)$$

At a potential E the ratio of concentrations of the two forms is determined by the Nernst equation

$$E = E_{In}^{o\prime} - \frac{RT}{nF} \ln \frac{[In_{red}]}{[In_{ox}]} \qquad (15\text{-}20)$$

where $E_{In}^{\circ\prime}$ is the formal potential of the indicator. Since the value of $E_{In}^{\circ\prime}$ varies to some extent with the composition of the solution, it should be determined in the actual titration medium. If the molar absorptivities of the two forms are comparable, a practical estimate of the color-change interval corresponds to the change in the ratio $[In_{red}]/[In_{ox}]$ from 1/10 to 10. (See also Section 3-10.) The corresponding transition interval of potential (in volts) is

$$E = E_{In}^{\circ\prime} \pm \frac{0.059}{n} \qquad \text{at 25°C} \qquad (15\text{-}21)$$

If one form is much more intensely colored than the other, the intermediate color is attained at a potential somewhat removed from $E_{In}^{\circ\prime}$.

If an indicator is to change color within 0.1% of the end point, the color change must occur between $X = 99.9$ and 100.1, and therefore in the potential region

$$E_1^{\circ\prime} + \frac{3 \times 0.059}{n_1} < E < E_2^{\circ\prime} - \frac{3 \times 0.059}{n_2} \qquad (15\text{-}22)$$

as is seen from Equations (15-6) and (15-16).

EXAMPLE 15-3   In the titration of Fe(II) with Ce(IV) in 0.5 M H_2SO_4, we have $n_1 = n_2 = 1$, $E_1^{\circ\prime} = 0.68$, $E_2^{\circ\prime} = 1.44$; and we calculate the interval

$$0.86 < E < 1.26$$

In the titration of Fe(II) with dichromate in 1 M HCl, we have $n_1 = 1, n_2 = 3$, $E_1^{\circ\prime} = 0.69$, $E_2^{\circ\prime} = 1.09$; and the useful indicator interval is

$$0.87 < E < 1.03$$

In the titration of Fe(II) with dichromate in 1 M HCl containing 0.25 M H_3PO_4, $E_1^{\circ\prime} = 0.50$, $E_2^{\circ\prime} = 1.05$; and the useful interval becomes

$$0.68 < E < 0.99 \qquad\qquad ////$$

15-5   EXAMPLES OF REDOX INDICATORS

Many substances have been studied for use as redox indicators. For detailed reports the reader is referred to appropriate monographs.[12] To exemplify redox indicator behavior, we limit our discussion to two classes of redox indicators: diphenylamine and 1,10-phenanthroline-iron(II).

Diphenylamine   This compound was introduced by Knop[13] as an indicator for the Fe(II)-dichromate titration. Since then the chemistry of this indicator has been studied extensively. Detailed review of the several possibilities for an oxidation mechanism was made by Bishop.[12] Bishop and Hartshorn[14] concluded that diphenylamine is oxidized directly to In_{ox} in a single two-electron step that we formulate as follows:

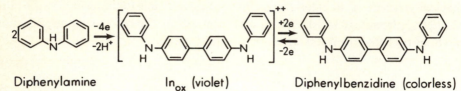

Diphenylamine In_{ox} (violet)   Diphenylbenzidine (colorless)

The colored oxidized form In_{ox} (traditionally called diphenylbenzidine violet) can exist as any of five possible products, depending on the particular intermolecular reaction between the two diphenylamine molecules; the p,p' oxidation product is shown here. The In_{ox} form can be reduced reversibly to the benzidine base ($E^{\circ\prime} =$ 0.76 V in 0.5 to 1 M sulfuric acid). Spontaneous decomposition of In_{ox} also takes place, with stability increasing with increasing concentration of sulfuric acid.

Sarver and Kolthoff[15] introduced diphenylaminesulfonic acid, which has the advantage of being water-soluble in the form of its sodium or barium salt. The color change in 0.5 M sulfuric acid occurs at a potential of 0.85 V, which is significantly higher than the potential for diphenylamine. As shown by the titration curves of Figure 15-2, the color change of diphenylamine in 0.5 M sulfuric acid occurs well before the equivalence point. The addition of phosphoric acid is beneficial, for it lowers the formal potential of the Fe(III)-Fe(II) system so that its equivalence potential coincides more nearly with that of the indicator. The higher potential of diphenyl-amine sulfonate is advantageous. The addition of phosphoric acid is usually recommended, although Stockdale[16] obtained good results, as well as a better warning of the approach of the end point, by omitting the phosphoric acid and titrating to the fully developed violet color of the indicator.

Ferroin   With the introduction of Ce(IV) as an oxidant and the evaluation of the formal potential of the Ce(IV)-Ce(III) couple,[17] the need for indicators with higher electrode potentials became evident. The indicator ferroin, tris(1,10-phenanthroline)-iron(II), was discovered by Walden, Hammett, and Chapman,[18] and its standard potential was evaluated at 1.14 V. Hume and Kolthoff[19] found that the formal potential was 1.06 V in 1 M hydrochloric or sulfuric acid. The color change, however, occurs at about 1.12 V, because the color of the reduced form (orange-red) is so much more intense than that of the oxidized form (pale blue). From Figure 15-1 it can be seen that ferroin should be ideally suited to titrations of Fe(II) and other reductants with Ce(IV), particularly when sulfuric acid is the titration medium. It has the further advantages of undergoing a reversible oxidation-reduction reaction and of being relatively stable even in the presence of oxidant.

The formal potential of ferroin can be modified greatly by the introduction of various substituents into the 1,10-phenanthroline nucleus. Brandt and Smith,[20] studying a series of methyl-substituted 1,10-phenanthrolines, found an interesting regularity in the values of the formal potentials. The effect of substitution was found to be additive in the sense that each methyl group in the 3- or 8-position lowered the formal potential by 0.03 V, in the 5- or 6-position by 0.04 V, and in the 4- or 7-position by 0.11 V. In this way, the formal potential could be varied in 0.01- to 0.03-V increments from 0.84 to 1.10 V.

The most important substituted ferroins are 5-nitro-1,10-phenanthroline and 4,7-dimethyl-1,10-phenanthroline.[3] The former ($E^{\circ\prime} = 1.25$ V) is especially suitable for titrations using Ce(IV) in perchloric or nitric acid solution,[21] where the formal potential of the oxidant is especially high. The 4,7-dimethyl derivative has a formal potential sufficiently low ($E^{\circ\prime} = 0.88$ V) to make it useful for the titration of Fe(II) with dichromate in 0.5 M sulfuric acid. It should be satisfactory also for Fe(II) titrations in hydrochloric acid, with either dichromate or Ce(IV).

In the foregoing discussion the indicator has tacitly been assumed to come rapidly to equilibrium at each point of the titration curve. That this is an over-simplification is evident from a number of experimental observations. Kolthoff and Sarver[22] found that the oxidation of diphenylamine with dichromate is induced by the Fe(II)-dichromate reaction. The direct oxidation is so slow that the indicator blank is best determined by comparison of the visual with the potentiometric end point. With ferroin, Smith and Brandt[10] and Stockdale[16] found that the reverse titration, dichromate with iron, gave satisfactory results at sufficiently high acidities, whereas the direct titration failed because the indicator could not be oxidized. Here the oxidation seems to be slow and the reduction rapid because of the irreversible nature of the oxidant and the reversible nature of the reductant.

With As(III) as the reductant and Ce(IV) as the oxidant in sulfuric acid, the situation is reversed. The oxidized indicator, ferriin, is reduced hardly at all by excess As(III) even in the presence of osmium tetroxide as a catalyst. If a drop of Ce(IV) solution is added, however, the red color of ferroin is rapidly developed, evidently because of an induced reaction. In hydrochloric acid the induced reaction does not occur, ferroin is oxidized by the first drop of Ce(IV), and so the titration fails. A small amount of chloride (for example, 0.1 M hydrochloric acid in 0.5 M sulfuric acid) does not interfere. Addition of excess Hg(II) perchlorate prevents the interference by complexation of the chloride.[23]

15-6 REDOX REACTIONS IN NONAQUEOUS SOLVENTS

The analytical chemistry of redox reactions in nonaqueous solvents has received less attention than acid-base reactions in these solvents. It should be a fruitful subject for future study. Thus far the Karl Fischer titration for water has been the most important example (Section 19-8). The developments and possibilities of analytical redox reactions in nonaqueous media have been reviewed by Kratochvil.[24]

Nonaqueous solvents have the advantage of enhanced solubility of organic reactants and products and avoidance of the leveling effects of aqueous solvent.

Useful solvents must themselves resist oxidation or reduction, should dissolve suitable ionic solutes and nonelectrolytes, and in addition should be inexpensive and obtainable in high purity. Kratochvil indicated that the most potentially useful solvents are those that have a dielectric constant greater than about 25 and have Lewis-base properties. Some solvents meeting these criteria are acetonitrile, dimethyl-sulfoxide, dimethylformamide, dimethylacetamide, propylene carbonate, ethylene carbonate, formamide, sulfolane, and γ-butyrolactone. Solvents of the Lewis-base type show specific solvation effects with many metal cations (Lewis acids). Thus acetonitrile functions as a Lewis base toward the silver ion. At the same time it reacts but little with the hydrogen ion.

When, as a result of strong interaction with a solvent, one oxidation state is strongly stabilized, much stronger oxidants or reductants are possible than with water as solvent. Thus copper(II) can be used as an effective oxidant in acetonitrile because of the stabilization of copper(I).

Change in the dielectric constant can have a notable effect on the rate of a reaction. For ion-ion reactions it can be estimated[25] that reducing the dielectric constant from 78 to 36 should decrease the rate of reaction[24] for singly charged ions by a factor of about 10^7. Specific solvent-solute interaction effects are important as well.

Because the coefficients of expansion for most nonaqueous solvents are high compared with water, the technique of gravimetric titration is attractive. Precise titrations can be performed in this way with small volumes of titrant and sample. End points can be obtained potentiometrically or with visual indicators[26,27] such as ferroin and diphenylamine.

The most important oxidants in glacial acetic acid as solvent are: lead tetra-acetate, which oxidizes mercaptans RSH to disulfide RSSR; cerium(IV), a strong oxidant, though its reactions tend to be slow; iodobenzene dichloride;[28] and bromine, which may be obtained conveniently by coulometric generation.

Copper(II) and cerium(IV) have been studied as oxidants in acetonitrile. The copper(II)-copper(I) couple has an estimated[29] electrode potential of 0.68 V relative to the silver reference electrode. It has been studied as an oxidant for substances such as iodide, hydroquinone, thiourea, potassium ethyl xanthate, diphenylbenzidine, and ferrocene. Cerium(IV) reactions are catalyzed by acetate ion. Copper(I) is a suitable reductant for chromium(VI), vanadium(V), cerium(IV), and manganese(VII) in the presence of iron(III). For details on many studies of redox reactions in non-aqueous solvents, the reader is referred to the summary by Kratochvil.[24]

15-7 CATALYZED REACTIONS

A redox reaction is particularly apt to be slow when unequal numbers of electrons are involved in the two half-reactions.[30] In that case the reaction cannot proceed through a single bimolecular step, but instead through either a succession of steps or a higher-order reaction. A succession of steps generally involves an unstable intermediate oxidation state; a higher-order reaction involves improbable higher-order collisions. As an example, consider the reaction between Ce(IV) and As(III), which cannot proceed through a single bimolecular step and is therefore slow:

$$2Ce(IV) + As(III) \rightarrow 2Ce(III) + As(V) \qquad (15\text{-}23)$$

There is some disagreement about the exact mechanism. Although earlier research had indicated third-order kinetics, the work of Csányi and Szabó[31] established the existence of As(IV) as an intermediate, implying the two-step mechanism

$$Ce(IV) + As(III) \xrightarrow{\text{slow}} Ce(III) + As(IV) \qquad (15\text{-}24)$$

$$Ce(IV) + As(IV) \xrightarrow{\text{fast}} Ce(III) + As(V) \qquad (15\text{-}25)$$

The intermediate species is a strong reducing agent, as indicated by the estimated potential:

$$As(V) + e^- \rightleftharpoons As(IV) \qquad E° \simeq 0.3 \text{ V} \qquad (15\text{-}26)$$

In the presence of iodide or iodine a pronounced catalysis occurs, probably through the sequence[32]

$$Ce(IV) + I^- \rightarrow I^0 + Ce(III) \qquad (15\text{-}27)$$

$$2I^0 \rightarrow I_2 \qquad (15\text{-}28)$$

$$I_2 + H_2O \rightleftharpoons HOI + H^+ + I^- \qquad (15\text{-}29)$$

$$H_3AsO_3 + HOI \rightarrow H_3AsO_4 + H^+ + I^- \qquad (15\text{-}30)$$

because all the reactions involving iodine are rapid. Dubravćić[33] found that, if only small amounts of iodide are present, no catalysis is observed unless chloride is present. Evidently, in the absence of chloride a side reaction occurs with the formation of iodate which is ineffective as a catalyst. In the presence of chloride, unipositive iodine is stabilized as iodine monochloride rather than as hypoiodous acid.

Sandell and others described determinations of ruthenium[34] and osmium[35] based on catalysis of the Ce(IV)-As(III) reaction. In both cases, the reaction rate is proportional to catalyst concentration. However, with ruthenium the rate is independent of [As(III)] and dependent on [Ce(IV)], whereas with osmium the rate is independent of [Ce(IV)] and dependent on [As(III)]. Although the complete reaction mechanisms have not been elucidated, one may infer that the rate-determining steps in the two cases are different, namely, oxidation of the reduced form of ruthenium and reduction of the oxidized form of osmium. In the osmium-catalyzed reaction, one crucial reaction may be[36]

$$As(III) + Os(VIII) \rightarrow As(V) + Os(VI) \qquad (15\text{-}31)$$

Another example of homogeneous catalysis is that of the decomposition of hydrogen peroxide by bromide ion. Here hydrogen peroxide acts as both an oxidant and a reductant, with two different reactions proceeding at characteristic rates under a given set of conditions:

$$H_2O_2 + 2Br^- + 2H^+ \rightarrow Br_2 + 2H_2O \qquad (15\text{-}32)$$

$$H_2O_2 + Br_2 \rightarrow O_2 + 2H^+ + 2Br^- \qquad (15\text{-}33)$$

If bromide or bromine is added to hydrogen peroxide, the reaction mixture reaches a steady state such that the rates of the two reactions are equal. If the reaction rates are initially unequal, either bromide or bromine will accumulate until the unequal concentrations cause the rates to become equal. Then the only observable reaction is the sum of (15-32) and (15-33), or

$$2H_2O_2 \rightarrow 2H_2O + O_2 \qquad (15\text{-}34)$$

Rates of reactions and standard potentials   For most redox reactions there is no simple relation between the equilibrium constant and the rate constant. There are many examples of reactions with favorable free-energy change but extremely slow reaction rate [$S_2O_8^=$ and As(III), H_2 and O_2, Ce(IV) and H_2O]. There is, however, one class of reactions, outer-sphere electron-transfer reactions, in which a relatively simple relation exists between free-energy change and reaction rate.

Marcus[37] derived an expression involving the forward rate constant k_{12} for an electron-transfer reaction:

$$ox_1 + red_2 \rightleftharpoons red_1 + ox_2 \qquad (15\text{-}35)$$

$$k_{12} = \sqrt{k_1 k_2 K_{12} f} \qquad (15\text{-}36)$$

Here k_1 and k_2 are rate constants for the electron exchange reactions of the two half-reactions, K_{12} is the equilibrium constant for the electron-transfer reaction, and f is a constant:

$$\log f = \frac{\sqrt{\log K_{12}}}{4 \log (k_1 k_2 / z^2)} \qquad (15\text{-}37)$$

where z is the collision frequency of two uncharged molecules in solution (often assumed to be 10^{11} 1/mole-s). Equation (15-36) is equivalent to

$$\Delta G_{12}^* = 0.5 \, \Delta G_1^* + 0.5 \, \Delta G_2^* + 0.5 \, \Delta G_{12}^0 - 1.15 \log f \qquad (15\text{-}38)$$

where ΔG^* is free energy of activation and ΔG^0 is standard free energy.

Table 15-1 RATE CONSTANTS[38] FOR THE OXIDATION OF IRON(II)-
PHENANTHROLINE COMPLEXES BY CERIUM(IV) IN
0.5 M SULFURIC ACID AT 25°C

Complex	$E°$ of Complex	k	log k
Tris(3,4,7,8-tetramethyl-1,10-phenanthroline)-iron(II)	0.83	1.6×10^6	6.20
Tris(5,6-dimethyl-1,10-phenanthroline)-iron(II)	0.99	4.3×10^5	5.63
Tris(5-methyl-1,10-phenanthroline)-iron(II)	1.04	2.2×10^5	5.34
Tris(1,10-phenanthroline)-iron(II)	1.08	1.42×10^5	5.15
Tris(5-phenyl-1,10-phenanthroline)-iron(II)	1.10	1.2×10^5	5.08
Tris(5-chloro-1,10-phenanthroline)-iron(II)	1.14	2.5×10^4	4.40
Tris(5-nitro-1,10-phenanthroline)-iron(II)	1.26	3.9×10^3	3.59

The important concept is that, if the reaction mechanism for a series of reactions does not change as the coordinating ligand is changed, then for outer-sphere electron transfers the relation between the formal potential and the log of the forward rate constant should be linear.

Sutin and others[38] showed that rates of reactions can be calculated reliably from Equation (15-36). For example, the observed and calculated rate constants for several reactions are

$$Ce(IV) - Fe(CN)_6{}^{4-} \qquad 1.9 \times 10^6 (obs) \text{ and } 6.0 \times 10^6 (calc)$$

$$Ce(IV) - Mo(CN)_8{}^{4-} \qquad 1.4 \times 10^7 (obs) \text{ and } 1.3 \times 10^7 (calc)$$

$$IrCl_6{}^{=} - Fe(CN)_6{}^{4-} \qquad 3.8 \times 10^5 (obs) \text{ and } 5.7 \times 10^5 (calc)$$

$$Fe(CN)_6{}^{3-} - W(CN)_8{}^{4-} \qquad 4.3 \times 10^4 (obs) \text{ and } 5.1 \times 10^4 (calc)$$

For a series of related complexes and a single redox reagent (Table 15-1), rate constants can be predicted from potentiometric data and rate measurements on a single member.

15-8 INDUCED REACTIONS

The phenomenon of chemical induction has been recognized for over a century.[39–41] Discussions of the role of induced reactions in analytical chemistry have been presented by Skrabel,[42] Kolthoff and Stenger,[43] and Medalia.[44]

A reaction between A and B is said to *induce* a reaction between A and C if the latter reaction under a given set of conditions does not occur at all, or proceeds only slowly, unless it is caused to proceed by the simultaneous occurrence of the reaction between A and B.

A distinction should be made between an induced and a catalyzed reaction. A catalyst is not altered permanently and so enters into a cyclic reaction or no reaction, whereas an inductor must take part in the primary reaction. A *side reaction* is distinguished from an induced reaction by the fact that its rate is unaffected by the occurrence of the primary reaction.

Schilow[45] recognized that induced reactions fall into two classes. The first class now is called an *induced chain reaction*, which can be described in terms of an initiation step, a propagation sequence, and a termination step. The other class is the *coupled reaction*, which can be distinguished from an induced chain reaction by the behavior of the *induction factor* defined by the ratio: equivalents of induced reaction/ equivalents of primary reaction.[46] In an induced chain reaction the induction factor increases without limit as the propagation chain length is increased. In a coupled reaction the induction factor approaches some definite small value such as 1, 2, or 1/2 as the induction reaction is favored.

In general, an induced reaction involves the formation, in the primary reaction, of an active intermediate[47] that then reacts with a substance that otherwise reacts slowly if at all. The active intermediate may be formed from either reagent in the primary reaction, and it may be either a free radical or a species of intermediate oxidation state.

Induced chain reactions   To illustrate an induced chain reaction, consider the oxidation of Fe(II) by hydrogen peroxide, which induces a chain decomposition reaction. The classical interpretation is that the decomposition proceeds through the Haber-Weiss mechanism,[48] as modified by Baxendale and others,[49] in which the chain decomposition is initiated by

$$H_2O_2 + Fe^{++} \rightarrow FeOH^{++} + \cdot OH \qquad (15\text{-}39)$$

The propagation chain is given by

$$\cdot OH + H_2O_2 \rightarrow \cdot HO_2 + H_2O \qquad (15\text{-}40)$$

$$FeOH^{++} + \cdot HO_2 \rightarrow O_2 + Fe^{++} + H_2O \qquad (15\text{-}41)$$

$$Fe^{++} + H_2O_2 \rightarrow FeOH^{++} + \cdot OH \qquad (15\text{-}39)$$

which adds to the induced reaction

$$2H_2O_2 \rightarrow 2H_2O + O_2 \qquad (15\text{-}42)$$

The termination step is

$$\cdot OH + Fe^{++} \rightarrow FeOH^{++} \qquad (15\text{-}43)$$

The initiation and termination steps are added to give the expected stoichiometric reaction

$$H_2O_2 + 2Fe^{++} \rightarrow 2FeOH^{++} \qquad (15\text{-}44)$$

The induction factor is determined by the number of times the chain reaction is propagated before it is terminated by Reaction (15-43). The induction factor therefore increases without limit as the ratio of hydrogen peroxide concentration to Fe(II) concentration is increased.

From the analytical viewpoint the stoichiometric reaction (15-44) can be observed only by adding a large excess of Fe(II) to accelerate the termination reaction (15-43). Additional complications are encountered in the presence of oxygen or organic substances such as alcohols.

Kremer and Stein[50] proposed a reaction pathway involving not $\cdot OH$ but instead the intermediates $FeOOH^{++}$ and Fe^{3+}. Kolthoff and Medalia[51] and Walling and Goosen,[52] however, interpreted the retardation of hydrogen peroxide decomposition by organic substances in terms of an extraction of H atoms from organic compounds RH to form radicals $R\cdot$, which react with oxygen to yield peroxide radicals $ROO\cdot$. These in turn are reduced by Fe(II):

$$ROO\cdot + Fe^{++} + H^+ \rightarrow ROOH + Fe^{3+} \qquad (15\text{-}45)$$

The hydroperoxide thus formed reacts with Fe(II) by a reaction analogous to (15-39), that is,

$$ROOH + Fe^{++} \rightarrow RO\cdot + Fe^{3+} + OH^- \qquad (15\text{-}46)$$

The reduction reactions of hydroperoxides are subject to the same types of chain mechanism as the reduction of hydrogen peroxide, and in addition the radical $RO\cdot$ can undergo various rearrangement reactions.

From the analytical viewpoint the important detail is that either low or high results for peroxide can be observed in the presence of organic matter and oxygen, depending on the relative rates of the competing reactions of radicals $R\cdot$ with Fe(III) or oxygen. The reaction of $R\cdot$ with Fe(III) leads to the induced reduction of peroxide and therefore to low results; Reaction (15-45) leads to an induced air oxidation of Fe(II),

$$O_2 + 4Fe^{++} + 4H^+ \rightarrow 4Fe^{3+} + 2H_2O \qquad (15\text{-}47)$$

and therefore to high results. Medalia[44] cited examples in which the apparent peroxide content of organic hydroperoxides determined by Fe(II) methods varied from 17 to 690% of the true peroxide content. The iodometric method, on the other hand, has been found to yield reliable results even in the presence of organic substances.

Similar induced reactions have been found in the reduction of peroxydisulfate by Fe(II). Kolthoff and Carr[53] expressed a mechanism involving the free-radical ion $\cdot SO_4^-$ as follows:

$$Fe^{++} + S_2O_8^= \rightarrow Fe(III) + SO_4^= + \cdot SO_4^- \qquad (15\text{-}48)$$
$$Fe^{++} + \cdot SO_4^- \rightarrow Fe(III) + SO_4^= \qquad (15\text{-}49)$$

In the presence of an organic substance RH (such as ethyl alcohol), a free radical can form by the reaction

$$\cdot SO_4^- + RH \rightarrow HSO_4^- + R\cdot \qquad (15\text{-}50)$$

The free radical $R\cdot$ then reduces Fe(III), leading to an induced oxidation of the organic substance. Various other induced reactions, including the induced air oxidation of

Fe(II), can take place. It is of analytical interest that the induced reactions are suppressed to some extent by the addition of bromide. The effect of bromide was interpreted by assuming that bromide destroys the free radical ion $\cdot SO_4^-$ by the reaction

$$\cdot SO_4^- + Br^- \rightarrow SO_4^= + Br\cdot \qquad (15\text{-}51)$$

in which bromine free radicals (atoms) are formed. These are reduced to bromide ions:

$$Fe^{++} + Br\cdot \rightarrow Fe(III) + Br^- \qquad (15\text{-}52)$$

Woods, Kolthoff, and Meehan[54] found for the induced oxidation of As(III) by the Fe(II)-peroxydisulfate or Fe(II)-H_2O_2 reactions that the induction factor can approach infinity in the presence of Fe(III) or Cu(II). With peroxydisulfate the initiation reaction is given by (15-48) and by

$$As(III) + \cdot SO_4^- \rightarrow As(IV) + SO_4^= \qquad (15\text{-}53)$$

The propagation chain is

$$As(IV) + O_2 + H^+ \rightarrow As(V) + \cdot HO_2 \qquad (15\text{-}54)$$

$$Fe^{++} + \cdot HO_2 + H^+ \rightarrow Fe(III) + H_2O_2 \qquad (15\text{-}55)$$

$$Fe^{++} + H_2O_2 \rightarrow Fe(III) + \cdot OH + OH^- \qquad (15\text{-}56)$$

$$As(III) + \cdot OH \rightarrow As(IV) + OH^- \qquad (15\text{-}57)$$

and the chain termination step is

$$\cdot SO_4^- + Fe^{++} \rightarrow SO_4^= + Fe(III) \qquad (15\text{-}58)$$

Various other induced air-oxidation reactions are important in analytical chemistry. A classical example, among the earliest observed,[40] is the air oxidation of Sn(II), which is induced by the reaction of Sn(II) and dichromate. Up to 98.3% of the oxidation of Sn(II) was found to be due to oxygen, and as little as 1.7% to dichromate. Several other examples of induced air oxidation, reported by Lenssen and Löwenthal and other early investigators, have been discussed by Bray and Ramsey.[55] In each of the following reactions an induced reaction occurs between the reductant and atmospheric oxygen: $SnCl_2$ + $KMnO_4$, O_3, H_2O_2, or ClO_2; H_2SO_3 + $K_2Cr_2O_7$, $KMnO_4$, $K_2S_2O_8$, or H_2O_2.

Boyer and Ramsey[56] found that the air oxidation of iodide is induced by the reaction of iodide with peroxydisulfate, Fe(III), ferricyanide, or V(V). The reaction with V(V) was investigated in detail and found to behave as a typical induced chain reaction. The proposed mechanism involves the formation of atomic iodine I^0 or the ion I_2^- as an intermediate that reacts with oxygen. The chain is initiated by the reaction

$$V(OH)_4^+ + I^- + 2H^+ \rightarrow VO^{++} + I^0 + 3H_2O \qquad (15\text{-}59)$$

The chain is propagated by the sequence

$$I_2^- + O_2 + H^+ \rightarrow I_2 + \cdot HO_2 \qquad \text{rate-determining} \qquad (15\text{-}60)$$

$$H^+ + \cdot HO_2 + I^- \rightarrow I^0 + H_2O_2 \qquad (15\text{-}61)$$

$$I^0 + I^- \rightleftharpoons I_2^- \qquad (15\text{-}62)$$

The hydrogen peroxide formed in (15-61) is reduced by iodide:

$$H_2O_2 + 2I^- + 2H^+ \rightarrow I_2 + 2H_2O \qquad (15\text{-}63)$$

The reaction is terminated by

$$I_2^- + I_2^- \rightarrow I_3^- + I^- \qquad (15\text{-}64)$$

In accordance with this mechanism, an induced air oxidation of iodide is to be expected whenever I^0 or I_2^- is formed as the primary oxidation product of iodine. Titrations of iodide with Ce(IV) or MnO_4^- are not sensitive to induced air oxidation, evidently because I_2^- is rapidly oxidized to I_2. The induction factor and therefore the relative error due to air oxidation (oxygen error, Section 19-2) are decreased by an increasing concentration of iodide and oxidant, because the importance of the termination reaction (15-64) increases.

Induced chain reactions can represent serious sources of error in analytical measurements. Though in principle such reactions can be used as the basis of sensitive qualitative tests, conditions are difficult to control closely enough that useful quantitative results can be obtained.

Coupled reactions A coupled reaction involves two oxidants with a single reductant, where one reaction taken alone would be thermodynamically unfavorable. The induction factor approaches some small definite value as the induced reaction is favored. An illustration is the oxidation of Mn(II) by dichromate[46,57] to give MnO_2:

$$2Cr(VI) + 3Mn(II) \rightarrow 2Cr(III) + 3Mn(IV) \qquad (15\text{-}65)$$

This reaction is thermodynamically unfavorable in the forward direction; in fact, the reverse reaction proceeds slowly. Yet, if the reaction of As(III) with dichromate is allowed to proceed in the presence of Mn(II), for every equivalent of Cr(VI) reduced by As(III), 0.5 equivalent is reduced by Mn(II), at least under suitable conditions of concentration.

The relation observed is satisfactorily accounted for by the following mechanism for the primary reaction between Cr(VI) and As(III) in the absence of Mn(II):

$$Cr(VI) + As(III) \rightarrow Cr(IV) + As(V) \qquad (15\text{-}66)$$

$$Cr(VI) + Cr(IV) \rightleftharpoons 2Cr(V) \qquad (15\text{-}67)$$

$$Cr(V) + As(III) \rightarrow Cr(III) + As(V) \qquad (15\text{-}68)$$

The sum of (15-66), (15-67), and twice (15-68) is the primary reaction

$$2Cr(VI) + 3As(III) \rightarrow 2Cr(III) + 3As(V) \qquad (15\text{-}69)$$

In the presence of Mn(II) the unstable Cr(IV) formed in (15-66) reacts according to

$$Cr(IV) + Mn(II) \rightarrow Mn(III) + Cr(III) \qquad (15\text{-}70)$$

The resulting Mn(III) is the final product if it is stabilized by the formation of a fluoride or metaphosphate complex.[58] In the absence of such stabilizing reagents the Mn(III) disproportionates to give Mn(IV), which then precipitates as MnO_2:

$$2Mn(III) \rightarrow Mn(II) + Mn(IV) \qquad (15\text{-}71)$$

If every Cr(IV) ion produced in (15-66) is used in (15-70), the overall induced reaction is the sum of twice (15-66) plus twice (15-70) plus (15-71), or

$$2Cr(VI) + 2As(III) + Mn(II) \rightarrow 2Cr(III) + 2As(V) + Mn(IV) \qquad (15\text{-}72)$$

which is the sum of the primary reaction (15-69) and the apparent induced reaction (15-65) taken in the ratio 2/1. Reaction (15-65) alone does not occur, and the paradoxical observation of a thermodynamically impossible reaction is due to the coupling of this reaction with a thermodynamically favorable reaction so that the overall induced reaction has a negative free-energy change.

Other examples of reactions coupled with the Cr(VI)-As(III) reaction are the oxidation of halides. With Cl^- the induction factor approaches 0.5, suggesting the induced reaction to be $Cr(IV) + Cl^- \rightarrow Cr(III) + Cl^0$, whereas with iodide and bromide the induction factor approaches 2, suggesting the induced reaction to be $Cr(V) + X^- \rightarrow Cr(III) + X^+$. The dichromate-As(III) reaction can also induce the oxidation of V(IV) to V(V) and Ce(III) to Ce(IV).

In alkaline solution the reaction between Cr(VI) and As(III) induces the air oxidation of arsenic(III). This reaction has been studied by Kolthoff and Fineman,[59] who found a limiting value of the induction factor of 4/3.

Other coupled reactions involving Cr(VI) have been discussed by Westheimer,[57] Medalia,[44] and Beattie and Haight.[60] If a 1-equivalent reductant such as Fe(II) reacts with Cr(VI), the first step presumably involves the formation of Cr(V):

$$Cr(VI) + Fe(II) \rightarrow Cr(V) + Fe(III) \qquad (15\text{-}73)$$

If iodide is present, an induced oxidation occurs, with the limiting value of the induction factor[61] of 2. The reaction probably proceeds through the steps[62]

$$Cr(V) + I^- \rightarrow Cr(III) + OI^- \qquad (15\text{-}74)$$
$$OI^- + I^- + 2H^+ \rightarrow I_2 + H_2O \qquad (15\text{-}75)$$

Reaction (15-74), however, is in competition with the reaction between Cr(V) and Fe(II):

$$Cr(V) + Fe(II) \rightarrow Cr(IV) + Fe(III) \qquad (15\text{-}76)$$

$$Cr(IV) + Fe(II) \rightarrow Cr(III) + Fe(III) \qquad (15\text{-}77)$$

so it would seem that the induction factor could approach 2 only if the ratio of iodide ion to Fe(II) concentration were large. But, since the rate constant of (15-74) is about six times that of (15-76),[63] the limiting value of the induction factor can be approached readily.

In summary, coupled reactions can be used in some cases to bring about induced reactions that, taken alone, would be thermodynamically impossible. In many cases, determination of the coupling index affords important clues to the mechanisms of complex redox reactions.

Oscillating reactions   In these coupled autocatalytic reactions, repeated undamped oscillations occur in concentrations of reactants. These reactions have been reviewed by Field.[64] For example,[65] in the cerium-catalyzed reaction of bromate with malonic acid, changes in concentration of bromide ion and in the Ce(IV)-Ce(III) ratio over several orders of magnitude occur repeatedly. These systems as yet are not well understood. For the bromate–malonic acid reaction a 10-step process for the mechanism has been proposed.[65]

REFERENCES

1   H. A. LAITINEN in "Physical Methods of Chemical Analysis," W. H. Berl (ed.), vol. 1, Academic, New York, 1950.

2   H. A. LAITINEN and I. M. KOLTHOFF: *J. Phys. Chem.*, **45**:1079 (1941).

3   G. F. SMITH: *Anal. Chem.*, **23**:925 (1951); G. F. SMITH and F. P. RICHTER: "Phenanthroline and Substituted Phenanthroline Indicators," G. F. Smith Chemical, Columbus, Ohio, 1944; "Cerate Oxidimetry," 2d ed., 1964.

4   P. K. WINTER and H. V. MOYER: *J. Amer. Chem. Soc.*, **57**:1402 (1935).

5   J. W. ROSS and I. SHAIN: *Anal. Chem.*, **28**:548 (1956).

6   H. GERISCHER: *Z. Elektrochem.*, **54**:366 (1950).

7   A. HICKLING: *Trans. Faraday. Soc.*, **41**:333 (1945).

8   I. M. KOLTHOFF and N. TANAKA: *Anal. Chem.*, **26**:632 (1954).

9   K. J. VETTER and J. W. SCHULTZE: *J. Electroanal. Chem.*, **34**:131, 141 (1972).

10   G. F. SMITH and W. W. BRANDT: *Anal. Chem.*, **21**:948 (1949).

11   N. HACKERMAN and R. A. POWERS: *J. Phys. Chem.*, **57**:139 (1953).

12   E. BISHOP in "Indicators," E. Bishop (ed.), chap. 8B, Pergamon, New York, 1972.

13   J. KNOP: *J. Amer. Chem. Soc.*, **46**:263 (1924).

14   E. BISHOP and L. G. HARTSHORN: *Analyst*, **96**:26 (1971).

15 L. A. SARVER and I. M. KOLTHOFF: *J. Amer. Chem. Soc.*, **53**:2902 (1931).

16 D. STOCKDALE: *Analyst*, **75**:150 (1950).

17 A. H. KUNZ: *J. Amer. Chem. Soc.*, **53**:98 (1931).

18 G. H. WALDEN, JR., L. P. HAMMETT, and R. P. CHAPMAN: *J. Amer. Chem. Soc.*, **53**:3908 (1931); **55**:2649 (1933).

19 D. N. HUME and I. M. KOLTHOFF: *J. Amer. Chem. Soc.*, **65**:1895 (1943).

20 W. W. BRANDT and G. F. SMITH: *Anal. Chem.*, **21**:1313 (1949).

21 G. F. SMITH and C. A. GETZ: *Ind. Eng. Chem.*, *Anal. Ed.*, **10**:304 (1938).

22 I. M. KOLTHOFF and L. A. SARVER: *J. Amer. Chem. Soc.*, **52**:4179 (1930); **53**:2906 (1931).

23 K. GLEU: *Z. Anal. Chem.*, **95**:305 (1933).

24 B. KRATOCHVIL: *Crit. Rev. Anal. Chem.*, **1**:415 (1971); *Rec. Chem. Progr.*, **27**:253 (1966).

25 G. SCATCHARD: *Chem. Rev.*, **10**:229 (1932).

26 G. P. RAO and A. R. V. MURTHY: *Fresenius' Z. Anal. Chem.*, **180**:169 (1961).

27 B. KRATOCHVIL and D. A. ZATKO: *Anal. Chem.*, **40**:422 (1968).

28 P. N. K. NAMBISAN and C. G. R. NAIR: *Talanta*, **18**:753 (1971).

29 J. K. SENNE and B. KRATOCHVIL: *Anal. Chem.*, **43**:79 (1971); **44**:585 (1972).

30 P. A. SHAFFER: *J. Amer. Chem. Soc.*, **55**:2169 (1933).

31 L. J. CSÁNYI and Z. G. SZABÓ: *Talanta*, **1**:359 (1958).

32 W. C. BRAY: *Chem. Rev.*, **10**:161 (1932).

33 M. DUBRAVČIĆ: *Analyst*, **80**:146, 295 (1955).

34 C. SURASITI and E. B. SANDELL: *Anal. Chim. Acta*, **22**:261 (1960).

35 R. D. SAUERBRUNN and E. B. SANDELL: *J. Amer. Chem. Soc.*, **75**:4170 (1953); *Mikrochim. Acta*, **1953**:22.

36 R. L. HABIG, H. L. PARDUE, and J. B. WORTHINGTON: *Anal. Chem.*, **39**:600 (1967); J. B. WORTHINGTON and H. L. PARDUE: *Anal. Chem.*, **42**:1157 (1970).

37 R. A. MARCUS: *Ann. Rev. Phys. Chem.*, **15**:155 (1964).

38 G. DULZ and N. SUTIN: *Inorg. Chem.*, **2**:917 (1963); R. J. CAMPION, N. PURDIE, and N. SUTIN: *Inorg. Chem.*, **3**:1091 (1964).

39 F. KESSLER: *Poggendorf's Ann.*, **95**:224 (1855); **119**:218 (1863).

40 E. LENSSEN and J. LÖWENTHAL: *J. Prakt. Chem.*, **86**:193 (1862).

41 C. F. SCHÖNBEIN: *Poggendorf's Ann.*, **100**:34 (1857); *J. Prakt. Chem.*, **75**:108 (1858).

42 A. SKRABEL: "Die Induzierten Reaktionen; ihre Gesichte und Theorie," F. Enke, Stuttgart, 1908.

43 I. M. KOLTHOFF and V. A. STENGER: "Volumetric Analysis," 2d ed., vol. 1, Interscience, New York, 1942.

44 A. I. MEDALIA: *Anal. Chem.*, **27**:1678 (1955).

45 N. SCHILOW: *Z. Phys. Chem.* (Leipzig), **42**:641 (1903); R. LUTHER and N. SCHILOW: *Z. Phys. Chem.* (Leipzig), **46**:777 (1903); *Z. Anorg. Allg. Chem.*, **54**:1 (1907).

46 R. LANG and J. ZWERINA: *Z. Anorg. Allg. Chem.*, **170**:389 (1928).

47 F. R. DUKE in "Treatise on Analytical Chemistry," I. M. Kolthoff and P. J. Elving (eds.), pt. 1, vol. 1, p. 637, Interscience, New York, 1959.

48 F. HABER and J. WEISS: *Naturwissenschaften*, **20**:948 (1932); *Proc. Roy. Soc., Ser. A*, **147**:332 (1934).

49 W. G. BARB, J. H. BAXENDALE, P. GEORGE, and K. R. HARGRAVE: *Trans. Faraday Soc.*, **47**:591 (1951).

50 M. L. KREMER and G. STEIN: *Trans. Faraday Soc.*, **55**:959 (1959); M. L. KREMER: *Trans. Faraday Soc.*, **58**:702 (1962); **59**:2535 (1963).

51 I. M. KOLTHOFF and A. I. MEDALIA: *J. Amer. Chem. Soc.*, **71**:3777, 3784, 3789 (1949).

52 C. WALLING and A. GOOSEN: *J. Amer. Chem. Soc.*, **95**:2987 (1973).

53 I. M. KOLTHOFF and E. M. CARR: *Anal. Chem.*, **25**:298 (1953).

54 R. WOODS, I. M. KOLTHOFF, and E. J. MEEHAN: *J. Amer. Chem. Soc.*, **85**:2385, 3334 (1963); **86**:1698 (1964).

55 W. C. BRAY and J. B. RAMSEY: *J. Amer. Chem. Soc.*, **55**:2279 (1933).

56 M. H. BOYER and J. B. RAMSEY: *J. Amer. Chem. Soc.*, **75**:3802 (1953).

57 F. H. WESTHEIMER: *Chem. Rev.*, **45**:419 (1949).

58 R. LANG and F. KURTZ: *Z. Anorg. Allg. Chem.*, **161**:111 (1929); *Z. Anal. Chem.*, **86**:288 (1931); R. LANG: *Z. Anal. Chem.*, **102**:8 (1935).

59 I. M. KOLTHOFF and M. A. FINEMAN: *J. Phys. Chem.*, **60**:1383 (1956).

60 J. K. BEATTIE and G. P. HAIGHT, JR.: in "Inorganic Reaction Mechanisms," J. O. Edwards (ed.), pt. 2, p. 93, Progress in Inorganic Chemistry, vol. 17, Wiley, New York, 1972.

61 W. MANCHOT and O. WILHELMS: *Justus Liebigs Ann. Chem.*, **325**:105 (1902).

62 R. LUTHER and T. F. RUTTER: *Z. Anorg. Chem.*, **54**:1 (1907).

63 C. WAGNER and W. PREISS: *Z. Anorg. Allg. Chem.*, **168**:265 (1928).

64 R. J. FIELD: *J. Chem. Educ.*, **49**:308 (1972).

65 R. J. FIELD, E. KOROS, and R. M. NOYES: *J. Amer. Chem. Soc.*, **94**:8649 (1972).

PROBLEMS

15-1   A titration of Fe(II) with Ce(IV) is carried out in H_2SO_4 medium, the end point being taken at 0.88 V. If the formal potentials of the Fe(III)-Fe(II) and Ce(IV)-Ce(III) couples are 0.68 and 1.44 V, what is the theoretical titration error?

Answer -0.04%.

15-2   Calculate the potentials for the following percentages of the equivalent amount of reagent added in the titration of Sn(II) with Fe(III): 9, 50, 91, 99, 99.9, 100, 100.1, 101, 110, 200%. Take the formal potentials of the Fe(III)-Fe(II) and Sn(IV)-Sn(II) couples to be 0.66 and 0.15 V.

Answer   0.12, 0.15, 0.18, 0.21, 0.24, 0.32, 0.48, 0.54, 0.60, 0.66 V.

15-3   Use the standard potentials in Table 12-1, and assume activity coefficients of unity:

(*a*) Estimate the equivalence potential at pH = 0 and at pH = 1 for the titration reaction

$$V(OH)_4^+ + Cr^{++} + 2H^+ \rightarrow Cr^{3+} + VO^{++} + 3H_2O$$

(*b*) At pH = 0, calculate the potential at 50, 99, 99.9, 100.1, 101, and 200% of the equivalent amount of Cr^{++} added.

(c) Sketch the titration curve for the stepwise reduction of V(V) to V(IV) to V(III) to V(II) with Cr(II) as the titrant.

Answer (a) 0.31, 0.25 V. (b) 0.999, 0.881, 0.822, −0.20, −0.26, −0.38 V.

15-4 For the titration of 0.01 M V(V) with relatively concentrated V(II) at pH = 0, neglecting dilution and assuming activity coefficients of unity, calculate the potential at the following points: (a) the point where half the V(V) has been titrated to V(IV); (b) the V(IV) equivalence point (*Hint:* calculate the equilibrium constant of the reaction $2VO^{++} + 2H_2O \rightleftharpoons V(OH)_4^+ + V^{3+}$); (c) the point where half the V(IV) has been titrated to V(III); (d) the V(III) equivalence point; (e) a point beyond the second equivalence point corresponding to the volume required to reach the first equivalence point.

Answer (a) 0.989 V. (b) 0.67 V. (c) 0.32 V. (d) 0.043 V. (e) −0.209 V.

15-5 Derive an expression for the minimum difference of formal potentials $E_2^{\circ\prime} - E_1^{\circ\prime}$ for which a redox indicator would show an essentially complete change in color within the interval −0.1 to +0.1% of the end point. Assume that $E_{ind}^{\circ\prime} = E_{equiv}$ and that the color-change interval is given by Equation (15-22).

Answer $E_2^{\circ\prime} - E_1^{\circ\prime} = (3/n_2 + 2/n_{ind} + 3/n_1)\,0.059$.

15-6 Derive Equation (15-18) for the equivalence potential of the Fe(II)-bromine reaction. Given the standard potentials $E_{Br_2,\,aq\,Br^-}^{\circ} = 1.08$ V and $E_{Fe^{3+},Fe^{++}}^{\circ} = 0.771$ V, calculate the equivalence potential for (a) the titration of 0.01 M Fe(II) with 0.01 M bromine, assuming activity coefficients of unity; (b) the same titration at 10-fold higher concentrations of sample and reagent; (c) the titration of (a) except with 0.1 M KBr initially present (*Hint:* A new equation is necessary).

Answer (a) 1.02 V. (b) 0.99 V. (c) 0.97 V.

15-7 Calculate $E^{\circ\prime}$ for the couple $Cu^{++} + e \rightleftharpoons Cu^+$, given that in acetonitrile $Cu^+ + e \rightleftharpoons Cu$ has an $E^{\circ\prime}$ against Ag, 0.01 M AgNO$_3$ = −0.45 V and $Cu^{++} + Cu \rightleftharpoons 2Cu^+$ has a $K_{eq} = 10^{21.3}$.

Answer 0.81 V.

15-8 The free radical diphenylpicrylhydrazyl can be oxidized or reduced in acetonitrile:

$$R^+ + e^- \rightleftharpoons R\cdot \qquad E^{\circ\prime} = +0.39 \text{ V against Ag, } 0.01\ M\ AgNO_3$$

$$R\cdot + e^- \rightleftharpoons R^- \qquad E^{\circ\prime} = -0.12 \text{ V against Ag, } 0.01\ M\ AgNO_3$$

Calculate the equilibrium constant for the reaction

$$R^+ + R^- \rightleftharpoons 2R\cdot$$

Answer 4.3×10^8.

PRIOR OXIDATION AND REDUCTION

In an analysis, prior to the actual determination, an oxidation or a reduction step may be necessary to bring the sample constituent quantitatively to a particular oxidation state. The following conditions must be fulfilled:

1 The oxidation or reduction must be rapid.

2 The equilibrium conditions, as calculated from standard or formal potentials, must represent a quantitative reaction.

3 It must be possible to completely remove excess oxidant or reductant by selective reaction in solution, physical separation of phases, or other means such as dilution or cooling.

4 The oxidation or reduction step must be sufficiently selective to avoid interference from other components of the sample.

Oxidants and reductants may be classified conveniently according to their physical state, that is, whether they are gases, insoluble solids, or in solution. A reagent is classified as a gas if it is removed by volatilization, even though it may have been added in the form of a solution. The following discussion presents a comparison of a number of representative reagents.

16-1   OXIDIZING AGENTS

Gaseous oxidants   *Ozone* is a strong oxidant, as evidenced by the electrode potential of the half-reaction

$$O_3 + 2H^+ + 2e^- \rightleftharpoons O_2 + H_2O \qquad E^\circ = 2.07 \text{ V} \qquad (16\text{-}1)$$

Mn(II) can be oxidized[1] quantitatively to permanganate in perchloric acid by the use of silver ion as a catalyst. In the presence of phosphate, Ce(III) is oxidized to Ce(IV) phosphate, which precipitates from a solution containing sulfuric and phosphoric acids. The precipitate can be dissolved in sulfuric acid. Other oxidations that can be performed quantitatively are V(IV) to V(V) in acid solution, hypophosphite and phosphite to phosphate, selenite to selenate, tellurite to tellurate, nitrite to nitrate, and in alkaline solution, iodide to periodate.

   Many oxidation reactions may be carried out by use of either *bromine* or *chlorine*. Chlorine water was used in the old Winkler method[2] of determining iodide by oxidation to iodate, which then was measured iodometrically. Bromine vapor was used by Sadusk and Ball[3] for iodide oxidation, and bromine water for oxidation of Tl(I) to Tl(III). The excess is readily removed with phenol.[4,5]

Homogeneous oxidants   *Peroxydisulfate* ion, $S_2O_8^=$ (often called persulfate), is a strong oxidant in acid solution in the presence of silver ion as catalyst. The reaction

$$S_2O_8^= + Ag^+ \rightarrow SO_4^= + \cdot SO_4^- + Ag^{++} \qquad (16\text{-}2)$$

is in agreement with experimental evidence[6-8] and appears to be the rate-determining step in peroxydisulfate oxidations. The sulfate radical ion, $\cdot SO_4^-$, and Ag(II) act as the active oxidants. The excess reagent is destroyed by boiling the solution, which results in liberation of oxygen from water. The activation energy for the decomposition decreases at low pH.[9] The decomposition is therefore autocatalytic, since HSO_4^- is a product of the decomposition. Kolthoff and Miller[10] found that the oxygen liberated comes from water in neutral solution and from peroxydisulfate at low pH. The proposed mechanism[7] for the silver-catalyzed decomposition is Reaction (16-2) followed by

$$\cdot SO_4^- + Ag^+ \rightarrow SO_4^= + Ag^{++} \qquad (16\text{-}3)$$

The Ag^{++} decomposes by reaction with water with the liberation of oxygen [see **Solid oxidants,** *silver(II) oxide*].

   Applications of the silver-catalyzed peroxydisulfate reaction are the oxidation of Ce(III) to Ce(IV),[11,12] Cr(III) to dichromate, W(V) to W(VI), and V(IV) to V(V).[12] In the oxidation of Cr(II) to Cr(III), the first step is formation[13] of $CrSO_4^+$ and $\cdot SO_4^-$. The $\cdot SO_4^-$ then reacts with Cr(II) in a second step. The oxidation of Mn(II) to permanganate is carried out in nitric or sulfuric acid solution containing orthophosphoric[14] or metaphosphoric[15] acid, which is added to prevent precipitation of manganese dioxide. The permanganate is titrated with Fe(II),[15] arsenic(III),[14] or a mixture of arsenic(III) and nitrite.[16] Arsenic(III) has the advantage over Fe(II) of permitting the selective reduction of permanganate in the presence of Cr(VI) or V(V), but suffers from the disadvantage that the permanganate is reduced only to an effective oxidation state of manganese of 3.3. By use of the arsenic(III)-nitrite mixture, permanganate is reduced to Mn(II), and the color change improved. Hillson[17] retarded decomposition of the permanganate by carrying out the oxidation in a buffered phosphate solution. After acidification the permanganate was titrated with arsenic(III) with osmium tetroxide as the catalyst. Under these conditions a stoichiometric reduction was achieved.

Permanganate has been used in the selective oxidation[18] of Ce(III) in the presence of fluoride or pyrophosphate. It has also been used in the selective oxidation of V(IV) to V(V) in the presence of Cr(III). The V(V) is then titrated with Fe(II). Cr(III) is oxidized slowly in acid solutions[19] and rapidly in alkaline solutions[20] to form Cr(VI). The excess permanganate can be reduced by adding sodium azide, which is oxidized to nitrogen. The excess hydrazoic acid is destroyed by boiling. Alternatively, the excess permanganate may be removed by adding sodium nitrite; the excess nitrite then is destroyed by adding urea.[21]

Periodate is used in the oxidation of Mn(II) to permanganate:

$$2Mn^{++} + 5IO_4^- + 3H_2O \rightarrow 2MnO_4^- + 5IO_3^- + 6H^+ \qquad (16\text{-}4)$$

The mechanism of this reaction has been found[22,23] to be autocatalytic. The rate is proportional to the concentrations of Mn(II) and periodate and also increases with permanganate concentration. In perchloric acid solution a violet-colored complex of Mn(III) is formed as an intermediate. The proposed mechanism may be written

$$Mn(II) + MnO_4^- \rightleftharpoons Mn(III) + MnO_4^= \qquad (16\text{-}5)$$

$$Mn(II) + MnO_4^= \rightarrow 2Mn(IV) \qquad (16\text{-}6)$$

$$Mn(IV) + Mn(II) \rightarrow 2Mn(III) \qquad (16\text{-}7)$$

$$Mn(III) + 2IO_4^- \rightarrow MnO_4^- + 2IO_3^- \qquad (16\text{-}8)$$

According to this mechanism, Mn(IV) cannot be oxidized unless it is first reduced by Mn(II), which is being removed. Therefore, some manganese may be trapped as Mn(IV) if its concentration ever exceeds that of Mn(II). For this reason, in the colorimetric method of Willard and Greathouse[24] the use of standard samples is necessary for color comparison.

By precipitating the excess periodate and most of the iodate as the Hg(II) salts $Hg_5(IO_6)_2$ and $Hg(IO_3)_2$, Willard and Thompson[25] were able to determine the permanganate by filtering the solution into excess standard Fe(II), which was back-titrated. Perhaps significant is that the addition of phosphoric acid is recommended to prevent the formation of manganese dioxide, particularly when larger amounts of manganese are to be oxidized. Phosphoric acid is not necessary for the oxidation, but undoubtedly plays a role as a complexing agent for Mn(III) and Mn(IV).

Perchloric acid is an effective oxidant (Section 16-3), when hot and concentrated, with the effectiveness improving as acid concentration and temperature increase.[26] It has long been used as an oxidant for chromium and vanadium[27] in steels, for the rapid dissolution of sulfide ores,[28] and in the wet oxidation of organic matter. On cooling and dilution the acid no longer has oxidizing properties. Smith[29] stressed the need for rapid cooling to avoid partial reduction of chromium, which he attributed to hydrogen peroxide formed from the hot, concentrated acid. After cooling and dilution, the solution should be boiled to remove chlorine before the determination is carried out. Knoeck and Diehl[30] quantitatively oxidized Mn(II) to Mn(III) by boiling with a 1:1 mixture of perchloric and phosphoric acids. The Mn(III) is stabilized by formation of a pyrophosphate complex,[31] $Mn(H_2P_2O_7)_3{}^{3-}$. Hayes, Diehl, and Smith[32] used perchloric acid to oxidize iodine directly to iodic acid.

Hydrogen peroxide is an efficient oxidizing agent (Section 16-3), particularly in alkaline solution. The excess peroxide is usually decomposed by boiling the alkaline solution; the process is hastened by a number of catalysts, including nickel salts, iodide, and platinum black. Schulek and Szakacs[33] removed the excess with chlorine water and then added potassium cyanide to destroy the excess chlorine. Examples of oxidation reactions are: the oxidation of Cr(III) to chromate in $2\,M$ sodium hydroxide,[33,34] Co(II) to Co(III) in bicarbonate solution,[35] Mn(II) to Mn(IV) in the presence of tellurate,[36] and Fe(II) to Fe(III) followed by titration with ascorbic acid.[37] *Sodium peroxide*, an even more vigorous oxidant, is applied in alkaline fusions. The fusion of chromite ore to form chromate has been critically studied,[38]

$$2Fe(CrO_2)_2 + 7O_2^= \rightleftharpoons 4CrO_4^= + Fe_2O_3 + 3O^= \qquad (16\text{-}9)$$

especially as to methods of decomposing the excess peroxide. Treatment with ammonium peroxydisulfate in acid solution in the presence of silver ion as a catalyst, or use of permanganate after acidification, is effective in destroying the excess peroxide. Permanganate is usually more convenient because manganese in the sample undergoes oxidation to manganate during the fusion. On acidification, manganate disproportionates to form permanganate and manganese dioxide. The latter largely escapes oxidation with peroxydisulfate, so that it is necessary to remove it in any case. Boiling with hydrochloric acid proved successful in removing excess of both permanganate and manganese dioxide.

Potassium dichromate has been used for some interesting oxidation reactions induced by the dichromate-arsenic(III) reaction. The mechanism is discussed in Section 15-8. Examples are the oxidation of Mn(II) to Mn(III) in the presence of fluoride[39] or metaphosphate[40] and of V(IV) to V(V) and Ce(III) to Ce(IV) in the presence of metaphosphate.[41] Since an excess of arsenic(III) is used, there is no excess dichromate to be removed. This method is of particular importance when the interference of chromium is to be avoided.

Solid oxidants   *Sodium bismuthate* is such a strong oxidant that it will oxidize Mn(II) to permanganate[42,43] at room temperature. Ce(III) is oxidized quantitatively to Ce(IV) in sulfuric acid solution,[44] and the excess bismuthate removed by filtration. Silver-catalyzed peroxydisulfate oxidation has largely supplanted the bismuthate method, thus avoiding the inconvenient filtration step required with solid oxidants.

   Lead dioxide is a moderately strong oxidant ($E° = 1.46$ V in acid solution) that can be used for oxidations such as that of Mn(II) to Mn(III) in the presence of pyrophosphate.[31] Under the same conditions, Cr(III), V(IV), and Ce(III) undergo oxidation. Here also, the excess is removed by filtration.

   The higher oxidation states of silver have been reviewed by McMillan.[45] *Silver(II) oxide* has been recommended[46] as an oxidant for Mn(II) to permanganate, Cr(III) to dichromate, and Ce(III) to Ce(IV) in acid solutions. It is an oxidant of considerable analytical potential. The oxide dissolves in nitric, perchloric, or sulfuric acids, forming Ag^{++}, which is such a powerful oxidant[47] ($E°'_{Ag^{++},Ag^{+}} = 2.0$ V in 4 M perchloric acid) that the above oxidation reactions occur rapidly at room temperature.

   The excess oxidant is removed by warming for a few minutes. Ag(II) undergoes reduction by the solvent at a rate proportional to the square of the Ag(II) concentration, and inversely proportional to the Ag(I) concentration.[48] The decomposition is believed[49,50] to proceed in a series of reactions involving Ag(III), H_2O_2, HO_2, and a Ag^{++}-HO_2 complex.[51] Alternatively, the solution can be titrated potentiometrically with Fe(II); two inflections are observed, the first due to reduction of excess Ag^{++} and the second to the reduction of permanganate, dichromate, or Ce(IV).

   Sodium perxenate, first prepared in 1963,[52] has a standard electrode potential estimated to be 3.0 V in acid solution. Without a catalyst and at room temperature it rapidly oxidizes[53] Mn(II) to Mn(VII). In neutral solution it oxidizes Np(IV) and Pu(IV) hydroxides to Np(VI) and Pu(VI). Sodium perxenate decomposes in acid solution:

$$2HXeO_6{}^{3-} + 6H^+ \rightarrow 2XeO_3 + O_2 + 4H_2O \qquad (16\text{-}10)$$

Xenon trioxide has been used[54] for the room-temperature oxidation of traces of alcohols.

OTHER OXIDANTS   Pickering[55] reviewed the use of solid oxidants in analysis. He included *Cu(II) oxide* for Dumas nitrogen, *iodine pentoxide* for oxidation and determination of carbon monoxide, and *Mn(IV) oxide* at 350°C for gas-solid reactions.

16-2   REDUCING AGENTS

Gaseous reductants   *Hydrogen sulfide* was formerly a popular reductant for Fe(III) because of its selectivity. The removal of the excess requires lengthy boiling.

Sulfur dioxide is a mild reducing agent:

$$SO_4^= + 4H^+ + 2e^- \rightleftharpoons H_2SO_3 + H_2O \qquad E° = 0.17 \text{ V} \qquad (16\text{-}11)$$

The mechanism of the reduction of Fe(III) is somewhat complicated,[56] involving the intermediate formation of a red material that appears to be a sulfite complex of Fe(III). The reaction rate is slow in the presence of excess sulfuric acid. It is accelerated by the presence of thiocyanate, which apparently replaces part of the coordinated sulfite in the intermediate. The reduction then proceeds quantitatively even in 1 M sulfuric acid when the cold solution is treated with sulfur dioxide and slowly heated to boiling. The excess reductant is removed with a stream of carbon dioxide.

Sulfur dioxide has also been used for the reduction of As(V) to As(III),[57] Sb(V) to Sb(III), Se(IV) and Te(IV) to the elements, Cu(II) to Cu(I) in the presence of thiocyanate, and V(V) to V(IV).

Homogeneous reductants   *Tin(II) chloride* rapidly reduces Fe(III) to Fe(II) in hot hydrochloric acid solution. The rate of the reaction was found[58] to be proportional to the concentrations of Fe(III) and Sn(II) and to increase rapidly with chloride ion concentration. A reevaluation[59] of the kinetic data led to the expression

$$k_2 = 8.7 \times 10^4 C^4 + 26 \times 10^4 C^5$$

where k_2 is the second-order rate constant in liters per mole per minute and C is the concentration of chloride ion not present as complexes of iron or tin. Except at the highest chloride concentrations, the rate was found to be proportional to the fourth power of chloride concentration. This may be taken to mean that the rate-determining step involves a total of four chloride ions coordinated to Fe(III) and Sn(II), that is, a reaction between Fe^{3+} and $SnCl_4^=$ or between $FeCl^{++}$ and $SnCl_3^-$. At higher chloride concentrations five chloride ions may be involved. From the analytical viewpoint it is important to keep the Fe(III) and hydrochloric acid concentrations high if quantitative reduction is to be achieved with only a small excess of Sn(II).

The excess Sn(II) is destroyed by adding quickly an excess of Hg(II) chloride:

$$Sn(II) + 2HgCl_2(\text{excess}) \rightarrow Sn(IV) + Hg_2Cl_2 + 2Cl^- \qquad (16\text{-}12)$$

Excessive Sn(II) chloride must be avoided, because a local excess causes the formation of metallic mercury,

$$Sn(II)(\text{excess}) + HgCl_2 \rightarrow Sn(IV) + Hg + 2Cl^- \qquad (16\text{-}13)$$

which will react with the oxidant. A large amount of Hg(I) chloride, resulting from a large excess of Sn(II) chloride, also is harmful in that it reacts to some extent with the oxidant. In any case, the titration should be completed without undue delay because Fe(III) slowly oxidizes Hg(I) chloride. During a normal titration the amount of interaction is negligible.

Wehber[60] used cacotheline as an indicator to detect the excess of Sn(II). Dichromate was then added, first to oxidize the excess Sn(II), which is evidenced by the change in the color of the cacotheline, and then to titrate the Fe(II) as usual, using diphenylaminesulfonate. Hume and Kolthoff[61] suggested the same indicator, but used Ce(IV) as titrant.

Tin(II) chloride has been used to reduce Mo(VI) to Mo(V)[62] and As(V) to As(III).[63] Main[64] used it for the reduction of U(VI) to U(IV), with Fe(III) chloride as a catalyst, and destroyed the excess as usual with Hg(II) chloride. An excess of Fe(III) chloride was added, and the resulting Fe(II) titrated with dichromate.

Chromium(II) chloride, a strong reducing agent ($E^0_{Cr^{3+},Cr^{++}} = -0.41$ V), was used[65] to reduce U(VI) to U(IV). The excess was removed by air oxidation, with phenosafranine as an indicator. This dye is reduced to a colorless compound by Cr(II). Upon air oxidation the indicator turns pink. Shatko[66] used Cr(II) to reduce As(V) to the element.

Iron(II) in concentrated sulfuric acid solution at a high temperature can be used as a reductant for perchlorate.[67]

Hydrazine, which is conveniently added as the hydrochloride, the sulfate, or the hydrate, is an interesting reductant in that its oxidation product, nitrogen, is innocuous. Sloviter, McNabb, and Wagner[68] reduced $HgI_4^=$ in alkaline solution with hydrazine sulfate, collected the mercury, and determined it bromometrically. Schulek and von Villecz[69] determined arsenic in organic compounds by destroying the organic matter and reducing the As(V) acid to As(III) acid with hydrazine sulfate, the excess being destroyed by heating with concentrated sulfuric acid. The As(III) was titrated with bromate. Arsenic and antimony can be reduced with hydrazine sulfate to the trivalent state and separated by distilling the arsenic as the trichloride.[70] Both distillate and residue are treated with bromate.

Bengtsson[71] studied the kinetics of the reduction of V(V) in perchloric acid solutions and concluded that a complex is formed as an intermediate with hydrazine or hydroxylamine as the reductant.

Hydroiodic acid, containing hypophosphorous acid, can be used to reduce sulfates quantitatively to hydrogen sulfide, which then can be distilled and determined by the iodine method.[72]

Hypophosphorous acid and *hypophosphites* are strong reducing agents. As(V) and As(III) are reduced to elemental arsenic,[73,74] Fe(III) to Fe(II),[75] Sn(IV) to (II),[76] Ge(IV) to (II),[77] and Se(IV) and Te(IV) to the elements.[78]

Metallic reductants   Metals are used as reducing agents in several forms (including sheet, wire, powder, and shot), as reductor columns and as liquid amalgams. Two important considerations in the choice of reductant are the selectivity of the reducing action, determined largely by the electrode potential of the metal–metal ion couple, and the method of removing excess reducing agent, determined largely by the physical form of the metal. Occasionally, as with powdered aluminum,[79] the excess is simply dissolved in acid. More often it is physically separated by filtration or by the use of reductor columns or liquid amalgams.

Reductor columns have been prepared from zinc, silver, lead, cadmium, bismuth, antimony, nickel, copper, tin, and iron.

The *zinc reductor*, commonly known as the *Jones*[80] *reductor*, generally is prepared from amalgamated zinc. The addition of mercury does not affect the standard potential of the Zn^{++}–Zn couple[81] (-0.7628 V) as long as solid zinc is present. The rate of reduction, however, depends on the concentration of zinc at the surface of the amalgam.[82] With relatively strong oxidants, such as Fe(III) and Ce(IV), which are reduced by mercury, a mercury content of 1 or even 5% may be used at high acid concentrations to control the rate of hydrogen evolution. With weaker oxidants the mercury content should be minimized so that the reduction reaction is not retarded.

Zinc is relatively nonselective as a reducing agent because of its strongly negative potential. Examination of a table of electrode potentials reveals that the following reductions are to be expected for some of the common metals: Fe(III) → Fe(II); Cr(III) → Cr(II); Ti(IV) → Ti(III); V(V) → V(II). The reduction of U(VI) to U(III) is only partially complete, as would be expected from the highly reducing character of the U(IV)–U(III) couple ($E° = -0.61$ V). Commonly, air oxidation to U(IV) is carried out before the titration to U(VI). It might be expected from their standard potentials ($E° = -0.44$, -0.28, and -0.25 V) that Fe(II), Co(II), and Ni(II) would be displaced as the metals. These reduction processes are highly irreversible, however, and do not proceed at appreciable rates. The more noble metals (Cu, Ag, Hg, Sb, Bi) are displaced from solution.

There have been reports of the formation of hydrogen peroxide by the reduction of atmospheric oxygen. Burdick[83] found peroxide only when water, not acid, was used as solvent in the zinc reductor. Lundell and Knowles[84] showed that hydrogen peroxide is destroyed by zinc reduction in acid solution. On the other hand, Sill and Peterson[85] found hydrogen peroxide, especially when the zinc was heavily amalgamated and air bubbles were passed rapidly through the column. With a lead reductor much greater amounts of peroxide were found. Traces of hydrogen peroxide were also noted when a silver reductor was used in the presence of air. Chalmers, Edmond, and Moser[86] showed that, when a zinc reductor (or other two-phase reductor) is used in the presence of air, a steady-state concentration of hydrogen peroxide is built up. This peroxide leads to a negative error of about 1% for iron determinations.

Particularly when small amounts of iron are being determined, the dissolved air should be removed with hydrogen[87] or carbon dioxide.[88]

The *silver reductor*, often called the *Walden reductor*,[89] is much more selective than zinc as a reducing agent. A hydrochloric acid solution is always used. The electrode potential varies with the concentration of chloride; therefore the acid concentration is more critical than with the zinc reductor. From

$$AgCl + e^- \rightleftharpoons Ag + Cl^- \qquad E° = 0.2223 \text{ V} \qquad (16\text{-}14)$$

it can be inferred that, in 1 M hydrochloric acid ($a_{Cl^-} \simeq 1$), the reductions Fe(III) → Fe(II) and U(VI) → U(IV) should be quantitative, whereas Cr(III) and Ti(IV) should not be reduced. These expectations are fully realized. From the standard potential of the V(IV)-V(III) couple (0.34 V), considerable reduction of vanadium to the trivalent state would be expected. Lingane and Meites[90] found that little V(III) is formed because the reaction is slow. For some reductions the product depends on the acid concentration. Mo(VI) can be reduced to either Mo(V) or Mo(III), depending on acid concentration and temperature.[91,92] For copper a high acid concentration favors formation of the soluble $CuCl_2^-$ rather than insoluble CuCl:

$$CuCl + Cl^- \rightleftharpoons CuCl_2^- \qquad K = 6.5 \times 10^{-2} \qquad (16\text{-}15)$$

The potential of the Cu(II)-Cu(I) couple

$$Cu^{++} + e^- + 2Cl^- \rightleftharpoons CuCl_2^- \qquad E° = 0.464 \text{ V} \qquad (16\text{-}16)$$

is increased by increasing the chloride concentration, as would be expected from the Nernst equation. Birnbaum and Edmonds[93] recommended reduction in 2 M hydrochloric acid and passing the reduced solution directly into a solution of Fe(III) to avoid air oxidation. For the reduction of U(VI) to U(IV) the acid concentration was 4 M, and the temperature was raised to 60 to 90°C.

In bromide solutions the formal potential of the silver reductor is 0.13 V lower,[94] and reductions take place more readily. It is used with 0.1 to 4 M hydrogen bromide.

The *lead reductor* ($E° = -0.126$ V) was proposed by Treadwell.[95] If hydrochloric acid more concentrated than 2.5 M is used, lead sulfate films do not form[96] even if highly concentrated sulfuric acid is present. The most important application of the lead reductor is in the reduction of U(VI) to U(IV).[85,96,97] Its advantages over the zinc reductor lie mainly in its ability to reduce uranium to a definite oxidation state and in avoidance of certain interferences.

The *cadmium reductor* has been used by Treadwell[95,98] as a substitute for the zinc reductor. One application was the reduction of chlorate to chloride. Perchlorate was reduced to chloride only in the presence of a small amount of titanium ion as a catalyst.

Mercury can be used as a reducing agent of about the same potential as silver in the presence of chloride ions. Furman and Murray[99] showed that hydrogen peroxide is formed by reduction of oxygen in the presence of chloride and therefore recommended rigorous exclusion of air. In 2 to 3.5 M hydrochloric acid the reductions $Mo(VI) \rightarrow Mo(V)$, $Fe(III) \rightarrow Fe(II)$, $V(V) \rightarrow V(IV)$, and $Sb(V) \rightarrow Sb(III)$ were performed.

The idea of using a *low-melting alloy*, which can be removed from solution after solidification, has been studied by several.[100-102]

Finely divided *copper*[103] has been used as a reducing agent for Fe(III). Activated copper,[104] prepared by reducing copper(II) oxide with hydrogen, has been used to displace cadmium but not zinc from a cyanide solution. In cyanide solution, copper is a strong reducing agent ($E^{\circ\prime} = -1.09$ V). Its formal potential lies between those of zinc (-1.26 V) and cadmium (-0.90 V) and several heavy metals (Pb, Bi, Sn, Ag, Hg) are displaced from solution.

Nickel shot has been used to reduce Fe(III) selectively in the presence of Sn(IV).[105] After the Fe(II) is titrated, the nickel is returned to the vessel, and Sn(IV) is reduced upon heating. After filtration the resulting Sn(II) is titrated with iodine.

Liquid amalgams For the liquid amalgams of zinc, cadmium, lead, and bismuth, the potentials do not differ widely from those of solid metals, for the following reasons.[106,107] If solid metal is added to mercury, a solid phase eventually is formed, and at equilibrium both phases of the saturated, two-phase amalgam must have the same electrode potential. The emf of the cell

$$M(\text{solid}) \,|M^{n+}|\, M(Hg) \qquad \text{two-phase}$$

depends on the free energy of interaction between the metal M and mercury and is small for most heavy metals (0.003 V for thallium, 0.006 V for lead, 0.051 V for cadmium, 0.000 V for zinc). The much greater emf of the alkali metals (0.780 V for sodium, 1.001 V for potassium[106]) indicates that the amalgams are much weaker reducing agents than the pure metals. The potential varies as the liquid amalgam is diluted, in accordance with the Nernst equation for a concentration cell. But the potential change on dilution is only $0.0591/n$ V for a 10-fold change in concentration (assuming activity coefficients of unity), which is negligible from the analytical viewpoint.

Applications of liquid amalgams have been suggested[108] in conjunction with EDTA titrations of metal ions. A liquid amalgam can be used essentially as a controlled-potential reductor. The amount of metal ion entering solution from the amalgam is determined by EDTA titration, which in effect substitutes for a coulometer in integrating the total amount of reduction that has occurred. The principle can be applied to both inorganic and organic systems. For example, *p*-nitrophenol was reduced to *p*-aminophenol by shaking a deoxygenated solution in an acetate buffer for 5 min with a liquid zinc amalgam. The Zn(II) produced, as shown by titration with EDTA, corresponded to 5.96 electrons per molecule of *p*-nitrophenol compared with the theoretical value of 6.

Smith and Kurtz[109] devised a titration flask with a raised bottom and a gas-entry tube. By addition of carbon tetrachloride or chloroform the reduced solution was separated from the liquid amalgam, and the titration could be carried out without draining off the amalgam. An inert atmosphere could be introduced through the gas-entry tube.

The most important liquid amalgams are those of lead and bismuth, though several other metals such as zinc, cadmium, and tin may be used.

Lead amalgam ($E = -0.13$ V) is best used in relatively high concentrations of hydrochloric acid to prevent undue separation of lead chloride.

Bismuth amalgam is a weaker reducing agent ($E = 0.32$ V), reducing V(V) to only V(IV). Mo(VI) is reduced to Mo(V) or Mo(III), depending on the acidity. W(VI) is reduced to W(V).

Sodium amalgam is a strong reducing agent ($E \simeq -1.9$ V) of interest mainly because it accomplishes rapid reduction at room temperature and does not introduce heavy metal cations into solution.

Cadmium amalgam has been studied[110] as a reductant for perchlorate, with molybdenum as catalyst. The catalytic cycle involves Mo(V) and Mo(VI), in which a Mo(V) dimer is oxidized by ClO_4^-.

16-3 OXIDATION OF ORGANIC MATTER

An analysis often must be made for an element in an organic compound or for a substance present with an organic matrix. The removal of the organic-matter interference frequently entails its complete destruction. Several reviews of methods have been made.[111-114] Removal of organic matter can be brought about by either wet or dry ashing.

In the separation of organic interferences several sources of potential error should be recognized. In wet ashing, large amounts of oxidant may be needed, and nonvolatile impurities in the reagents must be negligible or innocuous. If a precipitate forms during digestion, coprecipitation may be serious enough to take up some wanted materials. Although temperatures are normally lower in wet ashing, volatilization still may be a problem with mercury, ruthenium, and osmium under oxidizing conditions, selenium under reducing conditions, and volatile chlorides such as those of germanium and arsenic in chloride-containing samples. Silica from reaction vessels may combine with heavy metals[115] to form silicates that are difficult to bring into solution.

Wet ashing   *Iron(II)-H_2O_2, Fenton's reagent*, is an oxidant that acts through an induced chain mechanism involving hydroxyl radicals (Section 15-8).[116] It is nearly ideal as an oxidant for the destruction of organic matter because it can be used under mild experimental conditions, it is a reagent of high purity, and the decomposition products H_2O and O_2 are benign. At a concentration of 50%, H_2O_2 must be handled with adequate safety precautions.[117] With 50% H_2O_2 the rapid destruction of large amounts of even highly refractory organic matter can be brought about, and the reagent is especially suitable for biological materials.[118] Degradation of organic material to small volatile molecules probably proceeds by way of a series of steps involving formation of hydroperoxides,

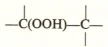

followed by cleavage of the carbon–carbon bond to give one molecule with a ketone and another with an alcohol functional group. H_2O_2 is sometimes used after treatment of a sample with sulfuric acid.[112]

Perchloric acid, alone or in mixtures with other reagents such as nitric, hydrochloric, or sulfuric acids, is extensively used for wet ashing.[119] Vanadium is a catalyst for the oxidations, evidently fluctuating between V(IV) and V(V) in the catalytic cycle. Cr(III) can act both as a catalyst and as an indicator to show when oxidation is complete. If dilute perchloric acid is heated in the absence of reducing agents, it is concentrated gradually to form an azeotrope (the dihydrate) containing 72% perchloric acid and having a boiling point of 203°C. Thus anhydrous perchloric acid, which is so unstable that it explodes spontaneously upon standing except at low temperatures, cannot be formed by simple boiling. Concentrated perchloric acid, however, when heated with hydroxyl-containing compounds, notably ethanol, cellulose, or polyhydric alcohols, becomes explosively hazardous owing to the formation of perchlorate esters such as ethyl perchlorate. Therefore nitric acid is added to the cold solution, which is then gradually heated. Most of the organic matter is destroyed by the nitric acid, which eventually is volatilized from the mixture. Only small quantities of organic matter should remain unoxidized when the last of the nitric acid has been expelled and the perchloric acid begins to become more concentrated with a corresponding rise in boiling point. Perchloric acid should be used only with adequate knowledge of its properties.[119] For details the original literature should be consulted, and the recommended procedures followed carefully.

When *sulfuric acid* is used as the primary oxidizing-dehydrating reagent, about 10 to 20 g is required for each gram of organic material, with sulfur dioxide as the main reduction product. In the Kjeldahl method for nitrogen, sulfuric acid is used to destroy organic matter, its boiling point being raised by the addition of potassium sulfate. The rate of reaction is speeded by catalysts such as selenium and mercury.

The disadvantages of sulfuric acid are the formation of insoluble sulfates and the difficulty in removal of excess acid.

Nitric acid is commonly used as the primary oxidant of organic matter. Although the excess of concentrated acid is removed easily by heating (bp 120°C), for this reason it cannot be used at higher temperatures except in conjunction with sulfuric acid. When nitration of organic matter occurs, nitro compounds are formed that are difficult to decompose. The rapid destruction of plant material through vapor-phase oxidation has been described.[120]

Dry ashing This procedure normally involves evaporating water and other volatile materials followed by heating to a high temperature in the presence of excess air or oxygen. When volatile materials of interest would be removed along with the oxidation products, closed systems such as those employing a combustion train with traps are appropriate. When little of critical importance is lost through volatilization, open systems are conventionally used, with a final temperature of about 500°C.

Low-temperature dry ashing can be accomplished by use of reactive oxygen[121] that has been electrically excited by a 300-W, 13.56-MHz radiofrequency oscillator. The metastable oxygen produced has enough energy to rupture all carbon-carbon and carbon-hydrogen bonds. High recovery rates have been obtained with samples containing such elements as mercury, selenium, lead, arsenic, and iodine.

16-4 REDOX RESINS

Redox resins were originated by Cassidy, and he reviewed the early developments.[122] They can be regarded as analogous to ion-exchange polymers (Section 25-1) except that electrons rather than ions are interchanged. Although redox resins are interesting in principle, few applications appear promising. Probably the most promising type of application is the quantitative oxidation or reduction of trace constituents without the introduction of oxidation or reduction products of the reagent.

Equilibrium often is attained slowly, and the potential varies with the ratio of the amounts of resin in the oxidized and reduced forms. In other words, not every site is equally strongly oxidizing or reducing.

Manecke[123] described the preparation and properties of several resins prepared by the condensation of hydroquinone, pyrogallol, resorcinol, or pyrocatechol with phenol and formaldehyde.

Another type of redox resin comprises those obtained by loading cationic resins[124] with ions such as Fe(III), Cu(II), or Sn(II) or by loading anionic resins with methylene blue, sulfite, or hydroquinone. Cerrai and Testa[125] prepared columns from Kel-F powder (a porous organic material) and tetrachlorohydroquinone with a redox capacity of 1.6 meq/g. An anion-exchange resin in $3 M$ hydrochloric acid strongly retains both Sn(II) and Sn(IV) (Figure 25-1). An anion column containing Sn(II) can be used as a reductor.[126]

REFERENCES

1   H. H. WILLARD and L. L. MERRITT, JR.: *Ind. Eng. Chem., Anal. Ed.*, **14**:486, 489 (1942).

2   First published by J. VON WESZELSZKY: *Z. Anal. Chem.*, **39**:81 (1900).

3   J. F. SADUSK, JR., and E. G. BALL: *Ind. Eng. Chem., Anal. Ed.*, **5**:386 (1933).

4   J. PROSZT: *Z. Anal. Chem.*, **73**:401 (1928).

5   J. F. REITH and K. W. GERRITSMA: *Rec. Trav. Chim. Pays-Bas*, **65**:770 (1946).

6   C. E. H. BAWN and D. MARGERISON: *Trans. Faraday Soc.*, **51**:925 (1955).

7   S. FRONAEUS and C. O. OSTMAN: *Acta Chem. Scand.*, **9**:902 (1955); **10**:320 (1956).

8   D. A. HOUSE: *Chem. Rev.*, **62**:185 (1962).

9   E. HAKOILA: *Ann. Univ. Turku., Ser. A.*, **66**:51p (1963).

10   I. M. KOLTHOFF and I. K. MILLER: *J. Amer. Chem. Soc.*, **73**:3055 (1951).

11   N. H. FURMAN: *J. Amer. Chem. Soc.*, **50**:755 (1928).

12   H. H. WILLARD and P. YOUNG: *J. Amer. Chem. Soc.*, **50**:1379 (1928); *Ind. Eng. Chem., Anal. Ed.*, **4**:187 (1932); **5**:154, 158 (1933).

13   D. E. PENNINGTON and A. HAIM: *J. Amer. Chem. Soc.*, **90**:3700 (1968).

14   H. A. BRIGHT and C. P. LARRABEE: *J. Res. Nat. Bur. Stand.*, **3**:573 (1929).

15   R. LANG and F. KURTZ: *Z. Anal. Chem.*, **85**:181 (1931).

16   E. B. SANDELL, I. M. KOLTHOFF, and J. J. LINGANE: *Ind. Eng. Chem., Anal. Ed.*, **7**:256 (1935).

17   H. D. HILLSON: *Ind. Eng. Chem., Anal. Ed.*, **16**:560 (1944).

18   V. I. BOGOVINA, Y. I. USATENKO, and V. F. MAL'TSEV: *Zh. Anal. Khim.*, **23**:1152 (1968); *J. Anal. Chem. USSR*, **23**:1012 (1968).

19   H. A. FALES and P. S. ROLLER: *J. Amer. Chem. Soc.*, **51**:345 (1929).

20   B. REINITZER and P. CONRATH: *Z. Anal. Chem.*, **68**:81 (1926).

21   R. LANG and F. KURTZ: *Z. Anal. Chem.*, **86**:288 (1931).

22   J. D. H. STRICKLAND and G. SPICER: *Anal. Chim. Acta*, **3**:517 (1949).

23   G. R. WATERBURY, A. M. HAYES, and D. S. MARTIN, JR.: *J. Amer. Chem. Soc.*, **74**:15 (1952).

24   H. H. WILLARD and L. H. GREATHOUSE: *J. Amer. Chem. Soc.*, **39**:2366 (1917).

25   H. H. WILLARD and J. J. THOMPSON: *Ind. Eng. Chem., Anal. Ed.*, **3**:399 (1931).

26   G. F. SMITH: *Talanta*, **15**:489 (1968).

27   H. H. WILLARD and R. C. GIBSON: *Ind. Eng. Chem., Anal. Ed.*, **3**:88 (1931).

28   W. C. HOYLE and H. DIEHL: *Talanta*, **18**:1072 (1971).

29   G. F. SMITH: *Analyst*, **80**:16 (1955).

30   J. KNOECK and H. DIEHL: *Talanta*, **14**:1083 (1967).

31   I. M. KOLTHOFF and J. I. WATTERS: *Ind. Eng. Chem., Anal. Ed.*, **15**:8 (1943).

32   J. M. HAYES, H. DIEHL, and G. F. SMITH: *Talanta*, **13**:1019 (1966).

33   E. SCHULEK and M. SZAKACS: *Acta Chim. Acad. Sci. Hung.*, **4**:457 (1954).

34   F. FEIGL, K. KLANFER, and L. WEIDENFELD: *Z. Anal. Chem.*, **80**:5 (1930).

35   H. A. LAITINEN and L. W. BURDETT: *Anal. Chem.*, **23**:1268 (1951).

36   I. M. ISSA and I. F. HEWAIDY: *Chemist-Analyst*, **44**:70 (1955).

37   L. ERDEY and E. BODOR: *Anal. Chem.*, **24**:418 (1952).

38   P. J. HARDWICK: *Analyst*, **75**:9 (1950); F. J. BRYANT and P. J. HARDWICK: *Analyst*, **75**:12 (1950).

39   R. LANG and F. KURTZ: *Z. Anorg. Allg. Chem.*, **161**:111 (1929); R. LANG: *Z. Anorg. Allg. Chem.*, **170**:387 (1928); R. LANG and J. ZWERINA: *Z. Anorg. Allg. Chem.*, **170**:389 (1928).

40   R. LANG: *Z. Anal. Chem.*, **102**:8 (1935).

41   R. LANG and E. FAUDE: *Z. Anal. Chem.*, **108**:181 (1937).

42   W. BLUM: *J. Amer. Chem. Soc.*, **34**:1379 (1912).

43   G. E. F. LUNDELL: *J. Amer. Chem. Soc.*, **45**:2600 (1923).

44   F. J. METZGER: *J. Amer. Chem. Soc.*, **31**:523 (1909).

45   J. A. MCMILLAN: *Chem. Rev.*, **62**:65 (1962).

46   J. J. LINGANE and D. G. DAVIS: *Anal. Chim. Acta*, **15**:201 (1956).

47   A. A. NOYES, D. DEVAULT, C. D. CORYELL, and T. J. DEAHL, *J. Amer. Chem. Soc.*, **59**:1326 (1937).

48   G. A. RECHNITZ and S. B. ZAMOCHNICK: *Talanta*, **11**:713, 1645 (1964); **12**:479 (1965).

49   J. B. KIRWIN, F. D. PEAT, F. J. PROLL, and L. H. SUTCLIFFE: *J. Phys. Chem.*, **67**:1617 (1963).

50   H. N. PO, J. H. SWINEHART, and T. L. ALLEN: *Inorg. Chem.*, **7**:244 (1968).

51   C. F. WELLS and D. MAYS: *Inorg. Nucl. Chem. Lett.*, **5**:9 (1969).

52   J. G. MALM, B. D. HOLT, and R. W. BANE: in "Noble Gas Compounds," H. H. Hyman (ed.), p. 167, University of Chicago, Chicago, 1963.

53   R. W. BANE: *Analyst*, **90**:756 (1965); **95**:722 (1970).

54   B. JASELSKIS and J. P. WARRINER: *Anal. Chem.*, **38**:563 (1966).

55   W. F. PICKERING: *Chemist-Analyst*, **53**:91 (1964).

56   F. BURRIEL-MARTI and F. L. CONDE: *Anal. Chim. Acta*, **3**:547 (1949).

57   A. KURTENACKER and I. FURSTENAU: *Z. Anorg. Allg. Chem.*, **212**:289 (1933).

58   F. R. DUKE and R. C. PINKERTON: *J. Amer. Chem. Soc.*, **73**:3045 (1951).

59   F. R. DUKE and N. C. PETERSON: *Iowa State Coll. J. Sci.*, **32**:89 (1957).

60   P. WEHBER: *Angew. Chem.*, **66**:271 (1954).

61   D. N. HUME and I. M. KOLTHOFF: *Anal. Chim. Acta*, **16**:415 (1957).

62   R. LANG and S. GOTTLIEB: *Z. Anal. Chem.*, **104**:1 (1936).

63   S. TRIBALAT: *Anal. Chim. Acta*, **1**:149 (1947).

64   A. R. MAIN: *Anal. Chem.*, **26**:1517 (1954).

65   W. D. COOKE, F. HAZEL, and W. M. MCNABB: *Anal. Chim. Acta*, **3**:656 (1949).

66   P. P. SHATKO: *Zh. Anal. Khim.*, **7**:242 (1952).

67   G. ARAVAMUDAN and V. KRISHNAN: *Talanta*, **13**:519 (1966).

68   H. A. SLOVITER, W. M. MCNABB, and E. C. WAGNER: *Ind. Eng. Chem., Anal. Ed.*, **13**:890 (1941).

69   E. SCHULEK and P. VON VILLECZ: *Z. Anal. Chem.*, **76**:81 (1929).

70   C. L. LUKE: *Ind. Eng. Chem., Anal. Ed.*, **15**:626 (1943).

71   G. BENGTSSON: *Acta Chem. Scand.*, **25**:2989 (1971).

72   C. L. LUKE: *Ind. Eng. Chem., Anal. Ed.*, **15**:602 (1943).
73   I. M. KOLTHOFF and E. AMDUR: *Ind. Eng. Chem., Anal. Ed.*, **12**:177 (1940).
74   J. HASLAM and N. T. WILKINSON: *Analyst*, **78**:390 (1953).
75   M. N. SASTRI and C. RADHAKRISHNAMURTI: *Z. Anal. Chem.*, **147**:16 (1935).
76   B. S. EVANS: *Analyst*, **56**:171 (1931); B. S. EVANS and D. G. HIGGS: *Analyst*, **69**:201 (1944).
77   B. N. IVANOV-EMIN: *Zavod. Lab.*, **13**:161 (1947); *Chem. Abstr.*, **42**: 480(1948).
78   B. S. EVANS: *Analyst*, **63**:874 (1938).
79   E. R. RIEGEL and R. D. SCHWARTZ: *Anal. Chem.*, **24**:1803 (1952); **26**:410 (1954).
80   C. JONES: *Trans. Amer. Inst. Mining Eng.*, **17**:411 (1889).
81   W. J. CLAYTON and W. C. VOSBURGH: *J. Amer. Chem. Soc.*, **58**:2093 (1936).
82   H. W. STONE and D. N. HUME: *Ind. Eng. Chem., Anal. Ed.*, **11**:598 (1939).
83   W. L. BURDICK: *J. Amer. Chem. Soc.*, **48**:1179 (1926).
84   G. E. F. LUNDELL and H. B. KNOWLES: *Ind. Eng. Chem.*, **16**:723 (1924).
85   C. W. SILL and H. E. PETERSON: *Anal. Chem.*, **24**:1175 (1952).
86   R. A. CHALMERS, D. A. EDMOND, and W. MOSER: *Anal. Chim. Acta*, **35**:404 (1966).
87   C. F. FRYLING and F. V. TOOLEY: *J. Amer. Chem. Soc.*, **58**:826 (1936).
88   C. C. MILLER and R. A. CHALMERS: *Analyst*, **77**:2 (1952).
89   G. H. WALDEN, JR., L. P. HAMMETT, and S. M. EDMONDS: *J. Amer. Chem. Soc.*, **56**:57 (1934).
90   J. J. LINGANE and L. MEITES, JR.: *J. Amer. Chem. Soc.*, **69**:277 (1947).
91   N. BIRNBAUM and G. H. WALDEN, JR.: *J. Amer. Chem. Soc.*, **60**:64 (1938).
92   J. BECKER and C. J. COETZEE: *Analyst*, **92**:166 (1967).
93   N. BIRNBAUM and S. M. EDMONDS: *Ind. Eng. Chem., Anal. Ed.*, **12**:155 (1940).
94   F. PANTANI: *Anal. Chim. Acta*, **31**:121 (1964).
95   W. D. TREADWELL: *Helv. Chim. Acta*, **5**:806 (1922).
96   W. D. COOKE, F. HAZEL, and W. M. MCNABB: *Anal. Chem.*, **22**:654 (1950).
97   T. W. STEELE: *Analyst*, **85**:55 (1960).
98   W. D. TREADWELL: *Helv. Chim. Acta*, **4**:551 (1921); **5**:732 (1922).
99   N. H. FURMAN and W. M. MURRAY, JR.: *J. Amer. Chem. Soc.*, **58**:429, 1689, 1843 (1936).
100 P. G. POPOV: *Ukr. Khim. Zh.*, **10**:428 (1935).
101 S. KANEKO and C. NEMOTO: *J. Soc. Chem. Ind. Jap.*, **35**:185 (1932).
102 R. A. EDGE and G. W. A. FOWLES: *Anal. Chim. Acta*, **32**:191 (1965).
103 K. KÜRSCHNER and K. SCHARRER: *Z. Anal. Chem.*, **68**:1 (1926).
104 A. BRYSON and S. LENZER-LOWY: *Analyst*, **78**:299 (1953); **79**:636 (1954).
105 A. C. SIMON, P. S. MILLER, J. C. EDWARDS, and F. B. CLARDY: *Ind. Eng. Chem., Anal. Ed.*, **18**:496 (1946).
106 J. J. LINGANE: *J. Amer. Chem. Soc.*, **61**:2099 (1939).
107 M. VON STACKELBERG: *Z. Elektrochem.*, **45**:466 (1939).
108 W. G. SCRIBNER and C. N. REILLEY: *Anal. Chem.*, **30**:1452 (1958).
109 G. F. SMITH and L. T. KURTZ: *Ind. Eng. Chem., Anal. Ed.*, **14**:854 (1942).
110 G. A. RECHNITZ and H. A. LAITINEN: *Anal. Chem.*, **33**:1473 (1961).
111 T. T. GORSUCH: "The Destruction of Organic Matter," Pergamon, New York, 1970.
112 E. C. DUNLOP in "Treatise on Analytical Chemistry," I. M. Kolthoff and P. J. Elving (eds.), pt. 1, vol. 2, chap. 25, Interscience, New York, 1961.
113 G. MIDDLETON and R. E. STUCKEY: *Analyst*, **78**:532 (1953).
114 ANALYTICAL METHODS COMMITTEE: *Analyst*, **85**:643 (1960).

115 E. B. SANDELL: "Colorimetric Determination of Traces of Metals," 3rd ed., p. 21, Interscience, New York, 1959.

116 N. URI: *Chem. Rev.*, **50**:375 (1952).

117 R. P. TAUBINGER and J. R. WILSON: *Analyst*, **90**:429 (1965).

118 B. SANSONI and W. KRACKE: *Fresenius' Z. Anal. Chem.*, **243**:209 (1968); "Rapid Methods for Measuring Radioactivity in the Environment," p. 217, IAEA, Vienna, 1971.

119 G. F. SMITH: *Anal. Chim. Acta*, **8**:397 (1953); *Talanta*, **11**:633 (1964); "The Wet Chemical Oxidation of Organic Compositions," G. F. Smith Chemical, Columbus, Ohio, 1965.

120 A. D. THOMAS and L. E. SMYTHE: *Talanta*, **20**:469 (1973).

121 C. E. GLEIT and W. D. HOLLAND: *Anal. Chem.*, **34**:1454 (1962).

122 H. G. CASSIDY: "Fundamentals of Chromatography," chap. 10, Interscience, New York, 1957.

123 G. MANECKE: *Z. Elektrochem.*, **57**:189 (1953); **58**:363, 369 (1954).

124 B. SANSONI: *Naturwissenschaften*, **39**:281 (1952); B. SANSONI and W. WIEGAND: *Talanta*, **17**:973 (1970); B. SANSONI and E. BAUER-SCHREIBER: *Talanta*, **17**:987 (1970).

125 E. CERRAI and C. TESTA: *Anal. Chim. Acta*, **28**:205 (1963).

126 L. ERDEY, J. INCZEDY, and I. MARKOVITS: *Talanta*, **4**:25 (1960); J. INCZEDY: *Acta Chim. Acad. Sci. Hung.*, **27**:185 (1961).

PROBLEMS

16-1 A sample containing Fe(III) and V(V) is made up to 250 ml. A 25-ml aliquot is passed through a Zn reductor into Fe(III) solution and titrated with standard dichromate, requiring 43.21 ml of 0.01667 M solution. Another 25-ml portion is titrated after passage through a Ag reductor, requiring 23.45 ml of the dichromate. Calculate the V and Fe content in the sample.

Answer V, 0.5035 g; Fe, 0.7576 g.

16-2 Prepare a chart showing oxidation states of Fe, Ti, V, and Cr formed when amalgams of Zn, Bi, Pb, and Hg are used as reducing agents in HCl solutions of appropriate concentrations. Show in principle how the four reductants could permit analysis of a sample containing Fe, Ti, V, and Cr.

16-3 Enlarge the chart prepared in Problem 16-2 by adding the oxidation states of U, Mo, W, Cu and adding Cd amalgam and Ag as reducing agents.

16-4 Describe how to perform each of the following operations, giving the appropriate equations:

(*a*) Reduce permanganate without reducing dichromate or V(V).

(*b*) Reduce V(V) to V(IV) without reducing V(IV), Ti(IV), or Cr(III).

(*c*) Oxidize V(IV) without oxidizing Mn(II) or Cr(III).

(*d*) Oxidize Fe(II) without oxidizing Cr(III) or V(IV).

(*e*) Oxidize Mn(II) to Mn(III) using dichromate.

17

PERMANGANATE AND DICHROMATE AS OXIDANTS

Permanganate and dichromate are among our earliest titrimetric reagents;[1] permanganate was introduced in 1846 by Margueritte, and dichromate independently by Schabus and by Penny in 1850. They were used for the titration of iron(II) produced by reduction with zinc. Both oxidants are strong, $E°$ values for the half-reactions in acid solution being:

$$MnO_4^- + 8H^+ + 5e^- \rightarrow Mn^{++} + 4H_2O \qquad E° = 1.51 \text{ V} \qquad (17\text{-}1)$$

$$Cr_2O_7^= + 14H^+ + 6e^- \rightarrow 2Cr^{3+} + 7H_2O \qquad E° = 1.33 \text{ V} \qquad (17\text{-}2)$$

A prominent aspect of the chemistry of permanganate and dichromate relates to their unstable intermediate oxidation states.

17-1 HALF-REACTIONS

Permanganate Reactions (17-1) and (17-2) are irreversible half-reactions, and their potentials are not subject to direct measurement. For permanganate, by taking proper precautions (especially by using properly prepared, pure manganese dioxide), it is possible to measure the potential corresponding to the following half-reaction in acid solution:[2,3]

$$MnO_2 + 4H^+ + 2e^- \rightleftharpoons Mn^{++} + 2H_2O \qquad E^\circ = 1.23 \text{ V} \qquad (17\text{-}3)$$

In alkaline solution the reversible potential[4] of the half-reaction

$$MnO_4^- + 2H_2O + 3e^- \rightleftharpoons MnO_2 + 4OH^- \qquad E^\circ = 0.588 \text{ V} \qquad (17\text{-}4)$$

can be measured directly. The standard potential in acid solution,

$$MnO_4^- + 4H^+ + 3e^- \rightleftharpoons MnO_2 + 2H_2O \qquad E^\circ = 1.695 \text{ V} \qquad (17\text{-}5)$$

can be calculated from the pH dependence of (17-4), but cannot be measured directly because manganese dioxide catalyzes the decomposition of permanganate. The standard potential of (17-1) can be calculated by combining (17-3) and (17-5).

In neutral, slightly acidic, or moderately alkaline solution, permanganate is reduced quantitatively to manganese dioxide [(17-4) and (17-5)]. Permanganate is a much weaker oxidant in alkaline than in acid solution, but since many reducing agents likewise become stronger in alkaline solution, permanganate is an effective and versatile oxidant.

In strongly alkaline solution, permanganate is reduced to manganate:

$$MnO_4^- + e^- \rightleftharpoons MnO_4^= \qquad E^\circ = 0.564 \text{ V} \qquad (17\text{-}6)$$

Stamm[5] took advantage of the insolubility of barium manganate to stabilize the manganate ion and to increase the electrode potential.

Mn(III) in the absence of a complexing agent is a strong oxidant. Latimer estimated[6]

$$Mn^{3+} + e^- \rightleftharpoons Mn^{++} \qquad E^\circ \simeq 1.5 \text{ V} \qquad (17\text{-}7)$$

Taube[7] estimated the value $E^\circ = 1.6$ V, and Vetter and Manecke[8] determined $E^{\circ\prime} = 1.488$ V in 7.5 M sulfuric acid, in which complexes of Mn(III) are formed.

In the absence of complexing agents, Mn(III) is unstable with respect to Mn(II) and manganese dioxide:

$$2Mn^{3+} + 2H_2O \rightarrow MnO_2 + Mn^{++} + 4H^+ \qquad \Delta G^\circ = -50 \text{ kJ} \qquad (17\text{-}8)$$

In the presence of pyrophosphate in acid solution, Mn(III) is greatly stabilized, as shown by the half-reaction[9]

$$Mn(H_2P_2O_7)_3{}^{3-} + 2H^+ + e^- \rightleftharpoons Mn(H_2P_2O_7)_2{}^= + H_4P_2O_7 \qquad E^\circ \simeq 1.15 \text{ V}$$
$$(17\text{-}9)$$

Fluoride ion also exerts a strong stabilizing action on Mn(III).

In 7.5 M sulfuric acid the formal potential of the couple Mn(IV)-Mn(III) was evaluated[10] as 1.052 V; the low value indicates that relatively strong Mn(IV) complexes must be present.

Dichromate   The unstable intermediate species in dichromate reductions involve Cr(V) and Cr(IV). Westheimer[11] calculated the $E°$ values for the Cr(VI)-Cr(V) couple to be 0.62 V; for Cr(V)-Cr(III), 1.75 V; For Cr(IV)-Cr(III), 1.5 V; and for Cr(VI)-Cr(IV), 1.3 V. Kolthoff[12] estimated the potential for the Cr(IV)-Cr(III) couple to be about 2 V.

The equilibria in solution are further complicated through polymerization; a dilute acid solution of chromate contains $CrO_4^=$, $Cr_2O_7^=$, and probably also small amounts of $Cr_3O_{10}^=$ and $Cr_4O_{13}^=$. In addition, the species $HCrO_4^-$ and $HCr_2O_7^-$ are present. In an acidic dichromate solution we have the equilibria[13,14]

$$HCrO_4^- \rightleftharpoons H^+ + CrO_4^= \qquad K_2 = 3.2 \times 10^{-7} \qquad (17\text{-}10)$$

$$2HCrO_4^- \rightleftharpoons Cr_2O_7^= + H_2O \qquad K = 98 \qquad (17\text{-}11)$$

The two most striking aspects of dichromate as an oxidant are the wide variation of the formal potential, depending on the nature and concentration of acid and complexing agents, and the variety of induced oxidation reactions that would be considered thermodynamically impossible on the basis of the $E°$ value. This para-doxical behavior is due to the transitory existence of active intermediates; an example is given in Section 15-8. Cr(IV) and Cr(V) are strong oxidants and are important active intermediates in the induced reactions of Cr(VI). Cr(VI) oxidations of inorganic substances have been reviewed,[15] with particular attention to the nature of the activated complexes and the fate of intermediates. It was concluded that the reduction Cr(V) → Cr(IV) or the oxidation Cr(IV) → Cr(V) is the slow step in the many reactions with 1-equivalent reagents.

17-2   THE PERMANGANATE-OXALATE REACTION

If permanganate solution is added drop by drop to an acidic oxalate solution, the first drop is decolorized slowly. During the course of the titration the reaction rate becomes progressively more rapid, owing to the catalytic effect of the Mn(II) produced in the reaction. Elucidation of the mechanism has been the object of many studies for over a century.[16] Undoubtedly the last study is yet to be made. The present picture appears to be as follows.

If Mn(II) ion is present initially, the mechanism involves the rapid oxidation of Mn(II) in the presence of oxalate to form oxalate complexes of Mn(III) according to the equation

$$MnO_4^- + 4Mn^{++} + 5nC_2O_4^= + 8H^+ \rightarrow 5Mn(C_2O_4)_n^{(3-2n)+} + 4H_2O \qquad (17\text{-}12)$$

which takes place in several steps. The first step is the transfer of one electron from a monooxalate complex of Mn(II):

$$MnO_4^- + MnC_2O_4 \rightarrow MnO_4^= + MnC_2O_4^+ \qquad (17\text{-}13)$$

The manganate ion, $MnO_4^=$, which would disproportionate in acid solution in the absence of reducing agent, is reduced rapidly to $Mn(III)$:

$$Mn(VI) + Mn(II) \rightarrow 2Mn(IV)$$

$$Mn(IV) + Mn(II) \rightarrow 2Mn(III) \qquad (17\text{-}14)$$

No significant concentrations of $Mn(IV)$, $Mn(V)$, or $Mn(VI)$ can be detected in solutions containing both permanganate and $Mn(III)$.

$Mn(III)$ forms several oxalate complexes—$MnC_2O_4^+$ (red), $Mn(C_2O_4)_2^-$ (yellow), and $Mn(C_2O_4)_3^{3-}$ (red)—and equilibrium among them is rapidly attained. The $Mn(III)$ complex decomposes slowly[17] to give $Mn(II)$ and CO_2.

Grobler and Berg[18] showed that the rate-limiting step in the reduction of $Mn(VII)$ to $Mn(II)$ is mainly the one involving protonation of the dioxalato complex of $Mn(III)$:

$$[Mn(C_2O_4)_2(H_2O)_2]^- + H^+ \rightarrow Mn^{++} + C_2O_4^= + CO_2 + \cdot COOH + 2H_2O$$
$$(17\text{-}15)$$

The carboxyl radical (or $\cdot CO_2^-$ or $\cdot C_2O_4^-$) then reacts rapidly with permanganate:

$$MnO_4^- + \cdot COOH \rightarrow MnO_4^= + H^+ + CO_2 \qquad (17\text{-}16)$$

$Fe(III)$ can catalyze the oxidation of oxalate, with iron replacing manganese in the rate-controlling step:[19]

$$[Fe(C_2O_4)_2(H_2O)_2]^- + H^+ \rightarrow Fe^{++} + C_2O_4^= + CO_2 + \cdot COOH + 2H_2O$$
$$(17\text{-}17)$$

In this case both the $Fe(II)$ and the carboxyl radical react rapidly with excess permanganate.

If the concentration of $Mn(II)$ is low, as it is during the earliest part of the titration, manganate can be reduced by oxalate to $Mn(IV)$:

$$Mn(VI) + C_2O_4^= \rightarrow Mn(IV) + 2CO_2 \qquad (17\text{-}18)$$

This is reduced further to $Mn(III)$:

$$2Mn(IV) + C_2O_4^= \rightarrow 2Mn(III) + 2CO_2 \qquad (17\text{-}19)$$

The $Mn(III)$ then decomposes slowly according to (17-15).

The long accepted explanation of the $Mn(II)$-catalyzed decomposition[20] of $H_2C_2O_4$ to give CO and CO_2 as a source of nonstoichiometry cannot be correct, because CO could not be detected[21] in the residual gases using gas chromatography after absorption of the CO_2.

The hydrogen peroxide problem   When oxygen of the air is present, the reactions are more complicated; H_2O_2 is formed during the permanganate-oxalate titration according to the overall reaction

$$O_2 + C_2O_4^= + 2H^+ \rightarrow H_2O_2 + 2CO_2 \qquad (17\text{-}20)$$

in which 2 equivalents of H_2O_2 appear for 2 equivalents of oxalate. Thus, if no decomposition of H_2O_2 occurs, there is no noticeable result in the stoichiometry. H_2O_2 formation has been studied by many. Some of the observations of Kolthoff, Meehan, and Kimura[22] were that: no peroxide is formed if no oxygen is present; the amount of H_2O_2 formed is the same whether MnO_4^-, Ce(IV), or dichromate is the oxidant; if no Mn(II) is present, a negligible amount of H_2O_2 is formed; the amount of H_2O_2 formed increases as the rate of addition of oxidant decreases; and much more H_2O_2 is formed than corresponds to the added oxidant. They concluded that the oxidizer reacts with Mn(II) to form Mn(III), which forms the complex with oxalate. Upon decomposition, according to (17-15), the radical $\cdot CO_2^-$ (or $\cdot COOH$ or $\cdot C_2O_4^-$) is formed. It then reacts rapidly with oxygen:

$$\cdot CO_2^- + O_2 \rightarrow \cdot O_2CO_2^- \qquad (17\text{-}21)$$

and $\qquad \cdot O_2CO_2^- + Mn(II) + 2H^+ \rightarrow H_2O_2 + CO_2 + Mn(III) \qquad (17\text{-}22)$

Thus the Mn(III) is regenerated. There is less probability of the radical being terminated when its concentration is lower. Hence the amount of H_2O_2 formed increases with decreasing rate of addition of oxidizer.

The amount of H_2O_2 formed is considerably decreased in the presence of Fe(III) or Cu(II), which act as chain terminators:

$$Fe(III) + \cdot CO_2^- \rightarrow Fe(II) + CO_2 \qquad (17\text{-}23)$$

$$Cu(II) + \cdot CO_2^- \rightarrow Cu(I) + CO_2 \qquad (17\text{-}24)$$

Cu(II) is the more effective since with it the rate constant is about six times larger than with Fe(III). With addition of Fe(II) no induced formation of H_2O_2 occurs as long as it is present in excess of the oxidant. In a practical analytical application[22] the inhibiting effect of Fe(III) is used in the determination of active oxygen in samples containing MnO_2. The sample is digested with an excess of oxalic acid in 2 M H_2SO_4. The excess oxalic acid then is determined by titration with standard permanganate. High results are eliminated if Fe(III) is present during the digestion, so that the induced formation (and decomposition during digestion) of H_2O_2 is prevented.

17-3   THE IRON(II)-DICHROMATE REACTION

By far the most important analytical reaction of dichromate is the oxidation of Fe(II):

$$6Fe(II) + Cr_2O_7^= + 14H^+ \rightarrow 6Fe(III) + 2Cr(III) + 7H_2O \qquad (17\text{-}25)$$

which has been the subject of a number of mechanism studies. The Fe(II)-dichromate reaction was studied long ago by Benson[23] and more recently by Espenson and King.[24] Under conditions where ionic strength is constant and most of the chromium is present as $HCrO_4^-$, the rate of disappearance[24] of Fe(II) can be formulated

$$-\frac{d[Fe(II)]}{dt} = \frac{[H^+]^3[Fe(II)]^2(k_1[HCrO_4^-] + k_2[HCrO_4^-]^2)}{[Fe(III)]} \qquad (17\text{-}26)$$

Interpretation of the data in terms of pH is complicated by equilibria involving formation of H_2CrO_4 as well as hydrolysis of Fe(III).

The mechanism of the reaction probably involves the monomeric form as the reactive species and the equilibrium

$$Cr(VI) + Fe(II) \rightleftharpoons Cr(V) + Fe(III) \qquad (17\text{-}27)$$

The Cr(V) intermediate formed reacts slowly according to

$$Cr(V) + Fe(II) \rightarrow Cr(IV) + Fe(III) \qquad (17\text{-}28)$$

followed by fast reaction of the Cr(IV) intermediate,

$$Cr(IV) + Fe(II) \rightarrow Cr(III) + Fe(III) \qquad (17\text{-}29)$$

According to (17-26) the rate of reaction increases with the square of the Fe(II) concentration and varies *inversely* with the Fe(III) concentration. This inhibition is interpreted in terms of the formation of an Fe(III)-Cr(VI) complex, probably $FeCrO_4^+$ or, more generally, $H_nFeCrO_4^{n+1}$. The importance of the Fe(III) concentration is further illustrated by the fact that fluoride, which forms stable complexes with Fe(III), *increases* the reaction rate.[25]

The potentiometric titration curves and indicators for the Fe(II)-dichromate reaction are discussed in Section 15-3. In general, from curves such as in Figure 15-2, the dichromate potential has been found to increase with acidity, as expected. The variations with the nature of the acid, however, have not yet been explained. In particular, the dichromate potential is so low in 0.1 M perchloric acid that the potential break is barely discernible.[26] From the practical viewpoint it should be emphasized that the rate of attainment of electrode equilibrium, particularly near the end point and beyond it, becomes slower with increasing dilution. Nevertheless, the reaction itself proceeds quantitatively and reasonably rapidly even at extreme dilution. As little as 1 μg of chromium in 100 ml of solution ($\sim 10^{-7}$ M dichromate) has been successfully titrated[27] with an accuracy of $\pm 1\%$ by use of an amperometric end point.[28]

17-4   APPLICATIONS OF PERMANGANATE

Standard solution   *Pure* solutions of permanganate are surprisingly stable. Solutions have been stored without much deterioration for 1 to 3 years.[29,30] In the presence of Mn(II), however, permanganate is inherently unstable because of the *Guyard reaction*,

$$2MnO_4^- + 3Mn^{++} + 2H_2O \rightarrow 5MnO_2 + 4H^+ \qquad (17\text{-}30)$$

which is slow in acid solution.[31,32] The rate of this reaction increases with decreasing acidity until, in a neutral solution, it is essentially instantaneous. The classical Volhard method for manganese,[33] in which Mn(II) is titrated with permanganate in the presence of zinc oxide, is based on the Guyard reaction. Once solid manganese dioxide is formed, the rate of decomposition of permanganate increases enormously, by a complex heterogeneous reaction. In view of the catalytic action of solid manganese dioxide toward decomposition of permanganate, it is essential to eliminate sources of this contaminant. Prior to standardization, common practice[34] is to heat a freshly prepared solution to boiling and digest it near the boiling point for about an hour before filtering the solution through a nonreducing filtering medium such as fritted glass. Alternatively, the solution may be allowed to stand for several days at room temperature before filtration. The storage vessel should be a glass-stoppered bottle carefully freed from grease and prior deposits of manganese dioxide. Solutions should be protected from unnecessary exposure to light. Diffuse daylight causes no appreciable decomposition, but direct sunlight readily decomposes even pure solutions.[35] Acidic and alkaline solutions are less stable than neutral ones.

Although sodium oxalate is commonly used as a primary standard, its use is recommended only when oxalate is to be determined.[36] For best absolute accuracy As(III) oxide is recommended.[36] The direct titration of As(III) in acid solution does not proceed readily without a catalyst, probably because of the stabilization of Mn(III) by complex formation with As(V). With potassium iodate (1 drop of 0.0025 M solution) as catalyst, the potentiometric end point was found to coincide with the visual end point, using ferroin, to within 0.01%, and the accuracy, tested against pure potassium iodide, was within 0.02%.

Iodine monochloride has the advantages[37] of being more efficient as a catalyst than iodate and of avoiding a blank or uncertainty about final oxidation state that might be encountered in using other iodine compounds. It was shown that relatively large amounts of iodine monochloride could be added without error and that the accuracy was within 1 part in 3000.

The mechanism of iodine monochloride catalysis has been explained by Swift[38] as being due to rapid oxidation of As(III) by iodine monochloride,

$$2ICl + H_3AsO_3 + H_2O \rightarrow I_2 + H_3AsO_4 + 2H^+ + 2Cl^-$$

followed by rapid reoxidation of iodine to iodine monochloride by permanganate,

$$5I_2 + 2MnO_4^- + 10Cl^- + 16H^+ \rightarrow 10ICl + 2Mn^{++} + 8H_2O$$

Other primary standards are: pure electrolytic iron; Fe(II) ammonium sulfate, $FeSO_4 \cdot (NH_4)_2SO_4 \cdot 6H_2O$; potassium ferrocyanide, $K_4[Fe(CN)_6] \cdot 3H_2O$; and potassium iodide. Permanganate often is used without an indicator since a 2×10^{-6} M solution has a discernible pink color.

Determination of iron Permanganate can be used for titration of Fe(II) in acid solution:

$$5Fe^{++} + MnO_4^- + 8H^+ \rightarrow 5Fe^{3+} + Mn^{++} + 4H_2O \qquad (17\text{-}31)$$

In sulfuric acid solution the reaction is rapid and quantitative.[39] Phosphoric acid usually is added to decolorize the Fe(III) by formation of a colorless complex. The complex $Fe(HPO_4)^+$, of formation constant 2.3×10^9, has been found to be the species involved.[40] Permanganate may be used as its own indicator, or ferroin may be added to provide a more sensitive end point. Whether phosphoric acid is present or not, the end point may be detected potentiometrically.

In many practical procedures, titration in sulfuric acid solution is inconvenient. For example, when Sn(II) chloride or the silver reductor is used in the preliminary reduction of Fe(III) to Fe(II), hydrochloric acid is necessarily present. The iron-permanganate reaction has long been known to give high results in hydrochloric acid solution, the deviation increasing with increasing hydrochloric acid concentration. (See also Section 16-2 concerning hydrogen peroxide and the zinc reductor.) The induced oxidation of chloride gives an error that is relatively smaller the larger the amount of iron being titrated and the slower the titration.[39] Zimmermann[41] showed that the error can be decreased by the addition of Mn(II) sulfate, and Reinhardt[42] added phosphoric acid, primarily to decolorize the Fe(III). In present-day practice the Zimmermann–Reinhardt (Z–R) *preventive solution* is prepared from Mn(II) sulfate, sulfuric acid, and phosphoric acid.

The action of the Z–R reagent is first to supply an adequate concentration of Mn(II), which reacts with local excesses of permanganate and ensures the reduction of intermediate oxidation states of manganese to Mn(III). The Mn(II) also depresses the potential of the *reversible* Mn(III)-Mn(II) couple. Phosphoric acid (and to a lesser extent, sulfuric acid) also lowers the Mn(III)-Mn(II) potential, so that Mn(III) is reduced by Fe(II) rather than by chloride. Schleicher[43] stressed the importance of the Mn(III)-Mn(II) couple and maintained that five Mn(II) ions should be present locally for each Mn(VII) ion, to ensure that no manganese oxidation state higher than Mn(III) can exist. For this purpose four Mn(II) ions should suffice:

$$Mn(VII) + 4Mn(II) \rightarrow 5Mn(III) \qquad (17\text{-}32)$$

Hydrogen peroxide The mechanism of the reaction

$$5H_2O_2 + 2MnO_4^- + 6H^+ \rightarrow 2Mn^{++} + 5O_2 + 8H_2O$$

cannot involve rupture of the O—O bond in hydrogen peroxide because the oxygen atoms appearing as molecular oxygen have been shown to come only from the hydrogen peroxide and not from water.[44] The reaction, therefore, involves removal of protons and electrons (for instance, $H_2O_2 \rightarrow HO_2^- \rightarrow HO_2 \rightarrow O_2^- \rightarrow O_2$).

The permanganate–hydrogen peroxide reaction shows an induction period similar to that observed in the oxalate reaction. There is no induced oxidation of hydrochloric acid, however, perhaps because of the formation of a peroxy complex of Mn(III), which stabilizes the acid. Nevertheless, it is good practice to use a high acid concentration (15 to 20% sulfuric acid) and a reasonably low rate of addition of MnO_4^- to minimize the danger of forming manganese dioxide, which is an active catalyst for the decomposition of hydrogen peroxide. The permanganate method suffers from the disadvantage, as compared with the Ce(IV) titration, that organic substances are likely to interfere and that pure permanganate must be used; iron, for example, brings about an induced chain reaction (Section 15-8). A fading end point is an indication of the presence of organic matter or other reducing agents.[45] For colored solutions or for titrations with dilute permanganate, the use of ferroin as an indicator is advantageous.

Other applications   The present treatment does not undertake a comprehensive survey of the applications. For details the reader is referred to the monograph by Kolthoff and Belcher[46] or to the original literature.

Permanganate is useful for the determination of all the substances mentioned as primary standards for its standardization. Other reducing agents that can be titrated are Sb(III), Mn(II), V(IV), W(V), U(IV), Tl(I), Cr(III), and Ce(III). Several metals can be reduced to a lower oxidation state by passage through a suitable reductor, but cannot be titrated conveniently because of the extreme sensitivity of the reduced solution to air oxidation. For example, the zinc reductor can be used to produce Cr(II), V(II), Ti(III), Nb(III), Mo(III), and Re(–I). Techniques of rigid exclusion of dissolved oxygen from the titrant often can be replaced by the simple operation of passing the reduced solution directly into an air-free solution of Fe(III), usually as Fe(III) ammonium sulfate. The Fe(II) thus produced is titrated with permanganate, Ce(IV), or dichromate.

Permanganate is used in two types of indirect determinations. In one type an excess of permanganate is added to reducing agents that react too slowly to enable a direct titration to be performed but undergo stoichiometric reactions with excess permanganate. The method of Stamm[5] involves the addition of excess permanganate in strongly alkaline solution in the presence of barium ion to form barium manganate. The excess permanganate is determined by titration with sodium formate. Examples of oxidation reactions are iodide to periodate, phosphite or hypophosphite to phosphate, cyanide to cyanate, thiocyanate to cyanate and sulfate, and formate or formaldehyde to carbonate. In the other type of indirect determination, excess reductant is added to oxidizing agents that do not react rapidly enough to permit direct titration with a reductant. The excess reductant is back-titrated with permanganate [or Ce(IV) or sometimes dichromate]. An example involves the determination of available oxygen in compounds such as manganese dioxide, lead dioxide, and hydrogen peroxide.

17-5 APPLICATIONS OF DICHROMATE

Standard solution Potassium dichromate can be prepared as a primary-standard chemical by recrystallization from water and drying at 150°C. Its standard solutions, usually prepared by direct weight, are extraordinarily stable. Carey[47] found that a 0.017 M solution did not change appreciably in titer in 24 years. Dichromate reacts less readily with organic matter than does permanganate and does not react with chloride in acid solutions in the cold at concentrations less than 3 M hydrochloric acid.

Determination of iron The most important applications of dichromate involve either directly or indirectly the titration of Fe(II). An excess of standard Fe(II) can be added to determine oxidants, or an excess of Fe(III) to determine reductants. These determinations usually can be carried out equally well with Ce(IV). For routine applications, however, the low cost and ease of preparation of standard solutions and the great stability of dichromate offer some advantages. Permanganate is at a disadvantage, expecially if hydrochloric acid solutions are to be used.

Dichromate reacts too slowly with some reducing agents to permit a direct titration. If the reductant does not react rapidly and quantitatively with Fe(III), excess dichromate can be added and back-titrated with Fe(II).

Chemical oxygen demand In the examination of water and wastewater an important measurement involves determination of the amount of oxygen required by aerobic organisms to consume the organic material available to them, and of the amount required by substances that consume oxygen, such as Fe(II), sulfide, and ammonia. This oxygen demand can be measured biochemically and is known as BOD (biochemical oxygen demand). A measurement of BOD takes 5 days of incubation. A more rapid test that is not completely equivalent to the BOD test is the measurement of chemical oxygen demand (COD).[48] In this test the oxidizable materials are destroyed by boiling the water for 2 h with a dichromate–sulfuric acid mixture with an Ag(I) catalyst. The amount of oxidizable material is measured in terms of the amount of potassium dichromate consumed and involves a back titration with Fe(II).

Reactions in concentrated acid In phosphoric acid solutions the formal potential of the Cr(VI)-Cr(III) couple increases with increasing concentration of acid. In contrast, for couples such as Mn(III)-Mn(II) and Ce(IV)-Ce(III) the potentials decrease. Rao[49] exploited these changes to develop several methods based on dichromate in concentrated phosphoric acid as a titrimetric reagent with potentiometric end points. As examples: Ce(III) is oxidized and can be titrated to Ce(IV)[50] rapidly at room temperature; Mn(II) to Mn(III);[51] V(III) to V(IV) and then to V(V);[52] and Mo(V) to Mo(VI).[53] A titration involving oxidation of V(IV) by Cr(VI) in 0.5 M sulfuric acid fails for kinetic reasons.[54]

Dichromate often is used in sulfuric acid solution at higher temperatures for oxidation of organic compounds[55] such as hydrocarbons, alcohols, ethers, carboxylic acids, and aldehydes. The end products are usually water and carbon dioxide. Ethanol gives acetic acid as the oxidation product. The dichromate-ethanol reaction is used widely for the rough quantitative determination of alcohol in blood and alveolar air. A device (called an Alcolyser) for a screening test[56] consists of a glass tube containing an appropriate solution of dichromate in concentrated sulfuric acid dispersed on an inert support. Expired air is blown through the tube to fill a plastic measuring bag. The dichromate-alcohol reaction produces a green chromium(III) stain whose length gives a measure of the alcohol level in the blood.

REFERENCES

1   F. SZABADVARY: "History of Analytical Chemistry," pp. 230, 232, Pergamon, New York, 1966.
2   A. W. HUTCHISON: *J. Amer. Chem. Soc.*, **69**:3051 (1947).
3   A. D. WADSLEY and A. WALKLEY: *Trans. Electrochem. Soc.*, **95**:11 (1949).
4   L. V. ANDREWS and D. J. BROWN: *J. Amer. Chem. Soc.*, **57**:254 (1935).
5   H. STAMM: "Die Reduktion von Permanganat zu Manganat abs Grundlage eines neuen Titrationsverfahrens," Akademische, Halle, 1927.
6   W. M. LATIMER: "The Oxidation States of the Elements and Their Potentials in Aqueous Solutions," 2d ed., p. 237, Prentice-Hall, Englewood Cliffs, N.J., 1952.
7   H. TAUBE: *J. Amer. Chem. Soc.*, **70**:3928 (1948).
8   K. J. VETTER and G. MANECKE: *Z. Phys. Chem.* (Leipzig), **195**:270 (1950).
9   J. I. WATTERS and I. M. KOLTHOFF: *J. Amer. Chem. Soc.*, **70**:2455 (1948).
10   K. J. VETTER and G. MANECKE: *Z. Phys. Chem.* (Leipzig), **195**:337 (1950).
11   F. H. WESTHEIMER: *Chem. Rev.*, **45**:419 (1949).
12   I. M. KOLTHOFF, private communication.
13   J. D. NEUSS and W. RIEMAN, III: *J. Amer. Chem. Soc.*, **56**:2238 (1934).
14   J. Y. TONG and E. L. KING: *J. Amer. Chem. Soc.*, **75**:6180 (1953).
15   J. K. BEATTIE and G. P. HAIGHT, JR., in "Inorganic Reaction Mechanisms," J. O. Edwards (ed.), pt. 2, Progress in Inorganic Chemistry, Vol. 17, Wiley, New York, 1972.
16   J. W. LADBURY and C. F. CULLIS: *Chem. Rev.*, **58**:403 (1958).
17   H. TAUBE: *J. Amer. Chem. Soc.*, **70**:1216 (1948).
18   A. H. GROBLER and J. A. VAN DEN BERG: *J. S. Afr. Chem. Inst.*, **18**:6 (1965).
19   G. E. MAPSTONE: *J. Appl. Chem. Biotechnol.*, **21**:238 (1971).
20   I. M. KOLTHOFF: *Z. Anal. Chem.*, **64**:185 (1924).
21   H. A. LAITINEN and R. P. DURBIN, unpublished experiments.
22   I. M. KOLTHOFF, E. J. MEEHAN, and M. KIMURA: *J. Phys. Chem.*, **75**:3343 (1971); *Talanta*, **19**:1179 (1972); **20**:81 (1973).
23   C. C. BENSON: *J. Phys. Chem.*, **7**:1, 356 (1903).
24   J. H. ESPENSON and E. L. KING: *J. Amer. Chem. Soc.*, **85**:3328 (1963).
25   R. A. GORTNER: *J. Phys. Chem.*, **12**:632 (1908).
26   G. F. SMITH: *Anal. Chem.*, **23**:925 (1951).

27 H. A. LAITINEN and A. S. O'BRIEN, unpublished experiments.

28 I. M. KOLTHOFF and D. R. MAY: *Ind. Eng. Chem., Anal. Ed.*, **18**:208 (1946).

29 G. BRUHNS: *Chem. Ztg.*, **47**:613 (1923).

30 J. O. HALVERSON and O. BERGEIM: *J. Ind. Eng. Chem.*, **10**:119 (1918).

31 M. J. POLISSAR: *J. Phys. Chem.*, **39**:1057 (1935); *J. Amer. Chem. Soc.*, **58**:1372 (1936).

32 F. C. TOMPKINS: *Trans. Faraday Soc.*, **38**:131 (1942).

33 J. VOLHARD: *Justus Liebigs Ann. Chem.*, **198**:318 (1879).

34 T. KATO: *J. Chem. Soc. Jap.*, **48**:408 (1927).

35 H. N. MORSE, A. J. HOPKINS, and M. S. WALKER: *Amer. Chem. J.*, **18**:401 (1896).

36 I. M. KOLTHOFF, H. A. LAITINEN, and J. J. LINGANE: *J. Amer. Chem. Soc.*, **59**:429 (1937).

37 D. E. METZLER, R. J. MYERS, and E. H. SWIFT: *Ind. Eng. Chem., Anal. Ed.*, **16**:625 (1944).

38 E. H. SWIFT: "Introductory Quantitative Analysis," p. 132, Prentice-Hall, New York, 1950.

39 I. M. KOLTHOFF and N. SMIT: *Pharm. Weekbl.*, **61**:1082 (1924); *J. Chem. Soc.*, **126**(Pt. 2): 786 (1924).

40 O. E. LANFORD and S. J. KIEHL: *J. Amer. Chem. Soc.*, **64**:291 (1942).

41 C. ZIMMERMANN: *Ber. Deut. Chem. Ges.*, **14**:779 (1881).

42 C. REINHARDT: *Stahl Eisen*, **4**:704 (1884); *Chem. -Ztg.*, **B**:323 (1884).

43 A. SCHLEICHER: *Fresenius' Z. Anal. Chem.*, **135**:259 (1952); **140**:321 (1953); **144**:100 (1955); **151**:413 (1956).

44 A. E. CAHILL and H. TAUBE: *J. Amer. Chem. Soc.*, **74**:2312 (1952).

45 J. S. REICHERT, S. A. MCNEIGHT, and H. W. RUDEL: *Ind. Eng. Chem., Anal. Ed.*, **11**:194 (1939).

46 I. M. KOLTHOFF and R. BELCHER (eds.): "Volumetric Analysis," vol. 3, Interscience, New York, 1957.

47 W. M. CAREY: *J. Amer. Pharm. Assoc.*, **16**:115 (1927).

48 "Standard Methods for the Examination of Water and Wastewater," 12th ed., p. 510, American Public Health Association, New York, 1965.

49 G. G. RAO: *Talanta*, **13**:1473 (1966).

50 G. G. RAO, P. K. RAO, and S. B. RAO: *Talanta*, **11**:825 (1964).

51 G. G. RAO and P. K. RAO: *Talanta*, **10**:1251 (1963).

52 G. G. RAO and P. K. RAO: *Talanta*, **13**:1335 (1966).

53 U. MURALIKRISHNA and G. G. RAO: *Talanta*, **15**:143 (1968).

54 K. SRIRAMAM: *Talanta*, **18**:361 (1971).

55 M. R. F. ASHWORTH: "Titrimetric Organic Analysis," pt. II, Interscience, New York, 1965.

56 T. P. JONES: Report to 5th International Conference on Alcohol and Traffic Safety, Freiburg, Germany, 1969.

PROBLEMS

17-1 From the pH dependence of the potential of Equation (17-4), calculate the standard potential of (17-5).

17-2 Combine (17-3) and (17-5) to derive the standard potential of (17-1).

17-3 Write balanced equations for the reactions in the following titrations with permanganate:

(a) Mn(II) to Mn(III) in the presence of pyrophosphate [Equation (17-9)];

(b) H_3AsO_3 to H_3AsO_4 in H_2SO_4 solution;

(c) Iodide to ICl, ICN, and CH_3COCH_2I.

17-4 Confirm the value of $\Delta G°$ for the reaction (17-8) by combining the half-reactions (17-3) and (17-7).

17-5 Calculate the standard free-energy change and the equilibrium constant of (17-30).

Answer -120 kJ, log $K = 21$.

17-6 Write balanced equations to describe the oxidation of the following substances with strongly alkaline permanganate in the presence of barium ion (Stamm reaction): iodide, phosphite ($HPO_3^=$), hypophosphite ($H_2PO_2^-$), cyanide, thiocyanate, formate.

17-7 A 1.234-g sample, containing lead as PbO and PbO_2, is treated with 20 ml of 0.25 M oxalic acid, which reduces the PbO_2 to Pb^{++}. The resulting solution is neutralized with ammonia to precipitate all the lead as lead oxalate. The filtrate is acidified and titrated with standard permanganate with 10.00 ml of 0.04 M $KMnO_4$. After acidification the precipitate requires 30.00 ml of permanganate for titration. Calculate the percentage of PbO and PbO_2 in the sample.

Answer 36.18, 19.38%.

17-8 A sample containing V, Cr, and Mn is oxidized to yield a solution containing V(V), dichromate, and permanganate, 40.00 ml of 0.1 M Fe(II) being required. The resulting vanadyl ion requires 2.5 ml of 0.02 M permanganate. After the addition of pyrophosphate, the resulting Mn(II) and the original Mn(II) are titrated to Mn(III), requiring 4.0 ml of the same permanganate solution. Calculate the milligrams of V, Cr, and Mn in the sample.

Answer 12.7, 41.6, 14.8 mg.

17-9 A 1.50-g sample containing MnO and Cr_2O_3 is fused with sodium peroxide, giving Na_2MnO_4 and Na_2CrO_4. After dissolution and decomposition of the excess peroxide the solution is acidified, whereupon the manganate disproportionates to MnO_4^- and MnO_2, which is filtered off. The filtrate is heated with 50.0 ml of 0.1 M Fe(II) sulfate. The excess Fe(II) requires 18.40 ml of 0.01 M permanganate. The precipitate is treated with 10 ml of 0.1 M Fe(II) sulfate, the excess requiring 8.24 ml of $KMnO_4$. Calculate the percentage of MnO and Cr_2O_3 in the sample.

Answer 4.17, 1.92%.

17-10 An 0.80-g sample of steel containing Cr and Mn is dissolved and treated to yield Fe(III), Cr(VI), and Mn(II). The Mn(II) is titrated in the presence of fluoride with 0.005 M $KMnO_4$, 20.0 ml being required. The resulting solution is titrated with 0.04 M Fe(II) sulfate, 30.0 ml being required. Calculate the percentage of Cr and Mn in the sample.

Answer 1.52, 2.75%.

Quadrivalent cerium was first used as a titrimetric oxidizing agent in 1927 by Martin.[1] Systematic studies of its uses were begun soon thereafter by Furman[2] and Willard.[3] The rate and extent of reaction with reductants is affected by the solvent, by pH, and by complex formation, and mechanisms of reactions often have been difficult to untangle.

18-1 HALF-REACTIONS AND ELECTRODE POTENTIALS

Smith and Getz[4] determined the formal electrode potentials in various concentrations of perchloric, nitric, and sulfuric acids, with the results given in Table 18-1. They postulated that Ce(IV) existed as anionic complexes $Ce(ClO_4)_6^=$, $Ce(NO_3)_6^=$, $Ce(SO_4)_3^=$, and $CeCl_6^=$ in perchloric, nitric, sulfuric, and hydrochloric acids. Although Ce(IV) is not particularly stable in hydrochloric acid, a value of 1.28 V in 1 M was reported. Note that Smith and Getz measured formal potentials using a saturated calomel electrode, and the measured values, therefore, include liquid-junction potentials that cannot be estimated accurately.

A study of the electrode potential of the Ce(IV)-Ce(III) couple in sodium perchlorate solutions was made by Sherrill, King, and Spooner,[5] who found that the potential varied with hydrogen ion concentration but was practically independent of perchlorate ion concentration. They concluded that neither Ce(IV) nor Ce(III) reacts with perchlorate ion and that Ce(III) is not hydrolyzed in 0.2 to 2.4 M perchloric acid, but that Ce(IV) is hydrolyzed in two stages involving one or two hydroxyl ions. Later, Heidt and Smith[6] offered photochemical evidence for dimeric species of Ce(IV) in perchloric acid solutions; they showed by recalculation of the Sherrill, King, and Spooner emf data that greater consistency could be obtained by assuming dimeric species. They estimated the following standard potentials:

$$Ce(OH)_2{}^{++} + 2H^+ + e^- \rightleftharpoons Ce^{3+} + 2H_2O \qquad E° = 1.7286 \text{ V}$$
$$(CeOCe)^{6+} + 2H^+ + 2e^- \rightleftharpoons 2Ce^{3+} + H_2O \qquad E° = 1.6652 \text{ V}$$
$$(HOCeOCeOH)^{4+} + 4H^+ + 2e^- \rightleftharpoons 2Ce^{3+} + 3H_2O \qquad E° = 1.6783 \text{ V} \tag{18-1}$$
$$(CeOCeOH)^{5+} + 3H^+ + 2e^- \rightleftharpoons 2Ce^{3+} + 2H_2O \qquad E° = 1.6628 \text{ V}$$

Supporting evidence for dimeric species was obtained by Hardwick and Robertson[7] from absorption spectra. They estimated that, in 2 M perchloric acid, 28% of the Ce(IV) is present as aquated Ce^{4+}, the remainder predominantly as $CeOH^{3+}$ and as a dimer such as $(CeOCe)^{6+}$ or $(CeOCeOH)^{5+}$. Duke and Parchen[8] found that they could most satisfactorily explain the kinetics of electron exchange between Ce(IV) and Ce(III) in perchloric acid solutions by assuming that Ce(IV) is present as $Ce(OH)_2{}^{++}$, $Ce(OH)_3{}^+$, and $(CeOCeOH)^{5+}$ in 5 to 6 M perchloric acid. At lower acid concentrations, more highly hydrolyzed and polymerized species seemed to be involved. Greef and Aulich[9] studied the kinetics of the Ce(IV)-Ce(III) couple at a platinum electrode and concluded that the postulate of exclusive participation of dimeric species is unjustified. Exchange currents vary with changes in the electrode surface over short periods of time. Their observations are in agreement with the postulate that reduction kinetics at the electrode are governed by the nature of the anion. Ce(III) is strongly adsorbed on the electrode, particularly when oxide-coated, and this adsorption retards the reduction of Ce(IV). Bonewitz and Schmid[10] found

Table 18-1 FORMAL POTENTIALS OF THE
Ce(IV)-Ce(III) COUPLE[4]

Acid concentration, M	$HClO_4$	HNO_3	H_2SO_4
1	1.70	1.61	1.44
2	1.71	1.62	1.44
4	1.75	1.61	1.43
6	1.82	...	...
8	1.87	1.56	1.42

the charge transfer coefficient on gold electrodes for the Ce(IV)-Ce(III) reaction to be independent of oxygen coverage.

An optical study[11] of Ce(III) indicated the existence of a complex, $CeClO_4^{++}$. The formation constant was evaluated as about 80 at zero ionic strength and was found to decrease rapidly with increasing ionic strength, in accordance with the Debye–Hückel theory. If this interpretation is correct, it would seem that Ce(IV) must also form a perchlorate complex of comparable stability to account for the lack of effect of perchlorate on the potential.

In nitric acid solution, Noyes and Garner[12] found standard potentials of 1.6085, 1.6096, and 1.6104 V for acid concentrations of 0.5, 1.0, and 2.0 M from measurements against a hydrogen electrode. From the slight variation with acid concentration Noyes and Garner concluded that neither hydrolysis nor complex formation was involved at the concentrations in question. Larsen and Brown[13] found by x-ray examination of a 54% solution of ammonium hexanitratocerate that the ion $Ce(NO_3)_6^{=}$ exists in solution. No evidence was found for Ce–Ce interaction. The low formal potential in nitric acid compared with that in perchloric acid appears to indicate complex formation between Ce(IV) and nitrate ions; qualitatively, the deepening of color on addition of nitrate to a Ce(IV) perchlorate solution is also an indication of complex formation. It appears, then, that anionic complexes, perhaps $Ce(NO_3)_5^-$ and $Ce(NO_3)_6^{=}$, exist in nitric acid solutions of concentrations greater than 6 M but that the interaction between aquated Ce(IV) and nitrate ions decreases rapidly at lower acidities. Danesi[14] concluded that Ce(IV) in nitric acid is complexed less strongly with the anion than in sulfuric acid and more strongly than in perchloric acid.

Kunz[15] measured the Ce(IV)-Ce(III) potential in sulfuric acid solution against the hydrogen electrode and obtained standard-potential values of 1.461 and 1.443 V in 0.5 and 1.0 M solutions of the acid. The fact that the potential is lower than in perchloric acid solutions is indicative of complex formation. Jones and Soper[16] found from electrical-migration experiments that, in 0.25 to 10 M sulfuric acid, Ce(IV) exists in the form of anionic complexes, which they believed to be largely $Ce(OH)(SO_4)_3^{3-}$. In a more extensive study, Hardwick and Robertson[7] studied both absorption spectra and electrical migration. In solutions of constant hydrogen ion concentration of 1 M (using perchloric acid) but variable sulfate concentration, they found that the electrical migration at a sulfate concentration of 0.01 M was mainly to the cathode. Migration proceeded in both directions at 0.05 M sulfate but only to the anode at 0.5 M. From optical data they evaluated the constants:

$$Ce^{4+} + HSO_4^- \rightleftharpoons CeSO_4^{++} + H^+ \qquad K_1 = 3500$$

$$CeSO_4^{++} + HSO_4^- \rightleftharpoons Ce(SO_4)_2 + H^+ \qquad K_2 = 200 \qquad (18\text{-}2)$$

$$Ce(SO_4)_2 + HSO_4^- \rightleftharpoons Ce(SO_4)_3^{=} + H^+ \qquad K_3 = 20$$

Ce(III) also forms complexes with sulfate. Newton and Arcand[17] estimated the formation constant of the species $CeSO_4^+$ to be 2.4×10^3 at zero ionic strength and found that it decreased rapidly with increasing ionic strength, until at unit ionic strength the formation constant was estimated to be about 18.

In hydrochloric acid solutions, Ce(IV) is not stable, but oxidizes chloride at a rate proportional to the concentration of Ce(IV) and to chloride.[18] The rate increases with increasing acidity and decreases in the presence of sulfate. The reaction is catalyzed by silver ion. The rate is controlled by reactions such as[19]

$$CeCl^{3+} + Cl^- \rightleftharpoons Ce^{3+} + Cl_2^-$$
$$CeCl_2^{++} \rightleftharpoons Ce^{3+} + Cl_2^-$$

(18-3)

A complication is introduced by the hydrolysis of Ce(IV); apparently the hydrolyzed ion $Ce(OH)^{3+}$ does not oxidize chloride at an appreciable rate.

Considering the reaction between Ce(IV) and chloride ion, it appears that the observed formal potential of 1.28 V in 1 M hydrochloric acid is actually a mixed potential[20] determined partly by the chlorine-chloride couple. Consequently, measured values of the potential cannot be used to calculate the formation constants of Ce(IV)-chloride complexes. From a practical analytical viewpoint, however, it is important that Ce(IV) can be used as a titrant for solutions containing up to 3 M hydrochloric acid[21] without loss of chlorine.

18-2   THE CERIUM(IV)-OXALATE REACTION

Sodium oxalate has been used[3,22] as a primary standard substance for Ce(IV) in sulfuric acid. In the absence of a catalyst a temperature of 70 to 75°C is necessary. Smith and Getz[23] found that in 1 to 2 M perchloric acid solution, sodium oxalate can be titrated at room temperature with Ce(IV) perchlorate or nitrate but not with sulfate. Rao, Rao, and Rao[24] carried out the titration at room temperature in the presence of barium chloride to remove sulfate, which retards the reaction between oxalate and Ce(IV) and between oxalate and oxidized ferroin. Alternatively, some Fe(III) was added, and the trace of Fe(II) produced photochemically then reacted with the indicator. Rao, Rao, and Murty[25] carried out the titration in 0.5 M HNO_3 with ammonium hexanitratocerate(IV) instead of the sulfate. With a small amount of KI and KIO_3, a satisfactory end point was obtained at room temperature with ferroin as indicator.

Several observations resulting from studies[26,27] of the Ce(IV)-oxalate reaction are: sulfate and phosphate decrease the rate of reaction; the rate of reduction of Ce(IV) increases with decreasing oxalate concentration (within limits); mixing Ce(IV) and oxalate produces a red color that develops rapidly and then fades as reaction occurs; when a small amount of cerium is added to a large amount of oxalate, an apparent induction period is observed; and at equimolar concentrations the reaction is first order with respect to Ce(IV) and oxalate. The following equilibria and mechanism appear to be consistent with the experimental observations for Ce(IV)-oxalate reaction in sulfuric acid solution:

$$Ce(SO_4)_2 + HSO_4^- \rightleftharpoons Ce(SO_4)_3^= + H^+ \qquad (18\text{-}4)$$

$$Ce(SO_4)_2 + HC_2O_4^- \rightleftharpoons Ce(SO_4)_2C_2O_4^= + H^+ \qquad (18\text{-}5)$$

$$Ce(SO_4)_2C_2O_4^= \rightleftharpoons Ce(SO_4)_2^- + \cdot C_2O_4^- \qquad (18\text{-}6)$$

$$Ce(SO_4)_2C_2O_4^= + \cdot C_2O_4^- \rightarrow Ce(SO_4)_2^- + 2CO_2 + C_2O_4^= \qquad (18\text{-}7)$$

Reaction (18-6) is considered to be the rate-limiting step. This is analogous to the oxidation of oxalate by permanganate (Section 17-2), where a critical intermediate is a complex between oxalate and Mn(III). The rate-limiting step is reversible in that the $\cdot C_2O_4^-$ radical can reoxidize the Ce(III) to Ce(IV). The concentration of sulfate controls the amount of the Ce(IV)-oxalate reactive intermediate formed, but not its specific rate of decomposition.[26] Thus the error in the Fe(II)-Ce(IV) titration in the presence of oxalate can be decreased by the addition of sulfate. Phosphate also inhibits the Ce(IV)-oxalate reaction.

Inhibition of the Ce(IV)-oxalate oxidation by oxalate itself is interpreted in terms of the formation of higher complexes at high oxalate concentrations. If the 1:1 complex is the most reactive, its concentration is decreased when higher complexes are formed. For 10^{-4} M Ce(IV) the maximum rate of reaction with oxalate occurs[28] with an oxalate concentration of 2×10^{-3} M. It seems reasonable to presume the formation of a Ce(IV) complex as the reactive intermediate in the oxidation of several other substances such as citric acid,[29] mandelic acid,[25] benzilic acid,[30] and polyamino-polycarboxylic acids.[31]

When oxalate is complexed with another metal ion such as Cr(III), Rh(III), or Co(III), a polynuclear complex becomes the critical intermediate.[32] The addition of Ce(IV) to solutions of $Co(C_2O_4)_3^{3-}$ brings about an accelerated rate of *reduction* of Co(III). If oxalate is assumed to function as a bidentate bridging ligand to form complexes between both the metal and Ce(IV), the rate-determining equilibrium for cobalt is

$$[Co(C_2O_4)_3^{3-}]Ce(IV) \rightarrow [Co(C_2O_4)_3^=]Ce(III) \qquad (18\text{-}8)$$

followed by

$$[Co(C_2O_4)_3^=]Ce(III) \rightarrow Co(II) + 2C_2O_4^= + 2CO_2 + Ce(III) \qquad (18\text{-}9)$$

18-3   THE CERIUM(IV)-HYDROGEN PEROXIDE REACTION

Ce(IV) reacts rapidly and completely with H_2O_2 in two one-electron steps:

$$Ce(IV) + H_2O_2 \underset{k_{r_1}}{\overset{k_1}{\rightleftharpoons}} Ce(III) + \cdot HO_2 + H^+ \qquad (18\text{-}10)$$

$$Ce(IV) + \cdot HO_2 \xrightarrow{k_2} Ce(III) + O_2 + H^+ \qquad (18\text{-}11)$$

The existence of the $\cdot HO_2$ radical has been confirmed[33] by spectroscopic means and isotopic exchange.[34] The reaction resulting in the formation of the $\cdot HO_2$ radical is reversible; that is, $\cdot HO_2$ can oxidize Ce(III). Samuni and Czapski[35] confirmed this reversibility and suggested the formation of a Ce(III)-HO$_2$ complex.[36] They modified (18-10) and (18-11) to

$$\cdot HO_2 + Ce^{3+} \rightleftharpoons Ce(III)\text{-}HO_2 \text{ complex} \qquad (18\text{-}12)$$

$$Ce(III)\text{-}HO_2 \text{ complex} + Ce^{4+} \rightarrow H^+ + O_2 + 2Ce^{3+} \qquad (18\text{-}13)$$

The forward rate constant of (18-10) for the formation of $\cdot HO_2$ has been measured,[37] and a value of 1.0×10^6 1/mole-s obtained. A value of 13 for the ratio k_2/k_{r_1} also has been obtained.

In the presence of oxalate[38] the rate of the reverse reaction is accelerated so that Ce(IV) is more stable toward H_2O_2.

18-4   OTHER CERIUM(IV) REACTIONS

Thallium   Dorfman and Gryder,[39] investigating the kinetics of the Ce(IV)-Tl(I) reaction, concluded that the mechanism involves two 1-equivalent oxidation steps with a Ce(IV) dimer, with Tl(II) as an intermediate. They pointed out that the reaction may be complicated by the formation of $\cdot OH$ or $\cdot NO_3$ free radicals that oxidize Tl(I). Sinha and Mathur[40] studied the Ce(IV)-Tl(I) reaction in the presence of the catalysts Ag(I) and Os(VIII). The rate was found to be first order with respect to both Ce(IV) and Tl(I) but unaffected by Ce(III) or Tl(III). Bisulfate and hydrogen ions acted as inhibitors. The slow step in sulfuric acid medium was considered to be

$$TlCe(SO_4)_2{}^+ + Os(VIII) \rightarrow Tl^{3+} + Ce(SO_4)_2 + Os(VI) \qquad (18\text{-}14)$$

The Os(VI) then is oxidized rapidly in two 1-equivalent steps. The inhibition by $HSO_4{}^-$ was presumed to be due to

$$Ce(SO_4)_2 + HSO_4{}^- \rightleftharpoons HCe(SO_4)_3{}^- \qquad (18\text{-}15)$$

In perchloric acid medium the rate-determining step was considered to be[41]

$$TlCeOH^{4+} + Os(VIII) \rightarrow Tl^{3+} + CeOH^{3+} + Os(VI) \qquad (18\text{-}16)$$

Chromium Tong and King[42] concluded that the Ce(IV)-Cr(III) reaction involves the equilibria

$$Ce(IV) + Cr(III) \rightleftharpoons Ce(III) + Cr(IV) \qquad (18\text{-}17)$$

$$Ce(IV) + Cr(IV) \rightleftharpoons Ce(III) + Cr(V) \qquad (18\text{-}18)$$

$$Ce(IV) + Cr(V) \rightleftharpoons Ce(III) + Cr(VI) \qquad (18\text{-}19)$$

Their observations indicate that (18-18) is rate-limiting. The equilibrium constant for (18-17) must be relatively small because Cr(IV) is strongly oxidizing. The reverse reaction, for example, is involved in the induced oxidation of Ce(III) in the Cr(VI)-As(III) reaction (Section 15-8).

Because ammonium hexanitratocerate(IV) in dilute nitric acid oxidizes Cr(III) quantitatively at room temperature in 10 to 15 min, it has found practical application in the determination of chromium.[43]

Arsenic and antimony Csányi and Szabo[44] postulated that As(IV) is the first product of the Ce(IV)-As(III) reaction. The later work of Woods, Kolthoff, and Meehan[45] supports this postulate. As(IV) would be oxidized rapidly by Ce(IV), so its steady-state concentration would be low and induced air oxidation should be negligible, especially when a catalyst is used for the primary reaction.

Mishra and Gupta[46] similarly postulated that, in perchloric acid, Sb(IV) is formed from Sb(III) in a first step by reaction with Ce^{4+}, $CeOH^{3+}$, or $Ce(OH)_2^{++}$. The Sb(IV) then reacts rapidly with additional Ce(IV). The rate is decreased in the presence of bisulfate, and the active Ce(IV) species then becomes $CeSO_4^{++}$.

The Ce(IV)-As(III) reaction is catalyzed by iodine and thereby furnishes the basis for the determination of traces of iodine.[47] A possible pathway involves the formation of I^0 or I^+ by reaction of I^- with Ce(IV).[48] The rate of the overall reaction (15-23) is determined by that of (15-27), which is proportional to the steady-state concentration of iodide ion. Because each iodide ion enters the catalytic cycle many times, the method is extremely sensitive. Chloride is beneficial in the iodide-catalyzed reaction, presumably through formation of ICl as an intermediate that inhibits formation of HIO_3 as an unreactive product.[49]

Phosphorus The rate of oxidation of HPO_2 with most oxidants is determined by the rate of transformation from an inactive form, $H_2PO(OH)$, to an active form, $HP(OH)_2$. The rate of oxidation is therefore largely independent of the nature of the oxidant. Carroll and Thomas[50] found, however, that with Ce(IV) the rate is only about $\frac{1}{200}$ of that with other oxidizing agents. An explanation for this may be in terms of the formation of Ce(IV) complexes with one, two, or three $H_2PO_2^-$ ions. The mechanism of the formation of H_3PO_4 must involve two Ce(IV) ions. It can be speculated that an internal electron transfer takes place, forming an unstable P(II) species that would then react rapidly with a second Ce(IV).

Mercury McCurdy and Guilbault[51] used a combined catalyst of Ag(I) and Mn(II) in 2 M $HClO_4$ for the Ce(IV) determination of mercury, phosphite, hypophosphite, tellurium, polyhydric alcohols, and metal chelates of 8-quinolinol. They found that the rate of reaction of Ce(IV) with Hg(I) is more rapid with both catalysts than with either alone. The rate increases with $HClO_4$ concentration, reaching a maximum in the region of 3 to 4 M. Variation of the concentration would affect both the hydrolysis of the aquated Ce(IV) ion and the dimerization equilibrium. They suggested that the rate-determining step in the mechanism is the breaking of the $(Hg-Hg)^{++}$ bond and transfer of an electron from Ce(IV). The catalytic effects arise through Ag(II) or Mn(III) as active intermediates that assist in the breaking of the $(Hg-Hg)^{++}$ bond. The rate is inhibited[52] by sulfate, and a proposed pathway for the reaction is

$$Ce(HSO_4)_3{}^+ \rightleftharpoons Ce(HSO_4)_2{}^{++} + HSO_4{}^- \qquad (18\text{-}20)$$

$$Ce(HSO_4)_2{}^{++} + Hg_2{}^{++} \rightleftharpoons Ce(HSO_4)_2{}^{++} \cdots Hg_2{}^{++} \qquad (18\text{-}21)$$

The transient complex decomposes in a slow irreversible step,

$$Ce(HSO_4)_2{}^{++} \cdots Hg_2{}^{++} \rightarrow Ce(HSO_4)_2{}^+ + Hg^{++} + Hg^+ \qquad (18\text{-}22)$$

The Hg^+ radical reacts rapidly with a second $Ce(HSO_4)_3{}^+$ entity.

The Ce(IV)-Hg(I) reaction is catalyzed also by iridium compounds[53] in which the active intermediate is Ir(IV), formed by reaction of Ce(IV) with Ir(III). The catalytic action of iridium in this system was proposed as a basis for the kinetic determination of iridium.

18-5   STANDARD CERIUM(IV) SOLUTIONS

Preparation   In the early work with Ce(IV), standard solutions were prepared from relatively crude preparations of Ce(IV) oxide, Ce(IV) sulfate, or Ce(IV) ammonium sulfate. Contamination from other lanthanides and phosphate led to the gradual precipitation of lanthanide phosphates from a sulfuric acid solution.

The preparation of Ce(IV) compounds free of foreign metals is facilitated by the fact that, when ammonium nitrate is added to a concentrated nitric acid solution of Ce(IV), the compound $(NH_4)_2Ce(NO_3)_6$ is precipitated in relatively pure form.[54] Because the Ce(IV) exists as the complex ion $Ce(NO_3)_6{}^=$, the ammonium salt is properly called ammonium hexanitratocerate(IV). In water and in dilute nitric acid, extensive dissociation occurs. Ammonium hexanitratocerate(IV) was suggested[54] as a primary standard, and drying it to constant weight at 100°C was recommended. Later it was recognized that slow decomposition occurs at this temperature, and Smith and Fly[55] recommended a procedure that involved drying a product of "primary-standard" grade at 85°C for 1 to 6 h. The accurate results of Smith and Fly have been confirmed.[56]

Accurate coulometric titrations[56] have indicated a purity of 99.96 to 99.98% for hexanitratocerate products of low thorium content. Two older preparations containing 0.22 and 0.79% thorium as $(NH_4)_2Th(NO_3)_6$ gave assays of 99.4 and 98.7%. Apparently the NH_3 content has never been accurately checked to determine whether compensation of errors could occur. Appreciable amounts (0.1 to 0.5%) of Ce(III) in "primary-standard" grade hexanitratocerate have been found[57] by direct titration of Ce(III) to Ce(IV) with ferricyanide in carbonate medium.

Stability Ce(IV) solutions in perchloric acid are sensitive to photochemical reduction by water. Weiss and Porret[58] postulated that a hydroxyl radical is formed from activated Ce(IV). The radical can either reoxidize Ce(III) or react to release oxygen:

$$\cdot OH + Ce(III) \rightarrow Ce(IV) + OH^- \qquad (18\text{-}23)$$

$$\cdot OH + \cdot OH \rightarrow H_2O + O \qquad (18\text{-}24)$$

$$O + O \rightarrow O_2 \qquad (18\text{-}25)$$

Hydrogen peroxide, formed from hydroxyl radicals, also can be involved. Photochemical reduction is negligible in sulfuric acid solution, because of the low concentration of Ce^{4+} ions. Furman[59] stated that 0.1 M solutions of Ce(IV) in sulfuric acid solutions are stable for at least 6 years, but was of the opinion that nitrate and ammonium ions are undesirable. Stored at room temperature, 0.1 M Ce(IV) in nitric or perchloric acid solutions showed a 0.01 to 0.03% decrease in concentration per day.[60] Protection from light is necessary for nitric and perchloric but not for sulfuric acid solutions.

Matthews, Mahlman, and Sworski[61] observed net Ce(III) oxidation induced by gamma radiation in 4 M sulfuric acid solution. The Ce(IV) concentration goes through a maximum, caused by formation of oxygen, which in turn removes H atoms to form $\cdot HO_2$. The $\cdot HO_2$ then reduces Ce(IV). Studies of this nature are particularly important in understanding the stability and in the storage of solutions under radioactive conditions.

18-6 APPLICATIONS OF CERIUM(IV) OXIDATIONS

The high $E°$ value for the Ce(IV)-Ce(III) couple makes possible the oxidizing of substances as resistant to the process as dithionite.[62]

Iron Ce(IV) has the advantage of being applicable to a wide variety of solutions of Fe(II). Perchloric or sulfuric acid can be used[4] in concentrations of 0.5 to 8 M. Hydrochloric acid solutions ordinarily are used in concentrations of 0.5 to 3 M. Ferroin is the usual indicator, although Cagle and Smith[63] used 2,2'-bipyridine in

1 M hydrochloric or sulfuric acid. A better indicator in hydrochloric acid is 5,6-dimethylferroin,[64] which has a lower transition potential. Nitroferroin, with its high potential, may be used to advantage in nitric or perchloric acid solutions.[65]

For many purposes, hydrochloric acid is the most convenient titration medium, as it permits the use of Sn(II) chloride or a silver reductor as a selective reducing agent (Section 16-2). Petzold[66] titrated small amounts of Fe(II) in the presence of arsenic, antimony, and Sn(II). Fe(II) can be titrated in the presence of V(IV) by using a silver reductor with 1 M hydrochloric acid and by making the solution 5 M in sulfuric acid to prevent the oxidation of V(IV) at the ferroin end point.[67]

Hydrogen peroxide The use of Ce(IV) for the titration of hydrogen peroxide,

$$H_2O_2 + 2Ce(IV) \rightarrow 2Ce(III) + O_2 + 2H^+ \qquad (18\text{-}26)$$

introduced by Atanasiu and Stefanescu,[68] has been shown to be applicable in sulfuric, hydrochloric, nitric, and acetic acid solutions.[22,69,70]

Hurdis and Romeyer[71] found nitroferroin more suitable than ferroin in titrations with nitric and perchloric acids, because it has less tendency to be oxidized locally by a transient excess of Ce(IV). The oxidized indicator was slow to return to its reduced state in the immediate vicinity of the end point. With sulfuric acid solutions, nitroferroin changes color at too high a potential, and the slow fading of the ferroin color made the use of ferroin preferable to titration without an indicator, using the color of Ce(IV) itself. Baer and Stein[72] found an exact stoichiometry using pure Ce(IV). With ordinary reagent-quality Ce(IV), however, they observed an error due to catalytic decomposition of hydrogen peroxide when hydrogen peroxide was in excess, although the results were exact when Ce(IV) was in excess.

Other direct titrations Oxalate, As(III), ferrocyanide, and iodide can be determined by direct titration. Other reducing agents that react rapidly and stoichiometrically and can be titrated directly are: Sb(III), Mo(V), Pu(III), Sn(II), $S_2O_3^=$, Tl(I), U(IV), and V(IV). Bromide[73] in 4 M HClO$_4$ and also thiosulfate,[74] thiourea, and related substances have been titrated by enthalpimetric methods.[75]

Ammonium hexanitratocerate(IV) as a standard oxidant for reactions at room temperature has been developed for several substances.[25,76-78] Oxalate[76] can be titrated with Ce(IV) in 0.5 M HNO$_3$ with an iodide catalyst; mandelic acid at a lower acid concentration, 0.1 M HNO$_3$; Mn(II) with excess Ce(IV) in 0.5 to 2 M HNO$_3$ with silver nitrate catalyst; hydrazine or isonicotinic acid[77] in HCl-KBr solution; and As(III) in HCl, HNO$_3$, or H$_2$SO$_4$ solution with a trace of iodine as catalyst.

Peroxydisulphate The direct titration of peroxydisulfate, $S_2O_8^=$, by Fe(II) is not possible because the reaction is too slow. Therefore, an excess of standard Fe(II) is added and then back-titrated with Ce(IV). The reaction takes place through the $\cdot SO_4^-$ radical ion, as shown by Equations (15-48) and (15-49). Accurate results are possible under the proper conditions, but certain precautions need to be taken.[79] When Fe(II) was added slowly[80] to peroxydisulfate with or without oxygen, a mole ratio of iron to peroxydisulfate of less than 2 was found. Presumably, an induced decomposition similar to that observed for hydrogen peroxide (Section 15-8) occurs, although the decomposition is much less extensive, and no oxygen was detected. If oxygen is present, induced air oxidation leads to the consumption of too much iron. The effect of oxygen is accentuated by the presence of organic matter. In the absence of oxygen the induced oxidation of organic substances causes the consumption of too much peroxydisulfate. From the analytical viewpoint it is important that bromide suppresses both induced reactions, presumably through the reaction

$$\cdot SO_4^- + Br^- \rightarrow SO_4^= + Br\cdot \qquad (18\text{-}27)$$

forming a bromine atom that rapidly oxidizes Fe(II):

$$Br\cdot + Fe(II) \rightarrow Br^- + Fe(III) \qquad (18\text{-}28)$$

The bromine atom no doubt is stabilized by the formation of the bromine molecule ion, Br_2^-.

Kolthoff and Carr[79] used the suppressing effect of bromide to permit the determination of peroxydisulfate even in the presence of organic matter. The optimum bromide concentration was of the order of $1M$; ferroin served as the indicator for the back titration, which was carried out with Ce(IV) in $0.5\ M\ H_2SO_4$.

Other indirect determinations A number of reducing agents react too slowly with Ce(IV) to be titrated directly; yet they react quantitatively with excess Ce(IV), which can be back-titrated with a reducing agent such as Fe(II) sulfate or arsenic(III). Such substances include[81] As(0), Cr(III), V(IV), HN_3, NH_2OH, NO_2^-, H_3PO_2, H_3PO_3, $H_4P_2O_6$, Hg(I), Re(III), and Te(IV).

Indirect determinations by use of iron(III) In determining strong reducing agents that are sensitive to air oxidation, it is advantageous to pass the reduced solution directly into an excess of Fe(III). The equivalent amount of Fe(II) then is determined. Ce(IV) has the advantage over permanganate of permitting the use of relatively concentrated solutions of hydrochloric acid as the reducing medium. Several applications include[81] copper, molybdenum, thorium, titanium, and niobium.

REFERENCES

1 J. MARTIN: *J. Amer. Chem. Soc.*, **49**:2133 (1927).

2 N. H. FURMAN: *J. Amer. Chem. Soc.*, **50**:755 (1928).

3 H. H. WILLARD and P. YOUNG: *J. Amer. Chem. Soc.*, **50**:1322 (1928).

4 G. F. SMITH and C. A. GETZ: *Ind. Eng. Chem., Anal. Ed.*, **10**:191 (1938); see also G. F. SMITH: "Cerate Oximetry," 2d ed., G. F. Smith Chemical, Columbus, Ohio, 1964.

5 M. S. SHERRILL, C. B. KING, and R. C. SPOONER: *J. Amer. Chem. Soc.*, **65**:170 (1943).

6 L. J. HEIDT and M. E. SMITH: *J. Amer. Chem. Soc.*, **70**:2476 (1948).

7 T. J. HARDWICK and E. ROBERTSON: *Can. J. Chem.*, **29**:818, 828 (1951).

8 F. R. DUKE and F. R. PARCHEN: *J. Amer. Chem. Soc.*, **78**:1540 (1956).

9 K. GREEF and H. AULICH: *J. Electroanal. Chem., Interfacial Electrochem.*, **18**:295 (1968).

10 R. A. BONEWITZ and G. M. SCHMID: *J. Electrochem. Soc.*, **117**:1367 (1970).

11 L. J. HEIDT and J. BERESTECKI: *J. Amer. Chem. Soc.*, **77**:2049 (1955).

12 A. A. NOYES and C. S. GARNER: *J. Amer. Chem. Soc.*, **58**:1265 (1936).

13 R. D. LARSEN and G. H. BROWN: *J. Phys. Chem.*, **68**:3060 (1964).

14 P. R. DANESI: *Acta Chem. Scand.*, **21**:143 (1967).

15 A. H. KUNZ: *J. Amer. Chem. Soc.*, **53**:98 (1931).

16 E. G. JONES and F. G. SOPER: *J. Chem. Soc.*, **1935**:802.

17 T. W. NEWTON and G. M. ARCAND: *J. Amer. Chem. Soc.*, **75**:2449 (1953).

18 A. A. ALEXIEV and P. R. BONTCHEV: *Z. Phys. Chem.* (Leipzig), **242**:333 (1969).

19 F. R. DUKE and C. E. BORCHERS: *J. Amer. Chem. Soc.*, **75**:5186 (1953).

20 E. WADSWORTH, F. R. DUKE, and C. A. GOETZ: *Anal. Chem.*, **29**:1824 (1957).

21 I. TSUBAKI: *Bunseki Kagaku*, **3**:253 (1954).

22 H. H. WILLARD and P. YOUNG: *J. Amer. Chem. Soc.*, **55**:3260 (1933).

23 G. F. SMITH and C. A. GETZ: *Ind. Eng. Chem., Anal. Ed.*, **10**:304 (1938).

24 V. P. RAO, P. V. K. RAO, and G. G. RAO: *Fresenius' Z. Anal. Chem.*, **176**:333 (1960); V. P. RAO and G. G. RAO: *Talanta*, **2**:370 (1959).

25 G. G. RAO, P. V. K. RAO, and K. S. MURTY: *Talanta*, **9**:835 (1962).

26 G. A. RECHNITZ and Y. A. EL-TANTAWY: *Fresenius' Z. Anal. Chem.*, **188**:173 (1962); **193**:434 (1963); *Anal. Chem.*, **36**:2361 (1964); Y. A. EL-TANTAWY and G. A. RECHNITZ: *Anal. Chem.*, **36**:1774 (1964).

27 P. S. PERMINOV, S. G. FEDOROV, V. A. MATYUKHA, V. B. MILOV, and N. N. KROT: *Russ. J. Inorg. Chem.*, **13**:245 (1968).

28 K. B. YATSIMIRSKII and A. A. LUZAN: *Russ. J. Inorg. Chem.*, **10**:1233 (1965).

29 K. K. SENGUPTA: *J. Proc. Inst. Chem.* (India), **36**:149 (1964).

30 V. K. GROVER and Y. K. GUPTA: *J. Inorg. Nucl. Chem.*, **31**:1403 (1969).

31 S. B. HANNA, R. K. HESSLEY, W. R. CARROLL, and W. H. WEBB: *Talanta*, **19**:1097 (1972).

32 M. W. HSU, H. G. KRUSZYNA, and R. M. MILBURN: *Inorg. Chem.*, **8**:2201 (1969); H. G. KRUSZYNA and R. M. MILBURN: *Inorg. Chem.*, **10**:1578 (1971).

33 E. SAITO and B. H. J. BIELSKI: *J. Amer. Chem. Soc.*, **83**:4467 (1961); *J. Phys. Chem.*, **66**:2266 (1962).

34 P. B. SIGLER and B. J. MASTERS: *J. Amer. Chem. Soc.*, **79**:6353 (1957).

35 A. SAMUNI and G. CZAPSKI: *Isr. J. Chem.*, **8**:551 (1970).

36 G. CZAPSKI and A. SAMUNI: *Isr. J. Chem.*, **7**:361 (1969); G. CZAPSKI, H. LEVANON, and A. SAMUNI: *Isr. J. Chem.*, **7**:375 (1969).

37 G. CZAPSKI, B. H. J. BIELSKI, and N. SUTIN: *J. Phys. Chem.*, **67**:201 (1963).

38 V. A. MATYUKHA, V. B. MILOV, N. N. KROT, and P. S. PERMINOV: *Russ. J. Inorg. Chem.*, **12**:1763 (1967).

39 M. K. DORFMAN and J. W. GRYDER: *Inorg. Chem.*, **1**:799 (1962).

40 B. P. SINHA: *Z. Phys. Chem.* (Leipzig), **233**:161 (1966); B. P. SINHA and H. P. MATHUR, *Z. Phys. Chem.* (Leipzig), **246**:342 (1971).

41 B. P. SINHA and H. P. MATHUR: *J. Inorg. Nucl. Chem.*, **33**:1673 (1971).

42 J. Y. TONG and E. L. KING: *J. Amer. Chem. Soc.*, **82**:3805 (1960).

43 G. G. RAO, K. S. MURTY, and M. GANDIKOTA: *Talanta*, **19**:65 (1972).

44 L. J. CSÁNYI and M. SZABO, *Talanta*, **1**:359 (1958).

45 R. WOODS, I. M. KOLTHOFF, and E. J. MEEHAN: *J. Amer. Chem. Soc.*, **85**:2385, 3334 (1963); **86**:1698 (1964).

46 S. K. MISHRA and Y. K. GUPTA: *J. Inorg. Nucl. Chem.*, **30**:2991 (1968); *J. Chem. Soc.*, **1970**:A260.

47 E. B. SANDELL and I. M. KOLTHOFF: *J. Amer. Chem. Soc.*, **56**:1426 (1934); *Mikrochim. Acta*, **1**:9 (1937).

48 P. A. RODRIGUEZ and H. L. PARDUE: *Anal. Chem.*, **41**:1369 (1969).

49 J. DEMAN: *Mikrochim. Acta*, **1964**:67.

50 R. L. CARROLL and L. B. THOMAS: *J. Amer. Chem. Soc.*, **88**:1376 (1966).

51 W. H. MCCURDY, JR., and G. G. GUILBAULT: *J. Phys. Chem.*, **64**:1825 (1960); **70**:656 (1966); *Anal. Chem.*, **32**:647 (1960); **33**:580 (1961); *Anal. Chim. Acta*, **24**:214 (1961).

52 Y. A. EL-TANTAWY, F. M. ABDEL-HALIM, and M. Y. EL-SHEIKH: *Z. Phys. Chem.* (Frankfurt am Main), **73**:277 (1970).

53 K. B. YATSIMIRSKII, L. P. TIKHONOVA, and I. P. SVARKOVSKAYA: *Russ. J. Inorg. Chem.*, **14**:1572 (1969).

54 G. F. SMITH, V. R. SULLIVAN, and G. FRANK, *Ind. Eng. Chem.*, *Anal. Ed.*, **8**:449 (1936).

55 G. F. SMITH and W. H. FLY: *Anal. Chem.*, **21**:1233 (1949).

56 J. KNOECK and H. DIEHL: *Talanta*, **16**:181 (1969).

57 D. N. HUME, private communication.

58 J. WEISS and D. PORRET: *Nature*, **139**:1019 (1937).

59 N. H. FURMAN in foreword to G. F. Smith, "Cerate Oxidimetry," G. F. Smith Chemical, Columbus, Ohio, 1942.

60 G. F. SMITH and C. A. GETZ, *Ind. Eng. Chem.*, *Anal. Ed.*, **12**:339 (1940).

61 R. W. MATTHEWS, H. A. MAHLMAN, and T. J. SWORSKI: *J. Phys. Chem.*, **72**:3704 (1968).

62 V. R. NAIR and C. G. R. NAIR: *Talanta*, **18**:432 (1971).

63 F. W. CAGLE, JR., and G. F. SMITH: *Anal. Chem.*, **19**:384 (1947).

64 W. W. BRANDT and G. F. SMITH: *Anal. Chem.*, **21**:1313 (1949).

65 G. F. SMITH and F. P. RICHTER: *Ind. Eng. Chem.*, *Anal. Ed.*, **16**:580 (1944).

66 A. PETZOLD: *Fresenius' Z. Anal. Chem.*, **149**:258 (1956).

67 G. H. WALDEN, JR., L. P. HAMMETT, and S. M. EDMONDS: *J. Amer. Chem. Soc.*, **56**:350 (1934).

68 J. A. ATANASIU and V. STEFANESCU: *Ber. Deut. Chem. Ges.*, **61**:1343 (1928).

69 N. H. FURMAN and J. H. WALLACE, JR.: *J. Amer. Chem. Soc.*, **51**:1449 (1929).

70   J. S. REICHERT, S. A. MCNEIGHT, and H. W. RUDEL: *Ind. Eng. Chem., Anal. Ed.*, **11**:194 (1939).

71   E. C. HURDIS and H. ROMEYER, JR.; *Anal. Chem.*, **26**:320 (1954).

72   S. BAER and G. STEIN: *J. Chem. Soc.*, **1953**:3176.

73   H. W. WHARTON: *Talanta*, **13**:919 (1966).

74   B. F. A. GRIEPINK and H. A. CYSOUW: *Mikrochim. Acta*, **1963**:1033.

75   W. A. ALEXANDER, C. J. MASH, and A. MCAULEY: *Analyst*, **95**:657 (1970).

76   G. G. RAO, K. S. MURTY, and P. V. K. RAO: *Talanta*, **10**:657 (1963); **11**:955 (1964).

77   G. G. RAO and P. V. K. RAO: *Talanta*, **11**:1489 (1964).

78   G. G. RAO, K. S. MURTY, and M. GANDIKOTA: *Talanta*, **19**:59 (1972).

79   I. M. KOLTHOFF and E. M. CARR: *Anal. Chem.*, **25**:298 (1953).

80   I. M. KOLTHOFF, A. I. MEDALIA, and H. P. RAAEN: *J. Amer. Chem. Soc.*, **73**:1733 (1951).

81   I. M. KOLTHOFF and R. BELCHER: "Volumetric Analysis," vol. 3, chap. 4, Interscience, New York, 1957.

Applications of iodine as a redox reagent are extensive[1] because (1) its $E°$ is intermediate and therefore it can act as either an oxidizing agent (as I_2) or a reducing agent (as I_3^-), and (2) its E^0 is nearly independent of pH (at values less than about 8), a useful property when conditions for favorable conditional equilibrium constants are being selected. We include in this chapter a few examples to illustrate the range of application.

19-1 ELECTRODE POTENTIAL

The iodine-iodide couple may be characterized by the half-reaction

$$I_2 \text{ (solid)} + 2e^- \rightleftharpoons 2I^- \qquad E° = 0.535 \text{ V} \qquad (19\text{-}1)$$

which shows that iodine is a relatively weak oxidant and iodide a relatively weak reductant. Thus iodine is quantitatively reduced to iodide by moderately strong reductants [such as tin(II), H_2SO_3, and H_2S in neutral solution], and iodide is quantitatively oxidized to iodine by moderate or strong oxidants (H_2O_2, IO_3^-, $Cr_2O_7^=$, and MnO_4^-).

The behavior of iodine is complicated by its relatively low solubility in water (1.18×10^{-3} M at 25°C) and by the formation of triiodide ion,

$$I_2 + I^- \rightleftharpoons I_3^- \qquad K = \frac{a_{I_3^-}}{a_{I_2}a_{I^-}} = 710 \qquad (19\text{-}2)$$

Equation (19-1) describes the behavior of a solution saturated with solid iodine. This half-reaction occurs, for example, toward the end of a titration of iodide with an oxidant such as permanganate, when the iodide ion concentration becomes relatively low. Near the beginning, and in most indirect determinations, an excess of iodide is present, and the half-reaction is more accurately written

$$I_3^- + 2e^- \rightleftharpoons 3I^- \qquad E^\circ = 0.545 \text{ V} \qquad (19\text{-}3)$$

The relation between (19-1) and (19-3) may be clarified by considering the potential for the couple in the absence of solid iodine:

$$I_2 + 2e^- \rightleftharpoons 2I^- \qquad E^\circ = 0.621 \text{ V} \qquad (19\text{-}4)$$

Since the activity of solid iodine may be taken to be unity, we may compare (19-1) and (19-4) and write the Nernst equation, either in a form that specifies constant activity for I_2 or in a form in which I_2 activity appears in the logarithm term

$$E = E^\circ_{I_2(\text{solid}),I^-} - \frac{RT}{2F} \ln a_{I^-}^{\ 2} = E^\circ_{I_2,I^-} - \frac{RT}{2F} \ln \frac{a_{I^-}^{\ 2}}{a_{I_2}} \qquad (19\text{-}5)$$

or $\qquad E^\circ_{I_2,I^-} = E^\circ_{I_2(\text{solid}),I^-} - \dfrac{RT}{2F} \ln S = 0.535 + 0.0866 = 0.622 \text{ V} \qquad (19\text{-}6)$

where S is the molar solubility of iodine. Writing the Nernst equation for (19-4) and considering (19-2), we have

$$E = 0.621 - \frac{RT}{2F} \ln \frac{a_{I^-}^{\ 2}}{a_{I_2}}$$

$$= 0.622 - \frac{RT}{2F} \ln K - \frac{RT}{2F} \ln \frac{a_{I^-}^{\ 3}}{a_{I_3^-}} \qquad (19\text{-}7)$$

$$= 0.545 - \frac{RT}{2F} \ln \frac{a_{I^-}^{\ 3}}{a_{I_3^-}} \qquad (19\text{-}8)$$

which is the Nernst equation for (19-3). The standard electrode potentials of (19-1) and (19-3) happen to be only 10 mV apart because the concentration of triiodide ion happens to be about 1 M in a solution saturated with iodine and containing 1 M free iodide ion. The potential for a solution unsaturated with iodine may be calculated by either (19-7) or (19-8), with care to use the actual equilibrium concentration of either triiodide ion or iodine. For the sake of simplicity and to emphasize the stoichiometric relations, we often write a reaction showing the formation of iodine even in a solution containing excess iodide ion.

19-2 STANDARD SOLUTIONS AND PRIMARY STANDARDS

Iodine can be purified by sublimation from potassium iodide and calcium oxide and weighed as a primary standard. Because of the limited solubility and volatility of iodine, it must be dissolved in concentrated potassium iodide solution and diluted to volume. Air oxidation of iodide should be minimized by preparing the solution with water free of heavy-metal ions and storing it in a cool, dark place. Because of the inconvenience of weighing iodine accurately, its solutions are commonly standardized against arsenic(III) oxide (primary standard) or thiosulfate.[2]

Two common sources of error in the quantitative use of iodine are (*1*) loss of iodine due to its volatility and (*2*) air oxidation of iodide. The first is most likely to be encountered if the concentration of iodide is so low that solid iodine is present. Sufficient iodide should be present to decrease the concentration of free iodine below the saturation value. Loss of iodine is enhanced by evolution of gases (such as carbon dioxide generated for deaeration) and by elevated temperatures. Determinations should be carried out in cold solutions.

Air oxidation, the second source of error, is negligible in neutral solutions in the absence of catalysts, but its rate increases rapidly with increasing acidity. The reaction is catalyzed by metal ions of variable oxidation number (particularly copper) and is photochemical in nature. Moreover, air oxidation may be induced by a reaction between iodide and the oxidizing agent, especially when the main reaction is slow.[3] Therefore, solutions containing an excess of iodide and acid must stand no longer than necessary before titration of the iodine. When prolonged standing is required, the air should be displaced by an inert gas such as carbon dioxide, and in extreme cases the solution should be protected from light. The formation of triiodide ion in extremely dilute aqueous solutions of iodine has been attributed to reduction of iodine by traces of impurities.[4]

Thiosulfate solutions are generally prepared from sodium thiosulfate penta-hydrate, $Na_2S_2O_3 \cdot 5H_2O$, which under ordinary conditions is not a primary standard. The solutions should be prepared from water free of heavy-metal impurities to avoid "catalytic" air oxidation. Ordinary air oxidation is negligible in rate and proceeds through the slow decomposition of thiosulfate to sulfite, which is rapidly air-oxidized to sulfate. "Catalyzed" air oxidation, on the other hand, proceeds through the reduction of metals such as copper(II) or iron(III), present as thiosulfate complexes, followed by air oxidation of the lower oxidation state:

$$2Cu(II) + 2S_2O_3^{=} \rightarrow S_4O_6^{=} + 2Cu(I) \qquad (19\text{-}9)$$

$$2Cu(I) + \tfrac{1}{2}O_2 + H_2O \rightarrow 2Cu(II) + 2OH^{-} \qquad (19\text{-}10)$$

Freshly boiled water is recommended, to prevent decomposition of the thiosulfa solution by bacteria. A few drops of chloroform[5] act as an effective preservative Boiling also removes dissolved carbon dioxide, which is harmful unless neutralized because thiosulfate decomposes to sulfite and sulfur in dilute acid solution:

$$H^+ + S_2O_3^= \rightarrow HSO_3^- + S \qquad (19\text{-}11)$$

With higher concentrations of acid a series of reactions has been postulated to occur[6,7] in which the primary reaction is:

$$H_2S_2O_3 \rightarrow H_2S + SO_3$$

followed by formation of a trithionate

$$H_2S + 2SO_3 \rightarrow H_2S_3O_6$$

which by reaction with excess thiosulfate forms higher thionates

$$S_3O_6^= + H^+ + S_2O_3^= \rightarrow S_4O_6^= + HSO_3^- \qquad (19\text{-}12)$$

A *small amount* (0.1 g/l) of sodium carbonate may be added to advantage to ensure a slightly alkaline solution; but the addition of sodium hydroxide or large amounts of carbonate or borax should be avoided, because there is evidence that such solutions are actually less stable.[8] If turbidity develops, the solution is best discarded and a fresh one prepared.

Several primary standards are available for the standardization of thiosulfate solutions.[2,9] *Potassium dichromate* liberates iodine according to

$$Cr_2O_7^= + 14H^+ + 6I^- \rightarrow 3I_2 + 2Cr^{3+} + 7H_2O \qquad (19\text{-}13)$$

Because the reaction is not instantaneous at moderate acidity, a relatively high acidity is used, with the concomitant danger of air oxidation of iodide. Accurate results can be obtained by regulating the concentrations of acid and potassium iodide closely (for example, using 0.2 M hydrochloric acid and 2% potassium iodide) and allowing the mixture to stand 10 min.

Other suitable primary standards for thiosulfate include potassium iodate (Section 20-1), potassium bromate (Section 20-2), potassium ferricyanide,[10] metallic copper, potassium acid iodate,[11] and iodine.

19-3 STARCH AS AN INDICATOR

The characteristic deep blue imparted to a dilute solution of iodine by soluble starch has been used as a sensitive indicator for 150 years. Iodine at a concentration of 10^{-5} M is detected readily. The color intensity decreases with increasing temperature, being 10 times less sensitive[12] at 50°C than at 25°C. The sensitivity decreases upon the addition of solvents such as alcohol. No color is observed in solutions containing at least 50% ethanol.

Starches can be separated into two major fractions, which exist in different proportions in various plants. Amylose, a straight-chain fraction abundant in potato starch, forms a definite blue complex with iodine.[13] The other major fraction, amylopectin, is branched in structure and interacts only loosely with iodine to form a red-purple product. Simple methods for preparation of a linear starch fraction in 10% acetic acid have been described.[14,15] The nature of the starch-iodine-iodide complex has been the object of continuing studies for decades. A recent study[16] indicates that the rate-determining step in the formation of the blue complex involves formation of a nucleus of I_{11}^{3-} polyiodine tetramer (such as $4I_2 + 3I^-$ or $I_2 + 3I_3^-$) inside the amylose helix. The monomers, dimers, and trimers tend to dissociate, while the tetramer acts as a polymerization nucleus for the formation of a polyiodine chain. The minimum length of the amylose chain for the blue complex is about 30 to 40 glucose units, providing a cavity corresponding closely in length to an 11-atom polyiodine chain.

Sodium starch glycollate[17,18] and polyvinyl alcohol[19] have been suggested as being superior to starch in forming water-soluble complexes. Pritchard and Akintola[20] suggested that the blue complex of iodine and polyvinyl alcohol is the result of helical envelopment of iodine molecules. They concluded that the linear I_3^- ion provides a matrix on which the polyvinyl alcohol helix can wind. Methylene blue or malachite green has been recommended as preferable to starch as an indicator for iodine in the presence of acetone, glycerol, ethanol, or high concentrations of electrolyte.[21,22]

Soluble starch, available from chemical supply houses, is readily dispersed in water. The iodine-starch complex has limited water solubility, and it is therefore important not to add the starch indicator until near the end point when the iodine concentration is low. Because starch is subject to attack by microorganisms, the solution usually is prepared as needed. Among the products of hydrolysis is dextrose, which can cause large errors because of its reducing action. Various substances have been recommended as preservatives, including mercury(II) iodide and thymol. With formamide[23] a clear solution containing 5% starch is obtained that is stable indefinitely.

Complexes of ruthenium(II) and methyl-substituted derivatives of 1,10-phenanthroline have been suggested[24] as iodine fluorescent indicators under conditions where a starch does not function, such as highly acid, dilute, or colored solutions.

19-4   THE IODINE-THIOSULFATE REACTION

When a neutral or slightly acidic solution of iodine in potassium iodide is titrated with thiosulfate, the following reaction occurs rapidly and stoichiometrically:

$$I_3^- + 2S_2O_3^= \rightarrow 3I^- + S_4O_6^= \qquad (19\text{-}14)$$

Although the reaction is rapid and some tetrathionate is formed immediately upon mixing, a colorless intermediate, $S_2O_3I^-$, is formed by the rapid reversible reaction

$$S_2O_3^= + I_3^- \rightleftharpoons S_2O_3I^- + 2I^- \qquad (19\text{-}15)$$

The reaction of the intermediate with iodide,

$$2S_2O_3I^- + I^- \rightarrow S_4O_6^= + I_3^- \qquad (19\text{-}16)$$

accounts for the reappearance of iodine near the end point in the titration of dilute iodine solutions.[25] The intermediate also reacts with thiosulfate[26,27] to provide the principal course of the overall reaction

$$S_2O_3I^- + S_2O_3^= \rightarrow S_4O_6^= + I^- \qquad (19\text{-}17)$$

The rate law for the formation of tetrathionate turns out to involve the square of the concentration of the ion $S_2O_3I^-$. At low concentrations of iodide (below $0.003\ M$) the stoichiometry is no longer exact, because some sulfate is formed by the reaction

$$S_2O_3I^- + 3I_3^- + 5H_2O \rightarrow 2SO_4^= + 10H^+ + 10I^- \qquad (19\text{-}18)$$

which involves the first power of the $S_2O_3I^-$ concentration in the rate law. Thus sulfate formation becomes more important toward the end of a titration, as the $S_2O_3I^-$ concentration becomes smaller. Tetrathionate undergoes slow oxidation by excess iodine to form sulfate,[27] though the reaction is too slow to be of serious consequence under ordinary analytical conditions. Nevertheless, the instability of tetrathionate must be considered if the solution is to be used for other determinations, because substances that consume iodine, such as thiosulfate and sulfite, may be slowly formed, particularly in neutral or alkaline solution.[28]

From a practical viewpoint the iodine-thiosulfate reaction is accurate under the usual experimental conditions, provided the pH is less than 5. In alkaline solution (pH > 8) iodine reacts with hydroxyl ion to form hypoiodous acid ($K_a \simeq 10^{-11}$),

$$I_2 + OH^- \rightleftharpoons HOI + I^- \qquad (19\text{-}19)$$

or hypoiodite ion,[29]

$$I_2 + 2OH^- \rightleftharpoons OI^- + I^- + H_2O \qquad (19\text{-}20)$$

either of which oxidizes thiosulfate partially to sulfate.[30] For example, Kolthoff[25] found that the consumption of thiosulfate was low by 2.5, 2.3, and 7.1% when equal volumes of $0.2\ M$ $NaHCO_3$, Na_2HPO_4, and $Na_2B_4O_7$ were added to $0.05\ M$ iodine solution prior to titration. The same salts had no effect, however, on the reverse titration, because the rapid thiosulfate reaction consumed the iodine before it could react with hydroxyl ion. In strongly alkaline solution, the reaction goes quantitatively to sulfate.[31] Acidification ensures proper stoichiometry. In extremely acid solution thiosulfate (even a local excess) must not be exposed to acid, or it may decompose to sulfurous acid and sulfur. Sulfurous acid reacts with 1 mole of I_2 per mole, compared with $\frac{1}{2}$ mole of I_2 per mole of thiosulfate:

$$S_2O_3^= + 2H^+ \rightarrow H_2SO_3 + S \qquad (19\text{-}21)$$

$$H_2SO_3 + I_2 + H_2O \rightarrow SO_4^= + 4H^+ + 2I^- \qquad (19\text{-}22)$$

Actually, a strongly acidic solution of iodine can be titrated if the thiosulfate is added slowly with vigorous stirring. In the reverse titration (thiosulfate with iodine), however, a weakly acidic solution must be used to avoid decomposition. A low concentration of acid tends to prevent appreciable air oxidation of iodide.

Strong oxidants generally oxidize thiosulfate to a mixture of tetrathionate, sulfate, and sulfur. The reaction between the oxidant and excess iodide must, therefore, be allowed to go to completion before titration of the iodine with thiosulfate.

19-5 THE IODINE-ARSENIC(III) REACTION

For an acid solution (pH < 2) we may write the equilibrium

$$H_3AsO_3 + I_3^- + H_2O \rightleftharpoons H_3AsO_4 + 3I^- + 2H^+ \qquad (19\text{-}23)$$

for which an equilibrium constant of 5.5×10^{-2} has been evaluated.[32] At higher pH values the ionization of arsenious acid ($pK_1 = 9.2$) and of arsenic acid ($pK_1 = 2.3$, $pK_2 = 4.4$, $pK_3 = 9.2$) must be considered. Thus the half-reactions for oxidation of arsenic(III) are written

$$H_3AsO_3 + H_2O \rightarrow H_3AsO_4 + 2H^+ + 2e^- \qquad \text{pH} < 2.3 \qquad (19\text{-}24)$$

$$H_3AsO_3 + H_2O \rightarrow H_2AsO_4^- + 3H^+ + 2e^- \qquad 2.3 < \text{pH} < 4.4 \qquad (19\text{-}25)$$

$$H_3AsO_3 + H_2O \rightarrow HAsO_4^= + 4H^+ + 2e^- \qquad 4.4 < \text{pH} < 9.2 \qquad (19\text{-}26)$$

$$H_2AsO_3^- + H_2O \rightarrow AsO_4^{3-} + 4H^+ + 2e^- \qquad 9.2 < \text{pH} \qquad (19\text{-}27)$$

Up to pH 9 the iodine-iodide half-reaction is independent of pH. From pH 9 to 11 the formation of hypoiodite is appreciable; but, since hypoiodite oxidizes arsenic(III) to arsenic(V) stoichiometrically,

$$H_2AsO_3^- + OI^- + 2OH^- \rightarrow AsO_4^{3-} + I^- + H_2O \qquad (19\text{-}28)$$

arsenic(III) can be titrated successfully with iodine up to pH 11.[33] Above this value the formation of iodate becomes marked, and iodate does not readily oxidize arsenic(III). The reverse titration of iodine with arsenic(III) is not successful above a pH of 9 because iodate is formed.

Arsenic(V) is a stronger oxidant than iodine in strongly acidic solution and, indeed, can be determined quantitatively by addition of excess iodide at a hydrogen ion concentration of about 4 M and titration of the resulting iodine. The more usual determination is based on the direct titration of arsenic(III) with iodine at a higher pH. To ensure adequate speed and quantitativeness, the titration should be carried out at pH values between 4 and 9.

Disodium phosphate or borax may be used conveniently to buffer a strongly acidic solution to the proper pH value. Sodium bicarbonate is less desirable because of the possible loss of iodine vapor with the escaping carbon dioxide.

19-6   DETERMINATION OF COPPER

When a slightly acidic solution of Cu(II) is treated with an excess of iodide, a precipitate of Cu(I) iodide is formed, and an equivalent amount of iodine is released. The iodine is titrated with thiosulfate. The reaction may be written

$$2Cu^{++} + 4I^- \rightarrow 2CuI + I_2 \qquad (19\text{-}29)$$

or preferably

$$2Cu^{++} + 5I^- \rightarrow 2CuI + I_3^- \qquad (19\text{-}30)$$

The iodide ion not only serves as a reducing agent, but also exerts an enormous influence on the potential of the Cu(II)-Cu(I) couple because of the slight solubility of Cu(I) iodide ($K_{sp} = 10^{-12}$). Cu(I), which is not stable in water at appreciable concentrations, owing to its disproportionation into Cu(II) and metallic copper, is stabilized by the formation of Cu(I) iodide. The half-reaction

$$Cu^{++} + I^- + e^- \rightleftharpoons CuI(\text{solid}) \qquad E° = 0.85 \text{ V} \qquad (19\text{-}31)$$

corresponds to a stronger oxidizing system than the iodine-triiodide couple:

$$I_3^- + 2e^- \rightleftharpoons 3I^- \qquad E° = 0.545 \text{ V} \qquad (19\text{-}32)$$

Although hydrogen ions do not appear in (19-29) or (19-30), the effect of pH on the experimental results is important. If the pH is above about 4, hydrolysis of the Cu(II) ion or precipitation of copper hydroxide causes the reaction to become sluggish. The end point then becomes fleeting, owing to a shift in the equilibrium to release more iodine as it is removed by reaction with thiosulfate. Complexing agents also inhibit the reaction by decreasing the concentration of Cu(II) ion. If the pH is below about 0.5, air oxidation of iodide becomes significant.[34] Actually, the lower limit of pH often is restricted more severely by the presence of As(V) or Sb(V), which oxidize iodide in acid solution. If the pH is 3.2 or greater, no oxidation occurs.[35]

Generally, the pH is adjusted to the proper value with the aid of a suitable buffer. Ammonium hydrogen fluoride is especially effective in that it acts as an equimolar buffer of HF and F^- ($K_a \simeq 7 \times 10^{-4}$) of the proper pH value[36] and also serves as a complexing agent to prevent the interference of Fe(III) by its strong complexation.[35] The effects of various buffer components have been studied and the permissible pH range has been found[37] to depend on the nature of the buffer components as well as on the expected impurities. Formic acid buffers were recommended for solutions free of iron. Acetate and phthalate buffers are commonly used, with the addition of fluoride if iron is present.

A source of error is the adsorption of iodine by Cu(I) iodide. The adsorbed iodine imparts a buff color to the precipitate and interferes with the sharpness of the end point. In addition, the results may be low[34] by as much as 0.3%, and fleeting end points may be observed.

Foote and Vance[38] suggested the addition of thiocyanate just before the end point. Cu(I) thiocyanate is less soluble than the iodide, and it is expected that at least the surface layers of the CuI will be transformed into CuSCN, which will have less tendency to adsorb iodine (as triiodide ion). Also, the last of the Cu(II) should produce the thiocyanate rather than the iodide of Cu(I), and the greater insolubility should bring about a more favorable equilibrium near the end point. Less interference from the formation of Cu(II) complexes should be experienced, owing to the more favorable equilibrium. The thiocyanate should not be added until most of the iodine has been reduced by thiosulfate, or appreciable reduction of iodine by thiocyanate may occur.[39]

If thiocyanate is not added, the thiosulfate should be standardized with copper metal under the same conditions as for the determination of copper.[40,41]

The error due to adsorption of iodine may be prevented also by the use of sufficiently high concentrations of iodide to dissolve the Cu(I) iodide as CuI_2^-:

$$CuI + I^- \rightleftharpoons CuI_2^- \qquad K = 6 \times 10^{-4} \qquad (19\text{-}33)$$

This method has been recommended by Meites,[42] but has the disadvantage that relatively large amounts (25 g) of potassium iodide are consumed. Hahn[43] used only enough potassium iodide (5 g) to dissolve the precipitate initially. Although some Cu(I) iodide precipitated, it was white, and the adsorption error was negligible.

Oxidizing agents, especially oxides of nitrogen remaining from the dissolution of the sample, must be completely removed. Boiling is effective in removing nitrogen oxides, but inconvenient. Moreover, a boiled nitric acid solution may develop appreciable amounts of nitrous acid on standing.[44] Urea, which is commonly used to remove nitrous acid, is effective in warmed solutions.[45] Brasted[44] recommended sulfamic acid as being more convenient. The reactions are

$$2HNO_2 + CO(NH_2)_2 \rightarrow 2N_2 + CO_2 + 3H_2O \qquad (19\text{-}34)$$

$$HNO_2 + HSO_3NH_2 \rightarrow N_2 + H_2SO_4 + H_2O \qquad (19\text{-}35)$$

The reaction between Cu(II) and iodide can be prevented or reversed by the addition of certain complexing agents. Kapur and Verma[46] used pyrophosphate to repress the reaction of copper and determined iodate in the presence of Cu(II). The use of pyrophosphate is limited by the relatively high pH necessary for stable complex formation, with the consequence of a slow reaction between iodate and iodide. By adding citrate, Hume and Kolthoff[47] reversed the Cu(II)-iodide reaction and therefore were able to determine strong oxidants in the presence of Cu(II). They found that, by adding excess mineral acid, the Cu(II)-citrate complex was destroyed and copper could be determined in the same sample. An alternative method of destroying the citrate complex without the necessity of titrating at low pH is to add cyanide, which also prevents the precipitation of Cu(I) iodide.[48]

A similar use of complexing is made in the titration of As(III) with iodine in the presence of Cu(II). Neutral tartrate is used to prevent the interaction of Cu(II) with iodide. After the iodine titration the copper may be determined by acidification to the proper pH and titration of the resulting iodine.[49]

Verma and Bhuchar[50] determined copper by reducing its tartrate complex with glucose to form insoluble Cu_2O, which was treated with an excess of standard iodine and back-titrated with standard As(III). Oxalate was added as a complexing agent to aid in the oxidation of the Cu_2O, and precautions were taken to avoid air oxidation. The method has the advantage of avoiding interference from V(V). For the determination of copper in alloys, Rooney and Pratt[51] separated copper by precipitation as its diethyldithiocarbamate from EDTA solution.

19-7 OXYGEN: WINKLER METHOD

The classical method of determining dissolved oxygen in water is that of Winkler,[52] which is based on the reaction between oxygen and a suspension of manganese(II) hydroxide in strongly alkaline solution. Upon acidification in the presence of iodide, the oxidized manganese hydroxide is reduced back to manganese(II) by iodide. The iodine equivalent to the dissolved oxygen present is titrated with standard thiosulfate. The important reactions can be written:

$$2Mn(II) + 4OH^- + O_2 \rightarrow 2MnO_2 + 2H_2O \qquad (19\text{-}36)$$

$$MnO_2 + 4H^+ + 2I^- \rightarrow I_2 + Mn(II) + 2H_2O \qquad (19\text{-}37)$$

Jones and Mullen[53] showed that (19-36) is quantitative at pH 9 or higher and that the half-time for the reaction is less than 2 s.

Potter and White[54,55] discussed critically the sources of error in the original Winkler method and various modifications, with special reference to the determination of low concentrations (below 0.01 ppm) of dissolved oxygen. The principal sources of error are: (1) reducing agents, such as sulfite, thiosulfate, iron(II), and organic matter, that react with oxygen or oxidized manganese in alkaline solution or with iodine in acid solution; (2) oxidizing agents that react with suspended manganese(II) hydroxide in alkaline solution or with iodide in acid solution (for example, chlorine, hypochlorite, nitrite, and peroxides); (3) failure to compensate properly for dissolved oxygen in the reagents; (4) errors inherent in the final determination.

In practical analysis it is common to run a blank in which the usual order of adding reagents ($MnSO_4$, KOH and KI, H_2SO_4) is reversed. Potter[54,55] pointed out that the preferred order in running the blank is KOH and KI, H_2SO_4, $MnSO_4$. Thus, if the blank is made alkaline initially, allowance is made for substances that interfere specifically in alkaline solution. An important interference not corrected for by running a blank is iron(II), which leads to low results by forming iron(II) hydroxide, which in turn reacts with oxygen. The resulting iron(III) ion does not produce an equivalent amount of iodine upon acidification.

19-8   WATER: KARL FISCHER METHOD

Measurement of the water content of a wide variety of materials is a problem of universal interest; a review of physical and chemical methods for its measurement has been made.[56,57] Though we are concerned here only with the Karl Fischer[58] titrimetric method, it should be mentioned that gas-chromatographic methods for the determination of traces of water in small samples would seem to be an attractive alternative.[59]

The Karl Fischer method is widely applicable not only to the determination of water in organic and inorganic systems but also to many indirect determinations based on quantitative reactions that consume or produce water.[56,60]

Stoichiometry   The Karl Fischer method usually consists of the titration of a sample in anhydrous methanol with a reagent composed of iodine, sulfur dioxide, and pyridine in methanol.[61] In the titration the end point corresponds to the appearance of the first excess of iodine detected visually or by electrical means.

The iodine and sulfur dioxide are present as addition compounds with pyridine. The reaction path has been shown[61] to involve two distinct steps. The first step involves the reaction between iodine and sulfur dioxide in the presence of pyridine and water to form a salt of pyridine-N-sulfonic acid. Pyridine acts as a base to form pyridinium iodide:

$$C_5H_5N \cdot I_2 + C_5H_5N \cdot SO_2 + C_5H_5N + H_2O$$

$$\rightarrow 2C_5H_5NH^+I^- + C_5H_5N \cdot SO_3 \quad (19\text{-}38)$$

This first step involves the reduction of iodine and the oxidation of sulfur dioxide, the water serving as a source of oxide ion. In the absence of methanol, the pyridine salt, "pyridine sulfur trioxide," can be isolated as a stable product. The second step of the reaction involves methanol, with the formation of pyridinium methyl sulfate:

$$C_5H_5N \cdot SO_3 + CH_3OH \rightarrow C_5H_5NHOSO_2OCH_3 \quad (19\text{-}39)$$

The methanol plays an important role; without it the second step could involve water instead of methanol, with the formation of pyridinium hydrogen sulfate:

$$C_5H_5N \cdot SO_3 + H_2O \rightarrow C_5H_5NHOSO_2OH \qquad (19\text{-}40)$$

This reaction, however, is not specific, because any active hydrogen compound could replace methanol. Therefore, practical titrations always are carried out in the presence of an excess of methanol. Under these conditions the overall reaction involves 1 mole of iodine per mole of water, and a simplified overall equation can be written

$$I_2 + SO_2 + H_2O + CH_3OH + 3py \rightarrow 2pyH^+I^- + pyHOSO_2OCH_3 \qquad (19\text{-}41)$$

where py represents pyridine.

Preparation of reagent   Even in the absence of water a reagent consisting of iodine, sulfur dioxide, pyridine, and methanol undergoes rapid initial loss in strength, followed by slower change.[61] The reactions involve the formation of quaternary pyridinium salts and pyridinium iodide, for example,

$$I_2 + SO_2 + 2CH_3OH + 3py \rightarrow 2pyH^+I^- + py(CH_3)OSO_2OCH_3 \qquad (19\text{-}42)$$

Various procedures[62] have been suggested to overcome at least partially the disadvantage of instability of the reagent. Of the various methods the following are noteworthy.

The *two-reagent method*[63,64] involves introduction of the sample into a reagent composed of pyridine, methanol, and sulfur dioxide. The titration reagent is iodine in methanol, which is relatively stable. This method is as precise as the usual one-reagent method, as long as no procedure is employed that involves the addition of excess iodine and back titration with water.

A *stabilized reagent* with methyl Cellosolve in place of methanol was proposed by Peters and Jungnickel.[65] A mixture of ethylene glycol and pyridine was recommended as the sample solvent. By avoiding the use of methanol, they minimized or eliminated interferences by ketones and acids. Dimethylformamide has been suggested[66] as a replacement for methanol in the reagent when traces of water are being determined.

Applications   The Karl Fischer reagent can be applied directly to the determination of water in a variety of organic compounds,[56,60] including saturated or unsaturated hydrocarbons, alcohols, halides, acids, acid anhydrides, esters, ethers, amines, amides, nitroso and nitro compounds, sulfides, hydroperoxides, and dialkyl peroxides. The use of sodium tartrate dihydrate for standardization of the response of the Karl Fischer reagent has been shown[67] to lead to a small error because of occlusion of about 2% (relative) water in the crystal structure.

Active carbonyl compounds interfere by forming acetals or ketals by reaction with methanol. The interference can be avoided by adding hydrogen cyanide to form cyanohydrins that do not interfere. Other interfering substances include mercaptans and certain amines, which react with iodine, and various oxidizing substances such as peroxy acids, diacyl peroxides, and quinones, which react with iodide to produce iodine.

With inorganic substances, a greater variety of interferences is encountered.[68] Both sulfur dioxide and iodide act as reducing agents, and therefore many oxidizing agents interfere. For certain oxidants that react exclusively with iodide, a correction can be made for the amount of iodine produced. For example, if $CuSO_4 \cdot 5H_2O$ is to be analyzed for water of hydration, 5 moles of iodine are consumed per mole of pentahydrate by the water of hydration, but 0.5 mole is produced by the reduction of copper(II) to copper(I) iodide, leaving a net consumption of 4.5 moles. Reducing agents that react with iodine cause erroneously high values for the water content of the sample. Oxides and hydroxides of several metals that react to form iodides behave like water toward the Karl Fischer reagent. Alkali and alkaline earth metal salts of weak acids may act like basic oxides to produce water.

Theoretically,[60] the Karl Fischer reagent can be used for the determination of any organic functional group that will undergo a quantitative and stoichiometric reaction to produce or consume water under conditions that do not interfere with the titration. These include alcohols, carboxylic acids, acid anhydrides, carbonyl compounds, primary and secondary amines, nitriles, and peroxides.

19-9 AMPLIFICATION REACTIONS

Amplification reactions have been reviewed.[69] A straightforward example involves the determination[70] of small amounts of CO_2 by passing it over alternate layers of heated carbon (at 900°C) and CuO (at 500°C), whereby the following transformations occur:

$$CO_2 \rightarrow 2CO \rightarrow 2CO_2 \rightarrow 4CO \rightarrow 4CO_2 \rightarrow 8CO \rightarrow 8CO_2 \cdots$$

The multiplication thus proceeds according to a geometric progression. A small amount of water or oxygen can be amplified by first converting it to CO with heated carbon.

The best known amplification reactions are used in connection with the determination of small amounts of iodide. By oxidation of iodide to iodate,

$$I^- + 3H_2O \rightarrow IO_3^- + 6H^+ + 6e^- \qquad (19\text{-}43)$$

followed by treatment of the iodate with excess iodide in acid solution,

$$IO_3^- + 5I^- + 6H^+ \rightarrow 3I_2 + 3H_2O \qquad (19\text{-}44)$$

an amplification factor of 6 is achieved. The factor may be increased to 36 or even 216; the iodine is extracted with carbon tetrachloride, extracted back into aqueous solution, and reoxidized with bromine water. Excess bromine is removed by boiling, excess iodide added, and the extraction repeated.[71] Another method involves use of silver iodide instead of potassium iodide.[72]

If iodide is oxidized with periodate[73]

$$I^- + 3IO_4^- \rightarrow 4IO_3^- \qquad (19\text{-}45)$$

followed by masking the excess periodate with molybdate,[74,75] the IO_3^- formed makes possible a 24-fold amplification in a single stage.

A micromethod has been described for organic sulfur determination by an amplification reaction involving barium bromate and sulfuric acid.[76] If amplification reactions are carried out in a chromatographic column, phosphate, ferrocyanide, chromate, zinc, iron, and silver can be amplified directly or indirectly.[77]

19-10 OTHER METHODS

Direct reaction with iodine A standard iodine solution as a weak oxidant can be used as a titrant for strong reductants. Its wide range of applicability may be illustrated by brief mention of some examples: determination of arsenic by titration of arsenic(III) in bicarbonate solution using starch as an indicator; determination of tin after reduction to tin(II) by the use of lead, antimony, aluminum, nickel, or iron; determination of thallium(III) after reduction to thallium(I); determination of sulfide, either by direct titration with iodine or indirectly after adding an excess of iodine followed by a back titration; determination of thioacetamide by titration with iodine as a basis for the microdetermination of heavy-metal ions; determination of sulfite by back titration; resolution of hypophosphite and phosphite by titrations at two different pH values; determination of cyanide by quantitative reaction with iodine in alkaline solution; and determination of many organic compounds by titration with iodine,[78] including polyphenols, ascorbic acid, mercaptans, uric acid, hydrazines, phenols, dithioglycolic acid, organometallic mercaptides, and aluminum alkyls. *Iodine numbers* are used as measures of unsaturation in fats and oils. Details of many methods can be obtained from Kolthoff and Belcher.[1]

Methods based on iodine titration with thiosulfate Iodide, being a weak reductant, can react with an enormous variety[1,79] of oxidants to liberate an equivalent quantity of I_2 that can be titrated with thiosulfate. Such oxidants include peroxides, peroxy compounds, peroxydisulfate, ozone, iron(III), chromate, selenium (as $SeO_3^=$), silver(II) oxide, xenon trioxide, iodate, and bromate. Bromide can be determined by oxidizing it to bromine, followed by extraction and determination of the bromine iodometrically. Metals such as barium, strontium, and lead can be determined after precipitation of their chromates, followed by determination of the chromate in the precipitate. Lithium can be precipitated as a complex periodate, and after filtration and washing the periodate can be determined iodometrically. Thorium can be separated from the lanthanides by precipitation as the iodate from solutions of relatively high nitric acid concentration, and the iodate then determined iodometrically.

REFERENCES

1   I. M. KOLTHOFF and R. BELCHER: "Volumetric Analysis," vol. 3, Interscience, New York, 1957.
2   S. POPOFF and J. L. WHITMAN: *J. Amer. Chem. Soc.*, **47**:2259 (1925).
3   KOLTHOFF and BELCHER,[1] p. 203.
4   J. H. WOLFENDEN: *Anal. Chem.*, **29**:1098 (1957).
5   J. L. KASSNER and E. E. KASSNER: *Ind. Eng. Chem., Anal. Ed.*, **12**:655 (1940).
6   M. SCHMIDT: *Z. Anorg. Allg. Chem.*, **289**:141 (1957).
7   F. H. POLLARD, G. NICKLESS, and R. B. GLOVER: *J. Chromatogr.*, **15**:518 (1964).
8   S. O. RUE: *Ind. Eng. Chem., Anal. Ed.*, **14**:802 (1942).
9   KOLTHOFF and BELCHER,[1] p. 213.
10 I. M. KOLTHOFF: *Pharm. Weekbl.*, **59**:66 (1922).
11 I. M. KOLTHOFF and L. H. VAN BERK: *J. Amer. Chem. Soc.*, **48**:2799 (1926).
12 I. M. KOLTHOFF, E. B. SANDELL, E. J. MEEHAN, and S. BRUCKENSTEIN: "Quantitative Chemical Analysis," 4th ed., p. 845, Macmillan, London, 1969.
13 F. L. BATES, D. FRENCH, and R. E. RUNDLE: *J. Amer. Chem. Soc.*, **65**:142 (1943); R. E. RUNDLE and D. FRENCH: *J. Amer. Chem. Soc.*, **65**:1707 (1943).
14 J. L. LAMBERT: *Anal. Chem.*, **25**:984 (1953).
15 J. L. LAMBERT and S. C. RHOADS: *Anal. Chem.*, **28**:1629 (1956).
16 J. C. THOMPSON and E. HAMORI: *J. Phys. Chem.*, **75**:272 (1971).
17 S. PEAT, E. J. BOURNE, and R. D. THROWER: *Nature*, **159**:810 (1947).
18 L. DEIBNER: *Chim. Anal.* (Paris), **33**:207 (1951).
19 S. A. MILLER and A. BRACKEN: *J. Chem. Soc.*, 1933 (1951).
20 J. G. PRITCHARD and D. A. AKINTOLA: *Talanta*, **19**:877 (1972).
21 E. KOLOS, E. BANYAI, K. ERÖSS, and L. ERDEY: *Mikrochim. Acta*, **1967**:333.
22 J. O. MEDITSCH: *Anal. Chim. Acta*, **31**:286 (1964).
23 A. C. HOLLER: *Anal. Chem.*, **27**:866 (1955).
24 B. KRATOCHVIL and M. C. WHITE: *Anal. Chim. Acta*, **31**:528 (1964).
25 I. M. KOLTHOFF: *Z. Anal. Chem.*, **60**:338 (1921).
26 G. DODD and R. O. GRIFFITH: *Trans. Faraday Soc.*, **45**:546 (1949).
27 A. D. AWTREY and R. E. CONNICK: *J. Amer. Chem. Soc.*, **73**:1341 (1951).
28 KOLTHOFF and BELCHER,[1] p. 214.
29 T. L. ALLEN and R. M. KEEFER: *J. Amer. Chem. Soc.*, **77**:2957 (1955).
30 J. H. BRADBURY and A. N. HAMBLY: *Aust. J. Sci. Res., Ser. A*, **5**:541 (1952).
31 E. ABEL: *Z. Anorg. Chem.*, **74**:395 (1912).

32   E. W. WASHBURN and E. K. STRACHAN: *J. Amer. Chem. Soc.*, **35**:681 (1913).

33   R. K. MCALPINE: *J. Chem. Educ.*, **26**:362 (1949).

34   E. W. HAMMOCK and E. H. SWIFT: *Anal. Chem.*, **21**:975 (1949).

35   B. PARK: *Ind. Eng. Chem., Anal. Ed.*, **3**:77 (1931).

36   W. R. CROWELL, T. E. HILLIS, S. C. RITTENBERG, and R. F. EVENSON: *Ind. Eng. Chem., Anal. Ed.*, **8**:9 (1936); W. R. CROWELL, S. H. SILVER, and A. T. SPIHER: *Ind. Eng. Chem., Anal. Ed.*, **10**:80 (1938).

37   W. R. CROWELL: *Ind. Eng. Chem., Anal. Ed.*, **11**:159 (1939).

38   H. W. FOOTE and J. E. VANCE: *J. Amer. Chem. Soc.*, **57**:845 (1935); *Ind. Eng. Chem., Anal. Ed.*, **8**:119 (1936).

39   H. BASTIUS: *Fresenius' Z. Anal. Chem.*, **250**:169 (1970).

40   H. W. FOOTE: *J. Amer. Chem. Soc.*, **60**:1349 (1938).

41   J. J. KOLB: *Ind. Eng. Chem., Anal. Ed.*, **11**:197 (1939).

42   L. MEITES: *Anal. Chem.*, **24**:1618 (1952).

43   F. L. HAHN: *Anal. Chim. Acta*, **3**:65 (1949).

44   R. C. BRASTED: *Anal. Chem.*, **24**:1040 (1952).

45   L. O. HILL: *Ind. Eng. Chem., Anal. Ed.*, **8**:200 (1936).

46   P. L. KAPUR and M. R. VERMA: *Ind. Eng. Chem., Anal. Ed.*, **13**:338 (1941).

47   D. N. HUME and I. M. KOLTHOFF: *Ind. Eng. Chem., Anal. Ed.*, **16**:103 (1944).

48   J. F. SCAIFE: *Anal. Chem.*, **29**:1224 (1957).

49   KOLTHOFF, SANDELL, MEEHAN, and BRUCKENSTEIN,[12] p. 855.

50   M. R. VERMA and V. M. BHUCHAR: *J. Sci. Ind. Res., Sect. B*, **15**:437 (1956).

51   R. C. ROONEY and C. G. PRATT: *Analyst*, **97**:731 (1972).

52   L. W. WINKLER: *Ber. Deut. Chem. Ges.*, **21**:2843 (1888).

53   M. M. JONES and M. W. MULLEN: *Talanta*, **20**:327 (1973).

54   E. C. POTTER; *J. Appl. Chem.*, **7**:285, 297 (1957).

55   E. C. POTTER and J. F. WHITE: *J. Appl. Chem.*, **7**:309, 317; 459 (1957).

56   J. MITCHELL, JR.: in "Treatise on Analytical Chemistry," I. M. Kolthoff and P. J. Elving (eds.), pt. II, vol. I, chap. 4, Interscience, New York, 1961.

57   C. HARRIS: *Talanta*, **19**:1523 (1972).

58   K. FISCHER: *Angew. Chem.*, **48**:394 (1935).

59   P. SELLERS: *Acta Chem. Scand.*, **25**:2295 (1971).

60   J. MITCHELL, JR., and D. M. SMITH: "Aquametry," Interscience, New York, 1948.

61   D. M. SMITH, W. M. D. BRYANT, and J. MITCHELL, JR.: *J. Amer. Chem. Soc.*, **61**:2407 (1939).

62   KOLTHOFF and BELCHER,[1] p. 413.

63   A. JOHANSSON: *Acta Chem. Scand.*, **3**:1058 (1949).

64   W. SEAMAN, W. H. MCCOMAS, JR., and G. A. ALLEN: *Anal. Chem.*, **21**:510 (1949).

65   E. D. PETERS and J. L. JUNGNICKEL: *Anal. Chem.*, **27**:450 (1955).

66   V. A. KLIMOVA, F. B. SHERMAN, and A. M. L'VOV: *Zh. Anal. Khim.*, **25**:158 (1970); *J. Anal. Chem. USSR*, **25**:127 (1970).

67   T. H. BEASLEY, SR., H. W. ZIEGLER, R. L. CHARLES, and P. KING: *Anal. Chem.*, **44**:1833 (1972).

68   W. M. D. BRYANT, J. MITCHELL, JR., D. M. SMITH, and E. C. ASHBY: *J. Amer. Chem. Soc.*, **63**:2924, 2927 (1941).

69 R. BELCHER: *Talanta*, **15**:357 (1968).

70 W. SCHÖNIGER: *Microchem. J.*, **11**:469 (1966).

71 H. SPITZY, H. SKRUBE, and F. S. SADEK: *Mikrochim. Acta*, **1953**:375.

72 U. FRITSCHE, *Mikrochim. Acta*, **1969**:1322.

73 H. H. WILLARD and L. H. GREATHOUSE: *J. Amer. Chem. Soc.*, **60**:2869 (1938).

74 D. BURNEL, M. F. HUTIN, and L. MALAPRADE: *Chim. Anal.* (Paris), **53**:230 (1971).

75 R. BELCHER, J. W. HAMYA, and A. TOWNSHEND: *Anal. Chim. Acta*, **49**:570 (1970).

76 Y. A. GAWARGIOUS and A. B. FARAG: *Talanta*, **19**:641 (1972).

77 H. WEISZ and M. GONNER: *Anal. Chim. Acta*, **43**:235 (1968).

78 M. R. F. ASHWORTH: "Titrimetric Organic Analysis," pt. I, Interscience, New York, 1964.

79 M. R. F. ASHWORTH: "Titrimetric Organic Analysis," pt. II, Interscience, New York, 1965.

20

OTHER OXIDANTS AND REDUCTANTS

The purpose of this chapter is to review several redox reagents of lesser importance. More details, particularly about applications, can be obtained from Berka, Vulterin, and Zyka[1] and Kolthoff and Belcher.[2]

20-1 POTASSIUM IODATE

Andrews[3] introduced the titration, further developed by Jamieson,[4] in which potassium iodate is used as a titrating reagent. In 3 to 9 M hydrochloric acid, iodate is reduced to iodine monochloride:

$$IO_3^- + 6H^+ + Cl^- + 4e^- \rightarrow ICl + 3H_2O \qquad (20\text{-}1)$$

In hydrochloric acid solution, iodine monochloride forms a stable complex with chloride:[5-7]

$$ICl + Cl^- \rightleftharpoons ICl_2^- \qquad K = 1.7 \times 10^2 \qquad (20\text{-}2)$$

The half-reaction is more properly written

$$IO_3^- + 6H^+ + 2Cl^- + 4e^- \rightarrow ICl_2^- + 3H_2O \qquad E° = 1.23 \text{ V} \qquad (20\text{-}3)$$

When free iodine forms during the course of a titration, even though a reducing agent is present, it can be regarded as resulting from the further reduction of the iodine monochloride complex:

$$2ICl_2^- + 2e^- \rightarrow I_2 + 4Cl^- \qquad E° = 1.06 \text{ V} \qquad (20\text{-}4)$$

When all the reducing agent has been consumed, the free iodine is titrated to form the iodine monochloride complex:

$$2I_2 + IO_3^- + 6H^+ + 10Cl^- \rightarrow 5ICl_2^- + 3H_2O \qquad (20\text{-}5)$$

and the Andrews end point is marked by the disappearance of the last of the iodine.

The optimum acidity varies from one reductant to another. For example, with Sb(III) the acidity is critical, correct results being obtained only over the range 2.5 to 3.5 M hydrochloric acid.[8] At lower acidity the rate of oxidation of iodine to iodine monochloride (20-5) is too low. The upper limit of hydrochloric acid concentration appears to be governed by the rate of reduction of iodine monochloride to iodine. For example, Penneman and Audrieth[9] found that at excessively high concentrations of hydrochloric acid (above 9 M) no visual end point could be observed in the titration of hydrazine because no iodine was formed. Yet the oxidation proceeded smoothly, as evidenced by the formation of nitrogen. Arsenic(III) behaves similarly. Evidently the complex ICl_2^- reacts only slowly with certain reducing agents with which iodine monochloride reacts rapidly.

Because the characteristic blue color of the starch-iodine complex is not formed at high concentrations of hydrochloric acid, starch is not a suitable indicator. Andrews used a few milliliters of an extraction solvent such as carbon tetrachloride or chloroform to obtain a highly sensitive end point. The main disadvantage is the inconvenience of shaking the reagent with the extraction solvent after each addition near the end point. As internal indicators, amaranth, brilliant ponceau,[10] and p-ethoxychrysoidin[11] have given the best results.

In addition to iodine [Equation (20-5)], several other reducing agents can be titrated accurately. The following overall reactions indicate the stoichiometry of some of the more important applications. For simplicity, the product is written ICl rather than ICl_2^-.

$$2I^- + IO_3^- + 6H^+ + 3Cl^- \rightarrow 3ICl + 3H_2O$$

$$2H_2SO_3 + IO_3^- + 2H^+ + Cl^- \rightarrow ICl + 2H_2SO_4 + H_2O$$

$$2H_3AsO_3 + IO_3^- + 2H^+ + Cl^- \rightarrow ICl + 2H_3AsO_4 + H_2O$$

$$2SbCl_4^- + IO_3^- + 6H^+ + 5Cl^- \rightarrow ICl + 2SbCl_6^- + 3H_2O$$

$$2Hg_2Cl_2 + IO_3^- + 6H^+ + 13Cl^- \rightarrow ICl + 4HgCl_4^= + 3H_2O$$

$$N_2H_4 + IO_3^- + 2H^+ + Cl^- \rightarrow ICl + N_2 + 3H_2O$$

$$2Tl^+ + IO_3^- + 6H^+ + 13Cl^- \rightarrow ICl + 2TlCl_6^{3-} + 3H_2O$$

$$2SCN^- + 3IO_3^- + 6H^+ + Cl^- \rightarrow ICl + 2ICN + 2H_2SO_4 + H_2O$$

Lang[12] found that, in the presence of cyanide, iodine is oxidized quantitatively to iodine cyanide. The standard potential of the half-reaction

$$2ICN + 2H^+ + 2e^- \rightleftharpoons I_2(\text{solid}) + 2HCN \qquad E° = 0.711 \text{ V} \qquad (20\text{-}6)$$

has been evaluated.[13] This method has the advantage over the iodine monochloride end point that the acidity level is lower (1 M in hydrochloric or sulfuric acid), so that starch can be used as the indicator. Some chloride should be present, because iodine monochloride evidently is necessary as an intermediate. The concentration of hydrochloric acid must not be excessive, or the rate of conversion of iodine chloride to iodine cyanide may occur too slowly; and then extreme caution must be exercised to avoid the fumes of hydrogen cyanide. The applications are similar to those of the Andrews method. Other strong oxidants can be used in place of iodate, provided iodine monochloride or iodine cyanide is added to serve with the indicator.

20-2 POTASSIUM BROMATE

Bromate is a strong oxidant, as indicated by the electrode potential:

$$BrO_3^- + 6H^+ + 5e^- \rightleftharpoons \tfrac{1}{2}Br_2 + 3H_2O \qquad E° = 1.52 \text{ V} \qquad (20\text{-}7)$$

During a titration, in the presence of excess reducing agent, bromine is reduced to bromide:

$$Br_2 + 2e^- \rightleftharpoons 2Br^- \qquad E° = 1.08 \text{ V} \qquad (20\text{-}8)$$

With the first excess of bromate, free bromine appears in the solution, owing to the reaction

$$BrO_3^- + 5Br^- + 6H^+ \rightleftharpoons 3Br_2 + 3H_2O \qquad K_{eq} = 2 \times 10^{38} \qquad (20\text{-}9)$$

Bromate itself reacts only slowly with reductants; the reactive substance[14] is Br_2 formed according to (20-9), BrCl, or a mixture of BrCl and Cl_2 formed in solutions containing chloride. The electrode potential is controlled by the halide concentration, and the reaction is slow in the absence of both chloride and bromide. In (20-9) the rate of reaction is proportional to bromate ion concentration, bromide ion concentration, and hydrogen ion concentration squared.[15]

Sometimes the reaction between the reducing agent and bromine is so slow that an excess of bromate must be added and then back-titrated after being allowed to stand. In this case bromate is essentially a convenient substitute for the less stable solution of elemental bromine.

Smith[16] showed that the addition of Hg(II) is beneficial in several direct bromate titrations. The higher rate of reaction inhibits the premature appearance of excess bromine and is assumed[14] to be due to the formation of Br^+ ions:

$$Hg^{++} + Br_2 \rightarrow HgBr^+ + Br^+ \qquad (20\text{-}10)$$

From the solubility of Hg(II) bromide and the formation constants of HgBr$^+$ and HgBr$_2$, we estimate

$$Br_2(0.1\ M) + Hg^{++}(0.1\ M) + 2e^- \rightleftharpoons HgBr_2(sat) \qquad E = 1.59\ V \qquad (20\text{-}11)$$

For the As(III)-KBrO$_3$ reaction, OsO$_4$ has been found to be of value as a catalyst.[17]

Thompson[18] found the rate of reaction of weak 1-equivalent reductants with bromate to be independent of the concentration of the reductant. After an induction period the rate was second order in bromate. This was explained in terms of a mechanism that involves HBrO$_2$ formation at one stage.[19]

The first use of a bromate-bromide mixture as a source of bromine was by Koppeschaar,[20] who devised the classical bromination method for phenol. The method is still in use today and has been applied to a wide variety of organic compounds. Three types of error can be recognized:[21] (1) oxidation of readily oxidized substances such as o- and p-aminophenols; (2) precipitation of incompletely brominated products, which occurs especially with parasubstituted compounds such as p-nitroaniline and p-iodoaniline; and (3) replacement of certain groups such as —COOH, —CHO, or —SO$_3$H by bromine.

The bromination of 8-hydroxyquinoline (oxine) is applicable to the indirect determination (after separation) of various metals.[22] Each mole of oxine consumes 2 moles of bromine, or 4 equivalents of oxidant:

$$C_9H_7ON + 2Br_2 \rightarrow C_9H_5Br_2ON + 2H^+ + 2Br^- \qquad (20\text{-}12)$$

so the method is sensitive.[23] Propylene carbonate is a useful solvent for bromine substitution reactions[24] in terms of rates of reactions and solubilities of reactants and products.

Other examples of the use of potassium bromate as an oxidant include: the determination of various metals through bromination of anthranilic acid;[25] the determination of unsaturation;[26] As(III);[27,28] Sb(III);[27] Fe(II) in the presence of organic substances;[29,30] mixtures of iodide, bromide, and chloride;[31] hypophosphite;[32] Cu(I); Tl(I); selenium; hydrogen peroxide; and hydrazine.

20-3 PERIODATE

The equilibria existing in solutions of periodic acid and its salts have been investigated,[33,34] with the conclusion that, of the possible hydrates of I(VII) oxide, only paraperiodic acid, H$_5$IO$_6$, exists as a solid in equilibrium with its aqueous solutions. Acidic periodate solutions contain chiefly the two species[35] IO$_4^-$ and H$_5$IO$_6$, the proportion of H$_5$IO$_6$ increasing with decreasing pH.

Potentiometric titrations[33] of the acid with base reveal two distinct end points: a sharp one at about pH 5 and a less distinct one at pH 10. Spectrophotometric data give evidence of a third ionization step. The ionization equilibria are complicated by a dehydration step; the observations can be interpreted[33] in terms of the equilibria;[34]

$$H_5IO_6 \rightleftharpoons H^+ + H_4IO_6^- \qquad K_1 = 1 \times 10^{-3} \qquad (20\text{-}13)$$

$$H_4IO_6^- \rightleftharpoons IO_4^- + 2H_2O \qquad K = 43 \qquad (20\text{-}14)$$

$$H_4IO_6^- \rightleftharpoons H^+ + H_3IO_6^= \qquad K_2 = 3 \times 10^{-7} \qquad (20\text{-}15)$$

The third step of ionization is so weak ($pK_3 = 15$) that it fails to give an end point in aqueous titrations.

Periodic acid is a strong oxidant in acid solution, as shown by the estimated standard potential:[36]

$$H_5IO_6 + H^+ + 2e^- \rightleftharpoons IO_3^- + 3H_2O \qquad E^\circ \simeq 1.6 \text{ V} \qquad (20\text{-}16)$$

The Malaprade[37] reaction is the basis of the most important analytical application of periodate: the determination of α-diols and α-carbonyl alcohols. If two adjacent carbon atoms have hydroxyl groups, periodate causes a cleavage of the carbon-carbon bond, with the formation of carbonyl groups:

$$RCH(OH)CH(OH)R' \rightarrow RCHO + R'CHO + 2H^+ + 2e^- \qquad (20\text{-}17)$$

If an alcohol has an α-carbonyl group, the carbonyl acts as though it were hydrated, with the formation of a carboxyl group:

$$RCH(O)CH(OH)R' \overset{H_2O}{\rightarrow} RCH(OH)_2CH(OH)R'$$

$$\rightarrow RCH(OH)O + R'CHO + 2H^+ + 2e^- \qquad (20\text{-}18)$$

Polyhydroxy compounds can be considered to be oxidized stepwise, the first step giving an α-hydroxy carbonyl compound, which is oxidized to a carboxylic acid. If each carbon atom has a hydroxyl group, the final products are formic acid and formaldehyde:

$$CH_2OH\text{—}(CHOH)_n\text{—}CH_2OH$$

$$\rightarrow 2HCHO + nHCOOH + 2(n + 1)H^+ + 2(n + 1)e^- \qquad (20\text{-}19)$$

Periodate also oxidizes α-amino alcohols in which the amino group is primary or secondary.[38,39]

The oxidations usually are carried out at room temperature with an excess of periodic acid or its salts. The analyses can be completed by several means. The IUPAC procedure[40] for glycerol involves an alkalimetric titration of the formic acid generated, using a pH of 8.0 at the end point. More commonly, the excess periodate is determined by adding a slight excess of standard As(III) to reduce periodate to iodate; then the excess As(III) is titrated with standard iodine.[41] Both iodate and periodate can be determined by first masking the periodate with molybdate while iodate is titrated and then demasking the periodate with oxalate, followed by titration of the periodate.[42]

A rapid test for glycol in motor oil involves testing colorimetrically for the formaldehyde product of the cleavage with *p*-rosaniline hydrochloride and sulfurous acid. A wine-colored, quinoid-type dye forms.

Potassium periodate has been used in place of iodate for various titrations involving the iodine-chloride, iodine-bromide, and iodine-cyanide end points.[43] Reaction of periodate with iodide involves transfer of an oxygen atom[44] rather than an electron, followed by rearrangement of the structure of the IO_4^- ion. In the oxidation of VO_2^{++} the active oxidant appears[45] to be IO_4^-. An intermediate $[IO_4^- \cdot VO_2^{++}]$ is formed, and the rate-limiting step is its reaction with further IO_4^-.

α-Hydroxyamino acids can be determined[46] by measuring the ammonia produced through oxidation with periodate in a carbonate medium. Of the amino acids occurring in proteins, only serine, threonine, hydroxylysine, and α-hydroxyglutamic acid have adjacent hydroxy and amino groups. Although various other amino acids react with periodate, only those with the α-hydroxyamino structure yield ammonia. The ammonia can be removed by aeration and determined acidimetrically.

20-4 OTHER OXYHALOGENS

Hypochlorite is a strong oxidant in alkaline solution, as shown by

$$OCl^- + H_2O + 2e^- \rightleftharpoons Cl^- + 2OH^- \qquad E° = 0.89 \text{ V} \qquad (20\text{-}20)$$

The potential is higher than those of the permanganate-manganate (0.57 V) and manganate–manganese dioxide (0.60 V) couples. The mechanisms of hypohalite oxidations have been reviewed by Edwards[47] and Taube.[48] The lack of isotopic oxygen exchange indicates that hypochlorite occurs in solution as the discrete species OCl^- and not, for example, as the hydrate $Cl(OH)_2^-$. Many hypohalite oxidations proceed by direct transfer of oxygen atoms, but alternative reaction paths frequently are possible.

Hypobromite, although a weaker oxidant than hypochlorite ($E° = 0.76$ V), often reacts more rapidly with reducing agents. Since hypobromite solutions are relatively unstable,[49,50] it is advantageous to produce hypobromite in solution by adding an excess of bromide to the sample[51] and then oxidizing it with hypochlorite.

Velghe and Claeys[52] developed a procedure for the analysis of the BrO^-, BrO_2^-, and BrO_3^- content of hypobromite solutions and studied its decomposition. They concluded that it should be stored in the dark, because light can bring about reactions that reduce the total oxidizing capacity and change the internal composition through formation of BrO_2^- and BrO_3^-.

In summary, it appears that, whenever the reaction of hypobromite with the reducing agent occurs rapidly in mildly alkaline solution, a solution of hypochlorite should be used in the presence of bromide. If the oxidation is too slow to permit a direct titration, an excess of hypochlorite can be added and determined iodometrically. Alternatively, an excess of standard As(III) can be added and back-titrated with hypochlorite. If a strongly alkaline medium is necessary and the reaction of hypochlorite is too slow, a standard solution of hypobromite should be used, with due precaution to avoid the presence of bromates.

20-5   OTHER OXIDANTS

Vanadium(V)   The standard potential of the V(V)-V(IV) couple is given by[53]

$$VO_2{}^+ + 2H^+ + e^- \rightleftharpoons VO^{++} + H_2O \qquad E° = 0.999 \text{ V} \qquad (20\text{-}21)$$

As an oxidant, V(V) is comparable to dichromate, but is less subject to interference from organic compounds than is dichromate or Ce(IV). Thus phenol and *o*-, *m*-, and *p*-cresols interfere with the titration of hydroquinone with Ce(IV), but V(V) gives the correct result.[54] It has been used as a substitute for iodate in the Andrews titration (Section 20-1) in 7 *M* hydrochloric acid, in the presence of iodine monochloride.[55]

Chloramine-T (sodium salt of *p*-toluenesulfochloramide) behaves much like hypochlorite; both may be considered derivatives of unipositive chlorine. Chloramine-T, however, has the advantages of being applicable in both acid and alkaline solution[56] and being more stable,[57] especially if prepared in a state of high purity and stored in a brown glass bottle. The reagent is used directly as a titrant in hydrochloric acid solutions with the Andrews iodine monochloride end point. In these applications it appears to offer little advantage over iodate. It has been used for the analysis of sodium borohydride[58] and for the determination of As(III), Sb(III), hydrazine, thiocyanate, and Tl(I), using iodine monochloride as a reaction intermediate.[59]

Dichloramine-T has been proposed as a titrant for use in solutions with glacial acetic acid as solvent.[60]

Ferricyanide, or hexacyanoferrate(III), is of interest because it can be applied in alkaline solutions and is its own primary standard. It has an $E°$ value[61] of 0.355 V. It has been used for the determination of Sn(II) in at least 6 *M* hydrochloric acid[62] and of Co(II) in 4 *M* ammonia–1 *M* ammonium citrate.[63] Ascorbic acid,[64] sulfide[65] at pH 9.4, As(III), Sb(III), Cr(III), and hydrazine have been determined with excess ferricyanide and back-titrated with vanadyl sulfate[66] ($VOSO_4$). Potassium molybdicyanide, $K_3Mo(CN)_8$, has an $E°$ of 0.73 V and is therefore a stronger oxidant[67] than $K_3Fe(CN)_6$.

Manganese(III) is a strong oxidant that must be stabilized by complex formation.[68]

Copper(III), stabilized by the formation of periodate or tellurate complexes, has been applied as an analytical reagent, mainly in KOH solutions.[69]

Iodine monochloride is a suitable reagent for oxidations in acid or weakly alkaline solutions. An excess of ICl is added, and a reaction such as the following takes place:

$$ZnS + 2ICl \rightarrow ZnCl_2 + I_2 + S \qquad (20\text{-}22)$$

Either the I_2 produced (after extraction) or the excess ICl can be determined. It has been suggested as a reagent for iodides, sulfides, xanthates, dithiocarbamates, and thiophene.[70]

Iodine cyanide in acid solution reacts according to

$$ICN + H^+ \rightleftharpoons I^+ + HCN \qquad (20\text{-}23)$$

Reaction of I^+ with a reductant such as I^-, As(III), Sb(III), Sn(II), Hg(I), ascorbic acid, or β-naphthol can make the reaction go to completion.[71]

Lead tetraacetate is a strong oxidant that often can be used to differentiate organic compounds containing hydroxyl groups on adjacent carbon atoms in a manner in which periodic acid or Ce(IV) cannot. It has been suggested for the quantitative determination of glycol, mannite, citric acid, malic acid, calcium gluconate,[72] hydroxy steroids,[73] and organic sulfides,[74] as a few examples. It can be used in glacial acetic acid.[75,76]

Iron(III) is a suitable oxidant for the determination of strong reducing agents such as Ti(III),[77] Mo(III),[78] V(II),[79] and W(III).[80]

20-6 TITANIUM(III)

Ti(III) is a moderately strong reducing agent:[35]

$$TiO^{++} + 2H^+ + e^- \rightleftharpoons Ti^{3+} + H_2O \qquad E^\circ \simeq 0.1 \text{ V} \qquad (20\text{-}24)$$

A formal potential of 0.056 V was reported by Diethelm and Foerster[81] for 0.5 M titanium in 2 M sulfuric acid. According to Kolthoff[82] the potential of the couple in 0.72 M hydrochloric acid solution (18°C) is given by

$$E = 0.03 - 0.058 \log \frac{[(TiIII)]}{[Ti(IV)][H^+]} \qquad (20\text{-}25)$$

The pH dependence suggests a half-reaction such as

$$TiO^{++} + H^+ + e^- \rightleftharpoons TiOH^{++} \qquad E^{\circ\prime} = 0.03 \text{ V} \qquad (20\text{-}26)$$

Mixtures of high ratios of [Ti(III)] to [Ti(IV)] showed a tendency toward hydrogen evolution, catalyzed by a platinum surface, as would be expected from the negative value of the potential calculated from (20-25). Any potential measured under such conditions is a mixed potential (Section 12-9) and therefore of no precise thermodynamic significance. The formal potential became more negative[82] with increasing chloride concentration at a constant hydrogen ion concentration, probably because Ti(IV) has some tendency to form chloride complexes. Kolthoff and Robinson[83] used citrate or tartrate as complexing agent to lower the potential of the Ti(IV)-Ti(III) couple, thereby causing Ti(III) to act as a stronger reductant. With citrate or tartrate, in spite of this change in potential, the danger of reduction of hydrogen ion is lessened by the higher pH. EDTA solutions might possess advantages because of the enormous stability of the Ti(IV)-EDTA complex (Section 11-2).

Standard solutions of Ti(III) can be prepared determinately by dissolving pure metal in hydrochloric or sulfuric acid.[84] Alternatively, standardization is conveniently carried out against dichromate.[85] Standard solutions of Ti(III) chloride and sulfate, being sensitive to air oxidation, must be stored under an inert atmosphere or else generated coulometrically.

Fe(III) can be titrated accurately with thiocyanate as an indicator if the reagent is added slowly near the end point[86] or the temperature raised to 50 to 60°C. Dissolved air should be removed with a stream of carbon dioxide. Cu(II) can be titrated in the presence of thiocyanate to give a white precipitate of Cu(I) thiocyanate.[87] Some Fe(II) is added as an indicator, and the end point detected by the disappearance of the red Fe(III)-thiocyanate color from the solution. Details for the determination of many oxidants (usually by adding an excess of Ti(III) and back-titrating with Fe(III), with thiocyanate as indicator) have been given.[2,88] Potentiometric end points often are used.

20-7   CHROMIUM(II)

Cr(II) is so strong a reductant that it is unstable with respect to hydrogen ions:

$$2Cr(II) + 2H^+ \rightarrow 2Cr(III) + H_2 \qquad (20\text{-}27)$$

Therefore, measurements with an electrode of low hydrogen overpotential are subject to error due to mixed potential behavior. Using an electrolytic tin electrode, Grube and Schlecht[89] found values of the formal potential to be -0.411 V in 0.0015 M sulfuric acid and -0.386 to -0.373 V in 0.1 M sulfuric acid.

Solutions of Cr(II) chloride or sulfate are prepared readily by reduction of Cr(III) with zinc. Determinate solutions can be prepared by reducing weighed quantities of potassium dichromate to Cr(III), using hydrogen peroxide.[90] The resulting solution is reduced further to Cr(II) and stored over amalgamated zinc. Cr(II) is the strongest reducing agent used as a titrant and must be rigorously protected from atmospheric oxygen.

Cr(II) may be used to carry out all the reactions of Ti(III), but usually under milder conditions. Applications of Cr(II) as a reductant have been reviewed.[1] The applications include Sn(IV) chloride[91] in the presence of catalysts such as Sb(V) or Bi(III), Sb(V)[91,92] in 20% HCl at elevated temperatures, Cu(II),[93,94] silver, gold, mercury, bismuth, iron, cobalt, molybdenum, tungsten, uranium, dichromate, vanadate, titanium, thallium, hydrogen peroxide, oxygen in water and gases, as well as organic compounds such as azo, nitro, and nitroso compounds and quinones. Excess Cr(II) in sulfuric acid solution reduces nitrate to ammonium ion.[95] The reduction is catalyzed by Ti(IV), which is rapidly reduced to Ti(III).

20-8   OTHER REDUCTANTS

Mercury(I) nitrate was introduced[96] as a reagent for Fe(III), with thiocyanate as indicator. The Fe(III)-Hg(I) reaction has been studied[97] for the direct determination of iron and the indirect determination of oxidants that react quantitatively with Fe(II). For example, hydroxylamine and Hg(I) can be determined by adding an excess of Fe(III) and back-titrating the excess. An unusual example is the reduction of Cu(II) to Cu(I) by Fe(II) in the presence of thiocyanate. In this reaction, complex formation causes the usual direction of reaction to be reversed.

Mercury(I) perchlorate also has been used as a reductant,[98] in conjunction with alkaline ferricyanide as oxidant. Reducing agents such as Cr(II), hydrogen peroxide, hydrazine, and As(III) are treated with an excess of standard ferricyanide and back-titrated. Merrer and Stock[99] found 0.01 M solutions of Hg(I) perchlorate to change concentration with time.

Vanadium(II) is intermediate between Ti(III) and Cr(II) as a reductant ($E° = -0.255$ V).

Chromium(II)-EDTA complex (formation constant 4×10^{13}) is a strong reductant that undergoes rapid reactions.[100]

Arsenic(III) has been recommended[101] for the determination of traces of hydroperoxide, since no free radicals or induced reactions are expected.

Ascorbic acid, which has been studied extensively,[102] is a strong reductant ($E° = +0.185$ V). It reacts readily with oxygen; with EDTA and formic acid[103] the solutions are stabilized, and so an inert atmosphere is unnecessary. With EDTA, traces of heavy metals no longer catalyze the oxidation.

Hydroquinone is of interest mainly as a reductant for Au(III) to the metal, with *o*-dianisidine[104] or 3-methylbenzidine[105] as indicators. Its value as a reductant has been reviewed, and it has been suggested[106] that it can replace Fe(II) in many applications.

Oxalate stabilized with Fe(III) has been recommended[107] for the determination of active oxygen in manganese dioxide.

Iron(II) in concentrated phosphoric acid has been developed[108] as a reagent for the direct titration of substances such as U(VI), Mo(VI), and V(IV).

Several metal ions of low oxidation number can be generated electrochemically for coulometric titrations (Section 14-10).

The *hydrated electron*, symbolized by e_{aq}^-, is stated by Hart[109] to be an ideal analytical reagent. It is highly specific in its reactions and intensely colored. It can be produced photochemically by pulsed radiolysis, electrochemically, and by reduction of water where it is the precursor to formation of hydrogen atoms. In water the hydrated electron has a half-life of about 8×10^{-4} s.

$$e_{aq}^- + H_2O \rightleftharpoons H_{aq} + OH_{aq}^- \qquad (20\text{-}28)$$

At a concentration greater than 10^{-8} M the following reaction becomes important:

$$e_{aq}^- + e_{aq}^- \rightarrow (H_2)_{aq} + 2OH_{aq}^- \qquad (20\text{-}29)$$

The hydrated electron has been suggested as a reagent for the determination of submicromolar concentrations of scavengers. It can reduce ions not reducible by hydrogen atoms and is a more effective reducing agent[109] by about 0.6 V. The hydrated electron and a hydrogen atom can be considered a conjugate acid-base pair in the Brønsted sense.[110]

$$e_{aq}^- + H_{aq}^+ \rightleftharpoons H_{aq} \qquad (20\text{-}30)$$

Using pulsed radiolysis, ions such as Zn^+, Pb^+, Cd^+, and Co^+ can be produced[111] in solution by reaction with e_{aq}^-.

REFERENCES

1 A. BERKA, J. VULTERIN, and J. ZYKA: "Newer Redox Titrants," translated by H. Weisz, Pergamon, New York, 1965.
2 I. M. KOLTHOFF and R. BELCHER: "Volumetric Analysis," vol. 3, Interscience, New York, 1957.
3 L. W. ANDREWS: *J. Amer. Chem. Soc.*, **25**:756 (1903).
4 G. S. JAMIESON: "Volumetric Iodate Methods," Reinhold, New York, 1926.
5 E. H. SWIFT: *J. Amer. Chem. Soc.*, **52**:894 (1930).
6 F. A. PHILBRICK: *J. Amer. Chem. Soc.*, **56**:1257 (1934).
7 J. H. FAULL, JR.: *J. Amer. Chem. Soc.*, **56**:522 (1934).
8 E. W. HAMMOCK, R. A. BROWN, and E. H. SWIFT: *Anal. Chem.*, **20**:1048 (1948).
9 R. A. PENNEMAN and L. F. AUDRIETH: *Anal. Chem.*, **20**:1058 (1948).
10 G. F. SMITH and R. L. MAY: *Ind. Eng. Chem., Anal. Ed.*, **13**:460 (1941); G. F. SMITH and C. S. WILCOX: *Ind. Eng. Chem., Anal. Ed.*, **14**:49 (1942).
11 R. BELCHER and S. J. CLARK: *Anal. Chim. Acta*, **4**:580 (1950).
12 R. LANG: *Z. Anorg. Allg. Chem.*, **122**:332 (1922); **142**:229, 280 (1925); **144**:75 (1925).
13 D. F. BOWERSOX, E. A. BUTLER, and E. H. SWIFT: *Anal. Chem.*, **28**:221 (1956).

14 E. SCHULEK, K. BURGER, and J. LASZLOVSZKY: *Talanta*, **7**:51 (1960).

15 J. R. CLARKE: *J. Chem. Educ.*, **47**:775 (1970).

16 G. F. SMITH: *J. Amer. Chem. Soc.*, **45**:1115, 1417, 1666 (1923); **46**:1577 (1924).

17 D. R. BHATTARAI and J. M. OTTAWAY: *Talanta*, **19**:793 (1972).

18 R. C. THOMPSON: *J. Amer. Chem. Soc.*, **93**:7315 (1971).

19 R. M. NOYES, R. J. FIELD, and R. C. THOMPSON: *J. Amer. Chem. Soc.*, **93**:7315 (1971).

20 W. F. KOPPESCHAAR: *Z. Anal. Chem.*, **15**:233 (1876).

21 A. W. FRANCIS and A. J. HILL: *J. Amer. Chem. Soc.*, **46**:2498 (1924).

22 R. BERG: *Pharm. Ztg.*, **71**:1542 (1926).

23 R. G. W. HOLLINGSHEAD: "Oxine and Its Derivatives," vols. 1–4, Butterworths, London, 1954–1956.

24 R. D. KRAUSE and B. KRATOCHVIL: *Anal. Chem.*, **45**:844 (1973).

25 H. FUNK and M. DITT: *Z. Anal. Chem.*, **91**:332 (1933); **93**:241 (1933); H. FUNK and M. DEMMEL: *Z. Anal. Chem.*, **96**:385 (1934).

26 A. POLGAR and J. L. JUNGNICKEL in "Organic Analysis," vol. 3, p. 203, Interscience, New York, 1956.

27 S. GYÖRY: *Z. Anal. Chem.*, **32**:415 (1893).

28 J. M. OTTAWAY and E. BISHOP: *Anal. Chim. Acta*, **33**:153 (1965).

29 G. F. SMITH and H. H. BLISS: *J. Amer. Chem. Soc.*, **53**:4291 (1931).

30 G. G. RAO and N. K. MURTY: *Fresenius' Z. Anal. Chem.*, **208**:97 (1965).

31 R. BERG: *Z. Anal. Chem.*, **69**:1, 342 (1926).

32 A. SCHWICKER: *Z. Anal. Chem.*, **110**:161 (1937).

33 C. E. CROUTHAMEL, H. V. MEEK, D. S. MARTIN, and C. V. BANKS: *J. Amer. Chem. Soc.*, **71**:3031 (1949); C. E. CROUTHAMEL, A. M. HAYES, and D. S. MARTIN: *J. Amer. Chem. Soc.*, **73**:82 (1951).

34 S. H. LAURIE, J. M. WILLIAMS, and C. J. NYMAN: *J. Phys. Chem.*, **68**:1311 (1964).

35 D. J. B. GALLIFORD, R. H. NUTTALL, and J. M. OTTAWAY, *Talanta*, **19**:871 (1972).

36 W. M. LATIMER: "The Oxidation States of the Elements and Their Potentials in Aqueous Solutions," 2d ed., p. 66, Prentice-Hall, Englewood Cliffs, N.J., 1952.

37 L. MALAPRADE: *C.R. Acad. Sci.*, **186**:382 (1928); *Bull. Soc. Chim. Fr.*, (4) **43**:683 (1928).

38 B. H. NICOLET and L. A. SHINN: *J. Amer. Chem. Soc.*, **61**:1615 (1939).

39 R. D. GUTHRIE and A. M. PRIOR: *Carbohyd. Res.*, **18**:373 (1971).

40 R. F. BARBOUR and J. DEVINE: *Analyst*, **96**:288 (1971).

41 P. FLEURY and J. LANGE: *J. Pharm. Chim.*, (8) **17**:107 (1933).

42 R. BELCHER and A. TOWNSHEND: *Anal. Chim. Acta*, **41**:395 (1968).

43 B. SINGH and A. SINGH: *J. Indian Chem. Soc.*, **29**:34 (1952); **30**:143, 786 (1953); *Anal. Chim. Acta*, **9**:22 (1953).

44 A. INDELLI, F. FERRANTI, and F. SECCO: *J. Phys. Chem.*, **70**:631 (1966).

45 D. J. B. GALLIFORD and J. M. OTTAWAY: *Analyst*, **97**:412 (1972).

46 D. D. VAN SLYKE, A. HILLER, D. A. MACFADYEN, A. B. HASTINGS, and F. W. KLEMPERER: *J. Biol. Chem.*, **133**:287 (1940).

47 J. O. EDWARDS: *Chem. Rev.*, **50**:455 (1952).

48 H. TAUBE: *Rec. Chem. Progr.*, **17**:25 (1956).

49 I. M. KOLTHOFF, W. STRICKS, and L. MORREN: *Analyst*, **78**:405 (1953).

50   O. TOMIČEK and M. JASEK: *Collect. Czech. Chem. Commun.*, **10**:353 (1938).

51   I. M. KOLTHOFF and V. ∴. STENGER: *Ind. Eng. Chem.*, *Anal. Ed.*, **7**:79 (1935).

52   N. VELGHE and A. CLAEYS: *Talanta*, **19**:1555 (1972); *Anal. Chim. Acta*, **60**:377 (1972).

53   J. E. CARPENTER: *J. Amer. Chem. Soc.*, **56**:1847 (1934).

54   G. G. RAO, V. B. RAO, and M. N. SASTRI: *Curr. Sci.*, **18**:381 (1949).

55   B. SINGH and R. SINGH: *Anal. Chim. Acta*, **10**:408 (1954); **11**:412 (1954); B. SINGH and S. SINGH, *Anal. Chim. Acta*, **13**:405 (1955).

56   O. TOMIČEK and B. SUCHARDA: *Collect. Czech. Chem. Commun.*, **4**:285 (1932).

57   W. POETHKE and F. WOLF: *Z. Anorg. Allg. Chem.*, **268**:244 (1952).

58   A. R. SHAH, D. K. PADMA, and A. R. V. MURTHY: *Analyst*, **97**:17 (1972).

59   E. BISHOP and V. J. JENNINGS: *Talanta*, **8**:22, 34 (1961); **9**:581, 593 (1962).

60   T. J. JACOB and C. G. R. NAIR: *Talanta*, **19**:347 (1972).

61   G. I. H. HANANIA, D. H. IRVINE, W. A. EATON, and P. GEORGE: *J. Phys. Chem.*, **71**:2022 (1967).

62   H. BASINSKA and W. RYCHCIK: *Talanta*, **10**:1299 (1963).

63   B. KRATOCHVIL: *Anal. Chem.*, **35**:1313 (1963).

64   G. S. SASTRY and G. G. RAO: *Talanta*, **19**:212 (1972).

65   G. CHARLOT: *Bull. Soc. Chim. Fr.*, (5) **6**:1447 (1939).

66   H. H. WILLARD and G. D. MANALO: *Anal. Chem.*, **19**:167 (1947).

67   B. KRATOCHVIL and H. DIEHL: *Talanta*, **3**:346 (1960).

68   R. BELCHER and T. S. WEST: *Anal. Chim. Acta*, **6**:322 (1952).

69   G. BECK: *Mikrochim. Acta*, **35**:169 (1950); **36**:245 (1950); **38**:1, 152 (1950); **39**:22, 149 (1952); **40**:258 (1953).

70   P. N. K. NAMBISAN and C. G. R. NAIR: *Anal. Chim. Acta*, **52**:475 (1970); **60**:462 (1972).

71   R. C. PAUL, R. K. CHAUHAN, N. C. SHARMA, and R. PARKASH: *Talanta*, **18**:1129 (1971).

72   A. BERKA, V. DVORAK, and J. ZYKA, *Mikrochim. Acta*, **1962**:541.

73   E. CSIZER, S. GÖRÖG, and T. SZEN, *Mikrochim. Acta*, **1970**:996.

74   L. SUCHOMELOVA, V. HORAK, and J. ZYKA: *Microchem. J.*, **9**:201 (1965).

75   O. TOMIČEK and J. VALCHA: *Chem. Listy*, **44**:283 (1950); *Collect. Czech. Chem. Commun.*, **16**:113 (1951).

76   G. PICCARDI, *Talanta*, **11**:1087 (1964).

77   L. GIUFFRE and F. M. CAPIZZI: *Ann. Chim.* (Rome), **50**:1150 (1960).

78   J. DOLEZAL, B. MOLDAN, and J. ZYKA: *Collect. Czech. Chem. Commun.*, **24**:3769 (1959).

79   M. MATRKA and Z. SAGNER: *Chem. Prum.*, **9**:526 (1959).

80   W. D. TREADWELL and R. NIERIKER, *Helv. Chim. Acta*, **24**:1067 (1941).

81   B. DIETHELM and F. FOERSTER: *Z. Phys. Chem.* (Leipzig), **62**:129 (1908).

82   I. M. KOLTHOFF: *Rec. Trav. Chim. Pays-Bas*, **43**:768 (1924).

83   I. M. KOLTHOFF and C. ROBINSON: *Rec. Trav. Chim. Pays-Bas*, **45**:169 (1926).

84   W. SCHWARTZ and H. J. BARZ: *Fresenius' Z. Anal. Chem.*, **223**:90 (1966).

85   R. H. PIERSON and E. ST. C. GANTZ: *Anal. Chem.*, **26**:1809 (1954).

86   I. M. KOLTHOFF: *Rec. Trav. Chim. Pays-Bas*, **43**:816 (1924).

87   E. L. RHEAD: *J. Chem. Soc.*, **89**:1491 (1906).

88   E. KNECHT and E. HIBBERT: "New Reduction Methods in Volumetric Analysis," 2d ed., Longmans, New York, 1925.

89   G. GRUBE and L. SCHLECHT: *Z. Elektrochem.*, **32**:178 (1926).

90   J. J. LINGANE and R. L. PECSOK: *Anal. Chem.*, **20**:425 (1948).

91 H. BRINTZINGER and F. RODIS: *Z. Anorg. Allg. Chem.* **166**:53 (1927); *Z. Elektrochem. Angew. Phys. Chem.*, **34**:246 (1928).

92 A. R. TOURKY and A. A. MOUSA: *J. Chem. Soc.*, **1948**:759.

93 E. ZINTL and F. SCHLOFFER: *Z. Angew. Chem.*, **41**:956 (1928).

94 W. U. MALIK and K. M. ABUBACKER: *Anal. Chim. Acta,* **23**:518 (1960).

95 J. J. LINGANE and R. L. PECSOK: *Anal. Chem.*, **21**:622 (1949).

96 F. R. BRADBURY and E. G. EDWARDS: *J. Soc. Chem. Ind. London,* **59**:96 (1940).

97 R. BELCHER and T. S. WEST: *Anal. Chim. Acta,* **5**:260, 268, 360, 472, 474, 546 (1951).

98 F. BURRIEL-MARTI, F. LUCENA-CONDE, and S. ARRIBAS-JIMENO: *Anal. Chim. Acta,* **10**:301 (1954); **11**:214 (1954).

99 R. J. MERRER and J. T. STOCK: *Analyst,* **96**:359 (1971).

100 R. N. F. THORNELEY, B. KIPLING, and A. G. SYKES: *J. Chem. Soc., A,* **1968**:2847.

101 I. M. KOLTHOFF, E. J. MEEHAN, S. BRUCKENSTEIN, and H. MINATO: *Microchem. J.,* **4**:33 (1960).

102 L. ERDEY and E. BODOR: *Anal. Chem.*, **24**:418 (1952); *Fresenius' Z. Anal. Chem.*, **133**:265 (1951); **137**:410 (1953); L. ERDEY and I. BUZAS: *Acta Chim. Acad. Sci. Hung.*, **4**:195 (1954); **8**:263 (1955).

103 L. ERDEY and E. BODOR: *Magy. Kem. Foly.*, **58**:295 (1952); *Fresenius' Z. Anal. Chem.*, **136**:109 (1952).

104 W. B. POLLARD: *Analyst,* **62**:597 (1937).

105 R. BELCHER and A. J. NUTTEN: *J. Chem. Soc.*, **1951**:550.

106 A. BERKA, J. VULTERIN, and J. ZYKA: *Chemist-Analyst,* **51**:88 (1962).

107 D. S. FREEMAN and W. G. CHAPMAN: *Analyst,* **96**:865 (1971).

108 G. G. RAO and S. R. SAGI: *Talanta,* **9**:715 (1962); **10**:169 (1963); G. G. RAO and L. S. A. DIKSHITULU: *Talanta,* **10**:295 (1963).

109 E. J. HART: *Rec. Chem. Progr.*, **28**:25 (1967); *Acc. Chem. Res.*, **2**:161 (1969).

110 D. C. WALKER in "Radiation Chemistry," p. 49, Advances in Chemistry Series, Vol. 81, American Chemical Society, Washington, D.C., 1968.

111 J. H. BAXENDALE, J. P. KEENE, and D. A. STOTT, *Chem. Commun.*, **1966**:715.

PROBLEMS

20-1 Write an equation for the oxidation of Cu(I) thiocyanate in the Andrews titration.

20-2 Write equations for the titration of iodine to iodine monochloride, using permanganate, dichromate, and periodate as oxidants.

20-3 From the standard potential of the half-reaction (20-3) and the equilibrium constant of (20-2), calculate the standard potential of (20-1).   *Answer*   1.20 V.

20-4 From the standard potentials of (20-3) and (20-4), calculate the equilibrium constant of (20-5).   *Answer* 3.2×10^{11}.

20-5 Verify the potential given for (20-11), given the standard potential of the half-reaction $Br_2(aq) + 2e^- \rightleftharpoons 2Br^-$, $E° = 1.087$ V; the formation constants $k_1 = 10^9$, $k_2 = 10^{8.3}$ for $HgBr_2$; and the solubility of $HgBr_2$, 1.7×10^{-2} M.

21

REACTION RATES IN CHEMICAL ANALYSIS

Generally, substances can be analyzed chemically by methods based on either thermodynamic or kinetic principles. Thermodynamic methods employ reaction conditions such that reactions other than the one of interest are thermodynamically unfavorable. The methods treated up to this point have been largely of this equilibrium type. Reaction-rate, or kinetic, methods make it possible to exploit a larger number of chemical reactions by utilizing information from the rate of reaction rather than from the position of equilibrium. For example, they may be of value when a reaction is inconveniently slow for use in an analysis that depends on equilibrium being attained. They may involve catalyzed as well as uncatalyzed reactions. The rate of catalyzed homogeneous reactions is normally proportional to the concentration of the catalyst and can be used with high sensitivity in the estimation of catalyst concentration. Kinetic methods can be applied also to systems that have unfavorable equilibrium constants or systems that involve side reactions. When a mixture contains two closely related substances that undergo the same type of reaction at different rates, measurements of reaction rates can permit analysis of the mixture.

More broadly, through the measurement of kinetic constants it is often possible to elucidate reaction mechanisms and thereby explain the causes of nonstoichiometry.

For example, induced reactions (Section 15-8), which usually are annoying sources of error but occasionally can be made to proceed as useful quantitative reactions, can be understood only through a detailed study of reaction mechanisms.

Reaction-rate methods are readily adaptable to automated techniques of data acquisition and processing. Although favorable precision can be obtained with repetitive runs and signal averaging, usually the results are not so precise as those from most equilibrium methods. With improved instrumentation, however, the quality of results obtainable is steadily improving, and the precision of single-component kinetic methods is comparable to equilibrium methods based on measurements such as absorbance, current, or voltage. When time is a reaction variable, both concentrations of reactants and temperature must be carefully controlled. Kinetic methods of analysis have been reviewed by Mark and Rechnitz,[1,3] Yatsimirskii,[2] and Guilbault.[4]

21-1   RATES OF REACTIONS

Consider a reaction in which a reactant R undergoes a reaction to products:

$$R \rightarrow \text{products} \qquad (21\text{-}1)$$

If (21-1) is irreversible or the rate of the reverse reaction negligible, the rate of disappearance of the reactant R or of appearance of a product P is given by

$$\frac{-d[R]}{dt} = k_R[R] = \frac{d[P]}{dt} \qquad (21\text{-}2)$$

where k_R is the first-order rate constant at the temperature of the reaction. In integrated form

$$\ln[R]_t = \ln[R]_0 - k_R t$$

or

$$C = C_0 e^{-kt} \qquad (21\text{-}3)$$

where $[R]_t = C =$ concentration of R at time t, and where $[R]_0 = C_0 =$ initial concentration of R. If a first-order rate law is obeyed, the rate of disappearance of R or of appearance of P is given by curves of the type shown in Figure 21-1 (left). According to Equation (21-2) the reaction rate is linearly dependent on $[R]$; if the concentration is doubled, the rate is doubled.

When $[R]_t = \frac{1}{2}[R]_0$, half the reactant has undergone reaction and the half-time $t_{\frac{1}{2}}$ is equal to $\ln 2/k$. This half-time is independent of the concentration, and thus it takes the same length of time for a reaction to go from half to three-quarters completion as it does to go to half completion (Figure 21-1, right).

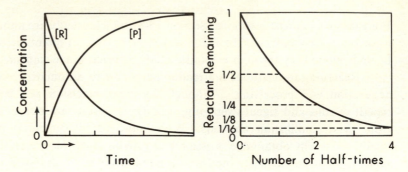

FIGURE 21-1   Left, variation with time of the concentrations of reactant R and product P.   Right, relation between half-time and fraction of reactant remaining.

For a rate-determining reaction

$$R_1 + R_2 \rightarrow \text{products} \qquad (21\text{-}4)$$

The rate of the forward reaction is equal to $k_f[R_1][R_2]$, where k_f is the rate constant for the forward reaction.   More generally for a rate-determining reaction

$$n_1 R_1 + n_2 R_2 + n_3 R_3 \cdots \rightarrow \text{products} \qquad (21\text{-}5)$$

the rate of the forward reaction is $k_f[R_1]^{n_1}[R_2]^{n_2}[R_3]^{n_3} \cdots$.   This reaction can be considered to be broadly general, and the reactants can include catalysts, hydrogen ions, and solvent molecules.   Their concentrations are included in the expression for the rate constant.   Further, it should be recognized that the value of the rate constant k_f increases exponentially with temperature according to the Arrhenius equation $d \ln k/dT = C/T^2$, where C is a constant for the particular reaction and T is absolute temperature.   The rate constant may be affected also by the intensity of radiant energy.

Reactions that are kinetically higher than first order are of little interest as such to chemical analysis.   Fortunately, concentrations of all reactants except one can normally be made sufficiently high (or constant as for a catalyst) that the rate is controlled by only the one reactant; the reactions are then termed pseudo first order. If one of the reactants is present at a large and therefore sensibly constant concentration, the reaction can be considered kinetically to be R → P as in (21-1), that is, now analogous to a first-order reaction.   For example, in the hydrolysis of cane sugar, $C_{12}H_{22}O_{11}$, to give a molecule of glucose and one of fructose, the reaction is bimolecular in that a water molecule is involved, but the reaction rate is first order in that in aqueous solution the concentration of water remains large and constant.

Another way of avoiding the complications of high-order reactions is to establish pseudo-zero-order conditions by making sufficiently sensitive measurements that no appreciable change in concentration occurs during the time of measurement. Under these conditions

$$\frac{-d[R]}{dt} = k_0 \qquad (21\text{-}6)$$

where k_0 is the constant for a zero-order reaction.

Studies of reaction rates become more complicated when it is realized that most reactions are *reversible*, so that while the rate of the forward reaction decreases that of the reverse reaction increases with time, and at equilibrium the overall net rate is zero. Other common complications involve *consecutive* reactions such as R → P → Q, *competing* reactions $R_1 + R_2$ → products and $R_1 + R_3$ → other products, and *chain* reactions such as described in Section 15-8. *Heterogeneous* reactions involve two or more phases and are difficult to formulate.

A slow reaction,

$$A + B \rightarrow X + Y \qquad (21\text{-}7)$$

can be accelerated by addition of catalyst C, which forms an intermediate compound with A that then reacts rapidly with B, regenerating the catalyst:

$$A + C \rightarrow AC \qquad (21\text{-}8)$$

$$AC + B \rightarrow X + Y + C \qquad (21\text{-}9)$$

Enzymatic reactions are examples of catalytic reactions; they can be formulated in a generalized way:

$$E + S \underset{k_2}{\overset{k_1}{\rightleftharpoons}} ES \overset{k_3}{\longrightarrow} E + P \qquad (21\text{-}10)$$

where E is enzyme, S is substrate, ES is the enzyme-substrate complex, and P is product. [There could be several of these complexes, and the enzyme kinetics might be more complex[5] than indicated by (21-10).] The equilibrium concentration of a complex depends on the rate constants k_1, k_2, and k_3. The *Michaelis constant* K_m, defined by $(k_2 + k_3)/k_1$, is an equilibriumlike constant for the formation of ES.

Under steady-state conditions,[6,7] when Michaelis-Menten kinetics (that is, kinetics applicable to systems with any number of complexes) are obeyed,

$$\frac{d[P]}{dt} = -\frac{d[S]}{dt} = \frac{k_3 C_E[S]}{K_m + [S]} \qquad (21\text{-}11)$$

where C_E is the total enzyme concentration. If K_m is small compared with [S], (21-11) becomes a zero-order reaction with respect to substrate:

$$\frac{d[P]}{dt} = k_3 C_E = k_{cat} C_E \qquad (21\text{-}12)$$

where k_{cat} is V_{max}/C_E, the reaction rate per unit of enzyme concentration when [S] → ∞ (Figure 21-7). On the other hand, if K_m is large compared with [S]

$(K_m > 100 \, [S]_0)$, Equation (21-11) can be rearranged to an equation analogous to (21-2):

$$\frac{d[P]}{dt} = \frac{k_{cat}C_E[S]}{K_m} \qquad (21\text{-}13)$$

Equation (21-13) is first order with respect to enzyme and substrate concentration, and with C_E constant the rate is directly proportional to $[S]$.

Several methods are used for obtaining concentrations from rate data. For a reaction occurring at a moderate rate the technique for providing the necessary kinetic data may include quenching the reaction by any of several methods—quick cooling, changing pH, adding a substance to combine with one of the reactants, or adding an inhibitor or other substance to interfere with the activity of a catalyst. After quenching, chemical or physical measurements on the system can be made at greater leisure. For measuring rates with half-times as low as about 0.001 s, continuous-flow or stopped-flow techniques have been devised. Some electrochemical and relaxation methods such as temperature or pressure jump are useful in the same range and for somewhat faster reactions. Active developments are taking place to increase both sensitivity and precision; reaction-rate methods have been reviewed by Malmstadt, Cordos, and Delaney,[8] Crouch,[9] Blaedel and Hicks,[10] and Pardue.[11]

Two general types of analysis can be performed: those in which a single component is determined and those in which multiple components are determined simultaneously without separation. Within each group, one of two types of measurements can be made: either (1) an instantaneous measurement of the slope of the response curve, which should be proportional to the reaction rate, or (2) measurements of changes in concentration over finite time intervals, in which the ratio $\Delta C/\Delta t$ under the correct conditions represents a reaction rate. The important differences between differential and integral types of measurements are discussed below.

The major disadvantage of kinetic methods is that the change in concentration of the indicator substance must be measured not only with little error but also as a function of time. As mentioned, the inclusion of time as an experimental variable means that temperature must be carefully controlled and also the solution composition (as it affects activities of the reacting species).

A summary[6] of commercially available reaction-rate instruments and their applicability has been made.

21-2   SINGLE-COMPONENT ANALYSIS

Differential methods: Initial rate   The derivative $-d[R]/dt$ (or $d[P]/dt$) of Equation (21-2) can be estimated directly at a time after the reaction has started but before a significant fraction of R has reacted or P has been produced. With knowledge of the rate constant k_R measured under the same conditions, the concentration of R

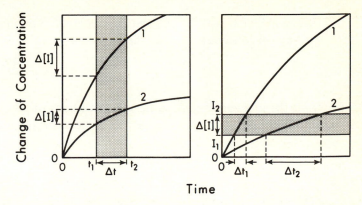

FIGURE 21-2 Fixed-time (left) and variable-time (right) integral methods of measurement of reaction rates.

can be calculated directly. With such initial-rate methods (sometimes called pseudo-zero-order methods) the concentration of R, P, or some other indicator substance giving a signal that is linear with change in concentration is plotted as a function of time, and the tangent to the curve at zero time† is directly proportional to the concentration.

Initial rates of reactions can be measured by electronic derivative circuitry.[12] High precision and accuracy are obtainable when the output signal from the measuring device is compared electronically with that from a standard-slope generator that produces a signal to match that of the unknown. The response time of the electronics can be extremely fast,[13,14] and precision in the range of 1 to 2% can be obtained.

Initial-rate methods have several fundamental advantages. Since the amount of product formed during the period of measurement is small, the reverse reaction decreases the overall net rate inappreciably. Complications from slower side reactions usually are minimal. The concentrations of reactants change but little, and pseudo-zero-order kinetics are obeyed. For reactions whose rates are in the useful range, measurements of initial rate should be more precise than measurements of rates at later times because the rate is highest at the beginning, where the slope of the curve is steepest and the relative signal-to-noise ratio is most favorable. Initial-rate methods permit the use of reactions whose formation constants may be too small for equilibrium methods.

Integral methods: Constant time In the fixed-time method of measurement the change in concentration of the indicator substance I [which could be [R] or [P] in Equation (21-2)] is measured twice to cover a preselected time interval (Figure 21-2).

---

† Strictly speaking, the rate of reaction begins at zero, and during some process (such as mixing reactants or changing temperature) that initiates the reaction, it rises to a maximum. Measurements should be made at a time somewhat after zero to avoid the effect of the rising rate.

The first measurement is at time t_1, and the second after an interval equal to Δt. The rate of change in concentration over the interval is given by

$$\frac{\Delta[I]}{\Delta t} = k'[R] \qquad (21\text{-}14)$$

where k' is the rate constant when the concentrations of all other reactants are held constant. With a constant time interval, Δt can be incorporated with the rate constant to give

$$\Delta[I] = k''[R]_0 \qquad (21\text{-}15)$$

that is, the concentration of the substance is directly proportional to the change in concentration of the indicator substance over the preselected time interval. The advantage of this type of measurement is that the fraction of the reaction occurring under specified conditions is invariant with the initial concentration, whether or not the two time values are on the linear portion of the curve. Therefore, a linear relation exists between $\Delta[I]$ and $[R]_0$. Ingle and Crouch[15] showed that the proportionality holds between $\Delta[I]$ and $[R]_0$ even for reversible reactions; they concluded that the fixed-time approach is theoretically and practically superior for pseudo-first-order reactions and those involving measurement of substrate concentrations with enzymes.

Fixed-time methods include those in which the reaction is quenched after a preselected period of time and the solution analyzed by chemical or physical techniques.

Parker and Pardue[16] described a miniature on-line, general-purpose digital computer that also has been applied to kinetic analysis in the constant-time mode.

Integral methods: Variable time   In the variable-time method of measurement of the initial slope, the concentration of the indicator substance I is measured twice, and the time interval Δt required to bring about a preselected change in concentration $\Delta[I]$ is the important quantity (Figure 21-2, right). Since the change in concentration is a fixed preselected value, it can be incorporated with the constant in Equation (21-5) to give

$$\frac{1}{\Delta t} = k'''[R]_0 \qquad (21\text{-}16)$$

that is, the initial concentration of the substance is directly proportional to the reciprocal of the time interval required to bring about the measured concentration change in the indicator material. For valid results the reaction must be essentially linear with time over the interval Δt. Figure 21-2 shows that at the lower concentration the relative change in rate over the interval Δt_2 is greater than over the interval Δt_1. The nonlinearity is a source of potential error. For first-order kinetics, Ingle and Crouch[15] calculated that the concentration-time relation can be considered linear only over a small fraction of the reaction; thus for 1% accuracy in the rate the concentration of R must change less than 2%. They further calculated that, if $[R]_0$ is 0.001 M and 0.01 M in two cases, and if $[I]_1 = 10^{-5} M$ and $[I]_2 = 2 \times 10^{-5} M$, then if Δt for the 0.01 M solution is 10 s, with first-order kinetics the value of Δt for the 0.001 M solution would be 101.3 instead of 100 s. Thus an error of -1.3% is introduced, even though the extent of reaction over the interval is only 1% for even the most dilute sample.

With variable-time methods, standards should be chosen that have concentrations near those of the samples. Variable-time techniques are most suited to measurements where the signal is nonlinear with concentration and to the measurement of concentrations of catalysts such as enzymes, where $\Delta[I]$ represents a constant fraction of a reaction (the catalyst being regenerated).

Parker, Pardue, and Willis[17] described a reaction-rate instrument including a reciprocal-time digital computer circuit designed for on-line variable-time analyses. Alkaline phosphatase could be determined satisfactorily with a relative standard deviation of about 1% if the times were in excess of 0.5 s. Their data indicate that, with proper calibration and temperature control, the limitations lie in the chemical and detection systems rather than in the rate-measuring system.

Integral methods: Constant signal Signal-stat[11] methods involve adding reagent at a rate that will maintain a constant signal from the indicator substance. The rate of addition of reagent is directly equal to the reaction rate and therefore directly proportional to the concentration to be measured. The most familiar example is the pH stat approach. For example, an enzyme reaction can produce or consume acid; by the continuous neutralization of the acid or base formed, the pH can be maintained at a predetermined level. An advantage of this technique is that the concentration of a critical reaction parameter is held constant during the measurement.

Matsen and Linford[18] maintained a constant amperometric signal from iron(III) during reaction by generating it electrolytically as needed. Thus they were able to follow low reaction rates for periods of up to a week. Malmstadt and Piepmeier[19] maintained constant pH to within 0.002 pH units in the kinetic analysis of urea and glucose by addition of sodium hydroxide solution from an automatic delivery pipet. Karcher and Pardue[20] described a pH stat apparatus that monitors pH spectrophotometrically with coulometric generation of acid or base. No errors from dilution are thereby introduced.

21-3 MIXTURES: SIMULTANEOUS ANALYSIS

In practice, a substance to be measured often must be separated from closely related substances unless kinetic techniques can be applied. For example, a sample containing two organic compounds having the same functional group may be difficult to handle by conventional methods. Often available are reagents that react with both compounds by the same mechanism but at different rates. Thus, the reaction of analytical significance is isolated kinetically whenever possible; enzyme methods can be highly selective in this regard. In cases where the difference in rates cannot be made large, differential-rate techniques are applicable.

Consider a reaction that is kinetically first order with respect to two sample constituents A and B. It could be either a reaction involving no other reactant or a pseudo-first-order reaction with a reagent R that is present in relatively large and therefore constant concentration:

$$A + R \rightarrow products$$

$$B + R \rightarrow products \qquad (21\text{-}17)$$

The kinetic order with respect to the reagent R is of no concern, as long as the concentration of R is held constant during all measurements. If k_A and k_B are the first-order or pseudo-first-order rate constants for the two reactions, the rate of disappearance of A is given by

$$\frac{-d[A]}{dt} = k_A[A] \qquad (21\text{-}18)$$

In integrated form

$$\ln [A]_t = \ln [A]_0 - k_A t \qquad (21\text{-}19)$$

where $[A]_t$ and $[A]_0$ are the concentrations at time t and at the initial time. Similar expressions can be written for the concentration of B.

Neglect of reaction of slower-reacting component   If component A is the faster reacting component, we can calculate[21] the ratio k_A/k_B so that the reaction of A is essentially complete in the period of time necessary for the reaction of a negligible amount of B. From (21-19) and the analogous equation for B

$$\frac{k_A}{k_B} = \frac{\ln ([A]_0/[A]_t)}{\ln ([B]_0/[B]_t)} \qquad (21\text{-}20)$$

For 99% reaction of A and 1% reaction of B

$$\frac{k_A}{k_B} = \frac{\log 100}{\log 1.01} = 463$$

Thus, if the ratio of rate constants is at least 463/1, it is possible to find a reaction time such that the faster reaction is within 1% of completion when the slower reaction has proceeded to less than 1% of completion. Both can then be measured by performing two analyses at appropriate times. A limitation is that the reaction rates must be within certain limits and should be controllable by simple means. For example, when both A and B are to be determined, if even the faster reaction does not reach 99% completion in a matter of minutes, the slower reaction will be impracticably

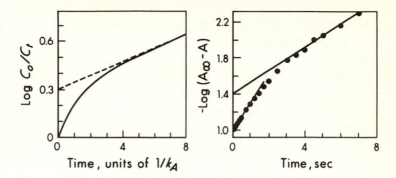

FIGURE 21-3 Left, plot of first-order differential rate, calculated for $k_A/k_B = 10$ and $[A]_0 = [B]_0$. Right, an experimentally observed curve for $k_{Ca}/k_{Mg} = 6.5$, $[Ca] = 2.0 \times 10^{-4}$ M, and $[Mg] = 1.11 \times 10^{-4}$ M. (*From Pausch and Margerum.*[22])

slow unless it can be speeded up by a simple change such as raising the temperature. At the other extreme, if even the slower reaction goes nearly to completion in a short time, determinations of concentration usually cannot be made rapidly enough to take advantage of the difference in reaction rates.

As an example, in the determination of alkaline earth metals in the presence of lanthanides by the methods involving CyDTA complexes described in Section 21-5, the rates of reaction of lanthanides are too slow to cause interference, and simple methods of measurement should be possible. The differences between the rates of reaction of the various alkaline earth metals are so small, however, that a simple procedure cannot be applied.

Logarithmic extrapolation method The sum of the total initial concentrations, $[A]_0 + [B]_0$, and the total concentration C_t at any time t, $[A]_t + [B]_t$, can be determined by titration or by instrumental means. If the rates of reaction of A and B are each independent of the presence of the other, C_t is given by

$$C_t = [A]_t + [B]_t = [A]_0 e^{-k_A t} + [B]_0 e^{-k_B t} \qquad (21\text{-}21)$$

If $\log (C_0/C_t)$ is plotted against t, a curve results, unless $k_A = k_B$, in which case a straight line is obtained. As time goes on the contribution of the first term of (21-21) becomes progressively smaller, and the semilogarithmic plot eventually becomes linear. By extrapolation of the linear portion to $t = 0$, the value of $\log [B]_0$ is obtainable. The value of $[A]_0$ then is obtained by difference. The advantage of this procedure is that a smaller ratio of k_A to k_B can be exploited than is acceptable by neglect of the reaction of the slower-reacting component. On the other hand, enough measurements are required to assure definition of the linear part of the curve.

Figure 21-3 (left) shows a calculated curve for the case of $k_A/k_B = 10$ and

$[A]_0/[B]_0 = 1$; at $t = 5/k_A$, the value of $[A]_t$ is only 0.67% of its initial value, whereas 60.5% of B remains. Figure 21-3 (right) shows an experimental curve for the differential kinetic analysis of a calcium and magnesium mixture,[22] in which the ratio of rate constants is 6.5.

Worthington and Pardue[23] described an analog circuit coupled with an XY recorder with which concentrations can be obtained directly from extrapolations of linear regions. Time-consuming plotting of data is eliminated. They pointed out that errors become large when concentration ratios are large. Nakanishi[24] described an analog computer circuit for the simulation of the reaction-rate curve, with a fit obtained by a trial-and-error procedure.

Reaction-rate method   A method independent of linear extrapolation but requiring more observations is to determine the rate of change in total concentration dC_t/dt, which approaches $k_B[B]_t$ as the concentration of A approaches zero:

$$\frac{-dC_t}{dt} = k_A[A]_t + k_B[B]_t \qquad (21\text{-}22)$$

The rate of change can be determined from the slope of the curve of a plot of C_t as a function of time. As constituent A disappears, the ratio of the negative slope to the concentration approaches a constant value of k_B:

$$\frac{-dC_t}{dt}\frac{1}{C_t} = k_B \qquad \text{for } [A]_t = 0 \qquad (21\text{-}23)$$

Then by use of k_B the value of $[B]_0$ can be calculated from points late in the run, where $C_t = [B]_t$, from the equation

$$\ln [B]_0 = \ln [B]_t + k_B t \qquad (21\text{-}24)$$

An alternative approach (the method of proportional equations)[25] is to use Equation (21-21) and measure C_t at two different reaction times to obtain two simultaneous equations that can be solved for $[A]_0$ and $[B]_0$.

For a three-component mixture, Equation (21-21) becomes

$$C_t = [A]_t + [B]_t + [C]_t = [A]_0 e^{-k_A t} + [B]_0 e^{-k_B t} + [C]_0 e^{-k_C t} \qquad (21\text{-}25)$$

The initial concentrations of each of the three species then can be calculated by measuring C_t at three reaction times and solving the resulting three simultaneous equations. A method for the kinetic determination of five peroxides by reaction with selected sulfides has been devised[26] that involves five equations and five unknowns.

In principle, if C_0 is known or measurable, and if k_A and k_B have been determined independently in separate experiments under carefully controlled conditions, then both $[A]_0$ and $[B]_0$ can be calculated from a single measurement of C_t. For such a kinetic measurement the highest precision is obtained at an optimum time that is neither too short nor too long. Lee and Kolthoff[27] showed that this optimum time

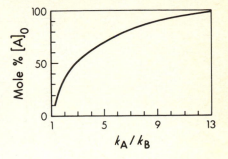

FIGURE 21-4 An experimentally determined curve of maximum mole percent $[A]_0$ tolerable (that is, amenable to
"complete" reaction) as a function of
k_A/k_B for determination of $[A]_0$ and $[B]_0$
in binary amine mixtures by graphical
extrapolation. (*From Papa, Mark, and
Rechnitz.*[31])

depends on the rate constants and is equal to $\ln (k_A/k_B)/(k_A - k_B)$. By this single-
point method and with the prior information required, they succeeded in determining
closely related pairs of esters such as ethyl acetate and isopropyl acetate by use of
their saponification rates, which at 25°C have a ratio of 4.2. The absolute error was
2%. They also analyzed mixtures of carbonyl compounds and measured the relative
amounts of the internal and external double bonds in polymers.[28]

Curve-fitting or regression method With advances in computer technology,
methods are being developed that can effectively use the entire signal-time response
curve. A simplified linear least-squares method has been developed and equipment
designed[29,30] for data acquisition and processing with an on-line computer for
multicomponent kinetic analysis. Data can be taken either at equally spaced time
intervals or at a decreasing rate as a reaction rate decreases. With the larger volume
of data, errors are markedly decreased (Section 21-4).

21-4 PRECISION OF MEASUREMENTS

Simultaneous analysis of mixtures Errors associated with methods for simultaneous analysis for pseudo-first-order reactions are considered briefly here. A more
complete analysis has been presented elsewhere:[31] With several simplifying assumptions, graphical and algebraic methods based on simultaneous equations were compared. Graphical extrapolation is limited in that one component must have essentially
completely reacted before data suitable for extrapolation for the second can be used.
Complete reaction is defined[31] by $k_A[A]_t/k_B[B]_t = 30$, at which time $[B]_t$ should be
large enough so that appreciable changes in concentration of the indicator substance
still can be observed (Figure 21-4).

Normally, for mixtures containing 50% of each component the minimum value
of k_A/k_B is about 4, and the accuracy improves as the ratio $[B]_0/[A]_0$ increases. In
the graphical-extrapolation method, because the reactions must be carried nearly to
completion, it is important that they be irreversible or have large equilibrium

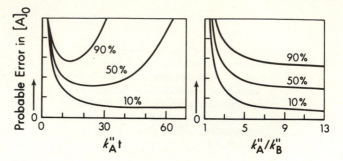

FIGURE 21-5   Left, variation of probable error in $[A]_0$ as a function of reaction time for $k_A''/k_B'' = 7.5$ and for 10, 50, and 90 percent of A in the mixture. Right, variation of probable error in $[A]$ as a function of k_A''/k_B'', for optimum reaction time. (*From Papa, Mark, and Rechnitz.*[31])

constants and that there be no side reactions. This method has the advantage that the rate constants need not be known and, though the reaction conditions must be carefully controlled within any one run, they need not be carefully controlled from run to run.

One algebraic method involves rates of reactions when the reagent concentration is large compared with $[A]_0 + [B]_0$. Two simultaneous equations are solved at two times of observation, one near the optimum (Section 21-3) and the other when the reaction is nearly complete. This method is less restricted than graphical extrapolation with respect to both $[A]_0/[B]_0$ and k_A/k_B, because $k_A[A]_t$ need not be negligible compared with $k_B[B]_t$ to make analytically useful observations feasible. With this method the error in analysis increases when the ratio k_A/k_B approaches unity and when the second observation is made at a time prior to complete reaction.

The widest range of applicability with respect to both $[A]_0/[B]_0$ and k_A/k_B is obtainable when $[A]_0 + [B]_0$ is large compared with the reagent concentration. With this condition,[32] a composite pseudo-first-order rate constant can be used for the mixture that is equal to $k_A''[A]_0 + k_B''[B]_0$, where k_A'' and k_B'' are the second-order rate constants for the reaction of A and B with the reagent R. In this method, $[A]_0$ is best calculated from the extent of reaction at a single optimum time when the probable error is smallest (Figure 21-5, left), and $[B]_0$ is obtained by subtracting $[A]_0$ from the prior determination of $[A]_0 + [B]_0$. At the optimum time the reaction should be about two-thirds complete.[33] The value of $[A]_0$ can be measured over a wide range of values of $[A]_0/[B]_0$ and k_A''/k_B'' as long as $k_A''[A]_0$ contributes significantly to the composite pseudo-first-order rate constant (Figure 21-5, right). Thus the method can be used for the determination of traces of $[A]_0$ when the ratio k_A''/k_B'' is large, as indicated by Figure 21-6. In this figure the line corresponds to the situation where the contribution of reactant A to the composite pseudo-first-order rate constant is 5%.

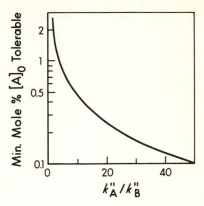

FIGURE 21-6 Minimum mole percent of A measurable when ($[A]_0 + [B]_0$) $\gg$ [R] as a function of k_A''/k_B''. (*From Papa, Mark, and Rechnitz.*[31])

Instrumental aspects For many methods of kinetic analysis, direct proportionality between signal and concentration is most easily obtained if the extent of reaction is kept small. This leads to instrumental problems, since small changes should be measured with high precision. The development of detectors of high stability to meet this need is proceeding.[34,35]

Because of the experimental work entailed, emphasis in early studies of kinetics in analysis was placed on retrieving precise information from a minimum number of observations. Emphasis at that time often was on analysis of two-component mixtures by, in the limit, a single measurement at an optimum time.[27] Continuing developments in electronics and instrumentation are revolutionizing kinetic methods, and in particular they are providing freedom to choose methods of gathering and processing data that eliminate concern over whether the calculations are cumbersome. An automated computer-controlled system for obtaining and processing data has been described.[36]

Pausch and Margerum,[22] in attempting to analyze differential kinetic data for a particular system using a two-point method including selection of optimum time, frequently encountered errors of disastrous magnitude (up to 500%). When 30 points were used, reduction of errors to 5% was routinely possible. Ingle and Crouch[37] discussed the effect of noise on the precision of rate measurements. They concluded that photocurrent shot noise for spectrophotometric rapid stopped-flow rate measurements limits the precision to about 1%. They indicated that repetitive measurements are required to obtain more precise data. In broad terms, the precision of an analysis depends on the intensity and character of the noise associated with a signal, the intensity of the signal, the prior information about a signal (say, that the model entails a linear change in signal with time), and the method of estimation (Section 26-1). Fundamentally, in kinetic analysis, high precision should be attainable with repetitive runs and signal averaging. On the other hand, the present feasibility of frequent sampling of an analog signal and data reduction by computers should not be a license for careless experimental work; instead, it should provide an opportunity for better results.

A regression analysis system has been described[29] that can utilize 200 data points taken at either regularly spaced time intervals or at decreasing rates. The data are analyzed on-line with a small computer[38] by a linear least-squares method. Rate constants and, in this case, molar absorptivities for each reactant system must be carefully measured as part of the prior information. The rate of data acquisition for a standard mixture is varied until a minimum in the least-squares error is attained. A weighting factor is included so that data taken near the end of the reaction with the slower-reacting component are deemphasized. The entire operation on mixtures with reaction half-lives up to a fraction of a minute takes about 3 min, including data processing. Binary mixtures of calcium and strontium in the concentration range 10^{-4} to 10^{-5} M with concentration ratios varying from 0.11 to 11 were analyzed, and standard deviations were obtained of less than 1% for the major component and somewhat more than 2% for the minor one. Ternary mixtures of magnesium, calcium, and strontium were analyzed with an error ranging up to 6%—an order of magnitude less than obtained earlier[22] without the on-line technique. Evidence is given that the precision of the method should be relatively independent of rate-constant ratios down to about 1.5 for binary mixtures. The upper limit of rate-constant ratios is limited by the necessity of collecting data for at least one half-life of the slower-reacting component.

Uncertainty of the temperature of reaction in stopped-flow reaction-rate measurements can be a serious source of error.[39] There is an advantage in having the temperature-measuring thermocouple positioned at the observation cell.

In the GeMSAEC[40] fast kinetic analysis system several samples are mixed simultaneously in a rotor that contains multiple assemblies for the samples. Many thousands of data points per minute can be obtained sequentially for samples, standards, and blanks.

21-5   APPLICATIONS

No attempt is made to list all reactions used for kinetic methods of analysis. Rather, a selection is made to typify several kinds of reactions that can be used. Mark[41] recommended that proposals for new methods include information on the mathematical basis, the reaction mechanism, the effects of trace impurities, instrumental factors, accuracy and precision attainable, and applicability.

For a reaction to be applicable in kinetic analysis, its rate must be neither too high nor too low. We may define fast reactions as those that approach equilibrium (several half-lives) during the time of mixing. For analytical measurements, reaction half-times for first-order reactions in the range of 0.1 to 1000 s are convenient, although stopped-flow mixing systems extend the range to a few milliseconds. If a reaction rate is outside this range, experimental conditions should be modified. Reaction rates can be modified by changing such variables as concentration, temperature, solvent, and addition of catalysts or inhibitors. In typical low-viscosity solvents at ordinary temperatures, the rate constant for diffusion-controlled bimolecular reactions that occur on every collision is of the order of 10^{10} l/mole-s. Although for unimolecular reactions there is no upper limit, vibrational frequencies are of the order of 10^{14} s. Most organic reactions are slow, whereas complexation reactions may be fast or slow.

Many kinetic methods of analysis involve reactions whose rates depend on catalysts in solution (Section 15-7); most of these involve redox systems. A *catalyst* may be defined broadly as an agent that alters the rate of a reaction without shifting the position of equilibrium. The catalyst itself undergoes no permanent change, although it may enter the reaction mechanism in a cyclic manner. The mechanisms and activation of catalytic reactions have been reviewed.[42]

From the analytical viewpoint, we are interested primarily in homogeneous catalysis, particularly where the rate of the catalyzed reaction is proportional to the concentration of catalyst. A catalyst may operate either by lowering the energy barrier for both the forward and back reactions or by introducing an alternative reaction path.

Analysis for traces of substances that act as catalysts for reactions can be both highly specific and sensitive. Calculations of the type carried out by Yatsimirskii[43] indicate that, if the rate of a catalytic reaction can be measured spectrophotometrically with a change in concentration equivalent to 10^{-7} moles/l, and if the catalytic coefficient amounts to 10^8 cycles/min, the minimum concentration of catalyst that can be measured by following the reaction for 1 min is about 10^{-15} moles/l. Such sensitivities can be attained by few other analytical techniques. Tölg[44] considers only mass spectroscopy, electron-probe microanalysis, and neutron-activation analysis methods applied to favorable cases to have better limits of detection than catalytic methods.

Catalytic reactions—Nonenzymatic Kinetic analysis through the use of catalyzed reactions is normally performed by the variable-time technique. This technique is appropriate because measured and constant quantities of A + B in (21-23) can be used in a reaction. Typically, the time for completion of the reaction is measured and then related to the concentration of the catalyst.

Several of the studies of catalyzed redox reactions involve hydrogen peroxide oxidations[2] of reducing agents such as thiosulfate, iodide, rubeanic acid, and methyl orange. Some substances that act as catalysts in acidic solution for the oxidation are Zr(IV), Hf(IV), Th(IV), Nb(V), Ta(V), Mo(VI), W(VI), Fe(III), Ti(IV), V(V), and

Cr(VI). In acting as catalysts, these substances form complexes or peroxides, though complex formation in itself is not sufficient to produce catalysis. In basic solution, catalysts involve substances that are easily oxidized and reduced, such as Fe(II), Co(II), and Cu(II). As an example, Svehla and Erdey[45] determined traces of molybdenum through its catalytic effect on the overall hydrogen peroxide–iodide reaction,

$$2I^- + H_2O_2 + 2H^+ \rightarrow 2H_2O + I_2 \qquad (21\text{-}26)$$

In the presence of a small amount of ascorbic acid, the I_2 formed is reduced back to iodide. When all the ascorbic acid is consumed, iodine is liberated, and the sudden appearance of iodine can be made visible with variamine blue (Landolt effect). The time to the appearance of iodine is measured. The reaction of hydrogen peroxide and iodide is much faster in the presence of molybdate, probably because of the reactions

$$H_2O_2 + H_2MoO_4 \rightarrow H_2O_2 \cdot H_2MoO_4 \qquad (21\text{-}27)$$

and $\qquad H_2O_2 \cdot H_2MoO_4 + I^- \rightarrow IO^- + H_2O + H_2MoO_4 \qquad (21\text{-}28)$

followed by, in acidic solution,

$$HIO + H^+ + I^- \rightarrow H_2O + I_2 \qquad (21\text{-}29)$$

The method is most suited to measurements in the range of 40 to 100 μg/ml of molybdenum, but can be used for concentrations as low as 0.1 μg/ml.

Many different redox reactions in acidic solutions are catalyzed by the same substances that catalyze hydrogen peroxide reactions. For example, Bognar and Jellinek[46] determined traces of V(V), Fe(III), and osmium tetroxide using a chlorate–bromide–ascorbic acid–o-tolidine system and the Landolt effect.

Catalysts have been used to make end points detectable; these have been reviewed by Mottola.[47] In such systems the titrant reacts rapidly with the substance titrated, but excess reagent reacts only slowly with an indicator in the absence of an appropriate catalyst. For example,[48] small amounts of complexing agents such as EDTA can be determined by titration with Mn(II). With malachite green as indicator in the presence of periodate, the excess Mn(II) catalyzes the indicator oxidation.

Yatsimirskii[2] tabulated about 50 different reactions that can be used for the measurement of catalytic amounts of specified ions, and Tölg[44] reviewed the problems associated with the measurement of traces. Reaction rates can be used also in the determination of an inhibitor through its retarding effect on a catalyzed reaction. An example is the determination of mercury by exploiting its inhibiting effect on the catalysis of the Ce(IV)-As(III) reaction by iodide.[49]

Margerum and Steinhaus[50] showed that, with a coordination chain-reaction system, metals that form complexes with EDTA could be measured at concentration levels of 10^{-7} M with limits of about 10^{-9} M. The method is based on the exchange reaction (charges omitted) between the nickel complex of triethylenetetraamine (T) and the copper complex of EDTA.

$$NiT + CuEDTA \rightarrow CuT + NiEDTA \qquad (21\text{-}30)$$

The reaction can be initiated by a trace of EDTA:

$$EDTA + NiT \rightarrow NiEDTA + T \qquad (21\text{-}31)$$

$$T + CuEDTA \rightarrow CuT + EDTA \qquad (21\text{-}32)$$

Thus the rate of chain propagation is proportional to the concentration of EDTA. A trace of a metal ion that forms a stable complex with EDTA decreases the [EDTA], and from the decreased rate the concentration of the trace metal can be deduced.

Catalytic reactions—Enzymatic An exceptional combination of selectivity and sensitivity occurs in the case of enzyme-catalyzed reactions. Enzymes are protein catalysts and may catalyze the reaction of a single substance in the presence of many others that might develop similar reactions. Thus glucose oxidase catalyzes the oxidation of β-D-glucose to gluconic acid with high specificity in the presence of many other oxidizable sugars. For example, α-D-glucose is oxidized at a rate less than 0.6% of the rate for β-D-glucose.[51] Enzyme selectivity applies as well to the particular products of a reaction, and therefore side reactions are of less concern.

Whereas some enzymes are highly specific to a single optical isomer (glucose oxidase, for example), other enzymes may catalyze reactions for several related substances. Thus D-amino acid oxidase reacts with several D-amino acid substrates.[52] The relative rates are tyrosine, 100; proline, 78; methionine, 42; alanine, 34; serine, 22; tryptophan, 19.5; valine, 18.4; phenylalanine, 13.7; isoleucine, 11.6; leucine, 7.4; and histidine, 3.3.

Enzyme-catalyzed reactions are used for the determination not only of substrates but also of the enzymes as well as activators and inhibitors of the enzymatic reaction. Substrate concentrations can be measured also by nonkinetic methods by allowing a reaction to proceed to completion before making measurements. Enzymatic methods of analysis have been reviewed by Guilbault.[4] He listed over 150 enzymes with their sources; many of these are now available in purified form with high specific activity. Urease was the first to be obtained in pure crystalline form.[53]

According to Equation (21-13), with a constant enzyme concentration the rate of reaction is linearly proportional to substrate concentration at low concentrations, and fixed-time methods are appropriate. Such methods are commonly used in clinical laboratories. At high concentrations of substrate the rate becomes independent of concentration[4] (Figure 21-7). For measurement of enzyme concentrations, variable-time procedures have advantages.

The rate of an enzymatic reaction increases as the value of k_3 increases (Equation 21-11) and as the value of the Michaelis constant decreases. For a constant substrate

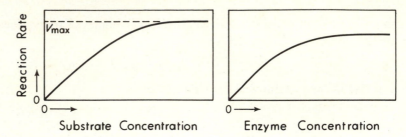

FIGURE 21-7 Relation between rate of an enzymatic reaction and concentration of substrate or enzyme.

concentration the rate of reaction is linearly proportional to enzyme concentration over a wide range (Figure 21-7). When substrate and enzyme concentrations both are constant, enzymatic reactions can be used to measure traces of activators or inhibitors with the high specificity inherent in enzyme reactions. Again, there is linearity up to some limiting concentration.

To specify the amount of an enzyme, an *international unit* has been adopted,[54] defined as the amount of enzyme that converts 1 μmole of substrate per minute at 25°C under optimum conditions of pH, ionic strength, and substrate concentration.

Reactions involving complexes A computer-controlled stopped-flow apparatus has been developed[30] and applied to the simultaneous determination of cysteine and thiolactic acid through an exchange reaction with the citrate complex of Ni(II). Fifty readings can be taken during the first 0.25 s; a computer program, using first-order equations for the two acids, processes the data to give best-fit results. Relative standard deviations of 1 to 2% were obtained.

Pausch and Margerum[22] used stopped-flow differential kinetic methods involving the reaction of Pb(II) ions with magnesium, calcium, strontium, and barium complexes of *trans*-1,2-diaminocyclohexane-*N,N,N',N'*-tetraacetate (CyDTA) to form the lead-CyDTA complex. The reaction rate constants have relative magnitudes of 1, 6.5, 96, and 1660 for magnesium, calcium, strontium, and barium. The rate-determining step is the reaction of the complex with hydrogen ions; a typical reaction could be (depending on the pH)

$$BaCyDTA^= + H^+ \rightarrow HCyDTA^{3-} + Ba^{++} \qquad (21\text{-}33)$$

The $HCyDTA^{3-}$ produced in turn reacts rapidly with lead ions, and there is no direct replacement of barium ions by lead ions. The rate of the reaction therefore can be adjusted by pH control. Some of the reactions are so fast that a pH as high as 8 is required for stopped-flow techniques to be applicable; at the same time, this high pH makes copper and cobalt (instead of lead) unsuitable as scavenging metals. The sensitivity is such that as little as 10^{-6} M metal ion can be measured. It is important that the CyDTA be pure, or other substances that react more rapidly may form complexes.

Lanthanide ions form stable complexes (K_{MY} 10^{15} to 10^{20}) with multidentate ligands. Kinetic studies have been carried out[55] with the lanthanide, transition, and Group III metals. Their reaction rates vary over a range of about 10^{12}, and their dissociation rates are dependent on the pH.[56] Reactions at pH 7.5 are suitable for determinations of alkaline earth metals; other metal ions react too slowly to interfere at this pH. At pH 4 the transition metals can be determined, since alkaline earth metal ions now react too rapidly to interfere. Analyses can be carried out for binary mixtures of lanthanum and cadmium with a rate-constant ratio as low as 1.4 and of samarium and europium with a rate-constant ratio of 1.7. Rate constants too close together for resolution of individual reactants often can be made more divergent by changes in such factors as pH, temperature, and anion composition of the solution. Thus the ratio of rate constants for cadmium and lead is 1.8 at 25°C, but 7.6 at 11°C. For copper and cobalt the ratio is ordinarily about 1.2 at 25°C, but 11 in 0.1 M perchloric acid. Masking agents also can be used to achieve selectivity in kinetic measurements.[57] Thus thiosulfate masks Ag(I) and Hg(II), and cyanide followed by chloral hydrate masks Ni(II), Co(II), and Fe(III).

Polyaminopolycarboxylic acids, alone or in mixtures, have been determined by measurements of the rate of reaction of cyanide with their Ni(II) complexes.[58] As little as 10 ppb of nitrilotriacetic acid (NTA) in water was detected, and binary, ternary, and even four-component mixtures were determined. The rate-determining step was shown to be

$$NiL(CN)_2{}^{n-} + CN^- \rightarrow Ni(CN)_3L^{(n+1)-} \qquad (21\text{-}34)$$

This reaction is rapidly followed by the addition of a fourth cyanide ion and loss of the NTA ligand L. Reagent-grade EDTA was found to contain 0.13% NTA by differential-rate measurements. The half-life of the NiL^- complex in 10^{-2} M cyanide was about 0.03 s.

The kinetics of the formation of complexes has been reviewed by Alimarin.[59]

Other reactions Some inorganic and many organic reactions are too slow or otherwise unsuitable for the reactions to be carried to completion, or they yield spurious end points in titrations.[60] Reactions of long half-life present a problem in that the procedure becomes tedious if many samples are analyzed. Kankare,[61] in a kinetic study of the reaction of benzyl chloride with dimethylaniline, used an automatic apparatus for continuous sampling and titration to take 72 data points at equal time intervals over a period of 12 h.

Lohman and Mulligan[62] resolved diethanolamide and monoethanolamide mixtures by a differential-rate technique involving saponification of the amides and the lauric esters. The reaction was quenched after 25 min by cooling, and the base formed was titrated with acid.

Hanna and Siggia[63] described the application of simultaneous analysis techniques to systems such as those containing primary and secondary hydroxyl groups, mixtures of amides and nitriles, and sugars and amino acids.

For faster reactions the speed with which reactants can be mixed is a limitation; the mixing time must be less than the half-time. Stopped-flow techniques have been developed that permit remarkably short kinetic analysis times. Beckwith and Crouch[64] described a stopped-flow kinetic analysis apparatus, with a mixing and dead time of less than 0.01 s, capable of analyzing 1000 phosphate samples per hour with a relative standard deviation of about 1%. Sample handling, mixing, and gathering and evaluation of data were automated with the help of on-line computer systems.

Ingle and Crouch[65] described a differential kinetic method for silicate and phosphate based on the faster rate of formation of heteropoly molybdenum blue from the yellow heteropoly acids in the presence of phosphate than in the presence of silicate. They found that silicon in the range of 1 to 10 ppm could be determined with 3% accuracy in the presence of 10 ppm of phosphorus, and phosphorus in the range of 1 to 10 ppm with 1% accuracy in the presence of 50 ppm of silicon. This system was also automated, with the analyses of mixtures being performed in less than 5 min.

In Chapters 15 to 20, reference to applications of kinetics to equilibrium systems is frequently made in connection with redox chemistry. Studying the kinetics of reactions can lead to better understanding of how chemical reactions proceed and thereby make clear how to select optimum conditions for a desired reaction and for minimum interference. In equilibrium methods of analysis, catalysts commonly are used to bring a reaction to completion in less time.

REFERENCES

1 H. B. MARK, JR., and G. A. RECHNITZ: "Kinetics in Analytical Chemistry," Interscience, New York, 1968.

2 K. B. YATSIMIRSKII: "Kinetic Methods of Analysis," translated by P. J. J. Harvey, Pergamon, New York, 1966.

3 H. B. MARK, JR.: *Talanta*, **19**:717 (1972).

4 G. G. GUILBAULT: "Enzymatic Methods of Analysis," Pergamon, New York, 1970.

5 L. PELLER and R. A. ALBERTY: *J. Amer. Chem. Soc.*, **81**:5907 (1959). See also V. BLOOMFIELD, L. PELLER, and R. A. ALBERTY: *J. Amer. Chem. Soc.*, **84**:4367 (1962); V. BLOOMFIELD and R. A. ALBERTY: *J. Biol. Chem.*, **238**:2804, 2811 (1963); M. TARASZKA and R. A. ALBERTY: *J. Phys. Chem.*, **68**:3368 (1964).

6 H. V. MALMSTADT, C. J. DELANEY, and E. A. CORDOS: *Crit. Rev. Anal. Chem.*, **2**:559 (1972); *Anal. Chem.*, **44**(12):79A (1972).

7 G. E. BRIGGS and J. B. S. HALDANE: *Biochem. J.*, **19**:338 (1925).

8 H. V. MALMSTADT, E. A. CORDOS, and C. J. DELANEY: *Anal. Chem.*, **44**(12):26A (1972).

9 S. R. CROUCH in "Spectroscopy and Kinetics," Computers in Chemistry and Instrumentation Series, vol. 3, J. Mattson and others (eds.), Dekker, New York, 1973.

10 W. J. BLAEDEL and G. P. HICKS in "Advances in Analytical Chemistry and Instrumentation," C. N. Reilley (ed.), vol. 3, p. 105, Wiley, New York, 1964.

11 H. L. PARDUE: *Rec. Chem. Progr.*, **27**:151 (1966).

12 H. L. PARDUE and W. E. DAHL: *J. Electroanal. Chem.*, **8**:268 (1964).

13 H. V. MALMSTADT and S. R. CROUCH: *J. Chem. Educ.*, **43**:360 (1966).

14 T. E. WEICHSELBAUM, W. H. PLUMPE, JR., R. E. ADAMS, J. C. HAGERTY, and H. B. MARK, JR.: *Anal. Chem.*, **41**:725 (1969).

15 J. D. INGLE, JR., and S. R. CROUCH: *Anal. Chem.*, **43**:697 (1971).

16 R. A. PARKER and H. L. PARDUE: *Anal. Chem.*, **44**:1622 (1972).

17 R. A. PARKER, H. L. PARDUE, and B. G. WILLIS: *Anal. Chem.*, **42**:56 (1970).

18 J. M. MATSEN and H. B. LINFORD: *Anal. Chem.*, **34**:142 (1962).

19 H. V. MALMSTADT and E. H. PIEPMEIER: *Anal. Chem.*, **37**:34 (1965).

20 R. E. KARCHER and H. L. PARDUE: *Clin. Chem.*, **17**:214 (1971).

21 J. S. FRITZ and G. S. HAMMOND: "Quantitative Organic Analysis," chap. 9, Wiley, New York, 1957.

22 J. B. PAUSCH and D. W. MARGERUM: *Anal. Chem.*, **41**:226 (1969).

23 J. B. WORTHINGTON and H. L. PARDUE: *Anal. Chem.*, **44**:767 (1972).

24 M. NAKANISHI: *Talanta*, **19**:285 (1972).

25 MARK and RECHNITZ,[1] p. 201.

26 J. P. HAWK, E. L. MCDANIEL, T. D. PARISH, and K. E. SIMMONS, *Anal. Chem.*, **44**:1315 (1972).

27 T. S. LEE and I. M. KOLTHOFF: *Ann. N.Y. Acad. Sci.*, **53**:1093 (1951).

28 I. M. KOLTHOFF, T. S. LEE, and M. A. MAIRS: *J. Polym. Sci.*, **2**:199, 220 (1947); I. M. KOLTHOFF and T. S. LEE: *J. Polym. Sci.*, **2**:206 (1947).

29 B. G. WILLIS, W. H. WOODRUFF, J. R. FRYSINGER, D. W. MARGERUM, and H. L. PARDUE: *Anal. Chem.*, **42**:1350 (1970).

30 D. SANDERSON, J. A. BITTIKOFER, and H. L. PARDUE: *Anal. Chem.*, **44**:1934 (1972).

31 L. J. PAPA, H. B. MARK, JR., and G. A. RECHNITZ, in Mark and Rechnitz,[1] chap. 7.

32 J. D. ROBERTS and C. REGAN: *Anal. Chem.*, **24**:360 (1952).

33 PAPA, MARK, and RECHNITZ,[31] p. 221.

34 W. J. BLAEDEL and G. P. HICKS: *Anal. Chem.*, **34**:388 (1962).

35 H. L. PARDUE and P. A. RODRIGUEZ: *Anal. Chem.*, **39**:901 (1967); H. L. PARDUE and M. M. MILLER: *Clin. Chem.*, **18**:928 (1972).

36 S. N. DEMING and H. L. PARDUE: *Anal. Chem.*, **43**:192 (1971).

37 J. D. INGLE, JR., and S. R. CROUCH: *Anal. Chem.*, **45**:333 (1973).

38 B. G. WILLIS, J. A. BITTIKOFER, H. L. PARDUE, and D. W. MARGERUM: *Anal. Chem.*, **42**:1340 (1970).

39 P. K. CHATTOPADHYAY and J. F. COETZEE: *Anal. Chem.*, **44**:2117 (1972).

40 C. D. SCOTT and C. A. BURTIS: *Anal. Chem.*, **45**(3):327A (1973).

41 H. B. MARK, JR.: *Talanta*, **20**:257 (1973).

42 P. R. BONTCHEV: *Talanta*, **17**:499 (1970); **19**:675 (1972).

43 YATSIMIRSKII,[2] p. 19.

44 G. TÖLG: *Talanta*, **19**:1489 (1972).

45   G. SVEHLA and L. ERDEY: *Microchem. J.*, **7**:206, 221 (1963).

46   J. BOGNAR and O. JELLINEK: *Mikrochim. Acta*, **1964**:317.

47   H. A. MOTTOLA: *Talanta*, **16**:1267 (1969).

48   H. A. MOTTOLA: *Anal. Chem.*, **42**:630 (1970).

49   P. J. KE and R. J. THIBERT: *Mikrochim. Acta*, **1973**:15.

50   D. W. MARGERUM and R. K. STEINHAUS: *Anal. Chem.*, **37**:222 (1965).

51   O. WARBURG: *Biochem. Z.*, **152**:51 (1924).

52   G. G. GUILBAULT and J. E. HIESERMAN: *Anal. Biochem.*, **26**:1 (1968).

53   J. B. SUMNER: *J. Biol. Chem.*, **69**:435 (1926).

54   J. COOPER, P. A. SRERE, M. TABACHNICK, and E. RACKER: *Arch. Biochem. Biophys.*, **74**:306 (1958).

55   D. W. MARGERUM, J. B. PAUSCH, G. A. NYSSEN, and G. F. SMITH: *Anal. Chem.*, **41**:233 (1969).

56   T. RYHL: *Acta Chem. Scand.*, **26**:3955 (1972); **27**:303 (1973).

57   R. H. STEHL, D. W. MARGERUM, and J. J. LATTERELL: *Anal. Chem.*, **39**:1346 (1967).

58   L. C. COOMBS, J. VASILIADES, and D. W. MARGERUM: *Anal. Chem.*, **44**:2325 (1972).

59   I. P. ALIMARIN: *Pure Appl. Chem.*, **34**:1 (1973).

60   P. W. CARR and J. JORDAN: *Anal. Chem.*, **45**:634 (1973).

61   J. J. KANKARE: *Acta Chem. Scand.*, **25**:3881 (1971).

62   F. H. LOHMAN and T. F. MULLIGAN: *Anal. Chem.*, **41**:243 (1969).

63   J. G. HANNA and S. SIGGIA in "Encyclopedia of Industrial Chemical Analysis," F .D. Snell and C. L. Hilton (eds.), vol. 1, p. 561, Interscience, New York, 1966.

64   P. M. BECKWITH and S. R. CROUCH: *Anal. Chem.*, **44**:221 (1972).

65   J. D. INGLE, JR., and S. R. CROUCH: *Anal. Chem.*, **43**:7 (1971).

66   Q. H. GIBSON, B. E. P. SWOBODA and V. MASSEY: *J. Biol. Chem.*, **239**:3927 (1964).

PROBLEMS

21-1   Calculate the ratio of first-order rate constants k_A/k_B necessary if 99.9% of A is to react in the time interval in which only 0.1% of B will react.

Answer   6900.

21-2   Plot $\log C_t/C_0$ as a function of time in units of $1/k_A$ when $k_A/k_B = 10$ and $[A]_0/[B]_0 = 10$. At the times when (*a*) 1% and (*b*) 0.1% of A remains, what percentage of B is left?

Answer   (*a*) 63.1%. (*b*) 50.1%.

21-3   For the case of Problem (21-2), plot C_t in arbitrary units as a function of time in units of $1/k_A$. For various arbitrary times, determine the ratio of slope to concentration and show that it approaches the value k_B as t increases.

21-4   For the glucose oxidase–catalyzed oxidation of glucose with oxygen at 0°C, pH of 5.6, and an enzyme concentration of 1.17×10^{-5} M the rate[66] of reaction is as follows:

$$\frac{1}{\text{Rate}} = \frac{k_2 + k_4}{k_2 k_4} + \frac{1}{k_1[\text{glucose}]} + \frac{1}{k_3[\text{oxygen}]}$$

where $k_1 = 2.1 \times 10^3/\text{mole-s}$, $k_2 = 650/\text{s}$, $k_3 = 1.3 \times 10^6/\text{mole-s}$, and $k_4 = 370/\text{s}$.

(a) What is the relation between rate and glucose concentration at an oxygen concentration of $10^{-3} M$?

(b) If the reaction rate is proportional to enzyme concentration what is the rate expression that includes both enzyme and glucose concentration?

(c) Draw a graph showing the relations: (1) between rate and glucose concentration at an oxygen concentration of $10^{-3} M$ and of enzyme concentration of $10^{-5} M$; and (2) between rate and enzyme concentration at an oxygen concentration of $10^{-3} M$ and a glucose concentration of $10^{-2} M$.

Answer (a) Rate $= 2.1 \times 10^3[\text{glucose}]/(1 + 10.5[\text{glucose}])$.

(b) Rate $= 1.8 \times 10^8[\text{glucose}][\text{enzyme}]/(1 + 10.5[\text{glucose}])$.

22

SEPARATIONS: INTRODUCTION

In Chapter 1 it is pointed out that analyses comprise several stages or operations: definition or recognition of the analytical aspects of a problem, sampling, separation, measurement, and evaluation of data. The preceding chapters have dealt mainly with the chemistry of the noninstrumental aspects of the measurement operation. Ideal methods of analysis would involve measurement techniques of such specificity that only the component of interest in a sample would give a reaction or response, without interference from other components. But in actual practice a sample may be composed of substances that are chemically similar, which undergo similar reactions and so are likely to interfere with the measurement operation. Therefore, analytical separations frequently are necessary to isolate or concentrate material before the measurement step can be carried out. The extent of such separation operations depends on the method of measurement selected or available, the nature and amount of sample, the time restrictions, and the precision required.

When an interfering element has several stable oxidation states, interference sometimes can be prevented simply by changing the oxidation number. Another simple approach to the problem of interferences is the technique of masking (Section 11-8), in which a reagent is added that decreases the effective concentration of a substance to a level sufficient to prevent certain chemical reactions. Masking techniques have been most successful in complexation analysis, where they eliminate the necessity for a separation.

Separation techniques encompass problems as diverse as isolating the several hundred components present in strawberry oil, isolating a few atoms of lawrencium quickly enough to observe its fission half-life, isolation of an insect pheromone, and the reduction of impurities in a material to a level low enough for use as a primary standard (Section 5-3). Ordinarily, a choice of several separation techniques is available. Two substances are separable only if they exhibit significant differences in such properties as size, density, vapor pressure, solubility, or rate of reaction. Although a satisfactory classification of the techniques is not available, for our purposes we can regard the separations as involving the formation of two phases and the transfer of matter from one phase to the other. One phase contains the material of interest, and the other the interference. The two phases are mechanically separated, and the one containing the material of interest normally is retained. Both the material of interest and the interference may be present initially as a sample of gas, liquid, or solid. The second phase formed from each of these three possibilities also may be gas, liquid, or solid. It is assumed that a phase is defined broadly enough to include gases of differing composition. Thus there are six major types of separation processes: solid–solid, solid–liquid, and so on. The theory of separation processes largely relates to heterogeneous equilibria involving these six types and to the rates at which such equilibria are attained. If both the equilibrium constant and the rate of attainment of equilibrium are favorable, the separation can be accomplished in a single stage. For less favorable systems, multistage operations are both feasible and attractive.

When the initial phase is a gas, a separation can entail formation of a second phase consisting of a solid (as in condensation, sorption, or chromatographic processes), a liquid (also as in condensation, sorption, or chromatography), or a gas phase of different composition (as in diffusion or magnetic separations). When the initial phase is a liquid, the second phase formed can be a solid (as in precipitation, sorption, electrodeposition, or chromatography), a liquid (as in distillation, extraction, dialysis, electrophoresis, electrodeposition, or chromatography), or a gas (as in volatilization). When the initial phase is a solid, the second phase can be another solid (as in sieving or magnetic separations), a liquid (as in solution processes), or a gas (as in sublimation or volatilization).

Separation techniques that involve two-phase equilibria selective and complete enough to accomplish a given separation in a single stage are most desired. When a single separation stage is not sufficiently selective or quantitative, increased complexity is encountered. Multistage techniques are advantageous when the components to be separated distribute themselves selectively between the phases. Thus in precipitation separations the amount of coprecipitation usually can be diminished by redissolving the precipitate in fresh solvent and repeating the process. Similarly, if in an extractive separation one component remains quantitatively in one of the two phases while another is distributed between both phases, then a repetitive extraction process can be advantageous. The Soxhlet extractor is a classical example; another application of the same principle is the use of fresh batches of mercury as a cathode in controlled-potential electrolytic separations of metals. Another level of complexity is reached when a moving phase containing the components to be separated is brought into contact with a stationary phase and both components are distributed between the two phases. The two phases can be divided into discrete units, as in the Craig countercurrent distribution method of extraction, in which one of the two phases remains in a set of extraction vessels and the other is moved stepwise from one vessel to the next. A still higher level of complexity is represented by true countercurrent methods, in which both phases are moving, but in opposite directions.

The separation techniques described here are those of most interest to analytical chemistry. The background and theory of separation by the single-stage processes of precipitation and electrodeposition are discussed in Chapters 7 and 14. In this chapter some general introduction to separations is presented along with a brief background for several separation processes.

22-1 RECOVERY AND ENRICHMENT FACTORS

The fundamental purpose of a separation operation is to obtain efficiently a component in a state of high purity. Sandell[1,2] defined a *recovery factor* R_i for a substance i by

$$R_i = \frac{Q_i}{Q_i^0} \qquad (22\text{-}1)$$

where Q_i is the quantity isolated and Q_i^0 is the amount in the sample. Recovery factors should approach unity for reliable analyses, except for measurement techniques such as tracer analysis, where the measurements are related to a control.

For the separation of a component A (desired) from B (unwanted) the *separation*[1,2] or *enrichment*[3] *factor* $S_{B/A}$ is defined by

$$S_{B/A} = \frac{Q_B/Q_A}{Q_B^0/Q_A^0} = \frac{R_B}{R_A} \qquad (22\text{-}2)$$

where Q_B and Q_A are the amounts of B and A in the recovered material and Q_B^0 and Q_A^0 are the original amounts. If the recovery factor is high, $Q_A \simeq Q_A^0$, and (22-2) is simplified to

$$S_{B/A} \simeq \frac{Q_B}{Q_B^0} \simeq R_B \qquad (22\text{-}3)$$

The values of enrichment factors should be small; how small depends on the nature of the analysis and the method of measurement. In trace analysis[2] the factor required may be typically of the order of 10^{-6}, whereas in macroanalysis approximately 10^{-3} is usually sufficient. Sandell[1] explained that the ratio of distribution ratios, often called the separation factor, does not allow the extent of separation to be calculated. He therefore recommended that it not be called a separation factor or index.

22-2 PRECIPITATION

Precipitation equilibria are described in part in Chapter 7. An important type of precipitation equilibrium considered here involves a coordination reaction of a metal ion with univalent anions to form an uncharged chelate that is insoluble. In general, it is necessary to consider stepwise equilibria:[4]

$$M^{n+} + L^- \rightleftharpoons ML^{(n-1)+} \qquad K_{eq} = K_1$$
$$ML^{(n-1)+} + L^- \rightleftharpoons ML^{(n-2)+} \qquad K_{eq} = K_2$$
$$\vdots \qquad\qquad \vdots$$
$$ML_{n-1}^+ + L^- \rightleftharpoons ML_n(soln) \qquad K_{eq} = K_n \qquad (22\text{-}4)$$

where $K_1, K_2, \ldots, K_n$ are the successive stepwise formation constants of the complexes with the ligand L^-. With each step of coordination the charge of the metal ion is lowered by 1 unit, and two coordination positions are occupied by the ligand.

For the moment we neglect the lower complexes and assume that the uncharged complex species ML_n is formed in one step:

$$M^{n+} + nL^- \rightleftharpoons ML_n(soln) \qquad K_{eq} = \beta_n = K_1 K_2 \cdots K_n \qquad (22\text{-}5)$$

where β_n is the overall formation constant (Section 11-1). Precipitation occurs when the solubility of the species ML_n is exceeded. In the behavior of organic precipitating reagents, the importance of intrinsic solubility S^0 (Section 7-1) has been stressed.[4] We write the solubility equilibrium

$$ML_n(solid) \rightleftharpoons ML_n(soln) \rightleftharpoons M^{n+} + nL^- \qquad (22\text{-}6)$$

With activity coefficients of unity, according to Equation (7-4) the expression for the solubility product is

$$K_{sp} = [M^{n+}][L^-]^n = K_d S^0 = \frac{S^0}{\beta_n} \qquad (22\text{-}7)$$

EXAMPLE 22-1 Christopherson and Sandell[5] estimated the solubility of nickel dimethylglyoxime, NiL_2, in water at an ionic strength of 0.05 to be 9.7×10^{-7} M and the solubility product to be 4.3×10^{-24}. The formation constant of NiL_2 is therefore $9.7 \times 10^{-7}/4.3 \times 10^{-24} = 2.3 \times 10^{17}$. ////

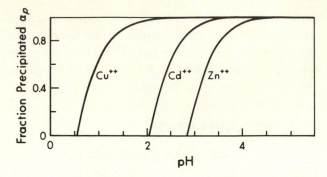

FIGURE 22-1   pH-precipitation curves for 0.001 M divalent metal ions with 0.003 M quinaldic acid. (*From Flagg,*[6] *p.* 63.)

Precipitation occurs when the ion product in (22-7) is equal to or greater than S^0/β_n; accordingly, the order of precipitation of two metals M_1 and M_2 of the same charge is given by the ratio $S_1^0\beta_{n,2}/S_2^0\beta_{n,1}$ rather than by the ratio of stability constants.

In Equation (22-6) the anion L is from a weak acid HL. This anion concentration is expressed conveniently in terms of the fraction of HL that is ionized [Equation (3-56)]:

$$[L^-] = \alpha_1 C_{HL} = \frac{K_a C_{HL}}{[H^+] + K_a} \qquad (22\text{-}8)$$

where C_{HL} is the total reagent† concentration $[HL] + [L^-]$ not coordinated to a metal ion. For the special case in which the reagent is in an acidic solution so that $K_a \ll [H^+]$ and in which the lower complexes are unimportant, we have $[L^-] \simeq C_{HL}K_a/[H^+]$, and

$$K_{sp} = [M^{n+}][L^-]^n = \frac{[M^{n+}]K_a^n C_{HL}^n}{[H^+]^n} \qquad (22\text{-}9)$$

If α_p represents the fraction precipitated, C_M^0 the initial concentration of metal, and C_{HL}^0 the initial concentration of reagent, then $[M^{n+}] = C_M^0(1 - \alpha_p)$, $C_{HL} = C_{HL}^0 - n\alpha_p C_M^0$, and (22-9) becomes

$$[H^+] = \frac{C_M^0(1 - \alpha_p)^{1/n} K_a (C_{HL}^0 - n\alpha_p C_M^0)}{K_{sp}^{1/n}} \qquad (22\text{-}10)$$

Equation (22-10) was derived by Flagg,[6] who showed it to be adequate to describe the precipitation of copper, zinc, and cadmium with excess quinaldic acid (Figure 22-1).

> † If the reagent HL is amphoteric, the cation H_2L^+ may be regarded as a divalent acid with ionization constants $K_1 = K_w/K_b$ and $K_2 = K_a$, where K_b and K_a are the base and acid dissociation constants. Then the anion concentration becomes $[L^-] = \alpha_2 C_{HL} = K_1 K_2 C_{HL}/([H^+]^2 + K_1[H^+] + K_1 K_2)$. See Equation (3-61).

Note that the pH of incipient precipitation, corresponding to $\alpha_p = 0$, for a given reagent depends first of all on the value of K_{sp} but also on the initial concentrations of metal ion and reagent. Therefore, these quantities need to be specified before quantitative meaning can be attached to the pH at which precipitation begins. The pH of quantitative precipitation can be calculated by taking $\alpha_p = 0.999$. The steepness of the precipitation curve or the change in pH to go from $\alpha_p = 0$ to $\alpha_p = 0.999$ depends on the value of n, the charge on the metal ion.

It is possible to calculate from (22-10) the theoretical degree of separation of two metals at a given pH value. The theoretical separation may not actually be achieved, however, owing to coprecipitation phenomena (Chapter 9).

Turn now to a more complete treatment of the equilibria, and consider the effect of lower complexes. The total metal concentration in solution C_M is given by

$$C_M = [M^{n+}] + [ML] + [ML_2] + \cdots + [ML_{n-1}] \qquad (22\text{-}11)$$

where the charges of the complexes have been omitted for the sake of simplicity and the concentration of ML_n (the intrinsic solubility) has been neglected on the assumption that it is negligible. If α_M is the fraction of the metal ion existing as the uncharged ion M^{n+}, we have [Equation (7-24)]

$$C_M = \frac{[M^{n+}]}{\alpha_M} = [M^{n+}](1 + \beta_1[L^-] + \beta_2[L^-]^2 + \cdots + \beta_{n-1}[L^-]^{n-1}) \qquad (22\text{-}12)$$

or $\quad C_M = [M^{n+}][1 + \beta_1\alpha_1 C_{HL} + \beta_2(\alpha_1 C_{HL})^2 + \cdots + \beta_{n-1}(\alpha_1 C_{HL})^{n-1}] \qquad (22\text{-}13)$

where α_1 is given by (22-8). From (22-7)

$$K_{sp} = [M^{n+}][L^-]^n = \alpha_M C_M \alpha_1^n C_{HL}^n \qquad (22\text{-}14)$$

or $$C_M C_{HL}^n = \frac{K_{sp}}{\alpha_M \alpha_1^n} \qquad (22\text{-}15)$$

The quantity $K_{sp}/\alpha_M \alpha_1^n$ may be regarded as a *conditional solubility product*, which is a function of the pH and of the reagent concentration. Equation (22-15) is also useful when one wishes to take into account the effect of hydrolysis or of masking agents that may be present. It is necessary only to make the appropriate calculation of the fraction α_M from the hydrolysis constants [Equation (7-17)] or from the formation constants of the secondary complexes [Equation (11-16)].

As a general rule, to rely on separation procedures that require control of pH over a narrow range appears to be unsafe, because the pH of incipient precipitation varies with the concentration of metal ion and of reagent. Thus it has been reported[7] that a quantitative separation of copper from cadmium can be made using quinaldic acid, with negligible coprecipitation of cadmium. Flagg,[8] however, preferred to make this separation using salicylaldoxime, because the cadmium salt of this reagent is so soluble in acid solution that the conditions need not be carefully controlled.

Coprecipitation can occur even though the pH of incipient precipitation of the second metal has not been reached. For example, Moyer and Remington[9] found that serious coprecipitation of magnesium occurred with zinc 8-hydroxyquinolate unless the pH was kept at least 2 units below the value at which magnesium alone begins to precipitate. Biefeld and Howe[10] found, however, that copper could be separated from nickel using salicylaldoxime with little coprecipitation when the pH was kept even 0.2 pH unit below the value at which nickel alone precipitated. In contrast, iron was seriously coprecipitated over a wide pH range.

22-3   ORGANIC REAGENTS

Complex-forming organic reagents are important in analytical chemistry because of the inherent sensitivity and selectivity of their reactions with metal ions. Much of the early work in the field was empirical, being directed toward a search for specific, or at least highly selective, reagents for metals. Selectivity often can be achieved for a particular purpose by controlling such variables as the pH and the reagent concentration and by taking advantage of masking agents that enhance the differences in behavior among various metals. The number of organic reagents is now so large that it is hardly possible even to compile a list of all the reagents for the various metal ions. They are used in two main ways in separations—in precipitation and in extraction. In addition, some are finding use in gas chromatography as volatile metalloorganic compounds. Although phase equilibria are involved in both precipitation and extraction, they are more complex in the case of extractions (Chapter 23).

Welcher[11] classified organic reagents according to the number of hydrogen ions displaced from a neutral molecule in forming one chelate ring.

Class I: Displacement of two hydrogen ions   The coordination reaction involves a metal ion and a divalent anion, with the result that with each step of coordination the charge of the complex becomes 2 units more negative than that of the metal ion.

If the coordination number of the metal ion for the reagent is equal to the charge on the metal ion, the resulting complex is a neutral molecule, generally insoluble in water, and the product usually can be extracted into an organic solvent. An example is α-benzoin oxime, $C_6H_5CHOHC{=}NOH(C_6H_5)$, which has two acidic hydrogen atoms and forms a 1:1 coordination compound with Cu(II).[12] The substituted arsonic acids, $RAsO(OH)_2$, which form 1:2 complexes with tetravalent metals, are sometimes considered to belong to this class. But with Ti(IV) and Zr(IV), which are present largely as hydrolyzed species such as TiO^{++} and ZrO^{++}, the bonding is more likely to involve displacement of only a single hydrogen atom of each of two molecules of reagent.[13]

If, as often happens, the coordination number of the metal ion for the reagent exceeds the charge on the metal ion, an anionic complex, usually soluble in water, is formed. Such complexes may be useful in a masking capacity; examples are the soluble complexes of metal ions with oxalate, citrate, and tartrate.

Class II: Displacement of one hydrogen ion The coordination reaction involves a metal ion and a univalent anion, with the result that each step of coordination lowers the charge of the metal by 1 unit and uses two coordination positions. If the coordination number of the metal ion for the reagent is twice the charge on the ion, a neutral species, usually insoluble in water, is formed. Note that coordination is often completed when the charge has become zero, even if the coordination positions are not fully occupied by the reagent, because further coordination would require ionization of the reagent and dissolution of the insoluble product. Since most reagents are only weakly acidic, such ionization is not energetically favorable.

For example, magnesium ion reacts with 8-hydroxyquinoline (oxine), which may be represented by HL, to form a dihydrate

$$Mg(H_2O)_6{}^{++} + 2HL \rightarrow MgL_2 \cdot 2H_2O + 2H^+ + 4H_2O$$

because the coordination number is 6 and the charge has become zero when two molecules of reagent have reacted with one magnesium ion. On the other hand, aluminum ion forms an anhydrous oxinate because the coordination number is only twice the ion charge.

By far the greatest number of analytically important organic reagents fall into class II. Examples are 1-nitroso-2-naphthol for Co(II), dimethylglyoxime and 1,2-cyclohexanedionedioxime (nioxime) for Ni(II), and oxine and diphenylthiocarbazone (dithizone) for various heavy metals.

Class III: Displacement of no hydrogen ions The coordination reaction involves the replacement of water by neutral molecules of reagent. The product is, therefore, a cation of the same charge as the original metal ion. Although the product is normally water-soluble, the unusually bulky cation together with appropriate anions can sometimes be extracted into organic solvents. For example,[14] salts of the Cu(I) and Fe(II) derivatives of substituted 1,10-phenanthrolines can be extracted into such solvents as the higher alcohols.

Reagents that do not fall into the above three classes are those forming more than one chelate ring per molecule of reagent (for example, EDTA) and those forming no chelate rings. As examples may be cited the ionic precipitates such as tetraphenylarsonium ion, which precipitates thallium as $(C_6H_5)_4AsTlCl_4$, and the tetraphenylborate(III) anion, which precipitates potassium as $KB(C_6H_5)_4$.

Structural factors that influence coordination A brief outline of the more important structural factors, with respect to the metal ion as well as the ligand, that determine the stability of coordination compounds should be helpful as a guide to qualitative predictions about the applicability of organic reagents for analytical purposes. At the outset it should be recognized that the stability of a coordination compound, expressed by its formation constant, is only one of several factors that determine the completeness of precipitation or extraction of a metal ion.

In precipitation, the intrinsic solubility (Section 7-1) of the complex and the dissociation constant of the reagent are involved. Freiser[15] observed that in 50% dioxane the Cu(II)-dimethylglyoxime complex is actually more stable than the corresponding Ni(II) complex, yet nickel is precipitated preferentially. The intrinsic solubility of the nickel complex is abnormally low, owing to metal-metal bonding between adjacent molecules in the solid planar structure.[16] Although copper dimethylglyoxime also has a planar structure with nearly the same lattice parameters, the arrangement of molecules in the lattice is such as to preclude a metal-metal interaction.[17]

For metal ions that form complexes of predominantly ionic character, stability increases with increasing charge and decreasing size of the (unhydrated) metal ion. This holds for metal ions possessing the electronic structure of the inert gases and for the lanthanide ions. In the alkali metal and alkaline earth families the stabilities are generally in the order: $Li^+ > Na^+ > K^+ > Rb^+ > Cs^+$ and $Mg^{++} > Ca^{++} > Sr^{++} > Ba^{++} > Ra^{++}$. In the lanthanide family the ionic size decreases and the stability of complexes increases with increasing atomic number.

For metal ions that form complexes of predominantly covalent character, valid generalizations can be made only within closely related groups. For the divalent ions of the first transition-metal series, Irving and Williams[4] found the order of stability to be $Mn^{++} < Fe^{++} < Co^{++} < Ni^{++} < Cu^{++} > Zn^{++}$ for many ligands. Anderegg[18] pointed out correlations involving entropy and enthalpy of association with atomic number for various complexes of transition metals.

A *steric factor*, reflecting that for each metal ion there is a particular preferred orientation of coordinated groups, is occasionally to be considered. This factor seems to be of secondary importance. For example, a square planar configuration is preferred over the tetrahedral configuration for tetracovalent complexes of Cu(II). Nevertheless, a reagent such as trimethylenetetramine, which could not form a planar complex, for steric reasons forces the tetrahedral configuration upon Cu(II) with only a small loss of expected stability.[19]

The basic dissociation constant K_b is a measure of the tendency of a ligand to coordinate with the hydrogen ion. Therefore, if *a series of ligands of structural similarity* are compared, a linear relation should be expected between log β (where β is the formation constant) and pK_b or pK_a for the ligand.[20] Examples are the Ag(I) complexes of primary amines and the Cu(II) complexes of β-diketones. There is a complication: Steric factors are apt to influence coordination with metal ions more than with hydrogen ions. Not only are metal ions larger, but they usually have larger coordination numbers than hydrogen ions, which at most can be bonded to two groups. Striking examples of steric effects are encountered in the use of high-molecular-weight amines as extraction reagents. Chain branching in the vicinity of the amine nitrogen causes interference with the coordinating tendency of the amine.

Irving, Cabell, and Mellor[21] found that 2-methyl-1,10-phenanthroline is a stronger base than 1,10-phenanthroline; yet it forms weaker complexes with Fe(II). The relative instability of the Fe(III) complex with 1,10-phenanthroline compared with the Fe(II) complex also seems to be due, at least in part, to steric hindrance. Steric hindrance plays an important role in the use of the reagent 2-methyl-8-hydroxy-quinoline, which precipitates zinc and magnesium but not aluminum.[22]

The chelate, or entropy, effect and the special stability of five- and six-membered rings is mentioned in Section 11-1.

22-4 SELECTED EXAMPLES OF ORGANIC REAGENTS IN PRECIPITATION

Oxine, or 8-hydroxyquinoline, is a precipitant for many metals. For the separation of heavy metals, extraction is generally a more attractive method than precipitation because difficulties due to coprecipitation can thus be avoided. Nevertheless, oxine has long been used[23,24] to separate aluminum from the alkali metals and alkaline earths, including magnesium and beryllium. An acetic acid–ammonium acetate buffer is used. The effect of the combination of a chelating reagent such as EDTA and of pH on selectivity has been examined.[25] Other separations have been described by Flagg.[6] Chalmers and Basit[26] showed that positive errors are attributable to coprecipitation of oxine and recommended heating the solution after mixing the reagents to ensure correct results. Magnesium may be separated from the alkali and alkaline earths by precipitation from ammoniacal buffers.[27] Since 2-methyl-8-hydroxy-quinoline (8-hydroxyquinaldine) does not precipitate aluminum,[22] zinc in the presence of aluminum and magnesium can be precipitated with 8-hydroxyquinaldine from an acetate buffer. At a pH higher than 9.3, magnesium forms a precipitate.

Dimethylglyoxime has long been used as a precipitant for nickel and palladium. Nickel[28] is usually precipitated from an ammoniacal tartrate buffer of pH about 8. Under these conditions iron and many other metals, even in large amounts, do not interfere. Palladium[29] is precipitated from hydrochloric or sulfuric acid solution. Nioxime[30] (1,2-cyclohexanedionedioxime) has the advantage of being more soluble in water than dimethylglyoxime and therefore less subject to coprecipitation with the metal chelate.

1-Nitroso-2-naphthol is important as a selective precipitant for cobalt.[31] The organic precipitate, however, is unsuitable as a weighing form, and so it must be ignited to Co_3O_4. For larger amounts of cobalt, metallic cobalt or $CoSO_4$ is recommended as the weighing form.

Cupferron (ammonium salt of N-nitrosophenylhydroxylamine) is of use principally as a group precipitant and extractant for several metal ions of higher charge from strongly acid solution. Several applications have been discussed by Clarke.[32] N-Benzoylphenylhydroxylamine, $C_6H_5CO(C_6H_5)NOH$, has been suggested[33] as a reagent similar to cupferron in its reactions but more stable.

The arsonic acids, notably phenylarsonic acid[34] [$C_6H_5AsO(OH)_2$], p-hydroxy-phenylarsonic acid,[35] and n-propylarsonic acid[36] are selective precipitants for quadrivalent metals in acid solution.

Tannin can hardly be classified with the usual organic reagents because it apparently acts as a negative colloid that is a flocculant for positively charged hydrous oxide sols such as those of WO_3, Nb_2O_5, and Ta_2O_5. For example, when a tungstate solution is treated with tannin and acidified, most of the tungsten is precipitated. A small amount remains colloidally dispersed and is flocculated with a tannin type of precipitant such as the alkaloid cinchonine.[37] In this way tungsten can be separated from many ions. The separation of tantalum from niobium also is of interest; tantalum is precipitated selectively from a slightly acidic oxalate solution.[38] The precipitation of germanium with tannin after distillation of the tetrachloride has been applied to the analysis of steel.[39] An exceptionally selective precipitant for tungsten is anti-1,5-di(p-methoxyphenyl)-1(hydroxylamino-3-oximino-4-pentene), which forms a 1:1 complex with tungstate in acid solution.[40]

Tetraphenylarsonium chloride has been recommended by Smith[41] as a precipitant for Tl(III) as $(C_6H_5)_4AsTlCl_4$, which is weighed as such. The principal interferences are cations that form insoluble chlorides and various anions other than chloride. A similar reagent is triphenylmethylarsonium iodide, which has been suggested for the determination of cadmium in the presence of zinc.[42] The weighing form is $[(C_6H_5)_3CH_3As]_2CdI_4$. Metals that form insoluble iodides or anionic iodide complexes interfere.

Benzidine, $H_2NC_6H_4$—$C_6H_4NH_2$, is used in slightly acid solution as the hydrochloride for the precipitation of sulfate as $C_{12}H_{12}N_2H_2^{++}SO_4^{=}$. After separation the analysis may be completed by weighing the precipitate as such or by titrating it with standard base.

Sodium tetraphenylborate, $Na^+B(C_6H_5)_4^-$, is an important ionic precipitant for potassium ion. The solubility product[43] of the potassium salt is 2.25×10^{-8}. Introduced by Raff and Brotz[44] in 1951, this reagent has many analytical uses. A review of the various applications has been given by Cluley,[45] who also described procedures for the separation and determination of potassium in silicates. In acid solution at pH 2 or pH 6.5 in the presence of EDTA, the separation is almost specific for potassium. Mercury(II) interferes in acid solution, but does not interfere at high pH in the presence of EDTA. The most important interferences are rubidium and cesium. Ammonium ion forms a slightly soluble salt that is removed by ignition. An acid solution should be used when possible, though a pH of much less than 2 or elevated temperatures tend to promote decomposition of the reagent,[46] leading to high results.

Myasoedova[47] reviewed the use of organic reagents for concentrating traces of metal ions by coprecipitation. Coprecipitated compounds can include salts of anionic metal complexes, salts of cations, chelate compounds, and colloidal suspensions.

22-5 DISTILLATION AND VOLATILIZATION

Separation by distillation involves differential vaporization of a liquid mixture, with subsequent collection of the vaporized material usually by cooling and condensation. Distillation comprises only a single stage in many analytically important applications; multistage techniques are more significant in preparative and industrial applications. Distillation may be used to give directly the quantitative composition of a sample, in which case complete separation of each component from the others is usually unnecessary. Our treatment of distillation techniques is abbreviated since, for the analysis of complex volatile mixtures, they have been largely supplanted by gas chromatography.

In volatilization we consider the principles of several types of analytical operations involving the separation of a gas from a solid or, less often, from a liquid sample. We are concerned primarily with methods that involve a chemical change such as a thermal decomposition or dehydration reaction or a reaction between a sample and a gas that removes a gaseous product from the sample. After separation, indirect determination can be based on the change in weight, or direct determination of the gaseous product can be carried out by measurement of its volume or weight, by titration or by instrumental means.

Distillation For separation to occur during distillation, the composition of the vapor must differ from that of the liquid. For example (Figure 22-2, left), the vapor from a solution of methanol and water contains a larger proportion of methanol than does the liquid. According to Figure 22-2, if 40% methanol is distilled, the first vapor to be condensed contains about 72% methanol. If the 72% vapor were condensed and a second distillation performed, the first part of the condensate would be about 88% methanol. To obtain near 100% methanol, a distillation with many such stages is required. In some systems the composition changes not toward the pure

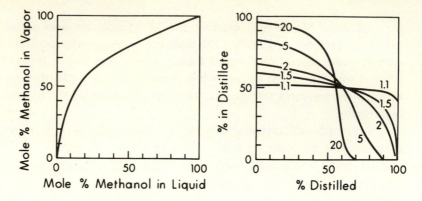

FIGURE 22-2   Left, relation between percentage of methanol in the liquid and in the vapor at equilibrium for methanol-water mixtures. Right, relation between vapor composition (as percentage of the more volatile component) and the fraction distilled for relative volatilities of 1.1, 1.5, 2, 5, and 20. An initial 50:50 mixture is assumed.

components but to a constant-boiling (azeotropic) mixture. For example, ethanol-water mixtures containing 89.4 mole % ethanol distill at 1 atm and 78.15°C without change in composition.

If we consider a binary liquid mixture, the volatility v_1 of the first component is defined by the ratio of its partial vapor pressure $\overline{p_1}$ to its mole fraction x_1 in the liquid:

$$v_1 = \frac{\overline{p_1}}{x_1} \quad (22\text{-}16)$$

If Raoult's law holds, $\overline{p_1} = x_1 p_1$, where p_1 is the vapor pressure of the pure liquid. Thus, for an ideal mixture the volatility of a component is simply proportional to the vapor pressure of the pure component.

The *relative volatility RV* is the ratio of the volatilities for two substances, or

$$RV = \frac{v_1}{v_2} = \frac{\overline{p_1} x_2}{\overline{p_2} x_1} \quad (22\text{-}17)$$

Normally, the ratio of partial vapor pressures $\overline{p_1}/\overline{p_2}$ can be replaced by the ratio of the mole fractions y_1/y_2 in the vapor (Dalton's law); so

$$\frac{y_1}{y_2} = RV \frac{x_1}{x_2} \quad (22\text{-}18)$$

where RV conventionally is set greater than 1 by specifying the components so that $v_1 > v_2$.

For a binary mixture, $x_2 = 1 - x_1$, and $y_2 = 1 - y_1$. It is customary to use

x and y as mole fractions of the more volatile component in the liquid and the vapor, so that

$$\frac{y}{1 - y} = RV\left(\frac{x}{1 - x}\right) \qquad (22\text{-}19)$$

Figure 22-2 (right) shows how distillate composition changes with fraction distilled in a simple one-stage distillation for varying values of relative volatility.[48]

Equation (22-19) is useful particularly for pairs of chemically similar liquids. If Raoult's law holds, relative volatility is equal to p_1/p_2. Therefore, it is possible to plot liquid-vapor composition diagrams for closely similar liquids, such as benzene-toluene, without further ado. Note that the value of RV is not strictly constant over the whole composition range, even for such a mixture, because p_1 and p_2 do not necessarily vary similarly with temperature (Clausius-Clapeyron equation).

By applying the Clapeyron equation and Trouton's rule to an ideal binary mixture, Rose[49] derived the relation

$$\log RV = 8.9\frac{T_2 - T_1}{T_2 + T_1} \qquad (22\text{-}20)$$

which is useful for estimating RV from the absolute normal boiling points T_1 and T_2 of the pure components. Equation (22-20) can be expected to hold only for chemically similar, normal liquids, to which Trouton's rule can be applied.

EXAMPLE 22-2 The normal boiling points of chlorobenzene and bromobenzene are 132 and 156°C. Estimate the relative volatility from (22-20) and compare this with the ratio of vapor pressures at 140°C (939.5 to 495.8 mm).

ANSWER From (22-20), $\log RV = 8.9(429 - 405)/(429 + 405) = 0.256$, or $RV = 1.80$ as compared with $939.5/495.8 = 1.89$.   ////

For nonideal liquid mixtures, more complex relations involving activity coefficients have been derived.[50,51] It is always necessary, however, to obtain the vapor-pressure data from experimental measurements.

Volatilization In dealing with thermal-decomposition and volatilization reactions, not only must the equilibria involved in reversible reactions be considered, but also the rates of both reversible and irreversible reactions. From thermodynamic data, accurate deductions often can be made as to the effects of conditions such as temperature and pressure on equilibrium behavior. The kinetic aspects, however, are more difficult to predict, and the conditions for carrying out the desired separation usually must be determined by experiment. Despite their practical importance, combustion analysis methods, such as those commonly carried out in elemental analysis of organic compounds, are arbitrarily omitted here owing to limitations of space:

The effect of temperature on chemical equilibria is expressed by the thermo-dynamic equations

$$\Delta G^0 = \Delta H^0 - T\Delta S^0 = -RT \ln K_p \qquad (22\text{-}21)$$

$$\frac{d \ln K_p}{dT} = \frac{\Delta H^0}{RT^2} \qquad (22\text{-}22)$$

$$\left(\frac{\delta \Delta G^0}{\delta T}\right)_p = -\Delta S^0 \qquad (22\text{-}23)$$

where ΔG^0 = standard free energy change; ΔH^0 = standard enthalpy change; ΔS^0 = standard entropy change (where for these three quantities the pressure is in atmospheres); K_p = equilibrium constant expressed in terms of pressure; R = gas constant; and T = absolute temperature. To carry out accurate calculations of ΔG^0 or K_p as a function of temperature, it is necessary to consider the variation of ΔH^0 and of ΔS^0, and therefore of the temperature coefficients given by (22-22) and (22-23), with changes in temperature. First, the heat of formation $\Delta H_T{}^0$ of each substance at the desired temperature T must be determined by carrying out the integration

$$\Delta H_T{}^0 = \Delta H_{298}^0 + \int_{298}^{T} \Delta C_p \, dT \qquad (22\text{-}24)$$

where ΔC_p is the difference between the heat capacities of the substance and its constituent elements. Similarly, the entropy of each substance is determined by the equation

$$S_T{}^0 = S_{298}^0 + \int_{298}^{T} \frac{\Delta C_p}{T} \, dT \qquad (22\text{-}25)$$

The integrations can be carried out graphically[52] from the values of $\Delta H_T{}^0$ and $\Delta S_T{}^0$ for the reaction, and the values of ΔG^0 and K_p then can be calculated by (22-21). As a first approximation, however, it often suffices to assume that ΔH^0 and ΔS^0 for the *reaction* are independent of temperature, even though the heats and entropies of formation of the individual reactants vary with temperature. This amounts to setting equal to zero the temperature coefficients $(\delta \Delta H^0/\delta T)_p = \Delta C_p$ and $(\delta \Delta S^0/\delta T)_p = \Delta C_p/T$, where ΔC_p is the difference between the heat capacities of the products and reactants, averaged over the desired temperature interval. Thus, using the values of $\Delta H°$ and $\Delta S°$ at 25°C, we write

$$\ln K_p = \frac{-\Delta H^0}{RT} + \frac{\Delta S^0}{R} \qquad (22\text{-}26)$$

This shows that, if the variations of ΔH^0 and ΔS^0 with temperature are neglected,

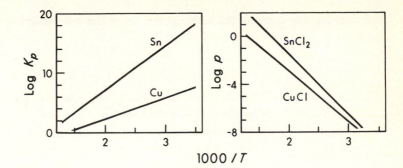

FIGURE 22-3 Variation of equilibrium constants with temperature. Left, $\log K_p$ for tin and copper reactions. (*From Trautzl and Treadwell.*[53]) Right, vapor pressure (mm Hg) of $SnCl_2$ and $CuCl$.

the logarithm of the equilibrium constant is a linear function of the reciprocal of the absolute temperature. From the slope and intercept, the values of ΔH^0 and ΔS^0 can be calculated.

EXAMPLE 22-3 Trautzl and Treadwell[53] investigated the selective oxidation and separation of tin from copper-tin alloys by means of a stream of dry hydrogen chloride. They determined the equilibrium constants $K_p = p_{H_2}/p_{HCl}^2$ for the reactions

$$2Cu + 2HCl(gas) \rightleftharpoons 2CuCl + H_2$$

and

$$Sn + 2HCl \rightleftharpoons SnCl_2 + H_2$$

and expressed them in the form of the equation

$$\log K_p = \frac{A}{T} + B$$

where A and B are constants.

For the copper and tin reactions the constant A was evaluated at 3489 and 7359 and the constant B at -4.74 and -7.60. From the values of A and B we calculate, for the copper and tin reactions: $\Delta H^0 = -2.3RA = -67$ and -141 kJ; $\Delta S^0 = 2.3RB = -91$ and -146 J/mole-deg; $\Delta G^0 = -2.3RT \log K_p = -40$ and -97 kJ at 25°C; and $\Delta G^0 = -5.8$ and -43 kJ at 400°C.

Figure 22-3 shows, for both reactions, the relation between temperature and K_p or vapor pressure p. The lines represent extrapolation to room temperature of data obtained in the temperature range 360 to 665°C for copper and 210 to 605°C for tin. The reaction rates at room temperature are too slow to attain equilibrium, especially in a flowing system. Greater selectivity of reaction exists at *low* temperatures than at high temperatures. However, since it is desired to vaporize the tin(II) chloride formed, it is necessary to consider the vapor pressures of the salts, given by equations of the form

$$\log P = \frac{C}{T} + D$$

where P is measured in millimeters of mercury, and where C has the values -4220 and -4490 and D the values 5.45 and 7.73 for CuCl and $SnCl_2$.

From Figure 22-3, greater selectivity with respect to vaporization apparently exists at *high* temperatures. A high temperature is favorable also with respect to the rate of the desired reaction. Trautzl and Treadwell chose a temperature of 400°C and showed that $SnCl_2$ could be vaporized quantitatively from copper-tin alloys containing 8.8 to 47.6% tin, leaving essentially pure copper as the residue.   /////

For decomposition reactions of substances in a condensed phase that result in the formation of a gas of the type

$$\text{M(liquid or solid)} \rightleftharpoons \text{N(liquid or solid)} + \text{Q(gas)} \qquad (22\text{-}27)$$

the standard entropy change ΔS^0 is a relatively large positive quantity, because the standard entropy values of gases[54,55] are in general much larger[56] than those of liquids or solids (Table 22-1). (According to Trouton's rule, the entropy of vaporization for an unassociated liquid is 90 J/mole-deg at the normal boiling point.) Therefore the standard negative free energy change $-\Delta G^0$ increases with rising temperature [Equation (22-23)]. Equation (22-26) contains two terms on the right side. The first, or enthalpy, term becomes increasingly important at low temperatures, whereas the second, or entropy, term is practically independent of temperature and so increases in relative importance at higher temperatures. With positive ΔS^0, reactions such as

Table 22-1   THERMODYNAMIC CONSTANTS† AT 25°C

Substance	H^0, kJ	G^0, kJ	S^0, J/mole-deg
H_2	0.0	0.0	130.59
H_2O(liquid)	-285.84	-237.19	69.940
H_2O(gas)	-241.83	-228.59	188.72
CO_2	-393.51	-394.38	213.64
HCl(gas)	-92.312	-95.26	186.68
BaO	-558.1	-528.4	70.3
$BaCO_3$	-1219	-1138	112
CaO	-635.5	-604.2	40
$CaCO_3$	-1207	-1127	88.7
$CaCl_2$	-795.0	-750.2	114
SiO_2	-859.4	-805.0	41.84
$CaSiO_3$	-1584	-1499	82.0

† From Latimer.[56]

(22-27) must become spontaneous at sufficiently high temperatures regardless of ΔH° unless vaporization of the other reactants intervenes; so there is no longer a net formation of gaseous products. Some other reaction also may intervene.

EXAMPLE 22-4 The following thermodynamic data are given for the dissociation of calcium carbonate at 25°C: $\Delta H^0 = 178$ kJ; $\Delta G^0 = 130$ kJ; $\Delta S^\circ = 161$ J/mole-deg. Estimate the temperature at which the dissociation pressure is 1 atm, ignoring the correction for heat capacity.

ANSWER From (22-21), $\ln K_p = 0$ when $\Delta H^0 = T \Delta S^0$; $T = 1106$K or 832°C. The actual temperature for $\ln K_p = 0$ is 882°C. The deviation of the estimated value was caused by the assumption that ΔH^0 and ΔS^0 were independent of temperature. ////

Examples of simple distillations and volatilizations A large number of applications involves characterization of organic materials;[48] no attempt is made to review this area. For some inorganic systems, relative volatility values are large, and quantitative separations can be made with a single-stage distillation. Examples[57] are: the separation of *chromium* as chromyl chloride, CrO_2Cl_2; *osmium* as OsO_4; *ruthenium* as RuO_4; *mercury* as the element; *boron* as methyl borate after reaction of boric acid with methanol; *carbonates* by carbon dioxide evolution upon treatment with a non-volatile acid; *carbon* (in steel, for example) by reaction with pure oxygen to produce carbon dioxide, *nitrogen* by conversion to an ammonium salt followed by distillation of ammonia from sodium hydroxide solution (Kjeldahl nitrogen, Section 6-4); *sulfur* as H_2S or SO_2; *water* by simple heating of the sample; *halogens* as either the free elements (often after selective oxidation) or the hydrogen halides; *fluorine* by steam distillation as H_2SiF_6 from perchloric or sulfuric acid solution in the presence of silica; *silicon* from silicates as SiF_4; *arsenic* as $AsCl_3$ or AsH_3; *antimony* as $SbCl_3$; and *tin* as $SnBr_4$ or SnI_4.

Distillation or sublimation techniques are commonly used to purify substances. For example, pure water as solvent is required in copious amounts; the removal of many impurities is best accomplished by distillation. Iodine as a standard substance is best purified by sublimation. An important method for concentrating trace elements from water prior to analysis is by the technique of freeze-drying.

REFERENCES

1 E. B. SANDELL: *Anal. Chem.*, **40**:834 (1968).
2 E. B. SANDELL: "Colorimetric Determinations of Traces of Metals," 3d ed., p. 24, Wiley, New York, 1959.

3 H. M. N. H. IRVING: *Pure Appl. Chem.*, **21**:109 (1970).

4 H. IRVING and R. J. P. WILLIAMS: *Analyst*, **77**:813 (1952); *J. Chem. Soc.*, **1953**:3192.

5 H. CHRISTOPHERSON and E. B. SANDELL: *Anal. Chim. Acta*, **10**:1 (1954).

6 J. F. FLAGG: "Organic Reagents," p. 55, Interscience, New York, 1948.

7 A. K. MAJUMDAR: *Analyst*, **68**:242 (1943); *J. Indian Chem. Soc.*, **21**:24 (1944).

8 FLAGG,[6] p. 251.

9 H. V. MOYER and W. J. REMINGTON: *Ind. Eng. Chem., Anal. Ed.*, **10**:212 (1938).

10 L. P. BIEFELD and D. E. HOWE, *Ind. Eng. Chem., Anal. Ed.*, **11**:251 (1939).

11 F. J. WELCHER: "Organic Analytical Reagents," vol. 1, chap. 4, Van Nostrand, New York, 1947.

12 F. FEIGL: *Ber. Deut. Chem. Ges.*, **56**:2083 (1923).

13 FLAGG,[6] p. 108.

14 G. F. SMITH and W. H. MCCURDY: *Anal. Chem.*, **24**:371 (1952); D. H. WILKINS and G. F. SMITH: *Anal. Chim. Acta*, **9**:538 (1953); A. A. SCHILT and G. F. SMITH: *Anal. Chim. Acta*, **15**:567 (1956).

15 H. FREISER: *Analyst*, **77**:830 (1952).

16 A. G. SHARPE and D. B. WAKEFIELD: *J. Chem. Soc.*, **1957**:281.

17 L. E. GODYCKI and R. E. RUNDLE: *Acta Crystallogr.*, **6**:487 (1953).

18 G. ANDEREGG in "Coordination Chemistry," A. E. Martell (ed.), vol. 1, chap. 8, Van Nostrand Reinhold, New York, 1971.

19 G. SCHWARZENBACH: *Helv. Chim. Acta*, **35**:2344 (1952).

20 R. W. PARRY and R. N. KELLER in "The Chemistry of the Coordination Compounds," J. C. Bailar, Jr. (ed.), chaps. 3–5, Reinhold, New York, 1956.

21 H. IRVING, M. J. CABELL, and D. H. MELLOR: *J. Chem. Soc.*, **1953**:3417.

22 L. L. MERRITT, JR., and J. K. WALKER: *Ind. Eng. Chem., Anal. Ed.*, **16**:387 (1944).

23 I. M. KOLTHOFF and E. B. SANDELL: *J. Amer. Chem. Soc.*, **50**:1900 (1928).

24 H. B. KNOWLES: *J. Res. Nat. Bur. Stand.*, **15**:87 (1935).

25 J. J. KELLY and D. C. SUTTON: *Talanta*, **13**:1573 (1966).

26 R. A. CHALMERS and M. A. BASIT: *Analyst*, **92**:680 (1967).

27 C. C. MILLER and I. C. MCLENNAN: *J. Chem. Soc.*, **1940**:656.

28 FLAGG,[6] p. 146.

29 F. E. BEAMISH and M. SCOTT: *Ind. Eng. Chem., Anal. Ed.*, **9**:460 (1937).

30 R. C. VOTER, C. V. BANKS, and H. DIEHL: *Anal. Chem.*, **20**:458, 652 (1948).

31 W. F. HILLEBRAND, G. E. F. LUNDELL, H. A. BRIGHT, and J. I. HOFFMAN: "Applied Inorganic Analysis," 2d ed., p. 421, Wiley, New York, 1953.

32 S. G. CLARKE: *Analyst*, **52**:466, 527 (1927).

33 S. C. SHOME: *Analyst*, **75**:27 (1950).

34 A. C. RICE, H. C. FOGG, and C. JAMES: *J. Amer. Chem. Soc.*, **48**:895 (1926).

35 C. T. SIMPSON and G. C. CHANDLEE: *Ind. Eng. Chem., Anal. Ed.*, **10**:642 (1938).

36 H. H. GEIST and G. C. CHANDLEE: *Ind. Eng. Chem., Anal. Ed.*, **9**:169 (1937).

37 D. A. LAMBIE: *Analyst*, **68**:74 (1943).

38 W. R. SCHOELLER: *Analyst*, **57**:750 (1932).

39 A. WEISSLER: *Ind. Eng. Chem., Anal. Ed.*, **16**:311 (1944).

40 J. H. YOE and A. L. JONES: *Ind. Eng. Chem., Anal. Ed.*, **16**:45 (1944).

41   W. T. SMITH, JR.: *Anal. Chem.*, **20**:937 (1948).

42   F. P. DWYER and N. A. GIBSON: *Analyst*, **75**:201 (1950).

43   W. GEILMANN and W. GEBAUHR: *Fresenius' Z. Anal. Chem.*, **139**:161 (1953).

44   P. RAFF and W. BROTZ: *Fresenius' Z. Anal. Chem.*, **133**:241 (1951).

45   H. J. CLULEY: *Analyst*, **80**:354 (1955).

46   H. W. BERKHOUT: *Chem. Weekbl.*, **48**:909 (1952).

47   G. V. MYASOEDOVA: *Zh. Anal. Khim.*, **21**:598 (1966); *J. Anal. Chem. USSR*, **21**:533 (1966).

48   A. ROSE in "Treatise on Analytical Chemistry," I. M. Kolthoff and P. J. Elving (eds.), pt. 1, chap. 29, Interscience, New York, 1961.

49   A. ROSE: *Ind. Eng. Chem.*, **33**:594 (1941).

50   H. C. CARLSON and A. P. COLBURN: *Ind. Eng. Chem.*, **34**:581 (1942).

51   A. M. CLARK: *Trans. Faraday. Soc.*, **41**:718 (1945).

52   G. N. LEWIS and M. RANDALL: "Thermodynamics," revised by K. S. Pitzer and L. Brewer, p. 121, McGraw-Hill, New York, 1961.

53   P. TRAUTZL and W. D. TREADWELL: *Helv. Chim. Acta*, **34**:1723 (1951).

54   W. F. GIAUQUE: *J. Amer. Chem. Soc.*, **52**:4808 (1930).

55   E. D. EASTMAN: *Chem. Rev.*, **18**:257 (1936).

56   W. M. LATIMER: "The Oxidation States of the Elements and Their Potentials in Aqueous Solutions," 2d ed., Prentice-Hall, Englewood Cliffs, N.J., 1952.

57   A. I. VOGEL: "A Textbook of Quantitative Inorganic Analysis," 3d ed., p. 137, Longmans, London, 1961.

58   I. M. KOLTHOFF and G. E. NOPONEN: *J. Amer. Chem. Soc.*, **60**:197 (1938).

PROBLEM

22-1   Kolthoff and Noponen[58] mixed 25 ml of 0.11 M barium nitrate, 10 ml of 0.1 M lead nitrate, and 25 ml of 0.1 M sodium sulfate. After precipitation and digestion for 1 h, 59% of the lead remained in the precipitate. Calculate the recovery and separation (enrichment) factors for barium.

Answer $R_{Ba} = 0.69$; $S_{Ba,Pb} = 0.85$.

23

LIQUID-LIQUID EXTRACTIONS AND SEPARATIONS

As pointed out in Chapter 22, of the enormous number of organic reagents that exist, many are suitable for making selective separations and analyses by way of precipitation. The effectiveness of many of these reagents for separations can be increased by use of the technique of liquid-liquid extraction. Such extractions are generally fast and are applicable to trace as well as macroquantities. Selectivity often can be enhanced by control of variables such as pH, reagent concentration, masking agents, choice of solvent, and even the rate of extraction.

Although Peligot observed in 1842 that uranyl nitrate is soluble in ether, it was not until materials of high purity were needed for nuclear reactors that extensive applications and developments, both industrial and analytical, were made. The literature on applications of liquid-liquid extraction (solvent extraction) is extensive; for details of the various procedures the reader is referred to the original papers and to compilations.[1-6] This chapter examines separations involving distribution of a solute between two immiscible phases and chemical equilibria of significance to the distribution ratio. Batch, countercurrent, and continuous liquid-liquid extractions are described in turn, followed by consideration of the factors governing the distribution ratio and finally by some illustrative applications.

23-1 DISTRIBUTION CONSTANT AND DISTRIBUTION RATIO

As indicated above, organic reagents are important not only in carrying out separations by precipitation but also in converting metal ions into forms that can be extracted readily from water into virtually immiscible organic solvents such as carbon tetrachloride, chloroform, ether, and benzene. Usually, though not always, the organic reagent is also preferentially extracted by the organic phase. Extraction has the advantages over precipitation of avoiding coprecipitation and of being applicable to smaller quantities of material. Moreover, the selectivity of a given reagent may be improved by the addition of another equilibrium step (extraction) in the separation.

For a nonionic solute that exists in the same molecular form in the two phases (if polymerization, ionization, and other secondary reactions are disregarded), the distribution equilibrium of a solute A in this single definite form between water and an organic phase is described by

$$A(\text{water}) \rightleftharpoons A(\text{organic}) \qquad K_{D,A} = \frac{[A]_{\text{org}}}{[A]_{\text{w}}} \qquad (23\text{-}1)$$

where $K_{D,A}$ is the *distribution constant* for A,[7] the subscript org denotes the organic phase, and the subscript w denotes water. Hereafter the subscript w usually is omitted.[7]

By applying Equations (4-1) and (4-2) to the distribution equilibrium (23-1), in general we have

$$[A]_{\text{org}}\gamma(A)_{\text{org}}\gamma_t(A)_{\text{org}} = [A]\gamma(A)\gamma_t(A)(\text{water}) \qquad (23\text{-}2)$$

where $\gamma(A)_{\text{org}}$ and $\gamma(A)$ are the usual salt or Debye–Hückel activity coefficients and $\gamma_t(A)_{\text{org}}$ and $\gamma_t(A)$ are the transfer activity coefficients (Section 4-1) in the organic and water phases. If the ionic strengths in the organic solvent and in water are low, $\gamma(A)_{\text{org}}$ and $\gamma(A)$ both approach unity; then, combining (23-1) and (23-2), we obtain $K_{D,A} \simeq \gamma_t(A)/\gamma_t(A)_{\text{org}}$, which indicates that, if the transfer activity coefficient of A in the organic phase is small compared with that in water, the distribution constant is large.

For some extractions, especially those involving ion pairs rather than neutral molecules, the assumption that salt-effect activity coefficients are unity may lead to serious error, as shown by the salt effect on the distribution constant. For example, Scott[8] took advantage of the salting-out effect (Section 2-7) of iron(III) nitrate and nitric acid to improve the extraction of uranyl nitrate from water into ether. The salting-out effects have been interpreted on the basis of changes in the activity coefficient of the uranyl nitrate.[9] Groenewald[10] recently illustrated the influence of these effects on liquid-liquid extractions, using ethyl ether and benzene as organic solvents. The complications due to salt activity-coefficient effects are beyond the scope of the present treatment, and salt activity coefficients are generally assumed to be unity. The limiting value of K_D at zero ionic strength is a true constant for a particular species under specified conditions.

If solid phase is present and *both* liquid phases are saturated with it and at equilibrium with each other, then the distribution constant is

$$K_D = \frac{S_{org}}{S} \qquad (23\text{-}3)$$

where S_{org} and S are the solubilities in the organic and water phases. Equation (23-3), however, is valid only if the same species is involved in both phases.

From the analytical viewpoint the most important type of extraction is that of uncharged molecules of chelate ML_n, which undergo no polymerization in the organic phase. In addition to the complexation equilibria in the aqueous phase, the distribution equilibria

$$ML_n(\text{water}) \rightleftharpoons ML_n(\text{organic}) \qquad K_{D,c} = \frac{[ML_n]_{org}}{[ML_n]} \qquad (23\text{-}4)$$

and

$$HL(\text{water}) \rightleftharpoons HL(\text{organic}) \qquad K_{D,r} = \frac{[HL]_{org}}{[HL]} \qquad (23\text{-}5)$$

must be considered, because both the uncharged reagent r and the chelate c are extracted to some extent into the organic phase.

The solute may be present in solution in several different chemical forms. Of more practical interest is the distribution of the total analytical concentration of a substance in the two phases. A *distribution ratio D* of a solute A is defined by $D = C_{A_{org}}/C_A$, where $C_{A_{org}}$ is the analytical concentration in the organic phase and C_A the analytical concentration in the water phase. The value of the distribution ratio varies with experimental conditions such as pH, whereas the value of the distribution constant at zero ionic strength is invariant for a system at a particular temperature.

If the solute exists in different states of aggregation or association in the two solvents, the equilibrium may be represented by

$$n(A_m)(\text{water}) \rightleftharpoons m(A_n)(\text{organic}) \qquad D = \frac{C_{A_{org}}^m}{C_A^n} \qquad (23\text{-}6)$$

and this distribution ratio varies with the analytical concentration of A. This complication is more common than at first might be expected. For example, osmotic measurements[11] indicate that extensive self-association of tetraalkylammonium and phosphonium salts occurs even at concentrations as low as 10^{-3} m in benzene. Complications from self-association are encountered more often in extraction of inorganic halides than in extraction of chelate complexes. For example, iron(III) is extracted from aqueous hydrochloric acid by diisopropyl ether as solvated $(H^+FeCl_4^-)_n$, where n varies from 2 to 4 depending on the iron concentration.[12,13]

A theoretical treatment of metal halide extractions, encompassing various cationic, anionic, and polymeric species, has been presented by Diamond,[14] who was particularly interested in molybdenum(VI) halides. Section 23-5 examines in more detail the questions of distribution ratios and chemical equilibria in the aqueous and organic phases.

23-2 BATCH LIQUID-LIQUID EXTRACTIONS

In general, with a sufficiently large equilibrium constant (distribution ratio) and a rapid rate of attainment of equilibrium, quantitative transfer from one phase to another can be made in a single stage. For such highly favorable systems, batch liquid-liquid extractions can be used in which one phase is equilibrated with several successive fresh portions of a second phase. Such batch separations are most effective when one component remains quantitatively in one phase while another distributes itself between the two phases.

Exhaustive extraction involves the quantitative removal of a solute, and *selective extraction*, the separation of two or more solutes from each other. One classical application of an exhaustive analytical extraction is the ether extraction of iron(III) chloride from hydrochloric acid solutions.[15] The extraction is not strictly quantitative in that a small fraction remains unextracted. Therefore the method is best suited to the removal of relatively large amounts of iron (several grams) from small amounts of such metals as nickel, cobalt, manganese, chromium, titanium, or aluminum.[16] It is of interest that iron(II) remains unextracted.[17]

In considering the theory of extraction procedures, it is useful to first treat the behavior of a single solute and then extend the discussion to separations with the assumption that two or more solutes undergo no interaction. Since this assumption is usually valid, we do not pursue the complication of cases where it may be unwarranted.

Consider, for the sake of convenience, the distribution of a solute between water and an organic liquid (with density less than that of water), although any two immiscible liquids could be taken. Also for convenience, assume the value of the distribution ratio D to be independent of concentration under given conditions of pH and independent of concentration of reactants including masking agents. If polymerization occurs in one of the two phases, the distribution ratio varies with concentration, and the treatment becomes more complicated, particularly for multistage separations. A more rigorous treatment would require the use of distribution-constant [Equation (23-1)] values for each molecular species present in the system to calculate the sum of their contributions to obtain the percentages extracted under specified conditions. We are interested here in the distribution of the solute between

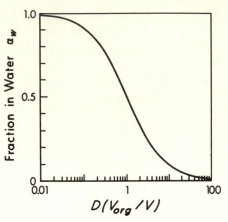

FIGURE 23-1 Fraction of solute in the aqueous phase as a function of the product of distribution ratio and relative volumes of the two phases DV_{org}/V.

the two phases, without concern for the particular chemical form in which it exists. Therefore the distribution ratio is the quantity of practical interest.

Consider, for instance, the use of V_{org} milliliters of organic solvent to extract a solution initially containing X_A moles of a solute A dissolved in V milliliters of water. From Equation (23-6) the distribution ratio D may be expressed by

$$D = \frac{C_{A_{org}}}{C_A} = \frac{(X_A - Y_A)/V_{org}}{Y_A/V} \qquad (23\text{-}7)$$

if Y_A moles of A remains in the water phase after a single extraction. From (23-7) the fraction remaining unextracted α_w is

$$\alpha_w = \frac{Y_A}{X_A} = \frac{1}{1 + D(V_{org}/V)} = \frac{V}{V + DV_{org}} \qquad (23\text{-}8)$$

This fraction remaining unextracted is assumed to be independent of the initial concentration. Figure 23-1 shows the relation between the amount of a solute in the aqueous phase and DV_{org}/V. When $V_{org} = V$, the quantity DV_{org}/V in Figure 23-1 can be replaced by the distribution ratio D. This figure shows that for efficient extractions a large value for the distribution ratio is favorable.

If n successive extractions are performed with fresh portions of solvent, after the nth extraction the fraction remaining unextracted α_n is

$$\alpha_n = \left(\frac{1 + DV_{org}}{V}\right)^{-n} \qquad (23\text{-}9)$$

The question of the limiting amount of solute remaining unextracted after n extractions as n approaches infinity has been considered by Griffin.[18] The limit of α_n in (23-9) is zero for a finite value of V_{org}/V. Such an extraction, however, does not represent a practical case in that the volume of extracting solvent would approach

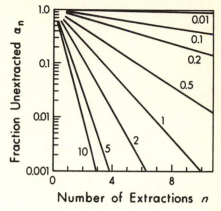

FIGURE 23-2 Fraction unextracted as a function of number of extractions for a system in which $V_{org} = V$, for solutes with distribution ratios D varying from 0.01 to 10.

infinity. For a finite volume V_t of extracting solvent divided into n portions, (23-9) becomes

$$\alpha_n = \left(1 + \frac{DV_t}{nV}\right)^{-n} \quad (23\text{-}10)$$

which in the limit, as n approaches infinity, approaches the value

$$\alpha_\infty = e^{-V_t D/V} \quad (23\text{-}11)$$

As the number of extractions approaches infinity, the volume of organic phase for each extraction approaches zero. In practice, little is gained by dividing the volume of extractant into more than four or five portions, because α_∞ is approached asymptotically. Equation (23-11) is useful in determining whether a given extraction is practicable, using a reasonable value of V_{org}/V, or whether an extractant with a more favorable distribution ratio should be sought.

EXAMPLE 23-1 For a system with $D = 2$, compare the fraction of solute remaining unextracted for the following situations: (a) single extraction with an equal volume of organic solvent, (b) five extractions with the same total amount of organic solvent, (c) limiting case of infinite number of extractions with same total amount of solvent.

ANSWER (a) $\alpha_1 = \frac{1}{3} = 0.333$. (b) $\alpha_5 = (1/1.4)^5 = 0.186$. (c) $\alpha_\infty = e^{-2} = 0.135$. ////

In practical separations where two or more solutes are involved, the conflicting goals of quantitative recovery and quantitative separation must be resolved. Figure 23-2 relates the number of extractions to the fraction unextracted. This figure shows

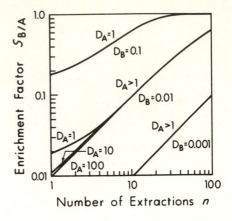

FIGURE 23-3 Enrichment factor for solute A from solute B as a function of number of extractions for a system in which $V_{org} = V$, for various values of D_A and D_B as shown.

that only a small number of extractions are required for quantitative recovery of a substance if its distribution ratio is much greater than unity. The enrichment factor for the separation of A from another solute B by liquid-liquid extraction is, from Equations (22-3) and (23-9),

$$S_{B/A} = \frac{1 - (\alpha_n)_B}{1 - (\alpha_n)_A} \qquad (23\text{-}12)$$

Figure 23-3 illustrates that this enrichment factor becomes progressively poorer (larger numerically) as the number of extractions increases. For quantitative recovery of A ($\alpha_n \leq 0.001$) and quantitative separation from B ($S_{B,A} \leq 0.001$), the distribution ratio[19] for A, D_A, must be at least about 10^4 times larger than D_B. This poor ability for selective separations is the principal disadvantage of the use of batch extractions for recovery or concentration of more than one solute.

23-3 COUNTERCURRENT LIQUID-LIQUID EXTRACTION

By the simple expedient of equilibration of a separated aqueous phase with fresh portions of organic phase, a powerful technique for selective separations is available. The method for carrying out such multiple liquid-liquid extractions is termed *countercurrent distribution*.[20]

An automatic apparatus permits the equilibration of two immiscible liquid phases in a series of specially designed tubes so arranged that one phase (say, the organic) is transferred from each tube to the next in sequence. After equilibrium is regained, another transfer occurs in the same direction. By adding fresh organic solvent, it is possible to carry out as many equilibrations and transfers as needed; several hundred is typical.

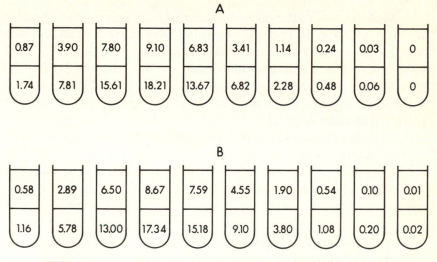

FIGURE 23-4   Distribution of 100 units of a solute with $D = 0.5$ (volumes of the two immiscible phases assumed equal) after (A) 10 equilibrations and 9 transfers and (B) 11 equilibrations and 10 transfers.

Distribution of a single solute   Consider a series of tubes numbered 0, 1, 2, 3, . . . , r, each containing (for the sake of simplicity) equal volumes of an aqueous phase and an immiscible organic liquid of lower density. A solute is placed in the aqueous phase of tube 0. After equilibrium is reached, the organic phase of tube 0 is transferred to tube 1 as that of tube 1 is transferred to tube 2, and so forth. The process of equilibration and transfer is repeated n times; each time, all upper organic layers are transferred to the next higher tube in sequence. The mechanics of connecting the vessels and of the transfer process need not concern us here.[20]

Suppose that 100 units of solute A with a distribution ratio of 0.5 is added to the first tube $[D = C_{A_{org}}/C_A(\text{water}) = 0.5$, or more generally, $DV_{org}/V = 0.5]$. If equilibrium is established, 66.7 units of the solute will be in the aqueous phase and 33.3 in the organic phase ($33.3/66.7 = 0.5 = D$). When the 33.3 units has been transferred to the second tube and equilibrium is again established, 22.2 units of solute will be in the organic phase of the first tube and 11.1 in the second. Figure 23-4A depicts the distribution of the 100 units of solute among the first 10 tubes after 10 such equilibrations and 9 transfers, and Figure 23-4B after 11 equilibrations and 10 transfers. Two aspects of the process of how a solute moves through a series of tubes are important from the point of view of separations: how completely components of varying values of distribution ratio are transferred from one tube to the next and how much spreading occurs. Observe that in Figure 23-4 the amount of solute reaches a maximum in one of the central tubes and decreases in both directions.

More generally, after the first transfer ($n = 1$) the fraction of solute remaining in tube 0 is $1/(D + 1)$, since the two phases were taken in equal volume [Equation (23-8)]. The fraction extracted and transferred in each step is $D/(D + 1)$. Therefore, upon equilibration the second tube (Tube 1, $r = 1$) contains $[1/(D + 1)][D/(D + 1)] = D/(D + 1)^2$ in the aqueous phase and $[D/(D + 1)][D/(D + 1)] = D^2/(D + 1)^2$ in the organic phase. In the second transfer the organic phase is moved to tube 2. Tube 1 then contains $D/(D + 1)^2$ transferred from tube 0 and an equal amount remaining after the second transfer, giving a total of $2D/(D + 1)^2$. Table 23-1 shows the distribution of solute during four extractions.[21]

Williamson and Craig[21] showed that the fraction of solute contained in the various tubes in the two phases can be expressed by the terms of the binomial expansion $[1/(D + 1) + D/(D + 1)]^n$, where the first term is the fraction unextracted and the second is the fraction transferred. A similar expansion had been used earlier by Martin and Synge[22] to represent the amounts of sample in succeeding theoretical plates in chromatography.

According to the general expression for the terms of a binomial expansion, the fraction $\alpha_{n,r}$ in tube r after n extractions is given by

$$\alpha_{n,r} = \frac{n!}{r!(n - r)!}\left(\frac{1}{D + 1}\right)^{n-r}\left(\frac{D}{D + 1}\right)^r \qquad (23\text{-}13)$$

Martin and Synge[22] pointed out that the successive terms of the binomial expansion approximate more and more closely the gaussian distribution of statistics (Section 26-3) as the number of terms increases. Craig[23] wrote for the gaussian curve

$$\alpha_{n,r} = \frac{1}{\sqrt{2\pi n D/(D + 1)^2}}\exp\left[\frac{-(r_{max} - r)^2}{2nD/(D + 1)^2}\right] \qquad (23\text{-}14)$$

Table 23-1   FRACTION OF SOLUTE IN SUCCEEDING TUBES IN LIQUID-LIQUID COUNTERCURRENT DISTRIBUTION

Number of extractions	Tube number				
	0	1	2	3	4
0	1	0	0	0	0
1	$\dfrac{1}{(D + 1)}$	$\dfrac{D}{(D + 1)}$	0	0	0
2	$\dfrac{1}{(D + 1)^2}$	$\dfrac{2D}{(D + 1)^2}$	$\dfrac{D^2}{(D + 1)^2}$	0	0
3	$\dfrac{1}{(D + 1)^3}$	$\dfrac{3D}{(D + 1)^3}$	$\dfrac{3D^2}{(D + 1)^3}$	$\dfrac{D^3}{(D + 1)^3}$	0
4	$\dfrac{1}{(D + 1)^4}$	$\dfrac{4D}{(D + 1)^4}$	$\dfrac{6D^2}{(D + 1)^4}$	$\dfrac{4D^3}{(D + 1)^4}$	$\dfrac{D^4}{(D + 1)^4}$

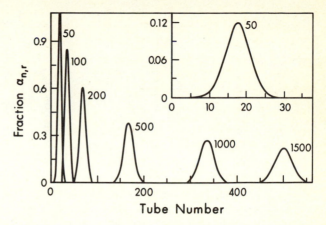

FIGURE 23-5 Distribution of a solute with $D = 0.5$ as a function of tube number, with the number of equilibrations and transfers varying from 50 to 1500. (*From Harris.*[26])

which is a reasonable approximation for $n > 25$. (Figure 23-4 exhibits a distribution not yet that of the gaussian curve.) The quantity $r_{max} - r$ represents the number of times that a particular tube r is removed from the tube of maximum concentration.

Equation (23-14) can be written more generally[24]

$$\alpha_{n,r} = \frac{1}{\sqrt{2\pi n X Y}} \exp\left[\frac{-(nX - r)^2}{2nXY}\right] \qquad (23\text{-}15)$$

where X is equal to the fraction transferred in the organic phase and is $D(V_{org}/V)/[1 + D(V_{org}/V)]$, and where Y is equal to the fraction remaining in the aqueous phase and is $1/[1 + DV_{org}/V]$. Note that, whereas (23-14) pertains to equal volumes of the two phases, (23-15) is valid for all relative volumes.

The approximation $r_{max} = nX$ was proposed by Craig[20] and derived by Nichols,[25] who pointed out that it is a reasonable approximation if n is large compared with r_{max}. This equation is analogous to that for the gaussian distribution [Equation (26-1)], where the population mean μ is replaced by nX, x is replaced by r, and the standard deviation σ by $\sqrt{nXY}$.

The fraction of solute in the tube of maximum concentration is

$$\alpha_{n,r_{max}} = \frac{1}{\sqrt{2\pi n X Y}} \qquad (23\text{-}16)$$

Thus, as was illustrated experimentally by Craig,[23] the concentration to be found in the tube r_{max} varies *inversely* with $\sqrt{n}$, where n is the number of transfers. Figure 23-5 illustrates the distribution[26] of a solute as a function of tube number for $D = 0.5$ (as in Figure 23-4). This figure shows the decrease in r_{max} with the inverse of $\sqrt{n}$. It

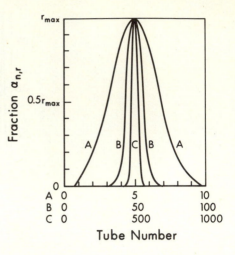

FIGURE 23-6 Comparative band widths as a function of the number of equilibrations and transfers. (*From Craig.*[23])

also indicates that the absolute spread of the band increases in proportion to $\sqrt{n}$. Critically significant, however, is that the *relative* spreading of the band decreases with an increasing number of transfers,[23] as is shown in Figure 23-6. Thus the solute as shown by curve *C* is retained in relatively fewer tubes than as shown by curves *A* and *B*.

So far we have considered only the spreading or movement of a solute *within* a series of tubes as a function of *n*. Spreading may also be considered from the point of view of elution *from* a series of tubes. The sample is spread within the series, as seen from Figures 23-4 and 23-5; also, the sample is spread as it passes any given point in the series. Figure 23-7 compares the two points of view. The term *band* or *zone* denotes the position and spread of a solute within a series of tubes; *peak* refers to a record of concentration of solute in the phase transferred forward as a function of volume while the solute is being eluted from a particular tube. One curve in

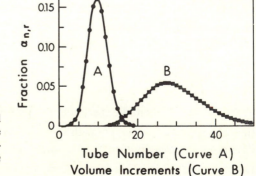

FIGURE 23-7 Comparison of a band or zone (*A*) and a peak (*B*) for a solute with a distribution ratio of 0.5. A volume increment is equal to V_{org}. (*From Harris.*[26])

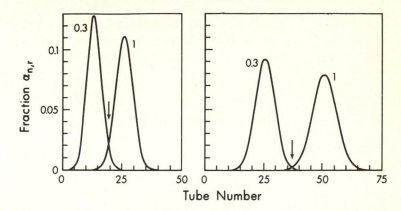

FIGURE 23-8 Band positions and widths illustrating overlap for a mixture of two solutes with distribution ratios of 1.0 and 0.333 for 50 (left) and 100 (right) equilibrations and transfers. (*From Harris.*[26])

Figure 23-7 shows the distribution of a solute in a series of tubes when the band maximum is at tube 10 (this maximum is attained after 28 equilibrations have been made with a solute of $D = 0.5$); the other curve shows the peak expected when the solute has been eluted from tube 10 (with a peak maximum at 28 volume increments). Both points of view are important, particularly when we come to consider chromatographic processes.

Separation of two solutes According to (23-15) the maximum in a distribution curve for a single solute moves along the tubes at a rate proportional to the fraction extracted and transferred in each step ($r_{max} \simeq nX$). Since the spreading of a band increases only with $\sqrt{n}$ (Figure 23-6), if two solutes act independently, in general each will move according to its value of X (fraction transferred), and the bands will separate from each other more rapidly than they spread. Figure 23-8 illustrates[26] this phenomenon for two solutes with distribution ratios of 1.0 and 0.333; overlap with 100 equilibrations and transfers is significantly less than that with 50.

Nichols[25] pointed out that the degree of overlap of two distribution curves can be determined by locating the tube corresponding to the intersection of the distribution curves (arrows in Figure 23-8). The area under the intersected portions of the distribution curves is a quantitative estimate of the degree of nonseparation. The area under any portion of a distribution curve can be determined from statistical tables, since the shapes of the curves closely approximate the gaussian distribution.

Nelson[24] wrote Equation (23-13) in terms of X and Y as follows:

$$\alpha_{n,r} = \frac{n!}{r!(n-r)!} X^r Y^{n-r} \qquad (23\text{-}17)$$

If X_1 and Y_1 refer to solute 1 and X_2, Y_2 to solute 2, and provided the starting amounts of the two solutes are the same, the point of intersection of the distribution curves corresponds to r_i, where

$$[\alpha_{n,r_i}]_1 = [\alpha_{n,r_i}]_2$$

or

$$X_1^{r_i} Y_1^{n-r_i} = X_2^{r_i} Y_2^{n-r_i} \qquad (23\text{-}18)$$

When solved for r_i, Equation (23-18) yields

$$r_i = \frac{n \log Y_2/Y_1}{\log (X_1 Y_2/X_2 Y_1)} = nA \qquad (23\text{-}19)$$

where A is defined as the ratio of the two logarithmic terms. The equation for a gaussian distribution [Equation (26-1)] can now be used to estimate the fraction of solute between the peak r_{max} and the point of intersection r_i of the distribution curves (Figure 23-8). From (23-15) and (23-18)

$$\alpha_{n,r_i} = \frac{1}{\sqrt{2\pi nXY}} \exp\left[\frac{-(nX - nA)^2}{2nXY}\right] \qquad (23\text{-}20)$$

This can be compared to the expression for the area under the tail of the gaussian-distribution curve

$$A_t = \frac{1}{\sqrt{2\pi}} \int_{-\infty}^{t} \exp \frac{-y^2}{2} \, dy \qquad (23\text{-}21)$$

where A_t is the area under one tail of the curve from $-\infty$ to t. The area under the one tail as a fraction of the area under the entire curve is 0.001 for $t = -3.090$, 0.005 for $t = -2.576$, 0.01 for $t = -2.326$, 0.025 for $t = -1.96$, 0.05 for $t = -1.645$, 0.10 for $t = -1.282$, and 0.20 for $t = -0.842$. (See Table 26-1, substituting t for y.)

From a table of areas under the gaussian-distribution curve (Table 26-1) corresponding to the variable

$$t_2 = \frac{nX_2 - nA}{\sqrt{nX_2 Y_2}} \qquad (23\text{-}22)$$

the fraction of solute 2 between the tube r_i and r_{max} can be determined directly. The quantity t_2 corresponds to the number of standard deviations along the gaussian-distribution curve that the position of maximum concentration of the second component is removed from the point of intersection of the two curves. Conversely, for any desired degree of separation the value of n (the number of necessary equilibrations and transfers) is determined from

$$n = \frac{t_2^2 X_2 Y_2}{(X_2 - A)^2} \qquad (23\text{-}23)$$

Similar calculations can be carried out for solute 1.

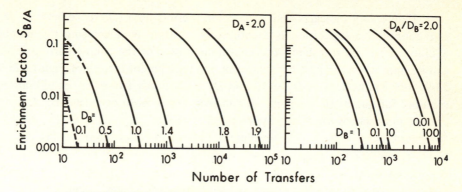

FIGURE 23-9   Enrichment factor $S_{B/A}$ as a function of number of equilibrations and transfers in liquid-liquid countercurrent extraction. Left, for $D_A = 2.0$ and values of the distribution ratio D_B varying from 0.1 to 1.9; right, for $D_A/D_B = 2$ and with distribution ratios of 2/1, 0.2/0.1, 20/10, 0.02/0.01, and 200/100.

EXAMPLE 23-2   Suppose that two solutes have distribution ratios $D_1 = 1$ and $D_2 = 2$ and that equal volumes of the two phases are used. Then $X_1 = 0.5$, $Y_1 = 0.5$, $X_2 = 0.667$, $Y_2 = 0.333$. From Equation (23-19), $A = 0.585$. Calculate the number of transfers required so that the fraction of solute 2 to the left of tube r_i is 2.5%. From the preceding data, this corresponds to $t_2 = 1.96$ (see also Table 26-3). Then, by substitution in (23-23), $n = 127$, the number of equilibrations and transfers required. Now $nX_2 = 84.5$, $nX_1 = 63.5$, and $r_i = 74.3$. The fraction of solute 1 contained in the tubes to the right of r_i is given by

$$t_1 = \frac{nX_i - r_i}{\sqrt{nX_1 Y_1}} = -1.81$$

Statistical t tables reveal that 3.5% of solute 1 is to be found at $r > r_i$. If a separation were made at $r = r_i$, the purity of solute 2 would be 96.5%, and of solute 1, 97.5%.

////

Figure 23-9 is a set of curves, calculated from (23-19) and (23-23), for enrichment factor [Equation (22-2)] as a function of the number of equilibrations and transfers under several conditions. The number of equilibrations and transfers for quantitative recovery (recovery factor = 0.999) of component A and for its quantitative separation (enrichment factor = 0.001) from component B is indicated by the intersections of the lines with the 0.001 line for $S_{B/A}$. The termination at the upper edge of the curves in Figure 23-9 ($S_{B/A} = 0.2$) represents resolution [Equation (24-30), Figure 24-7] slightly less than unity, a separation adequate for precise quantitative measurements.

Figure 23-9 indicates that, even when there is little difference between D_2 and D_1, quantitative separation is theoretically possible by increasing the number of equilibrations and transfers. The enormous power of the countercurrent technique can be appreciated in part by comparing the enrichment factors of Figures 23-3 and 23-9. For example, with $D_2 = 10$ and $D_1 = 0.1$, only five or six equilibrations are required for quantitative separation and recovery of both components. A matter of critical importance often not appreciated is that the quotient of D_2/D_1 must be favorable and *in addition* the absolute values of the distribution ratios D_2 and D_1 must be neither too large nor too small. When values of D_2 and D_1 are small, separation is inefficient, because little of either solute is transferred with each step; when the values are large, separation requires many stages, because neither solute is held back appreciably by the aqueous phase.

The countercurrent-distribution technique is of particular value whenever it is desirable to follow in detail the progress of a separation. Thus, in the resolution of a complex mixture an initial extraction may be performed using relatively few tubes of large size. Any desired fraction from the initial separation then may be subjected to further fractionation, by using, if necessary, an apparatus with a large number of tubes and performing a large number of transfers. An example of such a separation has been given by Bell,[27] who used up to 10,000 transfers in a 200-tube apparatus to isolate large peptide fragments from the degradation of ACTH.

Another special feature of countercurrent distribution is that it can be scaled up to permit the isolation of small quantities of components from large amounts of starting material. For example, Patterson and coworkers[28] used a 40-tube apparatus, consisting of stages with 1 l in each phase, to isolate protogen from 4 tons of mixed beef and pork liver.

23-4 CONTINUOUS LIQUID-LIQUID COUNTERCURRENT EXTRACTION

Continuous addition and transfer of the organic phase can be approached from the point of view of idealized continuous equilibrium, the limiting case of the addition of a series of increments δV, each small compared with the volume V_{org} in a single tube. Distribution equilibrium of the solute is assumed to be established between the two phases after the transfer and addition of each increment. Such incremental addition results in broader bands or peaks than would be obtained from complete transfer of the organic phase after each equilibration. If the fraction $\delta V/V_{org}$ of the amount at equilibrium in the organic phase is transferred with each increment, and if V_{org} is assumed equal to V, then the fraction of a solute transferred (extracted) with each increment of organic phase is $\delta V D/(D + 1)$, and the fraction remaining with the aqueous phase is $[1/(D + 1) + (1 - \delta V)D/(D + 1)]$. Figure 23-10 shows for two

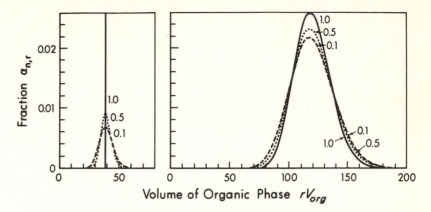

FIGURE 23-10 Effect of incremental addition and transfer of organic phase on peak widths in liquid-liquid countercurrent extraction for solutes with distribution ratios of ∞ (left) and 0.5 (right). (*From Harris.*[26])

solutes the peak widths resulting from such partial transfer.[26] Two aspects should be noted. First, the average rate of movement of a solute for equal volumes of organic phase used (in units of rV_{org}) is not affected by the incremental mode of operation and, second, the peaks are spread more. This spreading is most serious for solutes that are more strongly held in the organic phase (D large). Peak broadening increases significantly as the size of the increments δV is decreased until $\delta V/V \simeq 0.2$, and then further change is small.

The theory of continuous liquid-liquid countercurrent extraction leads directly to chromatographic theory. Martin and Synge[22] proposed a theory of chromatography based on the concept of theoretical plates (Section 24-2) that involved an equilibrium model analogous to continuous countercurrent liquid-liquid extraction but without physically discrete equilibrium stages. Instead of discrete equilibrium stages, their model involved visualizing a column with a series of zones, called plates, of such length that in each the two phases are at equilibrium. They paid particular attention to a point of fundamental concern in chromatography, the location of the maximum concentration of solute in the column. Adapting their derivation for chromatography to obtain the fraction $\alpha_{n,r}$ at this maximum of a solute with a distribution ratio D in zone (or plate) r after n successive increments δV of organic phase, we have

$$\alpha_{n,r} = \frac{1}{\sqrt{2\pi r}} \left[\frac{V_{org}}{r[V_{org} + (V/D)]} \right]^r \exp \left[r - \frac{V_{org}}{[V_{org} + (V/D)]} \right] \quad (23\text{-}24)$$

where V_{org} and V are volumes of organic solvent and water in one plate. According to (23-24) the maximum concentration of solute in the band varies inversely with the square root of the number of plates as it does in Craig's countercurrent liquid-liquid extraction. The band therefore broadens as it proceeds down the column. When r is large, (23-24) closely approximates the function for a gaussian-distribution curve:

$$\alpha_{n,r} = \frac{1}{\sqrt{2\pi r}} \exp\left(\frac{-\{V_{org}/[V_{org} + (V/D)]\}^2}{2r}\right) \qquad (23\text{-}25)$$

Equation (23-25) reaches a maximum value of α when $V_{org}/[r(V_{org} + V/D)] = 1$; this value is $1/\sqrt{2\pi r}$. The same result is obtained from (23-24).

Equation (23-24) is an example of a Poisson distribution, characteristic of a continuous-flow process, in contrast to the binomial-distribution characteristic of a batch process [Equation (23-13)]. Both approximate the gaussian-distribution curve when the number of stages becomes large.[29]

Apparatus for continuous addition and transfer of one phase, involving a helix of narrow-bore tubing, has been described by Ito and Bowman.[30] With a coil of 8000 turns and continuous phase separation provided by centrifugation they obtained separations equivalent to about as many as 5000 tubes used in the conventional Craig manner; this number would be expected under some conditions if each turn in the capillary acts as one equilibrium vessel under conditions of continuous transfer.

Mayer and Tompkins[31] adapted chromatographic theory to the problem of predicting the number of plates for a desired separation. Glueckauf[32] showed that their calculations were in error in the adoption of the discontinuous-flow equilibrium process (Craig countercurrent extraction) described in Section 23-3, in contrast to the continuous-flow equilibrium process described in Section 23-4. Figure 23-10 illustrates that the difference between the two models is particularly large when the distribution ratio in liquid-liquid extraction is such that the solute is mostly in the organic phase, and furthermore that it is far from negligible even when the solute is retained primarily by the aqueous phase. Continuous flow with continuous equilibrium is less efficient than the Craig countercurrent type of separation.

Material for analysis can be recovered or examined by determination of either the distribution of a solute in a series of tubes or its distribution as it is eluted from such a series. To indicate quantitatively the effectiveness of the selective separation of one substance from another, the term resolution may be used. Resolution of bands may be defined as the ratio of band separation to average band width, and similarly, resolution of peaks as the ratio of peak separation to average peak width (Section 24-6).

Johnson[33] discussed in detail band broadening in an extraction process involving the introduction of sample, all at once, into a continuously flowing countercurrent system where both phases are moving in opposite directions and continuously at equilibrium. This *countercurrent cascade* technique[34] is one of considerable industrial significance. With a given number of separation stages this technique can bring about more complete separation with high recovery factors than is possible with countercurrent extraction. The sample often is injected into one of the central stages rather than at one end, and the number of separation elements per unit length is decreased (cross section decreased) in both directions from the feed point, usually as a stepwise (squared-off) cascade. Two effluents are obtained from the two ends of the series of equilibration vessels if one solute has $DV_{org}/V > 1$ and another $DV_{org}/V < 1$. Multistage distillation is an example of a countercurrent cascade process. Digital computer simulations of the batchwise cascade process have been described.[35]

Cascade theory has been applied[34] to the separation of uranium 235 by gaseous diffusion of uranium hexafluoride; the ratio of the diffusion rates for uranium 235 and uranium 238 is 1.0043. With about 3900 equilibration stages a product containing 90 mole % uranium 235 can be obtained from natural uranium (0.71 mole %), giving a residue with 0.22 mole % uranium 235. The feed in this case is injected at the 550th stage.

Johnson[33] pointed out that isolation of components of mixtures is more efficient by way of elution from an extended system of vessels than by recovery from the vessels after the solutes have distributed themselves among them. Separation is more efficient when advantage is taken of the separating ability of all the vessels.

23-5 DISTRIBUTION RATIOS FOR COMPLEXES

The critical significance of the distribution ratio has become clear in the preceding sections. In this section several factors influencing the value of the distribution ratios of complexes are examined.

Irving, Rossotti, and Williams[36] considered the distribution ratios of inorganic compounds in a generalized way. They studied such effects as ion aggregation, solvation, stepwise complex formation between metal ion and inorganic anions or chelating reagent, and polymerization in the two phases. This general treatment is useful as a guide in planning experiments to determine the nature of the species present in both phases. For example, by determining the distribution ratio as a function of the metal concentration, it is possible to determine the difference in degree of association of the metal in the two phases.

Distribution ratio for a single chelate complex If we assume that the only form of the metal in the organic phase is the chelate ML_n, and if we neglect the formation of lower complexes in the aqueous phase, the distribution ratio of the chelate is

$$D = \frac{C_{M,org}}{C_M} = \frac{[ML_n]_{org}}{[ML_n] + [M^{n+}]} = \frac{K_{D,c}}{1 + [M^{n+}]/[ML_n]} \qquad (23\text{-}26)$$

where $K_{D,c}$ is the distribution constant for the chelate. Introducing the overall formation constant β_n of the complex ML_n in the aqueous phase,

$$\beta_n = \frac{[ML_n]}{[M^{n+}][L^-]^n} \qquad (23\text{-}27)$$

and the dissociation constant K_a of the reagent (Section 22-3) in the aqueous phase,

$$K_a = \frac{[H^+][L^-]}{[HL]} \qquad (23\text{-}28)$$

we have

$$\frac{[M^{n+}]}{[ML_n]} = \frac{[H^+]^n}{\beta_n K_a^{\,n}[HL]^n} = \frac{[H^+]^n K_{D,c}^{\,n}}{\beta_n K_a^{\,n}[HL]_{org}^n} \qquad (23\text{-}29)$$

From substitution in (23-26)

$$D = \frac{K_{D,c}}{1 + [H^+]^n K_{D,r}^{\,n}/\beta_n K_a^{\,n}[HL]_{org}^n} \qquad (23\text{-}30)$$

Equation (23-30) expresses the change in the distribution ratio of the metal in terms of the constants $K_{D,c}$, $K_{D,r}$, β_n, and K_a, which represent properties of the solvent and metal chelate system, and in terms of the variables $[H^+]$ and $[HL]_{org}$, which are subject to experimental variation for a given system. Equation (23-30) can be written

$$\frac{1}{D} = \frac{1}{D'} + \frac{1}{K_{D,c}} \qquad (23\text{-}31)$$

if D' is defined as

$$D' = \left(\frac{K_a[HL]_{org}}{K_{D,r}[H^+]}\right)^n K_{D,c}\beta_n \qquad (23\text{-}32)$$

When the second term of the denominator of (23-30) is large compared with unity, D' is a reasonable approximation to D. From (23-26), this condition is equivalent to $[M^{n+}] \gg [ML_n]$; that is, only a small fraction of the metal in the aqueous phase is present as the chelate. From (23-31) it appears that D' is generally a valid approximation to D when the chelate is preferentially extracted into the organic phase—unless $D \gg 1$, corresponding to a large percentage of extraction.

At the other extreme, if the metal in the aqueous phase is practically all present in the chelate, $[M^{n+}]/[ML_n] \ll 1$, and from (23-26) or (23-31), $D \simeq K_{D,c}$. This condition was encountered by Steinbach and Freiser[37] for acetylacetone extractions of metals (see also Example 23-4).

An equation of the form of (23-32) has been derived and verified experimentally[38] for the dithizone extraction of zinc. This equation may be derived by writing the equilibrium constant as the *extraction constant* K_{ex} of the reaction

$$M^{n+} + n(HL)_{org} \rightleftharpoons (ML_n)_{org} + nH^+ \qquad K_{ex} = \frac{[ML_n]_{org}[H^+]^n}{[M^{n+}][HL]_{org}^n} \qquad (23\text{-}33)$$

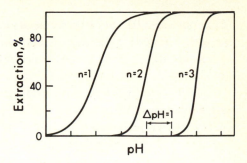

FIGURE 23-11 Extraction curves for chelates of metals with ionic charges of 1, 2, and 3.

This can be shown to yield

$$K_{ex} = \left(\frac{K_a}{K_{D,r}}\right)^n K_{D,c}\beta_n \qquad (23\text{-}34)$$

and

$$D' = K_{ex}\frac{[HL]^n_{org}}{[H^+]^n} = \left(\frac{K_a[HL]_{org}}{K_{D,r}[H^+]}\right)^n K_{D,c}\beta_n \qquad (23\text{-}35)$$

which is identical to (23-32).

Irving and Williams[39] defined the quantity $pH_{\frac{1}{2}}$ as the pH corresponding to $D = 1$, or to equal concentrations of metal in the two phases. When the volumes of the two phases are equal, $pH_{\frac{1}{2}}$ corresponds to the pH at which half the metal is extracted into the organic phase. In general, if E is the percentage of metal extracted into the organic phase and V_{org} and V are the volumes of organic and water phases, then

$$D = \frac{E}{(100 - E)}\frac{V}{V_{org}} \qquad (23\text{-}36)$$

Expressing (23-32) in logarithmic form gives

$$\log D' = n \log \frac{K_a}{K_{D,r}} + \log K_{D,c}\beta_n + n(pH + \log [HL]_{org}) \qquad (23\text{-}37)$$

Setting $pH = pH_{\frac{1}{2}}$ for $\log D' = 0$, and subtracting, we obtain

$$n(pH_{\frac{1}{2}} - pH) = -\log D' = \log\left(\frac{100 - E}{E}\frac{V_{org}}{V}\right) \qquad (23\text{-}38)$$

and

$$pH_{\frac{1}{2}} = \log \frac{K_{D,r}}{K_a} - \frac{\log K_{D,c}\beta_n}{n} - \log [HL]_{org} \qquad (23\text{-}39)$$

Figure 23-11 shows plots of several theoretical curves (see also reference 40) of percentage extraction against pH calculated from (23-38) for $V_{org} = V$. The curves become steeper with increasing values of n, the charge of the metal ion, and the value of $pH_{\frac{1}{2}}$ for a given system depends on the stability constant of the chelate and the concentration of excess reagent rather than on the concentration of metal ion

[Equation (23-39)]. For various metal ions, with the same reagent and solvent system, the variation in $pH_{\frac{1}{2}}$ is determined primarily by the formation constant of the chelate β_n, because $K_{D,r}$ and K_a are properties of the reagent alone and because $K_{D,c}$ is usually not significantly different for the chelates of different metals.

Irving and Williams[39] warned against uncritical acceptance of experimental extraction curves of the type in Figure 23-11, because (1) the concentration of excess reagent often is neither specified nor maintained constant as the percentage extraction is varied, and (2) equilibrium may not have been attained.

Particularly when the chelating reagent is present largely in the organic phase, its concentration in the aqueous phase may be so low that chelation equilibrium is attained only slowly. Morrison and Freiser[4] pointed out that two factors may affect the rate of achievement of equilibrium—the rate of formation of the extractable species and the rate of transfer of the species from one phase to the other. Oh and Freiser[41] found the rate of reaction of zinc, cobalt, and nickel ions with substituted diphenylthio-carbazones to be first order in metal ion and reagent concentrations but inversely proportional to hydrogen ion concentration. The formation of the chelate in the aqueous phase is the rate-determining step in the extraction. Irving and Williams[39] found the extraction of zinc dithizonate into carbon tetrachloride to be faster than into chloroform, in approximately the same ratio as the partition coefficients of dithizone between water and these two solvents. Thus the extraction rate appeared to be determined primarily by chelation rate, which in turn was proportional to the concentration of dithizone in the aqueous phase. Irving, Andrew, and Risdon[42] found a similar situation prevailing in the extraction of copper dithizonate, whereas mercury(II) dithizonate was extracted rapidly by both solvents at pH = 1. By using chloroform to retard the extraction of copper and by restricting the extraction time to 1 min, they were able to improve the separation of mercury(II) from copper beyond what could be achieved at equilibrium.

The following examples of calculations based on Equations (23-26) to (23-39) illustrate situations that may be encountered.

EXAMPLE 23-3   A metal ion M^{++} is 33% extracted as ML_2 from 100 ml of $10^{-5} M$ solution at pH = 5 with 20 ml of $10^{-3} M$ solution of a chelating reagent HL in an organic solvent. Calculate the percentage extraction expected for the same metal ion solution at pH = 6, using 50 ml of $5 \times 10^{-4} M$ reagent.

ANSWER   The extraction equilibrium constant [see (23-33) and (23-39)] is

$$K_{ex} = D \frac{[H^+]^2}{[HL]_{org}^2} = \frac{E}{100 - E} \frac{V[H^+]^2}{V_{org}[HL]_{org}^2}$$

$$= \frac{33}{67} \frac{100}{20} \left(\frac{10^{-5}}{10^{-3}}\right)^2 = 2.5 \times 10^{-4}$$

From this we calculate for the new conditions

$$\frac{E}{100 - E} = K_{ex} \frac{[HL]_{org}^2 V_{org}}{[H^+]^2 V} = 2.5 \times 10^{-4} \frac{(5 \times 10^{-4})^2}{(10^{-6})^2} \frac{50}{100} = 31.2$$

and $E = 96.9\%$. ////

EXAMPLE 23-4 Geiger and Sandell[43] reported for the extraction of Cu^{++} with dithizone from water into CCl_4 the following values: $K_a = 3 \times 10^{-5}$, $K_{D,r} = 1.1 \times 10^4$, $K_{D,c} = 7 \times 10^4$, $\beta_n = 5 \times 10^{22}$, $n = 2$. If 100 ml of an aqueous solution containing $10^{-7} M$ Cu^{++} in (a) 1 M HCl and (b) 0.1 M HCl is extracted with 10 ml of 0.01% dithizone in CCl_4 ($4 \times 10^{-4} M$ HL), calculate $pH_{\frac{1}{2}}$ and the percentage of the Cu^{++} remaining unextracted in each case.

ANSWER (a) At pH = 0, from (23-39), $pH_{\frac{1}{2}} = -5.21 - \log [HL]_{org}$ and, from (23-38), $2(pH_{\frac{1}{2}} - pH) = \log [(100 - E)V_{org}/EV]$. We find $pH_{\frac{1}{2}} = -5.21 - \log (4 \times 10^{-4}) = -1.81$ and, since $V_{org}/V = 0.1$, $\log (100 - E)/E = -2.62$ at pH = 0, $(100 - E)/E = 2.4 \times 10^{-3}$, and $(100 - E) = 0.24\%$ unextracted. To find whether the approximate expression (23-32) is applicable, we calculate the second term in the denominator of (23-30), which turns out to be 16.8 (considerably greater than unity). From (23-31), a better value of $D = K_{D,c}/17.8 = 3.9 \times 10^3 = EV/(100 - E)V_{org}$; then $100 - E = 0.256\%$ unextracted. (b) At pH = 1 the second term in the denominator becomes 0.168, and $D = K_{D,c}/1.168 = 6 \times 10^4$, from which $100 - E = 0.017\%$ instead of the 0.0024% that would be calculated from (23-32). At higher pH values, D approaches a limiting value of $D = K_{D,c} = 7 \times 10^4$, corresponding to $100 - E = 0.014\%$ unextracted. ////

Effect of competing complexing reactions An important limitation on the use of Equations (23-26) to (23-39) is that competing complexation reactions, including hydrolysis of the metal ion, have been neglected.

The effect of hydrolysis has been considered by Kolthoff and Sandell[38] and later in more detail by Connick and McVey.[44] The combined effects of hydrolysis, complexation with competing reagents, and formation of lower complexes with the chelating reagent have been discussed in the generalized treatment of Irving, Rossotti, and Williams.[36]

Including adduct species in the organic phase, the distribution ratio is

$$D = \frac{[ML_n]_{org} + \sum_q [ML_n \cdot B_q]}{[M^{n+}] + \sum_i [ML_i] + \sum_j [M(OH)_j] + \sum_p [MA_p]} \qquad (23\text{-}40)$$

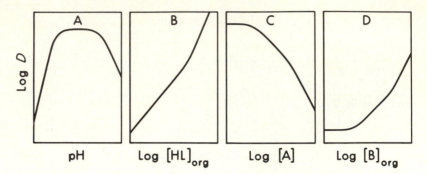

FIGURE 23-12   Schematic illustration of variation in distribution ratio as a function of pH, concentration of extracting reagent HL in the organic phase, concentration of masking reagent A in the aqueous phase, and concentration of an adduct former B in the organic phase. *A*, pH variable; *B*, [HL] variable; *C*, [A] variable; *D*, [B] variable. (*Adapted from Schweitzer.*[46])

where

$$\sum_i [ML_i] = \text{sum of concentrations of the complexes of the metal with the}$$
chelating reagent in the aqueous phase; similarly,

$$\sum_j [M(OH)_j] = \text{sum of concentrations of all hydroxyl complexes including}$$
hydrolyzed species and anionic complexes of amphoteric metals;

$$\sum_p [MA_p] = \text{sum of concentrations of the complexes of the metal with a masking}$$
reagent; and

$$\sum_q [ML_n \cdot B_q] = \text{sum of concentrations of complexes in the organic phase with an}$$
adduct B.

In (23-40) the polymeric species in both phases have been neglected. As mentioned above, such species may be of special importance in certain systems for metal halide extractions, but are of less importance in chelate extractions.

The general, minimizing computer program called LETAGROP has been modified[45] as DISTR and applied to the analysis of data for the distribution of a component between two phases when several equilibria and components are involved.

Equation (23-40) often can be simplified because of the virtual absence of many possible species. Figure 23-12 is a schematic diagram illustrating how the distribution ratio can be expected to vary with the concentration of hydrogen ion, extracting reagent, masking agent, and adduct.[46]

Case 1. Lower chelates in aqueous phase (hydroxyl complexes, masking effects, and adducts negligible)   Particularly with multivalent metals, an appreciable fraction of the metal ion in the aqueous phase may exist as lower complexes. Thus Rydberg,[47] in studying the extraction of thorium acetylacetonate from water into benzene, found it was necessary to consider all the lower complexes (ThL^{3+}, ThL_2^{++}, and ThL_3^+) but not the hydroxyl complexes, even though aquated thorium ion undergoes extensive hydrolysis in the absence of complexing agents. For such a case the distribution ratio is given by

$$D = \frac{[ML_n]_{org}}{[ML_n] + [M^{n+}] + [ML^{(n-1)+}] + [ML_2^{(n-2)+}] + \cdots + [ML_{n-1}^+]}$$

$$= \frac{[ML_n]_{org}}{[ML_n] + [M^{n+}]/\alpha} \qquad (23\text{-}41)$$

where, in analogy to Equation (7-24),

$$\frac{1}{\alpha} = 1 + \beta_1[L^-] + \beta_2[L^-]^2 + \cdots + \beta_{n-1}[L^-]^{n-1} \qquad (23\text{-}42)$$

This summation is carried out over the lower complexes, from ML to ML_{n-1} but not including the ultimate complex ML_n. Equation (23-41) can be applied readily only if the concentration of chelating anion L^- is known, that is, in alkaline solutions when a relatively large excess of chelating agent is present.

At lower pH values the reagent is present partly or largely as HL, which is extracted into the organic layer according to its distribution constant. Thus

$$[L^-] = \frac{K_a[HL]}{[H^+]} = \frac{K_a[HL]_{org}}{K_{D,r}[H^+]} \qquad (23\text{-}43)$$

and with substitution of (23-43) in (23-42)

$$\frac{1}{\alpha} = \sum_{i=0}^{i=n-1} \beta_i \left(\frac{K_a[HL]_{org}}{K_{D,r}[H^+]} \right)^i \qquad (23\text{-}44)$$

Comparing (23-41) with (23-26), we may write a modified equation analogous to (23-30),

$$D = \frac{K_{D,c}}{1 + [H^+]^n K_{D,r}{}^n/(\alpha \beta_n K_a{}^n[HL]_{org}^n)} \qquad (23\text{-}45)$$

and the approximate expression analogous to (23-32) will hold when the second term of the denominator is large compared with unity:

$$D' = \left(\frac{K_a[HL]_{org}}{K_{D,r}[H^+]} \right)^n \alpha K_{D,c} \beta_n \qquad (23\text{-}46)$$

Equation (23-46), however, is less convenient than (23-32) because α varies with $[H^+]$. This variation causes the extraction curves to be distorted from the symmetrical shapes shown in Figure 23-11, in accordance with the magnitudes of the formation constants of the lower chelates.

EXAMPLE 23-5   According to Rydberg[47] the following data describe the extraction of thorium acetylacetonate, ThL_4, from water into benzene: $K_a = 1.17 \times 10^{-9}$, $K_{D,c} = 315$, $K_{D,r} = 5.95$. The successive overall formation constants for the acetylacetone complexes of Th(IV) are $\beta_1 = 7 \times 10^7$, $\beta_2 = 3.8 \times 10^{15}$, $\beta_3 = 7.2 \times 10^{21}$, $\beta_4 = 7.2 \times 10^{26}$. Calculate the percentage extraction of thorium at a pH of 6, using a final equilibrium concentration of $10^{-3}\,M$ acetylacetone in the organic phase and a volume ratio $V_{org}/V = \frac{1}{5}$.

ANSWER   According to (23-44) we calculate $K_a[HL]_{org}/K_{D,r}[H^+] = 2 \times 10^{-7}$ and $1/\alpha = 1 + 7 \times 10^7 \times 2 \times 10^{-7} + 3.8 \times 10^{15} \times 4 \times 10^{-14} + 7.2 \times 10^{21} \times 8 \times 10^{-21} = 225$. The second term in the denominator of (23-45) is calculated to be 210, from which

$$D = \frac{E}{100 - E}\frac{V}{V_{org}} = \frac{315}{1 + 210} = 1.49$$

and $E = 23\%$.   ////

Case 2. Competing foreign complexing agents (formation of lower complexes, hydroxyl complexes, and adducts negligible)

We consider only the case of a reagent A (written without regard to its charge), which forms a series of complexes $MA, MA_2, \ldots, MA_p$.

The distribution ratio is given by

$$D = \frac{[ML_n]_{org}}{[ML_n] + [M^{n+}] + [MA^{n+}] + [MA_2^{n+}] + \cdots + [MA_p^{n+}]}$$

$$= \frac{[ML_n]_{org}}{[ML_n] + [M^{n+}]/\alpha} \tag{23-47}$$

where

$$\frac{1}{\alpha} = 1 + \beta_1[A] + \beta_2[A]^2 + \cdots + \beta_p[A]^p \tag{23-48}$$

Comparing (23-41) and (23-47), we may write expressions identical to (23-45) and (23-46), the only change being the different definition of α. Figure 23-12C indicates schematically how the distribution ratio would be expected to decrease with increasing concentration of A.

As shown by the following example, Cu(II) should be extracted with dithizone even in the presence of EDTA in acid solution.

EXAMPLE 23-6   Using the data in Example 23-4, calculate the distribution ratio for the extraction of Cu(II) dithizonate with $10^{-4}\,M$ dithizone in carbon tetrachloride, at (a) pH 4 and (b) pH 2, in the presence of $0.01\,M$ EDTA.

ANSWER   From Table 11-2, $\log K_{CuY}$ is 18.8. (a) From Table 11-1, $-\log \alpha_4$ for EDTA is 8.44 at pH 4. The conditional formation constant K'_{MY} of the Cu(II)-

EDTA complex is therefore $10^{10.4}$, whence $1/\alpha = 1 + 10^{10.4} \times 0.01 = 2 \times 10^8$. Equation (23-45) becomes $D = 7 \times 10^4/(1 + 538) = 130$. (b) At pH 2, $-\log \alpha_4 = 13.43$, and $1/\alpha = 2.3 \times 10^3$; so $D = 7 \times 10^4/(1 + 62) = 1.1 \times 10^3$.   ////

Nevertheless, when Friedeberg[48] carried out an extraction of silver using dithizone in carbon tetrachloride at pH 2, he found that copper was left in solution. Evidently, this separation is based on a slow rate of attainment of equilibrium of the reaction with copper in the presence of EDTA, for he noted that at higher concentrations the copper is extracted slowly. Friedeberg also found that EDTA prevents the extraction by dithizone of lead, zinc, bismuth, cadmium, nickel, cobalt, and thallium at any pH value.

The special case of hydroxyl ion as a competing complexing agent is encountered when hydrolysis or amphoteric behavior is to be considered. The formation of hydroxy complexes represents a competing reaction that would make the value of the distribution ratio smaller. If hydrolysis can be represented by the successive formation of the monomeric complexes $M(OH)^{(n-1)+}$, $M(OH)_2^{(n-2)+}$, ..., Equation (23-47) can be applied. Figure 23-12A indicates how pH would be expected to affect the value of the distribution ratio. At low values, D increases with pH because of increased ionization of the extracting reagent HL; at high pH values the extracting reagent is completely ionized, but the metal ion begins to form hydroxy complexes, and the conditional formation constant for the formation of the complex decreases (Figure 11-3). Hydrolysis is often complicated by the presence of species containing more than one metal atom per ion (Section 7-5), particularly when more highly charged metal ions are involved. Each metal ion-reagent-solvent system then represents a particular case that must be studied in detail to determine whether lower chelates or hydrolyzed species must be considered.

When a relatively large excess of hydroxyl ion is present and the metal is amphoteric, relatively simple anionic species (hydroxyl complexes) usually are formed. If the solution is sufficiently alkaline (pH $>$ pK_a + log $K_{D,r}$), the reagent is present largely in the aqueous phase as the anion L^-. If chelates lower than ML_n can be neglected and $M(OH)_j^{(n-1)+}$ is the highest hydroxy complex formed, we may write from (23-40)

$$D = \frac{K_{D,c}}{1 + [M^{n+}]/\alpha[ML_n]} \qquad (23\text{-}49)$$

where

$$\frac{1}{\alpha} = 1 + \beta_1[OH^-] + \beta_2[OH^-]^2 + \cdots + \beta_j[OH^-]^j \qquad (23\text{-}50)$$

But

$$\frac{[M^{n+}]}{[ML_n]} = \frac{1}{\beta_n[L^-]^n} \qquad (23\text{-}51)$$

so

$$D = \frac{K_{D,c}}{1 + 1/(\beta_n\alpha[L^-]^n)} \qquad (23\text{-}52)$$

Since most of the metal in the aqueous phase is present as hydroxy complexes, the second term of the denominator of (23-52) is large compared with unity, and

$$D' = K_{D,c}\beta_n\alpha[L^-]^n$$

Consider the case of an amphoteric metal that forms a single hydroxy complex $M(OH)_j^{(n-j)+}$ of formation constant β_j. Then

$$\frac{1}{\alpha} = 1 + \beta_j[OH^-]^j \simeq \beta_j[OH^-]^j \qquad (23\text{-}53)$$

and

$$D' = \frac{K_{D,c}\beta_n}{\beta_j}\frac{[L^-]^n}{[OH^-]^j} = \frac{K_{D,c}\beta_n[L^-]^n[H^+]^j}{\beta_j K_w^j} \qquad (23\text{-}54)$$

where K_w is the ion product for water. Whence,

$$j(pH - pH_{\frac{1}{2}}) = -\log D' = \log\frac{V_{org}(100 - E)}{VE} \qquad (23\text{-}55)$$

where

$$pH_{\frac{1}{2}} = \frac{1}{j}\log\frac{K_{D,c}\beta_n}{\beta_j K_w^j} + \frac{n}{j}\log[L^-] \qquad (23\text{-}56)$$

Equation (23-55) indicates that the value of E decreases with increasing pH; that is, the metal is extracted from the organic phase into the aqueous phase at sufficiently high pH values (Figure 23-12A). The value of $pH_{\frac{1}{2}}$ is determined primarily by β_n/β_j, the relative stabilities of the chelate and hydroxyl complexes, and varies with the concentration of the chelating ion. Examples of back extraction of amphoteric metals into an alkaline aqueous phase are extractions of aluminum, lead, and zinc with either oxine or dithizone as a reagent. At high pH, back extraction[49] of CH_3HgCl from an organic solvent takes place as CH_3HgOH.

Case 3. Adduct formation (lower chelates, masking effects, and hydroxy compound formation negligible)

Metal ion complexes retaining one or more molecules of hydrate water may react with ligands that can replace the coordinated water to produce a mixed-ligand chelate or ternary complex.[50] With water no longer part of the complex, the distribution ratio usually is increased strongly. For example, pyridine forms a ternary complex with zinc 8-hydroxyquinolate in chloroform with a formation constant[51] of 1.1×10^3. Frequently the organic solvent itself plays an active role as an adduct former.[2,52] For example,[2] thenoyltrifluoroacetone (TTA) complexes of cobalt(II) and uranium(VI) are more effectively extracted by alcohol, ketones, or esters than by hydrocarbons, whereas TTA complexes of iron(III) and thorium(IV) are coordination-saturated, and oxygen-containing solvents are no better than noncoordinating solvents.

More generally, solvents can promote extraction through promotion of ion-pair formation.[53] The dielectric constant of the solvent as well as specific solvent-ion interactions are important in this connection (Sections 4-2 and 4-3).

In other cases the adduct former may attach itself to the chelating reagent, and in still others self-adducts are formed that involve the extracting agent adding to itself.[54]

Equation (23-40) can be simplified to take account of adduct formation:

$$D = \frac{[ML_n]_{org} + [ML_n \cdot B]_{org} + [ML_n \cdot B_2]_{org} + \cdots + [ML_n \cdot B_q]_{org}}{[M^{n+}] + [ML_n]}$$

$$= \frac{[ML_n]_{org}/\alpha}{[M^{n+}] + [ML_n]} \tag{23-57}$$

where
$$\frac{1}{\alpha} = 1 + \beta_1[B]_{org} + \beta_2[B]_{org}^2 + \cdots + \beta_q[B]_{org}^q \tag{23-58}$$

where β_1, β_2, ... are the formation constants for the adducts of B with the complex ML_n.

23-6 SELECTED APPLICATIONS OF EXTRACTIONS

Diamond and Tuck[55] classified the extractable inorganic compounds under six headings:

1 Simple molecules. These involve the distribution of molecules such as iodine, bromine, $HgCl_2$, $AsCl_3$, OsO_4, fatty acids, and hydrocarbons between water and an organic solvent.

2 Pseudomolecular systems. These involve species that are partly ionized in water and are molecular in an organic solvent. Examples include weak acids such as phenols, dithizone, salicylic acid, cupferron, and many neutral metal-chelate complexes.

3 Coordinately unsolvated salts. These involve large ions that would extensively disrupt the structure of the solvent water and whose hydration energies are small and hence can exist as ions in an organic solvent. Examples are Ph_4As^+, Ph_4P^+, and Bu_4N^+. These cations can be extracted as ions with large unsolvated anions such as MnO_4^-, ClO_4^-, and IO_4^-.

4 Mineral acids. The tetrahydrated proton can be extracted into some basic organic solvents such as ethers and ketones. Large acid anions make the extraction more favorable. Thus $HClO_4$ is extracted more extensively than HCl.

5 Complex metal acids. The best-known example is the extraction of $HFeCl_4$ into solvents such as ether. The anion is large and has low charge. Similarly, the halides of ions such as Ga(III), Au(III), In(III), Tl(III), Mo(VI), Sb(V), As(III), Ge(IV), Hg(II), and Nb(V) have been found to be extractable from strong-acid solution.

6   Coordinately solvated salts. Tri-*n*-butyl phosphate, for example, can act as a solvating group to enhance the extractability of metal nitrates such as $UO_2(NO_3)_2 \cdot 2TBP$.

In the first two categories the distribution ratio should be relatively independent of the organic solvent, except for the influence of mutual solubility of the organic solvent with water on the distribution ratio. In the last four the organic solvent may play an active role in the extraction process.

Other factors being equal, the solvent[56] should be optically transparent (for later spectral measurements), have a density greater than water (for ease of separation of phases), be nontoxic, and have little tendency to form emulsions. Unfortunately, the commonly used solvents—chloroform, carbon tetrachloride, and benzene—are toxic. Propylene carbonate has been proposed[56] as a solvent having highly desirable properties.

To aid in deciding on the probable feasibility of a proposed extraction procedure, information on distribution ratios under a variety of conditions is useful. Distribution equilibria traditionally have been determined by the shake-test technique followed by separation and analysis of the phases. Reinhardt and Rydberg[57] developed the AKUFVE apparatus consisting of a mixing chamber, a small-volume centrifuge for continuous separation of phases, and detectors for continuous analysis of the phases. Phase equilibrium is attained in less than a minute, which means that hundreds of equilibrium points on extraction curves involving variation of pH, temperature, concentration of extracting reagent, and metal ion concentration as well as reaction kinetics can be obtained in a single day. Andersson and Spink[58] combined a dedicated computer with AKUFVE to control continuously the change in variables and to process the data obtained.

Metal chelate complexes   Kuz'min[59] summarized the developments in liquid-liquid extraction of metals and groups of metals with dithizone, dithiocarbonates, and 8-hydroxyquinolates, as well as other reagents and mixtures of reagents. He pointed out that extractions of traces of several elements can be carried out to advantage in several stages by use of pH control.

8-Hydroxyquinoline (oxine) and its derivatives have been used extensively as chelating and extraction reagents, taking advantage of the solubility of many oxinates in chloroform. The distribution of the reagent itself between water and chloroform has been studied in detail by Lacroix.[60] Since the reagent is amphoteric, it may be regarded as a cationic acid ($pK_1 = 5.09$ and $pK_2 = 9.82$). Only the neutral molecule is extracted appreciably into chloroform. Therefore, the distribution ratio of the reagent is given by

$$D_r = \frac{[HL]_{org}}{C_{HL}} = \frac{[HL]_{org}}{[HL]/\alpha_0} = K_{D,r}\alpha_0 \qquad (23\text{-}59)$$

where C_{HL} is the analytical concentration of the oxine in water and $\alpha_0 = K_1[H^+]/([H^+]^2 + K_1[H^+] + K_1K_2)$ is the fraction of the reagent in the aqueous phase present as the neutral molecule. The value of $K_{D,r}$ found by Lacroix was 720.

Starý[61] systematically studied the extraction of 32 metals with oxine, including measurement of their extraction constants. Conditions for the extraction of many metals have been summarized in several places.[1,4,5] Dyrssen[62] stressed the importance of the lower oxinates of thorium. His extraction data could not be explained by assuming that only Th^{4+} and ThL_4 existed in the aqueous phase. He also studied[63] extraction of strontium into chloroform using oxine and evaluated the following equilibrium constants:

$$Sr^{++} + L^- \rightleftharpoons SrL^+ \qquad \log K_1 = 2.39$$

$$SrL^+ + L^- \rightleftharpoons SrL_2 \qquad \log K_2 = 0.84$$

$$SrL_2 \cdot 2H_2O(solid) \rightleftharpoons SrL_2 + 2H_2O \qquad \log S = -5.42$$

Strontium oxinate is extracted into chloroform as $SrL_2 \cdot 2HL$, as shown by the fact that the solubility in the organic phase was found to be proportional to the square of the concentration of oxine in the organic phase, or

$$\log S_{org} = 2 \log [HL]_{org} - 3.41$$

Thus, a high concentration of organic reagent favors the extraction of strontium into the organic layer. As the following sample calculation shows, strontium is effectively extracted at pH 11 and $[HL]_{org} = 1 \, M$.

EXAMPLE 23-7 Calculate the distribution ratio of strontium between water and chloroform, using $[HL]_{org} = 1 \, M$ at pH 11.

ANSWER From Dyrssen's data, $\log S_{org} = -3.41$; $\log K_{D,c} = \log S_{org} - \log S = 2.01$, or $K_{D,c} = 102$. From the data of Lacroix[60] $D_r = 720\alpha_0 = 44.5$. Therefore $C_{HL} = [HL]_{org}/D_r = 0.0225$; $[L^-] = K_2C_{HL}/([H^+] + K_2) = 0.022$; $1/\alpha = 1 + K_1[L^-] = 6.4$; $D = K_{D,c}/(1 + 1/K_1K_2\alpha[L^-]^2) = 11.6$. /////

If the pH is raised further, the reagent is extracted largely into the aqueous layer, thus lowering the distribution ratio of the complex. On the other hand, if the pH is lowered, the strontium complex in the aqueous phase is weakened by competition from hydrogen ions. The conditions for quantitative extraction are therefore unusually critical.

Tungsten oxinate demonstrates unexpected behavior in that the curve for distribution ratio against pH shows two maxima.[64] At low pH values the reagent is in the protonated form; the competing reaction at higher pH probably involves formation of tungstates, $WO_4^=$.

Diphenylthiocarbazone (dithizone) extractions have been mentioned as examples of equilibrium calculations. Sandell[65] discussed many separations that take advantage of masking agents to increase the selectivity of the method. Stability constants with many metal ions have been measured.[66] Lead dithizone equilibria have been investigated,[67] and a pH of about 10.8 is recommended for the extraction with carbon tetrachloride. Conditions for the optimization of the extraction of zinc dithizonate have been considered.[68] Dithizone is widely recognized as a sensitive qualitative reagent as well as an important reagent for quantitative determinations. Colorimetric determinations are based on the intense green color of the reagent and the contrasting (usually orange or red) colors of the metal dithizonates in chloroform or carbon tetrachloride. Dithizone is not useful by itself with nickel because the complex forms too slowly; when pyridine or other nitrogen bases also are present, a mixed-ligand complex forms comparatively rapidly.[69] Math and Freiser[70] studied structures of metal dithizonates and concluded that the interatomic distance is unusually short in the N-N bond in the chelate ring and has double-bond character.

Acetylacetone is exceptional in that it is useful both as a solution in various organic solvents (carbon tetrachloride, chloroform, benzene) and as the pure liquid.[71] Rydberg[72] evaluated pK_a (8.82 in 0.1 M sodium perchlorate) as well as the distribution ratio of acetylacetone between chloroform ($K_{D,r} = 23.5$) or methyl isobutyl ketone ($K_{D,r} = 5.9$) and aqueous 0.1 M sodium perchlorate. The acetylacetonates of most metals are much more soluble in organic solvents than are other chelates used in analytical separations. Thus relatively large amounts of metals can be handled. Yet, small amounts also can be extracted, as illustrated by the isolation of the carrier-free radioisotope iron 59 from the cobalt from which it was prepared, using a xylene solution of acetylacetone.[73] The selectivity can be increased by use of EDTA as a masking agent.[71]

TTA is a fluorinated β-diketone[74] that is more acidic than acetylacetone and therefore permits extractions at lower pH values, in spite of its complexes being less stable than those of acetylacetone. For example, iron can be extracted from 10 M nitric acid.[75] Bolomey and Wish,[76] in developing a method for isolating carrier-free radioberyllium, studied the extraction of various metals using a dilute (0.01 M) solution of TTA in benzene. In certain cases, especially for beryllium and Fe(III) at low pH values, equilibrium was attained slowly.[77] The extraction *rate* can be increased by increasing the reagent concentration or by raising the pH. Thus in some instances equilibration time can be manipulated to improve separability. The difference in the extraction behavior of the lanthanides and the actinides is noteworthy.[78]

N-Nitroso-*N*-phenylhydroxylamine is used as its ammonium salt (cupferron) for many extractive separations. The equilibria involved in chloroform extractions have been considered by Furman, Mason, and Pekola,[79] who summarized in detail the properties of metallic cupferrates. Conditions for the extraction of several metals are listed in various places.[1,4,5]

Sodium diethyldithiocarbamate, $(C_2H_5)_2NCSS^-Na^+$, is effective for the extraction of about 20 metals[80-82] into various organic solvents such as ethyl acetate and carbon tetrachloride. The stabilities of the diethyldithiocarbamates are in general parallel to the insolubilities of the sulfides. Bode[81] and Starý and Kratzer[82] carried out systematic studies of various metals, using masking agents for selective extractions.

Amines and organophosphorus compounds   Long-chain amines and amine salts have been investigated[83,84] from both the preparative and the analytical viewpoints. Amines of high molecular weight, dissolved in solvents such as kerosene, benzene, or chloroform, are preferable to those of low molecular weight, primarily because of the lower solubility of the amines and amine salts in the aqueous phase compared with the nonaqueous. Amines can extract strong acids from aqueous solution to form the amine salt in the organic phase of the type $(R_3NHA)_n$[85] by the reaction

$$(R_3N)_{org} + H^+ + A^- \rightleftharpoons (R_3NH^+A^-)_{org}$$

where H^+ and A^- are in the aqueous phase. At higher acidities more than one acid per amine can be extracted. In alkaline solution the extraction is reversed. Examples of such extractions include not only perchlorate, nitrate, chloride, and sulfate, but also ions such as Cr(VI) as chromate,[86] Mo(VI) as polymolybdate, V(V) as vanadate or polyvanadate, and Tc(VII) or Re(VII) as pertechnetate or perrhenate.[87] An example of unusual practical importance is the extraction of U(VI) as the uranyl ion as an anion complex from acidic solutions containing sulfate, chloride, or phosphate.

Amine salts in the organic phase can also undergo anion exchange with an anion B^- in the aqueous phase:

$$(R_3NH^+A^-)_{org} + B^- \rightleftharpoons (R_3NH^+B^-)_{org} + A^-$$

It is interesting that, in general, the same order of anion extraction is observed as that found with anion-exchange resins, so that previous studies with resins can serve as a useful guide for extractions.

The structure of the amine and the nature of the organic diluent exert profound influences on the distribution ratios. Although the type of amine (primary, secondary, or tertiary) exerts an effect, this is strongly modified by the extent of branching of the alkyl groups and by the nature of the diluent. Highly branched chains generally interfere with efficient extraction, presumably owing to steric effects. But the branched-chain amines may be at the same time more compatible with the diluent, and so the net effect of chain branching on distribution ratio depends on the nature of the diluent. In general, a branched-chain secondary amine behaves more like a straight-chain tertiary than like a straight-chain secondary. Examples of analytical separations carried out by these liquid ion exchangers have been summarized by Green[88] and by De, Khopkar, and Chalmers.[89]

Organophosphorus compounds, in particular tri-*n*-butylphosphate, have long been used in nuclear processing as extractive reagents. Interestingly, the distribution ratio increases regularly in the sequence: trialkylphosphate, $(RO)_3PO$ < dialkylalkylphosphonate, $(RO)_2RPO$ < alkyldialkylphosphinate, ROR_2PO < trialkylphosphine oxide, R_3PO. This trend evidently is associated with the increasing polarity of the P—O bond as the number of such bonds decreases.[90] Of the various compounds, tri-*n*-octylphosphine oxide (TOPO) has received intensive study as an analytical reagent. Its remarkable properties as an extractant are illustrated by the fact that the distribution ratio of U(VI) is of the order of 10^5 times greater for TOPO than for tributylphosphate. It is often possible to carry out direct colorimetric determinations in the organic phase to achieve a rapid, sensitive, and selective determination. For example, U(VI) is extracted from nitric acid solution as $UO_2(NO_3)_2 \cdot 2TOPO$ by use of TOPO in benzene. Reaction of the extracted uranium with arsenazo(III) produces a compound with a sharp absorption band at 655 nm.[91]

REFERENCES

1 A. K. DE, S. M. KHOPKAR, and R. A. CHALMERS: "Solvent Extraction of Metals," Van Nostrand, New York, 1970.

2 H. FREISER: *Crit. Rev. Anal. Chem.*, **1**:47 (1970).

3 Y. MARCUS and A. S. KERTES: "Ion Exchange and Solvent Extraction of Metal Complexes," Wiley, New York, 1969.

4 G. H. MORRISON and H. FREISER: "Solvent Extraction in Analytical Chemistry," Wiley, New York, 1957.

5 G. H. MORRISON, H. FREISER, and J. F. COSGROVE in "Handbook of Analytical Chemistry," L. Meites (ed.), p. 10-5, McGraw-Hill, New York, 1963.

6 Y. A. ZOLOTOV: "Extraction of Chelate Compounds," Ann Arbor-Humphrey Science, Ann Arbor, Mich., 1970.

7 H. M. N. H. IRVING: *Pure Appl. Chem.*, **21**:109 (1970).

8 T. R. SCOTT: *Analyst*, **74**:486 (1949).

9 I. L. JENKINS and H. A. C. MCKAY: *Trans. Faraday Soc.*, **50**:107 (1954).

10 T. GROENEWALD: *Anal. Chem.*, **43**:1678 (1971).

11 I. J. GAL, I. PALIGORIC, and V. G. ANTONIJEVIC: *J. Inorg. Nucl. Chem.*, **32**:1645 (1970).

12 R. J. MYERS, D. E. METZLER, and E. H. SWIFT: *J. Amer. Chem. Soc.*, **72**:3767 (1950).

13 R. J. MYERS and D. E. METZLER: *J. Amer. Chem. Soc.*, **72**:3772 (1950).

14 R. M. DIAMOND: *J. Phys. Chem.*, **61**:69, 75 (1957).

15 J. W. ROTHE: *Stahl Eisen*, **12**:1052 (1892).

16 W. F. HILLEBRAND, G. E. F. LUNDELL, H. A. BRIGHT, and J. I. HOFFMAN: "Applied Inorganic Analysis," 2d ed., p. 134, Wiley, New York, 1953.

17 E. H. SWIFT: *J. Amer. Chem. Soc.*, **46**:2375 (1924).

18 C. W. GRIFFIN: *Ind. Eng. Chem., Anal. Ed.*, **6**:40 (1934); C. W. GRIFFIN and M. VON SAAF: *Ind. Eng. Chem., Anal. Ed.*, **8**:358 (1936).

19 I. M. KOLTHOFF, E. B. SANDELL, E. J. MEEHAN, and S. BRUCKENSTEIN: "Quantitative Chemical Analysis," p. 367, Macmillan, London, 1969.

20 L. C. CRAIG: *J. Biol. Chem.*, **155**:519 (1944); L. C. CRAIG and O. POST: *Anal. Chem.*, **21**:500 (1949).

21 B. WILLIAMSON and L. C. CRAIG: *J. Biol. Chem.*, **168**:687 (1947).

22 A. J. P. MARTIN and R. L. M. SYNGE: *Biochem. J.*, **35**:1358 (1941).

23 L. C. CRAIG: *Anal. Chem.*, **22**:1346 (1950).

24 E. NELSON: *Anal. Chem.*, **28**:1998 (1956).

25 P. L. NICHOLS, JR.: *Anal. Chem.*, **22**:915 (1950).

26 W. E. HARRIS: *Rev. Anal. Chem.*, **1**:7 (1971).

27 P. H. BELL: *J. Amer. Chem. Soc.*, **76**:5565 (1954).

28 E. L. PATTERSON, J. V. PIERCE, E. L. R. STOKSTAD, C. E. HOFFMANN, S. A. BROCKMAN, JR., F. P. DAY, M. E. MACCHI, and T. H. JUKES: *J. Amer. Chem. Soc.*, **76**:1823 (1954).

29 A. I. M. KEULEMANS: "Gas Chromatography," 2d ed., p. 118, Reinhold, New York, 1959.

30 Y. ITO and R. L. BOWMAN: *Science*, **167**:281 (1970).

31 S. W. MAYER and E. R. TOMPKINS: *J. Amer. Chem. Soc.*, **69**:2866 (1947).

32 E. GLUECKAUF: *Trans. Faraday Soc.*, **51**:34 (1955).

33 J. D. A. JOHNSON: *J. Chem. Soc.*, **1950**:1743.

34 H. R. C. PRATT: "Countercurrent Separation Processes," Elsevier, New York, 1967.

35 J. O. HIBBITS, *Anal. Chim. Acta*, **24**:113 (1961); K. K. STEWART, *Separ. Sci.*, **3**:479 (1968).

36 H. IRVING, F. J. C. ROSSOTTI, and R. J. P. WILLIAMS: *J. Chem. Soc.*, **1955**:1906.

37 J. F. STEINBACH and H. FREISER: *Anal. Chem.*, **26**:375 (1954).

38 I. M. KOLTHOFF and E. B. SANDELL: *J. Amer. Chem. Soc.*, **63**:1906 (1941).

39 H. IRVING and R. J. P. WILLIAMS: *J. Chem. Soc.*, **1949**:1841.

40 J. S. FRITZ, R. K. GILLETTE, and H. E. MISHMASH: *Anal. Chem.*, **38**:1869 (1966).

41 J. S. OH and H. FREISER: *Anal. Chem.*, **39**:295, 1671 (1967).

42 H. IRVING, G. ANDREW, and E. J. RISDON: *J. Chem. Soc.*, **1949**:541.

43 R. W. GEIGER and E. B. SANDELL: *Anal. Chim. Acta*, **8**:197 (1953).

44 R. E. CONNICK and W. H. MCVEY: *J. Amer. Chem. Soc.*, **71**:3182 (1949).

45 D. H. LIEM: *Acta Chem. Scand.*, **25**:1521 (1971).

46 G. K. SCHWEITZER: *Anal. Chim. Acta*, **30**:68 (1964).

47 J. RYDBERG: *Acta Chem. Scand.*, **4**:1503 (1950).

48 H. FRIEDEBERG: *Anal. Chem.*, **27**:305 (1955).

49 G. WESTÖÖ: *Acta Chem. Scand.*, **20**:2131 (1966); **21**:1790 (1967).

50 A. K. BABKO: *Talanta*, **15**:721 (1968).

51 F. CHOW and H. FREISER: *Anal. Chem.*, **40**:34 (1968).

52 I. P. ALIMARIN and Y. A. ZOLOTOV: *Talanta*, **9**:891 (1962).

53 T. HIGUCHI, A. MICHAELIS, and J. H. RYTTING: *Anal. Chem.*, **43**:287 (1971).

54 H. FREISER: *Pure Appl. Chem.*, **20**:77 (1969).

55 R. M. DIAMOND and D. G. TUCK in "Progress in Inorganic Chemistry," vol. 2, p. 109, Interscience, New York, 1960.

56 B. G. STEPHENS, J. C. LOFTIN, W. C. LOONEY, and K. A. WILLIAMS: *Analyst*, **96**:230 (1971).

57 H. REINHARDT and J. RYDBERG: *Chem. Ind.* (London), **1970**:488; J. RYDBERG: *Acta Chem. Scand.*, **23**:647 (1969).

58   S. O. S. ANDERSSON and D. R. SPINK: *Can. Res. Dev.*, **4**(6):16 (1971).

59   N. M. KUZ'MIN, *Zavod. Lab.*, **34**:395 (1968); *Ind. Lab. USSR*, **34**:471 (1968).

60   S. LACROIX: *Anal. Chim. Acta*, **1**:260 (1947).

61   J. STARÝ: *Anal. Chim. Acta*, **28**:132 (1963).

62   D. DYRSSEN: *Sv. Kem. Tidskr.*, **65**:43 (1953).

63   D. DYRSSEN: *Sv. Kem. Tidskr.*, **67**:311 (1955).

64   K. AWAD, N. P. RUDENKO, V. I. KUZNETSOV, and L. S. GUDYM: *Talanta*, **18**:279 (1971).

65   E. B. SANDELL: "Colorimetric Determination of Traces of Metals," 3d ed., Interscience, New York, 1959.

66   B. W. BUDESINSKI and M. SAGAT: *Talanta*, **20**:228 (1973).

67   O. B. MATHRE and E. B. SANDELL: *Talanta*, **11**:295 (1964).

68   O. BUDEVSKY, E. RUSSEVA, and R. STOYTCHEVA: *Talanta*, **19**:937 (1972).

69   B. S. FREISER and H. FREISER: *Talanta*, **17**:540 (1970).

70   K. S. MATH and H. FREISER: *Talanta*, **18**:435 (1971); *J. Chem. Soc., D*, **1970**:110.

71   J. F. STEINBACH and H. FREISER: *Anal. Chem.*, **25**:881 (1953); **26**:375 (1954); A. KRISHEN and H. FREISER: *Anal. Chem.*, **29**:288 (1957); J. P. MCKAVENEY and H. FREISER: *Anal. Chem.*, **29**:290 (1957); **30**:1965 (1958).

72   J. RYDBERG: *Sv. Kem. Tidskr.*, **65**:37 (1953).

73   A. W. KENNY, W. R. E. MATON, and W. T. SPRAGG: *Nature*, **165**:483 (1950).

74   J. C. REID and M. CALVIN: *J. Amer. Chem. Soc.*, **72**:2948 (1950).

75   F. L. MOORE, W. D. FAIRMAN, J. G. GANCHOFF, and J. G. SURAK: *Anal. Chem.*, **31**:1148 (1959).

76   R. A. BOLOMEY and L. WISH: *J. Amer. Chem. Soc.*, **72**:4483 (1950).

77   W. G. SCRIBNER, W. J. TREAT, J. D. WEIS, and R. W. MOSHIER: *Anal. Chem.*, **37**:1136 (1965).

78   L. B. MAGNUSSON and M. L. ANDERSON: *J. Amer. Chem. Soc.*, **76**:6207 (1954).

79   N. H. FURMAN, W. B. MASON, and J. S. PEKOLA: *Anal. Chem.*, **21**:1325 (1949).

80   Y. A. CHERNIKHOV and B. M. DOBKINA: *Zavod. Lab.*, **15**:1143 (1949); *Chem. Abstr.*, **44**:1358 (1950).

81   H. BODE: *Fresenius' Z. Anal. Chem.*, **144**:165 (1955).

82   J. STARÝ and K. KRATZER: *Anal. Chim. Acta*, **40**:93 (1968).

83   E. L. SMITH and J. E. PAGE: *J. Soc. Chem. Ind.* (London), **67**:48 (1948).

84   F. L. MOORE: *Anal. Chem.*, **29**:1660 (1957); C. F. COLEMAN, K. B. BROWN, J. G. MOORE, and K. A. ALLEN: *Proc. U.N. Int. Conf. Peaceful Uses At. Energy, 2d, Geneva*, **28**:278 (1958).

85   E. HÖGFELDT: *Sv. Kem. Tidskr.*, **76**:290 (1964); L. KUČA and E. HÖGFELDT: *Acta Chem. Scand.*, **25**:1261 (1971); E. HÖGFELDT, P. R. DANESI, and F. FREDLUND: *Acta Chem. Scand.*, **25**:1338 (1971).

86   J. ADAM and R. PŘIBIL: *Talanta*, **18**:91 (1971).

87   K. F. FOUCHÉ: *J. Inorg. Nucl. Chem.*, **33**:857 (1971).

88   H. GREEN: *Talanta*, **20**:139 (1973).

89   DE, KHOPKAR, and CHALMERS,[1] p. 201.

90   C. A. BLAKE, JR., C. F. BAES, JR., K. B. BROWN, C. F. COLEMAN, and J. C. WHITE, *Proc. U.N. Int. Conf. Peaceful Uses At. Energy, 2d, Geneva*, **28**:289 (1958).

91 J. A. PÉREZ-BUSTAMANTE and F. P. DELGADO: *Analyst*, **96**:407 (1971).
92 A. T. PILIPENKO: *Zh. Anal. Khim.*, **8**:286 (1953).
93 H. CHRISTOPHERSON and E. B. SANDELL: *Anal. Chim. Acta*, **10**:1 (1954).

PROBLEMS

23-1 Pilipenko[92] reported the values for the equilibrium constant

$$K = \frac{[M^{n+}][HL]_{org}^n K_a^n}{[ML_n]_{org}[H^+]^n K_{D,r}^n}$$

for the extraction of several metals into CCl_4 using dithizone. His values of K were: Hg(II), 7×10^{-45}; Bi(III), 1.1×10^{-37}; Cu(II), 1.1×10^{-27}; Ag(I), 2.3×10^{-18}; Co(II), 5×10^{-18}; Ni(II), 1.7×10^{-17}; and Sn(II), 4.5×10^{-16}.

(*a*) Show that the above equilibrium constant is related to the constants in the present treatment as follows:

$$K = \frac{1}{K_{ex}}\left(\frac{K_a}{K_{D,r}}\right)^n = \frac{1}{K_{D,c}\beta_n}$$

(*b*) Compare the value of K for Cu(II) with that calculated from the results of Geiger and Sandell (see Example 23-4).

(*c*) Derive the relation between pK and pH$_{\frac{1}{2}}$, using the values of the constants from Example 23-4. Calculate pH$_{\frac{1}{2}}$ for the above metals, taking $[HL]_{org} = 10^{-4}$.

(*d*) Taking the logarithms of the successive stepwise formation constants of $HgBr_4^=$ to be $\log K_1 = 9.0$, $\log K_2 = 8.3$, $\log K_3 = 2.4$, and $\log K_4 = 1.3$, calculate the value of pH$_{\frac{1}{2}}$ for the extraction of Hg(II), using 10^{-4} M reagent from (*i*) 1 M bromide solution and (*ii*) 0.1 M bromide solution.

Answer 0.9, -0.9.

(*e*) A solution of 10^{-5} M Ag^+ and 10^{-4} M Cu^{++} is treated to give a solution 0.01 M in EDTA and buffered at a pH of 6. If 100 ml of the mixture is shaken with 10 ml of 10^{-3} M dithizone in CCl_4, what percentage of Cu will remain in the aqueous phase?

Answer 99.9%.

23-2 Using Rydberg's data for the extraction of thorium acetylacetonate (Example 23-5), calculate the complete extraction curve of E against pH for $[HL]_{org} = 10^{-3}$, $V_{org} = V$.

23-3   Christopherson and Sandell[93] reported the following data for the extraction of nickel dimethylglyoxime, NiL_2, from water into chloroform at an ionic strength of 0.05%: $K_{D,c} = 410$, $\beta_n = 2.3 \times 10^{17}$, $K_a = 2.6 \times 10^{-11}$. The reagent HL has greater solubility in water ($S = 5.4 \times 10^{-3}$) than in chloroform ($S_{org} = 4.5 \times 10^{-4}$). Therefore, they wrote the equilibrium in terms of a water solution of reagent:

$$Ni^{++} + 2HL \rightleftharpoons (NiL_2)_{org} + 2H^+$$

(a)   Express the equilibrium constant of this reaction in terms of the above constants and find its value.

Answer   0.064.

(b)   Derive an expression for $D' = [NiL_2]_{org}/[Ni^{++}]$ and for $D = [NiL_2]_{org}/([Ni^{++}] + [NiL_2])$. Show that $1/D = 1/D' + 1/K_{D,c}$.

(c)   Calculate D' and D for the extraction of Ni^{++}, using a solvent system saturated with dimethylglyoxime at (i) a pH of 5 and (ii) a pH of 3.

Answer   (i) 1.9×10^4, 401; (ii) 1.86, 1.85.

23-4   Using the data of Example 23-7, calculate the distribution ratio of strontium between water and chloroform, using oxine at a concentration of 0.1 M in the chloroform layer, at pH 10.

Answer 7.1×10^{-5}.

23-5   Derive expressions for the percentage extraction of a metal M^{n+}, using V_{org} milliliters of *pure* acetylacetone as the extractant for V milliliters of aqueous solution as a function of pH, (a) considering that only the chelate ML_n is formed and (b) considering that lower chelates $ML, \ldots, ML_{n-1}$ are formed in aqueous solution but are not extracted.

23-6   Calculate the numbers in the lower set of tubes in Figure 23-4 from the numbers in the upper set.

23-7   Calculate the curves similar to those in Figure 23-9 for (a) $D_B = 1$, $D_A = 0.5$; (b) $D_B = 10$, $D_A = 5$; (c) $D_B = 100$, $D_A = 50$.

23-8   If the equilibrium constant is 10 for the reaction in the aqueous phase $2M \rightleftharpoons M_2$, and if the distribution constant $K_{D,M}$ is 5 and K_{D,M_2} is 25, (a) calculate the distribution ratio for overall concentration C_M in the aqueous phase equal to: 0.1 M, 0.01 M, 0.001 M, 0.0001 M. (b) For a system in which $V_{org} = V$, plot the distribution ratio as a function of $(C_M + C_{M_{org}})$.

Answer   (a) $D = 15, 7.9, 5.4, 5.0$.

CHROMATOGRAPHY

Section 23-4 should be regarded as serving a dual function: first, it is an essential part of liquid-liquid extraction theory, and second, it is a logical introduction to chromatography. Liquid-liquid chromatography is closely analogous to liquid-liquid extraction. If one liquid is rendered immobile by dispersal on a solid of large surface area and packed as a bed into a column, and if the other liquid is percolated through the bed, we have fundamentally changed little from continuous counter-current extraction. The main difference is that we no longer have physically discrete equilibrium stages or plates. The conclusions reached in Sections 23-3 and 23-4 regarding band and peak broadening and separation are applicable in large measure to chromatography.

The feature distinguishing chromatography from most other physical methods of separation is that one phase is stationary and the other mobile. The *moving phase* can be either liquid or gas; the *stationary phase* can be either liquid or solid. The four possible combinations lead to four broad types of chromatography: liquid-solid, liquid-liquid, gas-solid, and gas-liquid. The stationary phase is present in the form of a long bed and is either dispersed on or consists of a packing of large surface area. When a sample is added to one end of the column (broadly defined), components of the sample distribute themselves between the mobile and nonmobile phases, while the mobile phase percolates through the column, each solute proceeding down the column in a *band* or *zone*, at a rate slower than that of the mobile phase. This rate depends on the partition ratio (Section 24-1) of the solute. In some cases adsorption of a solute on a supposedly inert support may be a complicating factor, which is not considered in detail here. Suffice it to say that, if the amount adsorbed is proportional to the concentration (linear isotherm), the same theory holds for any level of adsorption. Recall that the Langmuir isotherm (Section 9-1) approaches a linear form in dilute solution, so that the theory applies here even if adsorption plays a role. In the more general case of a nonlinear isotherm, the elution bands undergo distortion.

With the continuous flow of the mobile phase in chromatography, not only do we have equilibrium spreading as the solute moves along a column, but in addition we must consider several nonequilibrium effects that bring about additional band and peak spreading. These nonequilibrium or dynamic effects have to do with the rate of the repeated transfer of a solute between the two phases, with diffusion processes of solute in the two phases, and with the so-called eddy effect. The equilibrium approach to the theory of chromatography assumes a discrete number of theoretical plates in a column, whereas the dynamic approach leads to an infinite number of available plates in the column when there is no dynamic spreading, since in the limit the separation stages become infinitesimal. An understanding of the chromatographic process should include an appreciation of spreading effects from both the equilibrium and dynamic viewpoints.

In elution chromatography the sample is introduced into the inlet end of the column as a short pulse, and the components are carried independently along the column at different speeds. In the alternative procedure of frontal chromatography, the sample is fed into the column continuously, and the various components break through at the outlet in a series of steps. Only the elution technique is considered here, since it is the one of importance in analytical chemistry.

24-1 SOME CHROMATOGRAPHIC TERMS

An adaptation of the distribution constant defined in Equation (23-3) could be made for solutes in chromatography. As in liquid-liquid extraction, however, solutes may be present in several chemical forms, and therefore a quantity analogous to the distribution ratio (Section 23-1), called the partition ratio, is preferred. The partition ratio must be a somewhat more broadly defined term than the distribution ratio in liquid-liquid extraction for two reasons. First, in chromatography, concentrations of solute in the two phases are usually unknown and may be unmeasurable, as when adsorption is important. Second, instead of the two phases being merely an aqueous phase and an immiscible organic solvent, in chromatography they can be any one of innumerable combinations of solid or liquid stationary phases and liquid or gas mobile phases. The *partition ratio k* is defined as the ratio of *total amount* of solute in the stationary phase to that in the mobile phase at equilibrium. Thus,

$$k = \frac{n_s}{n_m} \qquad (24\text{-}1)$$

where n_s and n_m are amounts of solute in the stationary and mobile phases. The definition in (24-1) often leads to an equilibrium-constant value that apparently involves the reciprocal of the distribution ratio used in liquid-liquid extraction; that is, $k = C_s V_s / C_m V_m = V_s / D V_m$. The partition ratio is called the mass distribution ratio† and has also been called the capacity factor.‡

The *retention volume V* for a solute is the volume of mobile fluid passed through a column from sample injection to its peak maximum. When the mobile fluid is a gas, temperature and pressure must be specified. Retention volume is the product of retention time t and flow rate F. The *dead space* or holdup volume V_m of a column is the volume of mobile fluid required to transport a nonabsorbed solute from injection to its maximum at the detector. It includes the effective volume contributions of the sample injector and detector as well as the column dead space.

The *net retention volume V'* is equal to the retention volume minus the column dead space, $V - V_m$, where both V and V_m are corrected for the pressure-gradient correction factor (see below). If the noncolumn contributions to the dead space are insignificant relative to the total dead space, and if the sample size is small enough so that the velocity of the mobile phase within the band is not increased by the solute in the mobile phase,[1] then the net retention volume is equal to kV_m, and the partition ratio is $(V - V_m)/V_m$. The *relative retention* for two solutes is the ratio of their net retention volumes $(V_2 - V_m)/(V_1 - V_m)$, or V_2'/V_1'; this is usually expressed by a number larger than unity and is often denoted by α. The *specific retention volume V_g* is the net retention volume per gram of stationary phase. In gas chromatography, V' is corrected to mean column pressure and to a standard temperature, 0°C, and V_g is equal to V'/w, where w is weight of stationary phase in grams.

One function of the mobile fluid is to serve as a medium for transporting solutes from the inlet to the outlet of a column. As previously noted, the retention volume is the volume of mobile fluid necessary to carry a solute band the full length of the column. When a noncompressible fluid is used, as in liquid chromatography, this

---

† The mass distribution ratio is given the symbol D_m, a symbol widely used to denote diffusion coefficient in the mobile phase. We choose to use the term partition ratio and the symbol k to minimize misunderstanding and in accord with extensive usage in chromatography.

‡ Unfortunately, the term retention ratio or retardation factor sometimes has been used instead of partition ratio. Retention ratio is the fraction (solute-band velocity/mobile-phase velocity) and is equal to $1/(1 + k)$. We prefer the term partition ratio because, in addition to its similarity to the distribution ratio in liquid-liquid extraction, it has the form of an equilibrium constant and is directly proportional to an important retention parameter, the net retention volume.

volume is unambiguous; but, when a gas is used as carrier fluid, the pressure gradient required for gas flow causes the volume at the band position to differ from the volume of gas measured at the column outlet. The true retention volume is the integrated sum of the gas volumes at the band positions along the column and is therefore smaller than the corresponding volume of gas observed at the column outlet. Thus, when the mobile phase is a gas, retention volumes are corrected for compressibility of the gas. The *pressure-gradient correction factor j*, derived by James and Martin,[2] is expressed by

$$j = \frac{3}{2} \frac{[(P_i/P_o)^2 - 1]}{[(P_i/P_o)^3 - 1]} \qquad (24\text{-}2)$$

where P_i is the inlet pressure and P_o the outlet pressure.

When multiplied by j, the retention volume measured at the column outlet is identical to the volume of carrier gas flowing past the solute band under the instantaneous pressure at the band position; it is the retention volume that would be observed with a noncompressible mobile phase. Values for the pressure-gradient correction factor have been tabulated;[3] a few selected values are: for $P_i/P_o = 1.1, j = 0.9517$; $P_i/P_o = 1.2, j = 0.9066$; 1.3, 0.8647; 1.4, 0.8257; 1.5, 0.7895; 2.0, 0.6429; 3.0, 0.4615; 4.0, 0.3571; 5.0, 0.2903.

24-2 BAND BROADENING AND THE THEORETICAL PLATE

Most chromatographic theory relating to band broadening was first developed in connection with, and substantiated by, gas-chromatographic techniques. Many of the findings were then broadened and applied more generally to other types of chromatography.

The important concept of the theoretical plate, which is taken from distillation theory, may be presented in terms of a column with a series of physically separate stages for vaporization and condensation. One *theoretical plate* is defined as the length of column required to produce a change in composition that corresponds to a single stage in the distillation of a liquid of one composition, producing a vapor and a condensate of different compositions in accordance with the temperature-composition diagram. More broadly, a theoretical plate is the portion of a multistage device that accomplishes the same amount of separation as a single-equilibration stage described in Sections 23-2 and 23-3. When, as in chromatography, there are no physically discrete separation stages, the concept of a theoretical plate is a mathematical, or intellectual, convenience and should not be construed as describing a physical process.

Martin and Synge[4] used a chromatographic model (Section 23-4) involving the hypothetical division of a column into a number of zones or plates, and the volume of one plate was such that the solute concentration in the mobile phase leaving the plate was in equilibrium with the average solute concentration in the stationary phase in that plate. Even though the conclusions arising from equilibrium-plate theory give an accurate description of chromatographic band broadening, Giddings[5] emphasized that an equilibrium-plate model conflicts conceptually with the actual processes occurring in a column and consequently the plate model should not be interpreted literally. He pointed out that equilibrium (actually a quasi equilibrium[6]) is attained only at the band maximum and that nonequilibrium exists at all other points. Nonequilibrium is the most important cause of band broadening. Therefore, chromatographic band broadening is treated here in terms of dynamic effects rather than of the equilibrium model of Martin and Synge. Plate terminology is not abandoned, however, because it is an effective way to express in simple terms the extent of band or peak broadening, including both equilibrium and dynamic effects.

Giddings viewed the highly complex chromatographic process from the point of view of a model that includes phenomena at the molecular, intraparticle, particulate, and column levels. Fundamental to all chromatographic separations is a differential rate of movement of different solutes that depends on their partition ratios, with bands and peaks separating from each other more rapidly than they spread. Even though a solute may be injected as a sharp band, a Poisson distribution (the limiting binomial distribution) of solute is obtained in the early stages of elution; after the solute has gone through the equivalent of 50 plates or so, this distribution closely approaches that given by a smooth gaussian-distribution curve (Figure 23-5). At the molecular level, movement is chaotic and seemingly a random stop-and-go process, a molecule sometimes moving forward in the mobile phase and at other times being essentially immobile in the stationary phase. The probability of a molecule being in the stationary phase is dependent on the partition ratio (proportional to k when k is small). Because at the band maximum there is inherent equilibrium, the average rate of movement of a large number of molecules is governed by the partition ratio.

With effects causing asymmetrical spreading assumed negligible, the *plate height h* (or HETP, height equivalent to a theoretical plate) for a gaussian distribution may be defined as variance per unit length of column:

$$h = \frac{\sigma^2}{L} \qquad (24\text{-}3)$$

where L is the length of the column and σ is the standard deviation of the band width expressed in the same units as L. For a uniformly packed column the value of σ or σ^2 is the resultant of all the random processes at the molecular level that cause band broadening. For a nonuniform column a local plate height h_{loc} is integrated[3] over the length of the entire column to give the observed plate height:

$$h = \frac{\int_0^L h_{loc}\, dz}{L} = \frac{\int_0^L \dfrac{d\sigma^2}{dz}\, dz}{L} \qquad (24\text{-}4)$$

where z is distance from column inlet. The complication of a nonuniform column is not considered in more detail.

Variances, rather than standard deviations, are statistically additive.[5] Thus, provided the individual effects are independent of each other, in Equation (24-3), σ^2 can be written

$$\sigma^2 = \sigma_1{}^2 + \sigma_2{}^2 + \sigma_3{}^2 + \cdots + \sigma_i{}^2 + \cdots + \sigma_n{}^2 \qquad (24\text{-}5)$$

where $\sigma_1{}^2, \sigma_2{}^2, \sigma_3{}^2, \ldots$ are variances arising from equilibrium spreading (Section 23-3), continuous flow (Section 23-4), dynamic effects such as resistance to mass transfer, diffusion processes, flow-pattern effects, sample injection spreading, and so on.

Since variances are additive, unless a band-broadening process has a σ_i value of at least 10 to 15% of the overall σ value, its contribution is slight. Thus, if $\sigma_i = 0.1\,\sigma$, the contribution of σ_i to the overall σ^2 is only 1%. Therefore, the observed plate height can be expected to result mainly from only one or two major band-broadening effects. Littlewood[7] pointed out that, since broadening is dominated by one or two major factors, the best designs bring about reduction of these factors, and chromatographic conditions should be chosen so that the broadening factors are of more-or-less equal magnitude.

EXAMPLE 24-1   The standard deviations in a chromatographic band resulting from several independent band-broadening processes are 0.013, 0.0024, 0.0067, and 0.047 cm.

(a)   What is the standard deviation of the band?
(b)   What percentage does each contribute to the total?

ANSWER   (a) $\sigma^2 = 0.013^2 + 0.0024^2 + 0.0067^2 + 0.047^2 = 0.00243$ cm^2. $\sigma = 0.049$ cm (that is, the standard deviation of the band is only slightly larger than that of the main component, 0.047). (b) $100\,\sigma_i{}^2/\sigma^2 = 7\%,\ 0.2\%,\ 1.8\%,$ and 91% of the total.   ////

For gaussian chromatographic *peaks*, as illustrated in Figure 24-1, the peak-base width measured from the intercepts of the tangents to the inflection points with the

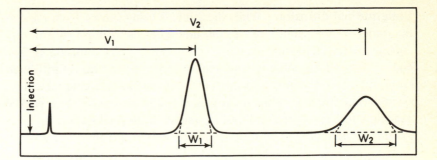

FIGURE 24-1 Two chromatographic peaks illustrating retention volume and peak-base widths.

peak base is equal to 4σ. Since the retention volume (Section 24-1) is proportional to column length, we obtain the following relation between the band width of a solute that has traveled the full length of a column and the peak width for that solute in the effluent:

$$\frac{4\sigma}{L} = \frac{W}{V} \qquad (24\text{-}6)$$

Combining Equations (24-3) and (24-6), we obtain

$$n = \frac{L}{h} = 16\left(\frac{V}{W}\right)^2 \qquad (24\text{-}7)$$

where n is defined as the number of theoretical plates, V is retention volume, and W is the peak-base width in the same units as V.

EXAMPLE 24-2 Estimate plate number and plate height h from Figure 24-1, assuming a column length of 0.5 m.

ANSWER In units of chart length, $V_1 = 4.29$ cm, $V_2 = 8.79$ cm, $W_1 = 0.88$ cm, and $W_2 = 1.62$ cm. Therefore n for the first peak is $16(4.29/0.88)^2 = 380$, and n for the second peak is $16(8.79/1.62)^2 = 471$. The plate heights for the first and second peaks are $\frac{50}{380} = 0.132$ cm and $\frac{50}{471} = 0.106$ cm. ////

Similarly, for a *band* or zone, the plate number is equal to 16 times the square of the ratio of distance moved to band width.

Equilibrium band spreading under conditions of batch or of continuous liquid-liquid extraction is treated in Sections 23-3 and 23-4. We now examine the more important dynamic effects.

Longitudinal diffusion   While the solute is being carried from the column inlet to the outlet by the mobile phase, the molecules can diffuse in any direction in the moving stream, including forward or backward. Net diffusion is from regions of high to low concentration, the amount of spreading increasing with time spent in the mobile phase. If the flow velocity of the mobile phase is u centimeters per second, the average time spent by solute molecules in this phase is L/u. In a packed column the packing interferes with free diffusion, and so diffusion distances are reduced. When, for this interference, an obstructive factor that is less than unity is introduced, the contribution arising from longitudinal diffusion in the mobile phase $\sigma_{d,m}^2$ is

$$\sigma_{d,m}^2 = \frac{2\gamma_m D_m}{L/u} \qquad (24\text{-}8)$$

where D_m is the diffusion coefficient in the mobile phase. Hence,

$$h_{d,m} = \frac{\sigma_{d,m}^2}{L} = \frac{2\gamma_m D_m}{u} \qquad (24\text{-}9)$$

When values for D_m are large (as in gas chromatography), longitudinal diffusion in the stationary phase can be expected to be small compared with that in the mobile phase. When D_m and D_s are comparable (as in liquid-liquid chromatography), longitudinal diffusion in the stationary phase may be larger than in the mobile phase. The average time spent in the stationary phase by solute molecules is kL/u, where k is the partition ratio. The contribution to plate height by longitudinal diffusion in the stationary phase $h_{d,s}$ is

$$h_{d,s} = \frac{2\gamma_s D_s k}{u} \qquad (24\text{-}10)$$

where γ_s is the obstructive factor in the stationary phase and D_s is the diffusion coefficient in the stationary phase.

Mass-transfer broadening   When a sample with a large partition ratio is injected into a column, it is retained almost totally by the stationary phase near the inlet. It is eluted from this region only with a large volume of mobile phase. Material eluted from this region is strongly retained in the next region. Thus nearly all molecules of such a solute must be transferred from the mobile to the stationary phase and from the stationary to the mobile phase for each region of the column equivalent to one plate. Transfer is not instantaneous, and in a continually flowing system equilibrium is not attained. The rate of transfer back and forth between the phases ultimately depends on diffusion and other rate processes in the two phases. The driving force for diffusion is a concentration gradient. The resistance to mass transfer is decreased by large diffusion coefficients, short diffusion distances, and large interfacial areas between the phases.

The plate-height contribution resulting from resistance to mass transfer in the stationary phase can be written

$$h_s = q \frac{k}{(1 + k)^2} \frac{d_s^2}{D_s} u \qquad (24\text{-}11)$$

where q is a configuration factor, d_s is the diffusion distance in the stationary phase (thickness of film of stationary phase), and D_s is the diffusion coefficient of the solute in the stationary phase. The configuration factor q depends on the shape of the pool of stationary phase and varies[5] from $\frac{2}{3}$ for a uniform liquid film to $\frac{2}{15}$ for a stationary phase composed of spheres such as ion-exchange beads. The factor $k/(1 + k)^2$ passes through a maximum of 0.25 when k equals unity.

The term for resistance to mass transfer in the mobile phase is expressed by

$$h_m = \frac{\omega_m d_p^2 u}{D_m} \qquad (24\text{-}12)$$

where d_p is the average particle diameter, D_m the diffusion coefficient in the mobile phase, and ω_m an overall parameter dependent on transchannel, transparticle, short- and long-range interchannel, and transcolumn effects. The magnitude of these ω terms depends on such variables as column and particle diameter and partition ratio; though they have not been evaluated exactly, estimates have been made.[8,9] The long-range interchannel effect probably has an ω term equal to about 2 and is likely the major contribution to the overall value of ω_m in (24-12).

Flow-pattern effects In a packed column there are innumerable pathways of various lengths along which mobile phase can travel from inlet to outlet. The average velocity in any one stream element or flow channel differs from that in the others. The net effect of differentials in flow velocity and in path length is that some molecules of solute are carried forward more rapidly and others less rapidly than the average. Owing to the packing in a column, the flow velocities change rapidly in any one flow channel from point to point. In addition, individual molecules diffuse back and forth between flow channels, some of which move rapidly and others slowly. The number of such transfers between flow channels is proportional to the time available (that is, inversely proportional to the flow rate). The plate-height contribution expected[5] for this diffusive exchange depends on the particle diameter, the dimensions of the column, the dimensions of the channels formed, and the flow differentials existing between flow channels, as well as on the diffusion coefficient of a solute in a mobile phase. At low flow velocities, diffusive exchange can nullify the spreading effects of different flow patterns; consequently, these spreading effects are not independent processes and so cannot be added statistically in the usual way.

The flow-pattern, or eddy, effect has been formulated in several ways. It was first expressed[10] as a term independent of flow rate and given the symbol A. Giddings formulated coupling between the interparticle structure-dependent and diffusion-dependent effects of the flow pattern. Littlewood[7] expressed the coupled term in the form

$$h_e = \frac{Ku}{D_m + \alpha u d_p} \qquad (24\text{-}13)$$

where K and α are constants that vary with the packing. Littlewood[11] reported values of α varying from 0.1 to 0.25 and values of K from about 10^{-4} to 4×10^{-4} for typical efficient packed columns. The factor K is equivalent to $\omega_e d_p^2$ in Giddings's formulation of the coupled term for eddy diffusion, where ω_e is the coupling constant. The coupled term at high flow rates approaches as a limit $K/\alpha d_p$ and at low flow rates becomes proportional to flow velocity.

When the terms contributing to band broadening are collected and equilibrium spreading is assumed to be negligible, the following general expression for overall plate height is obtained:

$$h = \frac{Ku}{D_m + \alpha d_p u} + \frac{2\gamma_m D_m}{u} + \frac{2\gamma_s D_s k}{u} + q\, \frac{k}{(1 + k)^2}\, \frac{d_s^2}{D_s}\, u + \frac{\omega_m d_p^2 u}{D_m} \qquad (24\text{-}14)$$

In this equation the first factor is called the eddy diffusion, the second and third are molecular diffusion, and the last two are called resistance-to-mass-transfer terms. All the terms include the mobile-fluid velocity as a variable that is proportional to the flow rate in some, and inversely proportional in others. The overall relation between plate height and flow velocity of the mobile phase is the statistical resultant of the five terms and is usually depicted in the form of a Van Deemter plot.[10] Such a diagram shows that an optimum flow velocity for minimum band spreading exists for a given chromatographic column.

The flow velocity at the optimum depends on the particular values of the constants γ_m, γ_s, q, K, α, and ω_m; the packing parameters d_p and d_s; the partition ratio k; and the diffusion coefficients in the mobile and stationary phases D_m and D_s. These diffusion coefficients are two of the most significant factors. Typically, diffusion coefficients of solutes in gaseous mobile phases are on the order of 0.3 cm²/s, whereas in liquid mobile phases they are on the order of 10^{-5} cm²/s (Figure 24-5). In liquid stationary phases, diffusion coefficients are typically in the range 10^{-6} to 10^{-10}cm²/s. Effective diffusion coefficients in solid stationary phases such as ion-exchange beads are probably of the same order of magnitude as in liquid stationary phases. (For example, at 25°C in sulfonated polystyrene with 10% divinylbenzene, sodium ions have a diffusion coefficient[12] of 2.8×10^{-7} cm²/s.) In gas-solid chromatography, mass-transfer broadening in the micropores with solid stationary phases may be independent of the carrier gas.[13]

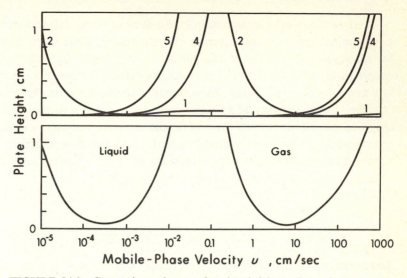

FIGURE 24-2   Comparison of curves for plate height against velocity for the individual terms (upper) and for the overall value (lower) of Equation (24-14). Left, the values for a liquid mobile phase; right, values for a gaseous mobile phase (liquid stationary phases in both cases). The numbered curves represent eddy diffusion (1), molecular diffusion in the mobile phase (2), and resistance to mass transfer in the stationary (4) and mobile (5) phases. The contribution of the term for molecular diffusion in the stationary phase is negligible at velocities near the optimum for both liquid and gas systems.

EXAMPLE 24-3   Calculate and plot the curves for plate height against velocity for each of the terms and the overall h in (24-14), using the following values for the constants: $K = 2 \times 10^{-4}$ cm^2; $\alpha = 0.2$; $\gamma_m = \gamma_s = 0.5$; $\omega_m = 2$; $d_p = 0.02$ cm. For a gaseous mobile phase take $q = \frac{2}{3}$; $D_m = 0.3$ cm^2/s; $D_s = 10^{-6}$ cm^2/s; $d_s = 10^{-4}$ cm. For a liquid mobile phase take $q = \frac{2}{15}$; $D_m = 10^{-5}$ cm^2/s; $D_s = 10^{-8}$ cm^2/s; $d_s = 10^{-3}$ cm. Use a value of unity for the partition ratio. The answer is given in the form of a set of curves in Figure 24-2. Note that the velocity u is given on a logarithmic scale rather than the more usual linear scale in Van Deemter plots.

////

Several of the main features of gas and liquid chromatography can be deduced from an examination of Figure 24-2. For example, the optimum flow velocity in liquid chromatography is several orders of magnitude lower than that in gas chromatography for d_p values of the same magnitude, owing primarily to the small diffusion coefficients in liquids compared with those in gases. If the stationary phase is retained on or is present as particles that are reasonably small, the plate height at the optimum is of the order of a millimeter, in agreement with experimental observations. Longitudinal-diffusion spreading in the stationary phase is probably insignificant for all reasonable values of the partition ratio. Although the magnitude of the coupled eddy-diffusion term is uncertain, its value increases at low velocity with flow rate and may be of the same order of magnitude as the terms for resistance to mass transfer. For the particular values chosen, the resistance to mass transfer has about the same magnitude in both the mobile and stationary phases, a situation commonly recommended[7] in chromatographic practice. In practice, effective separations are obtained through the use of long columns of small diameter packed with particles of uniform size and small diameter.

24-3 TEMPERATURE AND BAND BROADENING

Temperature, as an experimental variable, usually exerts a larger effect on the chromatographic process than any other single parameter under the experimentalist's control. Changes in temperature have profound effects, for instance, on retention volumes in gas chromatography and adsorption chromatography, on diffusion coefficients in the mobile and stationary phases, and on flow rate of the mobile phase.

Flow rate   To force the mobile fluid from inlet to outlet of a column, the inlet pressure must be higher than the outlet pressure. When the temperature is changed, the fluid viscosity changes, and then either the flow rate changes for the same inlet and outlet pressures or the pressure drop must be altered to maintain the same flow rate. The flow rate of a fluid through a packed or capillary column is inversely proportional to the viscosity. For example, from Darcy's law[14] the flow of a gas through a packed column is given by[3]

$$F_T = \frac{BA}{\eta L} \frac{p_o[(p_i/p_o)^2 - 1]}{2} \qquad (24\text{-}15)$$

where F_T is volumetric flow rate at temperature T and at pressure p_o, B is a constant dependent on the permeability of the bed, A is cross-sectional area of the empty tube, η is fluid viscosity, L is column length, p_i is inlet pressure, and p_o is outlet pressure. Similarly, the flow of a liquid through a capillary column is governed by Poiseuille's law:

$$F_T = \frac{\pi r^4}{8\eta L} \Delta P \qquad (24\text{-}16)$$

where r is the inside radius of the column and ΔP the pressure drop. Geometrical factors such as BA/L in (24-15) and r^4/L in (24-16) are primarily dependent on the column and affected little by changes in temperature. The viscosity η, however, is

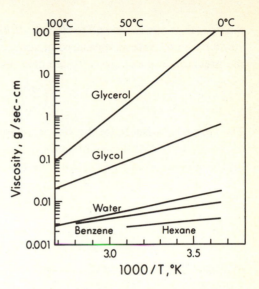

FIGURE 24-3   Change   in   viscosity
with temperature for several liquids.

strongly dependent on temperature for both gases and liquids. For gases the effective
flow rate also may change with temperature, owing to changes in the pressure-gradient
correction factor.

The decrease in the viscosity of liquids with increasing temperature is given by
an empirical equation of the form

$$\log \eta = \frac{B}{T} + C \qquad (24\text{-}17)$$

where B and C are constants for the liquid and T is absolute temperature. When η
is plotted on a logarithmic scale against the reciprocal of the temperature, a series of
straight lines for various liquids is obtained, as shown in Figure 24-3. The viscosity
of liquids commonly used as mobile fluids can be decreased by a factor of nearly 10
by increasing the temperature from 0 to 100°C. Therefore, for a constant pressure
drop ΔP, flow rates can be increased by this same factor.

The viscosities of gases suitable for chromatographic mobile phases increase
as the 0.7 power of the absolute temperature.[15] Under conditions of constant outlet
pressure, with neglect of the slight variation in column dimensions and permeability
with temperature, Equation (24-15) can be rewritten to give the ratio of flow rates at
two temperatures:

$$\frac{F_{T_1}}{F_{T_2}} = \left(\frac{T_2}{T_1}\right)^{0.7} \frac{[(p_i/p_o)^2 - 1]_{T_1}}{[(p_i/p_o)^2 - 1]_{T_2}} \qquad (24\text{-}18)$$

where the flow rates F_{T_1} and F_{T_2} are expressed at column temperature. In gas
chromatography, either the mass flow rate can be maintained constant while the
inlet pressure is varied, or the inlet and outlet pressures can be held constant while

the mass flow rate is allowed to change. With constant mass flow rate (that is, constant volumetric flow rate as measured at some reference condition), as a result of (24-18) the pressure drop changes with temperature according to:

$$\left[\left(\frac{p_i}{p_o}\right)^2_{T_2} - 1\right] = \left[\left(\frac{p_i}{p_o}\right)^2_{T_1} - 1\right]\left(\frac{T_2}{T_1}\right)^{1.7} \qquad (24\text{-}19)$$

The factor 1.7 arises from the viscosity dependence of 0.7 and the thermal expansion factor of 1.0. With constant inlet and outlet pressures, p_i/p_o is constant, and the outlet flow rate then changes with the 1.7 power of the temperature:

$$F = F_{T_1}\left(\frac{T_1}{T_2}\right)^{1.7} \qquad (24\text{-}20)$$

Here the flow rate F is expressed at temperature T_1.

For minimum pressure drop and least change in pressure drop with temperature, hydrogen is the best eluent gas in that its absolute viscosity is about half that of most other gases. For example, at 100°C the viscosity of hydrogen is 1.035×10^{-4} g/s-cm, while that of helium is 2.28×10^{-4} g/s-cm.

Retention volume Change in temperature can affect the distribution of a sample between the mobile and stationary phases and thereby affect retention behavior. In ion-exchange chromatography the effect of temperature is slight, but when the value of the partition ratio depends on adsorption, temperature variation is more significant. In gas-liquid chromatography the effects of temperature have been thoroughly studied, and the fundamental retention relation is

$$V - V_m = Ae^{\Delta H/RT} \qquad (24\text{-}21)$$

where A and ΔH are constants characteristic of the solute–stationary phase pair, the quantity ΔH is a measure† of the heat of solution of the solute in the stationary phase, and $R \ln A$ is a measure of the entropy of solution. Figure 24-4 illustrates the relation between the net retention volume and temperature for a particular system.[16]

Gaseous diffusion coefficients and temperature Chen and Othmer[17] proposed the following empirical equation to describe gaseous diffusion coefficients in terms of other measurable quantities:

$$D = \frac{0.43\left(\dfrac{T}{100}\right)^{1.81}\left(\dfrac{1}{M_1} + \dfrac{1}{M_2}\right)^{0.5}}{P\left(\dfrac{T_{c_1}T_{c_2}}{10{,}000}\right)^{0.1405}\left[\left(\dfrac{V_{c_1}}{100}\right)^{0.4} + \left(\dfrac{V_{c_2}}{100}\right)^{0.4}\right]^2} \qquad (24\text{-}22)$$

† Strictly speaking, ΔH in (24-21) is $-\Delta H°$, where $\Delta H°$ is the standard molar enthalpy of solution from perfect gas to infinitely dilute solution.

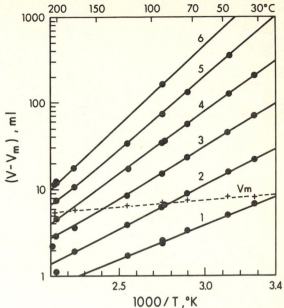

FIGURE 24-4   Dependence of specific retention volume on temperature for hydrocarbons ranging from propane (1) to n-octane (6), with Apiezon L as stationary phase. (*From Harris and Habgood.*[3])

where D is the diffusion coefficient in square centimeters per second, T is absolute temperature, P is pressure in atmospheres, M_1 and M_2 are molecular weights of the gas matrix and the diffusing solute, T_{c_1} and T_{c_2} are the critical temperatures, and V_{c_1} and V_{c_2} are the critical volumes. According to (24-22), gaseous diffusion coefficients increase strongly with temperature (to the 1.8 power) and are inversely proportional to the pressure.

The influence of temperature on diffusion coefficients of solutes in liquids has been studied in less detail. Diffusion coefficients often can be estimated from viscosity measurements. Hydrodynamic theory[18] relates self-diffusion coefficients to the viscosity by

$$D_{\text{liq}} = \frac{k}{2\pi} \left(\frac{N}{V}\right)^{\frac{1}{3}} \frac{T}{\eta} \qquad (24\text{-}23)$$

where D_{liq} is the self-diffusion coefficient in square centimeters per second, k is Boltzmann's constant (1.38×10^{-23} J/K), N is Avogadro's number (6.023×10^{23}), V is molar volume in milliliters, T is absolute temperature, and η is viscosity in g/s-cm. For calculation of the diffusion coefficient for a large molecule or particle diffusing through a solvent consisting of small molecules, (24-23) is modified by replacing 2π with 3π. For calculation of the diffusion coefficient for molecules whose size is smaller than that of the solvent, fragmentary evidence[3] indicates that 2π should be replaced by a smaller value, probably 1.4π. Figure 24-5 illustrates how

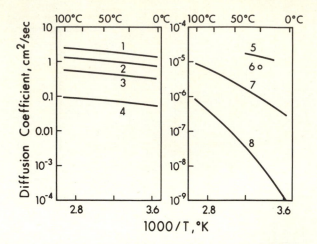

FIGURE 24-5   Diffusion coefficients of solutes in gases (left) and liquids (right) as a function of temperature at 1 atm pressure: (1) hydrogen in helium; (2) water in helium; (3) benzene in helium; (4) benzene in CO_2; (5) benzene in CCl_4 (*from Caldwell and Babb*[19]); (6) zinc ions in water; (7) nitrobenzene in ethylene glycol (*from Evanoff*[20]); (8) nitrobenzene in glycerol (*from Evanoff*[20]).

diffusion coefficients of solutes and typical gases and liquids vary with temperature. For this figure the diffusion coefficients in gases were calculated by (24-22), that of zinc ions in water was obtained from polarographic sources, the data for benzene in carbon tetrachloride were obtained from Caldwell and Babb,[19] and the data for nitrobenzene in glycol or glycerol from Evanoff.[20] The figure shows that, on a logarithmic scale, diffusion coefficients are in general roughly linear with the reciprocal of the absolute temperature. Also, diffusion coefficients in liquids are clearly many orders of magnitude smaller than those in gases.

Band broadening and temperature   The five terms of Equation (24-14) can be examined in the context of the influence of temperature on flow rates, retention volumes, and diffusion coefficients to obtain an estimate of the overall influence of temperature on band broadening. Through thermal expansion, temperature also influences such factors as thickness of a liquid film and particle and column diameters, and it may also influence slightly the empirical constants in (24-14). With a liquid mobile phase, flow velocity (with the same inlet and outlet pressures) is strongly dependent on temperature. But with flow velocity u maintained constant the first term of (24-14) becomes smaller as diffusion coefficients increase in the mobile phase. For flow rates near the optimum the first term is approximately inversely proportional to D_m. The second and third terms increase in direct proportion to the diffusion coefficients in the mobile and stationary phases D_m and D_s, whereas the fourth and fifth

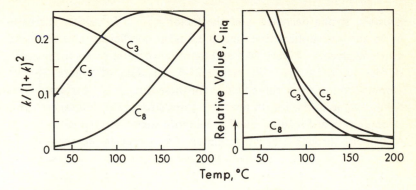

FIGURE 24-6 The factor $k/(1 + k)^2$ (left) or $[k/(1 + k)^2][d_s^2/D_s]$ (right) as a function of temperature in gas chromatography for propane, pentane, and octane, obtained from the retention data of Figure 24-4 and the diffusion-coefficient data corresponding to ethylene glycol of Figure 24-5. (*From Harris and Habgood.*[3])

terms are inversely proportional to D_m and D_s. The contribution to band broadening resulting from the fourth term includes a factor dependent on the partition ratio. Accordingly, any changes in the partition ratio with temperature also must be included.

In gas chromatography, influences of temperature are more complex. Under conditions of constant mass flow rate, but with variable temperature, the gas velocity at the column outlet is proportional to the temperature. The factor D_m/u in the second term of Equation (24-14) then increases according to the 0.81 power of the temperature rather than the 1.81 power of Equation (24-22). With constant inlet and outlet pressures the flow velocity at the outlet is proportional to $T^{-0.7}$, and D_m increases with the 1.8 power of temperature. Therefore the factor D_m/u increases with temperature according to the 2.5 power. The first and fifth terms have the inverse factor u/D_m.

In the fourth term the factor for thickness of the stationary-phase film d_s would be expected to show a slight increase with temperature, the factor $1/D_s$ a strong decrease, and the factor $k/(1 + k)^2$ either an increase or decrease depending on the value of the partition ratio. The factor $k/(1 + k)^2$ passes through a maximum at $k = 1$ and becomes smaller for either higher or lower values of the partition ratio, as shown in Figure 24-6. In normal chromatographic practice the partition ratio is somewhat greater than 1, and therefore the usual effect of increasing temperature is to increase this factor. According to Figure 24-5, the liquid diffusion coefficient D_s increases strongly with temperature. The term d_s in Equation (24-14) (thermal expansion of the liquid) shows a slight increase with temperature. (For example, d_s^2 for dinonylphthalate increases[21] by 20% between 0 and 150°C.) The factor $1/D_s$

shows a strong decrease, and the factor $k/(1 + k)^2$ either increases or decreases. With the assumption that liquid diffusion coefficients are the same as those for ethylene glycol in Figure 24-5 and that the partition ratios for propane, pentane, and octane are those given in Figure 24-4, the data of Figure 24-6 are obtained by combination of these three factors for these three hydrocarbons. In this instance the overall value of the fourth term decreases strongly with temperature for the two less strongly retained solutes, but for the more strongly retained solute the factor is seen to increase through a broad maximum with increasing temperature. For a complete description of the influence of temperature on the fourth term in Equation (24-14), changes in the flow rate and in the pressure-gradient correction factor with temperature must be included. Under conditions of constant mass flow, the flow rate increases in direct proportion to the absolute temperature, and the pressure drop increases so that the pressure-gradient correction factor decreases. The combined effect is that the effective flow rate increases with absolute temperature to a power somewhat less than 1 and thus somewhat offsets the tendency at a constant flow rate for the term for resistance to mass transfer to decrease with temperature. Under conditions of operation with constant inlet and outlet pressure, the pressure-gradient correction factor remains unchanged, and the outlet flow velocity expressed at column temperature decreases according to the 0.7 power of temperature.

Thus the effect of temperature on plate height involves a complex assessment of the interaction of many factors. At any single temperature, flow conditions cannot be chosen that give minimum plate height for all possible solutes. Within limits, plate height should usually decrease as temperature is increased.

24-4 SAMPLE-INJECTION EFFECTS

Sternberg[22] reviewed extracolumn contributions to band broadening, including sample-injection and detector effects. [As pointed out in Section 24-2, independent dispersion effects leading to band broadening are additive as their variances σ^2.] With the assumption[23,24] that the injected sample has a gaussian distribution, the variance due to finite sample width σ^2_{sample} can be combined with the variance due to column broadening processes σ^2_{col} to give the observed variance:

$$\sigma^2_{obs} = \sigma^2_{col} + \sigma^2_{sample} \qquad (24-24)$$

The value of σ_{sample} is one-quarter of the base width due to the sample band;[25] or, in terms of an increment in retention volume, σ_{sample} is equal to a quantity ΔV_{sample}. The retention volume V is itself proportional to column length. Hence

$$\frac{\sigma_{sample}}{L} = \frac{\Delta V_{sample}}{4V} \qquad (24-25)$$

Substitution of (24-24) and (24-25) in (24-6) yields

$$h_{obs} = h_{col} + \frac{L}{16}\left(\frac{\Delta V_{sample}}{V}\right)^2 \qquad (24\text{-}26)$$

where h_{col} is the plate height observed with limiting small samples. Equation (24-26) indicates that the plate-height increment due to the sample is proportional to sample volume squared. In terms of the relative broadening,

$$\frac{h_{obs}}{h_{col}} = 1 + \frac{n}{16}\frac{\Delta V_{sample}}{V} \qquad (24\text{-}27)$$

where n is plate number with $\sigma_{sample} \to 0$. Nearly universal advice to practicing chromatographers is to inject the sample onto the column with as little width as possible; for no band broadening from sample injections, sample volumes would have to be zero. Investigation[26] of the relative increase in plate height as a function of relative sample volume has confirmed the suggestion[10] that the maximum sample volume should be no greater than $V/2\sqrt{n}$. This value corresponds to an increase in plate height of about 2%. Retention volumes for various solutes are highly variable; accordingly, negligible effective sample volume depends on the particular solute. For example, the limit $V/2\sqrt{n}$ for $n = 300$ and $V = 100$ or 1000 ml is 2.9 or 29 ml. On the other hand, for $n = 3000$ and $V = 100$ or 1000 ml, the limit is 0.91 or 9.1 ml. Thus for solutes with small retention volumes and for columns with large plate numbers, impossibly small sample injection volumes are needed (the sample itself must occupy a finite volume).

Another factor to consider is that, even though samples may be injected within the above limits, their solutions in the stationary phase may have finite concentration, especially near the column inlet. Littlewood[27] indicated that sample size in gas chromatography may either increase or decrease retention volumes as a result of such variables as diffuse injection, finite concentration in the mobile phase, unique viscosity effects at the band position, nonlinear isotherms, and adsorption. Even though quantitative assessment of all these influences is not possible, let us briefly examine the nonlinear isotherm effect. With a standard state for the solute defined in terms of unit activity for mole fraction of unity, positive deviations from Raoult's law (that is, the vapor pressure of a solute is more than proportional to its mole fraction in solution in the stationary phase) lead to activity coefficients larger than unity for dilute solutions. Variation with concentration of the activity coefficients of volatile solutes from their infinite-dilution values results from two factors (Section 2-7): difference in size of solute and stationary-phase molecules and their interaction with each other. Variations can be expected to be largest when the solute strongly disrupts a strongly self-associated stationary phase. The net effect of the terms for entropy and for energy of interaction [Equation (2-39)] is most commonly that deviations from Raoult's law are positive; Cruickshank and Everett,[28] however, found negative deviations from Raoult's law to be unusually common.

The gas phase itself cannot be neglected when activity-coefficient effects are under consideration, because of interactions involving molecules of solute-solute, solute-gas, and gas-gas. Activity coefficients calculated according to Equation (2-37) can be corrected by the inclusion of additional terms known as second virial co-efficients. Usually this correction[29] to the activity coefficient for solute-solute inter-actions is made according to

$$\ln \gamma_i = \ln \gamma_{i_{\text{uncorrected}}} + \frac{(2B_{ig} - B_{gg} - \overline{V}_i^\infty)\overline{P}}{RT} - \frac{(B_{ii} - V_i^0)P_i^0}{RT}$$

$$\simeq \ln \gamma_{i_{\text{uncorrected}}} - \frac{P_i^0 B_{ii}}{RT} \tag{24-28}$$

where B_{ig}, B_{gg}, and B_{ii} are the second virial coefficients for solute–carrier gas, carrier gas–carrier gas, and solute-solute interactions; $\overline{V}_i^\infty$ is the partial molar volume of the solute at infinite dilution in the stationary phase and V_i^0 is the molar volume of pure solute; $\overline{P}$ is the mean column pressure; and P_i^0 is the vapor pressure of the pure solute at temperature T. Typical values of second virial coefficients are:[30] for benzene-benzene interactions at 25°C, -1326 cm³/mole; for benzene interaction with helium, nitrogen, and carbon dioxide, $+22$, -149, and -415; and for helium-helium, nitrogen-nitrogen, and carbon dioxide–carbon dioxide, $+15$, -8, and -124. Thus, with helium as the carrier gas, solute-solute interactions may be significant, and with other gases solute-gas and gas-gas interactions may be significant as well. At low temperatures the value of P_i^0 is small, and therefore the difference between γ_i and $\gamma_{i_{\text{uncorrected}}}$ also is small, but at higher temperatures the difference may be appreciable.

EXAMPLE 24-4   A column operated at 25°C contains 0.470 g of $C_{20}H_{42}$ as stationary phase. Inlet pressure is 888 mm, outlet pressure 740 mm, observed re-tention volume of a small sample of n-hexane 2884 ml, column dead space 6.0 ml, and vapor pressure of pure hexane 15 mm at 25°C. (a) What is the uncorrected activity coefficient of the hexane in the stationary phase? (b) If the second virial coefficient B_{ii} for hexane-hexane interaction at 25°C is -1468 cm³/mole, what is the corrected activity coefficient γ_i?

ANSWER   (a) Inlet/outlet pressure ratio = 888/740 = 1.2, pressure-gradient correction factor = 0.9066, net retention volume at 25°C = $(2884 - 6.0) \times$ 0.9066 = 2609 ml or 5551 ml/g of stationary phase. From Equation (2-37) the activity coefficient $\gamma_{i_{\text{uncorrected}}}$ = $(6.24 \times 10^4 \times 298.1)/(5551 \times 15 \times 282.6) = 0.791$. (b) The quantity $P_i^0 \times B_{ii}/RT = 15 \times -1468/(6.24 \times 10^4 \times 298) = -0.00119$. Hence, $\ln \gamma_i = \ln 0.791 - (-0.0012)$, and $\gamma_i = 0.792$.   ////

Application of the correction of Equation (24-28) is important when activity coefficients are of small or moderate size; when they are large, other sources of uncertainty (such as adsorption) make such corrections inappropriate.

Samples used in making chromatographic measurements must be finite even though they may be small. If a large sample is injected that consists of a major component with small proportions of minor components, the mole fraction of the major component in solution may approach unity, especially near the inlet where the band spreading is least and the mole fraction at a maximum. With positive deviations from Raoult's law (Figure 2-1) the activity coefficient increases toward the column outlet as the mole fraction decreases with band spreading. The overall effect of use of a large sample is on the average to decrease the activity coefficient relative to that for a small sample and thereby to increase the retention volume of the major component.[31] With negative deviations from Raoult's law, large samples should show decreased retention volumes. Cruickshank and Everett[28] found that deviations from Raoult's law caused skew in the peaks as well, and they measured skew ratios.

The effect of injection of large samples on the retention behavior of minor sample components can be either[31] to decrease or, more likely, to increase retention volumes, even though their mole fraction in solution may approach zero under all conditions, because at least at the column inlet the major sample component can act in part as the stationary phase. For example, Deans[32] observed substantially increased retention times for octane, nonane, and decane injected as minor components in large samples of hexane.

Another problem, that of adsorption at any of the several interfaces (such as gas-liquid and gas-solid), may arise with minute samples. Martin[33] found gas-chromatographic retention volumes to be markedly affected by such adsorptions and in some cases to be the cause of a larger contribution to retention volume than solution in the bulk stationary phase. He both measured and calculated the extent of such adsorptions for a variety of solutes. Adsorption at a gas-liquid interface is serious for polar stationary phases when surface areas are relatively high, when the ratio of stationary phase to solid support is low, and when the temperature is low. The effects of adsorption are minimized[34] by the use of relatively highly loaded columns and nonpolar stationary phases and by avoiding solute–stationary phase pairs in which the infinite-dilution activity coefficients deviate markedly from unity. Ottenstein[35] indicated that adsorption on a liquid surface can be considered negligible in packed columns when liquid loadings exceed 5% and activity coefficients are less than 10.

24-5 OTHER FACTORS AND PEAK BROADENING

Schmauch,[36] Johnson and Stross,[37] and Sternberg[22] studied the influence of the detection system on peak broadening, and Glenn and Cram[38] described a digital logic system for evaluation of instrumental contributions to peak broadening. Because the detection system (possibly including a recorder) lags behind the actual elution of a solute from the end of a column, the peak maximum may be decreased, the width increased, and asymmetry introduced. Quantitative measurements involving peak area are virtually unaffected by slow detector response time and peak broadening due to the detector. On the other hand, quantitative measurements dependent on peak height are affected. A time lag introduced by the detector generates errors in values of retention volume and partition ratio as well as those of plate number.

The variance due to the detection system is additive with the other variances contributing to peak broadening. The maximum effective volume of the detection system should be no greater than $V/2\sqrt{n}$, a limitation identical to that for sample injection.[22] Thus, in analogy to sample injection, the effects of detector volume are most significant when a peak with small retention volume is eluted.

With respect to column geometry and band broadening, the usual assumption is that for a uniformly packed column the average plate height is independent of the length of the column. Although for liquid mobile phases this assumption probably is correct, in gas chromatography an increased pressure drop is required for longer columns, and the pressure-gradient correction factor may change significantly; moreover, depending on the region of the Van Deemter curve and on whether the loading is high or low, the plate height may decrease, increase, or remain unchanged with increasing column length. Plate number tends to decrease as column diameter is increased because of greater flow differentials, with flow rates being higher in the packing near the column walls. Giddings[39] indicated that the resistance to mass transfer in the mobile phase and the term for eddy diffusion include a transcolumn parameter proportional to $(d_p/2r)^2$, where d_p is the particle diameter of the packing and r is the column radius. A plate number[40] of 880 was obtained for a gas-chromatographic column 0.4 cm in diameter, and of 310 for a column (of the same length) 6 cm in diameter.

For a capillary column the first term of Equation (24-14) makes a negligible contribution to band broadening, and the third term is negligible in any case. So, if the layer of stationary phase on the capillary wall is thin enough that resistance to mass transfer in the stationary phase is negligible, the plate-height equation need include only the appropriate modifications of the second and fourth terms and can be written in the following general form:

$$h = \frac{B}{u} + C_m u \qquad (24\text{-}29)$$

where $B = D_m$ [Equation (24-14)] since γ is now unity, and C_m, the coefficient of resistance to mass transfer in the mobile phase, becomes[41]

$$\frac{1 + 6k + 11k^2}{24(1 + k)^2} \times \frac{r^2}{D_m}$$

where r is the capillary radius. If plate height h is plotted against velocity u according to Equation (24-29), a hyperbola is obtained that has a minimum when $h = 2\sqrt{BC_m}$ and $u = \sqrt{B/C_m}$. When the partition ratio is small, the plate height at this minimum is equal to

$$2\sqrt{\frac{2r^2}{24}} = 0.57r$$

at a velocity of

$$\sqrt{\frac{2D_m(24)}{r^2}} = \frac{6.9D_m}{r}$$

When the partition ratio is large, the plate height at the minimum is $1.9r$ and the velocity $2.1D_m/r$. The important conclusion to be reached here is that, the smaller the diameter of the capillary, the smaller the optimum plate height and the higher the optimum flow velocity. This situation means more theoretical plates per unit length and the possibility of shorter analysis time for a given level of separation.

The effects of many other variables can be interpreted in terms of Equation (24-14). For example, a large proportion of stationary phase increases the resistance to mass transfer through the factor d_s^2 and may, depending on the solute, increase or decrease it through the term $k/(1 + k)^2$. With only a small proportion of stationary phase the resistance to mass transfer in the stationary phase may become small; then the first two terms in (24-14), eddy and molecular diffusion, may be the predominant causes of band broadening. Carrier gases of low molecular weight or liquids of low viscosity as mobile phases are favorable in terms of eddy diffusion and resistance to mass transfer but unfavorable with respect to molecular diffusion; consequently, the analysis time may be shorter because the optimum flow rate is higher (Figure 24-2). A factor rarely considered in apparent band broadening is that of isotope fractionation. Klein[42] concluded that isotope fractionation is an unusually common occurrence and that it may be observed with columns of relatively few plates (for example, 250). The effect is one of increased band width.

In summary, variances rather than standard deviations are additive, and therefore it is well to reduce the size of the factor contributing most to band broadening and to operate a column under conditions where the several broadening factors are about equal in magnitude.

24-6 RESOLUTION: ITS RELATION TO RELATIVE RETENTION, PARTITION RATIO, AND PLATE NUMBER

Theoretical-plate number is a useful concept with which to compare similar columns or various operating conditions for a single column. Nevertheless, the concept should not be overemphasized, and for broader comparisons resolution should be

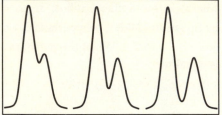

FIGURE 24-7   Three pairs of peaks with resolution values of 0.5, 0.75, and 1.0.

used. The most satisfactory way to compare different operating conditions and to indicate quantitatively how effective a separation has been is by measurement of resolution. *Resolution* is defined in chromatography as the ratio of peak separation to average peak width (Figure 24-1):

$$R = \frac{V_2 - V_1}{0.5(W_2 + W_1)} \qquad (24\text{-}30)$$

where V is retention volume and W is peak width. Figure 24-7 illustrates peak overlap for resolution values of 0.5, 0.75, and 1.0.

   Normally a resolution value much greater than unity denotes but little overlap; resolutions of 1.5 and 2.0 for two gaussian peaks of equal size represent overlap of only 0.13 and 0.003%. Table 24-1 indicates the relation between purity and resolution for a variety of binary mixtures.[3]

Table 24-1   PURITY AND RESOLUTION IN THE SEPARATION OF A BINARY MIXTURE[3]

Impurity in original mixture, %	Impurity in recovered peak, % (cut made midway between two peaks)				
	Resolution R				
	0.5	1.0	1.5	2.0	2.5
0.01	0.0014	0.0003	...	...	...
0.10	0.019	0.0024	0.0002	...	...
1.0	0.19	0.023	0.0014	...	...
10	2.05	0.26	0.015	0.000	0.0000
50	15.9	2.27	0.13	0.0032	0.00003
90	62.9	17.3	1.20	0.028	0.00026
99	94.9	69.7	11.8	0.31	0.003
99.9	99.5	95.9	57.5	3.07	0.03
99.99	99.95	99.6	93.1	24.0	0.3

Resolution can be related to theoretical-plate numbers and retention characteristics. From Equation (24-7) we can write

$$W = \frac{4V}{\sqrt{n}} \qquad (24\text{-}31)$$

Substitution in (24-30) yields

$$R = \frac{2(V_2 - V_1)}{(4V/\sqrt{n})_2 + (4V/\sqrt{n})_1} \qquad (24\text{-}32)$$

For a particular column the value of n in (24-31) varies with the nature of the solute. For two consecutive, closely spaced peaks the difference is likely to be small, however, so an average $(\sqrt{n})_{av}$, that is, $[(\sqrt{n_1} + \sqrt{n_2})/2]$, can be used to give

$$R = \frac{2(V_2 - V_1)\,(\sqrt{n})_{av}}{V_2 + V_1}\,\frac{}{4} \qquad (24\text{-}33)$$

For two closely spaced peaks we can use the approximations $V_1 \simeq V_2 \simeq V$, $k_2 \simeq k_1 \simeq k$, and $n_1 \simeq n_2 \simeq n$, so that (24-33) can be written

$$R \simeq \frac{(V_2 - V_1)\,\sqrt{n}}{V}\,\frac{}{4} \qquad (24\text{-}34)$$

This expression for resolution may be written in an alternative form in terms of the relative retention V_2'/V_1' (Section 24-1) and the partition ratio k. Thus it can be shown that (24-34) is equivalent to the following fundamental equation involving three nearly independent factors:

$$R \simeq \frac{(V_2'/V_1' - 1)}{V_2'/V_1'}\,\frac{k}{1 + k}\,\frac{\sqrt{n}}{4} \qquad (24\text{-}35)$$

On occasion the factor $\sqrt{n}$ is combined with the partition-ratio term to give a factor called *number of effective plates*, defined by $[k/(1 + k)]^2 n$. Hence

$$R = \frac{V_2'/V_1' - 1}{V_2'/V_1'}\,\frac{\sqrt{n}_{effective}}{4} \qquad (24\text{-}36)$$

In (24-35) the term $\sqrt{n}/4$ is influenced by many factors [Equation (24-14)], but mainly reflects the characteristics of a particular column and its mode of operation. This plate-number factor is described in the preceding sections (24-2 and 24-3). The factors of relative retention and partition ratio in (24-35) depend on the solutes, the nature of the mobile and stationary phases, and the temperature. The factor $k/(1 + k)$ also strongly depends on the dead-space volume in a column. Figure 24-8 illustrates how it varies according to the dead-space volume per gram of stationary phase. When the specific retention volume of a solute is relatively low and a column

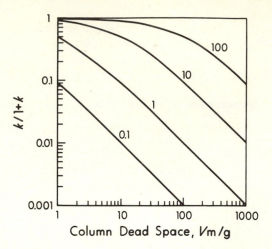

FIGURE 24-8 The factor $k/(1 + k)$ as a function of column dead space per gram of stationary phase for solutes with specific retention volumes of 0.1, 1, 10, and 100 ml/g of stationary phase.

has a large dead space, the factor $k/(1 + k)$ decreases sharply with increasing column dead space. To achieve the same resolution, a much larger number of theoretical plates would be required when column dead space increases. Thus an enormous amount in resolution can be lost through the use of columns with large dead space (capillary columns in gas chromatography, for instance).

EXAMPLE 24-5 Assume that two solutes have specific retention volumes of 0.9 and 1.0 ml/g of stationary phase. Estimate the plate number required to achieve a resolution of unity if the specific column dead space amounts to (a) 1 ml/g; (b) 100 ml/g.

ANSWER (a) The quantity $[(V_2'/V_1') - 1]/V_2'/V_1' = 0.11/1.1$, and the partition ratio of the second component is 1.0. Hence,

$$n = \left(4 \times \frac{1.111}{0.111} \times \frac{2.0}{1.0}\right)^2 = 6400$$

(b) The partition ratio is $\frac{1}{100}$, and

$$n = \left(4 \times \frac{1.111}{0.111} \times \frac{101}{1.0}\right)^2 = 1.6 \times 10^7$$

(an improbably large number!)   ////

Figure 24-9 indicates some of the interrelations between plate number, partition ratio, and relative retention required to achieve a resolution value of unity.[43] Other things being equal, resolution is proportional to the square root of the plate number, and therefore four times as many plates would be required for a resolution of 2.0 as for the 1.0 of Figure 24-9.

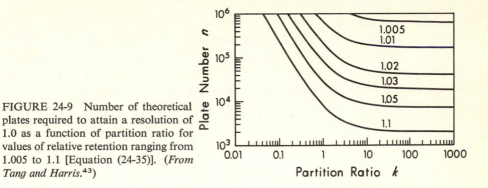

FIGURE 24-9 Number of theoretical plates required to attain a resolution of 1.0 as a function of partition ratio for values of relative retention ranging from 1.005 to 1.1 [Equation (24-35)]. (*From Tang and Harris.*[43])

In making estimates of the minimum number of plates required to achieve a desired level of separation it is important to realize that the often reproduced chart by Glueckauf[44] is misleading.[43] It is applicable with negligible error only when k is so large as to be of little chromatographic interest.

Figures 24-8 and 24-9 and Example 24-5 illustrate why capillary columns, lightly loaded columns, and columns operated under conditions where the solutes have low partition ratios have never entirely lived up to the promise of their sometimes enormous plate numbers or small plate heights. In fairness, it should be pointed out that capillary and lightly loaded columns are frequently used under conditions where high specific retention volumes partly compensate for the loss of separating efficiency resulting from the large dead space in these columns.

Comparisons of the effectiveness of chromatographic columns solely on the basis of plate numbers can be highly misleading. As pointed out by Littlewood,[7] resolution for a pair of peaks may be poorer with a column having hundreds of thousands of theoretical plates than with another having only 1000. The factor most often neglected is column dead space. A large dead space does nothing to increase the difference $V_2 - V_1$ in Expression (24-33), but increases the value of $V_2 + V_1$ in the denominator and thus decreases the resolution.

To maximize resolution, an assessment of all three factors in (24-35) is required— relative retention, partition ratio, and theoretical-plate number. For a particular column, separations are improved by operation at the minimum on the curve for plate height against velocity. If the dead space per gram of stationary phase is unchanging, then separations are improved by dispersal of the stationary phase as a thin film on packing of small-diameter particles [Equation (24-14)]. If the dead-space volume per gram is large, then the conditions should be such that the specific retention volume also is large. Often the most effective practical means of improving resolution is to select phases or conditions under which the relative retention factor is improved.

Although the principal criterion of the value of a chromatographic column or procedure should be the resolution achievable, the time required to achieve satisfactory resolution must also be considered. Optimum performance of a column means a satisfactory balance between the time required for analysis and the resolution achieved. The retention time is equal to $hn(1 + k)/u$. For a given relative retention the best resolution in the shortest time is obtained if the number of effective plates per second is at a maximum. If plate height is assumed to be independent of partition ratio k, then

$$\frac{n_{\text{effective}}}{t} = \frac{n[k/(1 + k)]^2}{t} = \frac{u}{h} \frac{k^2}{(1 + k)^3} \qquad (24\text{-}37)$$

The factor $k^2/(1 + k)^3$ passes through a maximum at a k value[45] of 2. When partition ratios are much less than 1 or much greater than about 10, the factor $k^2/(1 + k)^3$ becomes small, and much less than optimum performance can be attained.

When optimization is being considered, many parameters must be assessed. For example, one basic instrumental limitation has to do with the maximum column inlet pressure. In gas chromatography, when column length is increased to give a larger value of n, the temperature may have to be increased to obtain elution in a reasonable time. If a lightly loaded column is used, the unfavorable partition-ratio factor that may occur in the resolution equation can be offset by lowering the temperature. Changing the temperature probably affects the relative retention value and may enhance or decrease resolution. Comparisons of performances and determination of optimum conditions have been reviewed and studied by several investigators.[46-50]

24-7   COMPLEX MIXTURES: GRADIENT ELUTION AND TEMPERATURE PROGRAMMING

In chromatography an attempt is made to arrive at conditions that give maximum resolution of sample components without excessively large retention volumes for the most strongly retained component. In general, the aim is to obtain large differences in retention volumes and narrow peaks in a reasonable time. To accomplish this experimentally means compromises involving a number of factors such as choice of stationary phase, amount of stationary phase, temperature, mobile-phase velocity, and column length. For a single pair of chromatographic peaks the optimum conditions usually can be established by trial and error. Many chromatographic separations, however, are concerned with a large number of compounds and possibly several difficult separations. A single set of operating conditions may not allow even an approach to optimum resolution for all pairs.

In complex mixtures of unknown composition a sample may contain a variety of solutes with partition ratios ranging from small to large; for these the most efficient chromatographic separations are obtained through variation in the partition ratio as a band moves along a column. One technique involves increasing partition ratios down the length of the column through the use of a gradient bed. Materials with both high and low partition ratios thus tend to be separated as well-spaced bands. Another technique is that of gradient elution, or solvent programming,[51] in which the polarity or strength of the mobile phase is gradually increased to obtain a continual decrease in the partition ratios. This is accomplished by a continuous mixing process involving two or more liquid mobile phases and an external mixing chamber. Drake[52] concluded that, because sharpening of the bands is accompanied by a relative pushing together of the bands, improvement in resolution may or may not be obtained. Factors that affect resolution include variation in concentration with volume of eluent, motion of the gradient through the column, variation in partition ratio with concentration of eluent, and shape of the adsorption isotherm. Tiselius and co-workers[51,53] used gradient elution for the separation of amino acids, peptides, and sugars; Donaldson, Tulane, and Marshall[54] used the method for the separation of organic acids.

Chromathermography takes advantage of a moving temperature gradient in a column, using a gaseous mobile phase. This technique has been most highly developed by Turkel'taub and Zhukhovitskii and coworkers and has been reviewed by Tudge[55] and Ohline and DeFord.[56] In this technique a coaxial heater with a uniform temperature gradient is moved along a gas-chromatographic column during elution. When a solute band in the heated zone moves faster than the heater, its partition ratio increases and the band velocity decreases. When stabilized, each band moves with the same speed as the heater in a characteristic temperature zone that depends in part on the flow rate. Solutes with low partition ratios tend to stabilize at a position near the leading end of the heater, and those with large partition ratios at the trailing end. Each solute band tends to be narrow and to be continually sharpened as it moves along the column.

In gas chromatography the most successful solution to the resolution of complex mixtures has been that of programmed temperature gas chromatography. In this technique all partition ratios are decreased over the entire length of the column with time. When a separation is started at a relatively low temperature, the partition ratios of most components are so large that they are almost completely immobilized, or frozen, at the inlet of the column. The components with small partition ratios can move normally along the column. As the temperature is raised, partition ratios decrease, and the other solutes successively reach temperatures at which they have vapor pressures that enable them to be eluted at a reasonable rate. In effect, each compound tends to be eluted at its optimum temperature for the heating and flow rates chosen.

To illustrate the approach to the theory of retention and band broadening, we examine some aspects of programmed temperature gas chromatography.[3]

In a uniform column the isothermal retention volume V is related to eluent-gas flow rate F, column length L, and average band velocity $(dz/dt)_{av}$, by the expression

$$V = \frac{FL}{(dz/dt)_{av}} \qquad (24\text{-}38)$$

To speak of the retention volume is usually inappropriate in temperature programming; instead, the characteristic quantity is *retention temperature* T_R, the temperature reached by the column at the solute peak maximum. Retention temperature is related to heating rate, mobile-phase flow rate, and isothermal retention volumes. In temperature programming, as indicated above, the rate of band movement increases sharply with time. Nevertheless, even with isothermal elution, solute bands do not move at a uniform rate, but more rapidly as the carrier gas expands from the inlet to the outlet of the column. For a band at a distance z from the column inlet, the rate of movement is given by

$$\frac{dz}{dt} = F_z \frac{L}{V} \qquad (24\text{-}39)$$

where F_z is the volumetric flow rate of the gas at position z. Equation (24-39) is similar to (24-38) except that it refers to a particular position in the column rather than to an average over the whole column. Because the quantities F_z and V occur as a ratio, both should be specified under the same conditions of temperature. Under isothermal operation, with constant mass flow rate, the retention volume is

$$V = jFt \qquad (24\text{-}40)$$

where j is the pressure-gradient correction factor, F is flow rate, and t is time to peak maximum. On the assumption that within a column there is a uniform average flow rate equal to jF, equations describing retention behavior satisfactory for most conditions of temperature programming are obtained. For adverse conditions necessitating additional refinements, methods to allow for change in flow velocity along the column and for increase in the resistance to flow of gas with increasing temperature have been developed. We are not further concerned here with these refinements. With the flow rate assumed to be constant from the inlet to the outlet of a column, the variables F and z in (24-39) may be separated, and the equation written in the integral form:

$$\frac{1}{L}\int dz = \int \frac{F}{V} dt \qquad (24\text{-}41)$$

where V is the isothermal retention volume at the temperature corresponding to the time element dt. With a linear temperature program

$$T = T_0 + rt \quad \text{and} \quad dt = \frac{dT}{r} \qquad (24\text{-}42)$$

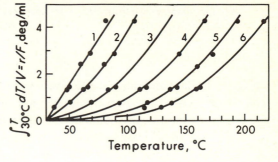

FIGURE 24-10 Characteristic curves of r/F against retention temperature for the hydrocarbons ranging from propane (1) to octane (6) obtained from the chromatographic system of Figure 24-4. Solid lines, calculated from isothermal data; points, experimental. (*From Harris and Habgood.*[3])

where T_0 is the initial temperature of the program and r is the rate of temperature increase. For a program in which the temperature is increased at a uniform rate, (24-41) can be written as the general integral

$$1 = F \int_0^{t_R} \frac{dt}{V} \qquad (24\text{-}43)$$

where t_R is the retention time for a solute. Substitution of (24-42) in (24-43) gives

$$\frac{r}{F} = \int_{T_0}^{T_R} \frac{dt}{V} \qquad (24\text{-}44)$$

This equation indicates that the retention temperature T_R may be related to the program in a characteristic way as defined by r/F. Note that the ratio of heating rate to flow rate is the significant variable, rather than either one alone, and that retention temperatures can be obtained from the isothermal retention volumes over the temperature range T_0 to T_R.

To calculate retention temperatures from isothermal retention data, first measure isothermal retention volumes at a few different temperatures within the range of interest. Then plot net retention volumes (all corrected to a single standard temperature) against the reciprocal of the temperature to obtain straight-line relations as in Figure 24-4. Use values of the dead-space volume V_m to obtain retention volumes at a series of temperatures, and plot the reciprocal $1/V$ as a function of temperature. The integral on the right side of (24-44) is the area under the curve of $1/V$ against temperature from a chosen lower temperature limit to any desired upper temperature. Evaluation of the integral is difficult by analytic methods; one simple method is graphical. The integrals for various upper temperature limits are plotted as a function of the upper temperature limit to obtain characteristic curves that designate the retention temperature for various linear programs given by the specified r/F values. Figure 24-10 shows a set of characteristic curves calculated from the data of Figure 24-4, to show how the retention temperature is affected by changing heating or flow rates.

To calculate plate number n in temperature programming, Equation (24-7) must be modified to

$$n = 16 \left(\frac{V_{T_R}}{W} \right)^2 \qquad (24\text{-}45)$$

where V_{T_R} is the isothermal retention volume at the retention temperature (obtainable from Figure 24-4). The isothermal retention volume at the retention temperature is reasonable because peak widths depend on two factors—the partition ratio at the time of elution and the band width a solute would have with the band maximum at the column outlet. When partition ratios are small, peak widths approach a limiting value equal to that for a solute with a partition ratio of 0, and thus this peak width depends almost solely on the limiting band width. For solutes having a large partition ratio the peak width depends mainly on the partition ratio, and so we must be interested in the partition ratio at the column exit as a peak emerges; the isothermal retention volume at the retention temperature reflects this partition ratio. As expected, plate numbers for programmed temperature and isothermal chromatography for the same column used under the same flow-rate conditions are in agreement.[57]

Resolution in temperature programming is calculated by means of Equation (24-30) without alteration. Equation (24-35), however, has no practical analog in temperature programming because relative retention has little direct significance and the partition ratio has meaning only at a particular temperature. Nevertheless, it can be concluded in a qualitative way that low partition ratios must be avoided if the best resolution is to be attained.

In the matter of band broadening and sample injection, temperature programming offers a significant advantage in gas chromatography. The sample injection limit of $V/2\sqrt{n}$ for isothermal chromatography (Section 24-4) becomes $V_{T_0}/2\sqrt{n}$, where V_{T_0} is the isothermal retention volume a solute would have if eluted isothermally at the initial temperature. That the expression for the maximum sample volume in programmed temperature chromatography includes V_{T_0} is highly significant. It means that in this type of chromatography the time for sample injection and the subsequent time for analysis can be controlled virtually independently.[3] Advantage can be taken of low-temperature injection without incurring the disadvantage of the inordinately large retention volumes that must follow in isothermal operation.

REFERENCES

1   M. JAFFAR and M. A. KHAN: *Anal. Chem.*, **45**:1842 (1973).
2   A. T. JAMES and A. J. P. MARTIN: *Biochem. J.*, **50**:679 (1952).
3   W. E. HARRIS and H. W. HABGOOD: "Programmed Temperature Gas Chromatography," Wiley, New York, 1966.

4 A. J. P. MARTIN and R. L. M. SYNGE: *Biochem. J.*, **35**:1358 (1941).

5 J. C. GIDDINGS: "Dynamics of Chromatography," pt. I, Dekker, New York, 1965.

6 S. WICAR, J. NOVAK, and N. RUSEVA-RAKSHIEVA: *Anal. Chem.*, **43**:1945 (1971).

7 A. B. LITTLEWOOD: "Gas Chromatography," 2d ed., p. 167, Academic, New York, 1970.

8 GIDDINGS,[5] p. 44.

9 R. H. PERRETT and J. H. PURNELL: *Anal. Chem.*, **35**:430 (1963).

10 J. J. VAN DEEMTER, F. J. ZUIDERWEG, and A. KLINKENBERG: *Chem. Eng. Sci.*, **5**:271 (1956).

11 LITTLEWOOD,[7] p. 203.

12 B. A. SOLDANO: *Ann. N.Y. Acad. Sci.*, **57**:116 (1953).

13 W. R. MACDONALD and H. W. HABGOOD: *Can. J. Chem. Eng.*, **50**:462 (1972).

14 P. C. CARMAN: "Flow of Gases Through Porous Media," chap. 1, Butterworths, London, 1956.

15 O. A. HOUGEN and K. M. WATSON: "Chemical Process Principles," pt. 3, "Kinetics and Catalysis," p. 871, Wiley, New York, 1947.

16 HARRIS and HABGOOD,[3] p. 54.

17 N. H. CHEN and D. F. OTHMER: *J. Chem. Eng. Data*, **7**:37 (1962).

18 J. C. M. LI and P. CHANG: *J. Chem. Phys.*, **23**:518 (1955).

19 C. S. CALDWELL and A. L. BABB: *J. Phys. Chem.*, **60**:51 (1956).

20 J. EVANOFF: M.Sc. Thesis, University of Alberta, Edmonton, Alberta, Canada, 1965.

21 E. R. ADLARD, M. A. KHAN, and B. T. WHITHAM in "Gas Chromatography 1960," R. P. W. Scott (ed.), p. 251, Butterworths, London, 1960.

22 J. C. STERNBERG in "Advances in Chromatography," J. C. Giddings and R. A. Keller (eds.), vol. 2, p. 205, Dekker, New York, 1966.

23 J. C. GIDDINGS: *Anal. Chem.*, **34**:722 (1962).

24 G. GUIOCHON: *Anal. Chem.*, **35**:399 (1963).

25 HARRIS and HABGOOD,[3] p. 116.

26 L. W. HOLLINGSHEAD, H. W. HABGOOD, and W. E. HARRIS: *Can. J. Chem.*, **43**:1560 (1965).

27 LITTLEWOOD,[7] p. 37.

28 A. J. B. CRUICKSHANK and D. H. EVERETT: *J. Chromatogr.*, **11**:289 (1963).

29 Y. B. TEWARI, D. E. MARTIRE, and J. P. SHERIDAN: *J. Phys. Chem.*, **74**:2345 (1970).

30 D. H. DESTY, A. GOLDUP, G. R. LUCKHURST, and W. T. SWANTON in "Gas Chromatography 1962," M. van Swaay (ed.), p. 67, Butterworths, London, 1962.

31 W. E. HARRIS: *J. Chromatogr. Sci.*, **11**:184 (1973).

32 D. R. DEANS: *Anal. Chem.*, **43**:2026 (1971).

33 R. L. MARTIN: *Anal. Chem.*, **33**:347 (1961); **35**:116 (1963).

34 D. E. MARTIRE and L. Z. POLLARA in "Advances in Chromatography," J. C. Giddings and R. A. Keller (eds.), vol. 1, p. 335, Dekker, New York, 1965.

35 D. M. OTTENSTEIN in "Advances in Chromatography," J. C. Giddings and R. A. Keller (eds.), vol. 3, p. 137, Dekker, New York, 1966.

36 L. J. SCHMAUCH: *Anal. Chem.*, **31**:225 (1959).

37 H. W. JOHNSON, JR. and F. H. STROSS, *Anal. Chem.*, **31**:357 (1959).

38 T. H. GLENN and S. P. CRAM: *J. Chromatogr. Sci.*, **8**:46 (1970).

39 GIDDINGS,[5] pp. 45, 52.

40 W. J. DE WET and V. PRETORIUS: *Anal. Chem.*, **32**:1396 (1960).

41   J. C. GIDDINGS: *Anal. Chem.*, **34**:1186 (1962); **35**:439 (1963).

42   P. D. KLEIN in "Advances in Chromatography," J. C. Giddings and R. A. Keller (eds.), vol. 3, p. 3, Dekker, New York, 1966.

43   S. H. TANG and W. E. HARRIS: *Anal. Chem.*, **45**:1977 (1973).

44   E. GLUECKAUF: *Trans. Faraday Soc.*, **51**:34 (1955).

45   B. L. KARGER in "Modern Practice of Liquid Chromatography," J. J. Kirkland (ed.), p. 3, Wiley, New York, 1971.

46   G. GUIOCHON: *Anal. Chem.*, **38**:1020 (1966); also in "Advances in Chromatography," J. C. Giddings and R. A. Keller (eds.), vol. 8, p. 179, Dekker, New York, 1969.

47   R. P. W. SCOTT in "Advances in Chromatography," J. C. Giddings and R. A. Keller (eds.), vol. 9, p. 193, Dekker, New York, 1970.

48   B. L. KARGER and W. D. COOKE: *Anal. Chem.*, **36**:985, 991 (1964).

49   E. GRUSHKA: *Anal. Chem.*, **43**:766 (1971); E. GRUSHKA and F. B. LO: *Anal. Chem.*, **45**:903 (1973).

50   J. PERRY in "Recent Advances in Gas Chromatography," I. I. Domsky and J. A. Perry (eds.), p. 1, Dekker, New York, 1971.

51   R. S. ALM, R. J. P. WILLIAMS, and A. TISELIUS: *Acta Chem. Scand.*, **6**:826 (1952).

52   B. DRAKE: *Ark. Kemi.*, **8**:1 (1955).

53   A. TISELIUS: *Endeavour*, **11**:5 (1952); L. HAGDHAL, R. J. P. WILLIAMS, and A. TISELIUS: *Ark. Kemi*, **4**:193 (1952); R. S. ALM: *Acta Chem. Scand.*, **6**:1186 (1952).

54   K. O. DONALDSON, V. J. TULANE, and L. M. MARSHALL: *Anal. Chem.*, **24**:185 (1952).

55   A. P. TUDGE: *Can. J. Phys.*, **40**:557 (1962).

56   R. W. OHLINE and D. D. DEFORD: *Anal. Chem.*, **35**:227 (1963).

57   H. W. HABGOOD and W. E. HARRIS: *Anal. Chem.*, **32**:450 (1960).

58   A. B. LITTLEWOOD, C. S. G. PHILLIPS, and D. T. PRICE: *J. Chem. Soc.*, **1955**:1480.

PROBLEMS

24-1   Calculate the pressure-gradient correction factor for inlet and outlet pressures in atmospheres of (*a*) 1.34 and 1, (*b*) 1.89 and 1, (*c*) 5 and 1, (*d*) 1.6 and 1.5.

Answer   (*a*) 0.8487. (*b*) 0.6708. (*c*) 0.2903. (*d*) 0.9674.

24-2   Suppose the standard deviations $\sigma_1, \sigma_2, \ldots$ in a chromatographic band resulting from several band-broadening processes are 0.009, 0.007, 0.020, and 0.037 cm. (*a*) What is the standard deviation of the band? (*b*) What percentage does each contribute to the total?

Answer   (*a*) 0.044. (*b*) 4%, 2%, 21%, 72%.

24-3   If the outlet flow rate, at column temperature, of helium is 40 ml/min with a column at 50°C, what is the flow rate at 150°C (and column temperature) with the assumption that (*a*) the inlet and outlet pressures remain constant, (*b*) the mass flow rate and the outlet pressure are held constant and the pressure drop is small?

Answer   (*a*) 33.1 ml/min. (*b*) 52.4 ml/min.

24-4 Calculate the gaseous diffusion coefficient of diethyl ether in hydrogen at 25°C and 1.3 atm. Use critical temperatures of 33 and 467°C and critical volumes of 65 and 281 ml for hydrogen and ether.

Answer   0.29 cm²/s.

24-5 If the flow velocity of an eluent gas is 50 cm/min and the diffusion coefficient of a solute is 0.6 cm²/s at 100°C, what is the value of D_m/u at 200°C (*a*) with constant inlet and outlet pressures and (*b*) with constant mass flow rate.

Answer   (*a*) 1.30/cm. (*b*) 0.87/cm.

24-6 A column operated at 80°C contains 2.64 g of $C_{24}H_{50}$ as stationary phase; column inlet pressure is 907 mm Hg and outlet pressure 726 mm. Observed retention volume of benzene is 285.0 ml, column dead space is 10.6 ml, and vapor pressure of pure benzene at 80°C is 760 mm. Calculate the activity coefficient for benzene, with and without the correction for the second virial coefficient B_{ii}. Assume a value of -1500 cm³/mole for B_{ii} for benzene-benzene interaction, neglecting solute-gas and gas-gas interactions.

Answer   0.93, 0.98.

24-7 Suppose the corrected net retention volume at 90°C for a limiting small sample of dodecane is 74.1 ml/g of triethylene glycol as stationary phase.

(*a*) Calculate the value of the infinite-dilution activity coefficient [Equation (2-37)], assuming the vapor pressure of pure dodecane at 90°C to be 10.0 mm. If a sample of moderate size has a net retention volume of 78.4 ml/g, what is the activity coefficient, assuming $x_i \simeq N_i/N_s$?

(*b*) If the standard state were defined in terms of activities and concentrations being equal in solution at infinite dilution, what would be the values of the activity coefficients in the above two cases? What would be the limiting value for large samples?

Answer   (*a*) 204, 192. (*b*) 1.00, 0.95, 0.005.

24-8 A gas-chromatographic column[58] using 2.82 g of silicone 702 as stationary phase was operated at 56.2°C and with inlet pressure 1230 mm and outlet pressure 781 mm, the atmospheric pressure being 760 mm. At a nitrogen flow rate of 11.0 cm³/min measured at 0°C and 760 mm, the retention time of methylpropionate was 58.2 min and that of hydrogen 2.7 min. Calculate the specific retention volume of methylpropionate, assuming that hydrogen was not retained by the liquid phase.

Answer   161 cm³.

24-9 Two solutes have specific retention volumes of 20.0 and 19.0 ml/g of stationary phase. For a column dead space of 1 ml/g and of 200 m/g, estimate the plate numbers required to achieve resolutions of (*a*) 0.75 and (*b*) 2.0.

Answer   (*a*) 3970; 440,000. (*b*) 28,000; 3 × 10⁶.

24-10   A gas-chromatographic column was operated under the following conditions: flow rate (50°C) 20 cm³/min, column temperature 50°C, inlet pressure 820 mm, outlet pressure 760 mm, and volume of stationary liquid phase (50°C) 3.12 cm³. The following data were obtained for time of peak maximum from time of injection: air, 0.50 min; n-hexane, 3.50 min; hexene-1, 3.86 min; n-heptane, 4.10 min.

    (a)   Calculate the partition ratios of n-hexane, hexene-1, and n-heptane.

    (b)   Calculate the relative-retention values for the pair of solutes n-hexane and hexene-1 and the pair hexene-1 and n-heptane.

    (c)   If the peak base width for n-hexane corresponds to 0.43 min, what is the maximum sample injection volume that should be used in isothermal chromatography at 50°C?

Answer   (a) 6, 6.72, 7.2. (b) 1.12, 1.07. (c) 1.1 ml.

24-11   Calculate and plot $k^2/(1 + k)^3$ as a function of k, and note the position of the maximum.

24-12   Show how to obtain Equation (24-35) from (24-34).

APPLICATIONS OF CHROMATOGRAPHY

Separations of the chromatographic type have been carried out for more than a century, and the process cannot be credited to any one person. The history of paper chromatography, in particular, extends into antiquity and includes observations of Pliny, Runge, Schoenbein, and Goppelsroeder. Day[1] was one of the first to use adsorption chromatography, and Tswett[2] made some of the most extensive early investigations and gave chromatography its name. Modern developments began with Kuhn and Lederer.[3]

In Chapter 24, chromatography is broadly defined as encompassing separations in which the mobile phase is either a gas or a liquid and the stationary phase a liquid or a solid. More generally, chromatography may be gas or liquid, according to the state of the mobile phase. In *gas chromatography* the stationary phase consists of either a thin film of liquid on a support or a solid of large surface area. The main types of *liquid chromatography* are: *ion exchange*, in which the mobile phase is usually aqueous and the stationary phase is an insoluble polymer possessing ionic sites; *adsorption*, in which the stationary phase is a solid of large surface area; *liquid-liquid*, in which the stationary phase is a thin film of a second immiscible liquid supported on a solid; *gel* or *exclusion*, in which the stationary phase is a gel or other porous material; *thin-layer*, in which the stationary phase is either a second liquid supported on a layer of finely divided solid or an adsorbent solid; *paper*, in which the stationary phase is a thin film of liquid on a paper as support. In *electrophoresis* the separation is made to take place under the influence of an electric field.

Broadly speaking, compounds that have significant vapor pressure and are stable at temperatures up to 100°C below their boiling points are separable by gas chromatography; ionizable nonvolatile substances by ion exchange; nonionic, nonvolatile organic compounds by adsorption, liquid-liquid, paper, or thin-layer chromatography; and solutes of high molecular weight by gel or exclusion chromatography. Electrophoretic separations can be applied to many substances such as proteins, emulsions, and colloidal particles.

Many books on the subject of chromatography have been published; a number of these are referred to in the following sections. A recent book that deals with both gas and liquid chromatography is the second edition of "Chromatography," edited by Heftmann,[4] which covers the subject in general; two recent publications that give a broad perspective of the two types of chromatography are those by Littlewood[5] and Perry, Amos, and Brewer.[6]

In this chapter, the special features of each of the types of chromatography are reviewed briefly. Because of its central role in the chromatographic process, we focus to a considerable extent on the stationary phase. Chromatography is an extremely broad subject, and therefore in a single chapter we can only touch on a few of its more important practical aspects. The topic of electrophoretic separations is defined only briefly, also because of limitations of space.

25-1   ION EXCHANGE

An ion-exchange medium consists essentially of a solid, insoluble phase that contains a number of positively or negatively charged sites. Adjacent to each site an equivalent amount of opposite charge in the form of counterions is present in the liquid phase. The ion-exchange equilibrium may be written

$$B^+ + A^+R^- \rightleftharpoons A^+ + B^+R^- \qquad \text{cation exchange} \qquad (25\text{-}1)$$

$$B^- + R^+A^- \rightleftharpoons A^- + R^+B^- \qquad \text{anion exchange} \qquad (25\text{-}2)$$

where R^+ or R^- represents the ion-exchange medium and A^+ or A^- represents the mobile counterions. A recent general reference book on the subject is one by Rieman and Walton;[7] some earlier works are by Kunin,[8] Samuelson,[9] and Helfferich.[10]

The most important types of ion-exchange materials are the synthetic resins, which are composed of networks of relatively loosely cross-linked polymers with attached cationic or anionic groups. For example, a copolymer of styrene and *p*-divinylbenzene may be cross-linked to the desired extent by choosing the fraction of divinylbenzene (typically 4 to 12%). The resulting copolymer can be sulfonated[11] to produce a strong-acid cation exchanger with about 1.1 sulfonate groups for each benzene ring, or it can be aminated[12] to give a strong-base anion exchanger with 1 exchange group for about every 2 benzene rings. In either type the concentration (in terms of univalent ions) of exchange sites is high, up to roughly 5 or 6 M.

Styrene-divinylbenzene cation-exchange resins are classified as strong (RSO_3H) or weak acids (RCOOH and ROH). A strong-acid type (Dowex 50 or Amberlite IR-120) is advantageous for most analytical applications because its exchange capacity does not vary with pH if the pH is above 3 or 4. The weak-acid types are useful for the selective takeup of strongly basic substances in the presence of weak bases. Similarly, anion-exchange resins are classified as strong (quaternary ammonium salts) or weak bases (primary, secondary, or tertiary amines). Like the acids, the strong-base types (Dowex 1 or 2 or Amberlite IRA-400) are preferable to the weak-base types (Dowex 3 or Amberlite IR-45) for most analytical applications. The weakly basic resins become inert at high pH values, whereas the strongly basic resins in the hydroxyl form can be used even at high pH values. Weakly basic resins can be used to retain strong-acid anions in the presence of weak acids, although this application is of limited use in analysis.

Chelating resin is a cross-linked polystyrene containing the active group $—CH_2N(CH_2COOH)_2$. Other ion-exchange materials of occasional interest to analytical chemistry include: inorganic exchangers;[13] silicates such as clays, aluminosilicates, and molecular sieves; liquid ion exchangers[14] (Section 23-6); redox resins (Section 16-4); and cellulose fibers treated to give ion-exchange sites.

Polystyrene ion-exchange resins are stable to about 150°C. They are resistant to strong acids and bases but not to strong oxidants or hydrogen peroxide. The ion-exchange capacity of dry strong-acid cation resins in the hydrogen form is typically of the order of 5 meq/g, and of dry strong-base anion resins in the chloride form, 2 meq/g. In water the resins swell to an amount depending on the extent of cross-linking.[7] Thus in 0.1 M hydrochloric acid the weight of water absorbed per gram of resin is 3.1, 1.5, 0.9, 0.6, and 0.4 g for 2, 5, 10, 15, and 25% divinylbenzene in the resin.

Since electrical neutrality must be maintained in ion-exchange chromatography, the distribution of ions between the mobile and stationary phases is complicated by the need for counterions. The use of partition or distribution ratios in the usual way is therefore not entirely appropriate; instead, a selectivity coefficient is used to denote the equilibrium distribution of ions. Thus for (25-1) and (25-2) the law of mass action gives[15]

$$\frac{a_A a_{BR}}{a_B a_{AR}} = K_{eq} \qquad (25\text{-}3)$$

where a_A and a_B represent the activities of two ions of the same charge in solution and a_{AR} and a_{BR} the activities in the immobilized state with the resin. If the activities are replaced by the products of the concentrations and activity coefficients, we have

$$\frac{[A]\gamma_A[BR]\gamma_{BR}}{[B]\gamma_B[AR]\gamma_{AR}} = K_{eq} \qquad (25\text{-}4)$$

or

$$\frac{[BR]}{[AR]} = K_{eq}\frac{\gamma_{AR}\gamma_B}{\gamma_A\gamma_{BR}}\frac{[B]}{[A]} = E_{B,A}\frac{[B]}{[A]} \qquad (25\text{-}5)$$

where $E_{B,A}$ is defined as the *selectivity coefficient* of the resin for B with respect to A. Usually the coefficient is written as a number larger than unity and depends on the counterion. Several listings of equilibrium data for ion-exchange reactions have been given.[7,15-21] For $E_{B,A}$ greater than unity the resin holds B more strongly than A. At a fixed pH and with the concentrations of complexing and other agents in solution also fixed,

$$[B] = \alpha_B C_B \qquad (25\text{-}6)$$

where α_B is the fraction of free ions and C_B the total analytical concentration. If C_B is small, as it normally is during analytical elutions, there is little variation in the activity coefficients in (25-5), the ratio $[BR]/[B]$, and the partition ratio (Section 24-1) $[BR]/C_B$. Hence, since $[AR]$ is approximately constant, as long as the two ions A and B in Equations (25-1) and (25-2) are of the same charge, the ratio between the amounts of an ion in the resin and in solution is independent of concentration. On the other hand, if an ion-exchange reaction involves ions of different charges, it follows that the resin phase becomes increasingly selective toward ions of higher charge as the dilution is increased.[17] This selectivity is called *electroselectivity*. If one ionic species is present in low concentration and another in large excess, the equilibrium for the dilute species can be represented by a constant partition ratio without regard to the charges of the ions.[22,23] Consequently, the concentration of a dilute species in the resin phase is proportional to its concentration in the solution phase, a condition analogous to linear isotherms in gas chromatography.

The activity coefficients in Equation (25-5), and hence the selectivity coefficient and partition ratio, depend on the concentration of electrolyte in solution and taken up by the resin. The fraction α_B for metal ions depends on factors such as hydrolysis, complexation, and masking (Section 23-5).

The rate of exchange of ions with a resin is usually regulated[7] by the diffusion rate through the resin or the thin film of solution surrounding the resin bead. Weak-acid resins in the hydrogen form often have a slow rate of exchange.[24] For ion-exchange particles of the order of 0.05 mm in diameter, flow rates of about 1 ml/cm^2-min are commonly used.

As particle sizes become small, the pressures required for maintenance of reasonable flow rate become high and usually a balance must be struck between flow rate and diffusion rate limitations. Pellicular ion-exchange resins are ones in which the exchange sites are limited to a thin surface layer. With beads of small diameter they are useful in high-pressure, high-speed separations of small samples.

25-2 ANALYTICAL APPLICATIONS OF ION EXCHANGE

Conversion of salt to acid If a salt solution is passed through a cation-exchange column in the hydrogen form, a quantitative reaction occurs:[25]

$$M^+ + A^- + H^+R^- \rightleftharpoons M^+R^- + H^+ + A^- \qquad (25\text{-}7)$$

The acid HA formed can be titrated with standard base. The column is regenerated by washing with 3 to 4 M hydrochloric acid and then with water. This procedure is applicable generally to the determination of total salt concentration, provided the acid HA can be washed from the column and titrated. If only a single anion is present, such as sulfate, perchlorate, acetate, or halide, the anion concentration can be determined by titration of the acid formed. The method affords a simple standardization of solutions of salts that, owing to hygroscopic behavior or an uncertain degree of hydration, are not conveniently prepared by direct weight. Another application is the preparation of standard solutions of acids by weighing an appropriate salt (HNO_3 from $AgNO_3$, HCl from NaCl, and so forth) (Section 6-1).

Sources of error have been reviewed by Rieman and Walton.[7] For example, correct results are not obtained with: acids whose ionization constants are much smaller than 10^{-5}; carbonates, bicarbonates, or bisulfites (owing to gaseous products); or salts that are adsorbed on the resin. Some organic compounds, especially ones containing aromatic rings or having high molecular weight, may be strongly adsorbed. Some organic acids that tend to be retained on the column can be eluted with alcohol. Phosphoric acid requires a considerable amount of washing, and the eluent must be concentrated by evaporation before titration.[25]

Wiesenberger[26] determined the saponification equivalent of esters by a method in which the ester is first saponified with an excess of base. The excess is then neutralized on passage through a bed of acidic resin, and the organic acid formed during saponification is titrated with standard base. In this way a relatively large excess of base can be used to ensure quantitative saponification. The determination is a direct titration; therefore only one standard solution is required, rather than the two that are customary in indirect determination.

The conversion of salt to hydroxide by use of a strongly basic anion-exchange resin, followed by titration with standard acid, is usually not so appropriate as the conversion of salt to acid because the resin-regeneration procedure is less convenient. Hydrolysis of the resin to give amines may occur, as well as absorption of carbon dioxide. The anion-exchange method can be used, for example, in the determination of alkali metal phosphates or sulfites.

Removal of interfering ions prior to analysis One application of ion exchange is in the conversion of alkali metal sulfates or phosphates to the chlorides. Thus cation exchange can be used for separating potassium from sulfate[27] or alkali metals from phosphate.[28] The procedure involves washing out the sulfuric or phosphoric acid while the alkali metals are retained by the hydrogen ion column, and then eluting the alkali metals from the column with hydrochloric acid. Similar separations can be effected with strongly basic anion-exchange resins,[29] a method that requires no elution to recover the alkali metal ion.

Another example is in the separation of sulfate or phosphate from various cations.[30] Samuelson devised a method for sulfur in pyrites that is based on the retention of iron(III) on a cation-exchange resin. The sulfuric acid that passes through the column can be determined by the usual gravimetric method as barium sulfate. In a similar way, phosphate in phosphate rock can be determined by retention of calcium, magnesium, iron, and aluminum on a cation-exchange resin followed by determination of the phosphate as magnesium pyrophosphate. The metal ions can be eluted from the column with 4 M hydrochloric acid.

Separation of metal ions   Metal ions can be separated as their hydrated cations on cation-exchange resin. Usually they can be separated more satisfactorily as complexed anions on an anion-exchange resin because the partition ratio can be controlled by variation of the anion concentration. Figure 25-1 shows how the partition ratio with cation- and anion-exchange resins varies with hydrochloric acid concentration for many of the elements in the periodic table.[31] The partition ratio of a metal is at a maximum when the chloro complex (possibly as the chloro acid) has a charge of 0. Thus[32] several of the transition metals can be separated by anion-exchange chromatography through variation of the concentration of HCl in the eluent. For example, a mixture of nickel(II), manganese(II), cobalt(II), copper(II), iron(III), and zinc(II) can be separated by elution first with 12 M HCl to remove nickel, then with 6 M HCl to elute manganese, 4 M HCl to elute cobalt, 2.5 M HCl to elute copper, 0.5 M HCl to elute iron(III), and finally 0.005 M HCl to elute zinc. Wilkins[33] developed separation procedures for the analysis of metals in high-temperature alloys using anion-exchange resin in the fluoride form and 1.2 M HF as eluent.

The alkali metals and to some extent the alkaline earth metals can be separated from each other through the use of cation-exchange resin and elution with HCl. Sodium ions are less strongly retained than either potassium ions or heavier alkali metal ions.

Calcium in lithium salts has been determined[34] by EDTA titration after a preliminary separation of the calcium from lithium using a chelating resin.

Separations based on complex formation   The relative retentions for closely similar metal ions often may be enhanced by use of a suitable reagent (such as EDTA) to take advantage of differences in formation constants of the complexes. A classic example is the separation of the lanthanides with a buffered citrate solution as the eluting agent. Especially effective separations are possible when one metal is converted to an anionic complex while another is present as a cation. Teicher and

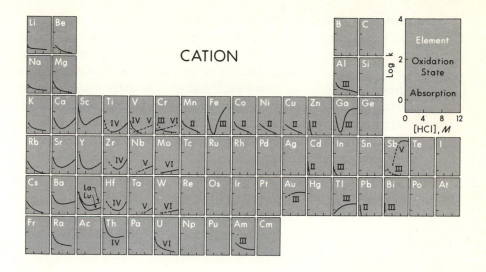

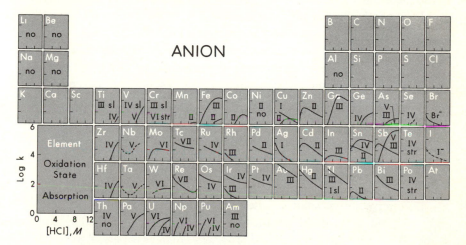

FIGURE 25-1 Variation in partition ratio with HCl concentration using Dowex strong-acid or strong-base cation- or anion-exchange resins. *No* denotes no absorption in the concentration range 0.1 to 12 *M* HCl; *sl*, slight absorption in 12 *M* HCl (*k* in the range 0.3 to 1); *str*, strong absorption (*k* large). Partition ratio *k* is given here as ratio of amount per kilogram of resin to amount per liter of solution. (*Adapted from Nelson and others.*[31])

Gordon[35] separated iron and aluminum by retaining the iron on an anion exchanger as a thiocyanate complex while the aluminum was being eluted. MacNevin and Crummett[36] separated palladium(II) and iridium(III) by treating the chlorides with ammonia to produce $Pd(NH_3)_4^{++}$, which was retained on a cation-exchange column while the iridium passed through as $IrCl_6^{3-}$.

Concentration of dilute electrolytes   An ion-exchange column is effective for collecting ionic substances from large volumes of dilute solutions. By elution employing a small volume of solution, a considerable concentration effect can be achieved. As examples may be cited the concentration of cations and anions in natural waters,[37-39] beryllium from bones,[40] copper from milk,[41] and silver in atmospheric precipitation.[42]

Ion-exchange beads have been proposed[43] as chemical microstandards since they can retain small measurable quantities (10^{-12} g) of ions such as sodium, potassium, calcium, and uranium. It is highly desirable that the beads be uniform in size and homogeneous in the trace constituent.

25-3   ADSORPTION CHROMATOGRAPHY

In this section we arbitrarily consider only adsorption chromatography with a liquid mobile phase; that with a gaseous mobile phase is considered with other gas-chromatographic techniques. A recent book on the subject of adsorption chromatography is one by Snyder.[44]

In adsorption chromatography in its original form a solid adsorbent such as alumina was placed in a tube, the column saturated with solvent, a small quantity of sample added to the top of the column, and solvent then passed through until colored bands of the solute components separated. The bed was then pushed out of the tube, and the bands were separated by cutting the bed. This method was limited to colored substances or to materials that could be observed in ultraviolet light or treated with suitable reagents to give colored bands.

Koschara[45] made the important contribution of continuing the elution procedure until each adsorbed substance in turn emerged at the end of the column. This method, although a great improvement, still suffered from a lack of a convenient technique for detecting colorless solutes. Tiselius[46] introduced the continuous measurement of refractive index for the detection of eluted materials and thereby enormously extended the usefulness of the method. In a series of later papers, Tiselius and coworkers described refinements involving interferometric methods for detecting small changes in refractive index and automatically registering the changes as a function of volume of solvent.[47]

A large number of stationary phases have been applied, the most widely used being silica, $SiO_2 \cdot XH_2O$, a polar adsorbent. Chromatographic silica typically has a surface area of about 500 m^2/g, a pore volume of about 0.4 ml/g, and an average pore diameter[44] of about 10 nm. The surface area may be decreased by aging in steam. Surface adsorption sites are composed of hydroxyl groups that are attached to silicon atoms and interact with adsorbing molecules through hydrogen bonding.[44] The maximum concentration of these surface hydroxyl groups is obtained by heating to about 200°C, where most of the adsorbed water is removed. At higher temper-

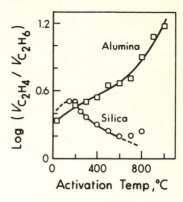

FIGURE 25-2 Variation of surface activity as a function of temperature, as indicated by the relative retentions of ethylene and ethane. (*From Snyder.*[44])

atures, hydroxyl groups interact with each other to liberate water, and the surface activity decreases (Figure 25-2). Distinction is made between several types of hydroxyl groups: *free*, in which the hydrogen ion is bonded only to the oxygen atom; *bound*, in which the hydrogen atom is bonded to two oxygen atoms; and *reactive*, in which the hydrogen atom is bonded to an oxygen atom, which in turn is hydrogen-bonded to a second hydrogen atom. Reactive sites are presumed to be the best for adsorption of polyfunctional solutes (the most common type), whereas monofunctional solutes (that is, those with one strong adsorbing group) adsorb equally well on reactive and free sites.

Next in importance to silica is alumina, an adsorbent having basic sites.[44] A problem with alumina is that base-catalyzed reactions may occur with many solutes, particularly when the alumina has been activated at a high temperature. Strong-acid solutes may be chemisorbed (energy of adsorption comparable to that in bond formation in a chemical reaction), esters and anhydrides may saponify, aldehydes and ketones may form condensation products, double bonds may change position, and hydrogen halide may be lost. Acetone in particular may react with alumina and therefore should not be used as a solvent with this adsorbent. Most commonly used is low-temperature (200°C), impure γ-alumina with a surface area of 100 to 200 m²/g. It is usually assumed that oxide ions ($O^=$) are present in the surface layer and aluminum ions (Al^{3+}) in the layer below. Water may be present as either hydroxyl groups or adsorbed water. Heating to about 300°C removes most of the adsorbed water and brings about a reaction to form hydroxyl groups, which in turn can be removed by heating to about 800°C. Surface activity increases with temperature, as indicated by Figure 25-2. At high temperatures (1100°C) α-alumina is formed, which has low surface area and is chromatographically inactive.

Charcoal historically has been an important adsorbent for a wide range of solutes. Surface areas are often in the range of 1500 m²/g. Graphitized charcoal prepared at high temperature is relatively nonpolar, whereas oxidized charcoal prepared by low-temperature oxidation is relatively polar.

Other adsorbents not discussed in detail here are magnesia, which somewhat resembles alumina; magnesium silicates (Florisil) formed by simultaneous precipitation of silica and magnesia, which have a surface that is usually highly acidic; bentones, derivatives of montmorillonite; polyamides; and diatomaceous earths. Group-selective, or group-specific, adsorbents also may be used. For example, adsorbents that specifically adsorb olefin compounds can be prepared by impregnating silica with silver nitrate. Weak compound formation occurs between the double bonds and silver.

In adsorption chromatography the choice of mobile phase (solvent) is probably more important than that of the stationary phase, because the mobile phase plays an active role in the adsorption of solutes and hence in the partition ratios. Frequently, silica is used as the stationary phase, and the solvent is chosen to meet the specific needs of the sample through balance of the interactions of the solute, mobile phase, and stationary phase to permit effective separations in a reasonable time. Solvents are described as weak or strong: *weak* solvents favor large, and *strong* solvents favor small partition ratios. Solvent molecules themselves compete for adsorption sites, and one that is strongly adsorbed is a strong solvent. When the interaction between solvent and solute molecules in solution predominates over other interactions, the solvent is stronger when there is increased tendency for the solute to dissolve in the solvent. When the interaction between adsorbed solvent and solute predominates, the solvent strength may be either increased or decreased, depending on whether the complex is more or less strongly adsorbed than the solute itself.

Tables of relative solvent strengths for various adsorbents have been developed. Table 25-1 lists relative solvent strength in terms of a solvent-strength parameter for several solvents with four adsorbents,[48] defining the adsorption energy from pentane solution to be zero. Solvent strength increases generally downward in the table.

Table 25-1   RELATIVE SOLVENT STRENGTHS
             WITH FOUR ADSORBENTS[48]

Solvent	Relative solvent strength			
	SiO_2	Florisil	MgO	Al_2O_3
Pentane	0.00	0.00	0.00	0.00
Carbon tetrachloride	0.11	0.04	0.10	0.18
Benzene	0.25	0.17	0.22	0.32
Chloroform	0.26	0.19	0.26	0.40
Methylene chloride	0.32	0.23	0.26	0.42
Methyl acetate	. . .	. . .	0.28	0.60
Ethyl acetate	0.38	. . .	. . .	0.58
Acetone	0.47	. . .	. . .	0.56
Dioxane	0.49	. . .	. . .	0.56
Acetonitrile	0.50	. . .	. . .	0.65
Ethanol	. . .	. . .	. . .	0.88
Ethylene glycol	. . .	. . .	. . .	1.11

Although much of the work is highly empirical, a number of rules have been formulated to aid in the selection of suitable adsorbents and eluents and in the prediction of the order of appearance of a related series of compounds on an adsorption column. Lederer[49] observed that the tendency toward adsorption in a related series of compounds increases with increasing numbers of double bonds and hydroxyl groups. A rough classification of organic compounds in decreasing order of adsorption tendency is: acids > alcohols, aldehydes, ketones > esters > unsaturated hydrocarbons > saturated hydrocarbons. Strain[50'] emphasized the importance of the nature of the adsorbent, the composition of the solvent, and the temperature in determining the relative order of adsorption zones. In general, in this field several descriptive terms have come into common use, with the result that meanings are not always entirely clear. For example, Martire and Pollara[51] recommended that the word polar be used with care.

LeRosen and coworkers[52] carried out calculations of retention behavior in terms of bond energies of interaction among the adsorbent, the solute, and the eluent. For instance, they considered the tendencies toward donation or acceptance of electron pairs and the hydrogen donor and acceptor behavior in hydrogen bond formation. The various eluting solvents were compared, with petroleum ether as standard, and the various adsorbents evaluated by experiment. Then the behavior of a solute could be predicted by a summation of the effects of its functional groups, together with a weighting factor taking into account the sum of the molecular weights of all the side chains.

Solvents should be purified sufficiently so that contaminants do not collect on the stationary phase and change retention characteristics or even order of elution. In other instances, contaminants must be removed to meet the needs of the detection system.

Snyder[44] reviewed the fundamental contributions leading to adsorption. He concluded that normally no more than monolayer adsorption is important in chromatography and that the probability of finding a solute adsorbed increases with increasing surface of the stationary phase. A major contribution to the free energy of adsorption is made by dispersion forces, particularly for nonpolar solutes with nonpolar adsorbents. A basic problem still present in adsorption chromatography is the nonreproducibility of successive batches of adsorbent, owing to differences both in preparation and in treatment after preparation. Exposure to traces of impurities may block some adsorption sites. The nonuniformity of the adsorption sites, even for a single solute, ordinarily leads to tailing during elution.

In the separation of multicomponent mixtures a single liquid may not be a suitable mobile phase for separating all components. The technique either of stepwise change in strength of the eluate or of gradient elution (Section 24-7) is employed. In the stepwise technique a succession of stronger and stronger solvents is used to elute the more strongly adsorbed solutes. In theoretical studies, Drake[53] concluded that the sharpening of elution bands must be accompanied by a relative pushing together of the bands, so that an improvement in resolution may or may not result. The degree of separation is affected by the shape of the gradient, the motion of the gradient through the column, the variation of partition ratio with concentration of eluent, and the shape of the adsorption isotherm.

In Section 24-2 it is shown that the flow rate at the optimum in the curve of plate height against velocity of mobile phase strongly depends on particle diameter and, because of small diffusion coefficients in liquids, the optimum velocity occurs at low flow rates. Recent trends have been toward the use of smaller particles (10 to 15 μm) and of longer columns (0.1 to 0.5 m) of small diameter (0.1 to 1.0 cm). Small-diameter packings allow considerably increased flow rates, which result in short separation times, without loss of efficiency [fifth term, Equation (24-14)]. With columns of moderate length, high inlet pressures (20 to 300 atm) are required for operation; rapid progress has been made in the development of instrumentation, techniques, and detection[54,55] systems for such high-resolution, high-pressure, high-speed chromatography. Theoretical-plate numbers in the range 1000 to 10,000 are obtainable with separations taking from a few minutes to an hour—a speed and efficiency comparable to that in gas chromatography. Plate heights of a fraction of a millimeter have been obtained with 20- to 30-μm and 5- to 10-μm silica gel at flow rates of about 0.1 cm/s[56] and 1.2 cm/s.[57] In high-pressure liquid chromatography the support for the stationary phase should be sufficiently rigid that it does not change dimensions significantly under high pressure. Porous glass beads meet this requirement. A collection of articles on modern liquid chromatography has been compiled.[58]

25-4   LIQUID-LIQUID CHROMATOGRAPHY

The technique and instrumentation for liquid-liquid chromatography are much the same as for adsorption chromatography; the main distinction lies in the nature of the stationary phase. In the original development of the method, Martin and Synge[59] effected a separation of acetylated amino acids by using silica gel as support for an aqueous stationary phase. The mobile phase was a mixture of chloroform, butanol, and water. The method was used for analysis of the amino acids of various proteins.[60]

The right combination of the two immiscible liquid phases, mobile and stationary, can lead to highly selective separations at ordinary temperatures for both volatile and nonvolatile solutes. An important characteristic is that the partition ratio ideally is independent of concentration; the elution bands are therefore more symmetrical and less subject to tailing than those observed in adsorption chromatography. Consequently, better resolution is usually possible. The principal problem is that of stabilizing the stationary phase; the stationary and mobile phases are not completely insoluble in each other, even if one is aqueous and the other a hydrocarbon.

As a result of solubility effects, stationary liquid phase can be lost from a column, particularly if lightly loaded, after a period of use. Stabilization can be achieved through presaturation of the mobile phase with the stationary phase. This requires both careful control of the temperature, with perhaps several hours allowed for equilibration before use of the mobile phase, and also the use of a small precolumn for removal or addition of a slight amount of stationary phase for the final adjustment of saturation. Stable stationary phases have been prepared for use in gas chromatography through chemical bonding to the support of long-chain silanizing reagents.[61,62] Similarly,[63,64] stable organic "liquid" phases have been prepared with up to 25% organic groupings, bonded to siliceous supports, that are not eluted with organic eluents. Supports[65] having a range of polarities and involving bonding of oxydipropionitrile, Carbowax 400, or n-octane to porous glass have been prepared. Reaction of the bonded phase with water, resulting in hydrolysis or loss of stationary phase, is a danger that should be avoided by the use of a dry mobile phase.

Support materials themselves are undergoing significant improvement. For instance, glass beads with porosity only in a thin surface layer[66] make it possible to have a minimum of band broadening from resistance to mass transfer in the stationary phase, a rigid support for applications involving high inlet pressure, and lower resistance to flow of the mobile liquid. Higher flow rates also are possible without inordinate reduction in the plate number.[67] Stationary phases, chemically bonded to glass beads of controlled surface porosity, may be present at a level of about 1% or less, and therefore the dead-space volume may be relatively large. In this situation lower sample capacity and a possible loss of resolution due to the partition-ratio factor of Equation (24-35) need to be kept in mind.

The mobile liquid phase is chosen on the basis of its two main effects—on band broadening and on the partition ratio of the solute. Since band broadening depends in part on the diffusion coefficient in the mobile liquid phase, with high-speed liquid chromatography the coefficient should be large. Hence, liquids of low viscosity are desirable; Snyder and Saunders[68] recommended the use of liquids with viscosities less than 0.004 g/s-cm (0.4 cP) at the temperature of the column. Some liquids with suitably low viscosities at 25°C are n-hexane, 0.0031; diethyl ether, 0.0023; methyl acetate, 0.0037; acetonitrile, 0.0037; carbon disulfide, 0.0037; diethylamine, 0.0038; benzene, 0.0065; methanol, 0.0060; acetic acid, 0.0126; chloroform, 0.0057; pyridine, 0.0094. The factor governing the partition ratio of the solute between the stationary and mobile phases is the resultant of the interactions of the mobile liquid phase with the stationary phase and the solute. The selectivity depends in part on the polarities of these three materials.

The types of compounds that can be separated by liquid-liquid chromatography include fatty acids, amino acids, organometallics, chelates, alcohols, amines, hydrocarbons, pesticides, herbicides, steroids, hormones, alkaloids, and antibiotics. A number of such separations have been illustrated by Schmit.[69]

25-5 GEL AND EXCLUSION CHROMATOGRAPHY

Several books and a recent review have been written on this subject.[70-73] This type of chromatography has two unusual characteristics: (*1*) both the mobile and stationary phases are of the same composition and (*2*) size, rather than specific chemical characteristics of the solute, is almost the sole basis for separations. Solutes distribute themselves between a mobile phase that is external to a porous packing within which is a stationary phase of the same composition. Solute molecules too large to enter the pores of the packing are eluted without holdup with a retention volume equal to the dead-space volume and to the volume of the mobile phase. Smaller molecules penetrate the stationary solvent in the porous packing and are eluted with a larger retention volume. A unique feature, then, is that the largest molecules have the smallest partition ratios and the smallest molecules have the maximum partition ratios. Two main fractions are obtained, one larger and one smaller than the pore size. Over a restricted range of sizes, approximately equal to the pore dimensions, more selective separations are possible.

In the absence of adsorption effects the maximum retention volume is V_m/V_s, where V_m is volume of mobile phase and V_s is volume of stationary fluid. The maximum value of the partition ratio is then V_s/V_m. With typical porous packings the values of V_s and V_m are comparable, and therefore the maximum partition ratios are of the order of unity. Thus, there is the restriction of having to operate within a limited range of values of partition ratios, between zero for the largest molecules and approximately unity for the smallest. In terms of Equation (24-35) and the middle term involving partition ratio, resolution is unfavorable compared with other chromatographic columns in which larger partition ratios can be used. As a result, only a few substances can be separated on a column relative to other types of chromatographic columns with the same number of theoretical plates.

Over much of the useful molecular-weight range, the relation between retention volume and the logarithm of the molecular weight has been found empirically to be linear.

An important practical problem is the determination of the useful molecular-weight range corresponding to a given porous packing. Some examples of proteins that have been used as standards for molecular weights[70] are: cytochrome C (molecular weight 13,000), myoglobin (17,800), trypsin (24,000), pepsin (35,500), egg albumin (45,000), serum albumin monomer (67,000), serum albumin dimer (134,000), catalase (230,000). Polystyrene fractions have been prepared as standards with average molecular weights ranging from 900 to 2,145,000.

Even when the stationary phase is of the same composition as the mobile phase, the porous support for the immobilized fluid is customarily referred to as "stationary phase." Porath and Flodin[74] used cross-linked dextrans as a molecular sieve for carrying out separations, and this material has been unusually effective. It is available as the gel Sephadex,[75] a cross-linked copolymer bead of epichlorohydrin [$CH_2CH(O)CH_2Cl$], and dextran, a polysaccharide consisting of glucose with mostly $\alpha,1,6$-glucosidic linkages. The amount of swelling varies with the solvent, the amount of cross-linking, and the degree and type of hydroxyl-group substitution. In water, the usual solvent, Sephadex has been used for separating globular proteins in the molecular-weight range of 100 to 500,000, polysaccharides in the range of 100 to 200,000, as well as[76] lower-molecular-weight lipids, sterols, steroids, and trimethylsilyl ethers of sugars.

Cross-linked polystyrene beads (Section 25-1) are most commonly used in conjunction with organic solvents for the separation and characterization of polymers. Like the rigid gel Styragel,[77] it is available with pore sizes that have exclusion limits for molecules ranging from 4 to 10^6 nm (expanded chain length), corresponding to molecular weights from 200 to 5×10^7.

The most rigid type of porous support consists of porous glass beads[78,79] made as spheres (Porasil)[77] of uniform size and with pore sizes ranging from 8 to 200 nm. For dextrans in distilled water, porous glass with a pore diameter of 7.5 nm has an exclusion limit of 28,000 and an operating range of 300 to 28,000; with a pore diameter of 24 nm, the exclusion limit is 95,000 and the operating range is 1150 to 95,000; and with a pore diameter of 200 nm, the exclusion limit is 1,200,000 and the range[80] is 120,000 to 1,200,000. Porous glass spheres can be packed into columns with high uniformity and with low resistance to the flow of mobile fluid. Since porous glass has a large surface area, adsorption may cause tailing of peaks. Deactivation by treatment of the surface with hexamethyldisilazane often is recommended. Other porous supports include agarose (the nonionic part of agar), cross-linked polymethacrylate, and polyacrylamide.

Retention behavior has been compiled[81] for organophosphorus pesticides, insecticides, steroids, gibberellins, pigments, esters, purines, sugars, monomers and oligomers in nylon, pyrimidines, phenols, aromatic acids, alcohols, alkaloids, amino acids, carboxylic acids, resins, carbonyls, amides, food preservatives, organic halides, iodotyrosines, and triglycerides. Other applications[68,69] include separation of proteins, enzymes, nucleic acids, carbohydrates, peptides, lipids, humic acid, crude oil, and polymers such as polyethylene, polybutadiene, and cellulose acetate. Gel chromatography can be used to desalt solutions, to separate lithium from a salt brine,[82] and also to remove low-molecular-weight compounds from solutions containing high-molecular-weight molecules.

Molecular sieve zeolites[83,84] constitute a class of stationary phase that combines exclusion with specific adsorption properties. These materials, which are crystalline aluminum silicates (commonly sodium or calcium aluminum silicates), have rigid, highly uniform three-dimensional porous structures containing up to 0.5 ml/g of free pore volume, resulting when water of crystallization is removed by heating. Although numerous natural zeolites are known,[83] most practical work is done with the synthetic zeolites A, X, and Y. Zeolite NaA or 4A has a pore size of about 0.4 nm; zeolite CaA or 5A, a pore size of about 0.5 nm; and zeolites X and Y (usually used in the sodium form), a pore size of about 0.9 nm. Molecular-sieve effects are most commonly found with the forms of zeolite A; 5A will adsorb straight-chain but not branched-chain hydrocarbons. For example, butane can be separated from isobutane because isobutane, being nonlinear, cannot enter the small pores.

25-6 THIN-LAYER AND PAPER CHROMATOGRAPHY

The history, theory, and applications of paper and thin-layer techniques are described in several books.[85-89] The first report on the subject of thin-layer chromatography was by Izmailov and Shraiber,[90] who applied the technique to the separation of pharmaceutical tinctures.

In thin-layer chromatography a stationary phase such as silica, instead of being packed into a column, is spread as a thin layer (up to 1 or 2 mm) on a backing of glass or flexible plastic support (typically 10×20 cm). In paper chromatography a sheet of paper acts as either the inert support or the stationary phase. Since paper may be made from short filaments of glass fiber and thin-layer plates from short fibers of a cellulose slurry, it is clear that the borderline between paper and thin-layer chromatography is difficult to define. Paper and thin-layer are sometimes called planar or open-column chromatography. They have the advantages of low cost, simplicity, speed, and wide range of applicability in terms of both types and amount of sample. Thin-layer- and paper-chromatographic techniques can be highly sensitive for traces of materials; with the ninhydrin reaction, amino acids can be separated and detected[91] at the 10^{-7}-g level, while fluorescent materials are detectable at the 10^{-9}-g level. Detection of solutes normally is carried out on the column rather than after elution; this necessary practice makes precise quantitative measurements difficult. Since many applications are essentially qualitative or semiquantitative rather than quantitative, our consideration of this topic is abbreviated.

A solution of a sample is placed as a spot or a line a centimeter or two from one end of the paper or thin-layer plate, and the solvent allowed to evaporate. Then the end of the paper or plate is dipped into a suitable developing solvent (mobile phase) in a closed tank saturated with the solvent vapor. Through capillary action, the solvent rises past the sample spots and continues upward. Individual solutes are carried forward at different rates until the solvent reaches at most the upper edge. It is inappropriate here to speak of retention volume—instead the amount of movement of a band is denoted by its R_f value, the ratio of the distance traveled by the center of the band to the distance traveled by the mobile phase.

The stationary and mobile phases used in thin-layer chromatography are the same as those used in liquid chromatography generally. Thus adsorbent solids, ion-exchange resins, porous glass beads, and supported liquids can act as stationary phase. Most commonly the stationary phase is either silica or powdered cellulose. Activated silica, when exposed to air, picks up water and so loses much of its activity. Although such deactivated silica may be suitable for polar solutes that tend to sorb strongly, in most cases it is not the most effective chromatographic substrate. Silica can be made more active (a stronger adsorbent) by heating the thin-layer plates to 110°C. Cellulose and paper are relatively weak adsorbents and are preferred for highly polar solutes that would be irreversibly adsorbed on silica or alumina. Thin-layer plates are made by preparing a slurry and coating a plate by spreading, pouring, dipping, or spraying. Uniformity of layer thickness is important for optimum separations and reproducibility of R_f values. The most uniform layers are prepared by use of an aligning tray to hold the slurry and a spreader to dispense it in a selected thickness, typically 0.25 mm. The plates are then allowed to dry. A binder of calcium sulfate usually is added to silica slurries; it hardens on drying to give stability to the layer.

Stewart[92] and Knight[93] reviewed the fundamentals of cellulose structure and ion-exchange cellulose papers. In paper, some water is integrally associated, chemically and physically, with the cellulose fibers. Stewart concluded that in paper chromatography the stationary phase is heterogeneous in a cross section involving the passages between crystallites and the glucosidic chains and that it resembles a solution of polyols. The stationary phase in paper chromatography then may be considered to be a cellulose-water complex, with the cellulose acting in addition as an adsorptive surface.

So-called standard papers are commercially available filter papers,† which tend to give the most reproducible results. Other papers are made for special purposes. For example, less dense papers are made for higher flow rates, and thicker papers for preparative purposes. Some are washed in acid to remove traces of metals, and others are free of lipid-soluble substances. The cellulose may be modified[93] chemically (esterification, hydroxylation) for specific purposes or impregnated with silicone oil (for reversed-phase chromatography).

Mobile phases (solvents) from eluotropic or other series (Section 25-3) are chosen to give optimum partition ratios; for single-pass development, R_f values in the range 0.2 to 0.8 are best. In paper chromatography, even though chemically or physically bound water is presumed to be present in the stationary phase, the mobile phase may be either miscible or immiscible with water. Choice of solvent can be substantially expanded and adapted to the needs of the sample by use of azeotropic mixtures as well as single-component solvents. Azeotropic mixtures behave like single solvents with respect to vaporization and do not change composition upon evaporation. Some useful azeotropic mixtures are (by weight): 4% water, 96%

† Whatman 1 and 2, Schleicher and Schull 2043B, Munktell CHR 100, Ederol 202, and Macherey-Nagel 260.

ethanol; 8% ethanol, 92% chloroform; 39% methanol, 61% benzene; 68% ethanol, 32% toluene; 87% acetone, 13% carbon tetrachloride.

Development of thin-layer or paper chromatograms can be through ascending, descending, horizontal, or radial movement of the solvent. The ascending mode tends to minimize problems of channeling of the solvent flow. *Continuous* development involves removal of the solvent as it reaches the end of the paper or thin layer. *Multiple* development involves vaporization of the solvent after the solvent front reaches the end of the sheet, followed by repetition of development in the same direction. *Stepwise* development involves development of a plate with a nonpolar solvent, followed by its evaporation and then development a second time with a more polar solvent. *Two-dimensional* development also involves two solvents used in succession; the chromatogram is first run in one direction, then dried to evaporate the first solvent, turned 90°, and run a second time with the second solvent. In *gradient elution* the solvent composition is changed as development takes place.[94] With this technique, however, much of the simplicity of thin-layer or paper chromatography is lost.

In *argentation thin-layer chromatography*, a major technique for lipid separations, application is made of the ability of silver to form complexes with lipids. Silver nitrate is incorporated in either the stationary or mobile phase, and separations are based on the type (such as *cis-trans*) and extent of unsaturation in the lipids. This type of chromatography has been reviewed by Morris.[95,96]

In *electrophoresis* a separation is carried out under an electric field, but the products are not permitted to migrate all the way to, and be discharged at, the electrodes. The rate of movement of a solute depends in part on its charge. The migration velocity is defined in terms of the centimeters per second of movement for a potential drop of 1 V/cm.

After development, location and identification of colorless solutes require either physical or chemical tests. Examination under ultraviolet light reveals most aromatic and many other compounds as darkened regions on a light background. Exposure of a chromatogram to iodine vapors reveals brown bands where iodine has been absorbed by the solute. Sulfuric acid or a mixture of sulfuric acid and dichromate is used with heating to carbonize organic solutes and give dark bands; these reagents are not suitable if the substrate is organic. Many reagents specific for given compounds or classes of compounds have been developed. For example, ninhydrin is used for α-amino acids, bromocresol green for carboxylic acids, bromothymol blue for lipids, antimony trichloride for steroids and glycosides, and anisaldehyde for sugars.

Applications of thin-layer and paper chromatography are associated with many aspects of experimentation in chemistry, biochemistry, pharmacy, medicine, and biology. Extensive tables of R_f values are summarized in the "Handbook of Chromatography."[81] They include data for alcohols, alkaloids, amines, amino acids, carboxylic acids, inorganic anions and cations, nitrogen heterocyclics, nucleic acid derivatives, aldehydes, ketones, flavonoids, peroxides, pesticides, phenols, pigments, purines, steroids, sulfur compounds, vitamins, carbohydrates, antibiotics, drugs, hydrocarbons, and lipids.

Janak[97] described the combination of thin-layer chromatography with gas chromatography as two-dimensional chromatography. Some mixtures may be subjected to a preliminary separation with thin-layer chromatography to remove undesirable materials. With others the eluted peaks from a gas-chromatographic separation are spotted on a thin-layer plate for further separation by thin-layer chromatography. This combination can be used only when a sample is sufficiently volatile for gas chromatography but not so volatile as to be lost from a thin-layer plate. It has been applied, for example, to esters of high-molecular-weight fatty acids and to trimethylsilyl ethers of steroids.

25-7 GAS CHROMATOGRAPHY

Because of the number and variety of books and monographs on the subject of gas chromatography and related topics, it is recommended that those seeking information consult one of the current compilations of the literature. Two such compilations have been made by Signeur[98] and Preston.[99]

The term *gas chromatography* describes all chromatographic methods in which the mobile phase is a gas. *Gas-liquid chromatography* describes methods in which the stationary phase is a liquid distributed on a solid support, and *gas-solid chromatography*, methods in which the stationary phase is a solid.

Although Martin and Synge,[59] as long ago as 1941, suggested specifically what is now known as gas chromatography, it remained for James and Martin[100] to make the first analytical applications in 1952. They used a column having a stationary phase of diatomite, impregnated with a silicone preparation containing 10% stearic acid, to separate a small quantity of a mixture of fatty acids by elution through the column with a relatively large amount of a carrier gas. The acids were determined initially by a manual, and later by an automatic, titration procedure that gave the total amount of acids as a function of time.

Probably no technique in analytical chemistry has been more quickly and widely adopted than gas chromatography. It is ideally suited to the separation and analysis of nonpolar volatile materials such as hydrocarbons. Factors that have contributed to this rapid development are: (*1*) the success of the method in performing quantitative separations of closely related substances, (*2*) the ease with which the operation can be made automatic, (*3*) the speed, precision, and accuracy with which quantitative determinations can be made, (*4*) the small quantities of sample required, and (*5*) the rapid development of sensitive detection devices. The principal limitation is that the method is restricted to materials that exert vapor pressures of at least 10 mm Hg at the temperature of the column (ranging up to 300°C or somewhat higher).

Several advantages accrue from the use of a gaseous as compared with a liquid mobile phase. The low viscosity of gases (several orders of magnitude less than that of liquids) means that it is possible to use long columns and at the same time high flow rates with only moderate column inlet pressures. Diffusion coefficients in a gas phase (Figure 24-5) are about 10^5 larger than those in a liquid, and therefore, in terms of band broadening, optimum flow rates are high when packings of moderate particle size are used (Figure 24-2). Separations and analyses combining high speed and moderate cost ordinarily can be performed in a matter of minutes, and sometimes in seconds. The low density of gases means that only a small mass of mobile fluid is associated with the sample constituents; this makes it convenient to monitor continuously the concentration in the carrier gas with a device such as a thermal conductivity cell or flame ionization detector.

A serious disadvantage of a gaseous mobile phase in terms of selectivity stems from its passive role (other than as a transport medium) in determining the value of the partition ratio. Unlike in liquid chromatography, interactions between mobile phase and sample and between mobile phase and stationary phase are normally insignificant. In gas chromatography, then, variation and optimization of interactions of solute with stationary phase are accomplished almost solely through the correct choice of stationary phase. Martire and Pollara[51] indicated that gas chromatography is a promising technique for studying interactions of solutes with liquid phases. The theory of solutions and thermodynamics has been applied to the optimization of choice of stationary phase by Martire and Locke.[101]

Variation of temperature plays but a minor role in the optimization of relative retention values. Desty[102] indicated that variation of the carrier gas can make small differences in relative retention values. Thus the relative retention values for benzene and 2,4-dimethylpentane are 1.009, 1.017, and 1.012 for helium, nitrogen, and carbon dioxide at an inlet pressure of 2.9 atm. Since the carrier gas can do little to optimize relative retention values for the more difficult separations in gas chromatography, the main recourse is to longer, more efficient columns.

Littlewood[103] classified stationary phases as paraffinic, dilute, concentrated, or specific, and listed many examples. He defined *dilute* stationary phases as those in which a large proportion of the molar volume is occupied by nonpolar groups, and *concentrated* stationary phases as those in which a large proportion of the molar volume is occupied by polar groups. Rohrschneider[104] developed a classification system for describing the gas-chromatographic polarity of stationary liquid phases on a 0 to 100 scale, where squalane is rated 0 and oxydipropionitrile 100. This scale is based mainly on the ability of a phase to retain nonparaffins more strongly than paraffins, and, in particular, retention ratios for butadiene and *n*-butane are used. McReynolds,[105] applying Rohrschneider's concepts, showed that many of the liquid phases in use are similar and suggested that the number could be reduced. A thermodynamic concept for the polarity of stationary phases based on the partial molar free energy of the methylene group has been proposed.[106]

Examples of stationary phases with some of their characteristics are: squalane (hexamethyltetracosane), a substance useful to a maximum temperature of about 100 to 150°C and suited to the analysis of hydrocarbons of low molecular weight; Apiezon grease and silicone oils, also suited to hydrocarbon analyses but having higher upper-temperature limits (about 200 to 250°C); dinonylphthalate, a somewhat polar stationary phase useful to about 100°C and showing good selectivity between aromatic and paraffinic hydrocarbons, with the activity coefficients of the aromatics being less than those of the paraffins; polyethylene glycol, useful to about 200°C, highly polar, and interacting strongly with polar solute molecules; silicone gum rubber SE 30, useful as a high-temperature stationary phase (to about 300°C) and relatively nonpolar; and silicone gum rubber XE 60, a moderately high-temperature (to about 250°C) material that is somewhat polar. Ethylene glycol–silver nitrate is an example of a specific stationary phase that retains unsaturates more strongly than saturates. Unfortunately, it is useful to only about 50°C. Poly-m-carborane siloxane polymers on etched or silylated glass capillaries have been used to 350°C.[107] In the form of a stationary-phase packing (Dexil 300) it can be used[108] to 450°C. Inorganic salts such as lithium or calcium chloride have been used to temperatures as high as 500°C for polyphenyl separations, and for a separation of metal halides.[109]

Hawkes and Mooney[110] studied volatility-temperature relations for several stationary phases. The history and background of literature dealing with chromatographic supports has been reviewed by Ottenstein.[111]

The solid supports for dispersal of the stationary liquid as a uniform thin film should be of moderate surface area and should be inert so that adsorption effects are absent at solid-gas interfaces. Adsorptive effects (with peak tailing) are most serious with lightly loaded columns. Sawyer and Barr[112] found lowest adsorption with supports of silanized Chromosorb W, glass beads, and Fluoropak (a fluorocarbon polymer). Silanizing the siliceous supports slightly reduces the surface area and sharply reduces the number of active adsorption sites. The support for the stationary phase can be simply the inner walls (porous or not) of a capillary. The use of open tubes of this sort means that column lengths can be large, with minimal resistance to flow of mobile gas phase. Their small sample capacity can be compensated for by the use of support-coated open tubular columns or highly sensitive detectors, such as ionization detectors.[113] The history and background of capillary columns have been reviewed by Desty.[102]

Chromosorb W is made from diatomite calcined with a small amount of sodium carbonate above 900°C. Larger agglomerates are formed during calcination, and much of the micropore structure of the diatom skeletons disappears. Urone and Parcher[114] indicated that supports are neither homogeneous nor inert and reviewed their effects on retention volumes.

Commonly used stationary phases in gas-solid chromatography are activated charcoal, silica gel, Fluorosil, and molecular sieves 5A, X, or Y (synthetic zeolites). These are useful to temperatures as high as 500°C. At high temperatures the adsorptive properties may change through loss of water by condensation of hydroxyl groups or by reactions of other oxygen-containing groups on the surface of charcoal. Copolymer beads of styrene-divinylbenzene with a surface area of about 300 m^2/g are useful at low and moderate temperatures for the separation of volatile organic compounds of low molecular weight and polar compounds such as water and ammonia. Molecular sieve 5A is especially useful for the separation of atmospheric gases from each other, for example, oxygen–nitrogen–methane–carbon monoxide mixtures.

Kiselev[115] reviewed and made fundamental studies of adsorbents in gas chromatography. He attempted to eliminate the heterogeneity of the surfaces of adsorbent solids to improve their selectivity. He classified adsorbents in terms of their specificity. Graphitized carbon black is a nonspecific adsorbent and silica and zeolite are specific adsorbents. The adsorptive properties and the tailing of peaks can be greatly modified by the addition of a small amount of liquid to a large-area solid. The maximum temperature limit is then that of the liquid.

Although the achievement of selective separations depends mostly on the proper choice of stationary phase, the control and variation of partition ratios to appropriate values is achieved primarily through the choice of the temperature of operation (Figure 24-4). For samples containing solutes with a wide range of boiling points, the technique of temperature programming (Section 24-7) serves the same purpose as gradient elution in liquid chromatography.

Interactions between solute and stationary phase have been considered in terms of the activity coefficient of the solute in the stationary phase.[116] For various homologous series of solutes, it has been shown that, if the first few members of a homologous series are excluded, the logarithm of the partition ratio is a linear function of the number of carbon atoms per molecule in the solute. This observation corresponds to a constant increment of free energy of vaporization per added methylene group. Similar regularities of contribution to log k were found for other structural groups, for example, methyl- or hydroxyl-group additions to paraffin hydrocarbons. Likewise, if the solute is held constant, and if the stationary phase is varied but kept within a particular homologous series, the partition ratio varies systematically with the carbon number of the stationary phase.

Applications   The most extensive sources of information on applications of gas chromatography are "Gas Chromatography Literature—Abstracts and Index"[99] and "Gas Chromatography Abstracts."[117] Compilations of retention data are available from several sources; two are "Gas Chromatographic Retention Data," by McReynolds[118] and "The Handbook of Chromatography."[81]

The number of compounds that can be volatilized at moderate temperatures is large; in general, all that can be volatilized without decomposition can be separated by gas-chromatographic techniques. Thus it is possible to separate many fatty acids, alcohols, aldehydes, amines, esters, ethers, halogenated hydrocarbons, hydrocarbons, ketones, phenols, sulfur compounds, metal complexes, inert gases, and even isotopes and isomers of hydrogen. Gas chromatography is used in studies of flavors and odors, pesticides, trace analysis, trace impurities, and polymers (by pyrolysis);[119] in biochemical problems; in the preparation of materials of high purity; and in automated analysis and control[120] of refinery operations.

25-8 IDENTIFICATION AND MEASUREMENT

Retention data for peaks and R_f values for bands give qualitative information about a solute, while area or intensity provides quantitative information. In this section some aspects of developments in identification and measurement in chromatography are reviewed briefly. In general, the developments have been most extensive for gas chromatography systems; textbooks on the subject usually contain substantial sections dealing with the identification problem, and several books[121-123] deal primarily with this topic. Identification of solutes in gas chromatography by retention and by response factors has been reviewed by Schomburg.[124] The correlation between retention behavior and chemical constitution of several solutes in paper chromatography systems has been reviewed by Green and McHale,[125] and a compilation on the quantitative aspects of paper and thin-layer chromatography has been published.[126]

Qualitative identification Relative retention values are used as evidence in characterization. The most highly developed system of relative retentions is that of the retention indices used in gas chromatography, due to Kovats.[127] In this system the n-paraffin hydrocarbons are used to provide a series of fixed points on a retention-index scale. Each is assigned a retention-index value of 100 times the number of carbon atoms in the hydrocarbon chain. Thus n-pentane, n-hexane, and n-heptane are assigned values of retention indices of 500, 600, and 700. Figure 24-4 shows that, in isothermal chromatography at a particular temperature, net retention volumes for consecutive homologs are equally spaced (over a moderate range) on a logarithmic scale; V' is a linear function of carbon number. Therefore, retention indices are best obtained by logarithmic interpolation:

$$I_i = 100N + 100 \left(\frac{\log V'_i - \log V'_N}{\log V'_{(N+1)} - \log V'_N} \right) \qquad (25\text{-}8)$$

where I_i is the retention index of the species i, N the carbon number of the n-paraffin eluted immediately before species i, and $N + 1$ the carbon number of the n-paraffin species eluted immediately after species i. In programmed temperature gas chromatography, retention temperatures are most appropriately displayed as a function of carbon number on a linear scale (Figure 24-10), although the relation between carbon number and retention temperature is not strictly linear. To calculate retention indices in this case, linear interpolation is appropriate:

$$I_{P,i} = 100N + 100 \left(\frac{T_{R,i} - T_{R,N}}{T_{R,(N+1)} - T_{R,N}} \right) \qquad (25\text{-}9)$$

where $I_{P,i}$ is the retention index under conditions of temperature programming, $T_{R,i}$ the retention temperature of species i, and $T_{R,N}$ the retention temperature of the n-paraffin with carbon number equal to N.

More broadly, the concentration of solutes in chromatographic effluents can be measured, and differential elution peaks recorded directly as a function of time at a constant rate of flow of mobile fluid. Under a given set of operating conditions, the retention time (often relative to some standard solute), which is proportional to the retention volume, serves as a method of qualitative identification of simple mixtures. Extensive tables of retention data are available for the several types of chromatography to aid in this type of identification. This method of comparison with known materials is adequate for well-characterized simple mixtures but not for more complex mixtures or those containing uncommon or unexpected components. A refinement of this simple procedure consists of measuring retention volumes with at least two different types of stationary phases in gas chromatography or two mobile phases in liquid chromatography. It is unlikely that the retention characteristics (activity coefficients) of two solutes with two different stationary or mobile phases will coincide. Thus the position of a peak, or the change in its position when another condition is introduced, provides more positive evidence for identity. Characterization evidence can be sought also in the regularities in retention for various homologous series. For example, in gas chromatography the linear relation between the log of the net retention volume and carbon number for the members of a homologous series can be used. Once the homologous series to which a compound belongs is established, the particular member of the series can be determined from a plot of log net retention volume against carbon number. Thus the retention characteristics of many homologs can be established by examination of only a few members of the series.

Mass, infrared, ultraviolet, and nmr spectrometry have been highly developed as aids to identification. In some gas-chromatographic techniques some of the sample corresponding to a peak must be condensed and examined; in others the effluent can be monitored continuously with an instrument such as the time-of-flight mass spectrometer, whereby mass spectra corresponding to each peak can be recorded as it is eluted. Developments in the mass-spectrometric analysis of gas chromatography effluents have been reviewed by McFadden.[128]

Another means of identification is a two-detector system. One detector should be nonspecific, such as the thermal conductivity cell in gas chromatography, or the refractive-index detector in liquid chromatography, while the other should have different and specific response characteristics, such as the electron-capture detector in gas chromatography or the ultraviolet detector in liquid chromatography. A comparison of chromatograms from the two detectors can provide information about functional groups and elemental composition. For example, Adlard, Creaser, and Matthews[129] proposed that the identification of the source of an oil spill be made through the use of two detectors selected to give the distribution of both hydro-carbons and sulfur-containing compounds in the oil. The characteristics of ionization detectors for gas chromatography have been reviewed by Karmen,[130] and nonioniza-tion detectors in gas chromatography by Winefordner and Glenn.[131]

Information about functional groups can be obtained by performing chemical tests on peaks or bands or by carrying out chemical reactions on a small precolumn at the inlet end of a column.

Quantitative measurements   Quantitative analysis is typically based on measure-ment of peak characteristics as defined by signals sent to a recorder by a detector or measurement of band characteristics as obtained by a recording densitometer. Thus, analysis does not depend on the column. Johnson[132] reviewed and compared methods for quantitative interpretation of gas-chromatographic data, including peak overlap, tailing, and unsteady base lines. For thermal conductivity detectors in gas chromatog-raphy, the signal is normally linearly proportional to the concentration of solute in the carrier gas. Under this condition the weight of the component is proportional to the peak area. Since the sensitivity of a detector varies for different substances, a proportionality constant must be determined for each substance by calibration.

Quantitative information can be extracted from peaks by either manual or automated methods. Laboratory integrators have been reviewed by Ewing.[133] Manual methods include measurements of height, measurements of the area by height times width, and perimeter methods such as planimetry and cut-and-weigh techniques; automated methods include disk and digital integration. For peaks whose width (standard deviation) is independent of sample size, quantitative measurements by way of height alone can be more precise than area measurements. To achieve high accuracy, however, it is imperative to keep constant all experimental variables that affect the retention volume or peak shape including column temperature, amount of liquid phase, volume of detector, and sample injection and size. On the other hand, provided the flow rate remains constant, the peak area is insensitive to several of these experimental variables that affect retention volume and peak height. Therefore, measurement of peak area is usually more satisfactory than measurement of peak height, because it permits variation of many experimental conditions without affecting the calibration.

The relative precision obtainable in quantitative measurements depends on the intensity and character of the base-line noise, the size and shape of the peak, the prior information, and the method of estimation.[134] In comparisons of integration

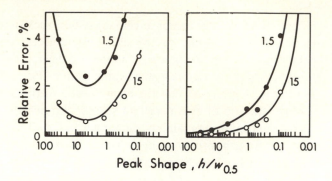

FIGURE 25-3   Relative error by height-width (left) and height only (right) measurements as a function of peak shape for noise-free peaks of 1.5-cm² and 15-cm² area. Points, experimental; lines, calculated from an analysis of the indeterminate errors associated with placement of the base line, measurement of the height from the base line, and positioning of a measuring instrument at a predetermined intermediate height and measurement of the width at that height. (*From Ball, Harris, and Habgood.*[135])

techniques, the influences especially of size and shape of peaks should be known and controlled. The areas of neither flat nor sharp peaks can be measured as precisely as those of optimum shape.

Figure 25-3 indicates how precision is affected by peak size and shape with two manual measurement techniques for noise-free peaks.[135] By height-width measurements the precision for peak areas improves with increasing area in proportion to the square root of the area, and by planimetry, in proportion to the $\frac{3}{4}$ power of the area.

Planimeter measurements are less sensitive to shape than are height-width measurements. The optimum shape for areas of noise-free peaks by height-width measurements is a height to width at half-height $h/w_{0.5}$ of about 3. The precision obtainable by height-width measurements is improved when the width is measured at a position below the half-height position; one-fourth of the height is recommended.[135]   Cut-and-weigh methods are subject to considerable error from variations in paper density and are recommended primarily for peaks of small area. For poorly resolved peaks, planimetry or cut and weigh, among the manual techniques, is indicated.

Integration of peak areas by automatic methods involves the use of integrator systems such as electromechanical integrators, voltage-to-frequency converters, and analog-to-digital converters with a digital computer. Automatic methods are less expensive than manual ones in terms of the operator time required, and digital integration with a well-adjusted instrument can give more precise measurements than manual methods.

Virtually all the information in a chromatographic peak lies in a section of the chromatogram centered at the peak mean and with a length of about eight standard deviations of the peak plus some allowance for base-line noise. The highest precision possible will be obtained with an estimator that uses all the information in a chromatogram including prior information.[134] The effects of sampling interval and number of samples needed to digitize various types of peaks have been studied.[136] Automatic and computer-assisted methods of data evaluation are developing rapidly,[137-142] involving off-line, on-line with a dedicated computer, or on-line with time-sharing systems.

REFERENCES

1 D. T. DAY: *Proc. Amer. Phil. Soc.*, **36**:112 (1897).

2 M. TSWETT: *Trav. Soc. Natur. Varsovic*, **1903**:14; *Ber. Deut. Bot. Ges.*, **24**:384 (1906).

3 R. KUHN and E. LEDERER: *Naturwissenschaften*, **19**:306 (1931); *Ber. Deut. Chem. Ges. B*, **64**:1349 (1931).

4 E. HEFTMANN (ed.): "Chromatography," 2d ed., Reinhold, New York, 1967.

5 A. B. LITTLEWOOD: "Gas Chromatography," 2d ed., Academic, New York, 1970.

6 S. G. PERRY, R. AMOS, and P. I. BREWER: "Practical Liquid Chromatography," Plenum, New York, 1972.

7 W. RIEMAN and H. F. WALTON: "Ion Exchange in Analytical Chemistry," Pergamon, New York, 1970.

8 R. KUNIN: "Ion Exchange Resins," 2d ed., Wiley, New York, 1958.

9 O. SAMUELSON: "Ion Exchange Separations in Analytical Chemistry," Wiley, New York, 1963.

10 F. HELFFERICH: "Ion Exchange," McGraw-Hill, New York, 1962.

11 K. W. PEPPER: *J. Appl. Chem.*, **1**:124 (1951).

12 R. M. WHEATON and W. C. BAUMAN: *Ind. Eng. Chem.*, **43**:1088 (1951).

13 C. B. AMPHLETT: "Inorganic Ion Exchangers," Elsevier, New York, 1964; V. PEKAREK and V. VESELY: *Talanta*, **19**:1245 (1972).

14 H. GREEN: *Talanta*, **11**:1561 (1964); **20**:139 (1973).

15 G. E. BOYD, J. SCHUBERT, and A. W. ADAMSON: *J. Amer. Chem. Soc.*, **69**:2818 (1947).

16 B. H. KETELLE and G. E. BOYD: *J. Amer. Chem. Soc.*, **69**:2800 (1947).

17 W. C. BAUMAN and J. EICHHORN: *J. Amer. Chem. Soc.*, **69**:2830 (1947).

18 T. R. E. KRESSMAN and J. A. KITCHENER: *J. Chem. Soc.*, **1949**:1190, 1201, 1208, 1211.

19 A. W. DAVIDSON and W. J. ARGERSINGER, JR.: *Ann. N.Y. Acad. Sci.*, **57**:105 (1953).

20 R. M. DIAMOND: *J. Amer. Chem. Soc.*, **77**:2978 (1955).

21 H. P. GREGOR, J. BELLE, and R. A. MARCUS: *J. Amer. Chem. Soc.*, **77**:2713 (1955).

22 E. R. TOMPKINS and S. W. MAYER: *J. Amer. Chem. Soc.*, **69**:2859 (1947).

23 SAMUELSON,[9] p. 37.

24 G. E. BOYD, A. W. ADAMSON, and L. S. MEYERS, JR.: *J. Amer. Chem. Soc.*, **69**:2836 (1947).

25 O. SAMUELSON: *Z. Anal. Chem.*, **116**:328 (1939).

26 E. WIESENBERGER: *Mikrochim. Acta*, **30**:241 (1942).

27 G. RUNNEBERG: *Sv. Kem. Tidskr.*, **57**:114 (1945).

28   G. RUNNEBERG and O. SAMUELSON: *Sv. Kem. Tidskr.*, **57**:91 (1945).

29   G. GABRIELSON and O. SAMUELSON: *Sv. Kem. Tidskr.*, **62**:221 (1950).

30   O. SAMUELSON: *Sv. Kem. Tidskr.*, **52**:115 (1940); **54**:124 (1942).

31   F. NELSON and D. C. MICHELSON: *J. Chromatogr.*, **25**:414 (1966); F. NELSON, T. MURASE, and K. A. KRAUS: *J. Chromatogr.*, **13**:503 (1964).

32   K. A. KRAUS and G. E. MOORE: *J. Amer. Chem. Soc.*, **75**:1460 (1953).

33   D. H. WILKINS: *Talanta*, **2**:355 (1959).

34   R. L. OLSEN, H. DIEHL, P. F. COLLINS, and R. B. ELLESTAD: *Talanta*, **7**:187 (1961).

35   H. TEICHER and L. GORDON: *Anal. Chem.*, **23**:930 (1951).

36   W. M. MACNEVIN and W. B. CRUMMETT: *Anal. Chim. Acta*, **10**:323 (1954).

37   F. NYDAHL: *Proc. Int. Ass. Theor. Appl. Limnol.*, **11**:276 (1951).

38   C. CALMON: *J. Amer. Water Works Assoc.*, **46**:470 (1954).

39   J. DINGMAN, JR., S. SIGGIA, C. BARTON, and K. B. HISCOCK: *Anal. Chem.*, **44**:1351 (1972).

40   T. Y. TORIBARA and R. K. SHERMAN: *Anal. Chem.*, **25**:1594 (1953).

41   H. A. CRANSTON and J. B. THOMPSON: *Ind. Eng. Chem., Anal. Ed.*, **18**:323 (1946).

42   J. A. WARBURTON and L. G. YOUNG: *Anal. Chem.*, **44**:2043 (1972).

43   D. H. FREEMAN: *Chem. Eng. News*, **46**(11):14 (1968).

44   L. R. SNYDER: "Principles of Adsorption Chromatography," Dekker, New York, 1968.

45   W. KOSCHARA: *Hoppe-Seyler's Z. Physiol. Chem.*, **239**:89 (1936).

46   A. TISELIUS: *Ark. Kemi Miner. Geol.*, **14B**(22):5p (1940).

47   S. CLAESSON: *Ark. Kemi Miner. Geol.*, **23A**(1):133p (1946).

48   SNYDER,[44] p. 197; L. R. SNYDER, *J. Chromatogr.*, **12**:488 (1963); **16**:55 (1964); **25**:274 (1966); **28**:300 (1967).

49   E. LEDERER: *Bull. Soc. Chim. Fr.*, (5) **6**:897 (1939).

50   H. H. STRAIN: *Ind. Eng. Chem., Anal. Ed.*, **18**:605 (1946).

51   D. E. MARTIRE and L. Z. POLLARA in "Advances in Chromatography," J. C. Giddings and R. A. Keller (eds.), vol. 1, p. 335, Dekker, New York, 1965.

52   A. L. LEROSEN, P. H. MONAGHAN, C. A. RIVET, and E. D. SMITH: *Anal. Chem.*, **23**:730 (1951); E. D. SMITH and A. L. LEROSEN: *Anal. Chem.*, **23**:732 (1951).

53   B. DRAKE: *Ark. Kemi*, **8**:1 (1955).

54   R. D. CONLON: *Anal. Chem.*, **41**(4):107A (1969).

55   H. VEENING: *J. Chem. Educ.*, **47**:A549, A675, A749 (1970).

56   R. AMOS: *Proc. Soc. Anal. Chem.*, **10**:6 (1973).

57   R. E. MAJORS: *Anal. Chem.*, **44**:1722 (1972).

58   J. J. KIRKLAND (ed.): "Modern Practice of Liquid Chromatography," Wiley, New York, 1971.

59   A. J. P. MARTIN and R. L. M. SYNGE: *Biochem. J.*, **35**:1358 (1941).

60   A. H. GORDON, A. J. P. MARTIN, and R. L. M. SYNGE: *Biochem. J.*, **37**:79, 86, 92, 313 (1943).

61   E. W. ABEL, F. H. POLLARD, P. C. UDEN, and G. NICKLESS: *J. Chromatogr.*, **22**:23 (1966).

62   W. HERTL and M. L. HAIR: *J. Phys. Chem.*, **75**:2181 (1971).

63   H. N. M. STEWART and S. G. PERRY: *J. Chromatogr.*, **37**:97 (1968).

64   I. HALASZ and I. SEBESTIAN: *Angew. Chem., Int. Ed. Engl.*, **8**:453 (1969).

65   PERRY, AMOS, and BREWER,[6] p. 86.

66   J. J. KIRKLAND: *Anal. Chem.*, **41**:218 (1969).

67 J. J. KIRKLAND: *J. Chromatogr. Sci.*, **7**:7, 361 (1969).

68 L. R. SNYDER and D. L. SAUNDERS: *J. Chromatogr. Sci.*, **7**:195 (1969).

69 J. A. SCHMIT in Kirkland,[58] p. 375.

70 H. DETERMANN: "Gel Chromatography," Springer-Verlag, New York, 1968.

71 K. H. ALTGELT and L. SEGAL (eds.): "Gel Permeation Chromatography," Dekker, New York, 1971.

72 J. F. JOHNSON and R. S. PORTER (eds.): "Analytical Gel Permeation Chromatography," Interscience, New York, 1968.

73 D. M. W. ANDERSON, I. C. M. DEA, and A. HENDRIE: *Talanta*, **18**:365 (1971).

74 J. PORATH and P. FLODIN: *Nature*, **183**:1657 (1959).

75 HEFTMANN,[4] p. 344.

76 J. SJOVALL, E. NYSTROM, and E. HAAHTI in "Advances in Chromatography," J. C. Giddings and R. A. Keller (eds.), vol. 6, p. 119, Dekker, New York, 1968.

77 WATERS ASSOCIATES, Framingham, Massachusetts.

78 W. HALLER: *Nature*, **206**:693 (1965).

79 A. J. DE VRIES, M. LEPAGE, R. BEAU, and C. L. GUILLEMIN: *Anal. Chem.*, **39**:935 (1967); C. L. GUILLEMIN, M. DELEUIL, S. CIRENDINI, and J. VERMONT: *Anal. Chem.*, **43**:2015 (1971).

80 D. J. HARMON in Altgelt and Segal,[71] p. 13.

81 G. ZWEIG and J. SHERMA (eds.): "Handbook of Chromatography," vol. 1, Chemical Rubber, Cleveland, 1972.

82 M. RONA and G. SCHMUCKLER: *Talanta*, **20**:237 (1973).

83 R. M. BARRER: *Endeavour*, **23**:122 (1964).

84 D. W. BRECK: *J. Chem. Educ.*, **41**:678 (1964).

85 J. G. KIRCHNER: "Thin-Layer Chromatography," Technique of Organic Chemistry, vol. XII, Interscience, New York, 1967.

86 K. RANDERATH: "Thin-Layer Chromatography," 2d ed., Academic, New York, 1968.

87 E. STAHL (ed.): "Thin-Layer Chromatography," Academic, New York, 1965.

88 I. M. HAIS and K. MACEK (eds.): "Paper Chromatography," Czechoslovak Academy of Sciences, Prague, 1963.

89 A. NIEDERWIESER and G. PATAKI (eds.), "Progress in Thin Layer Chromatography and Related Methods," 3 vols., Ann Arbor–Humphrey Science, Ann Arbor, Mich., 1970–1972.

90 N. A. IZMAILOV and M. S. SHRAIBER: *Farmatsiya* (Moscow), **3**:1 (1938), translated by N. Pelick, H. R. Bolliger, and H. K. Mangold in "Advances in Chromatography," J. C. Giddings and R. A. Keller (eds.), vol. 3, p. 85, Dekker, New York, 1966.

91 R. R. GOODALL: *Proc. Soc. Anal. Chem.*, **9**:270 (1972).

92 G. H. STEWART in "Advances in Chromatography," J. C. Giddings and R. A. Keller (eds.), vol. 1, p. 93, New York, 1965.

93 C. S. KNIGHT in "Advances in Chromatography," J. C. Giddings and R. A. Keller (eds.), vol. 4, p. 61, Dekker, New York, 1967.

94 A. NIEDERWIESER and C. C. HONEGGER in "Advances in Chromatography," J. C. Giddings and R. A. Keller (eds.), vol. 2, p. 123, Dekker, New York, 1966.

95 L. J. MORRIS: *J. Lipid Res.*, **7**:717 (1966).

96   L. J. MORRIS and B. W. NICHOLS in Niederwieser and Pataki,[89] vol. 1, p. 75, 1970.

97   J. JANAK in Niederwieser and Pataki,[89] vol. 2, p. 63, 1971.

98   A. V. SIGNEUR: "Guide to Gas Chromatography Literature," Plenum, New York, 1964.

99   "Gas Chromatography Literature—Abstracts and Index," Preston Technical Abstracts, Evanston, Illinois, monthly.

100   A. T. JAMES and A. J. P. MARTIN: *Analyst*, **77**:915 (1952); *Biochem. J.*, **50**:679 (1952).

101   D. E. MARTIRE: *Anal. Chem.*, **33**:1143 (1961); D. E. MARTIRE and D. C. LOCKE: *Anal. Chem.*, **43**:68 (1971).

102   D. H. DESTY in "Advances in Chromatography," J. C. Giddings and R. A. Keller (eds.), vol. 1, p. 199, Dekker, New York, 1965.

103   LITTLEWOOD,[5] p. 96.

104   L. ROHRSCHNEIDER in "Advances in Chromatography," J. C. Giddings and R. A. Keller (eds.), vol. 4, p. 333, Dekker, New York, 1967. See also L. J. LORENZ and L. B. ROGERS: *Anal. Chem.*, **43**:1593 (1971).

105   W. O. MCREYNOLDS: *J. Chromatogr. Sci.*, **8**:685 (1970).

106   J. NOVAK, J. RUŽIČKOVA, S. WICAR, and J. JANAK: *Anal. Chem.*, **45**:1365 (1973).

107   M. NOVOTNY, R. SEGURA, and A. ZLATKIS: *Anal. Chem.*, **44**:9 (1972).

108   R. W. FINCH: *Analabs. Res. Notes*, **10**(3):1 (1970); J. A. YANCY and T. R. LYNN: *Analabs. Res. Notes*, **14**(1):1 (1974).

109   R. S. JUVET, JR., and F. ZADO in "Advances in Chromatography," J. C. Giddings and R. A. Keller (eds.), vol. 1, p. 249, Dekker, New York, 1965.

110   S. J. HAWKES and E. F. MOONEY: *Anal. Chem.*, **36**:1473 (1964).

111   D. M. OTTENSTEIN in "Advances in Chromatography," J. C. Giddings and R. A. Keller (eds.), vol. 3, p. 137, Dekker, New York, 1966.

112   D. T. SAWYER and J. K. BARR: *Anal. Chem.*, **34**:1518 (1962).

113   A. KARMEN in "Advances in Chromatography," J. C. Giddings and R. A. Keller (eds.), vol. 2, p. 293, Dekker, New York, 1966.

114   P. URONE and J. F. PARCHER in "Advances in Chromatography," J. C. Giddings and R. A. Keller (eds.), vol. 6, p. 299, Dekker, New York, 1968.

115   A. V. KISELEV in "Advances in Chromatography," J. C. Giddings and R. A. Keller (eds.), vol. 4, p. 113, Dekker, New York, 1967.

116   G. J. PIEROTTI, C. H. DEAL, E. L. DERR, and P. E. PORTER: *J. Amer. Chem. Soc.*, **78**:2989 (1956).

117   "Gas Chromatography Abstracts," Institute of Petroleum, London, England.

118   W. O. MCREYNOLDS: "Gas Chromatographic Retention Data," Preston Technical Abstracts, Evanston, Ill., 1966.

119   S. G. PERRY in "Advances in Chromatography," J. C. Giddings and R. A. Keller (eds.), vol. 7, p. 221, Dekker, New York, 1968.

120   I. G. MCWILLIAM in "Advances in Chromatography," J. C. Giddings and R. A. Keller (eds.), vol. 7, p. 163, Dekker, New York, 1968.

121   D. A. LEATHARD and B. C. SHURLOCK: "Identification Techniques in Gas Chromatography," Wiley, New York, 1970.

122   V. G. BEREZKIN: "Analytical Reaction Gas Chromatography," Plenum, New York, 1968.

123 L. S. ETTRE and W. H. MCFADDEN (eds.): "Ancillary Techniques of Gas Chromatography," Wiley, New York, 1969.

124 G. SCHOMBURG in "Advances in Chromatography," J. C. Giddings and R. A. Keller (eds.), vol. 6, p. 211, Dekker, New York, 1968.

125 J. GREEN and D. MCHALE in "Advances in Chromatography," J. C. Giddings and R. A. Keller (eds.), vol. 2, p. 99, Dekker, New York, 1966.

126 E. J. SHELLARD (ed.): "Quantitative Paper and Thin-Layer Chromatography," Academic, New York, 1968.

127 E. KOVATS in "Advances in Chromatography," J. C. Giddings and R. A. Keller (eds.), vol. 1, p. 229, Dekker, New York, 1965.

128 W. H. MCFADDEN in "Advances in Chromatography," J. C. Giddings and R. A. Keller (eds.), vol. 4, p. 265, Dekker, New York, 1967.

129 E. R. ADLARD, L. F. CREASER, and P. H. D. MATTHEWS: *Anal. Chem.*, **44**:64 (1972).

130 A. KARMEN in "Advances in Chromatography," J. C. Giddings and R. A. Keller (eds.), vol. 2, p. 293, Dekker, New York, 1966.

131 J. D. WINEFORDNER and T. H. GLENN in "Advances in Chromatography," J. C. Giddings and R. A. Keller (eds.), vol. 5, p. 263, Dekker, New York, 1968.

132 H. W. JOHNSON, JR., in "Advances in Chromatography," J. C. Giddings and R. A. Keller (eds.), vol. 5, p. 175, Dekker, New York, 1968.

133 G. W. EWING: *J. Chem. Educ.*, **49**:A333 (1972).

134 P. C. KELLY and W. E. HARRIS: *Anal. Chem.*, **43**:1170, 1184 (1971).

135 D. L. BALL, W. E. HARRIS, and H. W. HABGOOD: *Anal. Chem.*, **40**:129 (1968); **40**:1113 (1968); *J. Gas Chromatogr.*, **5**:613 (1967).

136 P. C. KELLY and G. HORLICK: *Anal. Chem.*, **45**:518 (1973).

137 C. MERRITT, JR., J. T. WALSH, R. E. KRAMER, and D. H. ROBERTSON in "Gas Chromatography 1968," C. L. A. Harbourn (ed.), p. 338, Institute of Petroleum, London, 1969.

138 H. R. FELTON, H. A. HANCOCK, and J. L. KNUPP, JR.: *Instrum. Control. Syst.*, **40**(8):83 (1967).

139 R. D. MCCULLOUGH: *J. Gas Chromatogr.*, **5**:635 (1967).

140 F. BAUMANN, A. C. BROWN, and M. B. MITCHELL: *J. Chromatogr. Sci.*, **8**:20 (1970).

141 C. W. CHILDS, P. S. HALLMAN, and D. D. PERRIN: *Talanta*, **16**:629, 1119 (1969).

142 D. FORD and K. WEIHMAN in "Recent Advances in Gas Chromatography," I. I. Domsky and J. A. Perry (eds.), p. 377, Dekker, New York, 1971.

PROBLEMS

25-1 A 0.2567-g sample of a mixture of NaCl and KBr was passed through a Dowex 50 cation-exchange column, the effluent requiring 34.56 ml of 0.1023 *M* sodium hydroxide for its titration. What percentage of each salt was present in the mixture?

Answer NaCl 61.67%, KBr 38.33%.

25-2   A column of Sephadex of small pore size gives a retention volume of 46.7 ml for a high-molecular-weight polysaccharide, 63.2 ml for sucrose, and 75.7 ml for sodium chloride. Estimate the partition ratios for these three materials. State any assumptions made.

Answer   0.62, 0.35, 0.

25-3   The retention times under particular conditions in gas chromatography are: air, 1.72 min; *n*-heptane, 9.63 min; 2-methylheptane, 12.40 min; cycloheptane, 13.19 min; and *n*-octane, 14.21 min. Calculate the retention indices for 2-methylheptane and cycloheptane.

Answer   766, 781.

25-4   Compare gas and liquid chromatography in terms of the contribution made by the mobile phase to the separation process.

26
STATISTICS

Statistical inference is concerned with drawing conclusions from a number of observations in accordance with formalized assumptions and objective computational rules. Through the use of statistics, trends in data may be sought and tests performed to track down nonrandom sources of error. Through a statistical approach with properly designed experiments, the effects of experimental variables may be found more efficiently than through the traditional approach of holding all variables constant but one and systematically investigating each variable in turn. Quality-control charts are an important guide in evaluating day-to-day performance and in uncovering otherwise unsuspected variations or long-term trends.

Although the subject is much too large to treat in depth in a single chapter, it is hoped that a brief discussion of the more important statistical operations will encourage the reader to seek more detailed information.[1-5]

Sections 26-1, 26-2, and to some extent 26-3 constitute a brief philosophical background to the reasoned application of statistics. Those new to the subject may prefer to first become familiar with some of the formalized methods of calculation described in the remaining sections. Before drawing many interpretative conclusions, however, the first three sections should be read to help develop a philosophical basis for interpretation.

The techniques for data reduction discussed in this chapter are based largely on the assumption of a gaussian† distribution in the experimental observations. It should be recognized that such a distribution probably is not the usual one in chemical measurements; the simplicity gained, however, makes such an assumption attractive.

26-1 METHODS OF DATA REDUCTION

The data processing step in analysis has as its main function the condensation of otherwise undigested observations to furnish a final result free of irrelevant information. When but few measurements are made, the results often can be combined in an intuitively satisfactory way so that reasonable conclusions are obtained with little need for even simple statistical calculations. In Section 21-4 it is pointed out that the trend is toward use of more and more of the information potentially available from a system under study. The result is that correlation of the information for drawing conclusions about internal consistency or for making external interpretations becomes increasingly complex. Statistical evaluation of results is a tool that is becoming more essential, and experience in practical data handling is important in developing intuition; both mathematical and practical understanding are vital and complementary. Critical judgment supplemented with intuition frequently allows the experimentalist to avoid error, with failures being those of the experimentalist, not the statistics.

Although the favorable assumption of a gaussian distribution is usual practice, it should not be made blindly. Schmitt[5] states, ". . . no physical process or collection of observations has ever followed or will ever follow the (Gaussian) distribution exactly." In both lifetime and threshold types of studies the mean may have little significance. Here the Weibull distribution may be of more general importance;[6] it can closely approximate several other distributions, even though the physical basis is unlikely to be as general and valid as that for a gaussian distribution.

Parametric results are numerical, whereas nonparametric[7] results often aim for yes-or-no answers and often require no assumption about the underlying distribution of observations. The use of a median instead of the mean to estimate the location of a distribution is an example of the use of distribution-free statistics. The median is more robust than the mean in that it is more distribution-free, but the median is less efficient (requires more observations to achieve the same precision) than the mean for

---

† We prefer to avoid the more commonly used word *normal* for this type of distribution because of its unfortunate and misleading connotation.

a gaussian distribution. Thus the generality of distribution-free methods is not necessarily a recommendation for them.

With greater sophistication of methods of data acquisition, intuition can play a less important role, and a bayesian philosophical approach becomes more important.[5] In contrast to the classical approach to statistics, which is concerned with the distribution of possible measured values about a unique true value, the bayesian approach is concerned with the distribution of possible true values about the measured value at hand—a concept often greeted with hostility by traditional statisticians.

Assumptions may be made or models adopted (often by implication) about a system being measured that are not consistent with reality. The selection of the method of data reduction may be partly on the basis of the model adopted and partly on the basis of features such as computation time and simplicity. Kelly[8] classified data processing methods as direct, graphical, minmax, least squares, maximum likelihood, and bayesian. Each method has rules by which computations are made, and each produces an *estimate* (or numerical result) of reality.

A model commonly adopted is that the amount of a substance is directly proportional to the response obtained from a transducer. On the assumption that all conditions necessary for that proportionality to hold are valid, the estimate obtained is logically justified. In the *direct* method of estimation an observation is simply multiplied by a proportionality constant to obtain the estimate. Two different observations are likely to produce two different estimates, and a more complete model may enable a determinate error to be assigned to a specific cause. In the *graphical* method a straight line is drawn through several observations to obtain an estimate that is intuitively attractive. In the *minmax* method the "best" straight line is defined as one minimizing the maximum deviation. This method requires at least three points and is wasteful of observations in that the estimator uses mainly the observations with the highest deviations. In the *least-squares* method the sum of the squares of the absolute deviations is minimized; the observations are weighted according to the reciprocal of their standard deviations. The *maximum likelihood* method is more complex, but for a gaussian distribution of errors it gives the same results as the least-squares method. It is not limited to observations with a gaussian distribution. In the *bayesian* method, in which the numerical procedure is still more tedious, the essential feature, as already mentioned, is the distribution of true values around the measured result, not the distribution of measurements around the true value.[9] The choice of method involves a compromise between mathematical tedium and the detail of approach to reality that is desired.

Condensation of information is obtained through formal calculations of central values and dispersions without prejudice as to the type of distribution. *Interpretation* of the meaning of the information obtained is another matter. Here knowledge of statistical decision theory is helpful.[10] Statistical tools should be employed as aids to common sense.

26-2   ERRORS IN QUANTITATIVE ANALYSIS[11]

Two broad classes of errors may be recognized. Errors of the first class, called *determinate* or *systematic* errors, can be assigned to definite causes, at least in principle, even though a cause may not have been located. Such errors are characterized by being unidirectional. The magnitude may be constant from sample to sample, proportional to sample size, or variable in a more complex way. An example is the error caused by weighing a hygroscopic sample. This error is always positive in sign; it increases with sample size, but varies according to the time required for weighing, the humidity, and the temperature. An example of a negative systematic error is that caused by irreversible adsorption in a chromatographic column.

Errors of the second class, called *indeterminate* or *random* errors, are brought about by the effects of uncontrolled variables. Usually a relatively large number of experimental variables, each of which causes a small error, must be left uncontrolled. For example, if a correction for solubility loss of a precipitate is made to reduce the systematic error from this source, random errors due to fluctuations in temperature or volume of wash water, for instance, remain. Random errors are as likely to cause high as low results, and a small random error is more likely to occur than a large one. If observations were coarse enough, random errors would cease to exist. Every observation would give the same result, but the result would be less accurate than the average of a number of finer observations with random scatter.

The *precision* of a result is its reproducibility; the *accuracy* is its nearness to the truth. A systematic error causes a loss of accuracy; it may or may not impair the precision, depending on whether the error is constant or variable. A random error causes a lowering of precision, but with sufficient observations the scatter can (within limits) be overcome, so that the accuracy is not necessarily affected. Statistical treatment can be applied properly only to random errors. The objection might be raised that it is not known in advance whether the errors are truly random but, here again, the laws of probability can be applied to determine whether nonrandomness (trends, discontinuities, clustering, or the like) is a factor. If it is, an effort should be made to locate and correct or make allowance for the systematic causes.

Random errors may not follow a gaussian distribution, which is usually assumed for the analysis of data. Once more, statistical tests may be applied to determine whether serious deviation from a gaussian distribution exists and to interpret the data accordingly.

26-3   THE GAUSSIAN DISTRIBUTION

In statistics a finite number of observations of a given kind is considered to represent a sample of an *infinite population* or *universe* of data. The properties of the universe

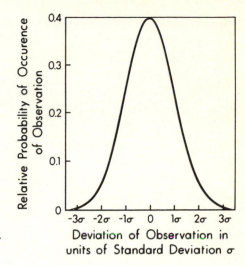

FIGURE 26-1   The gaussian-distribution or probability-density curve.

of random errors can be described by the gaussian distribution, which follows the
equation

$$p(x)\,dx = \frac{1}{\sigma\sqrt{2\pi}}\exp\left[-\frac{(x-\mu)^2}{2\sigma^2}\right]dx \qquad (26\text{-}1)$$

where $p(x)\,dx$ is the probability of a random error having a value in the interval x
to $x + dx$, μ is the average value of the entire population, and σ is the *standard
deviation* of the population.

The distribution of errors for a particular population of data is given by the
two *population parameters*† μ and σ. The *population mean* μ expresses the magnitude
of the quantity being measured; the standard deviation σ expresses the scatter and is
therefore an index of precision.

By introduction of the variable y as the deviation from the mean in units of the
standard deviation

$$y = \frac{x-\mu}{\sigma} \qquad (26\text{-}2)$$

a normalized expression of the gaussian distribution in terms of a single variable is
possible, giving the equation

$$p(x)\,dx = \frac{1}{\sqrt{2\pi}}\exp\frac{-y^2}{2}\,dy = f(y)\,dy \qquad (26\text{-}3)$$

which is plotted in Figure 26-1. Areas under the curve for the function $f(y)$ are

† Properties of a population, such as μ and σ, are known as *parameters*; analogous
properties of a finite sample are known as *statistics* (parameter estimates).

listed in mathematical tables under a title such as "cumulative normal distribution." The total area under the curve of Figure 26-1 from $-\infty$ to $+\infty$ corresponds to a total probability of unity for the entire population. Thus the area under the curve between any two values of y (in units of σ) gives the fraction of the total population having magnitudes of y between these two values. Table 26-1 lists the cumulative areas under the curve of Figure 26-1. The areas from $-y$ to $+y$ represent the probabilities for the absolute deviation $|x - \mu|$ to exceed a value $y\sigma$. Since σ is the standard deviation, $y\sigma$ [Equation (26-2)] is the deviation of a single observation from the population mean, measured in terms of the number of standard deviations.

From Table 26-1, for a gaussian distribution of errors the probability of an error greater than σ is 0.3174 (that is, $1 - 0.8413 + 0.1587$); of an error greater than 2σ, 0.0456; and of an error greater than 3σ, 0.0026. In each case, positive and negative deviations are equally probable.

The function $f(y)$ has the property of being a maximum for $y = 0$ and therefore for $x = \mu$. Thus the average value is the most probable value of the population. This probability forms the basis of one test for a gaussian distribution.

Table 26-1   CUMULATIVE AREAS UNDER THE PROBABILITY DENSITY CURVE FROM $-\infty$ TO y

y	Area	y	Area	y	Area
-4.265	0.00001	-1.282	0.10	$+1.645$	0.95
-3.719	0.0001	-1	0.1587	$+2$	0.9772
-3	0.0013	-0.524	0.30	$+2.326$	0.99
-2.576	0.005	0	0.50	$+2.576$	0.995
-2.326	0.01	$+0.524$	0.70	$+3$	0.9987
-2	0.0228	$+1$	0.8413	$+3.719$	0.9999
-1.645	0.05	$+1.282$	0.90	$+4.265$	0.99999

26-4   GAUSSIAN DISTRIBUTION APPLIED TO A FINITE SAMPLE

For a finite sample of n observations, the *sample mean* $\bar{x}$ is the arithmetic average of the n observations; $\bar{x}$ approaches the population mean μ in the limit as n approaches infinity. Correspondingly, the *sample standard deviation* s approaches the population standard deviation σ in the limit. The sample standard deviation is given by the equation

$$s = \sqrt{\frac{\sum_{i=1}^{n} (x_i - \bar{x})^2}{n - 1}} \qquad (26\text{-}4)$$

As n is increased, the quantity $n - 1$ approaches N (the entire population) relatively more and more closely.

The standard deviation of the sample is experimentally important as an estimate of the desired standard deviation of the population, which cannot actually be determined with a finite number of measurements. If a random sample is taken, the quantity s becomes a closer approximation to σ as the size of the sample is increased. Likewise, the sample mean $\bar{x}$ becomes a closer estimate of the population mean μ as the size of a random sample is increased. Although in many practical papers the standard deviation of a finite number of observations is represented by σ, strictly speaking the symbol σ should be reserved for the universe, or infinite population. For some purposes it is convenient to use the *variance*, which for the sample is s^2 or V.

From the relation

$$\sigma = \sqrt{\frac{\sum\limits_{i=1}^{N}(x_i - \mu)^2}{N}} \qquad (26\text{-}5)$$

the standard deviation σ of the population is the square root of the arithmetic mean of the squares of the deviations of the individual values x_i from the population mean μ. On the other hand, in (26-4) the quantity $n - 1$ rather than N appears when a finite sample is considered. For large values of n it is immaterial whether $n - 1$ or n is used, but for small numbers of observations—especially important in analytical chemistry—the distinction is significant, and the reason for it should be understood.

If a finite number n of observations is made with a sample mean $\bar{x}$, there are n individual deviations $x_i - \bar{x}$. The sum of the n deviations, however, is zero, and only $n - 1$ of the deviations are necessary to define the nth. Hence, only $n - 1$ independently variable deviations or *degrees of freedom* are left. We can regard the sample variance as the average of the square of the independently variable deviations. For small numbers of observations we want s to approximate σ as closely as possible, a condition that would be observed with an infinite number of observations. The sample mean $\bar{x}$ in general will not coincide with the population mean μ. It can be shown[12] that the use of $n - 1$ as a divisor, when averaged over all values of n, just compensates for the fact that the sample mean and the population mean are not identical. Because of the relative improbability of drawing large deviations in a small sample, the sample variance would otherwise underestimate the population variance.

26-5   MEASURES OF CENTRAL VALUE AND DISPERSION[1]

One of the objectives of data evaluation is to obtain values for the central value and dispersion that are as efficient as possible for a given expenditure of time and money. For a gaussian distribution the mean is the most efficient estimator for measuring the central value of a sample of data, it is also the most probable value (mode), and it is the middle value (median). When gross errors are present, the median and the probable deviation (the median, without regard to sign, of the deviations from the median) may be more efficient than the mean and the standard deviation. In general, the estimator with the smallest confidence interval (Section 26-8) is the most reliable.

Several measures of dispersion are the variance, the standard deviation, the relative standard deviation, the range, and the mean deviation. If the sampling distribution of an estimator has a mean equal to the corresponding universe parameter, the estimator is said to be an *unbiased* estimate of the parameter. The variance s^2 of a random sample is *on the average* equal to σ^2; thus the variance s^2 of a finite sample is an unbiased estimate of σ^2, although the standard deviation s is not an unbiased estimate of σ. The standard deviation is convenient because it is expressed in the same units as the measured value. The *relative standard deviation* is merely the standard deviation expressed as a fraction or percentage of the arithmetic mean. It is used mainly to show whether the relative or the absolute spread of values is constant as the values are changed. The *range* is the difference between the highest and lowest values of a sample of observations. It is of little use for large samples, but becomes increasingly useful as the number of observations decreases until for a pair of observations it, together with the mean, gives all the data. The *mean deviation* is the average deviation from the mean. It gives equal weight to large and small deviations, which are not equally probable.

The measures of dispersion partly reflect the estimator used, and an estimate of dispersion can be obtained either from the same series of observations used to obtain the central value or from a separate series. For a gaussian distribution the standard deviation and the mean are independent.

26-6   ERROR OF A COMPUTED RESULT[13]

The estimation of the error of a computed result R from the errors of the component terms or factors A, B, and C depends on whether the errors are determinate or random. The propagation of errors in computations is summarized in Table 26-2. The *absolute* determinate error ε or the variance $V = s^2$ for a random error is transmitted in addition or subtraction. (Note that the variance is additive for both a sum and a difference.) On the other hand, the *relative* determinate error ε_x/x or square of the relative standard deviation $(s_x/x)^2$ is additive in multiplication. The general case $R = f(A, B, C, \ldots)$ is valid only if A, B, $C, \ldots$ are independently variable; it is strictly true only for linear functions of A, B, $C, \ldots$, but only approximately so for other cases if the relative errors are smaller than about 20% of the mean values.[12]

Table 26-2 ERROR OF A COMPUTED RESULT

Computation	Determinate error	Random error
Addition or subtraction, $R = A + B - C$	$\varepsilon_R = \varepsilon_A + \varepsilon_B - \varepsilon_C$	$s_R{}^2 = s_A{}^2 + s_B{}^2 + s_C{}^2$
Multiplication or division, $R = AB/C$	$\dfrac{\varepsilon_R}{R} = \dfrac{\varepsilon_A}{A} + \dfrac{\varepsilon_B}{B} - \dfrac{\varepsilon_C}{C}$	$\left(\dfrac{s_R}{R}\right)^2 = \left(\dfrac{s_A}{A}\right)^2 + \left(\dfrac{s_B}{B}\right)^2 + \left(\dfrac{s_C}{C}\right)^2$
General, $R = f(A, B, C, \ldots)$	$\varepsilon_R = \dfrac{\partial R}{\partial A}\varepsilon_A + \dfrac{\partial R}{\partial B}\varepsilon_B + \cdots$	$s_R{}^2 = \left(\dfrac{\partial R}{\partial A}\right)^2 s_A{}^2 + \left(\dfrac{\partial R}{\partial B}\right)^2 s_B{}^2 + \cdots$

26-7 STANDARD ERROR OF THE MEAN

In Section 26-3 we considered the standard deviation σ, which is related to the probable error of a single observation. If a series of random samples of size n is drawn from an infinite population, the average value of the various sets of n observations will show a smaller scatter as n is increased. As n increases, each sample average approaches the population average μ in the limit, and the scatter approaches zero. It can be shown[14] that the standard error of the mean σ_m is inversely proportional to the square root of the number n, or

$$\sigma_m = \frac{\sigma}{\sqrt{n}} \qquad (26\text{-}6)$$

The precision of a measurement can be increased by increasing the number of observations. Nevertheless, because of the square root dependence expressed in (26-6), the improvement to be gained by replication has a practical limit. For example, to decrease the standard deviation by a factor of 10 requires 100 times as many observations. Systematic errors involved in a determination cannot be removed by replication. Consequently, the practical limit of useful replication is reached when the standard error of the random errors of the mean is comparable to the determinate error.

Information can be condensed into terms such as mean and standard deviation without prejudice as to the type of distribution. Kaiser,[15] however, pointed out the hazards in attempting to use such statistics for interpretive purposes (such as that the sample represents the population, the prediction of future events, the probability of error, or confidence in certain decisions).

26-8 MEASURE OF PRECISION OF A MEAN AND OF A STANDARD DEVIATION: CONFIDENCE INTERVALS

Precision of a mean If the mean $\bar{x}$ of n measurements is taken, the *population mean* μ lies within the limits

$$\mu = \bar{x} \pm 1.96 \frac{\sigma}{\sqrt{n}} \qquad \text{with 95\% confidence} \qquad (26\text{-}7)$$

$$\mu = \bar{x} \pm 2.58 \frac{\sigma}{\sqrt{n}} \qquad \text{with 99\% confidence} \qquad (26\text{-}8)$$

A difficulty is that the population standard deviation is not usually known and can only be approximated for a finite number of measurements by the sample standard deviation s, calculated from (26-4). This difficulty is overcome for gaussian distributions by use of the quantity t (sometimes known as *Student's t*),[16] defined by

$$\pm t = \frac{(\bar{x} - \mu)\sqrt{n}}{s} \qquad (26\text{-}9)$$

The quantity t takes into account the problem of a finite value for n (that is, both the possible variation of the value of $\bar{x}$ from μ and the use of s in place of σ). Values of t can be found in tables for any desired number of observations n or degrees of freedom $n - 1$ and for various desired confidence levels.

For independent measurements with a gaussian distribution the *confidence interval* of a mean may be written

$$\mu = \frac{\bar{x} \pm ts}{\sqrt{n}} \qquad (26\text{-}10)$$

For illustration, a few values of t are listed in Table 26-3 for various confidence levels

Table 26-3 VALUES OF t FOR ν DEGREES OF FREEDOM FOR VARIOUS CONFIDENCE LEVELS (TWO-TAILED VALUES)

Degrees of freedom	Confidence Level, %				
	50	90	95	99	99.5
1	1.000	6.314	12.706	63.657	127.32
2	0.816	2.920	4.303	9.925	14.089
3	0.765	2.353	3.182	5.841	7.453
4	0.741	2.132	2.776	4.604	5.598
5	0.727	2.015	2.571	4.032	4.773
6	0.718	1.943	2.447	3.707	4.317
7	0.711	1.895	2.365	3.500	4.029
8	0.706	1.860	2.306	3.355	3.832
9	0.703	1.833	2.262	3.250	3.690
10	0.700	1.812	2.228	3.169	3.581
20	0.687	1.725	2.086	2.845	3.153
∞	0.674	1.645	1.960	2.576	2.807

corresponding to v degrees of freedom, where $v = n - 1$. Note that the values for $v = \infty$ are reduced to the values of Equations (26-7) and 26-8).

Precision of a standard deviation   To obtain a measure of the precision of a standard deviation, we calculate[5] a value q.

$$q = \sqrt{\frac{s^2(n-1)}{n}} = \sqrt{\frac{\sum(x_i - \bar{x})^2}{n}} \qquad (26\text{-}11)$$

The value of q is multiplied by the tabulated values of the highest density regions for the degrees of freedom, and an appropriate confidence level (Table 26-4). The distribution curve (the posterior density of σ) is skewed so that the interval is larger on one side of the standard deviation than on the other, with low values of the standard deviation less probable than high values.

EXAMPLE 26-1   Ten replicate measurements of lead in a soil sample gave a mean of 0.1462% with a standard deviation of 0.0074%. Calculate the 95% confidence interval of the mean and of the standard deviation.

ANSWER   The 95% confidence interval of the mean $= 0.1462 \pm 2.262 \times 0.0074/\sqrt{10} = 0.1515$ to 0.1409. For the standard deviation

$$q = \sqrt{\frac{(0.0074)^2 9}{10}} = 0.00702$$

The 95% confidence interval for the standard deviation $= 0.0074 - (0.66 \times q)$ and $+ (1.77 \times q) = 0.0028$ to 0.0198.   ////

Table 26-4   VALUES[5] TO BE USED WITH q FOR v
DEGREES OF FREEDOM FOR VARIOUS
CONFIDENCE LEVELS OF THE
HIGHEST DENSITY REGIONS

Degrees of freedom	Confidence level					
	90		95		99	
1	0.38	11.29	...	...	...	...
2	0.49	3.81	0.45	5.44	0.38	12.25
4	0.59	2.21	0.55	2.69	0.48	4.13
9	0.70	1.59	0.66	1.77	0.59	2.22
19	0.78	1.35	0.75	1.44	0.69	1.64
39	0.84	1.22	0.81	1.27	0.76	1.39
99	0.89	1.13	0.87	1.16	0.84	1.22

26-9   COMBINATION OF OBSERVATIONS

A frequent situation in an investigation is a need to calculate the variance of a series of observations that can be arranged logically into subgroups. For example, suppose that the precision of a standardization of a solution is being evaluated for a new primary standard. On each of several days a set of values is observed, and it is desired to exclude the possible deterioration of the standard solution from day to day from the estimate of precision. Suppose that there are k subgroups, not necessarily containing the same number of observations, and N total observations. The variance is calculated by summing the squares of the deviations, $\sum (x_{ij} - \bar{x}_i)^2$, for each subgroup and then adding all the k sums and dividing by $N - k$, to give

$$s^2 = V = \frac{1}{N - k} \sum_i \sum_j (x_{ij} - \bar{x}_i)^2 \qquad (26\text{-}12)$$

The significance of dividing by $N - k$ is that, of the N deviations, only $N - k$ are independent, because 1 degree of freedom is lost in each subgroup.

For two groups of observations consisting of n_1 and n_2 members, of standard deviations s_1 and s_2, the variance is given by

$$V = s^2 = \frac{(n_1 - 1)s_1{}^2 + (n_2 - 1)s_2{}^2}{n_1 + n_2 - 2} \qquad (26\text{-}13)$$

26-10   TESTS OF SIGNIFICANCE

Statistical methods frequently are used to give a yes-or-no answer to a particular question concerning the significance of data. The answer is qualified by a confidence level indicating the degree of certainty of the answer. This procedure is known as *hypothesis testing*.[17]

A common procedure is to set up a *null hypothesis*, which states that there is no significant difference between two sets of data or that a variable exerts no significant effect.[18] To enable a yes-or-no answer to be given, a significance level, say 95 or 99%, is chosen to express the probability that the answer is correct. Careful judgment must be exercised in selecting the confidence level. If it is chosen too severely (such as 99.9%), a significant effect may be missed, and a true hypothesis rejected. Such an error is called α or type I. On the other hand, if too much latitude is allowed (such as 80%), an insignificant difference may be judged important, and a false hypothesis accepted. This error[19] is called β or type II.

The t test   An example of hypothesis testing is the t test (or Student's t test) based on the definition of t in Section 26-8. In this application the t test is used to test the hypothesis that two means do not differ significantly. Equation (26-10) is applicable directly to the comparison of the mean $\bar{x}$ of n observations drawn at random from a gaussian population of mean μ. The quantity $t = |\bar{x} - \mu| \sqrt{n}/s$ is compared with the values found in a table of values of t (Table 26-3) at the desired confidence level and corresponding to v degrees of freedom, where $v = n - 1$. If the value in question exceeds the tabular value, the null hypothesis is rejected, and a significant difference is indicated.

EXAMPLE 26-2   The following values were obtained for the atomic weight of cadmium: 112.25, 112.36, 112.32, 112.21, 112.30, 112.36. Does the mean of these values (112.30) differ significantly from the accepted value 112.40?

ANSWER   The standard deviation s is $\sqrt{\sum(x_i - \bar{x})^2/(n - 1)}$, or 0.060. The quantity t is $|112.30 - 112.40| \sqrt{6}/0.060 = 4.1$. From Table 26-3, for $v = 5$, t is 4.032 at the 99% confidence level and 4.773 at the 99.5% confidence level. A significant deviation is indicated; therefore a systematic source of error is highly probable.   ////

Such a test is limited in practical applicability because in it the population mean μ is regarded as known. If the deviation from a theoretical value is desired, and *if there is a gaussian distribution of error around the theoretical value*, the theoretical value is the population mean μ. Again, if a relatively large sample of data ($n > 30$) is available, its mean can be taken to be a measure of μ, and the average of a smaller set can be adequately compared with it. It is often desirable, however, to compare the means of two relatively small sets, of n_1 and n_2 observations with means $\bar{x}_1$ and $\bar{x}_2$, when *the variances within the sets can be regarded as the same* within the limits of random sampling. To test for homogeneity of variance, the F test discussed later in this section is applied. The variance based on both samples can be calculated from Equation (26-13). The variance of $\bar{x}_1$ is s^2/n_1, and that of $\bar{x}_2$ is s^2/n_2. The variance of the difference $\bar{x}_1 - \bar{x}_2$ is the sum of the two, or

$$V_{\bar{x}_1 - \bar{x}_2} = s^2 \left(\frac{1}{n_1} + \frac{1}{n_2} \right) = s^2 \left(\frac{n_1 + n_2}{n_1 n_2} \right) \qquad (26\text{-}14)$$

The quantity t is defined as the difference between the two means divided by its standard deviation [Equation (26-9)]:

$$t = \frac{\bar{x}_1 - \bar{x}_2}{s[(1/n_1) + (1/n_2)]^{\frac{1}{2}}} = \frac{\bar{x}_1 - \bar{x}_2}{s} \left(\frac{n_1 n_2}{n_1 + n_2} \right)^{\frac{1}{2}} \qquad (26\text{-}15)$$

which may be compared with tables of t corresponding to $n_1 + n_2 - 2$ degrees of freedom.

EXAMPLE 26-3   Two methods of analysis were applied to the same sample: the results for percentage of component x are shown below.

Test	Method A	Method B	Difference
1	4.68	4.81	$d_1 = 0.13$
2	4.64	4.70	$d_2 = 0.06$
3	4.69	4.74	$d_3 = 0.05$
4	4.55	...	...
	$\bar{x}_A = 4.64$	$\bar{x}_B = 4.75$	$\bar{d} = 0.08$

With our attention confined to the first three columns, the standard deviation is 0.06; t then is given by

$$t = \frac{|4.64 - 4.75|}{0.06} \sqrt{\frac{4 \times 3}{7}} = 2.4$$

The value of $t = 2.4$ for 5 degrees of freedom may be compared with values of 2.015 for 90%, 2.571 for 95%, and 4.032 for 99% confidence levels (Table 26-3). A result giving a positive test at the 95% level usually is regarded as significant, and one at the 99% level is highly significant. In this instance the t test suggests an affirmative result, but to decide with more assurance whether methods A and B are significantly different requires that a greater number of tests be run.   ////

In the special case where $n_1 = n_2 = n$, (26-15) is reduced to

$$t = \frac{\bar{x}_1 - \bar{x}_2}{s} \sqrt{\frac{n}{2}} \qquad (26\text{-}16)$$

Also, in the special case where $n_1 = n$, $n_2 = \infty$, $\bar{x}_2$ becomes the population mean μ, and (26-15) becomes (26-9).

Another type of application of the t test is to *individual* differences between sets of observations. In Example 26-3 the same sample was analyzed by two methods. But we may wish to determine whether there is a systematic difference between methods A and B irrespective of the sample. Both methods are assumed to have essentially the same standard deviation that does not depend on the sample. The fourth column lists the differences between the two methods for the three tests common to both methods. The standard deviation of the differences is

$$s_d = \sqrt{\sum_i \frac{(d_i - \bar{d})^2}{n - 1}} \qquad (26\text{-}17)$$

The value of t is the mean difference divided by its standard deviation, or

$$t = \frac{\bar{d}\sqrt{n}}{s_d} \qquad (26\text{-}18)$$

where the quantity $\sqrt{n}$ appears because the standard deviation of the *mean difference* is $s_d/\sqrt{n}$ if s_d is that of a single difference.

In the above example, $t = (0.08/0.0436)\sqrt{3} = 3.2$. From the table, corresponding to $n - 1 = 2$ degrees of freedom, $t = 2.920$ at the 90% confidence level and 4.303 at the 95% level. Thus a value of $t = 3.2$ or greater would occur by chance alone less than once in 10 trials, and therefore the differences between methods A and B may be judged to be real. Actually, it is risky to draw conclusions from a group of data as limited as the example presented here for the sake of simplicity; on the other hand, valid statistical conclusions are not necessarily based on an enormous body of data.

The F test In contrast to the t test, which is a comparison of means, the F test is a comparison of variances. The ratio between the two variances to be compared is the *variance ratio* F (for R. A. Fisher), defined by

$$F = \frac{s_1{}^2}{s_2{}^2} = \frac{V_1}{V_2} \qquad (26\text{-}19)$$

Values of F are available in statistical tables at various levels of significance. The values depend on the number of degrees of freedom v_1 and v_2 for the variances V_1 and V_2. The tables are arranged with F values greater than unity; therefore the two variances are compared in the order $V_1 > V_2$. Table 26-5 is a small section of a table of F values taken at the 95% confidence level. The entries correspond to a probability of 0.95 that the variance ratio will not exceed the value in the table. Extensive tables are available for various confidence levels and degrees of freedom.

EXAMPLE 26-4 Suppose that two kinds of observations are made, one of six observations, of standard deviation $s_1 = 0.055$, and another of four observations, of standard deviation $s_2 = 0.022$. We are to test whether s_1 is significantly greater than s_2. We have $v_1 = 5$, $v_2 = 3$, and from the table, $F = 9.01$. The experimental

Table 26-5 VALUES OF F AT THE
95% CONFIDENCE LEVEL

v_2	v_1					
	2	3	4	5	6	∞
2	19.00	19.16	19.25	19.30	19.33	19.50
3	9.55	9.28	9.12	9.01	8.94	8.53
4	6.94	6.59	6.39	6.26	6.16	5.63
5	5.79	5.41	5.19	5.05	4.95	4.36
6	5.14	4.76	4.53	4.39	4.28	3.67
∞	3.00	2.60	2.37	2.21	2.10	1.00

variance ratio is

$$\frac{V_1}{V_2} = \frac{s_1{}^2}{s_2{}^2} = \frac{0.0030}{0.00048} = 6.25$$

On the basis of chance, 5% of the time the value of F will exceed 9.01. We conclude that the divergence is not statistically significant and therefore the null hypothesis is valid.

As applied here, the F test is *one-sided*, testing the null hypothesis that $\sigma_1{}^2$ and $\sigma_2{}^2$ (as estimated by $s_1{}^2$ and $s_2{}^2$) are equal, the alternative hypothesis being $\sigma_1{}^2 > \sigma_2{}^2$. The F test may also be applied as a *two-sided* test, in which the alternative to the null hypothesis is $\sigma_1{}^2 \neq \sigma_2{}^2$. This doubles the probability that the null hypothesis is invalid and has the effect of changing the confidence level in the above example from 95 to 90%.   ////

The chi-square test   This test is a useful tool for determining to what extent data expressed quantitatively in terms of probabilities, or expected frequencies, correspond to theoretical values.

For example, if we were to examine the third digit of a five-place logarithm table, we would expect to find an equal probability, 0.1, of finding each of the 10 digits. On the other hand, if we examine (Example 26-5) the last digit of students' buret readings estimated to 0.01 ml, we find *number bias* in favor of certain digits and against others. For a limited number of observations a certain fluctuation is statistically expected, and the chi-square test makes allowance for this expectation.

The quantity χ^2 is defined by

$$\chi^2 = \sum \frac{(f_i - F_i)^2}{F_i} \qquad (26\text{-}20)$$

where f_i is observed frequency, F_i is expected frequency, and the summation is carried out over all classes of observations. In Table 26-6 the column headings give the probability P that the tabular value will be exceeded at v degrees of freedom.

Table 26-6   CHI-SQUARE DISTRIBUTION

v	P								
	0.995	0.99	0.95	0.90	0.50	0.10	0.05	0.01	0.005
1	0.00004	0.00016	0.0039	0.0158	0.455	2.71	3.84	6.63	7.78
2	0.0100	0.0201	0.1026	0.211	1.39	4.61	5.99	9.21	10.6
3	0.0717	0.115	0.352	0.584	2.37	6.25	7.81	11.3	12.8
4	0.207	0.297	0.711	1.064	3.36	7.78	9.49	13.3	14.9
5	0.412	0.554	1.15	1.61	4.35	9.24	11.1	15.1	16.7
6	0.676	0.872	1.64	2.20	5.35	10.6	12.6	16.8	18.5
7	0.99	1.239	2.17	2.83	6.35	12.0	14.1	18.5	20.3
8	1.34	1.65	2.73	3.49	7.34	13.4	15.5	20.1	22.0
9	1.73	2.09	3.33	4.17	8.34	14.7	16.9	21.7	23.6
10	2.16	2.56	3.94	4.87	9.34	16.0	18.3	23.2	25.2
15	4.60	5.23	7.26	8.55	14.3	22.3	25.0	30.6	32.8
20	7.43	8.26	10.85	12.44	19.3	28.4	31.4	37.6	40.0
30	13.79	14.95	18.49	20.60	29.3	40.3	43.8	50.9	53.7

EXAMPLE 26-5 An examination of the last digit of buret and balance readings reported by students of introductory quantitative analysis during a practical examination gave the following results.

Terminal digit	Buret reading		Balance reading	
	f_i	$\lvert f_i - F_i \rvert$	f_i	$\lvert f_i - F_i \rvert$
0	212	62	120	20
1	212	62	92	8
2	229	79	95	5
3	166	16	85	15
4	124	26	112	12
5	107	43	132	32
6	110	40	77	23
7	81	69	95	5
8	134	16	107	7
9	125	25	85	15
	1500		1000	
Sum of squares	23,952			2710

For the 1500 buret readings sampled, the expected frequency F_i is 150 in each class, with the assumption of no number bias. The calculated value of χ^2 is $23{,}952/150 = 160$, a value that at 9 degrees of freedom (the number of classes minus 1) lies far above the 99.9 percentile probability level and indicates pronounced number bias. The bias in this case is for small numbers and against large ones (a common type of bias for this type of reading). Another indication of number bias is obtained by comparing the observed standard deviation $s = \sqrt{23{,}952/9} = 52$ with that calculated from the binomial distribution (Section 27-3) $s = \sqrt{np(1 - p)}$, which for $n = 1500$, $p = 0.1$, and $1 - p = 0.9$ gives $s = 12$ for 10 *equal* classes of probability 0.1.

Making a decision about the terminal digit of a weighing from a balance with a vernier scale is more straightforward than that for a buret reading, and the likelihood of number prejudice is less. For the sample of 1000 balance readings the χ^2 value is $2710/100$ or 27, indicating number prejudice with a high probability but of a much smaller magnitude than for the buret readings. In this set the prejudice appears to be for the digits 0 and 5—another common type of bias. ////

Number bias varies considerably from observer to observer. The commonest bias is for the digits 0 and 5, but prevalent also is bias toward even numbers and against odd ones, and for low numbers and against high ones. Such number bias imposes a limitation on the accuracy of readings by an individual.

Tests for gaussian distribution The chi-square test can be used to find out whether an experimental distribution of errors follows the gaussian curve. The method is described here only in principle because only in unusual cases are enough data available. For a detailed example, see a paper by Nelson.[20] The steps are as follows:

1 Prepare a frequency table showing the numbers of observations falling into a series of classes.

2 Calculate the sample mean $\bar{x}$ and the variance V.

3 Define a gaussian-distribution curve with the same mean, the same variance, and the same total number of observations.

4 Take differences between the observed and expected frequencies, and calculate χ^2 for each class.

5 Add the values of χ^2 to obtain a total value, and compare this value with tabular values to find the probability level corresponding to the value of χ^2. This probability level expresses the likelihood that a random choice of values from a gaussian distribution will give the observed distribution. A probability level below 1 or 0.1% is generally regarded as necessary for a conclusion from a chi-square test that a population is nonrandom. The existence of abnormalities in the frequency distribution is indicated by abnormal contributions of certain classes to the total value of χ^2.

Another and more simple qualitative test for a gaussian distribution is the use of *probability paper* for plotting a distribution curve. If the cumulative probability of an observation is plotted as a function of the value of the observation, an *S*-shaped curve is observed, with the cumulative probability reaching 0.5 at the median value. By proper distortion of the ordinates of the graph paper, the cumulative probability can be a straight line from a gaussian distribution. Probability paper should be used with caution, because a straight line can be obtained from data that are far from gaussian. Also, deviations are difficult to judge, because points near the center of the graph carry more weight than those near the ends.

26-11 ANALYSIS OF VARIANCE

The technique of analysis of variance developed by Fisher[18] is a powerful tool in determining the separate effects of different sources of variation in experimental data.

In chemical analysis a frequently important question is whether a particular method gives the same precision in the hands of several analysts, or whether analysts within a laboratory agree more closely among themselves than with another group in a separate laboratory, or whether the same relative precision is obtained by a given group of analysts regardless of variations in the sample.

Analysis of variance is most useful when applied to a set of experiments planned with statistical evaluation in mind. *Factorial* experiments, in which several factors are changed in all possible combinations in a single integrated experiment, allow the estimation of *interaction effects*, the simultaneous effects of two or more variables. Such interactions may be extremely important; yet they may escape detection completely by the classical method of experimentation in which the variables are changed one at a time, with all other conceivable sources of variation held constant.

When several factors are involved, the statistical design and the analysis of variance become complicated. Therefore, we examine in detail here only the simplest type of example to illustrate the principles involved. More elaborate situations can be understood by an extension of the same principles.

The simplest application is that of a *one-way classification with equal numbers*, such that a total number of $N = nk$ observations are classified into k classes of n observations each. As an illustration, consider n standardizations of a single solution by k analysts. If, in a set of observations, $i = 1, 2, \ldots, k$, and if $j = 1, 2, \ldots, n$, an observation x_{ij} is then the jth observation of the ith class. There is a single *assignable* source of variation (the different analysts) in addition to the random error inherent in the method, and evaluation of the magnitude of the variance from this source is desired.

If the overall mean of the $N = nk$ observations is $\bar{x}$, and if $\bar{x}_i$ is the mean of the n observations of the ith class, we can write

$$x_{ij} - \bar{x} = (x_{ij} - \bar{x}_i) + (\bar{x}_i - \bar{x}) \qquad (26\text{-}21)$$

Squaring both sides and summing over i and j, it can be shown that

$$\sum_{ij} (x_{ij} - \bar{x})^2 = \sum_{ij} (x_{ij} - \bar{x}_i)^2 + n \sum_{i} (\bar{x}_i - \bar{x})^2 \qquad (26\text{-}22)$$

In (26-22) the quantity on the left side is the *total sum of squares S*, obtained by summing the squares of each individual deviation from the overall mean. For simplicity of calculation, use can be made of the identity

$$S = \sum_{ij}(x_{ij} - \bar{x})^2 \equiv \sum_{ij}(x_{ij})^2 - \frac{\left(\sum_{ij} x_{ij}\right)^2}{N} \qquad (26\text{-}23)$$

The second term on the right side of (26-22) is S_b, the sum of squares of deviations *between classes*, obtained by summing the squares of deviations of the class mean from the overall mean:

$$S_b = n \sum_{i}(\bar{x}_i - \bar{x})^2 \equiv \sum_{i} \frac{\left(\sum_{j} x_{ij}\right)^2}{n} - \frac{\left(\sum_{ij} x_{ij}\right)^2}{N} \qquad (26\text{-}24)$$

The sum of squares *within classes* S_w, the first term on the right side of (26-22), is obtained directly by subtracting the calculated value of S_b in (26-24) from the calculated value of the total sum of squares S in (26-23).

If all classes do not contain n members but do contain a variable number n_i, then $S_b = \sum_{i} n_i(\bar{x}_i - \bar{x})^2$, which is the weighted sum of squares. In each class the multiplier n_i is used because it is desired to express the variance for a single observation, which is n_i times the variance of the mean of n_i observations. The calculations are made more convenient by arranging all classes to contain an equal number of observations, as assumed above.

From (26-22)

$$S = S_w + S_b \qquad (26\text{-}25)$$

which states that the total sum of squares can be separated into a term S_w resulting from deviations *within* classes and a term S_b resulting from deviations *between* classes.

Variances (mean squares) are calculated by dividing the sum of squares in each case by the number of degrees of freedom. Since there are k classes each containing n observations, the total number of degrees of freedom is $nk - 1 = N - 1$. For S_b there are $k - 1$ degrees of freedom, because S_b is computed from the deviations of the k class means $\bar{x}_i$ from the overall mean $\bar{x}$. For S_w there are $N - k = nk - k = k(n - 1)$ degrees of freedom, representing the difference between the total number of observations and the k-class means used in the calculations. Note that, just as the sum of squares is additive, so is the number of degrees of freedom; that is, the number $N - 1$ of total degrees of freedom is equal to the number $k - 1$ of degrees of freedom "between classes" plus the number $N - k$ of degrees of freedom "within classes."

These observations may be summarized conveniently in an *analysis-of-variance table*; Table 26-7 illustrates this type of table for the above case. The overall variance (total mean square) $S/(N - 1)$ contains contributions due to variances within as well as between classes. The variation between classes contains both variation within classes and a variation associated with the classes themselves and is given by the expected mean square $\sigma_w^2 + n\sigma_b^2$. Whether $n\sigma_b^2$ is significant can be determined by the F test. Under the null hypothesis, $\sigma_b^2 = 0$. Whether the ratio

$$F = \frac{[S_b/(k - 1)]}{[S_w/(N - k)]}$$

Table 26-7   ANALYSIS-OF-VARIANCE TABLE FOR k CLASSES OF n OBSERVATIONS

Source of variation	Sum of squares	Degrees of freedom	Mean square	Expected mean square
Between classes	S_b	$k - 1$	$\dfrac{S_b}{k - 1}$	$\sigma_w^2 + n\sigma_b^2$
Within classes	S_w	$N - k$	$\dfrac{S_w}{N - k}$	σ_w^2
Total	S	$N - 1$	$\dfrac{S}{N - 1}$	$\cdots$

for $k - 1$ and for $N - k$ degrees of freedom is significant is determined by comparison of tabulated values of F at the desired significance level.

EXAMPLE 26-6 Suppose that five different analysts ($i = 5$) A, B, C, D, E reported triplicate results ($j = 3$) in parts per million for iron in water. We wish to determine whether the values of the means differ among analysts.

Determination	Analyst				
	A	B	C	D	E
1	10.3	9.5	12.1	7.6	13.6
2	9.8	8.6	13.0	8.3	14.5
3	11.4	8.9	12.4	8.2	15.1
Mean $\bar{x}_i$	10.5	9.0	12.5	8.0	14.4

The following analysis-of-variance table for the above table is obtained by summing squares and arranging as shown in Table 26-7. It is noted that $k = 5$, $n = 3$.

Source of variation	Sum of squares	Degrees of freedom	Mean square	Expected mean square
Between analysts	$S_b = 80.39$	4	20.10	$\sigma_w^2 + 3\sigma_b^2$
Within analysts	$S_w = 3.61$	10	0.36	σ_w^2
	$S = 84.00$	14	6.00	—

The ratio $F = 20.1/0.36 = 56$; this is compared with the value $F_{0.01} = 5.99$ found in statistical tables of F values at 4 and 10 degrees of freedom, at the 99% confidence level. Therefore, F is highly significant and the means among laboratories differ significantly. ////

26-12 CONTROL CHARTS[21]

A control chart is a sequential plot of some quality characteristic. It might be a day-by-day plot of the average moisture content of grain samples or the concentration of a standard solution, or the percentage of a constituent in successive production lots. The chart consists of a central line and a pair of limit lines, or in some cases two pairs of limit lines for the *inner* and *outer control limits*. Plotting a sequence of points in order makes available a continuous record of the quality characteristic. Trends in data or sudden lack of precision can be made evident so that the causes may be sought. Although control charts may be plotted for many statistics, the most common are for averages and ranges of observations.

The control chart is set up to answer the question of whether the data are in *statistical control*, that is, may be regarded as random samples from a single population of data. Because of this feature of testing for randomness, such a chart may be useful in searching out systematic sources of error in laboratory research data as well as in evaluating data on plant production or control analysis.

A control chart might be set up by plotting individual observations in sequential order and comparing them with control limits established from sufficient past experience. For example, if the mean $\bar{x}$ and standard deviation s of a supposedly constant quantity have been established from, say, 20 past observations, these quantities may be regarded as valid estimates of μ and σ for the population. Limits of $\pm 1.96\sigma$, corresponding to a confidence level of 95%, might be set as control limits. The probability that a future observation will fall outside these limits as a result of chance is only 1 in 20. A greater proportion of scatter might indicate a nonrandom distribution, that is, a systematic error. If the control limits are set up with a limited sample (20 in the above example), there is some probability that excessive scatter is caused by setting the initial control limits too narrow because of inadequate estimates of μ and σ. To check this possibility, a new calculation based on a larger number of observations should be made. It is common practice in some industries to set inner control limits, or warning limits, at $\pm 1.96\sigma$, and outer control limits at $\pm 3.09\sigma$. The outer limits correspond to a confidence level of 99.8%, or a probability of 0.002, that a point will fall outside the limits. One-half of this probability corresponds to a high result and one-half to a low result. Special attention should be paid to a one-sided deviation from the control limits, because systematic errors cause deviation in one direction more often than abnormally wide scatter. Two systematic errors of opposite sign would cause scatter, but it is unlikely for both to enter at the same time. The control chart need not be plotted in a time sequence. In any situation where relatively large numbers of units or small groups are to be compared, this chart is a simple means of indicating whether any unit or group is out of line. Thus laboratories, production machines, test methods, or analysts may be put arbitrarily into a horizontal sequence.

For small groups of observations it is usually preferable to plot means rather than individual observations. The random scatter of pairs of observations is $1/\sqrt{2} = 0.71$ as great as that of single observations, and the likelihood of two "wild" observations in the same direction is vanishingly small. The groups of two to five observations should be chosen in such a way that only chance variations operate within a group, whereas assignable causes are sought for variations between groups. If duplicate analyses are performed each day, the pairs form logical groups.

Some measure of dispersion of the subgroup data should also be plotted, as a parallel control chart. For small groups of data the range may be used as a measure of scatter, and it is usually a simple matter to plot the range as a vertical line and the mean as a point on this line for each group of observations.

26-13   REGRESSION ANALYSIS[22]

In the analysis of data it is often desirable to determine whether two variable quantities are interdependent and to express this relation quantitatively. The present treatment is limited to the case of a linear relation, which is experimentally the most important.

If two variables are proportional, then a linear plot should have an intercept of zero, and a log-log plot of one against the other must yield a straight line with a *slope of unity*.

In some instances the interest may not be in predicting the dependent variable y from the independent variable x but in determining whether they are associated. In these cases the correlation coefficient (the degree of dependence between two variables) may be calculated by either parametric or nonparametric methods, depending on the data.[1]

For regression analysis a nonlinear relation often can be transformed into a linear one by plotting a simple function such as the logarithm, square root, or reciprocal of one or both of the variables. Nonlinear transformations should be used with caution because the transformation will convert a distribution from gaussian to nongaussian. Calculations of confidence intervals usually are based on data having a gaussian distribution.

If two variables x and y are related, either of two different situations may be recognized: (*1*) both variables are subject to comparable experimental error; (*2*) one variable may be regarded as being determinable to so high a degree of precision that its uncertainty can be ignored. The second is much more frequently encountered in analytical chemistry; usually we are interested in determining whether a statistically significant trend in results exists with some variable (such as temperature, pH, or sample size) that exerts only a small effect and therefore can be fixed as accurately as needed.

We consider here only the simplest case of *simple linear regression*, in which x is considered the accurately determinable independent variable and y the dependent variable subject to experimental uncertainty. The data are to be fitted to the straight line

$$y = a + bx \qquad (26\text{-}26)$$

where a is the intercept and b is the slope, or *coefficient of regression of y on x*. The procedure is to fit the best straight line to the data by the *method of least squares*.

If (x_i, y_i) is any given observation, the deviation from the line is measured in the y direction since the error of x is insignificant. The deviation is given by

$$y_i - (a + bx_i) \qquad (26\text{-}27)$$

The values of a and b are selected to make the sum of the squares of the deviations Q a minimum, where

$$Q = \sum [y_i - (a + bx_i)]^2 \qquad (26\text{-}28)$$

This can be done by differential calculus, setting the derivatives of Q with respect to a and b equal to zero and solving for a and b. The results are

$$a = \bar{y} - b\bar{x} \qquad (26\text{-}29)$$

$$b = \frac{\sum (x_i - \bar{x})(y_i - \bar{y})}{\sum (x_i - \bar{x})^2} \qquad (26\text{-}30)$$

where $\bar{y}$ is the mean of y's and $\bar{x}$ is the mean of x's. From (26-29) the equation for the straight line may be written

$$y - \bar{y} = b(x - \bar{x})$$

which indicates that it is a line drawn through the point represented by the coordinates $(\bar{x}, \bar{y})$, with a slope equal to b.

The minimum value of squares of deviations is obtained by substituting the values of a and b in the expression for Q:

$$Q_{min} = \sum (y_i - \bar{y})^2 - b^2 \sum (x_i - \bar{x})^2 \qquad (26\text{-}31)$$

This sum is useful for estimating the precision of the regression equation. The term *variance about regression* means that the deviations of y are measured not from the mean (as in the usual definition of variance) but from the regression line. The variances of the deviation of y from the straight line is Q_{min} divided by the number of degrees of freedom ($n - 2$, since 2 degrees of freedom corresponding to the two constants have been used up in finding the regression equation). Since the sum of squares of deviations was minimized, the straight line corresponds to a *minimum variance* of vertical deviations—which is the justification of the method of least squares.

We may express the result as an analysis of variance, given in Table 26-8. From the expression for Q_{min} the total sum of squares of y is diminished by the amount $b^2 \sum (x_i - \bar{x})^2$ when regression is taken into account.

Table 26-8   ANALYSIS OF VARIANCE OF REGRESSION

Source of variation	Sum of squares	Degrees of freedom	Variance
Due to regression	$b^2 \sum (x_i - \bar{x})^2$	1	V_1
About regression	$\sum (y_i - \bar{y})^2 - b^2 \sum (x_i - \bar{x})^2$	$n - 2$	V_2
Total	$\sum (y_i - \bar{y})^2$	$n - 1$	

In regression analysis we can recognize two extreme cases: (*1*) The sum of squares due to regression $b^2 \sum (x_i - \bar{x})^2$ is equal to the total sum of squares $\sum (y_i - \bar{y})^2$, the sum of squares about the regression is then zero, and all points fall on the straight line. (*2*) The sum of squares due to regression is zero, corresponding to $b = 0$, or a horizontal straight line passing through the mean value of y. In the first extreme a straight line without error is observed; in the second, y is independent of x.

The F test can be applied to the variance ratio V_1/V_2. For a straight line, V_1/V_2 is infinite; for no dependence of y on x, V_1/V_2 is zero.

The confidence interval of the slope can be estimated by the t test. It is important to realize that the quantities a and b in the regression line $y = a + bx$ are statistics that are estimates of the population parameters α and β. As the number of observations is increased without limit, a and b approach α and β.

For a finite number of observations x_n, y_n, the variance of b is given by[23]

$$V_b = \frac{\sigma^2}{\sum (x_i - \bar{x})^2} \qquad (26\text{-}32)$$

where σ^2 is the population variance of y about the line, with the assumption that the same variance σ^2 applies for all values of x. Normally, the population variance σ^2 is unknown, and it is therefore necessary to use the estimate s^2 from a finite number n of observations, that is, $s^2 = Q_{\min}/(n - 2) = V_2$. This requires the introduction of the Student's t, in a manner analogous to its use in testing a hypothesis concerning the mean of an unknown population.

The confidence interval for the slope β is given by

$$\beta = b \pm \frac{ts}{\sqrt{\sum (x_i - \bar{x})^2}} \qquad (26\text{-}33)$$

The confidence interval of the intercept α is given by

$$\alpha = a \pm ts \sqrt{\left[\frac{1}{n} + \frac{\bar{x}^2}{\sum (x_i - \bar{x})^2} \right]} \qquad (26\text{-}34)$$

where in both (26-33) and (26-34) the value of t corresponds to $n - 2$ degrees of freedom and to the desired confidence level.

EXAMPLE 26-7   To illustrate a control chart and regression analysis, consider the following set of replicate titration volumes arranged in the order in which the determinations were run: 41.41, 41.30, 41.59, 41.47, 41.53, 41.20, 41.33, 41.32, 41.51, 41.26, 41.58. The mean is 41.41 ml; the standard deviation is 0.135 ml.

The control chart in Figure 26-2 (left) seems to indicate that the observations are in control, and they appear to exhibit random fluctuations when plotted in

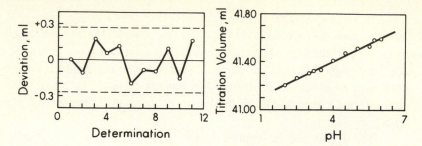

FIGURE 26-2   Left, control chart for titration volumes. Dashed lines are $\pm 2s$ limits. Right, regression line for titration volumes. Solid line is least-squares line.

sequence. It turns out, however, that the titrations were run at various pH values, which in sequence were 4.0, 3.0, 6.0, 4.5, 5.5, 2.0, 3.5, 3.2, 5.0, 2.5, 5.7. When the observations are plotted as a function of pH, the chart clearly shows a trend, which appears to be linear with pH. To show the significance of the trend, a regression analysis is carried out.

The mean values of x_i (pH) and y_i (titration volume) are

$$\bar{x} = \sum \frac{x_i}{n} = \frac{44.9}{11} = 4.082$$

$$\bar{y} = \sum \frac{y_i}{n} = \frac{455.5}{11} = 41.41$$

$$\sum (x_i - \bar{x})(y_i - \bar{y}) = 1.8269$$

$$\sum (x_i - \bar{x})^2 = 18.455$$

$$\sum (y_i - \bar{y})^2 = 0.18249$$

$$b = 1.8269/18.455 = 0.09899$$

$$a = \bar{y} - b\bar{x} = 41.41 - 0.09899 \times 4.082 = 41.006$$

The required least-squares line for the titration volume is $41.006 + 0.09899$ pH.

The analysis of variance of regression gives (Table 26-8) the following results: Variance V_1 due to regression (1 degree of freedom)

$$= b^2 \sum (x_i - \bar{x})^2 = 0.1808$$

Variance V_2 about regression (9 degrees of freedom)

$$= \frac{\sum (y_i - \bar{y})^2 - b^2 \sum (x_i - \bar{x})^2}{9} = 0.000183$$

To test for significance, the F value $V_1/V_2 = 990$ tells us that the regression is significant to a high level of confidence.   ////

A regression line should be interpreted with caution. First, its validity should be inferred beyond the experimentally determined values of x only when there is sound physical or theoretical justification. For example, in the above instance of regression of titration volume with pH, the quantity a does not necessarily have physical reality at pH $= 0$, because no experiments were run at that value. Another example is a regression plot in which the theoretically expected relation is $y = bx$ but the least-squares straight line gives an intercept a at $x = 0$. The intercept may be wrongly interpreted to be a "blank" value of y, when in fact it may have resulted merely from there being a finite uncertainty in the value of b for a limited set of data. A value of y should actually be determined for $x = 0$.

Second, the quantity x does not necessarily *cause* the regression of y, for another factor z may vary in a regular way with x and so be the actual cause. For example, the rate of an air-oxidation reaction could vary with pH and be the actual cause of a regression of titration volume with pH. Again, the slope of a least-squares plot of absorbance against concentration often is interpreted directly as a molar absorptivity, whereas the slope may in fact be affected by a third variable, such as the slit width of a spectrophotometer. Sometimes the calculation of simple correlation coefficients[1] can elucidate such problems.

Third, the foregoing regression analysis is carried out under the assumption that the absolute variance of y is independent of x. This may not be true. An important situation in analytical chemistry is one in which the *relative* variance or coefficient of variation is independent of x. Thus, an instrument reading may be theoretically proportional to concentration with the same relative precision over a range of concentrations. With appropriate changes, this case can be handled.[24]

More complicated situations of regression analysis, such as linear regression with more than one independent variable, are encountered less frequently in analytical chemistry and are not considered here.

26-14 STATISTICAL DESIGN OF EXPERIMENTS[25,26]

Mention has been made in connection with variance analysis that, if maximum information is to be obtained from a given amount of experimental work, experiments should be planned with statistical analysis in mind.

Table 26-9 FACTORIAL DESIGN FOR TWO FACTORS AT TWO LEVELS

	A_1	A_2
B_1	(1)	a
B_2	b	ab

Suppose that an experiment is to test the effect of n variables or *factors* at two values, a low level and a high level. The factors are designated A, B, C, ..., and the levels are designated A_1 and A_2, B_1 and B_2, ... To determine all the effects requires 2^n *treatments* (experiments), which can be set up according to a diagram, or *factorial design*. Table 26-9 represents a factorial design for two factors at two levels. The entries in the table represent the effects studied in each treatment, beginning with (1) as the low level of each factor. The entry a, for example, identifies a particular treatment and also the numerical result of the test.

An extension to three factors at two levels is shown in Table 26-10. The eight treatment combinations provide a means for estimating the effects of the three main factors, A, B, and C, the three first-order interactions between the main factors AB, BC, and AC, and the second-order three-factor interaction ABC.

As the number of factors increases, the number of treatments 2^n increases so rapidly that an excessive amount of experimental work may be involved to complete the entire factorial plan. Here the exercise of judgment on the part of the investigator may allow a considerable saving of effort with negligible loss of information. Let us say that five factors are judged to be of sufficient importance to merit testing at two levels. A complete factorial design would require $2^5 = 32$ treatments and would involve determining the 5 *primary* effects due to a single variable, 10 *first-order* interactions, 10 *second-order* interactions, 5 *third-order* interactions, and 1 *fourth-order* interaction. To the experimenter the higher-order interactions become decreasingly important the greater the number of variables involved, and the corresponding experiments may be omitted. Usually, one or two of the five variables may be considered on the basis of prior information to have little if any effect. This may be deduced from the primary effect. Accordingly, treatments corresponding to higher-order interactions may safely be omitted, because the likelihood of significance of a second- or higher-order interaction is even less than that of a first-order interaction. In short, the higher-order interactions turn out to measure the basic error of the experimentation, and it is unnecessary to make many estimates of this basic error. Such omission of selected interactions is known as *fractional replication*.

An example of a complete factorial design of $2^4 = 16$ treatments, together with

Table 26-10 FACTORIAL DESIGN FOR THREE FACTORS AT TWO LEVELS

	A_1		A_2	
	B_1	B_2	B_1	B_2
C_1	(1)	b	a	ab
C_2	c	bc	ac	abc

a discussion of the effect of half-replication, has been given by Box.[27] Important in fractional replication is that the proper experiments be omitted, for one incorrect omission can lead to a situation in which there is no estimate for the effect of one factor and there are unnecessary estimates of higher-order interactions. If any doubt exists about whether the primary effect of a specific variable is high enough to merit inclusion of its interactions, this variable can be tested early in the sequence of experiments, and the planned design altered accordingly. An elementary discussion has been given by Youden.[28]

A word of caution should be interjected as to the order in which statistically designed experiments should be run. With an ordered array of experimental variables such as that displayed in a factorial design, it is tempting to set up a systematic order of experimentation to ensure the completion of the design. Such a procedure may lead to incorrect conclusions if some unsuspected variable (such as time dependence) is exerting an effect. For example, suppose that an analyst has set up a factorial design to test the effects of two variables, say the temperature and acid concentration in a titration. Each titration is to be run in triplicate. Now, suppose that for the sake of convenience each set of replicates is run concurrently and that a gradual change occurs in the concentration of the reagent. The replicates might show satisfactory agreement, being run in close succession, and a later set run, say, at a higher temperature might agree internally but differ significantly from the first set. Then the effect might be falsely attributed to the temperature change. If the order of running the individual titrations had been randomized, divergence among supposed replicates would have revealed the presence of an unsuspected error.

Wernimont[29] reported an example of a nested design, which is an example of fractional replication. It was desired to evaluate an interlaboratory study of a method of acetyl determination. Results obtained by two analysts in each of eight laboratories were compared by having each analyst perform two tests on each of 3 days. The design was

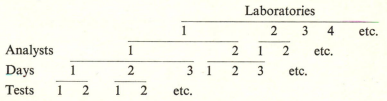

The results indicated a greater variance among laboratories than between two analysts within a laboratory or among tests run by the same analyst on the same day or different days.

Another form of fractional replication often applicable is the *latin square*. Suppose that three factors are to be considered at four levels. We may be interested, for example, in a comparison of results obtained in four laboratories on four samples by four methods of analysis. A complete factorial design would require 4^3, or 64, observations. By distributing the four samples to each of the laboratories, we could set up the following design in which the samples are denoted by A, B, C, and D:

Method	Laboratory			
	L_1	L_2	L_3	L_4
M_1	A	D	C	B
M_2	C	B	A	D
M_3	D	C	B	A
M_4	B	A	D	C

The arrangement shown above is one of 576 possible ways of arranging the samples so that each column and each row contains all the samples. The actual pattern to be used should be selected at random. Since only 16 of the possible 64 treatments are included, this design corresponds to a *quarter-replication*.

The analysis-of-variance table indicates how the total sums of squares of deviations can be divided into four groups, the *residual* sources being an estimate of the error against which the assigned sources are tested by the F test. Thus the variances arising from differences in methods or laboratories may be tested for statistical significance. We can compare the factorial and latin-square designs, both involving 16 treatments. The 2^4 factorial design permits a single estimate of the effect of each of four variables and of six interactions between two variables. The other 5 degrees of freedom may be regarded as error estimates. The latin-square design permits three estimates of the effects of each of three variables, but does not allow estimation of the effects of any of the interactions.

Table 26-11 ANALYSIS OF VARIANCE FOR LATIN-SQUARE DESIGN

From square	Source of variation	Degrees of freedom
Rows	Between methods	3
Columns	Between laboratories	3
Letters	Between samples	3
	Residual	6
	Total	15

26-15 REJECTION OF OBSERVATIONS

A question that often arises is the statistical justification for rejection of a divergent observation. The question is not serious if enough data are at hand to establish a reasonably valid estimate of the standard deviation. The t test is available as a criterion and, in any event, the effect of a single divergent result on the mean value is relatively small. For small groups of three to eight replicates, however, the question is a more difficult one.[30] Objective criteria for rejection or retention of a discrepant value from small numbers of observations often may be of the type that either rejects data too easily or retains highly suspect data.

The Q test[31] is one that is relatively critical and statistically sound. Q is defined as the ratio of the divergence of the discordant value from its nearest neighbor to the range of the values. If the value of Q exceeds tabular values (Table 26-12), which depend on the number of observations, the questionable value may be rejected. In Table 26-12 the tabular values of Q correspond to a 90% confidence limit that an error of the first kind, the rejection of valid measurements, has been avoided. For small numbers of observations, say three to five, the Q test allows rejection only of grossly divergent values. It therefore increases the probability of an error of the second kind, the retention of an erroneous result. The only valid justification for rejection of one discrepant observation from a group of three or four appears to be the location of an assignable cause of determinate error. To reject an intuitively doubtful observation is a dangerous practice, and the determination is best repeated until a statistically valid basis for retention or rejection is obtained. Otherwise the doubtful value should be retained.

The median, though less efficient than the mean for observations with a gaussian distribution, often is recommended because of its insensitivity to a divergent value especially when dealing with small numbers of observations.

Table 26-12 Q-TEST VALUES FOR REJECTION AT THE 90% CONFIDENCE LEVEL

n	3	4	5	6	7	8	9	10
$Q_{90\%}$	0.94	0.76	0.64	0.56	0.51	0.47	0.44	0.41

REFERENCES

1 W. J. DIXON and F. J. MASSEY, JR.: "Introduction to Statistical Analysis," 3d ed., McGraw-Hill, New York, 1969.
2 E. B. WILSON, JR.: "An Introduction to Scientific Research," McGraw-Hill, New York, 1952.
3 G. W. SNEDECOR and W. G. COCHRAN: "Statistical Methods," 6th ed., Iowa State University, Ames, Iowa, 1967.
4 R. LANGLEY: "Practical Statistics," Pan Books, London, 1970.
5 S. A. SCHMITT: "Measuring Uncertainty. An Elementary Introduction to Bayesian Statistics," Addison-Wesley, Reading, Mass., 1969.
6 See F. H. STEIGER: *Chem. Technol.*, p. 225, April 1971.
7 S. SIEGEL: "Nonparametric Statistics for the Behavioral Sciences," McGraw-Hill, New York, 1956.
8 P. C. KELLY: *Anal. Chem.*, **44**(11):28A (1972).
9 H. KAISER: *Anal. Chem.*, **42**(4):26A (1970).

10   C. W. HELSTROM: "Statistical Theory of Signal Detection," 2d ed., Pergamon, Oxford, 1968.

11   For a critical discussion, see E. B. SANDELL in "Treatise on Analytical Chemistry," I. M. Kolthoff and P. J. Elving (eds.), part I, vol. 1, chap. 2, Interscience, New York, 1959.

12   See O. L. DAVIES: "Statistical Methods in Research and Production," 3d ed., p. 50, Oliver and Boyd, London, 1957.

13   A. A. BENEDETTI-PICHLER: *Ind. Eng. Chem., Anal. Ed.*, **8**:373 (1936).

14   WILSON,[2] p. 252.

15   H. KAISER: *Anal. Chem.*, **42**(2):24A (1970).

16   W. S. GOSSETT: *Biometrika*, **6**:1 (1908). Gossett used the pseudonym Student.

17   WILSON,[2] chap. 3.

18   R. A. FISHER: "The Design of Experiments," 8th ed., p. 16, Hafner, New York, 1966, states, "Every experiment may be said to exist only in order to give the facts a chance of disproving the null hypothesis."

19   DIXON and MASSEY,[1] p. 84.

20   L. S. NELSON: *J. Chem. Educ.*, **33**:126 (1956).

21   E. L. GRANT: "Statistical Quality Control," 3d ed., McGraw-Hill, New York, 1964; G. WERNIMONT: *Ind. Eng. Chem., Anal. Ed.*, **18**:587 (1946); J. A. MITCHELL: *Ind. Eng. Chem., Anal. Ed.*, **19**:961 (1947).

22   DAVIES,[12] chap. 7.

23   C. A. BENNETT and N. L. FRANKLIN, "Statistical Analysis in Chemistry and the Chemical Industry," p. 227, Wiley, New York, 1954.

24   BENNETT and FRANKLIN,[23] p. 243.

25   W. J. YOUDEN: "Statistical Methods for Chemists," chaps. 8–10, Wiley, New York, 1951.

26   WILSON,[2] chap. 4.

27   G. E. P. BOX: *Analyst*, **77**:879 (1952).

28   YOUDEN,[25] chap. 10.

29   G. WERNIMONT: *Anal. Chem.*, **23**:1572 (1951).

30   W. J. BLAEDEL, V. W. MELOCHE, and J. A. RAMSAY: *J. Chem. Educ.*, **28**:643 (1951).

31   R. B. DEAN and W. J. DIXON: *Anal. Chem.*, **23**:636 (1951).

PROBLEMS

26-1   The following values were observed for the molarity of a particular solution of $KMnO_4$ with pure KI and As(III) oxide as primary standards:

Molarity versus KI	Molarity versus As_2O_3
0.44109	0.44118
0.44125	0.44122
0.44107	0.44127
0.44128	0.44117
0.44119	0.44124
0.44112	

Calculate the standard deviation of each set of standardizations. Is either method significantly more precise? Are the molarity values obtained by the two methods significantly different?

Answer $s_1 = 0.000086$; $s_2 = 0.000042$;
$s_1^2/s_2^2 = 4.36$, not significant at 95% level.
For $s^2 \cong s_2^2 = 17.3 \times 10^{-10}$ (the smaller variance), t is calculated to be only 1.96, which indicates that no significant difference exists between the means (0.44117 and 0.44122).

26-2 Suppose that the following numbers denote the relative frequency of occurrence of each of the digits in the last significant figure of an instrument reading:

Digit	Frequency	Digit	Frequency
0	324	5	308
1	267	6	313
2	292	7	284
3	264	8	311
4	319	9	268

(a) Estimate the probability that this distribution could have occurred by chance. (b) Is the excess of even over odd digits significant?

Answer (a) $\chi^2 = 16.2$, which for 9 degrees of freedom would be exceeded about 5% of the time. (b) $\chi^2 = 9.57$, which is highly significant. For 1 degree of freedom, $\chi^2 = 7.78$ at the 0.5% level.

26-3 Each of four analysts A, B, C, D carried out replicate sets of four determinations, with the following results:

Determination	Analyst			
	A	B	C	D
1	20.13	20.14	20.19	20.19
2	20.16	20.12	20.11	20.15
3	20.09	20.04	20.12	20.16
4	20.14	20.06	20.15	20.10

Carry out an analysis of variance to determine whether the variance between analysts is significantly higher than that among determinations by a single analyst.

Answer Sum of squares "between analysts" = 0.0085, "within analysts" = 0.0175, $V_b = 0.00283$, $V_w = 0.00146$. Therefore F is insignificant.

26-4 The following values were recorded for the potential E of an electrode, measured against the saturated calomel electrode, as a function of concentration C (moles per liter).

$-\log C$	E, mV	$-\log C$	E, mV
1.00	106	2.10	174
1.10	115	2.20	182
1.20	121	2.40	187
1.50	139	2.70	211
1.70	153	2.90	220
1.90	158	3.00	226

What is the least-squares straight line for these data?

Answer $E = 49.6 - 58.95 \log C$.

26-5   The following represent control-analysis data on successive production batches: 11.7, 10.9, 11.3, 11.5, 11.1, 11.3, 10.8, 11.5, 11.2, 10.7, 11.2, 10.8, 11.3, 10.4, 10.9, 10.6, 10.7. Prepare a control chart, using 95% confidence limits based on the data at hand. Perform a regression analysis to determine whether the apparent downward trend is statistically significant.

To the statistician the process of sampling consists of drawing from a population a finite number of units to be examined. From sample statistics, such as mean and standard deviation, estimates are made of the population parameters. By appropriate tests of significance, confidence limits are placed on the estimates. Sampling for chemical analysis is an example of statistical sampling in that conclusions are drawn about the composition of a much larger bulk of material from an analysis of a limited sample.

The process of sampling may involve an elaborate array of operations, such as crushing, grinding, and subdivision. Each operation makes a contribution to the overall variance. It is important to understand the basic principles underlying the sampling process so that the accuracy of the sampling operation as well as that of the laboratory analysis can be made appropriate to the problem at hand. Otherwise, too much or too little effort may be expended on the sampling operation, thereby either increasing the cost unnecessarily or failing to achieve the desired level of accuracy.

27-1 STATISTICAL CRITERIA OF VALID SAMPLING

A sampling scheme is set up with several objectives, which can be stated in statistical terms as follows:

1 The sample mean should provide an unbiased (Section 26-5) estimate of the population mean. This objective is achieved only if all members of the population have an equal chance of being drawn into the sample.

2 The sample should provide an unbiased estimate of the population variance, so that tests of significance may be applied.[1] This objective is achieved only if *every possible* unit of a preselected size has an equal chance of being drawn.

Consider a population consisting of packages of a chemical moving along a conveyor belt. Each package has an equal chance of being drawn into the sample if the packages are chosen at *random*, as with the aid of a table of random numbers. Even if samples are chosen at regular intervals, a random choice of the first unit to be chosen fulfills the requirement of equal chance. Suppose, for example, that every 10th package is taken in regular sequence. If the first to be chosen is picked according to first appearance of a particular digit in a column of random numbers, each package is given a probability of 0.1 of appearing in the sample. This procedure, however, can produce only 10 *different* samples of a given size out of many possible samples; therefore the second objective is not fulfilled. On the other hand, if the entire sample is picked according to the column of random numbers, every possible 1-in-10 sample has an equal probability of occurring.

3 A third objective may be stated in terms of efficiency. For a given expenditure of money or time, the sampling procedure should lead to estimates of the central value and dispersion that are as accurate as possible. Alternatively, the cost or effort of sampling should be minimized for a given accuracy. To accomplish this objective, it is often necessary to resort to a nonrandom sampling procedure, at least in part. If the population can be divided by a random procedure into a number of subdivisions or sections, and if the variance among sections is large compared with the variance within sections, a more accurate result will on the average be obtained by accurately sampling each section rather than by using a completely randomized procedure to draw the same number of units. Such a *stratified* procedure involves a risk of bias that cannot be estimated by the usual tests of significance, for these tests are based on probability theory, which in turn is based on the assumption of random selection.

Randomness has been violated to some extent by the use of an orderly procedure. The risk can be minimized by making the procedure as truly representative as possible.

27-2 SAMPLING UNITS

It is useful to distinguish between type-A materials such as piles of coal or ore, which contain no unique subdivisions, and type-B materials such as bottles of chemicals, which occur in discrete lots that can be specified as sampling units.[2] Type-A materials present a special problem, for it is not possible to specify a routine procedure that will ensure a properly unbiased sample.

To illustrate the problem, consider a conical pile of coal (a material whose sampling problems have been extensively studied) produced by unloading freight cars with a conveyor belt. Such a pile is subject to segregation because larger chunks tend to roll down the sides of the cone, fine dust tends to be blown by the wind, and small particulate matter tends to settle under the source. Once the pile has been formed, it is well-nigh impossible to specify a representative sampling procedure that does not involve subdividing the entire pile. On the other hand, if the coal is sampled directly from the freight cars, it can be regarded as type-B material. Then, owing to segregation during transportation, careful sampling *within* each freight car is necessary. (The question of the optimum number of freight cars to be selected at random, as compared with the number of samples to be taken within each car, is considered in Section 27-5). Better yet, the coal can be sampled directly from the conveyor belt, where sections taken across the belt can be regarded as sampling units.

The problem of sampling from a moving stream, such as a conveyor belt or pipeline, is not so simple as it might at first appear. A segment across a moving stream is not a representative sample unless the entire cross section of the stream is moving at a uniform velocity. The proper sampling unit, therefore, is not necessarily that which is present at any one instant in a cross-sectional segment of stream but rather that which passes a given cross-sectional plane in a given interval of time. Often, if it is not practical to take the total effluent for a selected time interval as a sample, it is possible to design a moving orifice that can sample a cross-sectional plane.

A *random* sampling procedure is one in which each portion of the whole is given an equal chance of appearing in the sample according to a procedure based completely on chance and involving no periodicity or exercise of judgment.

In practice, a truly randomized selection procedure is seldom used in sampling a sequential series. Instead, samples usually are taken at regular intervals. The danger always exists that a cyclic variation in quality could fall into phase with the sampling operation and thereby lead to a biased sample. For example, regular cyclic variations that are shorter than the sampling frequency can give a spurious indication of a lower frequency; this is known as *aliasing*.[3] At the other extreme, if the fluctuation cycle is longer than the sampling period, the uniform sampling procedure is analogous to graphical integration to find the area under a curve. If the total area is divided into *uniform* strips, the heights of which are measured at their centers, the integration is more accurate than if strips of random width are used. Likewise, a uniform sampling procedure that is frequent enough can give the average composition more accurately than a random procedure. If coincidence of sampling and fluctuation cycles is suspected, the error can be minimized by increasing the sampling frequency.

It is of interest to compare several procedures for the analysis of a material composed of essentially uniform sampling units, say headache tablets or bottles of peroxide, that can be sampled randomly as such:

1   Each unit is analyzed separately, and the results are averaged. Assume that the analysis of a single unit has a standard deviation σ_1. To the variance σ_1^2 must be added σ_2^2, arising from *differences* among the units. If n such units are analyzed, the variance of the mean is $(\sigma_1^2 + \sigma_2^2)/n$.

2   The n units are thoroughly mixed and divided into n parts before analysis. The mixing process averages the composition and thus changes the variance σ_2^2 to σ_2^2/n per unit. Now, if only one unit is analyzed, the variance is $\sigma_1^2 + \sigma_2^2/n$. If n analyses are carried out, the variance of the mean is $\sigma_1^2/n + \sigma_2^2/n^2$, which is smaller than $(\sigma_1^2 + \sigma_2^2)/n$. Therefore, *a more precise result is obtained by mixing groups of units and performing* n *analyses on the mixed groups than by analyzing the* n *units separately.*

3   The pooled mixture is analyzed by a method in which the standard deviation is proportional to the magnitude being determined. This method corresponds to a constant relative precision, a situation frequently encountered in chemical analysis. The standard deviation of determination for the whole sample is $n\sigma_1$, or σ_1 per unit. Alternatively, the variance of a determination for the whole sample is $n^2\sigma_1^2$; and, since the variance is proportional to the *square* of the sample size, the variance per unit is σ_1^2. The total variance is $\sigma_1^2 + \sigma_2^2/n$, which is the same as in the second procedure, indicating that, *with a method of constant relative precision, the analytical precision is independent of the amount analyzed.*

4   A single determination is carried out on the whole mixture by a method in which *the standard deviation is independent of the magnitude being determined.* The standard deviation σ_1 for the whole sample corresponds to a standard deviation of σ_1/n per unit and to a variance of only σ_1^2/n^2 per unit. The total variance is $\sigma_1^2/n^2 + \sigma_2^2/n$, which is less than $(\sigma_1^2 + \sigma_2^2)/n$. This situation might occur, for example, in a trace analysis, in which the absolute, rather than the relative, error is essentially independent of sample size. In this case *a single determination on the combined sample will give a more precise result than the average of* n *separate unit analyses.*

EXAMPLE 27-1 A random sample of 10 pills is taken for analysis. Each pill contains 5 mg of active material, with a gaussian distribution and an estimated standard deviation of 0.1 mg. Given that a single determination of active ingredient has a *relative* estimated standard deviation of 1% of the amount present, compare the standard deviations of the following analytical schemes:

(a) Analyze each pill separately and report the mean of the 10 determinations.
(b) Pool the sample, and run a single determination on $\frac{1}{10}$ of the pooled sample.
(c) Pool the sample, run a single determination on each of three pools of 10 pills, and report the mean.
(d) Pool the sample, and run a single determination on the entire sample.

ANSWER We have $\sigma_1 = 0.01x$, where x = active ingredient in milligrams, and $\sigma_2 = 0.1$ mg.

(a) $\sigma^2 = \dfrac{\sigma_1^2 + \sigma_2^2}{n} = \dfrac{(0.05)^2 + (0.1)^2}{10} = 0.00125$ $\qquad \sigma = 0.035$ mg

(b) $\sigma^2 = \sigma_1^2 + \dfrac{\sigma_2^2}{n} = (0.05)^2 + \dfrac{(0.1)^2}{10} = 0.0035$ $\qquad \sigma = 0.06$ mg

(c) $\sigma^2 = \dfrac{\sigma_1^2}{3} + \dfrac{\sigma_2^2}{3n} = \dfrac{0.0035}{3} = 0.00117$ $\qquad \sigma = 0.034$ mg

(d) $\sigma^2 = \dfrac{\sigma_1^2}{n^2} + \dfrac{\sigma_2^2}{n} = \dfrac{(0.5)^2}{100} + \dfrac{(0.1)^2}{10} = 0.0035$ $\qquad \sigma = 0.06$ mg

////

27-3 ESTIMATION OF REQUIRED SAMPLE SIZE OF PARTICULATE MATERIAL

For solid particulate matter the necessary weight of sample taken at random from a bulk must be increased in relation to the increase in each of the following: (1) variation in composition among the particles, (2) desired accuracy of analysis, and (3) particle size. For the general case of a complex mixture of several components, each containing the desired constituent at a different level and each existing in a wide range of particle sizes, rigorous statistical evaluation of indeterminate error is impractical. With several simplifying assumptions, some guiding conclusions can be drawn as to the minimum amount of sample for which the sampling error is less than a required limit. The order of magnitude of sample size can be estimated by a procedure given by Baule and Benedetti-Pichler.[4,5]

Consider a system in which the material to be sampled is a random binary population that consists of n_1 units of A and n_2 units of B. The two kinds of units might be black and white marbles, satisfactory and unsatisfactory items from a production line, or particles of ore mineral and gangue. The probability of drawing a unit of A is given by $p = n_1/(n_1 + n_2)$, and the probability of drawing a unit of B is $1 - p = n_2/(n_1 + n_2)$. Thus p and $1 - p$ represent the fractions (by number) of the two kinds of units in the population, so that $p + (1 - p) = 1$. If n units are removed at random from an infinite population, then the expected values for the numbers of the two kinds of units in the sample are pn and $(1 - p)n$. If a series of samples of n units are drawn at random, the number of A units will fluctuate around the expected value pn with a standard deviation σ_n that is obtainable from the properties of a binomial distribution:[6,7]

$$\sigma_n = \sqrt{np(1 - p)} \qquad (27\text{-}1)$$

Similarly, the number of B units will fluctuate around its expected value $n(1 - p)$ with the same variance.

EXAMPLE 27-2   If the incidence of defective units coming from a production line is 2% and a random sample of 10,000 units is drawn, what is the expected number of defective units in the sample, and what is the standard deviation?
   The *expected number* is $0.02 \times 10,000 = 200$; the standard deviation is

$$\sigma_n = \sqrt{10,000 \times 0.02 \times 0.98} = 14 \qquad ////$$

   The *relative* standard deviation (in percent) or sampling error σ_s of the units of A is given by

$$\frac{\sigma_n}{n_1} = \frac{\sigma_n}{np} = 100 \sqrt{\frac{1 - p}{np}} = \sigma_s \qquad (27\text{-}2)$$

and the corresponding relative standard deviation of the units of B is

$$\frac{\sigma_n}{n_2} = \frac{\sigma_n}{n(1 - p)} = 100 \sqrt{\frac{p}{n(1 - p)}} \qquad (27\text{-}3)$$

where $n_1 = np = $ number of A units drawn into the sample and $n_2 = n(1 - p) = $ number of B units drawn into the sample.
   Equation (27-2) allows the calculation of the sampling error σ_s as a function of the number of units in a sample. If A is 100% and B is 0% in the component of interest, Figure 27-1 indicates the relation[8] between the number of units (or particles) n in a sample and relative standard deviation, for the fraction p ranging from 0.0001 to 0.999 (0.01 to 99.9%). This figure shows that, when a constituent sought is present as only a small fraction of the total, for negligible sampling error the number of units or particles required in the sample becomes enormous.

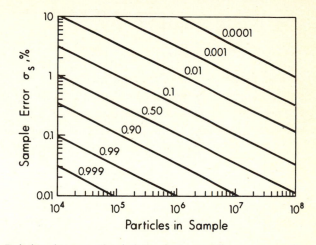

FIGURE 27-1   Relation between the indeterminate sampling error σ_s (in percentage) and the number of particles n for the fraction p ranging from 0.0001 to 0.999. (*From Harris and Kratochvil.*[8])

The weight of sample to be taken for analysis is the matter of practical concern in the laboratory. Figure 27-2 gives the approximate relation between the mesh size of spherical particles and the number of particles present per gram of material. It can be seen, for example, that if 10^6 particles are needed to keep within the required limit of sampling error for 1-g samples the material should be ground fine enough to pass a 200-mesh screen when the density is about 3. For high-precision analysis of traces it would be virtually impossible to grind matter to the required fineness for samples of reasonable weight.

More broadly, let P_{av} represent the average percentage of the component to be determined in the population, which is assumed to consist of uniform particles of two kinds A and B in the ratio $p/(1 - p)$. The A particles have density d_1 and contain $P_1\%$ of the component; the B particles have density d_2 and contain $P_2\%$. The average density is d, and all particles have the same volume. Benedetti-Pichler[5] showed that an error Δn_1 in the number of units of A contained in a sample of n units causes an error ΔP_{av} in the percentage of the component, where ΔP_{av} is given by

$$\Delta P_{av} = \frac{\Delta n_1}{n} \frac{d_1 d_2}{d^2} (P_1 - P_2) \qquad (27\text{-}4)$$

If σ_n from (27-1) is substituted for Δn_1 in (27-4), the *absolute* standard deviation of P_{av}, $\sigma_{P_{av}}$, is

$$\sigma_{P_{av}} = \sigma_n P_{av} = \frac{d_1 d_2}{d^2} (P_1 - P_2) \sqrt{\frac{p(1 - p)}{n}} \qquad (27\text{-}5)$$

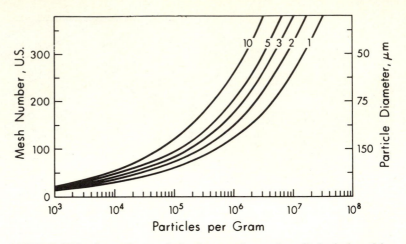

FIGURE 27-2 Approximate relation between the number of spherical particles per gram of sample and mesh size (U.S. Sieve Series ASTM E-11-61) or particle diameter, for densities from 1 to 10. (*From Harris and Kratochvil.*[8])

The number of particles n required in the sample is given by

$$n = p(1 - p)\left(\frac{d_1 d_2}{d^2}\right)^2 \left(\frac{100(P_1 - P_2)}{\sigma_s P_{av}}\right)^2 \qquad (27\text{-}6)$$

Thus the number of particles must be increased sharply as the *relative difference* in percentage of the desired constituent in the two kinds of particles approaches 100%.

EXAMPLE 27-3 A sample is composed of uniform particles of a mineral of density $d_1 = 5.2$ and a gangue of density $d_2 = 2.7$. The percentage P_1 of metal in the mineral is 75, and the percentage P_2 in the gangue is 5. If the sample contains $P_{av} = 60\%$ metal, how many particles must be taken to ensure a relative standard deviation of less than 1 ppt (0.1%) in P_{av}, due to sampling error?

ANSWER The fraction p of mineral particles is calculated from the following relation: $p/(1 - p) = d_2(P - P_2)/d_1(P_1 - P) = 1.903$, from which $p = 0.656$. The average density is given by $pd_1 + (1 - p)d_2 = 4.34$. The value of $P_{av}\sigma = (60 \times 0.1) = 6$, and n calculated from (27-6) is 1.7×10^5 particles. For spherical particles this corresponds to a sample weight of 0.39 g for particles 0.01 cm in diameter or 386 g for particles 0.1 cm in diameter.   ////

If we make the simplifying assumption that the density factor in (27-6) is unity, and if (for broad applicability) we consider the *relative*-percent difference in composition of the two kinds of particles, $100(P_1 - P_2)/P_1$, we can calculate the curves

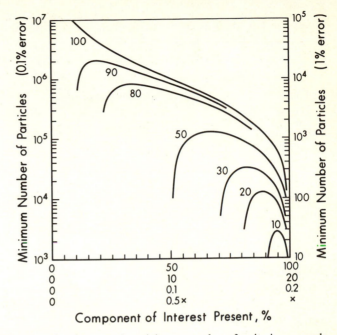

FIGURE 27-3 Relation between the minimum number of units in a sample required for sampling errors (relative standard deviations in percentage) of 0.1 and 1% (y axis) and the overall composition of a sample (x axis), for perfectly random mixtures having two types of particles with a *relative* difference in composition ranging from 100% to 10%. (*From Harris and Kratochvil.*[8])

of Figure 27-3. This figure indicates that sampling errors can be made insignificant more easily when the two kinds of particles differ but little in composition. From a practical viewpoint, the largest particle sizes consistent with negligible sampling error are the most convenient. When the relative difference in composition between A and B is 100%, it is difficult to grind samples adequately corresponding to regions to the left of the midpoint of the horizontal axis (Figure 27-3, left). An enormous advantage results from the point of view of sampling error when the relative difference in composition between A and B is small. For example, if a series of samples containing chloride is to be prepared, there is great advantage in using mixtures of NaCl and KCl instead of mixtures of NaCl with an inert diluent such as Na_2SO_4. When a sample contains only NaCl (60.66% chloride) or only KCl (47.55% chloride), sampling error is zero and is not a function of particle size. Maximum sampling error occurs with an approximately 40:60 NaCl-KCl mixture; if in this case a sampling error of 0.5 ppt is acceptable, a sample containing only about 60,000 particles need be taken for analysis. For a 1-g sample this corresponds to grinding the mixture to pass a 50-mesh sieve. On the other hand, if Na_2SO_4 is used as diluent, sampling error exceeds

1 ppt for 1 g of 50-mesh material when the Na_2SO_4 level is greater than 5% of the total composition of the sample.

For precise analyses it is just as serious if A is 0.1% and B is 0% in the constituent of interest as if A is 100% and B is 0% (Figure 27-3), because their relative difference in composition is the same.

Estimation of sample size by Equation (27-6) is idealized in two ways. In the first, only two components A and B are assumed to exist in the sample. If more than two components are present, it is possible to compute the necessary characteristics of an equivalent sample of just two components, one fraction rich and one fraction poor in the desired constituent. In the second simplification all particles are assumed to be the same size. This leads to a more serious difficulty, because rarely is detailed information available about the distribution of particle sizes in a sample. If the calculation of Example 27-3 were carried out with the assumption that all particles are spheres of diameter equal to the mesh spacing of a screen that will pass all the sample, the calculated sample size would be greater than that actually required. Benedetti-Pichler[5,9] showed that the standard deviation remains between 0.5 and 1.5 times that computed for uniform particles if any or all of the particles have volumes between 0.25 and 2.25 times as large as those assumed in the calculation. Within the same precision limits, up to 75% of the particles can be smaller without size limit, but only 18% by weight can be twice the normal diameter, or 8 times the normal weight. Thus it is important to avoid abnormally large particles or, to be on the safe side, to base the calculation on the largest particles.

Evaluation of the results of mineral analysis with a variety of sample sizes has shown [9a] that analytical error and sampling error can be distinguished.

27-4 STRATIFIED SAMPLING OPPOSED TO COMPLETELY RANDOM SAMPLING

When the material to be sampled can be subdivided into logical sampling units, there are two fundamentally different approaches to random sampling. A number of samples can simply be chosen at random from the whole bulk of material. Alternatively, the sampling can be stratified by first choosing from among the sampling units and then sampling within the units, again by a random procedure. As shown below, the stratified procedure always yields results at least as precise as the simple random scheme and gives superior results whenever the variance among sampling units is appreciable compared with the variance within such units.

In statistical terms, if the whole population to be sampled is homogeneous, that is, can be described by a single set of parameters, simple random sampling is as efficient as stratified sampling. But, if the population consists of a set of appreciably different subpopulations and so cannot be described by a single set of parameters, then sampling the subpopulations by a stratified procedure to estimate their parameters is preferred.

When the subpopulations or strata are unequal in size and in variance, it can be shown[10,11] that, if an estimate of the population mean is to be unbiased and the variance of the estimate is to be minimal, the number of samples taken from each stratum should be proportional to the size of the stratum and also to its standard deviation, or

$$\frac{n_r}{n} = \frac{w_r(\sigma_0)_r}{\sum w_r(\sigma_0)_r} \qquad (27\text{-}7)$$

where n_r is the number of samples from the rth stratum, n the total number of samples desired, w_r the weight of stratum, and $(\sigma_0)_r$ the standard deviation within the rth stratum. In calculation of the population mean, the stratum means must be weighted in proportion to the size of the strata, so that the estimate of the population mean may be unbiased. Thus

$$\bar{x} = \frac{\sum w_r \bar{x}_r}{\sum w_r} \qquad (27\text{-}8)$$

where $\bar{x}_r$ is the mean of the determined values in the rth stratum and $\bar{x}$ is the estimate of the population mean. If the strata differ in size but not in variance, σ_r is the same for all, and (27-7) becomes

$$\frac{n_r}{n} = \frac{w_r}{\sum w_r} \qquad (27\text{-}9)$$

This simply means that the numbers of samples within strata should be proportional to the sizes of the strata. This procedure is commonly called *representative sampling*; it gives an unbiased estimate of the population mean, but leads to a larger variance of the estimate than the procedure represented by Equations (27-7) and (27-8) *unless the variance is uniform in all the strata.*

EXAMPLE 27-4 A shipment of oil consists of equal numbers of 1-l and 5-l containers. A previous study has led to the expectation that the sulfur content of the 1-l containers is 10 ppm with a standard deviation of 2 ppm, and that of the 5-l containers is 15 ppm with a standard deviation of 3 ppm. If the analytical error is negligible, what is the optimum sampling scheme for a total of 24 samples to give an unbiased estimate of the sulfur content of the shipment, with minimum sampling error?

ANSWER The two strata are the two sizes of containers, with $w_1 = 1$, $w_2 = 5$. The number of samples n_1 to be taken from the 1-liter containers is given by $n_1 = (n_1 + n_2)(2w_1)/(2w_1 + 3w_2)$, from which $n_1 = 3$, $n_2 = 21$. If the difference in the stratum variance is neglected, the simple representative sampling scheme would call for $n_1 = 24w_1/(w_1 + w_2) = 4$ and $n_2 = 20$. In either case the weighted average $\bar{x} = (1x_1 + 5x_2)/6$, where x_1 and x_2 are the average values from the two strata, gives the sulfur content of the shipment.   ////

A simpler case, but one frequently encountered in analytical chemistry, is that in which the strata are equal in size. To judge the effect of stratification, suppose that a variance-analysis study has been carried out. If there are k strata containing n observations in each, the analysis-of-variance table will be identical to Table 26-7.

Let σ_0 and σ_i be the standard deviations of a single determination within strata and among strata. To calculate the variance of the mean of kn determinations, recall that the mean is calculated by (27-8), which for equal strata is

$$\bar{x} = \frac{\sum \bar{x}_i}{k} \qquad (27\text{-}10)$$

The variance of $\bar{x}$ is $1/k^2$ times the variance of $\bar{x}_i$ (Table 26-2), so

$$V_{\bar{x}} = \frac{1}{k^2} V_{\bar{x}_i} = \frac{1}{k^2} \frac{\sigma_0{}^2}{n^2} = \frac{\sigma_0{}^2}{N^2} \qquad (27\text{-}11)$$

where $N = nk =$ total number of determinations.

It is important that *the variance of the mean is independent of the variance among strata* and depends only on the variance within strata and on the number of determinations.

Now suppose that the same number of samples had been drawn at random from the population as a whole. The variance of a single determination would be $\sigma_0{}^2 + \sigma_i{}^2$, and the variance of the mean of N determinations would be $(\sigma_0{}^2 + \sigma_i{}^2)/N$. This formula would be identical to (27-11) if the variance among sections were equal to zero. Therefore, the stratified procedure must yield a result at least as good as the completely random procedure, and *it will yield a superior result if the variance among strata is appreciable compared with the variance within strata.* A qualification must be made that the *relative sizes* of the strata must be known. If there should be an appreciable error in sizes, the resulting bias in the estimate of the mean might more than compensate for the advantage in precision gained by stratification and, what is worse, introduce an unsuspected systematic error.

27-5   MINIMIZATION OF COST OR VARIANCE IN STRATIFIED SAMPLING

Suppose that the bulk of material to be analyzed is sampled by taking n_1 strata, each of which provides n_2 samples, and that n_3 determinations are to be carried out on each sample. The various strata are assumed to be equal in size and in variance within strata.

A practical question that arises is how to minimize the cost of determining an estimate of the population mean to within a desired variance. Alternatively, it may be desired to minimize the variance for a given allocation of funds, taking into account the relative costs of selecting the strata c_1, sampling within the strata c_2, and performing a determination c_3. The total cost of the procedure is

$$c = n_1 c_1 + n_1 n_2 c_2 + n_1 n_2 n_3 c_3 \qquad (27\text{-}12)$$

Suppose that an analysis of variance has been carried out and that the quantities $s_1{}^2$, $s_2{}^2$, and $s_3{}^2$ have been determined as estimates of the variance components $\sigma_1{}^2$, $\sigma_2{}^2$, and $\sigma_3{}^2$ for the two stages of sampling and the determination. These components contribute to the variance of the population mean that is being estimated. Thus

$$\sigma^2 = \frac{\sigma_1{}^2}{n_1} + \frac{\sigma_2{}^2}{n_1 n_2} + \frac{\sigma_3{}^2}{n_1 n_2 n_3} \qquad (27\text{-}13)$$

It can be shown[11] that, to minimize the cost c for a fixed value of σ^2, the values of n_1, n_2, and n_3 are given by

$$n_1 = \frac{\sqrt{\sigma_1{}^2/c_1}}{\sigma^2} \left(\sqrt{\sigma_1{}^2 c_1} + \sqrt{\sigma_2{}^2 c_2} + \sqrt{\sigma_3{}^2 c_3} \right) \qquad (27\text{-}14)$$

$$n_2 = \sqrt{\frac{\sigma_2{}^2 c_1}{\sigma_1{}^2 c_2}} \qquad (27\text{-}15)$$

$$n_3 = \sqrt{\frac{\sigma_3{}^2 c_2}{\sigma_2{}^2 c_3}} \qquad (27\text{-}16)$$

The significant result is that *the optimum allocation of sampling after the first stage is independent of the desired overall variance* σ^2. In other words, for different values of σ^2 the modification of the optimum sampling scheme consists of changing the number of strata n_1 sampled while maintaining constant the treatment of the various sections.

Similarly, if the total cost c is fixed, it can be shown[12] that the optimum values of n_1, n_2, and n_3 are

$$n_1 = \frac{c \sqrt{\sigma_i{}^2/c_1}}{\sqrt{\sigma_1{}^2 c_1} + \sqrt{\sigma_2{}^2 c_2} + \sqrt{\sigma_3{}^2 c_3}} \qquad (27\text{-}17)$$

$$n_2 = \sqrt{\frac{\sigma_2{}^2 c_1}{\sigma_1{}^2 c_2}} \qquad (27\text{-}18)$$

$$n_3 = \sqrt{\frac{\sigma_3{}^2 c_2}{\sigma_2{}^2 c_3}} \qquad (27\text{-}19)$$

showing that *the optimum allocation beyond the first stage is the same* for fixed total cost as for fixed total variance. The same principles can be extended to any number of stages in a nested sampling design. Also, similar relations are available for cases in which certain of the numbers $n_1, n_2, n_3, \ldots$ are fixed and either the total cost or the total variance is fixed.

EXAMPLE 27-5   If the standard deviation of sampling strata is 0.07, that of sampling within strata is 0.10, and that of a single determination is 0.21, and if the relative costs are in the ratio 4/2/1, calculate the optimum sampling scheme and minimum cost that will give an overall standard deviation of the mean of 0.08.

ANSWER $n_1 = \sqrt{(0.07)^2/4}(\sqrt{(0.07)^2 \times 4} + \sqrt{(0.10)^2 \times 2}$

$$+ \sqrt{(0.21)^2 \times 1})/(0.08)^2 = 2.7$$

$$n_2 = \sqrt{(0.10)^2 \times 4/(0.07)^2 \times 2} = 2.02$$

$$n_3 = \sqrt{(0.21)^2 \times 2/(0.1)^2 \times 1} = 2.97$$

Taking $n_1 = 3$, $n_2 = 2$, and $n_3 = 3$, we calculate $\sigma_0 = 0.076$ and $c = 3 \times 4 + 6 \times 2 + 18 \times 1 = 42$ on the relative scale.   ////

EXAMPLE 27-6   Given the same standard deviations and relative costs as in Example 27-5, calculate the minimum standard deviation that can be achieved for a cost of 30 times that of a single determination.

ANSWER $n_1 = 30\sqrt{(0.07)^2/4}(\sqrt{(0.07)^2 \times 4}$

$$+ \sqrt{(0.10)^2 \times 2} + \sqrt{(0.21)^2 \times 1}) = 2.14.$$

$$n_2 = 2.02 \qquad n_3 = 2.97$$

Taking $n_1 = 2$, $n_2 = 2$, and $n_3 = 3$, we calculate $c = 28$ and $\sigma_0{}^2 = (0.07)^2/2 + (0.10)^2/4 + (0.21)^2/12 = 0.00862$ so $\sigma_0 = 0.093$.   ////

27-6   SAMPLING PROCEDURES

From the statistical aspects of the sampling problem that have been considered, we may draw certain conclusions. First, the bulk of material to be sampled should be subdivided into real or imaginary sampling units, which might range from individual molecules in homogeneous gases or liquid solutions to carloads of coal. Next, it is helpful to know the relative variations to be expected among units and within units. (In some cases segregation is difficult to avoid—native gold in a sand deposit, for example). By carrying out analysis-of-variance studies on stratified sampling schemes it can be decided whether it is justifiable to continue stratification at several levels (nested sampling scheme) or to simplify the procedure. The decision usually will be based on cost and convenience as well as on the desired level of accuracy. As a rule, some stratification will turn out to be desirable, if strata are chosen that are known or suspected to have variations among them and samples are taken proportional to the sizes of the strata.

For particulate material the size of the sample should be commensurate with the maximum size of the particles, depending on the variations in composition to be found among particles (Section 27-3). If the same heterogeneity of composition exists at all stages of subdivision, then the calculations described in Section 27-3 will lead to the requirement that the same total number of particles be present in the sample at all stages of subdivision. This would mean that the sample weight could be decreased as the cube of the particle diameter. Actually, as the sample is ground to a finer mesh size, a greater number of particles usually are required, because coarse granules often are made up of aggregates of finer particles, which differ in composition more than the agglomerates do. In general, it is advisable not to carry out more grinding than essential for efficient sampling and subsequent chemical treatment, to avoid the danger of changes in composition (such as dehydration or oxidation) during the grinding operation.[13]

The overall sampling procedure may be divided into three steps: (*1*) collection of the gross sample, (*2*) reduction of the gross sample to a suitable size for the laboratory, and (*3*) preparation of the laboratory sample. The details of the steps differ considerably, according to the physical character of the material to be sampled. Often the initial sample that must be taken to ensure sufficient precision of sampling is so large that considerable reduction in size is necessary. For particulate matter the decrease in sample size must be accompanied by a corresponding decrease in particle size. For this purpose a variety of crushing, grinding, mixing, and dividing machines has been devised, the details of which are beyond the range of the present discussion. To prepare the laboratory sample requires packaging in a suitable form to ensure protection from changes in composition. For example, it is sometimes desirable to determine moisture immediately, before packaging.

General requirements have been outlined for the sampling of solids,[14] liquids,[15] and gases.[16] Some of the methods used for collecting samples are discussed here in a brief general way for several types of materials.

Material present in homogeneous solution For gases or liquids that contain no suspended matter and that can be regarded as homogeneous solutions, the sampling unit can be small.

EXAMPLE 27-7   What size of sample is required from a homogeneous aqueous solution of radioisotope present at a concentration of 10^{-8} M, if the standard deviation due to sampling is not to exceed 0.1% (relative)?

ANSWER   This is an example of sampling from a population of two types, solute molecules A and solvent molecules B. In 1 liter of solution there are $N_1 = 10^{-8} \times 6 \times 10^{23} = 6 \times 10^{15}$ solute molecules and $N_2 = 55.5 \times 6 \times 10^{23} = 3.3 \times 10^{25}$ solvent molecules. Accordingly, $p = N_1/(N_1 + N_2) = 1.8 \times 10^{-10}$, and $(1 - p) \simeq 1$. From Equation (27-2), $\sigma_n/N_1 = \sqrt{(1 - p)/pN} = 10^{-3}$, which yields $N = 10^6/p = 5.5 \times 10^{15}$ molecules in the total sample. This corresponds to $5.5 \times 10^{15}/55.5 \times 6 \times 10^{23} = 1.6 \times 10^{-10}$ l.   ////

Liquids containing suspended matter   Nonhomogeneity can be largely compensated for by taking a series of samples at various levels, meanwhile providing sufficient agitation to keep the solid matter suspended as uniformly as possible. A "sampling thief," a device that can be lowered to the desired depth and opened temporarily to gather a sample, often is used. In some instances the ratio of weights of the two phases can be determined independently, for example, from the material balance. Then it is not necessary to draw a representative sample of the mixture, but only to carry out an analysis on each of the phases after they have been separated.

Solids in particulate form   As discussed in Section 27-2, material consisting of discrete lots is sampled by taking a random selection of such lots. When large variations exist within a lot, for example, when segregation has occurred during transport, it is usually possible to resort to a representative sampling scheme within the lot or else to convert the lot into a flowing sample, which then can be sampled at random. If neither of these procedures is accessible, it may be possible to take samples from the composite pile while it is being formed, for instance, by taking samples at evenly spaced increments along lines drawn from several points around the base of the conical pile to the apex, after each lot is dumped at the apex.

Solids in compact form   Materials of this type are often in the form of discrete lots such as ingots, slabs, sheets, or bales that can be subjected to random sampling procedures. The procedure of sampling within the lot depends on the physical properties and the geometry of the material.

Sheets of metal often can be sampled conveniently and nondestructively by clamping a number of sheets together with edges flush and milling across the edges to obtain an edge sample.

Billets or ingots of nonferrous metals may be sampled by sawing completely across the specimen at several regularly spaced intervals along its length. The "sawdust" thus collected is combined to form the sample. Drilling or punching holes at regular intervals can be used only if it has been demonstrated to give valid samples. We have found,[17] for example, that drilling holes in carnallite-NaCl cores gives drillings that are seriously high in potassium.

It is beyond the scope of this book to consider the details of the special sampling procedures that have been worked out for various types of materials. Studies of the sampling of many types of materials have been carried out by testing organizations and government agencies. For metals, nonmetallic construction materials, paper, paints, fuels, petroleum products, and soils, the "Book of Standards" and other publications of the American Society for Testing and Materials[18] are recommended. The *Journal of the Association of Official Agricultural Chemists* regularly publishes tentative procedures for sampling and analysis of soils, fertilizers, foods, water, drugs, and so forth, and at 5-year intervals releases new editions of "Official Methods of Analysis."[19] Similar methods for vegetable fats, oils, soaps, and related materials are published and revised periodically by the American Oil Chemists' Society.[20] "Standard Methods of Analysis"[21] includes many references to the original literature.

REFERENCES

1   G. W. SNEDECOR and W. G. COCHRAN: "Statistical Methods," 6th ed., Iowa State University, Ames, Iowa, 1967.

2   L. TANNER and W. E. DEMING: *Amer. Soc. Test. Mater. Proc.*, **49**:1181 (1949).

3   H. V. MALMSTADT, S. R. CROUCH, C. ENKE, and G. HORLICK: "Optimization of Electronic Measurements," Module 4, Benjamin, Menlo Park, Calif., 1974.

4   B. BAULE and A. A. BENEDETTI-PICHLER: *Z. Anal. Chem.*, **74**:442 (1928).

5   A. A. BENEDETTI-PICHLER in "Physical Methods in Chemical Analysis," W. M. Berl (ed.), vol. 3, p. 183, Academic, New York, 1956.

6   M. R. SPIEGEL: "Schaum's Outline of Theory and Problems of Statistics," p. 122, McGraw-Hill, New York, 1961.

7   W. J. DIXON and F. J. MASSEY, JR.: "Introduction to Statistical Analysis," 3d ed., p. 413, McGraw-Hill, New York, 1969.

8   W. E. HARRIS and B. KRATOCHVIL: *Anal. Chem.*, **45**:313 (1974).

9   A. A. BENEDETTI-PICHLER: "Essentials of Quantitative Analysis," Ronald, New York, 1956.

9a C. O. INGAMELLS and P. SWITZER: *Talanta*, **20**:547 (1973); C. O. INGAMELLS, J. C. ENGELS, and P. SWITZER: 24th International Geological Conference, sect. 10, p. 405, 1972.

10 J. NEYMAN: *J. Roy. Statist. Soc.*, **97**:558 (1934).

11 C. A. BENNETT and N. L. FRANKLIN: "Statistical Analysis in Chemistry and the Chemical Industry," pp. 61, 482, Wiley, New York, 1954.

12 S. MARCUSE: *Biometrics*, **5**:189 (1949).

13 W. F. HILLEBRAND, G. E. F. LUNDELL, H. A. BRIGHT, and J. I. HOFFMAN: "Applied Inorganic Analysis," 2d ed., pp. 819, 907, Wiley, New York, 1953.

14   W. W. ANDERSON in "Standard Methods of Chemical Analysis," F. J. Welcher (ed.), 6th ed., vol. 2, part A, p. 28, Van Nostrand, Princeton, N.J., 1963.

15   W. V. CROPPER in Welcher,[14] vol. 2, pt. A, p. 39.

16   C. W. WILSON in Welcher,[14] vol. 2, pt. B, p. 1507.

17   W. E. HARRIS, unpublished results.

18   AMERICAN SOCIETY FOR TESTING AND MATERIALS: "Book of ASTM Standards," ASTM, Philadelphia, annual.

19   ASSOCIATION OF OFFICIAL AGRICULTURAL CHEMISTS: "Official Methods of Analysis of the Association of Official Agricultural Chemists," 10th ed., AOAC, Washington, D.C., 1965.

20   AMERICAN OIL CHEMISTS' SOCIETY: "Official and Tentative Methods of the American Oil Chemists' Society," AOCS, Chicago, 1951–, updated with supplements.

21   "Standard Methods of Chemical Analysis," 6th ed., 3 vols., Van Nostrand Reinhold, New York, 1962–1966.

PROBLEMS

27-1   A powdered sample consists of uniform particles of ore of density 4.5 and gangue of density 2.2, containing 60 and 15% metal M. If a sample of 0.20 g containing about 40% M is taken for analysis, and if the particles are assumed to have a uniform diameter of 0.01 mm, what is the expected standard deviation due to sampling?

Answer   0.0020% M.

27-2   A material occurs in a particular range of particle size. Two kinds of particles are present, one composed 100% and the other 0% of the component of interest, both with a density of 3. What is the minimum weight of sample that should be taken so that the relative sampling error is no greater than 1 ppt if the constituent of interest is present at the 25% level and the range of particle sizes is: (*a*) 150 to 210 μm, (*b*) 100 to 150 μm, (*c*) 75 to 100 μm, and (*d*) 45 to 75 μm?

Answer   (*a*) 63 g (on basis of largest diameter). (*b*) 23 g. (*c*) 7 g. (*d*) 2.9 g.

27-3   Repeat Problem 27-2 for a case where two kinds of particles are present that contain 0.5 and 0% of the component of interest, which is present at the 0.125% level.

Answer   (*a*) 63 g. (*b*) 23 g. (*c*) 7 g. (*d*) 2.9 g.

27-4   How large a sample of an equimolar mixture of two gases, measured at 0°C and 10^{-6} mm pressure, must be taken if the sampling error is not to exceed 0.1%?

Answer 2.8×10^{-5} ml.

INDEX